TAIPO.
SOURCE: DAVID PAYNE.

WALTER REEKS AT THE HELM OF THELMA (1909).
SOURCE: ROYAL SYDNEY YACHT SQUADRON.

WALTER REEKS

NAVAL ARCHITECT, YACHTSMAN AND ENTREPRENEUR

NICOLE MAYS & DAVID PAYNE

Navarine Publishing

Navarine Publishing

FIRST PUBLISHED IN 2023 BY NAVARINE PUBLISHING
GPO BOX 2178, HOBART, TASMANIA, AUSTRALIA 7001
WWW.NAVARINE.NET

TYPESETTING BY NICOLE MAYS

PRINTED IN AUSTRALIA BY INGRAMSPARK

ENQUIRIES SHOULD BE MADE TO NICMAYS@GMAIL.COM

ISBN: 978-0-6457086-2-2

NATIONAL LIBRARY OF AUSTRALIA

A catalogue record for this book is available from the National Library of Australia

PLEASE NOTE THAT WHILE WE HAVE INTENDED TO PROVIDE A DEFINITIVE AND ACCURATE HISTORY AND PROFILE OF WALTER REEKS' LIFE AND THE VESSELS HE WAS INVOLVED WITH, SOME EVENTS, NAMES AND DATES MAY HAVE UNINTENTIONALLY BEEN INCORRECTLY STATED OR OMITTED. VESSEL MEASUREMENTS MAY ALSO BE DIFFERENT OR INCORRECT, PARTICULARLY BETWEEN THOSE DIMENSIONS REPORTED IN THE PRESS VERSUS THOSE LISTED IN REGISTRATION PAPERS. WE APOLOGISE FOR ANY ERRORS OR OMISSIONS.

COVER IMAGES

ENA ON SYDNEY HARBOUR (2016).
SOURCE: AUSTRALIAN NATIONAL MARITIME MUSEUM COLLECTION.

LADY EDELINE AND LADY FERGUSON AT BALMAIN FERRY BASE (1975).
SOURCE: GRAEME ANDREWS WORKING HARBOUR PHOTOGRAPH COLLECTION, CITY OF SYDNEY ARCHIVES.

WALTER REEKS WITH HIS CHILDREN THELMA AND KENT (LATE 1890S).
SOURCE: RAE FIDOCK COLLECTION.

LAUNCH OF THE LADY MARY AT BERRYS BAY (1982).
SOURCE: AUSTRALIAN NATIONAL MARITIME MUSEUM COLLECTION.

THELMA AND IDUNA SAILING ON SYDNEY HARBOUR (LATE 1880S).
SOURCE: AUSTRALIAN NATIONAL MARITIME MUSEUM COLLECTION.

JENNY WREN ON SYDNEY HARBOUR (2008).
SOURCE: JOHN JEREMY.

CONTENTS

WITH THANKS ...

THE AUTHORS WOULD LIKE TO ACKNOWLEDGE AND THANK, WITH MUCH GRATITUDE, THE FOLLOWING INDIVIDUALS AND ORGANISATIONS FOR THEIR HELP AND SUPPORT IN PRODUCING THIS MANUSCRIPT: ROBERT ALBERT; BILL ALLEN; GRAEME ANDREWS; THE AUSTRALIAN NATIONAL MARITIME MUSEUM, INCLUDING SNR CURATOR DAINA FLETCHER AND DR PETER HOBBINS; RON BERTLESEN; GRAEME BROXAM; PETER CASUBIN; PETER COLE; JOHN DARROCH; JULIE DUELL; GWEN DUNDON; MORI FLAPAN; MEMBERS OF THE FOX ASHWIN FAMILY, INCLUDING RAE FIDOCK; BRIAN GALE; JOEL, OWEN AND LIAM GEOGHEGAN; DAVID GLASSON; COLIN GRAZULES; MICHAEL GREGG; PHIL HELMORE; THE HERRESHOFF MARINE MUSEUM AND AMERICA'S CUP HALL OF FAME; KELLY HOLDER; TONY HUNT; CHERRY JACKAMAN; JERVIS BAY MARITIME MUSEUM; HAROLD KIDD; STEPHEN KNIGHT; EDDIE LEWIS; MIKE LONGWORTH; PATRICIA LOWN; ALEX MACCORMICK; ANN MACINTOSH; JOHN MACLURCAN; ROBYN MAYS; SCOTT MILSON; MOSMAN LOCAL STUDIES; STEPHEN MULLINS; THE MUSEUM OF APPLIED ARTS & SCIENCES 'POWERHOUSE MUSEUM'; THE NATIONAL ARCHIVES OF AUSTRALIA; NEW YORK YACHT CLUB; NEVILLE AND MADELEINE OLLIFFE; SAM PAYNE; NIGEL PECK; TONY PRESCOTT; THE QUEENSLAND MARITIME MUSEUM; NOEL RILEY; THE ROYAL SYDNEY YACHT SQUADRON; SIMON SADUBIN; MIKE SCOTT; JOE SKRYZNSKI; JOHN STANLEY; RON STANNUS; OLIVER STREETON; THE SYDNEY HERITAGE FLEET; WARWICK TURNER; LAURENCE TURTLE; BILL WRIGHT; SYDNEY WATER; AND ANDY WHITE.

WE ADDITIONALLY THANK ALL OF THE ORGANISATIONS AND INDIVIDUALS THAT SUPPLIED IMAGES AND HAVE ENDEAVOURED TO CORRECTLY CITE THEIR SOURCES. REASONABLE STEPS HAVE BEEN TAKEN TO CONTACT COPYRIGHT OWNERS OF IMAGES USED IN THIS BOOK. PLEASE EMAIL NICMAYS@GMAIL.COM WITH ANY ERRORS OR OMISSIONS.

UNITS AND ABBREVIATIONS

BRAKE HORSEPOWER: B.H.P.
CP: PRISMATIC CO-EFFICIENT
FEET/FOOT: FT
HORSEPOWER: H.P.
INDICATED HORSEPOWER: I.H.P.
JUNIOR: JR
NOMINAL HORSEPOWER: N.H.P.
REVOLUTIONS PER MINUTE: RPM
SENIOR: SR
SIC: NOTES A SPELLING OR GRAMMATICAL ERROR

MAP OF ENGLAND, WALES AND SCOTLAND (CIRCA 1823), SHOWING PLACES OF INTEREST.
SOURCE: LIBRARY OF CONGRESS (ITEM 2018588029).

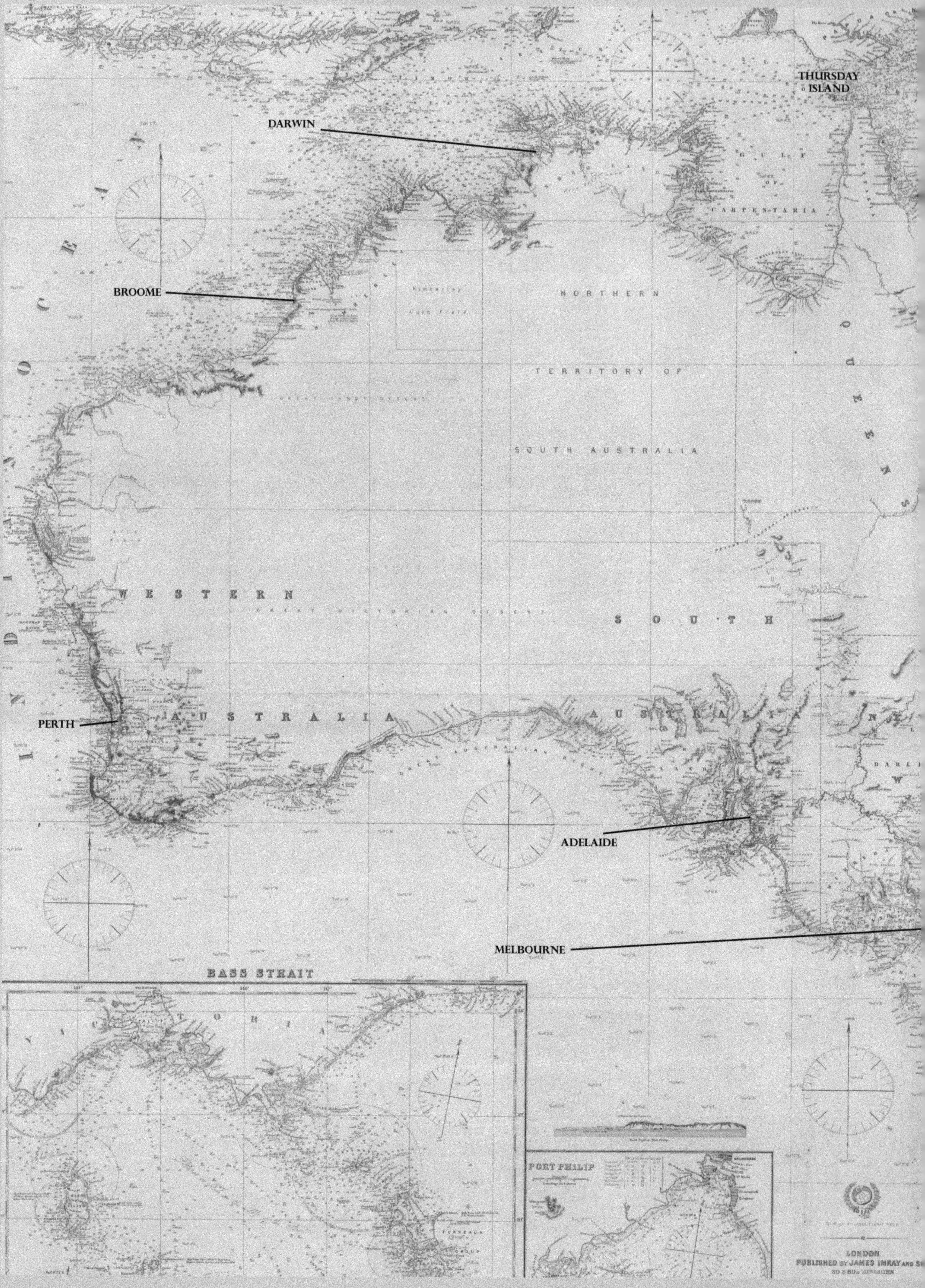
THURSDAY ISLAND
DARWIN
GULF OF CARPENTARIA
BROOME
NORTHERN
TERRITORY OF
SOUTH AUSTRALIA
WESTERN
AUSTRALIA
SOUTH
AUSTRALIA
PERTH
ADELAIDE
MELBOURNE
BASS STRAIT
PORT PHILIP
LONDON
PUBLISHED BY JAMES IMRAY AND S

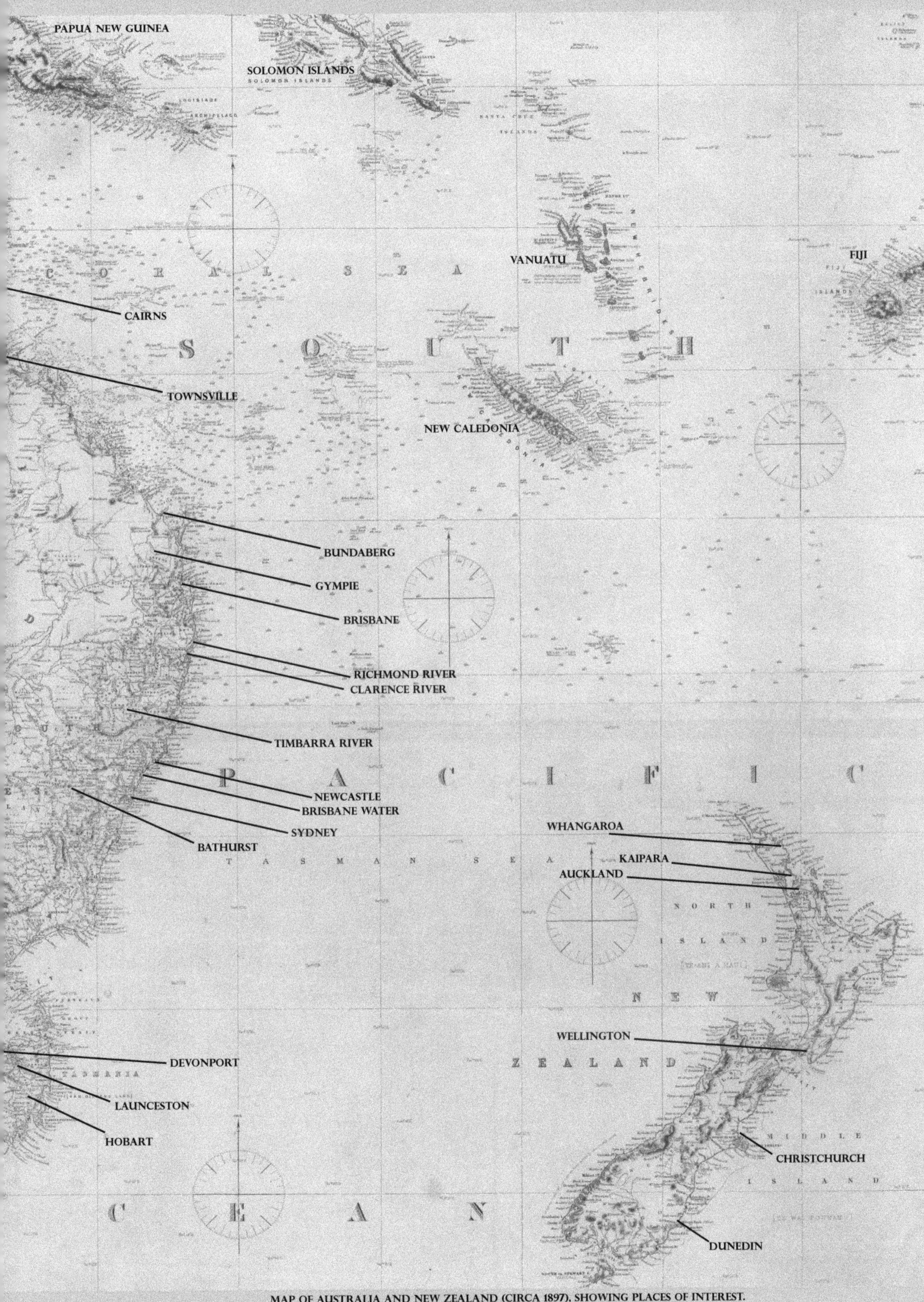

MAP OF AUSTRALIA AND NEW ZEALAND (CIRCA 1897), SHOWING PLACES OF INTEREST.
SOURCE: NATIONAL LIBRARY OF AUSTRALIA (NLA.OBJ-231517298).

WALTER REEKS (1906).
SOURCE: AUSTRALIAN NATIONAL MARITIME MUSEUM COLLECTION.

INTRODUCTION

Walter Reeks (1861 - 1925) lived a life of outstanding service to Australian naval architecture, yacht racing and administration, as well as to the communities in which he lived and worked, and to his family. In 1910 Sir William Bower Forwood (1840 - 1928), an English merchant, shipowner and politician, who is mentioned several times in this book, published a biography in which he wrote, *'As each life has its ambitions—small or great—its conquests, its trials, and its failures, so each day has to bear its own burden of trials and anxieties; and as the daily life is lived, and the daily task accomplished, so will our life's work be fulfilled; but how few there are who can look back and say their lives have been a success, and that they have accomplished all they should or all they might have done'.*[1] With charming clarity, as we reflect back on the life, relationships and individual and collective accomplishments of Reeks through a lens approaching 100 years after his death, this is profoundly the case.

William Bower Forwood further stated, '*The want of imagination prevents many men from making use of their opportunities'*. This was decidedly not the case for Reeks, naval architect, yachtsman and entrepreneur: a man who appears to have combined a far-reaching imagination with a tenacity to say '*Yes*' to new opportunities, '*Yes*' to trying new technologies and innovations, and '*Yes*' to building and forming new relationships. His success, in terms of business, social and community involvement, perhaps is only parlayed with those of his family tragedies. The loss of his first wife Mary at a young age and the murder of his son Kent in 1914 at the age of 24 reminding us that life is to one and all rather tenuous.

[1] Sir William B. Forwood (1910). *Recollections of a Busy Life: being the Reminiscences of a Liverpool Merchant*.

Born in Christchurch, Hampshire, England, in 1861, Reeks' life story is the product of fortunate and not so fortunate opportunities and circumstances, and local, national and international events, that took place in somewhat of a gilded age in terms of technological advancements. All of these factors combined favourably for Reeks, allowing his penchant for industriousness, inquiry, steady application and risk-taking to propel a life of outstanding involvement, output and achievement.

As with many people, we can look back on Reeks' formative years and see people, situations and experiences that became a foundation or that provided a strong influence on how his life would progress. His working-class father set a good entrepreneurial example, taking opportunities as they became available to move upwards beyond his class. From trading as a corn dealer, Charles Reeks created or took advantage of other business opportunities. He also saw the value of education and community involvement, supporting learning institutions and serving on various public committees and associations. Both of these themes feature strongly in his son's life.

The practical side of engineering and its strident advances in technology were also visible and on offer to Reeks from an early age, being demonstrated on his family's farms, in his town, in local industries and in improvements in transport. Another obvious clue to the quality of design, engineering and drafting abilities Reeks later developed comes from his attendance as a teenager at a college in nearby Southampton that was at the time a leading source of technical education.

Reeks began his career in shipbuilding and naval architecture in the late 1870s, during an era manifested with continuous development. He made good use of the opportunities in design and business that this presented. It was a pivotal time to join the field, professional vessel design and engineering was increasingly becoming an affordable necessity for many projects where once the actual builder had served as the designer. Technology was moving ahead too, driven by commercial competition and social pressures, bringing with it increased complexity, along with early regulations that were propagating an increased awareness of safety considerations. The need for full-time professional practitioners in vessel design became firmly established. This was echoed in the pastime of yachting as well: ever-expanding opportunities for organised competition at the local, regional, national and international levels driving the need for faster yet safer craft.

When Reeks arrived in Sydney in 1885 at the age of 24, only a relatively small number of engineers were operating in the colonies and few had any background or experience in vessel design. The healthy number of ship and boat building yards carrying out work on the shores of Port Jackson, the regional waterways of New South Wales, and further afield in Tasmania and other areas of Australia and New Zealand, were almost all operating on a steady, step-by-step evolution of their vessels: from one job to the next, backed by the shipwrights' experience and a 'rule of thumb' for calculation. The engineering profession, after it was initially called upon, became involved with the machinery as steam took over from sail. However, it was an industry that was no longer slowly advancing. Instead, boat and ship building were steadily growing and developing in new ways thanks in part to steam power, then propellers and then oil engines, as well as the introduction and use of iron and steel in construction. Australia's waterways were the highway for trade and movement, and a fast-growing field for recreation.

Reeks arrived in Australia from Liverpool, England, with a background of knowledge and practical work, an impressive resume, despite his young age, and an eagerness to prove himself. These factors combined to give him a head start in a new city and colony. He was fortunate to have grown up living close to and then being trained in a strong region of yachting in the United Kingdom. Arriving in Sydney, Reeks initially applied his skills to this pastime and quickly demonstrated success, but importantly he was making contacts and connections at many social levels and learning how to navigate as a respected member of this new society and as a professional. Reeks maintained this respect throughout his life, developing his network and becoming a national and international figure in his field. His client list shows he made time to work with projects big and small, for the well-off to those with lesser means, from the mainstream to intriguingly unique projects. He appears to have said '*Yes*' almost always, and most often without hesitation, his eagerness and energy revelling in disparate projects from a potential Australian challenge to the America's Cup to mining for gold in the outback. Reeks' involvement in society beyond work encompassed a similar range of interests and causes, and he was proud to be an Australian.

Sadly, it seems his family life probably fell short of these achievements. The tragic events of losing his first wife quite young and then his only son, who was following in his maritime engineering footsteps, were obviously significant setbacks. In contrast his only daughter Thelma chartered her own path, settling in western New South Wales and having a more stable home life that perhaps eluded her father. Thankfully, however, she passed on to her family the story of her father's deeds.

We cannot know Reeks' ambitions, big or small, but his eventual '*conquests, trials and failures*', including the design of more than 300 commercial and recreational vessels, many of them pioneering, and more than 40 years of community service and volunteer work, mapped and delivered a life that had a positive impact on many people, events, organisations and happenings throughout Australia and New Zealand. It was a life full of imagination and accomplishments, and it is one worth recounting to all.

WALTER REEKS (1861 - 1925).
SOURCE: THE SALON, VOL. 2, NO. 4 (1 NOVEMBER 1913).

1870 ORDNANCE MAP OF CHRISTCHURCH (HAMPSHIRE, ENGLAND) AND SURROUNDING AREAS.
SOURCE: NATIONAL LIBRARY OF SCOTLAND (HTTPS://MAPS.NLS.UK/VIEW/102343182).

CHAPTER ONE - CHRISTCHURCH

Walter Reeks was born on a cold February day at his family's residence in Castle Street, Christchurch, in the county of Hampshire, England. Exactly one month later, on 25 March 1861, he was christened at the local priory, a building of great significance to his paternal and maternal families having served as the conduit of life, marriage and death for many generations.

Reeks was the seventh child and third son born to Charles and Elizabeth Reeks (nee Kent). Twins, a son and daughter, followed his birth just over two years later. In an era indicative of high infant mortality, of the nine children born to Charles and Elizabeth two would not survive infancy. Another death would occur when the couple's eldest child John died in 1860 at the age of nine from an undisclosed illness.

Surviving childhood was often a tough ask during the reign of Queen Victoria. Scarlet fever, tuberculosis, measles, influenza and other contagious diseases periodically impacted families regardless of status or location.[1] Death from contaminated drinking water was also a possibility and a known common occurrence in the Christchurch area until at least 1902 when the town's sewerage system was established. Up until this point in time, cholera and other waterborne diseases were frequently prevalent. The adults seem to have overcome this problem through the consumption of beer, the most popular drink of the period. In 1855 Christchurch alone hosted six breweries, 16 inns and 26 alehouses.[2] Not bad for a local population of less than 4,000 inhabitants. In the decades that followed, tea consumption would grow rapidly across England, in part to alleviate concerns over ill-sourced drinking water, as well as a rising temperance movement.

Having been born a year after his oldest brother's death, Walter Reeks would not have known the family's intended heir. Instead, it was his second oldest brother William who would be drawn into and eventually take over the family business, leaving Reeks to determine his own fate. Their father Charles was a corn factor and auctioneer of good and honourable standing in the local Christchurch community. He also served as an insurance and valuation agent, including for the Royal Farmers' & General Insurance Company.[3] In addition to these professional activities, for many decades Charles was honorary secretary of the Avon and Stour Farmers' Club, as well as a member of the committee of the South Avon and Stour Agricultural Society and the Borough of Christchurch Conservative Association.[4] Charles was also secretary of the Tradesmen's Benefit Society, a waywarden and member of the Vestry Highway Board, a participant in activities organised by the Working Man's Institute, and a donor and supporter of the local poorhouse.[5]

This notion to contribute to the local community and be active within its societies and organisations would transfer well to the next generation, including to both of Charles' surviving sons William and Walter. In terms of professional provenance, the Reeks family was somewhat working class. Charles' father and Walter's paternal grandfather John Reeks was a butcher of Christchurch. Walter's maternal grandfather Moses

[1] *Christchurch Times*, 20 October 1860.
[2] en.wikipedia.org/wiki/History_of_Christchurch,_Dorset.
[3] *Hampshire Telegraph*, 20 May 1854, 23 May 1857.
[4] *Hampshire Telegraph*, 23 May 1857; *Christchurch Times*, 26 January 1867; *Hampshire Advertiser*, 5 May 1880.
[5] *Christchurch Times*, 28 December 1861, 22 March 1862, 2 March 1867, 1 April 1876; *Hampshire Advertiser*, 31 March 1877.

Kent was a grocer of nearby Ringwood. There is no indication of maritime pastimes nor related occupations taking preference in either of these familial lines. Instead trade and agricultural pursuits appear to be the norm, combined with a steadfast resolve to learn and be educated, as well as champion improvements and growth at the local level.

The corn factoring business served the growing Reeks family well. Charles traded in the grain wholesale and acted as an agent for its sale. He also established a seed store in Castle Street, Christchurch, selling grains, crops, manures and fertilisers to the local farming community.[6] Another income source was the auctioning of sheep and cattle, including at a fat cattle market established close by at Ringwood.[7]

The railway line first opened in Christchurch in 1847. Charles took full advantage of this new mode of transport, selling large quantities of manures and fertilisers at the local station direct to consumers, the product having been transferred from manufacturers in London and Bristol.[8] During this time Charles also began entering stock he had reared at annual shows and fairs. In 1858 he won second prize for the best sow.[9] He would continue to enter local agricultural events and win prizes in the succeeding decades.

[6] *Christchurch Times*, 10 April 1858.
[7] *Christchurch Times*, 10 April 1858.
[8] *Christchurch Times*, 29 May 1858.
[9] *Hampshire Independent*, 27 November 1858.

From 1859 Charles also became a real estate agent, a rather 'new' profession for the time, selling and leasing large farming and residential properties and their contents at auction. His business enterprises were not without risk, however. In December 1859 one of his employees was killed after falling from a wagon while loading hay.[10] Stock and crops were also lost due to weather events, pests and diseases. Still, it was a time of innovation and modernisation. Steam-powered farming equipment was starting to become available, and crop rotation and stock selection and breeding were resulting in increased yields. Animal husbandry practices were also improving. For Charles and his fellow farmers in the Christchurch area, turnips, kohl rabi and swedes were grown in abundance, particularly for use as a food source for stock. Potatoes and corn were also staples. In terms of animals, cattle and sheep were the most common stock, as well as pigs for the procurement of bacon and hams.

The 1861 Census notes the Reeks family as living at Castle Street in Christchurch. The family unit comprised Charles (aged 42), his wife Elizabeth (aged 38) and their surviving children: William (aged 8), Ellen (aged 7), Marian (aged 5), Annie (aged 3) and Walter (less than a year old).[11] A 17-year-old house servant named Ellen Jack was also part of the household.

A decade later, the 1871 Census notes the family as still living at Castle Street, Christchurch. Present were Charles,

[10] *Hampshire Advertiser*, 31 December 1859.
[11] 1861 England Census (Class: RG 9; Piece: 666; Folio: 66; Page: 34).

CASTLE STREET, CHRISTCHURCH, HAMPSHIRE, ENGLAND (CIRCA 1855).
SOURCE: CHRISTCHURCH HISTORY SOCIETY (A HISTORY OF CHRISTCHURCH DORSET IN BRIEF).

Elizabeth and their six children: William, Ellen, Marian, Annie, Walter and Louisa.[12] In addition to a house servant (Fanny Sandy), a governess had also joined the fold (Janet Ings), indicating Charles' commitment to the education of all of his children.

Foretelling some upward mobility, by the late 1860s Charles was also farming his own parcel of land on the outskirts of Christchurch.[13] Known as Wick Farm, the property provided a source of local employment to agricultural labourers and their families for several decades.[14] It was later taken over by Reeks' older brother William who continued to farm the property into the next century.[15]

Historically, the Reeks family had long been residents of Christchurch (originally referred to as Twynham), a town known to exist since at least 650 and officially recognised by name from 1094 owing to the construction of the large Augustinian priory church that stands today as one of the biggest parish churches in England. Situated at the convergence of the rivers Avon and Stour, and just over 30 miles from Southampton and 110 miles from London, the small town offered a seemingly idyllic place to grow up, providing a mix of rural and residential flavours. A small harbour entered the sea less than two miles from the town centre.

Despite its advantageous setting, the town's population was somewhat stagnant for most of the nineteenth century. It was not until Christchurch joined the main railway line in 1883, and a new station was built, that the town became popular, particularly as a holiday and retirement destination. By the 1890s the population had increased to 15,000 people, a doubling in just over a decade.

Today more than 45,000 people call Christchurch home with tourism the mainstay of the local economy, far-exceeding trade, agriculture and manufacturing. Continued improvements in trains, motor coaches and automobiles, as well as a general increase in the standard of living resulted in the area becoming a well-established holiday destination leading up to World War I. The town has retained this notoriety and, as part of the county of Dorset since 1974, is now a haven for retirees, with Britain's largest percentage of people over 65 years of age calling the region home.

With the scent of Christchurch becoming an oasis for retirees far off in the future, the location offered a more working class beginning for Reeks. However, by his teens the family appears to have progressed to more middle-class ranks. Education was regarded highly not only by the Reeks family but by members of the local community more generally. There were several free and private grammar schools in the town, providing ample opportunity for learning and development. Charles Reeks appears to have placed a priority on education, particularly for his

[12] 1871 England Census (Class: RG10; Piece: 1176; Folio: 126; Page: 47).
[13] *Hampshire Advertiser,* 14 December 1867.
[14] *Christchurch Times,* 19 October 1872, 27 October 1883.
[15] *Bournemouth Guardian,* 26 September 1903.

CHRISTCHURCH PRIORY (CIRCA 2010).
SOURCE: ANDREW MATTEWSON VIA DORSETS.CO.UK/ATTRACTIONS/CHRISTCHURCH-PRIORY-CHURCH.HTM.

two sons. A newspaper article published during Reeks' teenage years notes Charles presiding over a spelling bee held in the Christchurch school room.[16] Reeks obviously enjoyed learning too and performed well enough in school to earn a position at the prestigious Hartley Institution in Southampton. Founded in 1862 by the City of Southampton using proceeds from Henry Robinson Hartley's will, the higher-learning institution evolved over the coming decades into Hartley College (1896), the Hartley University College (1902), the University College of Southampton (1914) and, finally, the University of Southampton (1952). Between 1863 and the end of the nineteenth century, instruction was offered in a range of scientific, technical, engineering, art and literary subjects. The Hartley Institution also housed a museum, library and a venue for public lectures. As a leading source of technical education for scholars in Southampton and surrounding regions, examination results were routinely published in local newspapers. Of significance, in May of 1877, Reeks, by now 16 years of age, received a prize for his model drawing as part of the school's science examinations.[17]

HARTLEY UNIVERSITY COLLEGE, SOUTHAMPTON (CIRCA 1900S).
SOURCE: WWW.FACEBOOK.COM/HISTORICALSOTON/PHOTOS.

[16] *Christchurch Times*, 18 March 1876.
[17] *Hampshire Advertiser*, 25 July 1877.

In addition to trade, community service and education, kin also appear to have been at the forefront of the Reeks family's activities with aunts, uncles and a plethora of cousins living within walking distance to their home, either in town or in the surrounding countryside. There was also socialising and perhaps some frivolity too. In February of 1876 William Reeks and three of his sisters (likely Ellen, Marian and Annie) were presented at a Masonic Ball in nearby Bournemouth; Walter and Louisa were both too young to attend. The event was described in the press as the most spirited ball of the season, with 25 dances on the programme and over 200 persons present.[18] That same year William and his sisters performed at a concert for the local fire brigade, singing songs and playing the pianoforte.[19] It was one of many concerts that the Reeks family would participate in over the years.

In February of 1879 a performance was held at the Christchurch Town Hall in aid of the Christchurch Working Man's Institute Sports Club. The programme included two solo songs performed by William and a recitation on electricity given by Walter.[20] In January of 1880 the two brothers attended a ball held at the Bournemouth Town Hall in honour of the Hampshire Yeomanry Cavalry.[21] The local newspaper reported that dancing commenced at 10 pm and continued with great energy until 5 am the next morning, a rather substantial event considering it was held on a Wednesday night.

Another outlet for recreational hours was sport and it featured prominently in the Reeks family's pursuits. From 1870 William was involved in the Christchurch Rowing Club.[22] Reeks also appears to have been a good athlete. In September of 1876 he finished third in a quarter-mile flat race for boys under 16 years as part of athletic events associated with the Shirley, Millbrook and Fremantle Horticultural Society.[23] In late 1878, as a member of the Bournemouth Bicycle Club, he won a one-mile flat race at the organisation's annual sports day.[24] Cycling, however, proved problematic for him on at least one occasion. In February of 1880 Reeks was summoned to the Lymington Borough Bench for riding after sunset without a light. A fine of 6d and 4s 6d in costs was inflicted in lieu of seven days' hard labour.[25] The Lymington location is interesting as it confirms that Reeks had moved to the area for work by this time.

The first inclination of the Reeks family's participation in yachting as a pastime comes in 1877. William, sailing *The Freak,* competed in a race as part of a regatta associated with the Christchurch and Mudeford Rowing Club. With prize monies exceeding £1 on offer, unfortunately the craft failed to finish the race owing to an accident, an altogether disappointing outcome.[26]

[18] *Hampshire Advertiser*, 5 February 1876.
[19] *Christchurch Times*, 1 April 1876.
[20] *Christchurch Times*, 1 March 1879.
[21] *Hampshire Advertiser*, 21 August 1880.
[22] *Hampshire Advertiser*, 7 May 1870.
[23] *Hampshire Advertiser*, 2 September 1876.
[24] *Christchurch Times*, 15 June 1878.
[25] *Hampshire Advertiser*, 25 February 1880.
[26] *Hampshire Advertiser*, 8 September 1877.

The following year Reeks competed in a race for open boats under 20 feet (ft) in length at the Christchurch and Mudeford Rowing Club Regatta. It was an event that piqued brother against brother, with William competing in *The Freak* and Reeks sailing in the *Oder.* The prizes were significant: £2 for first, £1 5s for second, and £1 for third. Disappointingly, both boats finished well back in the fleet.[27]

The *Oder* returned to competitive sailing two years later, competing in several summer evening sailing matches between Christchurch's pleasure boats. '*Mr. W. Reeks*' was noted at the helm.[28] This descriptive likely refers to William. The races took place between the Quay and a boat stationed at Mudeford. The first event, held in August of 1880 and featuring seven starters, was won by the *Oder*.[29] Proving a considerable success, the vessel won the same event the following week.[30]

By now Reeks was entering his early 20s, formative years in terms of developing a career and establishing and supporting himself. The 1881 Census confirms that he had moved out of home and was living as a boarder with the Hodder family at 12 St Thomas Street, Lymington. His occupation was recorded as '*Articled pupil to shipwright*'.[31]

[27] *Christchurch Times*, 10 August 1878.
[28] *Hampshire Advertiser*, 11 August 1880.
[29] *Hampshire Advertiser*, 11 August 1880.
[30] *Christchurch Times*, 14 August 1880.
[31] 1881 England Census (Class: RG11; Piece: 1189; Folio: 89; Page: 3).

In an interview published over 30 years later Reeks stated that upon leaving school he first gained employment with Christopher Crabb Creeke, a prominent Bournemouth-based architect and surveyor. This was likely in 1878 when he was known to compete with the Bournemouth Bicycle Club.[32] Though the position would have been considered quite distinguished for a recent graduate, particularly in terms of networking and future prospects, Reeks took little time to realise that being involved with the design and construction of buildings was not his passion. Undeterred by little practical experience within the maritime realm, he turned to the water and was fortunate to gain an apprenticeship with George Inman & Son, shipbuilders of Lymington.[33]

Located on The Solent facing the Isle of Wight, and just 12 miles east of Christchurch, Lymington offered an ideal environment for Reeks to learn and become acclimatised to the shipbuilding industry and the increasingly popular pastime of yachting.

[32] *The Salon*, Vol. 2, No. 4 (1 November 1913).
[33] *The Salon*, Vol. 2, No. 4 (1 November 1913).

THOMAS STREET, LYMINGTON (CIRCA 1920S).
SOURCE: CINEMA TREASURES (WWW.CINEMATREASURES.ORG/THEATERS/46927/PHOTOS/285731).

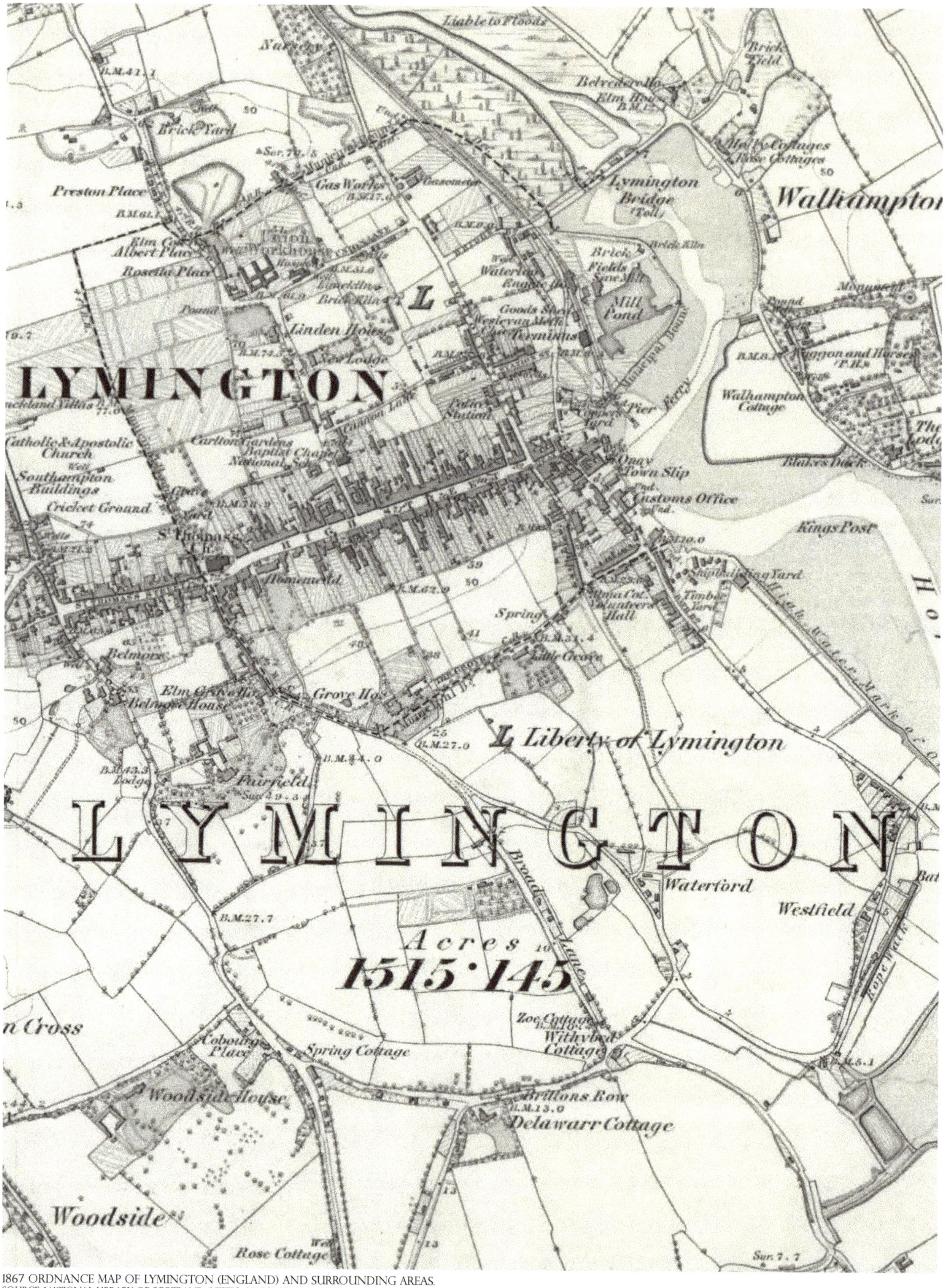

1867 ORDNANCE MAP OF LYMINGTON (ENGLAND) AND SURROUNDING AREAS.
SOURCE: NATIONAL LIBRARY OF SCOTLAND (HTTPS://MAPS.NLS.UK/VIEW/102343188).

CHAPTER TWO - LYMINGTON

With a population approaching 4,500, Lymington was considered somewhat of a modest market town. For Walter Reeks it was significantly smaller than his home town of Christchurch. However, dating back to Roman times, the area had supported a shipbuilding industry that by the mid-nineteenth century was thriving, helped in part by its close proximity to England's major yacht clubs, including the Royal Yacht Squadron based at Cowes on the Isle of Wight, as well as the success of Lymington-built yachts in prestigious sailing events.

In the late 1870s and into the early 1880s Reeks was articled to George Inman whose shipbuilding yard was situated along the banks of the Lymington River and is the present site of the Berthon Boat Company in Bath Road.[34] It was a noteworthy position. He was joining a family-based operation that over a period of six decades had earned worldwide acclaim for quality craftsmanship and innovative designs. Unfortunately the continuation of the Inman legacy would change abruptly towards the end of Reeks' tenure. The firm would cease operation following the sudden death of its owner George Inman in October of 1883 and the death of George's only surviving son Edwin less than two years later.[35] Still, the history of the Inman shipbuilding business is beyond impressive and, looking back, closely parallels with the trajectory of yacht design and development in Great Britain.

The tradition and high-profile reputation of the Inman family was very likely what drew Reeks to pursue employment with the firm. At the time of his appointment, it was managed by George Inman who had been born in the coastal town of Hastings, Sussex, in 1817 where his father Thomas was a local boat builder.[36] In 1819 Thomas moved his growing family along the coast to Lymington, apparently making the 100-mile trip in a boat he built.[37] Starting at a small scale, Thomas went on to establish himself as one of Lymington's premier shipbuilders, fast gaining national attention for building trading vessels and government revenue cutters, before transitioning to large-scale gentleman's yachts where his reputation was catapulted to the international level. It was a fortuitous move for the Inman family.

Thomas Inman's boat building enterprise was helped profoundly by the fact that it was established in a period of relative peace and prosperity. Britain was no longer at war with France nor with America. The post-Napoleonic depression had also subsided, supported in part by good agricultural harvests. Instead of being spent on defence and associated international efforts, industrial and financial resources were directed towards domestic improvements such as the development of railways, ports, bridges and other modes of transport, as well as long-distance communication systems. Improvements to manufacturing and farming practices were also significant during this period. For those with more robust personal finances and some free time, there were increasing opportunities for seaside vacations.

The Hampshire coast in particular, as well as the nearby Isle of Wight, soon found favour as summer resort locations. With groups of people gathering for recreation, they increasingly looked for activities and amusements to idly pass spare hours. Given largely unimpeded

[34] www.berthon.co.uk/about-berthon/berthon-history/#tab2Tab.

[35] *Shipping and Mercantile Gazette*, 26 October 1883; *Hampshire Advertiser*, 14 March 1885.

[36] 1881 England Census (Class: RG11; Piece: 1189; Folio: 54; Page: 3).

[37] J. Chitty (1983). *The River is Within Us: Maritime History of Lymington*.

access to sheltered waterways and The Solent, as well as predominantly good weather, aquatic pursuits began to prevail with boating and yachting growing in popularity. What initially began as private challenge races between owners of a few yachts soon turned into organised races between craft of similar types with not only substantial prizes on offer but, perhaps more importantly, bragging rights. The establishment of the Royal Yacht Squadron in 1815 by a kindred group with a shared interest further spurred significant development of the pastime.

Within a short period of time, annually spending the month of August holidaying in Cowes became a mainstay of 'The Ton'. Post-sailing entertainment, such as dinners, balls and parties, further enhanced the location's social appeal. The presence of royal family members within the milieu, as well as their interest in yachting, very quickly resulted in Cowes being considered the only fashionable place to be seen in late summer.

While his entrée into yachting was still a few years off, one of the first vessels Thomas Inman launched from his newly-established Lymington-based shipyard was the 80-ton cutter *Lion,* built for the Customs Service in 1822 for use in thwarting smugglers, predominantly those coming to England's shores from France looking to off load contraband spirits, tobacco, spices, lace, silk and tea.[38] Given increases in taxation on foreign goods that occurred following the end of the Napoleonic Wars, the smuggling industry was flourishing.

Four years later Inman completed the schooner *Sincerity*, built to the order of Robert Badcock for the London trade.[39] This and similar commercial vessels, along with luggers and fishing boats, were indicative of the type of craft turned out by shipbuilders of the southern English ports at the time.

A big break for Inman came in 1826 with an order for a pleasure boat. The *Ann*, a 42-ton cutter yacht built to the order of the Earl of Listowell, was launched in June of that year.[40] A year later the vessel performed extremely well at the inaugural Cowes Regatta, taking out the coveted Gold Cup valued at £100.[41] The Inman name was beginning to lay a foundation into the realm of yacht building.

Though only six miles apart, Lymington and Cowes were considered serious competitors within shipbuilding circles with the two locations historically producing recreational vessels quite distinct from one another. Through the Inman family, the Lymington-built yachts soon began to assert their dominance in regattas and other increasingly-popular sailing events. This success only vexed the Cowes-based yacht builders and owners even more with a growing rivalry leading to further growth and development in the pastime.

Following the success of the *Ann,* Inman soon found a patron in Joseph Weld, Esq., of the nearby Pylewell Park Estate, as well as the further afield Lulworth Castle.[42] A member of a very eminent Catholic family, Weld had stumbled into wealth and property after his elder brother joined the priesthood. With a penchant for yachting that was well supported by his newly-claimed disposable riches, Weld had previously superintended the design and build of his own racing yachts at a slip located near Lymington. These included the 60-ton cutter *Charlotte,* launched in 1815, and the very successful 85-ton cutter *Arrow,* launched in 1822.[43] Weld was also one of the founders of the Royal Yacht Squadron and one of the major advocates of the cutter rig. Still today his name remains synonymous with the history of modern-day yachting in Great Britain.[44] As Weld prophesied, for over 100 years the cutter would enjoy prominence amongst leading yachtsmen. The basic configuration of mainsail, topsail and two headsails produced a set of individual sail sizes that were easier for crew to handle. For racing it gave the best performance to windward until the sloop became almost universal with modern racing yachts during the twentieth century. The cutter rig did however maintain its popularity with cruising yachtsmen and is still common in contemporary designs.

In April of 1828 Inman launched the *Lulworth*, a 127-ton cutter racing yacht built for Weld.[45] The vessel was designed along the same lines as the *Arrow* and went on to enjoy similar success.[46] Also in 1828 Inman built a 110-ton cutter, the *Owen Glendower,* to the order of F. R. West, M.P. for Denbigh, based on a plan by Captain John Symonds.[47] The vessel raced competitively at Cowes against the likes of *Lulworth*, among others.[48]

In May of 1830 another fast-sailing racing yacht built by Inman to the order of Weld was launched, the 195-ton *Alarm*.[49] Completed at a cost of £20,000 (equivalent to nearly £4 million today), the craft exceeded all other cutters worldwide in terms of tonnage and was considered the epitome of racing yacht design and development at that time.[50] Living up to expectations, the *Alarm* decimated its competitors over a period of many decades.

Following the success of the *Alarm*, Inman continued to expand his business operations and offerings.[51] Vessels he built also won major races, amassing large monetary prizes.[52] Additionally, government and commercial vessels remained a significant output of his yard. In 1837 the revenue cruiser *Camelion* was completed, used to capture French vessels involved in smuggling activities along the British coast.[53] Soon followed two more revenue cutters: the 155-ton *Prince Albert* and the 150-ton *Royal George*.[54] The former was considered far superior to similar vessels in the government's fleet of revenue cruisers.[55]

[38] *Public Ledger and Daily Advertiser*, 18 November 1822.
[39] *Hampshire Advertiser*, 9 October 1826.
[40] *Hampshire Advertiser*, 5 June 1826.
[41] *Hampshire Chronicle*, 6 August 1827; *Morning Chronicle*, 20 August 1827.
[42] *Dorset County Chronicle*, 7 February 1828.
[43] *Salisbury and Winchester Journal*, 14 August 1815; 10 June 1822.
[44] www.maritimeviews.co.uk/british-yachts-yachtsmen/chapter-4-organisation-1815-1851/.
[45] *Hampshire Advertiser*, 3 May 1828.
[46] *Hampshire Advertiser*, 13 October 1827.
[47] *Hampshire Advertiser*, 13 October 1827; *Hampshire Chronicle*, 5 May 1828.
[48] *Hampshire Advertiser*, 9 August 1828.
[49] *Hampshire Advertiser*, 15 May 1830.
[50] rlymyc-history.org.uk/chronicles/times-gone-by-archivists-notes/.
[51] *Hampshire Advertiser*, 15 May 1830; *Morning Post*, 12 July 1831.
[52] *Salisbury and Winchester Journal*, 12 August 1833.
[53] *Morning Post*, 30 January 1837.
[54] *Sherbourne Mercury*, 1 June 1840.
[55] *Hampshire Advertiser*, 25 September 1841.

THE 195-TON YACHT ALARM (1830).
SOURCE: NATIONAL MARITIME MUSEUM, GREENWICH, LONDON (BHC4182).

More substantial yachts were built in the late 1830s and early 1840s. These included the 101-ton *Active,* the 53-ton *Rose,* and the 63-ton cutter yacht *Stormfinch.*[56] Next came a 126-ton schooner the Inman yard built on speculation that was subsequently purchased by John Greaves who intended to take a multi-year cruise to New South Wales and the East Indies.[57] Named *Osprey*, the vessel was launched in March of 1841.[58]

With Inman's customer base expanding amongst the English gentry and associated social class, in 1842 he built the 143-ton schooner *Resolution* for the Duke of Rutland.[59] Other vessels soon followed, including the cutter *Tar* for the Reverend D. George.[60]

Around this time an interesting article was published in the British press regarding substantial alterations required to make the Queen's yacht, the *Victoria and Albert,* more safe and serviceable. The opinion put forth was that the Royal Navy's architects, deemed '*crotchety and costly*', would be put to shame compared to the likes of '*Inman of Lymington*', among others.[61] It was another feather in Thomas Inman's cap.

Continuing to add to his repertoire, Inman launched several more schooners and cutters, including the 95-ton schooner yacht *Kate.*[62] Late in 1845 Inman's two surviving sons, George and James, took over the shipbuilding yard.[63] The duo's first commission appears to have been the schooner *Magic* built to the order of Lord Chetwydne.[64] A few weeks later the 67-ton schooner yacht *Fairy Queen* was launched.[65]

George Inman was the fourth of nine children, and second of three sons, born to Thomas and his wife Eliza. The couple's eldest son, Thomas (Jr), had died in 1837 at the age of 24.[66] As the next in line George and, to a much lesser extent his youngest brother James (1821 - 1904),

[56] *Shipping and Mercantile Gazette*, 8 March 1839; *Hampshire Advertiser*, 23 May 1840.
[57] *Shipping and Mercantile Gazette*, 1 January 1841.
[58] *Salisbury and Winchester Journal*, 5 April 1841.
[59] *Hampshire Independent*, 12 March 1842.
[60] *Shipping and Mercantile Gazette*, 4 May 1844.
[61] *Oxford Chronicle and Reading Gazette*, 2 November 1844.
[62] *Hampshire Independent*, 22 January, 12 March 1842; *Sun (London)*, 23 May 1842; *Shipping and Mercantile Gazette*, 4 May 1844; *The Era*, 20 April 1845; *Morning Post*, 15 September 1847.
[63] *Shipping and Mercantile Gazette*, 27 September 1845.
[64] *Shipping Mercantile Gazette*, 27 September 1845; *Hampshire Advertiser*, 9 May 1846.
[65] *Hampshire Chronicle*, 16 May 1846.
[66] *Hampshire Independent*, 15 July 1837.

were tasked with managing the family's shipbuilding business upon their father's retirement.

With George securely at the helm, the Inman shipyard grew significantly over the next three decades, both in terms of prominence and vessel output. This crescendo began formulating in the late 1840s with the launch of a suite of yachts of distinction. These include the 79-ton schooner yacht *Cayman,* the 20-ton yacht *Lizzy*, the 50-ton cutter yacht *Water Kelpie,* and the 44-ton schooner yacht *Aurora*.[67] Also launched during this period were two more government revenue cutters, *Sylvia* and *Defence*.[68]

In 1850 George completed the 120-ton schooner yacht *Claymore*, built to the order of Archibald Campbell, as well as the 60-ton *Magician* for Joseph Parker.[69] That same year the 78-ton schooner yacht *Ruby Queen* was launched, the second vessel the firm had built for T. L. Bridge.[70]

Business was more than brisk for the Inman yard in the early 1850s. In fact, never before had so many commissions been received. Another fateful catalyst that was to have historically-significant consequences not only for George Inman and his father Thomas but the sport of yachting more generally came in August of 1851 when Joseph Weld's *Alarm* went on to famously race against the American yacht *America* (of 164 tons) and others, including the *Arrow,* in a fleet race off the Isle of Wight. The event was staged at the invitation of the Earl of Winton of the Royal Yacht Squadron as part of the Great Exhibition of 1851. Now considered the forerunner to the America's Cup, the race at the time was dubbed the '*Championship of the World*' in some references, while the trophy was initially known as the '*100 Guinea's Cup*'.[71] However, once in the hands of the New York Yacht Club it soon took on its common name, the America's Cup, and as their success with its defence grew, the trophy developed the overtone of belonging exclusively to this mighty nation with an aura matching the challenge of extracting Excalibur's sword from the stone.

Given its previous wins, the *Alarm* was put forth as invincible in the press leading up to the event. Disappointingly, at least for the British, the race was handsomely won by the American yacht with both the *Alarm* and *Arrow* failing to finish.[72] Still the public interest and speculation surrounding the event only further propelled the Inman name to the national and international shipbuilding platform, significantly for the firm resulting in more new orders, many of which were from high-profile customers.

Immediately following the race for the '*Championship of the World*', Weld had the *Alarm* towed to Lymington. The Inman shipyard was then entrusted with lengthening it by 21ft aloft and 15ft by the keel.[73] The now 248-ton vessel was re-rigged as a schooner on the same plan as *America*. In hindsight the race result and the game of one-upmanship that followed between the two countries benefitted the sport of international as well as domestic yacht racing immensely, including propagating significant gains in yacht design from this point onwards. A sequence of gradual then more significant developments took place throughout the history of the event, and the pace of change sped up remarkably in the twenty-first century iterations of the America's Cup.

It was envisaged in the original Deed of Gift penned in 1852 and forwarded to the New York Yacht Club in 1857, as the holder of the America's Cup, that the event would become a demonstration of the best yachts of the two participating countries, i.e., those challenging for and defending the trophy, in terms of design and construction. The rakish lines of *America* had thrown a cat amongst the English pigeons in 1851 and this can be seen as the starting point. However, back then the contemporary sight of monohulls, 75ft in length, flying across the water on foils in New Zealand in 2021 would have been far beyond anyone's imagination.

The next development in the history of the America's Cup came in 1870 when British railway tycoon James Lloyd Ashbury challenged for the trophy. Competing in United States waters in his topsail schooner *Cambria*, of 188 tons, the solitary race followed the original 1851 format with the challenging yacht sailing against a fleet of 17 yachts from the defending yacht club. Despite losing the event, *Cambria* placed eighth, Ashbury issued another challenge to the New York Yacht Club in 1871, this time for a best of seven match series, which was duly accepted. Ashbury's new yacht *Livonia* (of 264 tons) was beaten at this event too by two different yachts: *Columbia* (races 1 and 2) and *Sappho* (races 4 and 5).

Another challenge for the America's Cup came several years later from the Royal Canadian Yacht Club and was the first to be undertaken solely between two yachts. The format of one-on-one match racing contests thus became established. From 1885 through to 1887 this developed as a series of challenges from the United Kingdom between two distinct styles of yacht, the English cutters and the American centreboard sloops. The latest examples of both were on show and as this was occurring Walter Reeks was taking the first steps toward his eventual career as a vessel and yacht designer. Little did he know that at the end of the 1880s he would also come very close to being involved in this auspicious international event and challenging for the America's Cup.

For the Inman shipyard, capitalising on the notoriety gained from its involvement in the 1851 '*100 Guinea's Cup*', the firm continued to steadily expand its operations, going on to launch a suite of vessels. These included the 63-ton schooner *Petrel*, the 150-ton schooner *Mavis*, a cutter named *Doris* of 43 tons, and a cutter of 50 tons named

[67] *Hampshire Telegraph*, 22 May 1847; *Hampshire Independent*, 17 February, 6 October 1849; *Liverpool Standard and General Commercial Advertiser*, 2 September 1851.
[68] *Dorset County Chronicle*, 13 January 1848.
[69] *Bell's Life in London and Sporting Chronicle*, 3 February 1850; *Shipping and Mercantile Gazette*, 1 May 1850.
[70] *Dorset County Chronicle*, 8 August 1850.
[71] en.wikipedia.org/wiki/1851_America%27s_Cup; *The Hampshire Independent*, 2 August 1851.
[72] en.wikipedia.org/wiki/1851_America%27s_Cup.
[73] *Portsmouth Times and Naval Gazette*, 22 May 1852.

Pandora.[74] 1852 additionally saw completion of the 74-ton schooner yacht *Vestal* and the 74-ton schooner yacht *Flying Cloud*.[75] As well, 1852 coincided with George Inman being elected mayor of Lymington.[76] He would go on to serve for many years as alderman and town councillor.[77]

Another small yacht launched from the Inman shipyard in 1852 and one with a remarkable history was the 20-ton cutter *Spray*, built to the order of a Mr Burrard of Lymington though quickly passed to W. E. Cox of the Royal Thames Yacht Club.[78] In 1854 the vessel became the property of J. A. Longmore of Edinburgh's Royal Eastern Yacht Club.[79] In his possession *Spray* twice circumnavigated Great Britain, paid two visits to the Orkney and Shetland Islands, made a voyage to Norway, and another to France.[80] In 1860 *Spray* was purchased by Duncan McPherson of Scotland. Its adventures then continued. With Captain Wyse at the helm, *Spray* sailed from Greenock in September of 1860 to Hobart, Tasmania, making the trip in 120 days with a crew of six.[81] As the property of Duncan McPherson (Jr), commodore of the Derwent Yacht Club, *Spray* then proved a handy addition to Hobart's racing fleet, also undertaking several coastal cruises. Advertised for sale in late 1862, it was eventually purchased by E. Luttrell and placed in the local river trade.[82] In 1864 *Spray* was transferred to Messrs Heath and Hendy of Geelong, Victoria, for use as a trading vessel amongst the Bass Strait Islands.[83] The following year it was purchased by Kong Meng, a well-known Melbourne merchant.[84] Another change of ownership took place in 1867 with the vessel being deployed on a cruise to Fiji as part of a shelling expedition. Sadly, it was on a reef at Rarotonga, the largest of the Cook Islands, in September of that year that *Spray* was wrecked.[85]

Returning to our profile of the Inman shipbuilding yard, at least four vessels were launched in 1853 including the 180-ton schooner yacht *St. Ursula* intended to compete against the famed *America* which still remained in English waters.[86] Another schooner yacht, the 100-ton *Rosalind* commissioned by Lord Alfred Paget, went on to successfully race with the Royal Thames Yacht Club of which its owner was commodore.[87]

1854 saw the launch of the 95-ton schooner yacht *Eugenie,* as well as the schooner *Petrel* of 64 tons, and the 30-ton schooner *Mona*.[88] Next came the *Mariquita,* a 105-ton schooner yacht.[89] The following year the 66-ton *Alma* was completed.[90] Also launched was the 80-ton schooner yacht *Harlequin*.[91]

THE 100-TON YACHT ROSALIND (1853).
SOURCE: NATIONAL MARITIME MUSEUM, GREENWICH, LONDON (PAH8722).

The mid-1850s, however, was a rather slow period, particularly in terms of new racing yacht commissions. The Crimean War took precedence. Owners opted not to put existing yachts into commission, with the intention that crews might be rendered available for service with the Royal Navy. Many regattas and sailing matches were dutifully postponed. The Inman yard instead joined the cause, constructing gun boats for the government.[92]

It was not until 1856 that yachting began to recover from a rather stymied few seasons. With the resumption of racing fixtures, new craft also entered the waters with the hope of new victories. One such vessel, a racing yacht for Joseph Weld was completed in July of that year. The cutter *Lulworth* was designed by Weld and constructed under his strict superintendence.[93] At 80 tons it was much smaller than the *Alarm* though, like its predecessor, proved an instant success. In 1857 *Lulworth* won no less than five major prizes during the yachting season and continued to gain success in subsequent seasons.[94]

1857 saw more vessels launched from the Inman yard. These included the schooner *Zouave* and the small cutter yacht *Oriole*.[95] Soon followed the 160-ton schooner yacht *Zoe*.[96] Completed in 1858 was the 110-ton yawl *Ursuline*.[97] Next came the *Oak Apple*, a roomy 30-ton cutter, and the *Ondine*, a 55-ton schooner purchased by Mortimer Ricardo part-way through construction.[98]

The schooner yacht *Imogene* of 80 tons was launched in January of 1859.[99] A month later the *Egret*, an 80-ton schooner was completed, designed by George Inman with

74 *Salisbury and Winchester Journal*, 13 March 1852; *Bell's Life in London and Sporting Chronicle*, 14 March 1852; *Hampshire Advertiser*, 8 May 1851.
75 *Hampshire Independent*, 17 July 1852; *Hampshire Advertiser*, 14 August 1852.
76 *Hampshire Independent*, 13 November 1852.
77 *Hampshire Chronicle*, 27 October 1883.
78 *Bell's Life in London and Sporting Chronicle*, 13 January 1861.
79 *Bell's Life in London and Sporting Chronicle*, 13 January 1861.
80 *Edinburgh Evening Courant*, 26 March 1861.
81 *Bell's Life in London and Sporting Chronicle*, 13 January 1861; *The Mercury*, 23 January 1861.
82 *The Mercury*, 21 November 1862, 23 December 1863, 6 May 1864.
83 *Launceston Examiner*, 14 June 1864; *The Cornwall Chronicle*, 2 July 1864.
84 *The Mercury*, 9 December 1865.
85 *The Argus*, 12 April 1867; *The Herald*, 20 August 1867; *Wellington Independent*, 5 November 1867.
86 *Field*, 19, 22 January, 7 May 1853; *Salisbury and Winchester Journal*, 30 April 1853.
87 *Salisbury and Winchester Journal*, 5 March 1853.
88 *Shipping and Mercantile Gazette*, 5 April 1854.
89 *Hampshire Advertiser*, 3 June 1854.
90 *Hampshire Advertiser*, 7 April 1855.
91 *Hampshire Advertiser*, 7 April 1855.
92 *Salisbury and Winchester Journal*, 6 October 1855.
93 *Hampshire Advertiser*, 26 July 1856.
94 *Dorset County Chronicle*, 3 December 1857.
95 *Bell's Life in London and Sporting Chronicle*, 17 May 1857; *Field*, 25 September 1858.
96 *Isle of Wight Observer*, 31 October 1857; *Sun (London)*, 25 November 1857; *Poole & Dorset Herald*, 8 April 1858.
97 *Poole & Dorset Herald*, 26 November 1857; *Salisbury and Winchester Journal*, 1 May 1858.
98 *Salisbury and Winchester Journal*, 1 May 1858; *Bell's Life in London and Sporting Chronicle*, 1 August 1858.
99 *Poole & Dorset Herald*, 20 January 1859; *Hampshire Independent*, 22 January 1859.

a light draft to adapt it for navigating shoal waters.[100] In April of that year the *Leonara,* a 116-ton schooner yacht, was launched.[101] Soon followed a 25-ton cutter named *Kingfisher*, a 78-ton schooner yacht named *Gannet*, and a schooner of 91 tons that was also named *Kingfisher*.[102]

In 1860 George Inman launched the 156-ton schooner *Albertine* for Lord Londesborough. It was considered a rakish-looking craft featuring a hollow razor-like entrance, characteristics of an Inman-build.[103] The vessel went on to make its debut on the Thames in 1861 famously beating the *Alarm* on time allowance in a smart breeze.[104]

Also completed that year was a 130-ton schooner for the Earl of Durham named *Beatrix*.[105] Thereafter came the 50-ton yacht *Kittiwake*.[106] Another craft by the name of *Rosalind*, this one slightly smaller than the first, at 94 tons, was launched in June of 1860.[107] A commission for Arthur Kavanah of the Royal Yacht Squadron was next off the slip in the form of the 150-ton schooner *Eva*.[108]

At least four more vessels were launched from the Inman shipyard in 1861. These were the cutters *Dolphin* and *Scout*, both of 80 tons and built to the order of the Coast Guard; the 110-ton schooner *Zelia*; and the 45-ton schooner *Endeavour*.[109] The latter named vessel was completed by George Inman with great generosity and likely at a financial loss. The cause was worthy, however, for the *Endeavour* was to embark on an Arctic voyage in search of Sir John Franklin's lost expedition to Canada's Northwest Passage.[110] HMS *Erebus* and HMS *Terror* had departed England in 1845 but no news of their progress had been received since 1848.

Another highlight of 1861 for the Inman shipyard was a visit by Prince Edward that occurred in October of that year.[111] He later became King Edward II. Also in 1861 *America* was purchased by Henry Decie and renamed *Camella.* The 215-ton schooner was soon pitted against the larger schooner *Alarm* in a rematch of the great race that had taken place a decade earlier.[112] With 100 sovereigns at stake and, perhaps more importantly an international reputation, the race took place off Cowes as the first event of the Royal Yacht Squadron Regatta with much expected from the veteran yachtsman Joseph Weld and his now 31-year-old vessel. It proved a leisurely romp for *Alarm*, the craft claiming victory from the New York-built yacht by a margin of more than 40 minutes.[113] *Alarm* was thus declared the fastest vessel afloat. For Weld, however, the victory was to be one of his last. He died two years later at Lulworth Castle, aged 86.[114] Both *Alarm* and *Lulworth* were advertised for sale as part of his estate.[115]

After the hype of the late 1850s, the early 1860s proved to be a rather quiet period. Times were somewhat sombre.

[100] *Hampshire Independent*, 12 February 1859.
[101] *Field*, 9 April 1859.
[102] *Hampshire Independent*, 22 January 1859; *Field*, 9 April 1859; *Poole & Dorset Herald*, 26 May, 22 September 1859.
[103] *Field*, 6 July 1895.
[104] *Field*, 6 July 1895.
[105] *Poole & Dorset Herald*, 22 September 1859; *Hampshire Independent*, 3 March 1860.
[106] *Poole & Dorset Herald*, 26 April 1860; *North British Daily Mail*, 30 April 1860.
[107] *Isle of Wight Observer*, 2 June 1860.
[108] *Isle of Wight Observer*, 2 June 1860; *Salisbury and Winchester Journal*, 22 September 1860.
[109] *Field*, 30 March 1861; *Newcastle Journal*, 15 July 1861; *Morning Herald*, 21 December 1869; *Shipping and Mercantile Gazette*, 31 January 1870.
[110] *John Bull*, 13 July 1861; *Newcastle Journal*, 15 July 1861.
[111] *Hampshire Chronicle*, 5 October 1861.
[112] *Sun (London)*, 1 August 1861.
[113] *Hampshire Advertiser*, 10 August 1861.
[114] *Sun (London)*, 22 October 1863.
[115] *Bell's Life in London and Sporting Chronicle*, 6 February 1864.

THE 156-TON SCHOONER YACHT ALBERTINE (1860).
SOURCE: NATIONAL MARITIME MUSEUM, GREENWICH, LONDON (PAH8750).

THE 130-TON SCHOONER YACHT BEATRIX (1860).
SOURCE SOURCE NATIONAL MARITIME MUSEUM, GREENWICH, LONDON (PAH8748).

The American Civil War was in progress, a scarlet fever epidemic gripped Great Britain with an estimated 30,000 lives lost, and Queen Victoria and the Empire as a whole mourned the death of Prince Albert. As such, many yacht racing fixtures were abandoned or truncated as a mark of respect. Yachting took a back seat for several seasons during this period of melancholy. George Inman also suffered personally with the loss of his wife Harriet in 1862, and his son Walter in 1863 at the age of 12.[116]

Few new vessels were launched from the Inman shipyard during these years. However, those that were built remained of substantial proportions and epitomised the Inman reputation for quality design and craftsmanship. There was also innovation in the launch of steam boats. New vessels included the 64-ton schooner yacht *Clytie* (1862), the 140-ton schooner *Red Gauntlet* (1862), the 140-ton schooner yacht *Pilgrim* (1862), the steam boat *Prince Albert* for the Portsmouth and Ryde Steam Packet Company (1863), the steam boat *Solent* for the Solent Sea Steam Navigation Company (1863), the 70-ton schooner *Reindeer* (1863), the 195-ton *Zelia* (1864), the 81-ton yacht *Fenella* (1864), the 33-ton yacht *Free Lance* (1865), the 78-ton cutter *Maria* (1865), and the 25-ton cutter *Era* which was notably sent to New South Wales in frame for James Milson (Jr) of the Royal Sydney Yacht Squadron (1865).[117]

Another Inman build, the *Red Gauntlet,* was also to make its way to Australia, having been purchased by C. C. Balfour of Sydney in 1880.[118] Following his death two years later, the vessel was transferred to C. and E. Millar of the Victoria Yacht Club.[119] Not intended as a racer, the craft made several extensive cruises of the Australian coast before being wrecked near Darwin in 1887.[120]

The remaining years of the mid-to-late 1860s were particularly quiet in terms of new Inman builds, largely indicative of a slowing British economy and the end of the Industrial Revolution. Trade unions were established and workers' rights became political issues with significant laws passed. The Inman shipyard, however, remained busy with refitting, repair and modification work, including adding length to the *Albertine.*[121] New builds launched during this period included the 58-ton schooner *Viking*, the schooner *Cuckoo*, the 9-ton cutter yacht *Waterwitch*, the substantial 300-ton schooner yacht *The Dream*, the 197-ton schooner *Mirage*, and the 43-ton screw steam yacht *Iole.*[122]

The 1870s began with great sadness for George Inman. His father Thomas died on 29 January 1870 at the age of 83.[123] His eldest son Herbert, aged 21 years, died just

[116] *Hampshire Advertiser*, 21 June 1862; *Hampshire Advertiser*, 15 August 1863.
[117] *Bell's Life in London and Sporting Chronicle*, 9 March 1862, 14 May 1864, 25 November 1865; *Hampshire Advertiser*, 11 October 1862; *Hampshire Independent*, 9 May 1863, 14 May 1864; *Portsmouth Times and Naval Gazette*, 24 October 1863; *Isle of Wight Observer*, 2 April 1864; *Salisbury and Winchester Journal*, 17 June 1865; *Hampshire Telegraph and Naval Chronicle*, 27 June 1866; *The Sydney Morning Herald*, 21 September 1865, 18 April 1873.
[118] *The Sydney Morning Herald*, 28 September 1880.
[119] *Illustrated Australian News*, 22 March 1882.
[120] *The Argus*, 4 November 1887.
[121] *Bell's Life in London and Sporting Chronicle*, 8 June 1867.
[122] *Greenock Advertiser*, 9 June 1866; *Banffshire Journal and General Advertiser*, 23 October 1866; *New York Daily Herald*, 18 February 1868; *The Hampshire Advertiser*, 30 May 1868; *Field*, 28 October 1871; *Hampshire Advertiser*, 5 February 1881; *Shipping and Mercantile Gazette*, 9 June 1882; www.maritimeviews.co.uk/byy-biographies/ormonde-marquis-of/.
[123] *The Hampshire Advertiser*, 2 February 1870.

THE 140-TON SCHOONER YACHT RED GAUNTLET (1862).
SOURCE: ILLUSTRATED AUSTRALIAN NEWS, 22 MARCH 1882.

two days later from consumption (tuberculosis).[124] Much sympathy was expressed by the populace of Lymington during this time of bereavement.[125] For the country as a whole, however, it was a period of increasing prosperity. The economy began bouncing back at a brisk pace, fuelled by the development and implementation of new technologies and innovations, including those associated with steel making, mass production and assembly lines, the electrical grid, large-scale manufacture of machine tools, and the use of machinery in steam-powered factories.

In terms of new builds, the Inman yard launched several vessels during the early 1870s, some of which were built using these developing technologies. First off the slip was the cutter yacht *Nann.*[126] Next came the 52-ton schooner *Ita.*[127] In 1871 there was launched the 20ft yacht *Irene* which was successfully sailed by George Inman's son Edwin in local yacht races.[128] A new steam yacht was also completed in 1871, as was the 110-ton schooner *Foam.*[129] In 1872 a 42-ton racing cutter was launched, in addition to a 12-ton yacht.[130]

The 110-ton steam yacht *Carlotta* was launched from the Inman shipyard in May of 1873.[131] It was followed several days later by the cutter yacht *Mars.*[132] Also under construction in 1873 was a 250-ton iron steam yacht.[133] This particular vessel was likely named *Whim* and completed the following year.

In 1874 came the launch of a 210-ton schooner by the name of *Ianira,* as well as the 230-ton schooner yacht *Castalia.*[134] Both vessels were designed by Edwin Inman.[135] A 175-ton schooner, the *Australia,* for W. W. Hughes, who was also its designer, was launched in 1875, as was the *Corsair,* a 48-ton cruising yawl, and the 137-ton schooner yacht *Ayacanora.*[136]

THE 210-TON SCHOONER YACHT IANIRA (1874).
SOURCE: NATHANIEL L. STEBBINS PHOTOGRAPHIC COLLECTION, HISTORIC NEW ENGLAND (PC047).

[124] *Salisbury and Winchester Journal*, 5 February 1870.
[125] *Hampshire Advertiser*, 5 February 1870.
[126] *Hampshire Advertiser*, 11 May 1870.
[127] *Greenock Telegraph and Clyde Shipping Gazette*, 7 April 1871.
[128] *Hampshire Advertiser*, 13 May 1871.
[129] *Hampshire Advertiser*, 23 August 1871, 4 October 1871.
[130] *Hampshire Advertiser*, 21 February 1872; *Bell's Life in London and Sporting Chronicle*, 16 March 1872.
[131] *Shipping and Mercantile Gazette*, 10 May 1873.
[132] *Hampshire Advertiser*, 14 May 1873.
[133] *Hampshire Advertiser*, 11 October 1873.
[134] *Hampshire Advertiser*, 11 October 1873, 18 February 1874; *Shipping and Mercantile Gazette*, 4 May 1874; *Poole & Dorset Herald*, 21 May 1874.
[135] *Hampshire Advertiser*, 18 February 1874, 20 May 1874.
[136] *Field*, 13 February 1875; *Hampshire Advertiser*, 8 May 1875; *Hampshire Independent*, 21 July 1875; *Isle of Wight Observer*, 24 July 1875; *Field*, 22 April 1876.

For yacht racing more generally, the mid-1870s coincided with the introduction of a yachting governing body and the implementation of standardised rules. Specifically, 1875 saw the establishment of the Yacht Racing Association, now considered one of the foundations of yacht racing. It came as a natural progression from the birth of a small number of yacht clubs into a collective of similar organisations spread throughout the United Kingdom. As the overarching authority, the Yacht Racing Association was charged with administering yachting on a national basis and helping to coordinate the ever-growing sport. Its first rules were associated with handicapping or fleet racing. Initially developed from commercial origins and based on a formula using internal length and beam measurements, these rules calculated the volume of cargo space or capacity in a vessel, expressed as its tonnage. Widely referred to as the 'Tonnage Rules', the outcome overall, for handicapping purposes, was a time allowance. This was then applied at the completion of a race, thus altering the elapsed time for each competing vessel. The winner was then declared as the yacht with the fastest time after the time allowance had been deducted from the elapsed time.

For yacht designers, including Alexander Richardson, George Lennox Watson, and a very young Walter Reeks who was about to start on this career path, the 'Tonnage Rules' resulted in a significant evolution in yacht design. Suddenly smaller craft were not only able to compete with the larger vessels, but also succeed. Another evolution would occur during the second half of the 1880s when rules governing waterline length and sail replaced beam and length.

For the Inman shipyard, it was a time of adjustment not just in the design and building of racing yachts but also in the development of pleasure vessels. In 1876 the firm completed the 230-ton schooner *Fortuna*, one of the largest vessels constructed to date.[137] Though never a racing yacht, the craft was much admired as a splendid cruising vessel in which it was solely intended.[138] *Fortuna* was wrecked in 1906 off the Falkland Islands after being purchased by the Falkland Islands Company over a decade prior.[139]

SCHOONER FORTUNA.
SOURCE: FALKLAND ISLANDS MAGAZINE, JUNE 1906.

[137] *Hampshire Advertiser*, 11 October 1873, 4 March 1876; *Shipping and Mercantile Gazette*, 25 April 1883.
[138] *Shipping and Mercantile Gazette*, 26 October 1883.
[139] www.facebook.com/groups/202600596533824/posts/3773632856097229.

The long list of well-built Inman vessels continued with the launch of *Nore,* a 190-ton cruising schooner.[140] A 160-ton auxiliary screw steam yacht named *Foam* was also launched, a replacement for a schooner of the same name which the firm had completed several years before.[141] Both vessels were launched in 1877, as was a 12-ton cutter built on speculation.[142] The auspicious *Lulworth*, built in 1856 for Joseph Weld and laid up for over a decade, was also given major repairs during this year having been purchased by Adrian Hope.[143]

In 1878 the firm began construction of the 25-ton cruising yawl *May Queen*.[144] Being built on speculation was a 112-ton cruising schooner that would take several years to complete.[145] Lengthening of the 145-ton schooner *Pilgrim* was also undertaken in 1879.[146]

It was during this period that Walter Reeks was privileged to be articled to George Inman & Son, celebrated shipbuilders of Lymington. It was to be a steep immersion into the realm of yacht design and building for the young Reeks, but one that was invaluable in terms of his future career path. It was also a prominent position given the decades-long standing that the Inman name enjoyed in the upper echelon of yachting circles, as well as the historical and continuing success of vessels laid down by the yard.

With Reeks watching on, in 1880 the Inman shipyard completed a 90-ton cutter to the order of Henry Thomson (Sr) of Itchen ferry.[147] Also at the yard awaiting its fate was the *Alarm,* the old vessel hauled up on the hard likely giving Reeks a welcome sight of intrigue and fascination.[148]

Of significance, under construction in early 1880, was a 94-ton cutter built to the order of John Jameson (Jr) of Dublin's famous Irish whisky family. The vessel's designer was Alexander 'Alex' Richardson, naval architect of Liverpool, and also creator of the famous 'fly to windward' craft *Lily* and *Challenge*. With naval architecture an emerging occupation, Jameson's vessel likely stands as one of the first instances of a professional designer being charged with creating a sailing craft that was not also the owner nor an employee of the builder.[149] The utilisation of naval architects would soon become the norm not the anomaly, a fortuitous development for Reeks.

The interest surrounding Jameson's new vessel was such that 'spies' in the incarnation of skippers of competing vessels from the Clyde and other regions manufactured themselves as distinguished gentlemen and began showing up at the Inman yard with the hopes of sneaking a look at the vessel and its plans.[150] George Inman eventually caught on to this charade and learned to increase security and keep a more aware and tight-lipped workforce.[151]

[140] *Field*, 27 May 1876; *Bell's Life in London and Sporting Chronicle*, 7 April 1877.
[141] *Field*, 30 September 1876; *Hampshire Advertiser*, 26 September 1877.
[142] *Hampshire Advertiser*, 31 January 1877.
[143] *Hampshire Advertiser*, 16 June 1877.
[144] *Field*, 1 February 1879; *Hampshire Advertiser*, 10 May 1879.
[145] *Field*, 4 January, 5 July 1879.
[146] *Poole & Dorset Herald*, 13 March 1879; *Field*, 22 May 1880.
[147] *Hampshire Advertiser*, 25 February 1880.
[148] *Field*, 3 January 1880.
[149] *Field*, 12 March 1880.
[150] *Field*, 13 March 1880.
[151] *Field*, 13 March 1880.

Jameson's racing yacht was launched in April of 1880 and bequeathed the name *Samaena*. The vessel was 81ft (on load line) with an extreme beam of 16ft 2in. It had not only been built to Richardson's plans and specifications, but he had also superintended its laying off and construction.[152] The London press reporting on the launch noted, *'never - at any rate since the days of the Alarm and Arrow - has any vessel exercised the local talent so much, or excited such curiosity and criticism in the town as this cutter'.*[153]

The *Samaena* certainly lived up to expectations, claiming victory in its class at regattas and events in the months immediately after launch. The press relayed that a match against its nearest rival, the famous steel-built Clyde flier *Vanduara*, was much anticipated.[154] This race took place in August of 1880 as part of the Royal Yacht Squadron Regatta. The *Samaena* went on to claim victory by a long margin.[155] It was one of many important victories for this vessel, and likely one eagerly followed by Reeks.

Another large schooner was launched from the Inman yard in mid-1881. Built on speculation over a period of several years, the 112-ton vessel was purchased by W. O'Connor of Dublin and subsequently named *Wild Rose*.[156] It went on to sail with the Royal Irish Yacht Club. Next came a second collaboration between Alex Richardson and George Inman: the 20-ton cutter yacht *Amethea* designed and built by the parties to the order of Henry Lamont and launched in May of 1881.[157] Finally the old *Alarm* received some attention. It was re-caulked and launched to lie as a sheer hulk for what the press expected was at least another decade.[158] In relaying this information, a Cowes-based newspaper opined that the Lymingtonians should give up the *'grotesque fiction that the owner of the Alarm annually receives a large sum, subscribed by the various yacht builders, not to fit out the Alarm, as she would wrest the prizes from every schooner afloat'.*[159]

In July of 1881 the Inman yard launched an 11-ton steam launch named *Elfin*, built to the order of G. W. Anderson of Liverpool, again to a design by Richardson.[160] Later that year the keel of the *Silver Star*, a 40-ton racer was laid down, the lines of which were also furnished by Richardson.[161] The vessel was built to the order of Malcolm Murray of Dundalk, Ireland, and launched in May of 1882.[162] A racing yawl of 46 tons designed by Richardson was launched from the yard one month later, named *Leander* and built to the order of William Bower Forwood, former mayor of Liverpool and then rear-commodore of the Royal Mersey Yacht Club.[163] Curiously both vessels carried the majority of their ballast internally (as handy-sized moulded pieces), this system of construction preferred by Richardson to allow the centre of gravity to be significantly lower than if the ballast was placed outside of the vessel.[164] The *Leander* in particular carried iron as ballast instead of the usually more expensive lead, resulting in the cost of its build being similar to that of a racing yacht half its size.[165]

Next from the Inman shipyard came a 10-ton racing cutter, *Katie*, built to the order of Francis Coddington of Blackburn to sail with the Royal Cheshire Yacht Club. It was launched a week after *Leander*.[166] This vessel marked the sixth Inman build to have been designed by Richardson.[167] Most if not all of the vessels were also built under his personal superintendence.[168] The *Katie*, however, proved somewhat uncompetitive. It was of extreme dimensions and had a very large displacement.[169]

As 1882 neared its end, the Inman yard was busy building a 5-ton racer for William Grunsell of Lymington, a well-known Solent yachtsman; a 40-ton cruising yawl for J. K. Soames; and a large steam yacht to the order of Philip H. George of Bristol.[170] A new shed was also under construction to support the ever-expanding business.[171] The press reluctantly stated that the old *Alarm* was once more languishing in the mud near the yard.[172]

The three vessels mentioned above were all completed in mid-1883. First came William Grunsell's 5-ton racer *Abeona* (later named *Dorothy*), launched in May of that year and, rather significantly, stated to have been designed by Walter Reeks.[173] Next came the 40-ton yawl *Thalia*, built to the order of J. K. Soames.[174] Of note this particular vessel was also designed in-house, in all likelihood by Reeks or at the very least with his assistance.[175] Finally, in July of 1883, the 108-ton screw steamer *Wild Rose* was launched, built to the order of Philip H. George of Bristol, and the seventh Inman vessel designed by Richardson.[176]

For Reeks, as well as being positioned within the epicentre of yacht design and building, the Lymingon location offered close proximity to his family in Christchurch. In the summer of 1882 he designed and built the 16ft centreboard yacht *Snowflake* which successfully competed in fortnightly sailing matches of the Christchurch Sailing Club, often helmed by his brother William.[177] Reeks and his older brother were obviously close. In June of the following year Reeks travelled to Leicester to be a witness at William's wedding to Sarah Pagett. Mid-1883 was not without sadness for the Reeks

[152] *Field*, 1 May 1880; *Hampshire Advertiser*, 1 May 1880.
[153] *Field*, 1 May 1880.
[154] *Shipping and Mercantile Gazette*, 7 July 1880.
[155] *Hampshire Independent*, 7 August 1880.
[156] *Hampshire Advertiser*, 11 December 1880, 15 January 1881; *Isle of Wight Observer*, 12 February 1881.
[157] *Isle of Wight Observer*, 12 February 1881; *Greenock Advertiser*, 3 May 1881.
[158] *Isle of Wight Observer*, 19 February 1881.
[159] *Isle of Wight Observer*, 19 February 1881.
[160] *Hampshire Advertiser*, 11 June 1881, 30 July 1881, 27 August 1881.
[161] *Greenock Telegraph and Clyde Shipping Gazette*, 28 November 1881; *Poole & Dorset Herald*, 20 April 1882.
[162] *North British Daily Mail*, 29 November 1881; *Field*, 6 May 1882.
[163] *Field*, 1 April 1882; *Poole & Dorset Herald*, 22 June 1882.
[164] *Field*, 1 April, 6 May 1882.
[165] *Poole & Dorset Herald*, 20 April 1882.
[166] *Hampshire Advertiser*, 28 December 1881; *Hampshire Independent*, 24 June 1882; *Salisbury and Winchester Journal*, 26 May 1883.
[167] *Poole & Dorset Herald*, 20 April 1882.
[168] *Poole & Dorset Herald*, 20 April 1882.
[169] www.maritimeviews.co.uk/british-yachts-yachtsmen/chapter-8-the-yra-tonnage-rule/.
[170] *Hampshire Independent*, 16 September 1882; *Hampshire Advertiser*, 29 November 1882.
[171] *Hampshire Advertiser*, 13 December 1882.
[172] *Isle of Wight Observer*, 21 October 1882.
[173] *Salisbury and Winchester Journal*, 19 May 1883; *Lloyd's Register of Yachts* (1903).
[174] *Hampshire Advertiser*, 23 June 1883, 10 September 1887.
[175] *Hampshire Advertiser*, 13 December 1882.
[176] *Field*, 28 July 1883; *Lloyd's Register of Yachts* (1903).
[177] *Christchurch Times*, 23 September 1882, 2 June 1883; *Hampshire Advertiser*, 4 July 1883; *Bournemouth Guardian*, 31 May 1884.

family, however, with the death of William and Walter's sister Annie in July at the age of 25.[178]

Upon moving to Lymington early in 1880, Reeks continued to be actively involved in sports and community organisations. In the summer of that year he joined the Lymington Rowing Club where he performed well in races.[179] Reeks was also a successful competitor in local swimming races.[180] Music was another outlet. In late 1880 he performed at a concert for the Lymington Mutual Improvement Society.[181]

Throughout Reeks' tenure, the Inman yard was additionally involved in overhauling, refitting, maintaining and repairing vessels, as well as modifying existing craft. The experience would have provided an invaluable starting point for Reeks' future career, a sound foundation from which he could potentially pivot to another position. This opportunity likely came earlier than expected. The death of George Inman in October of 1883, at the age of 66, after a short illness was the beginning of the end for the illustrious shipyard.[182]

In eulogising George Inman the Hampshire press stated that he carried with him, '*a sense of uprightness and conduct, as well as charity, benevolence and unostentatiousness*'.[183] It was hoped that those who followed in his footsteps would continue to set this example. He was also noted as a considerate and generous employer.[184] Undisputedly one of Britain's premier yacht builders, ironically George Inman was not at all interested in the pastime. The year before he died he plainly stated that he would only ever sail in vessels that were '*over 200 tons*' that also had '*a chaplain on board*'.[185]

At the time of Inman's death the shipyard had in the course of construction a 60-ton cutter being built for Mr Foster Connor of Liverpool who had previously commissioned the *Silver Star*.[186] The firm was also adjusting to changes within management. William Shilcock, who had been a valued employee for many years had left in mid-1883 to establish his own shipyard at East Cowes. He later returned to take on the role of working partner.[187] The firm was also actively pursuing a new era of composite construction with the purchase of plant and equipment.[188]

Designed by Richardson, Mr Foster Connor's substantial yacht *Marguerite* was launched in May of 1884.[189] Sadly the craft was to be the last big build launched from the Inman shipyard. Despite efforts made to expand the business following the death of its patriarch George Inman, another tragic event occurred in March of 1885 with the sudden death of Edwin Inman at the age of 32.[190] As the only surviving son of George Inman, Edwin's death marked the end of the family business.

In November of 1885 Inman & Son incorporated to become Inman & Co. with two new partners from a London-based ironwork company joining the enterprise.[191] Along with William Shilcock as a director, the partners planned to add an engineering department and make arrangements for the building and repair of iron and steel vessels.[192] A new 4-ton racing yacht, *Kitty,* was also launched in July of 1886, built for R. M. Tinkler.[193] The following year a 40-ton cruising yawl, *Maid Marian*, was built to the order of Myles Kennedy.[194]

In 1888 a 30-ton yawl was launched, the second vessel the company built for R. M. Tinkler.[195] The business also continued to undertake overhaul work, including at a second yard based at Cowes.[196] These attempts proved futile, however. In late 1888 part of Inman's Lymington-based premises was offered for lease, while the company's assets were beginning to be sold off.[197] The old *Alarm* was auctioned off for a measly £41, to be broken up.[198] Still, a few vessels were completed, including two 25ft steam launches for a telegraph company.[199] A 30-ton yawl built on speculation was also purchased by C. J. Morgan of London.[200] Named *Gracie*, this vessel was completed in April of 1889.[201] Six months later the company entered liquidation with the process continuing into 1891.[202] The yard's entire plant, gear, machinery, yachts and boats, etc., were sold off by public auction under execution by the Sheriff of Hampshire.[203] The affairs of the business were compulsorily wound up in 1892.[204]

It was the end of an era. For many employees, however, like Walter Reeks who had left the shipyard several years prior, they could reflect proudly on their time working at the Inman shipyard, relishing in its prominent place in yachting history and being nurtured in the field of yacht design and building.

178 Somerset, England, Church of England Burials, 1813-1914 for Annie Wiltshire.
179 *Hampshire Advertiser,* 14 August 1880.
180 *Hampshire Independent,* 21 August 1880.
181 *Hampshire Independent,* 18 December 1880.
182 *Shipping and Mercantile Gazette*, 26 October 1883.
183 *Hampshire Advertiser,* 27 October 1883.
184 *Hampshire Independent*, 27 October 1883.
185 *Poole & Dorset Herald,* 20 April 1882.
186 *Salisbury and Winchester Journal*, 27 October 1883; *Greenock Herald*, 22 March 1884.
187 *Hampshire Advertiser,* 14 November 1883; *Field,* 19 April 1884.
188 *Field,* 19 April 1884.
189 *Field,* 17 May 1884.
190 *Hampshire Advertiser*, 14 March 1885.
191 *Field,* 21 March 1885; *Salisbury and Winchester Journal,* 27 June 1885; *Hampshire Advertiser*, 21 November 1885.
192 *Field*, 20 February 1886.
193 *Western Gazette*, 9 July 1886.
194 *Hampshire Advertiser*, 16 October 1886, 30 April 1887; *Field,* 26 February 1887.
195 *Hampshire Advertiser,* 29 October 1887.
196 *Field,* 21 March 1885; *Hampshire Advertiser*, 21 November 1885.
197 *Hampshire Advertiser,* 8 September 1888.
198 *Hampshire Advertiser*, 8 September 1888.
199 *Poole & Dorset Herald,* 17 January 1889.
200 *Poole & Dorset Herald,* 14 February 1889.
201 *Poole & Dorset Herald,* 4 April 1889.
202 *Poole & Dorset Herald,* 3 October 1889.
203 *Hampshire Advertiser,* 28 February 1891.
204 *Salisbury and Winchester Journal,* 19 December 1891.

1851 ORDNANCE MAP OF LIVERPOOL (ENGLAND). RUMFORD STREET IS SHOWN TWO BLOCKS UP FROM ST. NICHOLAS' CHURCH.
SOURCE: NATIONAL LIBRARY OF SCOTLAND (HTTPS://MAPS.NLS.UK/VIEW/102344096).

CHAPTER THREE - LIVERPOOL

With his continuing employment placed into jeopardy by the sudden death of his supervisor and mentor George Inman in October of 1883, Walter Reeks likely found himself in somewhat of a difficult situation. Thankfully a new opportunity was not far off with Alexander 'Alex' Richardson, naval architect and vessel broker of Liverpool, offering Reeks a position with his firm. The pair had previously worked together on at least seven Inman-built vessels. These were the 94-ton cutter *Samaena* (1880), the 20-ton cutter *Amethea* (1881), the 11-ton steam launch *Elfin* (1881), the 40-ton racer *Silver Star* (1882), the 46-ton racing yawl *Leander* (1882), the 10-ton racing cutter *Katie* (1882), and the 108-ton screw steamer *Wild Rose* (1883).

For Reeks the move to Liverpool was significant. Approaching 23 years of age, he had been based in Lymington for several years, a location that was only a dozen miles away from family and friends in Christchurch. Whether he was excited to be moving 250 miles to the north, away from the Hampshire coast and away from all that was known and familiar is not recorded. Regardless of his apprehensiveness or eagerness, it was a brave move but one that would benefit his career immensely.

In the early 1880s Liverpool was a flourishing city with a population exceeding some 600,000 residents. For several decades manufacturing had been booming, particularly in areas such as shipbuilding, rope making, metal working and machine making.[205] The city had also become famous for its cotton mills. The development of railways and port facilities had enhanced both international and domestic trade, as well as the growing overseas passenger trade, helped by the introduction of steam-power, creating somewhat of a monopoly for the port of Liverpool and its neighbours, like Birkenhead on the opposite side of the River Mersey. Many of the city's merchants grew in wealth and prosperity during this period. The American Civil War (1861 - 1865), however, had cast a dark cloud over the region's cotton mills with blockades of cotton grown in the American south leading to much speculation and risk taking. During its height traders gained and lost fortunes overnight.

With the end of the Civil War, the cotton market had steadied with Liverpool buoyed by additional worldwide trade of produce, rice and sugar. With this increased affluence came enlightenment with the construction of libraries, museums, theatres and art galleries. Liverpool relished in a cosmopolitan flavour. Many new buildings were constructed by wealthy philanthropists.

It was in this context that Reeks took up employment with Richardson at his office on Rumford Street. Only a few blocks from the port district and around the corner from the Royal Exchange and Town Hall, the location was in the heart of Liverpool's financial and administration centre, populated by brokers, accountants, merchants, shipowners and agents.

Like that of Reeks, Richardson's biography is rather unique and it is surprising that no body of work exists profiling his life and life's work. Born in Rock Ferry (now considered part of Birkenhead) in 1844, his father Mervyn was a retired army captain of Irish descent, while his mother Harriet hailed from Leeds, Yorkshire. Mervyn Richardson had served in the 4th Light Dragoons from July of 1815 to February of 1826, during somewhat of a relatively peaceful time.[206] Thereafter he purchased an

[205] www.historic-uk.com/HistoryMagazine/DestinationsUK/Liverpool/.

[206] *Naval & Military Gazette and Weekly Courier of the United Service*, 1 January 1879.

unattached captaincy, allowing him to draw the half pay of the commission for more than 50 years without having to actively serve.

With Mervyn's pay seemingly adequate to support his large and ever-expanding family (Alex was the sixth of nine children), the large Richardson family lived comfortably at 100 Russell Road in Rock Ferry. Mervyn spent his time studying and writing books on horsemanship.[207] He died late in 1868 in Rock Ferry at the age of 81.[208]

In 1861, around the age of 16, Alex Richardson became apprenticed to a cotton broker, following in the footsteps of his older brother James.[209] It was likely a tumultuous early few years of introduction to the trade given the American Civil War had a dramatic effect on the import of cotton to Liverpool, with prices fluctuating to extremes. Still he managed to endure the speculation surrounding the commodity during these years such that by 1871, at the age of 26, Richardson was a single, fully-qualified cotton broker still living at home.

Likely looking for an outlet to spend some free time and his disposable income, as well as make business connections, around 1868 Richardson and his younger brother William (by now also a cotton broker) became members of the nearby Cheshire Yacht Club, an organisation with around 100 members that had only recently been formed.[210] The pair initially competed in the 6-ton *Dudu*.[211] In 1870 Richardson began successfully competing in the 10-tonner *Naiad*.[212] The vessel had been built by William Dickenson of Birkenhead, who had built several other 10-ton boats and was considered a champion of the class' development on the Mersey.[213]

Respected and well liked, the following year Richardson was elected vice-commodore of the Cheshire Yacht Club, retaining this position for a number of seasons.[214] During this period and into the succeeding years, he was also active with the Liverpool Model Yacht Club, as well as the Royal Mersey Yacht Club.[215]

Also a member of the Cheshire Yacht Club was James Leitch, one of Liverpool's wealthy sugar refiners who was originally from Scotland, and two of his sons James (Jr) and Robert. What very likely began as a friendship between Richardson and the Leitch trio of yachtsmen, who raced against one another but also took cruises of the Irish and Scottish coasts together, soon became a more formal union when Richardson married Leitch's eldest daughter Lydia Helen at Bidston Church, Birkenhead, in June of 1874.[216] The couple moved to Rock Ferry where twins, a boy and girl, were born in June of the following year.[217]

Richardson's marriage coincided with his entrée into yacht design, perhaps supported by the Leitch family. Two vessels he designed, the *Lily* and *Challenge*, soon began dominating local yacht races on the River Mersey.[218] The 10-tonner *Lily* was designed for his father-in-law and launched in May of 1875.[219] It was built under Richardson's strict superintendence by Messrs Buckley and Sherlock of Tranmere, Birkenhead, and went on to successfully compete with the Cheshire Yacht Club for several seasons, also cruising to Ireland and Scotland.[220]

In May of 1876 the 20-ton *Challenge* was launched, designed by Richardson and built by Messrs Buckley and Sherlock to the order of Frederick Thompson of Aston Hall, Cheshire.[221] The craft was stated to exhibit more originality of design than many of the newer vessels of its class. The *Challenge* was also considered a refreshing departure from the stereotyped mode of construction, with a bold but not too full midship section, considerable rise of floor and a moderate hollow in its garboards.[222]

THE 20-TON YACHT CHALLENGE (1876). THE VESSEL WAS LOST AFTER BEING RUN DOWN BY THE SCREW STEAMER OSSIAN IN SEPTEMBER 1883.
SOURCE: SIR EDWARD SULLIVAN (1894). YACHTING: VOLUME I.

With Richardson at the helm, the *Challenge* won handsomely. In its first year alone the craft won ten races and collected more than £320 in prize money.[223] Combined, the *Lily* and *Challenge* amassed nearly £700 in prize money in their first seasons.[224]

Encouraged by these successes, in February of 1878 Richardson adopted the profession of naval architecture, establishing his business at Rumford Street, Liverpool, where he soon advertised himself as '*Naval Architect and*

[207] *Horsemanship or the Art of Riding* (published 1853).
[208] *Naval & Military Gazette and Weekly Courier of the United Service*, 1 January 1879.
[209] 1861 England Census (Class: RG 9; Piece: 2646; Folio: 94; Page: 10).
[210] *Chester Courant*, 25 August 1869; *Liverpool Courier and Commercial Advertiser*, 1 August 1870.
[211] *Chester Courant*, 25 August 1869.
[212] *Liverpool Courier and Commercial Advertiser*, 1 August 1870.
[213] Sir Edward Sullivan (1894). *Yachting: Volume II*.
[214] *Bell's Life in London and Sporting Chronicle*, 27 May 1871.
[215] *Liverpool Mercury*, 18 November 1873.
[216] *Chester Courant*, 17 June 1874; *Morning Post*, 17 June 1874.
[217] *Liverpool Daily Post*, 24 June 1875.
[218] *Field*, 16 February 1878.
[219] *Field*, 5 June 1875.
[220] *Bell's Life in London and Sporting Chronicle*, 22 May 1875, 2 March 1878; *Field*, 5 June 1875, 25 August 1877.
[221] *Field*, 8 April 1876; *Liverpool Mercury*, 22 May 1876; *Bell's Life in London and Sporting Chronicle*, 17 June 1876.
[222] *Field*, 8 April 1876.
[223] *Field*, 12 August 1876; *Bell's Life in London and Sporting Chronicle*, 23 December 1876.
[224] *Bell's Life in London and Sporting Chronicle*, 16 March 1878.

Broker for the Designing, Sale and Purchase of Steamers, Sailing Vessels and Yachts'.[225]

One of Richardson's first professional commissions was a 5-tonner built by a Mr Willis of Birkenhead to the order of J. E. and G. W. Anderson to compete with the Cheshire Yacht Club.[226] Richardson also superintended the building of this vessel. Named *Urchin,* the craft's design was characteristic of his preference for a great rise of floor, as well as less hollow on the garboards and a powerful but easily-turned bilge.[227] Instead of the usual lead keel, the *Urchin* had an iron one and, without unduly extending it fore and aft, Richardson managed to get about 2½ tons into it.[228] In terms of scantlings, the vessel had alternate sawn and steamed frames, with a heavy angle iron floor. This plan was adopted in favour of the usual proportion of three steamed to each sawn frame.[229] The *Urchin* was considered a pioneer of the new 5-ton class which was intended to replace the famous Mersey 10-tonners that had dominated local yacht racing for a number of years.[230]

Another 5-tonner designed by Richardson was also completed around the same time. Named *Loro,* the vessel was built at Rutherglen near Glasgow to the order of J. Eadie.[231] While resembling the *Urchin* in general appearance, the *Loro* was said to have had more beam and less length. The craft went on to sail with the Royal Clyde Yacht Club.[232]

An iron steamer for use in conveying goods across the River Mersey between Woodside and Liverpool was the next vessel designed by Richardson.[233] The craft was launched in mid-1879, built by William Simons and Co. of Renfrew, Scotland, to the order of the Birkenhead Improvement Commission.[234] Named *Oxton,* the vessel was considered somewhat novel in that twin screws were fitted at each end.[235] The *Oxton* also had a rudder at each end such that it could turn around in its own length. Overall, the vessel measured 150ft in length and 50ft in breadth, and was fitted with two pairs of compound surface condensing engines of 140 h.p.[236] A spacious deck allowed for carriages, horses and carts to be driven onboard without the necessity of unyoking the horses.[237]

Another double-ended paddle steamer named *Primrose,* this time built to the order of the Wallasey Local Board in Egremont and intended for the passenger ferry trade between Egremont, Liverpool and New Brighton, was launched around the same time as the *Oxton.*[238] Constructed by T. B. Seath and Co. of Rutherglen, it was designed by Richardson and measured 150ft in length, 25ft in breadth with a depth of 10ft 6in.[239] The *Primrose* was also considered unique in construction in that the vessel was divided by seven cross and several longitudinal bulkheads to provide a dozen watertight compartments, making it almost unsinkable.[240]

225 *Field,* 2 February 1878.
226 *Field,* 16 February 1878, 6 July 1878.
227 *Field,* 16 February 1878, 6 July 1878.
228 *Field,* 16 February 1878.
229 *Field,* 16 February 1878.
230 *Field,* 16 February 1878.
231 *Field,* 16 February 1878, 15 June 1878.
232 *Field,* 15 June 1878.
233 *Liverpool Mercury,* 1 July 1879; openjournals.library.sydney.edu.au/index.php/EANSW/article/view/3148/3692.
234 *Liverpool Mercury,* 1 July 1879; www.clydeships.co.uk/view.php?ref=21322.

THE STEAMER OXTON.
SOURCE: SCOTTISH BUILT SHIPS (WWW.CLYDESHIPS.CO.UK).

235 *Liverpool Mercury,* 1 July 1879.
236 *Liverpool Mercury,* 1 July 1879.
237 *Liverpool Mercury,* 1 July 1879.
238 *Shipping and Mercantile Gazette,* 8 July 1879.
239 *Shipping and Mercantile Gazette,* 8 July 1879; *Liverpool Mercury,* 15 September 1879.
240 *Shipping and Mercantile Gazette,* 8 July 1879.

THE PADDLE STEAMER PRIMROSE.
SOURCE: SCOTTISH BUILT SHIPS (WWW.CLYDESHIPS.CO.UK).

Two more designs for paddle steamers were completed by Richardson for the Wallasey Local Board around the same time as the *Primrose*.[241] These were the *Sunflower* and *Daisy*.[242] The former was a luggage steamer used to convey goods, horses and cargo, while the latter was a saloon passenger steamer of a design very similar to the *Primrose*.[243]

The *Sunflower* unfortunately listed dangerously when carrying cargo and was rebuilt as a passenger ferry to accommodate some 950 people in 1882.[244] All three vessels remained in service for many decades.

THE PADDLE STEAMER SUNFLOWER.
SOURCE: SCOTTISH BUILT SHIPS (WWW.CLYDESHIPS.CO.UK).

[241] *Liverpool Mercury*, 15 September 1879.
[242] *Glasgow Herald*, 4 August 1879, 15 September 1879.
[243] *Shipping and Mercantile Gazette*, 7 February 1879.
[244] www.clydeships.co.uk/view.php?ref=20832.

Business continued to be brisk for Richardson. In the first of what would be many design commissions to be constructed by professional yacht builders in distant counties, in late 1879 he was tasked with designing a 90+ ton racing cutter to the order of John Jameson (Jr) of Dublin to be built by Inman & Son of Lymington.[245] Travelling to the Hampshire coast to superintend the building process, it was in this backdrop that Richardson first met Walter Reeks, then articled to George Inman, proprietor of the shipbuilding firm. With a long bow and bold-looking midship section, the resulting vessel *Sameana* was launched in April of 1880.[246]

THE 94-TON SCHOONER YACHT SAMAENA (1880).
SOURCE: SIR EDWARD SULLIVAN (1894). YACHTING: VOLUME II.

[245] *Field*, 29 November 1879, 12 March 1880.
[246] *Field*, 13 March 1880; *Bell's Life in London and Sporting Chronicle*, 1 May 1880.

THE PADDLE STEAMER DAISY.
SOURCE: NATIONAL MUSEUMS LIVERPOOL (MCR/90/298).

THE STEAMBOAT BANSHEE IN BERLIN, GERMANY (CIRCA 2009).
SOURCE: RAINER RADOW (WWW.STEAMBOATING.DE/STEAMBOAT/STEAMBOAT-BANSHEE.PHP).

It was the start of a busy summer period for Richardson. Appointed as official measurer of the Yacht Racing Association, he soon publicised that he would visit Erith, Gravesend, Birkenhead, Rothesay, Kingstown and Southampton in May and July of 1880 to measure yachts for tonnage with certificates then provided.[247]

Another diversion from his initial foundation of yacht designing also came in the summer of 1880 with Richardson designing the *Banshee,* an 11-ton iron screw steamer built by H. M. McIntyre at Paisley, Scotland.[248] The vessel was commissioned by William Bower Forwood for use as part of activities of the Windemere Yacht Club in the famous Lakes District.[249]

As previously stated, in 1881 two more vessels built by George Inman and Son at Lymington to designs by Richardson were launched. These were the 20-ton cutter *Amethea*, built for Henry Lamont of Glasgow, and the *Elfin*, an 11-ton steam launch built to the order of G. W. Anderson of Liverpool.[250] Also constructed during this year to a design by Richardson was the 85-ton yawl *Lorna*, built by Messrs Camper and Nicholson of Gosport (Hampshire) to the order of Samuel Hope Morley of Bristol.[251] Though initially built for cruising, the vessel was converted into a racing yacht, enjoying much success in its class.[252]

THE YAWL LORNA.
SOURCE: BRITISH YACHTS AND YACHTSMEN (WWW.MARITIMEVIEWS.CO.UK).

During this period Richardson also designed the 800-ton iron paddle steamer *Bonnie Princess,* built at Glasgow by T. B. Seath and Co. for the Liverpool, Llandudno and Welsh Coast Steamboat Company.[253] The 240ft vessel was launched in April of 1882, remaining in service for nearly two decades.[254]

Richardson spent more time at Lymington in 1882, superintending the construction of at least three vessels he had designed. These were the 10-tonner *Katie* (for Francis Coddington, a member of the significant Blackburn cotton milling family), the 46-ton yawl *Leander* (a second

[247] *Bell's Life in London and Sporting Chronicle*, 8 May 1880.
[248] *Lloyd's Register of Yachts* (1903).
[249] *Field*, 3 July 1880.
[250] *Hampshire Advertiser*, 11 June, 30 July, 27 August 1881; *Field*, 5 February, 7 May, 2 July 1881; Lloyds Register of Yachts (1903).
[251] *Hampshire Advertiser*, 16 February 1881; *Field*, 18 June 1881.
[252] *Hampshire Independent*, 1 October 1884.
[253] *Liverpool Daily Post*, 28 February 1882.
[254] www.clydeships.co.uk/view.php?ref=20857.

THE IRON PADDLE STEAMER BONNIE PRINCESS.
SOURCE: SCOTTISH BUILT SHIPS (WWW.CLYDESHIPS.CO.UK).

vessel built for William Bower Forwood) and the 40-ton racer *Silver Star* (for Malcolm Murray). This experience was very likely career-changing for Walter Reeks, eager to enter the world of naval architecture and learn from the mentorship offered by Richardson.

It was also a period of experimentation and advancement for yacht design more generally with Richardson leading the way initially in designing vessels with almost all of their ballast carried inside, thereby reducing the risk attendant with carrying a monstrous lead keel. However, this preference was considered obsolete just a few years later with designers, including Richardson, opting for more ballast on the keel and not inside of the hull.

With regards to the *Leander,* William Bower Forwood (a Liverpool-born merchant, shipowner and politician who would be knighted the following year) published a rather glowing letter in the *Field* in September of 1882 in reference to Richardson's tenacity not only to be innovative, but also to meet the needs of his customers.[255] Given Forwood was a co-founder of the Yacht Racing Association, as well as commodore of the Mersey and Windermere yacht clubs, his words wielded much weight. In the letter he stated that he was '*in want of a small cruising yacht to accommodate a party of four or five, with good headroom and plenty of elbow room, and yet that could be easily worked by a crew of five, all told, including a cook steward. I visited all the principal yacht depots for the purpose of buying such a vessel; my search was, however, fruitless. I could find no vessel with the accommodation I required, under 65 to 70 tons, and few vessels of even this tonnage had the roominess I desired, and, as these were all too large to be manned by a crew of five, I decided to ascertain if it was possible for a naval architect to design the yacht I wanted. I gave a note of my requirements to Mr. Alex Richardson, with the further instruction that she must be ballasted wholly with iron. Mr. Richardson designed for me a vessel, which has been built and launched this summer by Inman and Son, of Lymington; and, as I think, she marks a new departure in the construction of small cruising yachts*'.[256] Unfortunately there are no known images of the vessel.

In January of 1883 Richardson returned to Hampshire, this time superintending modifications to *Sameana* at the shipyard of J. G. Fay & Co., including fitting of a new and stouter mast.[257]

Several months later the Birkenhead boat and shipyards were beginning to present busy scenes in anticipation of the upcoming yachting season. Messrs Buckley and Sherlock were noted as building a cutter-rigged yacht, measuring 20ft by 6ft by 3ft 6in, to the order of S. Crowley of Southport to a design by Richardson.[258]

Even more noteworthy, this period coincided with the first instance of Walter Reeks being mentioned in the press for vessels he had designed. Specifically, two of his yachts were in the course of construction and nearing completion. The first of these was the 3½-ton yawl *Millie,* built by Gruer McGruer of Birkenhead to the order of Albert Coddington (older brother of Francis Coddington who only the previous year had commissioned the 10-tonner *Katie,* built by Inman's shipyard to a design by Richardson).[259] Stated to be exceedingly handsome, *Millie* measured 21ft by 7ft by 3ft 9in, with 1½ tons of iron on its keel, flush-decked and tastefully fitted below, all of

[255] *Field*, 9 September 1882; en.wikipedia.org/wiki/William_Bower_Forwood.

[256] *Field*, 9 September 1882.

[257] *Hampshire Independent*, 27 January 1883.

[258] *Field*, 31 March 1883.

[259] *Field*, 31 March 1883.

pine. Its beams were of half-round iron and apparently gave to the overheads a remarkably neat appearance and more headroom than would be the result if wood was used. *Millie* was in the water racing with the Cheshire Yacht Club by June of 1883.[260] The second vessel built to a design by Reeks was a 21ft cutter constructed by George Inman & Son at Lymington for William Grunsell.[261] The *Abeona* (later *Dorothy*) was launched in May of 1883.[262]

Substantial and significant commissions also continued to come Richardson's way. In late 1883 he received a contract from Mr Forster Connor of Belfast (who had previously commissioned the *Silver Star*) for a 60-ton composite racing yacht to be built at the Inman shipyard in Lymington.[263] The 74ft yacht *Marguerite* was launched in May of 1884. Of composite construction, the vessel was planked with teak and pitch pine and noted in the press, in reference to the popular Scottish naval architect George Lennox Watson who in 1873 had founded the world's first yacht design office dedicated to small craft, as somewhat of a '*sly-looking craft as she sits on the stocks, with quite a Watsonian look about the sheer*'.[264] *Marguerite* was also considered straighter in sheer than any vessel Richardson had previously designed, with a heavy rake to stern post, well-rounded off forward.[265] The craft solidified Richardson's reputation as one of Britain's leading modern yacht architects, though he was chided in the press at the time for not participating in ship and yacht model exhibitions. Richardson, however, was not the only one. Builders like Camper and Nicholson of Gosport, Inman of Lymington, and John White of Cowes were also on the naughty list.[266]

For Richardson the close of 1883 saw another customer return to the needs of his drawing board. John Jameson of the *Sameana* had tasked him with designing an 85-ton composite racing cutter to be built by J. G. Fay & Co. at Southampton.[267] The *Sameana* was subsequently advertised for sale.[268] The new vessel, *Irex,* was launched in mid-1884. It had a steel frame and the heaviest lead keel that had yet been placed on a yacht, weighing close to 72 tons.[269]

THE 85-TON RACING YACHT IREX (1884).
SOURCE: SIR EDWARD SULLIVAN (1894).
YACHTING: VOLUME II.

[260] *Birkenhead News*, 20 June 1883.
[261] *Field*, 31 March 1883.
[262] *Salisbury and Winchester Journal*, 19 May 1883; *Lloyd's Register of Yachts* (1903).
[263] *Field*, 27 October 1883.
[264] *Field*, 19 April 1884; en.wikipedia.org/wiki/George_Lennox_Watson.
[265] *Field*, 19 April 1884.

[266] *Field*, 6 May 1882.
[267] *Hampshire Advertiser*, 26 December 1883.
[268] *Field*, 2 February 1884.
[269] *Field*, 29 March 1884.

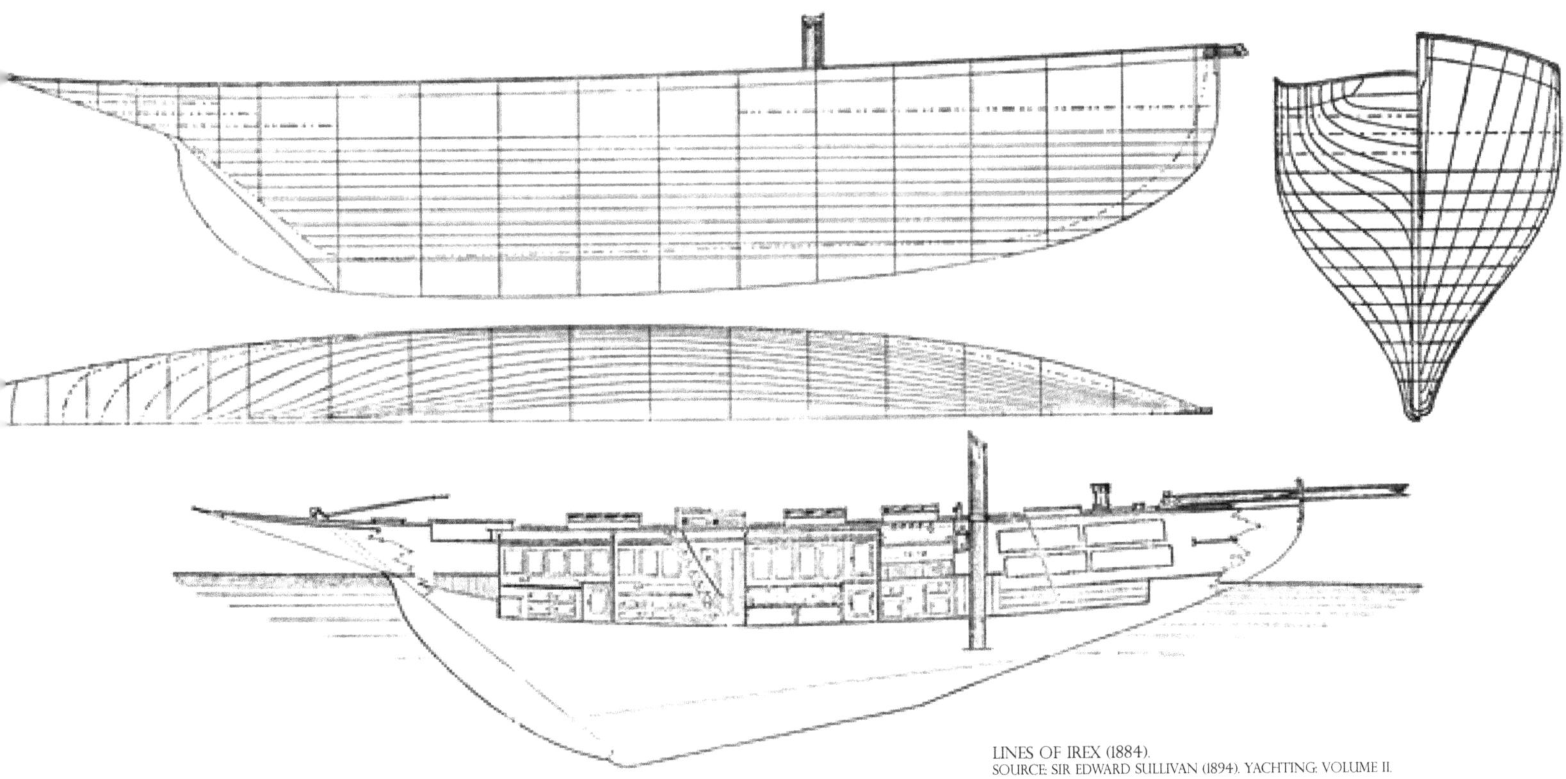
LINES OF IREX (1884).
SOURCE: SIR EDWARD SULLIVAN (1894). YACHTING: VOLUME II.

In March of 1884 there was launched from the shipyard of W. B. Thompson in Dundee, Scotland, a twin-screw tug steamer believed to be the largest and most powerful tug yet built.[270] Of 206ft in length, the vessel was constructed for the Retriever Steamship Company for use on the Hooghly River, India, the peculiar nature of the local currents requiring a tug specifically adapted for the trade.[271] With the hull designed by Richardson, the tug was smartly rigged as a fore-and-aft schooner with surface condensing engines of 1,500 h.p. collectively capable of propelling the vessel at 15 knots per hour.[272]

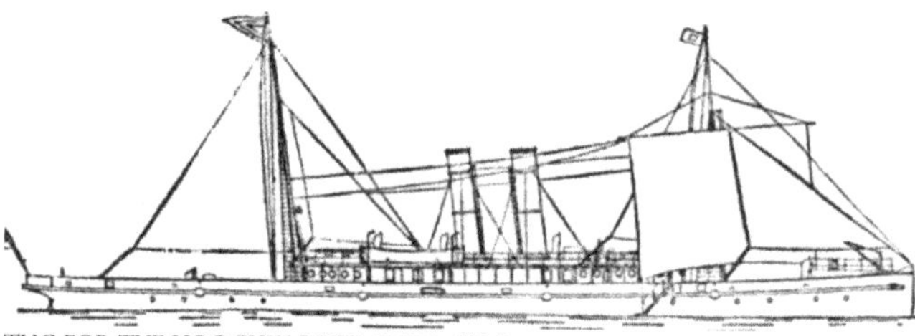

TUG FOR THE HOOGHLY RIVER, INDIA (1884).
SOURCE: DUNDEE ADVERTISER, 11 MARCH 1884.

It was likely in the summer of 1884, following completion of *Marguerite* at Inman's yard, that Walter Reeks moved to Liverpool, with newspaper reports noting him present at Rumford Street by this time.[273] Given the association between Richardson and the Inman yard, it was not a far off notion that Reeks would actively seek or perhaps have been sought by Richardson, particularly in the months after the death of George Inman in October of 1883. The move also allowed Reeks to be reunited with one of his first designs, the 21ft yacht *Millie.* The vessel had proved far more successful in local races on the Mersey than the *Katie*, though given the difference in their sizes, they did not compete in the same class. In fact, *Millie* showed the best record of the Southport-based yachts in 1884, sailed with consummate skill throughout the entire season by none other than Walter Reeks.[274] Deciding to enter the field of smaller yacht racing, Francis Coddington subsequently sold the *Katie* and instead commissioned the 3-ton *Snag*, built by Buckley and Sherlock of Birkenhead and launched in June 1884 to a design by Richardson.[275] This vessel also proved not up to the same calibre as *Millie.*

In terms of *Millie*'s subsequent history, the rather sudden death of Albert Coddington at the age of 43 in May of 1887 saw the vessel pass to his brother Francis.[276] Competing with the Southport Corinthian Sailing Club and the Lytham Yacht Club, the vessel distinguished itself as being '*a queen even among swift sailers*'.[277] *Millie* was still successfully racing in 1890 when Richardson was noted to be at the helm on at least one occasion.[278] In August of 1891 *Millie* won a handicap race at the Royal Anglesey Yacht Club Regatta in Wales with Francis Coddington at the helm.[279] However, the craft disappears from the records a few years later.

With regards to Richardson's professional activities, during mid-to-late 1884 his firm was charged with designing a composite 40-ton cruising cutter to be built by J. G. Fay & Co. to the order of George Moir of Manchester.[280] This vessel was launched in March of the following year and named *Challenge.*[281] Also designed towards the end of 1884 by Richardson and launched early the following year was a 47-ton composite racing cutter built by the same firm to the order of George Warren of Liverpool.[282] This craft, *Stranger,* was considered to exhibit more beam and less depth than its English counterparts and was built with the specific purpose of being sent to New York to race against the best of the American yachts, though not for the America's Cup itself.[283]

Two smaller vessels were additionally under construction during the end of 1884 to designs by Richardson. These were the *Launce,* a 3-ton cutter-rigged racing yacht built by Samuel Bond of Birkenhead to the order of J. H. Burton, commodore of the Southport Corinthian Sailing Club.[284] The vessel measured 27ft by 6ft 8in and was 21ft on the waterline.[285] Built nearby by J. Sherlock, who had just taken up a new yard, was *Lynette,* a 4-ton racing yacht of 31ft by 6ft 1in also designed by Richardson.[286] This particular vessel was commissioned by H. A. Graves to race with the Cheshire Yacht Club.[287]

The period, however, marked the start of a series of melancholy events for Richardson. On a professional level the newly-launched *Marquerite* was openly criticised in the press at the end of the 1884 yachting season, being described as the '*failure of the season*'.[288] Out of 20 starts the vessel had only managed to score two victories.[289] The article went on to describe *Marquerite* as a '*low uncanny-looking craft for sea going, and a sorry specimen of the modern type of racing vessel*'.[290] The *Irex* was also chided as being of extreme dimensions and only of modest success as a racing vessel, claiming nine wins out of 29 starts.[291] Still, the vessel had won the Royal Alfred Champion Cup making it somewhat credible. In contrast, the same article gave a glowing description of John Clark's steel-yawl *Wendur* of 125 tons that had been designed by George Lennox Watson, stating that it was '*unquestionably the fastest sailing yacht ever built*'.[292] The craft had won eight races out of 15 starts. For Richardson such negative publicity surely meant his reputation took a hit, while that of Watson's was very much on the rise.

[270] *Dundee Advertiser,* 11 March 1884.
[271] *Dundee Advertiser,* 11 March 1884.
[272] *Dundee Advertiser,* 11 March 1884.
[273] *Field,* 22 November 1884.
[274] *Field,* 22 November 1884.
[275] *Field,* 14 June 1884.
[276] *Preston Herald,* 18 May 1887; *Blackburn Standard,* 29 June 1889.
[277] *Blackburn Standard,* 29 June 1889.
[278] *Preston Herald,* 6 August 1890.
[279] *Liverpool Mercury,* 26 August 1891.
[280] *Field,* 31 May 1884.
[281] *Field,* 13 December 1884; *Hampshire Independent,* 28 March 1885.
[282] *Hampshire Independent,* 25 October 1884, 19 November 1884.
[283] *Hampshire Advertiser,* 8 November 1884; *Hampshire Independent,* 13 December 1884.
[284] *Field,* 14 February 1885; *Field,* 30 May 1885.
[285] *Field,* 14 February 1885.
[286] *Field,* 14 February 1885.
[287] *Field,* 14 February 1885; *Chester Courant,* 17 June 1885.
[288] *Hampshire Advertiser,* 4 October 1884; *Field,* 25 October 1884.
[289] *Hampshire Advertiser,* 4 October 1884.
[290] *Hampshire Advertiser,* 4 October 1884.
[291] *Hampshire Advertiser,* 4 October 1884.
[292] *Hampshire Advertiser,* 4 October 1884.

The end of 1884 also coincided with the end of Reeks' time in Liverpool and his employment with Richardson. Declining health and a recommendation to take a trip abroad saw Reeks take a passage on board the *Lancaster Castle* bound for Australia. The vessel left Liverpool on 21 December 1884.[293] Still, the overlap in their time working together was monumental in terms of both of their future careers. For Reeks, it set him up in good stead for the decades to come, while Richardson, only establishing himself as a professional naval architect less than eight years prior, would continue to receive important commissions in the coming years.

The departure of Reeks was followed by even more disastrous news for Richardson. His wife Lydia died on 26 March 1885 aged just 31 years at their house in Oxton, Cheshire, leaving him with four children between the ages of 8 months and 9 years to care for.[294]

—

Though this chapter profiling Alex Richardson is intended to provide context with regards to the time Walter Reeks spent working for him, some service is due to Richardson's ongoing career as a naval architect. As stated previously herein, the authors are unsure why so little has been written of Richardson and his life's work.

After weathering the loss of his wife, Richardson carried on working from his Rumford Street office, though there were very few known commissions received during the first half of 1885. Whether this was by choice or unintentional is not known. Richardson was, however, back sailing during the summer of that year with the Cheshire Yacht Club, on one occasion noted to be at the helm of the 4-ton *Lynette* which he had designed in 1884.[295]

In July of 1885 J. Sherlock of Birkenhead was announced as building a missionary schooner to a design by Richardson.[296] The 52ft vessel named *Nassau* was launched in December of that year intended for service along the west coast of Africa.[297] Also of note during this year, despite receiving criticism owing to its small size and light weight, George Warren's 78ft racing cutter *Stranger* made it safely across the Atlantic Ocean after a month-long journey. The vessel was then being prepped to race against the New York-based fleets.[298]

Richardson's reputation also earned redemption in 1885. The *Irex* was back in favour after undergoing a series of improvements. The vessel even earned the moniker '*Pride of the Season*' after claiming 13 victories, three second-place finishes and one third-place finish.[299]

In 1886 came the launch of the 91-ton composite schooner *Asteroid*, later *Astraea*, built for J. H. Burton by J. G. Fay & Co. of Southampton to a design by Richardson.[300] The vessel marked the second craft he had designed for the commodore of the Southport Corinthian Sailing Club in three years and was launched in May of that year.[301]

Continuing his involvement with the Yacht Racing Association, during this period Richardson joined forces with other naval architects, including Watson, Fife, Dixon Kemp and Clayton, to propose that the Tonnage Rules be altered to that of a rating.[302] Richardson expressed disfavour with the former, stating that because it favoured length '*its consequence had been an increase in the expense of racing yachts such that it was strangling the building altogether of vessels and practically killing the spirit of yacht racing*'.[303]

However, there were many differences of opinion regarding what alternate system should be implemented amongst Richardson and his cohort with some advocating for the sail area only to be the taxed dimension, including Richardson who believed this system would encourage international yacht racing. Others advocated for the sail area and length to be taxed, or for the sail area, length and breadth to be taxed.[304] It would take nearly a year of meetings for the changes to be agreed upon by the Yacht Racing Association with the final rule taxing waterline length and sail area only.[305]

Despite criticism from those that only wanted length taken into account, the new rule would have a lasting impact on the history of yacht design. It shepherded a fundamental change in direction toward assessing factors that affected speed, rather than those used to assess a vessel's commercial or cargo capacity.

Overall, however, it was unanimously agreed that rules were needed to enable yachts of a different size to race together. Historically, as noted in the previous chapter, an individual yacht's final position in a race amongst a mixed-sized fleet was calculated by a time allowance based on its size, which was then applied to the actual time taken to sail the course.[306] To assess size, yachts were given a rating as a means of handicapping.

The first rating rules were based on a formula that reflected only length and beam, and ignored the dominant speed-producing factors of waterline length and sail area. Implemented in 1843, the Builders Old Measurement, was simply the registered tonnage of the day. It had beam factored in three times and length only once. In 1855 this was replaced by another similar rule called the Thames Measurement but once again beam was factored in three times and length only once. Yacht designers were already awake to exploiting loopholes and whilst they played with the location of the points at which length was measured (essentially stem to sternpost), it was obvious that reducing beam reduced an individual yacht's rating or tonnage more significantly, resulting in a better time allowance for the same length. An alternate option was that for the same

[293] *London Evening Standard*, 23 December 1884.
[294] *Liverpool Mercury*, 30 March 1885; www.ancestry.com (Richardson family tree).
[295] *Chester Courant*, 17 June 1885.
[296] *Field*, 18 July 1885.
[297] *Field*, 5 December 1885.
[298] *Field*, 8 August 1885.
[299] *Hampshire Independent*, 23 September 1885.
[300] *Field*, 21 November 1885; *Hampshire Advertiser*, 15 May 1886; *Lloyd's Register of Yachts* (1903).
[301] *Hampshire Advertiser*, 15 May 1886.
[302] *Field*, 24 October 1885, 20 November 1886.
[303] *Hampshire Advertiser*, 11 December 1886.
[304] *Field*, 20 November 1886; *Hampshire Advertiser*, 25 December 1886.
[305] *Hampshire Advertiser*, 25 December 1886.
[306] The following discourse on rules is taken from various sources, but an excellent summary is found in C. A. Machaj (1996). *Seaworthiness: The Forgotten Factor.*

tonnage or rating, reducing beam further would allow an increase in length and therefore a faster yacht.

Whichever view was taken, gradually beam decreased relative to length and with it came a loss of hull-form stability. This loss was then countered by increased ballast, placed lower down and mostly inside the vessel, producing deeper hulls, primarily dependent on the ballast's centre of gravity for stability. All this combined to give a deep and narrow hull that ultimately became dangerous.

Concurrently, developments in vessel design theory were evolving worldwide through the work of prominent naval architects such as J. Scott Russell of Scotland, Colin Archer of Norway, and William Froude of England, amongst others. The nature and relationship of resistance, stability and sail power to hull shape were being studied and established. Some of the theories can now be seen as misguided in terms of correct hydrodynamics but they still gave a better design than what was previously on offer. Other theories were a preliminary step and even a foundation in the right engineering direction. However, all showed that a much better method of assessing a yacht's potential speed beyond beam and length factors was required.

Another significant figure in the United Kingdom during the mid-to-late nineteenth century was the British naval architect Dixon Kemp, a founding member of the Yacht Racing Association and, along with Richardson, a member of the committee which eventually established the final rule endorsed by the organisation in 1886. A key moment leading up to the rule's adoption occurred in May of that year with the capsize of the English cutter *Oona* and the loss of all crew, including its designer W. E. Patton. The yacht had the most extreme narrow and deep hull built to date, along with a huge ballast ratio that had yet been designed. Whilst *Oona*'s capsize highlighted the dangerous loss of hull-form stability that had occurred through the taxing of beam, another underlying problem was that older boats were being outdated too quickly by newer boats, and yachtsmen wanted some stability in the rule as well as better longevity of their craft. In November of 1886 the Yacht Racing Association adopted what was colloquially called the 'Dixon Kemp Rule', otherwise known as the 'Length & Sail Area Rule'. The formula was Waterline Length by Sail Area (imperial measurements) divided by 6,000 giving an answer in units known as 'sail tons'.

Internationally, the Dixon Kemp Rule was not the first rule of this nature. Yachting was an equally important pastime in the United States. The Seawanhaka Corinthian Yacht Club of New York first adopted a rating rule to govern all of its races in 1882. Referred to as the 'Seawanhaka Rule', the calculation involved load waterline length plus the square root of sail area, divided by two. The rule's success saw it serve as a rating for all eastern-seaboard races from 1887 on, including the America's Cup between 1893 and 1903.[307] Interestingly, the calculation had been guided by Dixon Kemp's initial proposal for a formula combining length and sail area in a rating rule, which was subsequently modified by the Seawanhaka Corinthian Yacht Club and has in fact been kept in use up to contemporary times. The British were also garrison to the new rule, it just took longer for the argument to take hold and be implemented.

Whilst the Dixon Kemp Rule began to respond to the problem of yacht stability, as well as assessing some of the speed-producing factors, subsequent history shows it did not satisfy the problem of older boats being outdated by newer ones. Indeed, this has been an eternal problem created by advances in technology and a designer's never-ending quest for improvement. Over time many new rules and even one-design classes have been established to try and solve this second problem, but it is one that frustratingly just won't quite go away.

In 1886 though, the Dixon Kemp Rule was an important first step in the right direction. Whilst Richardson can take

[307] en.wikipedia.org/wiki/Seawanhaka_Corinthian_Yacht_Club.

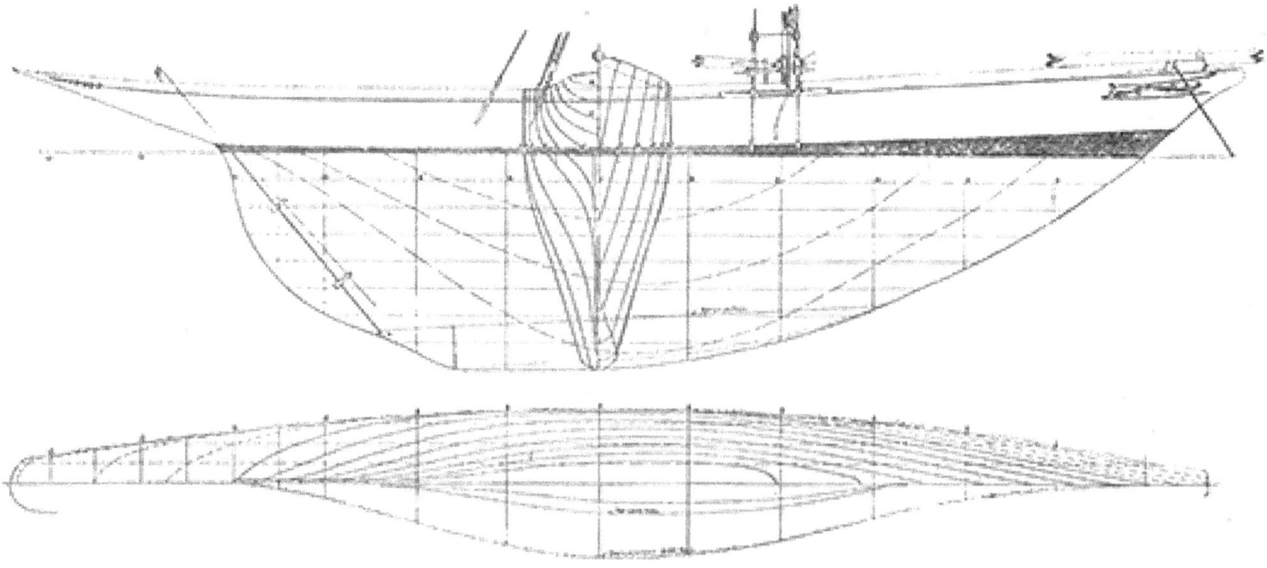

LINES OF OONA (1886).
SOURCE: THEO RYE (WWW.CLASSICBOAT.CO.UK/ARTICLES/GREAT-YACHT-DESIGNS-II-OONA/).

some credit for helping this change to occur, likely without knowing it, a short time later it would be his pupil Walter Reeks who in 1887, working on the opposite side of the globe, probably became the first designer to launch a new yacht designed to respond to the new rule.

For Richardson, however, he initially remained committed to his existing design practices and of service to his predominantly Liverpool-based customer base. In mid-1887 the 12-ton cutter *Brena* was launched, constructed by J. Sherlock of Birkenhead to a design by Richardson.[308] It was built to the order of a Mr Cox to be moored at Beaumaris.[309] By the same firm and designer combination, 1887 also saw the launch of a 24ft cutter with a large roomy cockpit for a Mr King of Beaumaris, named *Lucinda*, and a half-decked 18ft square-sterned boat for a customer based in Cork, possibly the *Surprise*.[310] That same year Richardson was on board the *Irex* during a Royal Mersey Yacht Club Regatta race against the *Thistle*, perhaps somewhat startled by how easily the latter vessel slipped through the lee and won the race.[311] *Thistle* was one of the first yachts George Lennox Watson had designed to the newly-implemented Dixon Kemp Rule and was built in Scotland under strict secrecy for a local syndicate and launched in April of 1887.[312] Soon thereafter the craft was sailed to New York where it unsuccessfully challenged the American yacht *Volunteer* for the America's Cup.[313]

Richardson continued to partner with J. G. Fay & Co. of Southampton during 1888 with the launch in August of the 60-ton rating composite racing cutter *Petronilla* to his design.[314] The vessel was built to the order of the Earl of Dunraven who had also been on board *Irex* when it unsuccessfully raced against *Thistle*.[315]

Also in 1888, from the yard of J. Sherlock of Birkenhead, came the *Redwing*, a 36ft by 9ft lugger-rigged centreboard boat Richardson had designed for W. G. Jameson to cruise the Irish coast.[316] A third vessel for J. H. Burton was also constructed at this yard to a design by Richardson: the 10-ton cutter *Odille* was launched in late May just in time for the opening of the Royal Mersey Yacht Club season.[317] Another Richardson-designed vessel launched in 1888 was the cutter *Waterwitch*, built by Samuel Bond of Birkenhead to the order of a Mr Potter of Manchester for use on Lake Windermere.[318]

In September of 1889, constructed by C. Hansen & Sons of Cowes, came the 128-ton composite schooner *Asterope*. Designed by Richardson it marked the fourth vessel he had designed for J. H Burton.[319] *Asterope* was intended for long distance sea cruising.[320] That same year Richardson designed two 5,000-ton merchant ships with the London press reporting that '*in consequence,* [he has] *little time to devote to his favourite pursuit of yacht designing*'.[321] Unfortunately, despite their size, there is little additional information known about these two vessels.

Richardson returned to his drawing board in late 1889, however, at the request of John Jameson charged with designing a new composite racing cutter.[322] The *Irex* was subsequently sold to Count Stroganoff, a Russian Prince, with the new 84ft vessel eagerly anticipated by the yachting fraternity. It was launched from the yard of J. G. Fay & Co. in May of 1890 as the *Iverna*.[323]

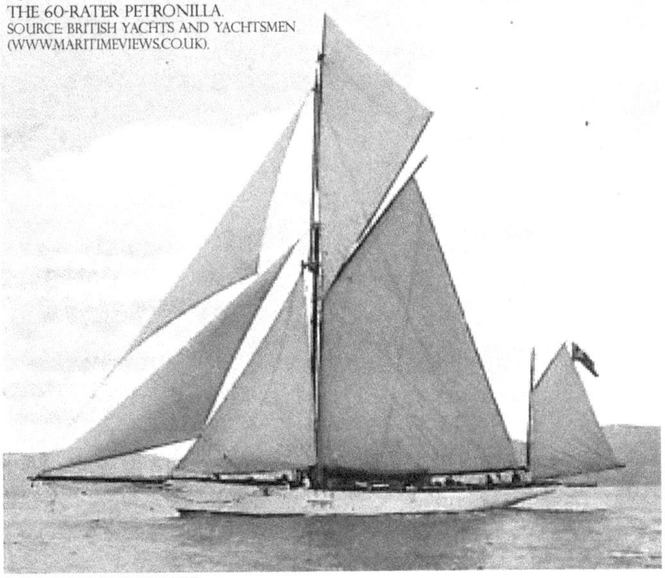

THE 60-RATER PETRONILLA.
SOURCE: BRITISH YACHTS AND YACHTSMEN (WWW.MARITIMEVIEWS.CO.UK).

THE 84-TON IVERNA (1890).
SOURCE: R. T. PRITCHETT, ET AL (2013). GUTENBERG EBOOK OF YACHTING VOL. 2.

[308] *Birkenhead News*, 9 July 1887; *Lloyd's Register of Yachts* (1903).
[309] *Birkenhead News*, 14 May 1887, 9 July 1887.
[310] *Liverpool Daily Post*, 9 April 1887; *Birkenhead News*, 14 May 1887; *Lloyd's Register of Yachts* (1903).
[311] *Field*, 25 June 1887.
[312] *Field*, 25 June 1887.
[313] en.wikipedia.org/wiki/Thistle_(yacht).
[314] *Field*, 21 January 1888; *Hampshire Advertiser*, 2 May 1888.
[315] *Hampshire Advertiser*, 2 May 1888; Motorboating - ND, January 1946.
[316] *Field*, 19 May 1888; *Lloyd's Register of Yachts* (1903).
[317] *Liverpool Daily Post*, 21 May 1888.
[318] *Liverpool Daily Post*, 21 May 1888.
[319] *Field*, 28 September 1889; *Lloyd's Register of Yachts* (1903).
[320] *Field*, 13 April 1889.
[321] *Field*, 13 April 1889.
[322] *Hampshire Independent*, 21 September 1889.
[323] *Hampshire Advertiser*, 18 January 1890; *Field*, 26 April 1890.

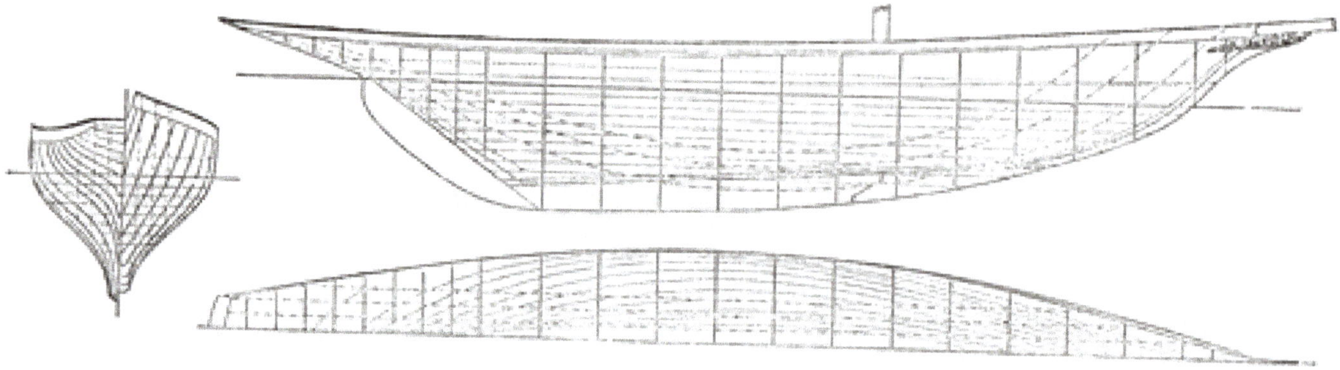

LINES OF THE IVERNA (1890).
SOURCE: R. T. PRITCHETT, ET AL (2013). GUTENBERG EBOOK OF YACHTING VOL 2.

The 1890s was another busy decade for Richardson. It was also the era of small yacht building and the development of innovative steam yachts and launches. The pastime of sailing was growing in popularity and extending its reach to the middle class. With this came the need for more recreational vessels and smaller racing yachts. For Richardson, with several prominent exceptions, it was a decade where he primarily designed smaller craft to meet the needs of a growing local customer base. During the early part of this decade he also moved to newly-constructed offices at 5 Castle Street in Liverpool.[324]

RECENT IMAGE OF 3 - 5 CASTLE STREET, LIVERPOOL. BUILT IN 1889.
SOURCE: JOHN BRADLEY (WIKIMEDIA COMMONS).

[324] *Field*, 9 January 1892.

In 1891 from Richardson's design board came the 20-rater cruising yacht *Nesta,* built by J. G. Fay & Co. at Southampton to the order of G. R. Cox of Liverpool.[325] It was launched in July of that year and was a popular and successful addition to the Royal Mersey Yacht Club's fleet, often sailed by Richardson in races.[326] Also in 1891 Richardson designed a steam yacht named *Ilma* for George W. Anderson of Liverpool of 58ft 6in (waterline) by 11ft beam.[327] The craft was built of steel supplied by Messrs Allsup and Sons of Preston and was fitted with Willans' patent compound engines.[328] *Ilma* was intended for general cruising on the lakes of the West Highlands.[329]

A 25-ton composite centreboard cutter named *Bronwen*, built by Hansen & Sons of Cowes to a design by Richardson was launched in 1892.[330] Also launched this year from the yard of Samuel Bond of Birkenhead to a Richardson design were two identical one-raters, the *Crane* and *Lady Maud.*[331] The former was built to the order of Colonel Gamble, commodore of the Royal Mersey Yacht Club, while the latter was built for S. G. Sinclair, vice-commodore of the same club.[332]

Late in 1892 came the launch of the pilot boat *George Holt,* a composite schooner of 101ft built by Philips & Son of Dartmouth (Devon) to the order of the Mersey Docks and Harbour Board to a design by Richardson.[333] During this period he was also publicly thanked, along with the likes of Watson, Fife, Nicholson, Clayton and Payne, for the submission of opinions to the Yacht Racing Association regarding the Dixon Kemp Rule which was set to expire at the end of 1893.[334] This auspicious group of yacht designers, after meeting at the Langham Hotel in London in October of that year, jointly drafted a letter in which they condemned the current system of measurement and instead offered an alternate system that would tend towards producing a bigger-bodied vessel, with more compact dimensions, adding that beam and

[325] *Field*, 29 August 1891.
[326] *Field*, 11 July 1891, 27 August 1892.
[327] *Field*, 28 November 1891.
[328] *Field*, 28 November 1891.
[329] *Field*, 28 November 1891.
[330] *Lloyd's Register of Yachts* (1903).
[331] *Field*, 25 June 1892.
[332] *Field*, 25 June 1892.
[333] *Liverpool Mercury*, 23 November 1892; *Lloyd's Register of Yachts* (1903).
[334] *Hampshire Advertiser*, 19 November 1892.

depth, or beam and girth, should be considered in the measurement formula.[335] It would take three years before a new rule, known as the 'International Rule' or 'Metre Rule', was implemented, in which length, beam, three-fourths of the underwater girth at 0.6 of the loadline from the bow, and one-half the square root of sail area, were all added together and divided by two.[336]

GEORGE HOLT, PILOT BOAT (1892).
SOURCE: NATIONAL MUSEUMS LIVERPOOL (MCR/90/61).

In 1893 came the launch of another one-rater designed by Richardson to the order of S. G. Sinclair to compete with the Royal Mersey Yacht Club.[337] Named *Minuet*, the vessel made its first appearance in June of that year being defeated by *Unit* (ex *Crane*).[338] The following year the 12-ton yacht *Irene* was built by Samuel Bond of Birkenhead to a design by Richardson.[339]

1895 saw Richardson spend more time in London discussing proposed changes to the Dixon Kemp Rule, including advocating for the International Rule.[340] In addition, in April of that year from the yard of Hansen & Sons of Cowes came the 39ft (waterline) cruising cutter *Rhoda,* built to the order of F. S. Schwabe of Manchester to a design by Richardson.[341] He also personally attended the launch of this vessel.[342] Another vessel completed in 1895 to his design was *Bojum*, a 3-ton centreboard lugger built by Samuel Bond of Birkenhead.[343] Two more one-raters were launched this year from the yard of Samuel Bond (the *Eira* and *Morwyn*), both were designed by Richardson.[344]

The following year Hansen & Sons launched the *St. Agnes*, a composite 65-ton cruising cutter built to the order of Vincent S. Caroll to a design by Richardson, as well as the *Alruda,* a comfortable 30-ton cutter for C. McIvor of Liverpool that was jointly designed by Richardson in coordination with Charles Livingstone.[345] Also launched in 1896 was the steam yacht *White Friar*, of 82 tons, built by J. Reid and Co. of Glasgow to the order of J. H. Burton to a design by Richardson.[346]

In 1897 Samuel Bond launched from his Birkenhead yard the 3-ton *Spindrift,* built to the order of H. R. Davies to a design by Richardson.[347] Competing with the Menai Straits centreboard class, the vessel was a standout that season, undoubtedly the fastest boat of the fleet, especially in light and moderate breezes.[348]

Also launched in 1897 was the 7-ton cutter *Myfanwy*, built by Samuel Bond to a design by Richardson. Of note, this vessel is one of 1,600 currently listed on the National Historic Ships UK Register.[349]

In February of the following year the steel steam yacht *Skipjack* was completed, built by W. H. Potter & Sons of Liverpool to the order of Sir Richard Williams Bulkeley of Beaumaris.[350] The 84ft (waterline) vessel was designed by Richardson.[351]

The close of the decade saw the launch of the rather substantial steel screw steam yacht *Ziska*, built by the Ailsa Shipbuilding Company in Troon to the order of S. Taylor Chadwick of Anglesey.[352] Designed by Richardson, the vessel measured 101ft 8in (waterline) with a moulded breadth of 18ft 6in and a depth of 11ft 6in.[353] *Ziska's* triple-expansion machinery were supplied by Messrs Muir and Houston of Glasgow.[354] In 1901 the craft was purchased by the Lancashire Sea Fisheries Joint Committee and renamed *John Fell*.[355] Another name change, to *Majesta*, occurred in 1910, along with several subsequent changes of ownership.[356] *Majesta* remained active for many decades, seeing service as a harbour defence patrol craft on the Thames during World War Two.[357] It was broken up in 1952.[358]

The new century saw Richardson entering his mid-50s and potentially contemplating retirement. At the time he was living in a rather comfortable house at 4 Kingsmead Road in Oxton, Cheshire, with his 22-year-old daughter Constance and 16-year-old son Foster.[359] His older two children, twins Lydia and Alexander (Jr), had both left home with the former moving to St Andrews, Scotland, with her husband Joseph Bowman Grimond, a wealthy merchant and jute manufacturer.[360] His son, Alexander (Jr), remained in Cheshire, embarking on a career as a manufacturer of disinfectants.

335 *Field*, 19 November 1892.
336 www.maritimeviews.co.uk/british-yachts-yachtsmen/chapter-10-modern-yachting-1896-1900/.
337 *Field*, 17 June 1893.
338 *Field*, 17 June 1893.
339 *Lloyd's Register of Yachts* (1903).
340 *Field*, 9 February 1895.
341 *Portsmouth Evening News*, 13 April 1895.
342 *Field*, 13 April 1895.
343 *Lloyd's Register of Yachts* (1903)
344 *Lloyd's Register of Yachts* (1903).
345 *Field*, 2 May 1896.
346 *Lloyd's Register of Yachts* (1903).
347 *Field*, 16 October 1897.
348 *Field*, 16 October 1897.
349 www.nationalhistoricships.org.uk/register/2449/myfanwy.
350 *Sporting Gazette*, 26 February 1898.
351 *Lloyd's Register of Yachts* (1903).
352 *Lloyd's List*, 15 April 1899.
353 *Lloyd's List*, 15 April 1899.
354 *Lloyd's List*, 15 April 1899.
355 www.clydeships.co.uk (vessel Ziska).
356 www.clydeships.co.uk (vessel Ziska).
357 www.clydeships.co.uk (vessel Ziska).
358 www.helensburgh-heritage.co.uk.
359 *1901 England Census* (Class RG13; Piece 3399; Folio 79).
360 *Scotland, National Probate Index*, 1929 for Joseph Bowman Grimond.

ZISKA (AS MAJESTA IN THE 1930S).
SOURCE: FIONA HOWARD, HELENSBURGH HERITAGE TRUST, 19 OCTOBER 2019.

In terms of Richardson's professional activities, in 1900 *Eira*, a small wooden centreboard cutter of 22ft which he designed was built by Samuel Bond of Birkenhead to the order of S. Taylor Chadwick.[361] The vessel competed with the Royal Anglesey Yacht Club in Beaumaris.[362] The following year came the 40ft steam launch *Novice* built by W. Roberts of Chester, the fourth vessel Richardson had designed for J. H. Burton.[363] Also completed to his design in 1901 was the 37-ton steam yacht *Zella,* another vessel built by Samuel Wood.[364] During this year Richardson was additionally noted as superintending the conversion of the cutter *Stranger* to a yawl.[365]

In June of 1902 the *James J. Bibby,* a 253-ton steel brigatine was launched, built by the Garston Graving Dock and Shipbuilding Co.[366] Designed by Richardson, the vessel was intended for the Liverpool Training School for Boys as a sail training ship in which occupation it spent several years.[367] Under a few different names, including HMS *Dargle*, the vessel saw service in World War I, after which it remained in use as a privately-owned commercial vessel for a number of decades.[368] Unfortunately no known photo exists of this vessel, despite its significant history and large size.

Between 1906 and 1908 three vessels were built by Philip & Son Ltd. of Dartmouth to designs by Richardson. These are the last vessels known to have been designed by him before retirement. First, in May of 1906, was launched the steel-composite schooner yacht *Dwyn Wen*, again built to the order of his most familiar customer, J. H. Burton of Beaumaris.[369] At 128ft, the considerable vessel remained in existence until quite recently, having sailed the world many times over.[370]

[361] *Lloyd's Register of Yachts* (1903).
[362] *North Wales Chronicle*, 11 August 1900.
[363] *Lloyd's Register of Yachts* (1903).
[364] *Field*, 11 July 1903; Theo Rye, pers. comm.
[365] *Field*, 23 March 1901.
[366] shippingandshipbuilding.uk.
[367] www.shipsnostalgia.com/threads/hms-dargle.36521/.
[368] www.shipsnostalgia.com/threads/hms-dargle.36521/.
[369] *Field*, 19 May 1906.
[370] classicyachtinfo.com/yachts/dwyn-wen/.

COMPOSITE SCHOONER DWYN WEN.
SOURCE: CLASSICYACHTINFO.COM/YACHTS/DWYN-WEN/.

In September of 1906 Richardson was tasked by the Lancashire and Western Sea Fisheries Committee with designing a new twin-screw steamer.[371] Launched a year later as the *James Fletcher* of 330 tons from the yard of Philip & Son Ltd., the vessel spent several years carrying out police work as well as scientific investigations before being requisitioned for war service and, following, being sold for commercial use.[372]

JAMES FLETCHER (CIRCA 1920S).
SOURCE: STEVEN BROWN (2016). PATROL VESSEL HISTORY
(WWW.NW-IFCA.GOV.UK/APP/UPLOADS/SB-PATROL-VESSEL-HISTORY-10-3-16.PDF).

Finally, the 12-metre yacht *White Heather* built to the order of Harter K. Glazebrook of Liverpool was launched from the Philip & Son Ltd. yard in mid-1909.[373] Designed by Richardson, it was one of the first of its class built in Britain.[374] Fittingly, as the last known vessel Richardson was involved with, it remains in existence as the *Varuna.*[375]

Shortly after completion of *White Heather*, at the age of 65, Richardson retired. He spent his later years living with his youngest daughter Constance, by now in her early 30s, at a smaller house at 29 Curzon Road, Prenton, in Birkenhead. Richardson died at Birkenhead in November of 1915.[376] A short eulogy published in the London press announcing his death chiefly reported his career accomplishments as the *Sameana*, *Irex* and *Iverna*.[377]

[371] *Blackpool Gazette & Herald*, 25 September 1906.
[372] *Bury Times*, 14 September 1907; *Burnley Express*, 11 April 1908; www.nw-ifca.gov.uk/app/uploads/SB-Patrol-Vessel-History-10-3-16.pdf.
[373] *Dartmouth & South Hams Chronicle*, 30 April 1909; classicyachtinfo.com/yachts/varuna/.
[374] *Field*, 7 May 1910.
[375] classicyachtinfo.com/designer/a-richardson/.
[376] *Illustrated Sporting and Dramatic News*, 27 November 1915.
[377] *Illustrated Sporting and Dramatic News*, 27 November 1915.

WHITE HEATHER (AS VARUNA).
SOURCE: CLASSICYACHTINFO.COM/YACHTS/VARUNA/.

With regards to his children, Richardson's daughter Constance remained in the Cheshire area, in 1926 at the age of 48, marrying Edward Beattie.[378] His youngest son Foster left home around 1910 and moved permanently to St Andrews (Scotland) after a few years spent in London. He was employed as a director of Jute Industries of Dundee (a company in which his brother-in-law Joseph Bowman Grimond was partner) and travelled overseas regularly throughout his career.[379] Foster Richardson also fought in World War I and was a prominent and successful member of the Royal and Ancient Golf Club. Playing off scratch, he won the Calcutta Cup in 1910 and 1924.[380]

Of Richardson's grandchildren, his daughter Lydia's youngest son, Jo Grimond, would become a prominent British politician, including being elected leader of the Liberal Party between 1956 and 1967 and again briefly on an interim basis in 1976.[381] Lydia's only daughter, Gwyn, married Sir Thomas Godfrey Polson Corbett who in 1931 became the 2nd Baron of Rowallan.[382] Of note Baron Rowallan, following a distinguished career in the military, was appointed Governor of Tasmania in 1959, serving for four years in this role.[383] During this period he also took an active role in sailing, purchasing the Derwent Class yacht *Nymph*.[384]

Overall, after initially gaining employment as a cotton broker, Richardson spent more than 30 years practicing as a successful naval architect in Liverpool, one of the earliest of his profession to do so. Over 60 vessels, small and large, recreational and commercial, are known to have been constructed to his designs.

Richardson's career benefitted immensely from repeat customers and high-profile patrons. He, in turn, also loyally served the boat and shipyards of Inmans of Lymington, J. G. Fay of Southampton, and J. Sherlock and Samuel Bond, both of Birkenhead. It was a period of innovation and design changes, including three separate Yacht Racing Association rules being implemented during his career, singularly and collectively having a huge impact on the types of vessels Richardson produced. Technological changes in terms of steam propulsion, and the implementation of new materials like iron and steel, also saw him capitalise on the growing need for commercial and working vessels of these types, including those to service the port of Liverpool's growing passenger and cargo trade.

Unfortunately, as far as we can tell at least, history has done Richardson little favours. His work has been largely overshadowed by designers considered more prominent, particularly Watson and Fife. In a short biography published about Richardson in the *Encyclopedia of Yacht Designers* (2005), he was stated to have been rather reclusive and not one to give interviews to the press. This shyness, whether real or perceived, may be why his impressive resume has been cloaked by the work of others. We hope this chapter provides him with some much-needed credence and credit, as well as does justice to our main objective of putting Walter Reeks' time spent working with Richardson in Liverpool in the early 1880s into some context.

ALEXANDER RICHARDSON (1844 - 1915).
SOURCE: THE YACHTSMAN, 16 MAY 1892.

[378] www.ancestry.com (Richardson family tree).
[379] *St Andrews Citizen*, 6 March 1948.
[380] *St Andrews Citizen*, 6 March 1948.
[381] en.wikipedia.org/wiki/Jo_Grimond
[382] en.wikipedia.org/wiki/Thomas_Corbett,_2nd_Baron_Rowallan.
[383] en.wikipedia.org/wiki/Thomas_Corbett,_2nd_Baron_Rowallan.
[384] adb.anu.edu.au/biography/rowallan-second-baron-11571.

1890 MAP OF SYDNEY AND ADJACENT MUNICIPALITIES.
SOURCE: NATIONAL LIBRARY OF AUSTRALIA (NLA.GOV.AU/NLA.OBJ-229943868).

CHAPTER FOUR - SYDNEY

We now return to the subject of our book, Walter Reeks, with the hope that the previous chapter provided context to his time working with and being mentored by Alex Richardson, naval architect of Liverpool. The two men communicated regularly throughout their later careers, and would even work together on other noteworthy projects, albeit via correspondence. For Reeks, however, a life-changing and surprising adventure was about to begin.

The fully-rigged ship *Lancaster Castle* was launched in October of 1884 from Workington, Cumberland, England, built by R. Williamson and Son.[385] With a net register of 2,095 tons and a length of 279ft, it was considered the largest vessel yet built in the small port town located 140 miles to the north of Liverpool. The *Lancaster Castle* was constructed entirely of iron with its masts and yards of steel and its upper deck of wood. The ship was built to the

[385] *Carlisle Journal*, 21 October 1884.

LANCASTER CASTLE AT PORT ADELAIDE IN THE EARLY 1900S (AS THE MARGRETHA).
SOURCE: STATE LIBRARY OF SOUTH AUSTRALIA (PRG 1373/8/28).

SCHEDULE A.

PASSENGERS' LIST.

REFERRED TO IN THE 8TH SECTION OF "THE COLONIAL PASSENGERS ACT 186A"

LIST of the PASSENGERS who have arrived at the Port of *Melbourne* on the *10th* day of *Mch*, 18*85*, from *Liverpool* on board the *Lancaster Castle* Ship, *John Sharpe* Master, *2095* Tons Register.

	Names of Passengers.	Ages. Adults. Married. M.	F.	Single. M.	F.	Children. 12—1. M.	F.	Infants. M.	F.	State whether British Subjects or Foreigners.
Number of Superficial Feet in Passengers' Deck.	*Mr W. Reeks*			*24*						
Greatest Number of Statute Adults, exclusive of Master, Crew, and Cabin Passengers, the Ship can legally carry.										

WALTER REEKS' ARRIVAL IN MELBOURNE ON BOARD THE LANCASTER CASTLE IN MARCH OF 1885. HE WAS THE ONLY PASSENGER.
SOURCE: VICTORIA, AUSTRALIA, ASSISTED AND UNASSISTED PASSENGER LISTS 1839 - 1923.

order of James Chambers and Co. of Liverpool and was to be commanded by Captain John Thorpe.[386] There is a reference, admittedly published in 1913, stating that Reeks was involved with the vessel's design. If true, this was more than likely through his employment with Richardson.[387] Perhaps this connection was how Reeks was able to secure a place on board. For the *Lancaster Castle* was not a passenger vessel by any means, instead the ship was to be involved in international trade.

The *Lancaster Castle* left Liverpool on its maiden voyage to Australia on 21 December 1884 having only arrived in port the week before, under tow from Workington.[388] The cargo manifest reveals a mixture of iron and chemical products, as well as machinery, glass, rum, whisky, beer, salt and paper goods, totalling 3,750 tons.[389] The vessel crossed the equator on 15 January, thus introducing Reeks to King Neptune and the line-crossing initiation ceremony, and after an 89-day direct passage, arrived in Melbourne well after sunset on 20 March 1885.[390] Reeks was the only passenger noted as disembarking.[391] He remained in Victoria for several weeks, probably visiting the city's yacht clubs, before travelling to Sydney via the A.S.N. steamer *Cahors.* Having left Melbourne on 11 April, Reeks arrived in New South Wales two days later, beginning his lifelong association with Australia.[392] We don't know why and when he made the decision to stay and then settle in Sydney. There is nothing known that indicates he left England with this intention, other than a report that Reeks had come '*to the colony in search of a congenial climate*', suggesting it was for health reasons.[393] However, within weeks of his arrival in Sydney, it was apparent to the newspapers that Reeks was here to stay.

The Sydney press soon got wind of Reeks' presence in Sydney, and his professional pedigree. He was very likely the first internationally-qualified naval architect to relocate to the city and practice this profession exclusively. Though the respected and very successful engineer Norman Selfe (1839 - 1911) was well established in the civil field in Sydney at the time, he included naval architecture as one aspect of several engineering-related works he was capable of undertaking.

With Reeks' credentials more than welcome, it was not long before he received invitations from Sydney's upper echelon and his experience and expertise was called upon. It is probable that he also carried letters of introduction to prominent members of Sydney's maritime community, and sought out introductions to specific members. James Milson (Jr) of the Royal Sydney Yacht Squadron was likely at the top of his list, having commissioned the small cutter *Era* built by the Inman yard at Lymington two decades prior.

On 30 May 1885, just seven weeks after Reeks' arrival, an opinion piece published in *The Sydney Mail and New South Wales Advertiser* by 'Bobstay', believed to be a prominent member of the Royal Navy's 'Flying Squadron', the flagship being HMS *Bacchante*, which visited Australia in the early 1880s, discussed the merits of naval architects and the immense strides that the profession had made in yacht designing.[394] He specifically stated that '*No man thinks of building nowadays from designs other than those supplied by a recognised and qualified naval architect, and, to digress for a moment, I find one of these, Mr. Reeks, of long experience in England, is about to practice in Sydney'*.[395] Given the author's high standing in Sydney society and amongst its yachtsmen, this ode to Reeks was something of a novelty. It also served as a pivotal launch pad for not only his admittance to a new community but also his access to its members. As well, it made clear that naval architects were the personification of modern yacht design and that no one should consider building a yacht without one.

Sydney in 1885 was a thriving city of some 250,000 people.[396] Approaching 100 years since the arrival of white Europeans in search of a place to banish Great Britain's criminals, the metropolis had shed its convict past with an energy and eagerness to situate itself as Australia's leading commercial, economic and cultural centre. Railway lines connected Sydney to Melbourne, telegraph lines increased intra and inter-colonial communication, and the gold rush of the 1850s had brought thousands and thousands of industrious free settlers to the colonies. Singularly and collectively, New South Wales, Victoria, Queensland, Tasmania, South Australia and Western Australia were becoming more independent from the 'mother country', slowly coalescing to become a federated nation.

With this increased mobility and upward wealth, recreational pursuits and pastimes came into vogue. Sydney Harbour offered bountiful opportunity for both. Amongst the city's earliest yacht and sailings clubs were the Royal Sydney Yacht Squadron and the Prince Alfred Yacht Club, both established in the 1860s. From sporadically-organised races between small open boats, naval and working craft beginning in the 1820s, local yachting had blossomed into a competitive and well-supported sport, with organised races being particularly popular amongst those well-known and well-resourced to splurge on ever-increasingly faster and larger yachts. Everyone wants to win. It is much easier to do so when you can afford to have the latest racing yacht custom designed and built for you.

Despite being welcomed with open arms to the new colony, Reeks was more than likely in need of an immediate income. Hence, it was not long before he applied his entrepreneurial skills and advertised for pupils that he could teach '*naval architecture and engineering*' to at a moderate premium out of his newly-established office at 139 Pitt Street.[397] Reeks obviously recognised that there

386 *Carlisle Journal*, 21 October 1884.
387 *The Salon*, Vol. 2. No. 4 (1 November 1913).
388 *London Evening Standard*, 23 December 1884.
389 *The Age*, 23 March 1885; *The Argus*, 23 March 1885.
390 *The Argus*, 21, 23 March 1885.
391 *The Age*, 23 March 1885.
392 *The Argus*, 13 April 1885; *The Daily Telegraph*, 14 April 1885.
393 *Sydney Mail and New South Wales Advertiser*, 6 November 1907.

394 *The Sydney Mail and New South Wales Advertiser*, 30 May 1885.
395 *The Sydney Mail and New South Wales Advertiser*, 30 May 1885.
396 en.wikipedia.org/wiki/Timeline_of_Sydney.
397 *The Sydney Morning Herald*, 13, 14 July 1885.

was a market for naval architects and engineers in the city, and that he could be remunerated for passing on his expertise. In an effort to further qualify his presence, Reeks also began venturing his professional opinion with regard to various local yachting matters.

The presence of Reeks in Sydney, however, appears, to have been inadvertently used to fuel a notion that yacht builders were to kowtow to those of naval architects and yacht designers. It began in relation to the surprising sale of the Victorian yacht *Assegai*, a 6-ton vessel designed and built in Melbourne in 1882 by Henry R. Murray, a rather prickly member of Victoria's yachting scene with a bull-headed tendency to overstate both his opinions and qualifications.[398] Originally built for C. H. Boyd, a solicitor's clerk who was sensationally arrested for fraud in association with providing worthless bills of payment for the vessel, *Assegai* was transferred to Sydney in June of 1885 after being purchased by G. F. Murnin.[399]

Even in this time period, there was a distinct divide between Victoria and New South Wales, with the eternal 'Melbourne versus Sydney' contest well underway. A newspaper report from 4 July 1885 refers to this bickering. '*The novelty of a Victorian designed and built yacht being purchased by a Sydney yachting gentleman, with the view of being put in racing commission for next season's matches, will come with somewhat of a surprise to most Victorian yachting men, and especially to those who have hitherto looked to the Sydney yacht fleet to supply their wants of out-classed racers for Victorian match sailing and cruising*'.[400] It continued in the same vein further on. '*Knowing how locally jealous Sydney people are of anything Victorian, and especially in view of the fact that hitherto they have inclined, and not without some reason, to regard Victorian yacht-building skill as bordering on something like Egyptian darkness.*'[401]

Victorian pride was probably not improved when later in July of 1885 a tussle between yacht builders and naval architects began playing out in the local press, relating to the fact that once in Sydney *Assegai* was to undergo changes by a local builder under Reeks' direction to improve its performance. An article published in the Melbourne-based *Leader* written under the pseudonym 'Mainstay' made its way to Sydney when quoted in the *Australian Town and Country Journal* of 25 July. In it 'Mainstay' publicly scolded the un-named Sydney yacht builder supposedly on Reeks' behalf for not opting to pad the garboards of *Assegai* with outside lead as had been suggested. In a forthright rebuke of the builder, 'Mainstay' exclaimed '*I was not aware that any Sydney yacht builders, since poor* [William] *Langford died, was entitled to make such a suggestion, especially in the presence of one whose opinion in such matters is unimpeachable, and who no doubt has frequently seen it done with the usual good results, viz., increased stability, power and speed*'.[402]

Obviously not impressed with these remarks and the fact that his name had been used to spur a fairly weighty criticism, Reeks wrote to the *Australian Town and Country Journal*. '*Referring to a paragraph in your issue of the 25th quoting "Mainstay" in the Leader, will you allow me to qualify his remarks. If "Mainstay" had confined himself to stating the simple fact that the builder in question did not approve of the padding garboards, it would not have been so bad, but even that seems to me quite uncalled for, and not of general interest; but he says no Sydney builder has any right to make such a statement, and that my opinion is unimpeachable. This I regret very much, and beg to say it is his opinion, not mine. There are more ways than one of getting lead lower down, and so increasing "stability, power and speed," and for anything I know to the contrary this builder may have had some such a plan in mind*'.[403]

The unnamed builder was later revealed as Walter 'Watty' M. Ford of Berrys Bay, one of Sydney's very best. Reeks had only been in the colony a matter of months and he was already establishing a connection with the right people, and this was a partnership that would become profitable for both men over many years. At this point in the relationship, however, there was no space for any disagreement, especially one created from an external source. Lead was added, and much more work was done such that when *Assegai* raced a second time against the 6-ton Dixon Kemp-designed yacht *Sao*, which had won the previous outing, *Assegai* prevailed. Noted at the helm was none other than Reeks himself. His skills were beginning to be recognised not only as a naval architect, but also as an expert yachtsman.

ASSEGAI SAILING ON SYDNEY HARBOUR (CIRCA 1880S).
SOURCE: STATE LIBRARY OF NEW SOUTH WALES, ROBERT HUNT PHOTOGRAPH ALBUM, 1858-1889.

Despite his young years, Reeks was well aware of the need to remain respectful, humble and well-mannered to one and all, regardless of profession, class, status or situation. He was also new to Sydney and the colonies and had yet to receive his first commission. It was not the time nor the place to make enemies, particularly of local

[398] *The Sydney Daily Telegraph*, 26 January 1882.
[399] *The Herald*, 12 December 1881, 3 July 1882; *The Australasian*, 20 June 1885.
[400] *Australian Town and Country Journal*, 4 July 1885.
[401] *Australian Town and Country Journal*, 4 July 1885.
[402] *Evening News*, 24 July 1885.
[403] *Australian Town and Country Journal*, 1 August 1885.

highly-regarded yacht builders that he could potentially form business relationships with. Reeks' brave issuance of a clarification also reveals his dislike of public and reputation-harming criticism and comments. Instead, his belief that everyone was entitled to their own opinion was shining through, likely passed down through his hardworking father. Also visible was Reeks' respect for the tradesmen of the maritime community. These attributes of fairness, reverence and civility would remain with Reeks for his entire professional life.

With his somewhat polite remonstration of 'Mainstay', and his firm defence of Ford, Reeks' notoriety was on the rise, especially amongst Sydney's leading yachtsmen. It was not long before he welcomed his first clients. However, Reeks' first two commissions appear to have been brokering the sale of the small local cutter yacht *Electra* to a Mr Bevau who intended to have the vessel fitted out at Ford's yard for use in the beche-de mer (commercial sea cucumber) fishery off the coast of New Guinea, as well as brokering the sale of the 10-ton yacht *Guinevere* on behalf of J. H. 'Jack' Want.[404]

Soon it was announced that Reeks was advising on extensive alterations to Alfred Milson's New Zealand-built gaff-rigged cutter yacht *Waitangi*, its owner looking for ways to increase the vessel's speed and steering and generally make it more competitive. The modifications, undertaken by George Ellis, boat builder of Balmain, involved rounding off large pieces of the forefoot and making changes to the rake of the sternpost, bringing the heel of the rudder closer to the midship section.[405] Lead was also removed from the fore end of the keel. In addition, Reeks developed a new sail plan for the vessel, including increasing the mainsail by 155 square ft. Collectively, the changes would prove favourable with the *Waitangi* going on to enjoy great success.

Alfred Milson was another worthy ally with whom Reeks was to establish a working and social relationship that lasted their entire lives. The Milson family was synonymous with the establishment, growth and administration of the Royal Sydney Yacht Squadron, as well as the development of Sydney's North Shore area. The latter distinction can be traced back to James Milson, a retired army officer who arrived in New South Wales in 1806 as one of the colony's first free settlers. Milson Precinct and Milsons Point are named after him and his son, James (Jr), who were pioneers of the area and of the North Shore ferry run. Well respected for his commercial and business activities, James (Jr) was also a director of the Bank of New South Wales.

In terms of yachting, James Milson (Sr) began participating in local regattas and sailing matches during the 1830s and has often been referred to as the 'Father of Australian yachting', while James (Jr) was one of the founding members of the Royal Sydney Yacht Squadron, and also its first vice-commodore. In time Alfred Milson, the son of James (Jr), would carry on this legacy. In 1882 he too was appointed vice-commodore of the squadron. Alfred Milson was also a leading figure in the commercial and philanthropic activities of Sydney, including being involved in administration of the Bank of New South Wales. For Reeks, being tasked to help improve *Waitangi*, at the time one of the best-known yachts on Sydney Harbour, was a huge coup.

Reeks' good fortune continued a few weeks later when he was commissioned by J. F. Fitzhardinge, a solicitor of Sydney, to design a yacht of 24 tons Thames measurement.[406] Not intended to be a racing craft, the vessel was considered purely for use as a wholesome cruiser. Construction was subsequently put out to tender with Ford duly being awarded the contract.[407]

With tenders seeming to be his preferred method of selecting boat builders, on 1 September 1885 Reeks invited quotes for the construction of another vessel he had designed, this time a 10-ton yacht.[408] This particular vessel was to join the 'ten-ton class' with the press noting that, '*if she comes off her advent will be hailed with joy and gladness, as the ranks in this class are all too thin*'.[409] Disappointingly there is no evidence of a 10-ton yacht being built to a design by Reeks during this period. Perhaps the on-going deliberations regarding the Yacht Racing Association's proposed new rules of measurement swayed the potential new owner's decision to delay or cancel the craft's construction. There were however many more opportunities for Reeks to meet new clients. September of 1885 saw him become a member of both the Prince Alfred Yacht Club and the Royal Sydney Yacht Squadron.[410] He was quickly confirming his place amongst a new fraternity.

While Reeks continued to act as a yacht broker in the months and years that followed, advertising various craft for sale on a regular basis in *The Sydney Morning Herald,* it was his design work that soon took precedence.[411] Reeks also began writing letters to the editors of various local papers earnestly and succinctly stating his opinions regarding numerous topics, including in October of 1885 offering his support for the proposed new rule of measurement, the Dixon Kemp Rule, that was adopted the following year.[412] This particular editorial and series of responses ran over successive issues of the paper and are not easy to follow in parts. Still, they illustrate the difficult opinions to be handled, as well as the expressions of the period. Reeks' measured replies also showcase his grasp of the old and new rule, no doubt assisted by him being present in England and in Alex Richardson's employ when the discussions were being held.

Despite some resistance, Reeks' support for the Dixon Kemp Rule was unwavering, and a few weeks later the

[404] *The Daily Telegraph*, 8 August 1885; *The Sydney Morning Herald*, 28 August 1885.
[405] *The Sydney Mail and New South Wales Advertiser*, 29 August 1885.
[406] *The Sydney Mail and New South Wales Advertiser*, 22 August 1885.
[407] *The Sydney Mail and New South Wales Advertiser*, 22 August 1885.
[408] *The Sydney Morning Herald*, 1 September 1885.
[409] *The Sydney Mail and New South Wales Advertiser*, 22 August 1885.
[410] *The Sydney Morning Herald*, 4, 5 September 1885.
[411] *The Sydney Morning Herald*, 19, 23, 30 September 1885; 24 February 1886.
[412] *The Sydney Mail and New South Wales Advertiser*, 3 October 1885.

ELECTRA PASSING SYDNEY HEADS.

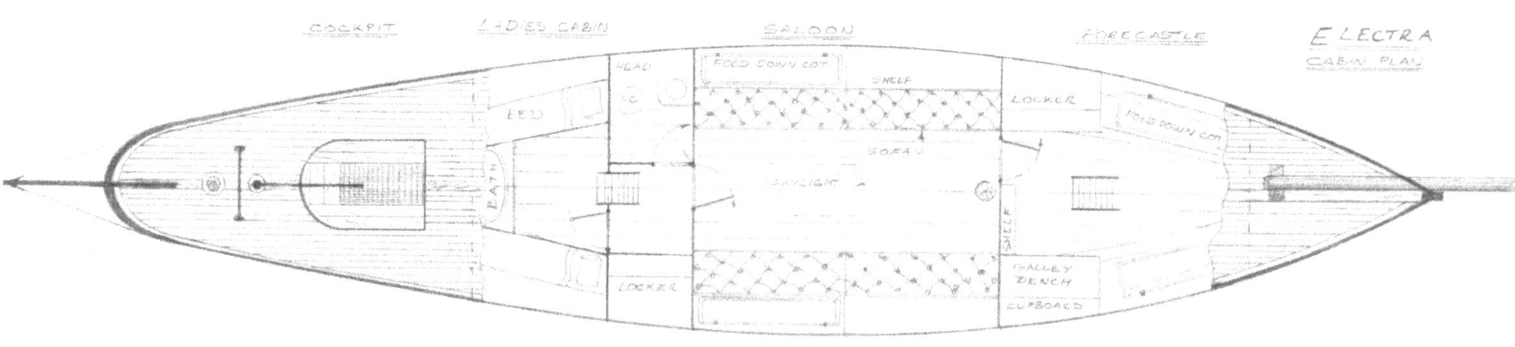

DRAWINGS OF THE CRUISER ELECTRA PASSING SYDNEY HEADS AND ELECTRA'S CABIN PLAN BASED ON CONTEMPORARY REPORTS.
BY DAVID PAYNE.

Prince Alfred Yacht Club appointed him its official measurer, tasked with implementing the new rule, a position he held until September of 1887.[413] Reeks was also making in-roads at the Royal Sydney Yacht Squadron, that same week proposing that the club sail its races as three classes denoted by letters, 'A', 'B' and 'C', under regulations of the Yacht Racing Association.[414] The Prince Alfred Yacht Club soon followed this lead.

Concurrently, a big step in Reeks' career was taken. On 21 November 1885 the first vessel designed by him in Australia, and also the first yawl built in the colony, was launched: J. F. Fitzhardinge's yacht *Electra*, built by Ford of Berrys Bay.[415] Early in construction the vessel had been heralded with the following words, '*Perhaps the most important item of news is the building of a new yacht, as it adds to the number, and that is just what we are all looking anxiously for*'.[416] The article continued with a detailed and admiring description of the craft's layout that included a main cabin with six berths, a ladies cabin, a water closet and a bath.

With registered measurements of 47ft 5in by 11ft 4in by 8ft 4in, *Electra*'s first trial was prefaced with the expectation that it would cause some sensation in the harbour, being the largest yacht built for many years. '*Mr. Ford, her builder, and Mr. Walter Reeks will sail her, in order to see justice done to the first craft in Australian waters which he has designed. She is a very handsome, high-wooded, roomy craft, essentially a sea boat, and fine cruiser, but bids fair to be fast as well*'.[417]

As 1885 neared its end, Reeks was busy superintending more alterations to G. F. Murnin's yacht *Assegai*.[418] The local press reported on the work. '*I understand that Assegai's new mast has been shortened and that an extra ton and three-quarters of lead has been put on her keel, under the advice and supervision of Mr. Reeks, who, by the way, appears to be actually reforming many of our racing yachts, to the great improvement of their speed and general behaviour. In addition to the extra lead, Mr. Reeks as architect and Mr. Ford as builder have retimbered and new-decked Assegai from stem to stem, strengthened her frame throughout, and given her a general overhaul.*'[419]

Assegai resumed competing in local yacht races in early January of 1886, notably with Reeks as skipper.[420] The modifications obviously made a positive impact as it won its first race on 9 January, beating *Sao*, by now considered an arch rival. The Sydney press was elated, '*Mr. Murnin, the spirited owner is to be congratulated on his first victory, for he has spared no expense in bringing out his Melbourne purchase as fit as a fiddle, Mr. Reeks sailed Assegai in a manner which does him very great credit, seeing that it takes some time for a newcomer, however cunning at the tiller, to get himself familiar with the racing wrinkles of Port Jackson, where the tide runs hardest, and where an eddy tide or an eddy wind should be looked for, which side of the harbour to work, and what courses to adopt*'.[421]

[413] *The Sydney Morning Herald*, 6 November 1885.
[414] *The Sydney Morning Herald*, 6 November 1885.
[415] *The Daily Telegraph*, 1 October, 23 November 1885.
[416] *Sydney Mail and New South Wales Advertiser*, 22 August 1885.
[417] *Evening News*, 19 December 1885.
[418] *Globe*, 6 January 1886.
[419] *The Sydney Mail and New South Wales Advertiser*, 16 January 1886.
[420] *Globe*, 9 January 1886.
[421] *The Sydney Mail and New South Wales Advertiser*, 16 January 1886.

Also in January of 1886 Reeks began presiding over meetings of the Prince Alfred Yacht Club, as well as being appointed official measurer of the Royal Sydney Yacht Squadron.[422] The latter position Reeks would hold continuously for 39 years, until his death in 1925; a remarkable achievement of volunteer work, dedication and service to this particular organisation.

January of 1886 additionally saw Alfred Milson's 20-tonner *Waitangi* make its reappearance on Sydney Harbour, almost looking like a new vessel. The changes Reeks had proposed and superintended resulted in the craft being transformed from a rather sluggish vessel to one of the fastest yachts on Port Jackson.

Though built as a cruising yawl, *Electra* also exceeded expectations within the first few months of launch such that the craft was soon encouraged to don racing spars and sails. *Electra*'s first race was at the Anniversary Regatta on 26 January 1886 where, competing against the first-class yachts *Waitangi, Magic* and *Sirocco*, it went on to claim third prize, its chance of winning hindered by a light breeze. *Waitangi* finished first.[423]

For Reeks there was much to celebrate from his inaugural attendance at the Anniversary Regatta. Nearing the end of his first year in Sydney, vessels he had designed or superintended significant modifications to were successful on the day. Another bonus came when he helmed *Assegai* in the race for yachts under 10 tons to another victory against *Sao*.[424]

Come February of 1886, Reeks had been elected treasurer of the Prince Alfred Yacht Club.[425] He had also volunteered his time to measure the club's yachts and allocate time allowances prior to the staging of a race to be sailed under the 'A', 'B' and 'C' classification of the Yacht Racing Association.[426]

It was not too long before *Electra* showed increased speed as well. In March of 1886, sailing in an offshore race to Barrenjoey and return, the vessel won handsomely by over 30 minutes against *Waitangi* and *Sao*.[427] Only a month earlier, sailing in a match race on a harbour course against *Waitangi,* the latter had defeated *Electra* by 14 minutes, even less on time allowance.[428] For Reeks, the craft's win was a strong statement of his design ability that was backed up by the success of his alterations to *Assegai* and *Waitangi*, with both yachts exhibiting significantly-improved performances. The risks were high, a new talent taking on the challenge of improving established yachts and then designing the biggest yacht seen for some time, and he passed the test comprehensively on all accounts.

In a confident yet measured fashion, Reeks was setting a good foundation in place within only a year of his arrival in Sydney and people were impressed. The accolades were numerous. '*When Electra was still on Mr. Ford's stocks I gave*

[422] *The Daily Telegraph*, 7 January 1886; *Globe*, 8 January 1886.
[423] *Globe*, 27 January 1886.
[424] *Globe*, 27 January 1886.
[425] *Globe*, 5 February 1886.
[426] *Globe*, 5, 12 February 1886.
[427] *Evening News*, 29 March 1886.
[428] *Evening News*, 15 February 1886.

some of her dimensions; ... and the opinion was then hazarded that she would make an excellent cruiser (for which, indeed, she was designed), and that her lines promised considerable speed. Those have now been thoroughly fulfilled, Electra is a very fast boat at every point, and a most comfortable cruiser.'[429]

Of social interest, at this point in time, it is evident that Reeks was part of something with wide public recognition. Local and colonial newspapers all included significant coverage of yacht and open boat races, as well as developments in racing rules, design improvements and the status of new builds and alterations. The public, at least in Sydney, was additionally heavily invested in the outcome of yacht races and sailing matches. There were many spectators both onshore or following the races on board a steamer. Sometimes, given bad weather, this was obviously an uncomfortable experience, but it did not deter people from continuing to patronise the sport and the steamers. Reports for the Barrenjoey race involving *Electra, Sao* and *Waitangi*, for example, state '*the weather was beautifully fine and most favourable for an outside trip; although the motion of the steamer was too much for some passengers'*.[430] It was no better for those following the harbour race. '*On the two occasions when it was necessary to pass the Heads a pretty stiff roll was encountered, and some of the passengers felt a tinge of regret they had not remained on terra firma.'*[431]

Returning to Reeks' professional activities of early 1886, there was much work on hand with him being tasked by Dr Mackeller, owner of the 15-tonner *Pethona*, with getting the vessel ready for racing. Reeks was fast developing a reputation as one who could turn '*a hopeless failure into the best boat in her class'*.[432] These efforts were not only centred on recreational craft. In March of 1886 an article written by Reeks on the adaptability of yachts for commercial fishing purposes was published in *The Sydney Mail and New South Wales Advertiser*. In it he opined '*It seems a very strange thing that an industry which represents over £60,000,000 annually in Great Britain, and an equal amount in Belgium, to say nothing of Sweden and Denmark, should be so neglected in New South Wales'*.[433] It was a truthful statement. The local fishing industry was in a lacklustre state, with catches predominantly coming from the local waters of Sydney Harbour and Botany Bay; an inability to transport live or fresh fish caught further away to markets in Sydney being a major obstacle to its development.

Reeks continued. '*Yachting pure and simple is a luxury, and means considerable expense to keep it up; but when it can be made to pay its own expenses, and perhaps something more, it puts a different phase on the whole thing.'*[434] An accompanying sketch of a vessel 70ft (waterline), 82ft (overall), with a beam of 15ft and a draft of 10ft was provided. There was a forcastle of 25ft in length, giving ample room to accommodate five to six men. Reeks proposed a main cabin, two steerage berths and a ladies cabin aft, also a water closet and lavatory. A chief feature was an ice chamber, 8ft long, 14ft wide and 10ft deep, around which were shelves with fenders in the front to store broken ice (obtained on shore). It was suggested that the ice chamber be lined with asbestos or another non-conductor to prevent rapid melting. The fish caught could then potentially remain fresh for a number of days. All told, there was the possibility of storing 400 to 500 fish. Rigged as a yawl, Reeks declared this prototype '*a very fine sea boat'*, with a probable cost of £3,500. He also gave calculations for wages and crew, wear and tear, and interest on the original outlay, estimating that such an enterprise could achieve a net weekly profit of £36. With half of its time spent fishing commercially and the other half spent as a recreational yacht, the hybridised vessel could then realise a yearly profit of some £312.

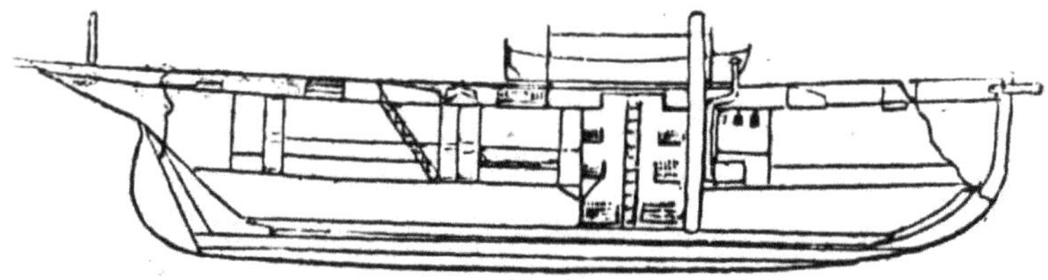

PLAN OF HYBRID YAWL-RIGGED FISHING YACHT (1886).
SOURCE: THE SYDNEY MAIL AND NEW SOUTH WALES ADVERTISER, 6 MARCH 1886.

This piece of design work shows lateral thinking, in terms of its dual-purpose concept, but also self-confidence too in the manner in which Reeks, still relatively new to the colony, was prepared to put forward publicly the idea in the first place, supported by strong argument, including costings. He was also ahead of his time in this instance, particularly with regards to the industry itself. It would be well over a decade before commercial fishing in New South Wales grew in productivity, and endemic species became not only acceptable to the largely-British menu but also palatable. The industry was additionally aided by the development of catch-specific vessels with wells for keeping fish fresh and, later, steam and auxiliary engines as a means of propulsion. Of course Reeks would be involved.

Continuing to embrace new projects and new opportunities, on 19 May 1886 *The Sydney Morning Herald* carried an advertisement placed by Reeks offering a reward for a '*Book containing figures headed Steamer Calculations'* believed lost on the Newtown tram the Friday before.[435] This book was likely associated with another divergence he had taken from yachting. A week later, on 26 May, Reeks was a guest on board the new passenger steamer *Leichhardt*, built by James Pashley and Sons of Balmain to the order of the Balmain Steam Ferry Company.[436] Intended to run between Leichhardt and Iron Cove on the Parramatta River, the vessel measured 85ft (overall) with

[429] *The Sydney Mail and New South Wales Advertiser*, 9 January 1886.
[430] *The Sydney Morning Herald*, 29 March 1886.
[431] *The Sydney Morning Herald*, 15 February 1886.
[432] *Globe*, 10 February 1886.
[433] *The Sydney Mail and New South Wales Advertiser*, 6 March 1886.
[434] *The Sydney Mail and New South Wales Advertiser*, 6 March 1886.
[435] *The Sydney Morning Herald*, 19 May 1886.
[436] *The Sydney Morning Herald*, 8 March, 27 May 1886.

a breadth of 18ft 2in and depth of 6ft 2in.[437] *Leichhardt* had a straight stem and elliptical stern, with machinery supplied by Mort's Dock and Engineering Company consisting of a steel boiler carrying a working pressure of 1001b to the square inch.[438] It is extremely likely that Reeks designed *Leichhardt*, though he appears to have been incorrectly referred to as '*Mr W. Kucks, marine architect*' in one newspaper report.[439] There was no such person by this name living in Sydney and Reeks was the only naval architect working in the city at the time. Moreover, he was one of only a handful of people invited on board for the *Leichhardt's* trial trip whereupon he gave a speech on behalf of its builder, James Pashley.[440]

Leichhardt presents as a relatively traditional looking ferry, featuring a plumb stem, counter stern, with a combination of enclosed and open spaces including a generous upper deck. Built for a new company needing vessels to form a fleet, it may be that when responding to a presumed call for tenders to design and build the craft, Reeks took a careful and orthodox path to ensure his design was considered favourable by the directors.

A second steamer built by James Pashley and Sons to a design by Reeks was soon under construction, commissioned by Henry Jeanneret and John Pinto and intended for the central coast trade.[441] Named *Gosford,* the craft measured 96ft (overall), with a 19ft 8in beam and a 7ft 4in depth of hold. It was expected to attain a speed of just over 11 miles per hour, driven by compound surfacing type engines.[442] *Gosford* was officially launched on 1 September 1886 with Reeks heartily congratulated during a vibrant launch party involving numerous toasts and speeches. In his custom humbleness, he briefly replied that '*it would be time for him to speak when the vessel had proved her part of the work a success*'.[443]

437 *The Sydney Morning Herald*, 8 March, 27 May 1886.
438 *The Sydney Morning Herald*, 27 May 1886.
439 *The Sydney Morning Herald*, 27 May 1886; 11 June 1886; *Evening News*, 11 June 1886.
440 *The Sydney Morning Herald*, 27 May 1886.
441 *The Sydney Morning Herald*, 11 June, 19 July 1886.

SS LEICHHARDT.
SOURCE: GRAEME ANDREWS WORKING HARBOUR PHOTOGRAPH COLLECTION, CITY OF SYDNEY ARCHIVES (A-00076173).

442 *The Sydney Morning Herald*, 11 June 1886.
443 *The Sydney Morning Herald*, 2 September 1886.

OIL PAINTING OF THE SS GOSFORD (1886).
SOURCE: AUSTRALIAN NATIONAL MARITIME MUSEUM COLLECTION (OBJECT NO. 00000053).

Reeks and Pashley, having initially formed an association with *Leichhardt*, were certainly recorded as working together during this project, and were able to branch out a little. For example, while on the stocks in Pashley's yard, the hull was glowingly reviewed twice, with Reeks' name attached to the design. Above the waterline *Gosford* was a conventional straight-stemmed vessel with an overhanging counter, but the differences below were noteworthy. *The Sydney Morning Herald* reported. '*She differs in design from the steamers usually built here, in that instead of the usual ram bow she is cut away or rounded in the forefoot. There is just the suspicion of hollow in the fore end, swelling away to a full round middle body, and lapsing again into a clean tail, thus giving free grip to the propeller.*'[444] At the time it was common for craft to continue the stem vertically to the keel line or even curve forward slightly in a ram profile. Here, however, Reeks had introduced a rounded profile blending the stem into the keel line. Another glowing review was published a month later. '*She is a study in symmetry. Bow, stern, topsides and bottom blend beautifully one into the other, round and clean. Fore-and-aft she is a perfect picture in the way of a model. If she is not fast it will be a puzzle to everybody, for she looks like speed, every inch of her.*'[445]

The design of *Gosford* was relatively conventional otherwise. '*On deck the vessel will be fitted something like the steamer Iolanthe, so far as bridge and wing houses go, but forward there is to be a spar deck, below which a first class saloon and ladies' cabin will be provided, well ventilated with ports and skylights, and with seats on it overhead. This spar deck carries the windlass. Aft there is to be a very spacious deck, covered in with an awning, and capable, if necessary, of carrying a large number of horses and light cargo.*'[446]

Gosford had to make a passage between Brisbane Water, on the New South Wales central coast, and Sydney, so this was an ocean-going coastal vessel; a big contrast in comparison to the yachts Reeks had designed thus far. It also highlights the range of craft he was capable of designing, even this early in his career, and the trust Sydney's commercial maritime operators were willing to place in him.

Another steamer built by the same company to a design almost duplicate to that of *Leichhardt* (so we assume designed by Reeks), was soon under construction, commissioned by the Atlas Engineering Company.[447] With registered dimensions of 75ft 6in in length, with a beam of 17ft 7in and a depth of hold of 6ft 10in, it was launched as the *Woolwich* in October of 1886 for use between the Atlas Works (in Woolwich) and Sydney.[448]

Though busy with a succession of steamers, Reeks remained resolute to his work regarding competitive yachts. In August of 1886 he was tasked by Dr Fred Milford, owner of *Sao*, with improving its sailing qualities. The work was carried out at Ford's yard under Reeks' supervision and included padding the bow on both sides below the waterline, cutting away several inches of the forefoot, reducing the spars, and raking the stern post.[449]

During this period Reeks also began teaching naval architecture at the Sydney Technical College, first delivering a lecture titled '*Ships: Their Design and Construction*' at the School of Arts, Sussex Street, to an audience of more than 300 people.[450] Though clearly stating that it was impossible to provide the merest outline of the subject in an hour, Reeks' lecture gave an overview of water and its relationship to ships, and details on resistance due to friction and a ship's bottom.[451] He also discussed the use of copper for sheathing purposes, as well as floatation and stability. Reeks went on to give more lectures as part of bi-weekly evening classes in naval architecture over several years, including two in October of 1886 titled '*Types of Vessels with Lantern Illustrations*' and '*Ships, their Design and Construction*'.[452]

Concurrently Reeks was tasked by William Cooper of Point Piper with designing a small steam yacht to be fired with liquid fuel (kerosene) stored in tanks under the fore deck.[453] Built by George Ellis of Balmain, the vessel measured 40ft (overall) by 7ft and was constructed of kauri pine with cedar fittings.[454] Launched as the *Bettina* in early 1887, the craft's engines were manufactured by Shipman's of America and on its trial trip the vessel achieved a speed of 8 knots.[455]

Bettina deserves more description than we can find, not just because it was the first steam yacht that Reeks designed under his name, but its kerosene-fired boiler arrangement was unique to the colony. This once again supports the theme common to Reeks where he was quite prepared to take a risk and to do something different if he felt it was right. He obviously carried his arguments well, given how often his unorthodox proposals were implemented on client's vessels.

With further reference to the *Bettina*, the press reported, '*Mr. W. Cooper of Point Piper, has commissioned Mr. Reeks, naval architect of this city, to design him a small steam yacht, which in many ways will be a novelty. She is to be engined by Messrs. Shipman, and will be fired with liquid fuel stored in tanks under the fore-deck. This system of firing is now engaging the attention of the engineering world, and will probably in a very few years, to a large extent, supersede coal in marine boilers, being far less expensive and infinitely cleaner*'.[456]

In hindsight it was optimistic to predict the demise of coal for steamers, and kerosene did not become the '*fashion on the harbour*' until later when oil-fuelled combustion engines began to make their appearance. Meanwhile *Bettina* must have been a pretty sight too, with an overhanging bow, probably clipper style, and finished at

444 *The Sydney Morning Herald*, 11 June 1886.
445 *The Sydney Morning Herald*, 19 July 1886.
446 *The Sydney Morning Herald*, 19 July 1886.
447 *The Sydney Morning Herald*, 25 August 1886.
448 *The Sydney Morning Herald*, 25 August 1886; *The Daily Telegraph*, 14 October, 7 December 1886.

449 *The Daily Telegraph*, 6 August 1886.
450 *The Sydney Morning Herald*, 31 August 1886.
451 *The Sydney Morning Herald*, 31 August 1886.
452 *The Daily Telegraph*, 29 September, 22 October 1886, 7 January 1888; *The Australian Star*, 2 February 1889.
453 *The Sydney Morning Herald*, 19 July 1886.
454 *The Sydney Morning Herald*, 19 July 1886, 23 March 1887.
455 *The Sydney Morning Herald*, 23 March 1887.
456 *Sydney Mail and New South Wales Advertiser*, 24 July 1886.

the stern with a long counter. The timber work combined beech, rosewood and teak. Even though it was only 40ft in length, *Bettina* was given the phrase of having a '*very light, rakish appearance*'. Reeks was certainly flaunting his style for bigger things to come.

George Ellis was the builder of *Bettina* and, having worked with Reeks on *Waitangi's* successful alterations, this commission cemented a strong and fruitful relationship between the pair over a small number of quite significant craft in the next few years.

There was more than enough work on hand for Reeks to advertise for an articled pupil in September of 1886.[457] That same month he was elected to the membership of the Engineering Association of New South Wales.[458] He also continued to be actively involved in administration, starting line and measuring efforts of the Prince Alfred Yacht Club and the Royal Sydney Yacht Squadron, being once again elected to the committee of both organisations.[459]

The year 1886 concluded with Reeks designing a new mast for *Waitangi* which had recently undergone additional extensive alterations under his superintendence in anticipation of the vessel competing in the upcoming 'Intercolonial Yacht Race'.[460] *Assegai* was also expected to compete in this event. However, the mere mention of this particular vessel's Sydney-based alterations in the press, as well as its new-found success on Port Jackson, saw matters of colonial rivalry combined with Henry R. Murray's ego resurface. He wrote a lengthy diatribe, published in November of 1886, not only discussing details of the Intercolonial Yacht Race and yachting in Victoria and New South Wales more generally, but perhaps even more revealing, taking credit for the recent improvements made to *Assegai*, the vessel he had designed and built some four years prior.

Murray put forth from his South Melbourne location. '*Unlike Sydney, the national importance and practical character of the pastime receives no consideration or support from our public men, or more wealthy citizens, who, with the exception of perhaps two or three, attach more importance to bowls and the effeminate pastime of lawn tennis; while the press, in every instance, incites the general craze for horse-racing and the brutalities of football, with their contingent betting agony and general demoralisation of the rising generation. As the older colony, and one whose yachting prestige can well afford a victory or two, by way of encouragement, to her younger sister, where the pastime is only in an incipient stage, I would humbly suggest through your columns the selection of a representative yacht to meet the Janet in the proposed series of matches. I do so, without any thought of the latter yacht scoring a victory over your representative. As the designer and builder of Janet, I can say that no claim has yet been put forward on the past of Janet to be considered anything more than a tolerably fast cruiser. Had my first design been approved of, your yachts would have had a 30-ton Assegai to compete with, embracing all the improvements to her that were effected at my suggestion when placed in Mr. Reeks's hands by her spirited owner, Mr. Murnin. To him I feel much indebted for releasing her from her isolation here in Victoria, and fitting her out with more regard to racing eventualities.*'[461] After an additional paragraph discussing the state of yachting overseas *vis a vis* Australia, Murray signed himself off at the end of the letter as '*H. R. Murray, N. A*'. Of note, here are two items of interest. First, Murray's claim of credit for the improvements made to *Assegai,* whereas the Sydney press had unanimously and quite rightly given credit to Reeks. Second, his self-promotion to naval architect.

Born in Ireland in 1832, though trained in Liverpool as an apprentice turner to an iron shipbuilding company, Murray moved to Victoria in the early 1850s becoming well-known as a boat builder of Geelong, then Ballarat, and from the 1870s of Melbourne.[462] This vocation, however, was not as fruitful as he needed such that over this same time period he oscillated between building boats and the more secure work of wood turning and ornamental carving. Though he designed many of the craft he built, as was customary at the time, Murray lacked formal qualification as a naval architect such that his self-elevation to this profession, at the age of 54, perhaps indicates that he was feeling somewhat threatened by Reeks' sudden rise within Australia's yacht designing ranks.

For Reeks, however, the reappearance in the local press of Murray's gripes about *Assegai* appear to be more of a nuisance rather than a personal criticism. Following several more letters published by Murray in *The Sydney Mail and New South Wales Advertiser* in the weeks and months that followed, all of them unequivocally stating that he had proposed the alterations made to *Assegai* and that Reeks had merely only superintended them, the latter sought to end the discourse once and for all.[463] In the same newspaper on 11 December 1886 and then again on 8 January 1887 Reeks published two brief letters to the editor, the final one proving enough of a directive for Murray to cease his relentless crusade.

TO THE EDITOR OF THE SYDNEY MAIL.

Sir,—Mr. H. R. Murray, writing under the heading "Intercolonial Yacht Racing," having, for the second time, dragged my name, in connection with the Assegai's alterations, forward, which, like the flowers that bloom in the spring, "*have nothing to do with the case,*" *I beg to* say those alterations were done, not by me, but by Mr. Ford, of Berry's Bay, under my supervision and from my designs, without reference to Mr. Murray or anybody else.

No doubt had Mr. Murray carried out the thing himself it would have been done as well; but the fact remains that he had nothing whatever to do with it.—Yours, &c.,

WALTER REEKS.

SOURCE: THE SYDNEY MAIL AND NEW SOUTH WALES ADVERTISER, 11 DECEMBER 1886.

TO THE EDITOR OF THE SYDNEY MAIL.

Sir,—I regret much having to trespass again on your space. I had my little say on this subject in your issue of December 11, and am content to leave it to my yachting friends to decide betwixt Mr. Murray and myself.—Yours, &c.

WALTER REEKS.

SOURCE: THE SYDNEY MAIL AND NEW SOUTH WALES ADVERTISER, 8 JANUARY 1886.

[457] *The Sydney Morning Herald*, 15 September 1886.
[458] *The Sydney Morning Herald*, 17 September 1886.
[459] *The Sydney Morning Herald*, 2 September 1886; *The Daily Telegraph*, 3 September 1886.
[460] *The Sydney Mail and New South Wales Advertiser*, 18 December 1886, 22 January 1887; *Referee*, 6 January 1887; *The Sydney Morning Herald*, 10 January 1887.
[461] *The Sydney Mail and New South Wales Advertiser*, 13 November 1886.
[462] N. Mays, D. Payne & C. Grazules (2021). *Little Boats with Sails: The History of Australia's 21 Foot Restricted Class.*
[463] *The Sydney Mail and New South Wales Advertiser*, 4 December 1886, 1 January 1887.

As Reeks entered his second year of living in Sydney, he had more than solidified himself as the city's premier naval architect, proving his worth with the design of both commercial and recreational vessels. He had also earned a reputation out on the water as an outstanding yachtsman, helming several of the harbour's showcase yachts.

In January of 1887 Reeks was invited to crew on board Alfred Milson's *Waitangi* in the Intercolonial Yacht Race, the first event of its type to be held in Sydney.[464] Competing against Sir William Clarke's *Janet* of Melbourne and James Fairfax's *Magic* of Sydney, the *Waitangi* went on to claim a well-earned victory in the first heat, much to the amazement of local pundits who expected *Magic* to eclipse the other two competitors, while Victorians were steadfast in their resolve that *Janet* (designed and built by Murray in 1880 reportedly at a cost nearing £3,000) was invincible.

In what was described in the press as a '*scientific display of seamanship*', *Waitangi's* win was followed a week later by another victory, making it the undisputed intercolonial champion and giving Milson the Intercolonial Cup (made by Elkington and Co. of London) and £100 in prize money.[465] It was then decided that the Cup should be an annual event until the same owner had been awarded the trophy three years in a row.[466]

[464] *The Sydney Mail and New South Wales Advertiser*, 22 January 1887.
[465] *The Sydney Morning Herald*, 24 January 1887;
The Sydney Mail and New South Wales Advertiser, 6 August 1887.
[466] *The Sydney Mail and New South Wales Advertiser*, 6 August 1887.

With his trust now firmly placed in Reeks, Milson commissioned the design of a new vessel with the firm intention of retaining the Intercolonial Cup.[467] Meanwhile, *Maysie,* the 10-ton cruising yawl Reeks had designed for J. M. Curtis, another Sydney-based solicitor, was launched from Ford's yard at Berrys Bay in late April of 1887.[468] With registered dimensions of 31ft 4in by 9ft 2in by 6ft 5in, *Maysie* featured a clipper bow profile, a novelty for the colony.[469] This was an overhang forward that ran in a gently-reversed curve line, most likely finished at the tip with scroll work; supposedly the first of this bow type built in Australia. Otherwise, although much smaller and with correspondingly less accommodation, *Maysie* exhibited similar design features to *Electra* and was expected to perform well.[470]

The report of *Maysie*'s launch carried a prophetic note at the end. '*This is the second boat from the design of Mr. Walter Reeks that we have in Port Jackson, the Electra being his first attempt. Though not intended for racing, I do not doubt she will well uphold the reputation already achieved by Mr. Reeks as an architect. Perhaps some day we may see what he can do when put on his mettle and ordered to produce a racing craft.*'[471] This call was soon answered with a significant response by Reeks in the form of Alfred Milson's yacht.

[467] *Referee*, 3 March 1887.
[468] *The Sydney Mail and New South Wales Advertiser*, 5 March 1887;
The Australian, 30 April 1887.
[469] *The Sydney Mail and New South Wales Advertiser*, 5 March, 2 April 1887.
[470] *The Sydney Mail and New South Wales Advertiser*, 5 March 1887.
[471] *The Sydney Mail and New South Wales Advertiser*, 23 April 1887.

LEFT TO RIGHT: WAITANGI, JANET AND MAGIC COMPETING IN THE INTERCOLONIAL YACHT RACE OF 1887.
SOURCE: SUPPLEMENT TO THE ILLUSTRATED SYDNEY NEWS, FEBRUARY 1887.

MAYSIE.
SOURCE: AUSTRALIAN NATIONAL MARITIME MUSEUM (WILLIAM HALL COLLECTION).

Meanwhile, in June of 1887, Reeks submitted a design for a steamer with two screws, one at each end, fitted with compound surface condensing engines, estimated to cost £4,300 to build.[472] It was one of nine submissions put forth as part of a competition initiated by the Balmain Steam Ferry Company for a new steamer to run between Balmain and Sydney.[473] A prize of £40 was on offer for the best design, while second place was to receive £10.[474] Unfortunately Reeks failed to win the competition. Instead a design submitted by Andrew Key, manager of the A. S. N. Company for a double-ended screw was awarded first prize, while second prize went to Messrs Humphrey and Cracknell for a wooden paddle boat with double engines.[475] Reeks was undeterred. There would be many more commissions for steamers in his future.

June of 1887 also coincided with more progress on the social front, with Reeks attending the Mayor's Jubilee Ball.[476] In addition he carried on with his entrepreneurial ways, in September of 1887 opening an exchange at his Pitt Street office for those wanting to dispose of or purchase yachts, steam launches or boats.[477] It was the first of its type in Sydney with the local press reporting that the '*inconvenience of going all over the harbour to look for a boat will thus be avoided, and, instead, all necessary information can be obtained right in the heart of the city*'.[478]

The semantics of yacht administration continued to play a major part in Reeks' daily life. At a meeting of the Royal Sydney Yacht Squadron in September of 1887, Alfred Milson proposed an amendment to 'Rule 39' such that yachts with centreboards, sliding keels or fins be allowed to sail in any match, but with a provision requiring dimensions be applied to ensure reasonable depth or draft of the hull, and therefore avoid lightly-built and shallow yachts known widely as 'skimming dishes', an apt description. The primary reason for bringing this issue to the squadron's attention was the Reeks-designed 40-ton centreboarder then under construction for an esteemed member of the organisation, J. H. 'Jack' Want. The press remarked on the delicate situation, '*The Prince Alfred Yacht Club have decided to admit centreboard yachts into their races; but it is uncertain how far the Squadron will go in that direction. Doubtless an unwillingness to shut out Mr. J. H. Want's new boat will have weight*'.[479]

Though most members of the Royal Sydney Yacht Squadron were in favour, the proposal met with some opposition and in a mistaken move, an amendment was proposed that if adopted Reeks noted would not be suitable for yachts rating 10 tons or less. Eventually a solution was reached that gave better definition with the required depth proportions varying with the tonnage and length of the boat. The following resolution was carried: '*That no yacht having a centreboard, sliding keel or fin shall be allowed to sail in any match unless the draught of such a vessel be to her length on waterline or not less than the following proportions: For 10 tons, 1 to 7; for 20 tons, 1 to 8; for 40 tons, 1 to 9; for 60 tons, 1 to 10; and an increase proportionately for higher tonnage, and shall further be subject to the approval of the committee*'.[480] The outcome avoided skimming dishes and permitted centreboarders of quite reasonable depth. No doubt Reeks' experience and expertise helped in this instance, recognising that factors in the design of yachts meant that a ratio of dimensions for one size was not necessarily a correct fit for all sizes.

The 40-tonner that caused the rule change discussion, and ultimately its implementation, was at the time being built by William Dunn at Careening Cove.[481] Named *Miranda*, the vessel measured 75ft (overall), 58ft (waterline) with an extreme beam of 13ft 6in.[482] Its hull was described in the press as somewhat along the lines of the famous American centreboard yachts *Puritan* and *Volunteer*.[483] The two similar craft had successfully defended the America's Cup in 1885 and 1887, respectively. Whilst they were single masted and cutter rigged for racing, *Miranda* was rigged with two masts as a yawl for cruising. *Miranda*'s centreboard, fitted in lieu of a deep keel, also allowed for entry to shallow waters such as Port Hacking and Broken Bay.[484] Built under Reeks' superintendence, the craft was launched on 24 October 1887 and in appearance exhibited a long elegant counter, similar to *Puritan*, and a rakish, overhanging bow, finished with a graceful carving, like *Volunteer*.[485]

Dunn was another of Sydney's long-standing and respected shipwrights, having first established a yard at Lavender Bay in the late 1850s before moving to Berrys Bay.[486] A prominent solicitor, Jack Want was a member

[472] *The Daily Telegraph*, 7 June 1887.
[473] *The Sydney Morning Herald*, 4 June 1887.
[474] *The Daily Telegraph*, 7 June 1887.
[475] *The Sydney Morning Herald*, 20 June 1887.
[476] *The Sydney Morning Herald*, 23 June 1887.
[477] *The Sydney Morning Herald*, 3 September 1887.
[478] *The Sydney Morning Herald*, 3 September 1887.
[479] *The Sydney Mail and New South Wales Advertiser*, 4 June 1887.
[480] *The Sydney Morning Herald*, 8 September 1887.
[481] *Referee*, 7 April 1887; *The Sydney Morning Herald*, 25 October 1887.
[482] *Referee*, 7 April 1887; *The Sydney Morning Herald*, 25 October 1887.
[483] *Referee*, 7 April 1887; *The Sydney Morning Herald*, 25 October 1887.
[484] *Newcastle Morning Herald and Miners' Advocate*, 27 April 1887.
[485] *The Sydney Morning Herald*, 25 October 1887.
[486] R. Svensen (2021). *A Changing Tide: the History of Berrys Bay.*

of the New South Wales parliament and, at the time of *Miranda*'s launch, was also serving as the colony's Attorney General.[487] The following year he would be appointed a Queen's Counsel.[488] Want was also a senior member of the Royal Sydney Yacht Squadron. Clearly he was a very high profile client for Reeks to attract.

Miranda marked the second vessel Want had commissioned Dunn to build; the 10-ton racing yacht *Daphne* had been launched in April of 1876.[489] Another 10-tonner, *Guinevere*, had been built for Want by William Langford a few years later.[490] However, *Miranda* was the outcome of Want's need for a cruising yacht that could take him to Tasmania or New Zealand, rather than relying on steamers and hotel accommodation. Both had given him poor experiences in the past. Want was after a vessel not only to reach these destinations, but also to stay on board in, somewhat akin to his own personal floating hotel. It was, overall, a unique commission for both the designer and the builder. Together the partnership of Reeks and Dunn not only attempted to deliver on these requirements, but also successfully achieved them. An article titled '*2000 Miles in a Forty Tonner*', published in *The Australian Town and Country Journal* and written by Want takes the reader through a trip on board *Miranda* to Tasmania in early 1890.[491] Of note, at the beginning of the article, he discusses the yacht's origins. '*I determined upon building a yacht large enough to make the voyage across (The Tasman Sea) with sufficient accommodation to enable me to be my own hotel-keeper upon arrival in port. Mr. Walter Reeks, the well-known naval architect, was entrusted with the designing; his primary instructions being to give me a fast boat, with plenty of room, not to draw too much water, so that I could visit shallow harbours, and above all bear in mind she must be a good sea-boat. The result was a forty (40) ton yawl, built on the American centre board system, and it is only fair to her designer to say that he gave me everything I asked for. How fast she is in comparison to other 40-tonners I do not know, but she is as fast a boat as I have ever been in; and as a sea-boat, I don't believe there is her equal in*

[487] adb.anu.edu.au/biography/want-john-henry-8979.
[488] adb.anu.edu.au/biography/want-john-henry-8979.
[489] *Evening News*, 1 May 1876.
[490] *The Sydney Mail and New South Wales Advertiser*, 27 March 1880.
[491] *Australian Town and Country Journal*, 12 April 1890.

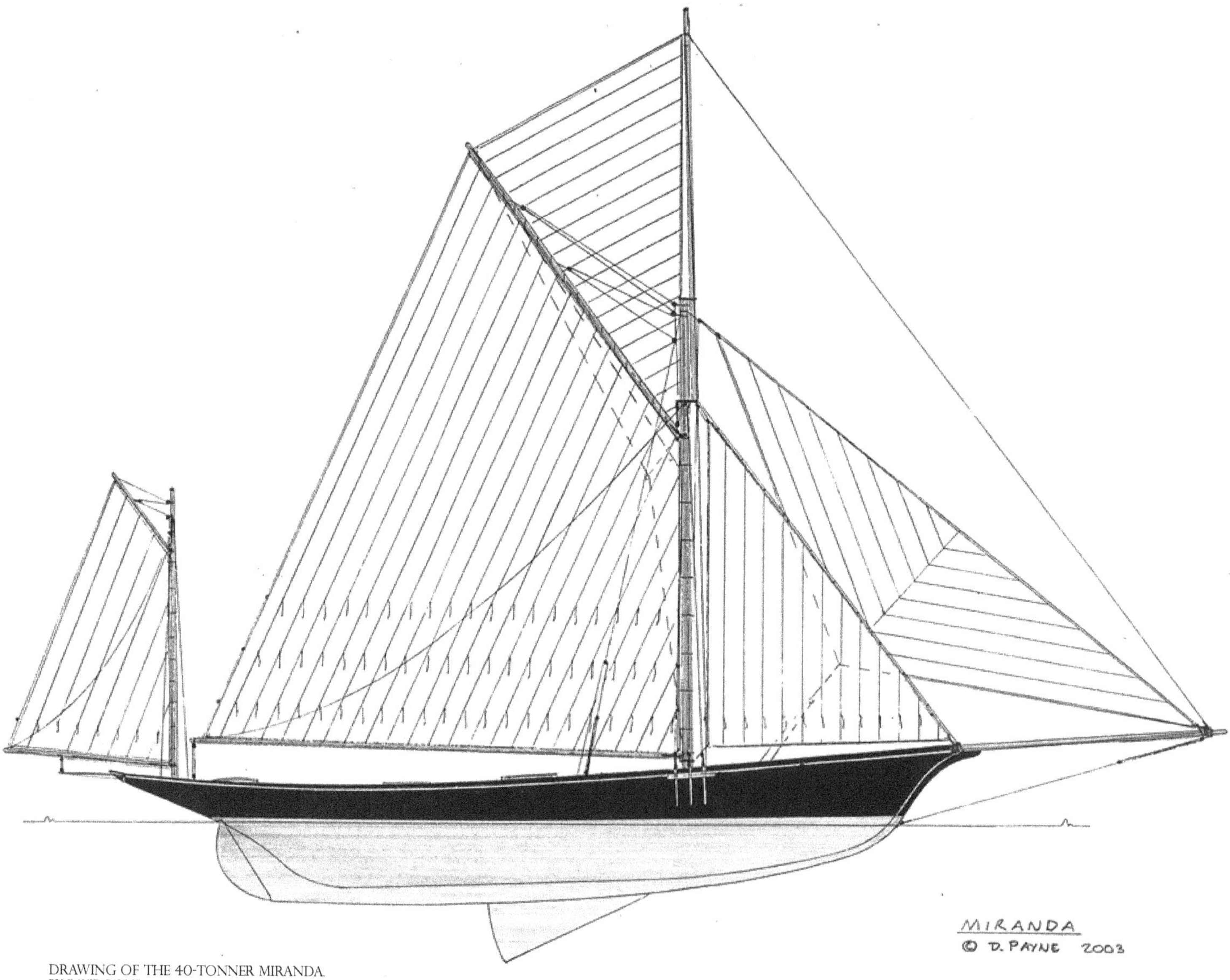

DRAWING OF THE 40-TONNER MIRANDA
BY DAVID PAYNE

MIRANDA.
SOURCE: AUSTRALIAN NATIONAL MARITIME MUSEUM COLLECTION (OBJECT NO. 00002633).

the world. I mention this in justice to Mr. Reeks, as some would-be critics have, without knowing anything about the yacht given their opinions (generally of an unfavourable character) respecting her merits. She is all I ordered, and exactly what I wanted.'

Construction of *Miranda* began in April of 1887, with the vessel launched six months later, on 24 October, christened by Want's niece Hilda. There were many interested and associated parties in attendance, including those perched on board a number of vessels moored out in the bay. Fitted out with a 14ft long saloon, an owner's cabin, a ladies' cabin, an open lounge, a foc'sle for a pantry, a galley and crew accommodation, with 7ft of headroom in the main areas, *Miranda* was a well-appointed yacht ideal for its dual purpose. However, despite Want's later high praise, the months immediately after the vessel's launch were not initially very successful. The craft's first sailing trips were reported in the Sydney press in early December of 1887 and detailed a number of small problems that plagued the trials. These issues continued. 'Backstay' reported in the *Referee* on 9 February the following year that, '*The Miranda is again in the doctor's hands. This time T. S. Rountree and Co. are entrusted with the job of taking out the centre board box and fitting a new one*'.

Eventually the problems were resolved and *Miranda* was underway. As the 1887/88 yachting season came to a close, and whilst the vessel had yet to race, it was finally ready for a passage offshore and cruised south with others, including *Electra,* to Jervis Bay.[492] More sailing took place that same year when the yachting season opened in October of 1888. However, although *Miranda* was seen out on Port Jackson during several race days, the vessel did not compete.

Miranda's long passage to Tasmania took place almost two years later, setting off on 8 January 1890. At South Head, just as the voyage began, the crew could see their first day was '*not going to be all beer and skittles; the indications of a southerly buster being unmistakable*'.[493] In fact, Want notes in his novella that the weather was bad enough to '*apparently demolish a hotel in George Street*'. Still the crew (and their vessel) weathered the blow and by evening were making progress, with many more adventures to come.

For Reeks, concurrent with *Miranda*'s construction, he remained busy with several other craft and activities on hand. This may be why he did not seek re-election to the committee of the Prince Alfred Yacht Club in September of 1887 at the organisation's annual meeting.[494] Instead he remained a member of a subcommittee tasked with developing a new code of rules to guide the measuring of tonnage with regards to club racing.[495]

The final months of 1887 saw Reeks superintend alterations to T. Marshall's decked boat *Asteroid*, including raising the vessel by seven inches, installing a new deck, changing its rig, and lengthening the counter. The work was undertaken by Morland of Berrys Bay.[496]

THE 40-TONNER MIRANDA SAILING ON PORT JACKSON.
SOURCE: MARITIME MUSEUM OF TASMANIA (P_OM_L_19A).

The most significant milestone to date for Reeks took place on 5 December 1887 with the launch of Alfred Milson's 40-tonner *Era* (taking its name from his father's earlier racing yacht), built by George Ellis at Balmain after a construction period of some five months.[497] However, the superstitious must have been extremely nervous as they witnessed the event that day, with two elements of bad luck taking place. The press reported on both. '*After a false start, the usual bottle of champagne was dashed against the bow by Miss I. Taylor, and the Era began to slide towards her native element. But when she had gone about half her own length she came to a standstill. It appeared that the ways* [the slipway rails] *were not set at a sufficient incline. A hawser was passed round her bow, and two small launches began to pull, while a ram was applied to the cradle. At last, at half-past 12, when all expected that the launch would have to be postponed, after a few preliminary slides the Era took the water in safety*.'[498]

Measuring 72ft (overall), 61ft 7in (waterline) with a beam of 12ft 8in and a depth of hold of 8ft 8in, *Era* was built strong enough to be classed 16 years by Lloyd's and, at the time of completion, was considered the largest racer in the country.[499] This latter point was only one of many significant features and facts attributed to *Era.* The most significant accolade, however, was not fully known at the time, but established later on. '*It is noteworthy that*

[492] *Sydney Mail and New South Wales Advertiser*, 14 April 1888.
[493] *Australian Town and Country Journal*, 12 April 1890.
[494] *The Sydney Morning Herald*, 9 September 1887.
[495] *The Daily Telegraph*, 10 September 1887.
[496] *The Sydney Morning Herald*, 16 November 1887.

[497] *Evening News*, 6 December 1887.
[498] *The Sydney Morning Herald*, 6 December 1887.
[499] *The Sydney Mail and New South Wales Advertiser*, 23 July 1887; *Adelaide Observer*, 3 September 1887.

the Era was the first boat built to the "length and sail area rule" in New South Wales, and it is believed, in any part of the world. The Mohawk, the first English yacht designed under this rule, was built subsequently to the Era.'[500] While this statement is undoubtedly correct for the New South Wales reference, and also true on a wider colonial view, it is quite possible that it is also accurate on a worldwide scale; a point that is still debated in English yachting circles to this day.

The recognition continued for *Era.* It was described in the local press as the '*beau ideal of a racing cutter',* while other references of the time state it was '*pretty*' and '*quite the regulation racing cutter'.*[501] *Era* was also unique in that it carried over 30 tons of lead, secured in one piece in the keel, with the casting said to be the largest ever made in Australia.[502] A measure of the keel's novelty was that a number of people were invited to witness the event. Even more noteworthy, Reeks had enlisted the experience gleaned from his time with Alex Richardson in England, where they had evolved their designs to hang almost all of the lead required for ballast outside of the hull from the

[500] *The Sydney Mail*, 6 November 1907.

[501] *The Sydney Mail and New South Wales Advertiser*, 29 October 1887; *Illustrated Sydney News*, 20 March 1890.

[502] *The Sydney Mail and New South Wales Advertiser*, 23 July 1887; *The Sydney Morning Herald*, 19 August 1887.

DRAWING OF THE 40-TONNER ERA.
BY DAVID PAYNE.

wooden keel, with only trimming ballast placed internally. In the case of *Era*, this was 4 tons of lead. The vessel demonstrated this 'outside lead construction' emphatically for the first time on a yacht in Australia, and set a pattern to be followed.

A contrast in appearance to *Miranda*, the *Era* was a long, rakish craft with low bulwarks and a long fine tail.[503] The local press also published an interesting point regarding its features, describing its tail or aft end as a '*long, graceful counter running right out to a "cheese cutter cap" end*'.[504] This is a very descriptive phrase, as the *Era* had an elegant rounded ending seen in top view. The open 'taff rail' that connected the two bulwarks from either side also created an image of a neat sharp cheese slice. Reeks used this design on many subsequent craft.

Importantly, however, as the first racing design undertaken to the new rule of measurement, *Era* showed the benefits of this change. It was deep keeled but with less draft than would have been the case under the old rule, and with more beam in comparison to its length than previous designs. These were the first steps toward a more wholesome and balanced combination of ballast and hull form to provide stability. The craft also had a straight stem and long running bowsprit.[505]

Internally, like *Miranda*, the *Era* was divided into cabins and had 7ft of headroom. Walnut and oak were used in the fit out, and the hatches, skylights and coamings were teak. Upholstery was of the highest standard; Alfred Milson spared no expense, though nothing was wasted or overly extravagant. Overall, this was a yacht that had it been launched in England would have drawn the same high praise and compared more than favourably with anything else of the period.

Era's first sail was to Manly, in January of 1888.[506] Though its sails were said to be fine, the mast was not up to standard and a new one was expected. It was fitted by the week's end, just in time for the craft to compete in its first race; a drifter where they apparently spent the night at sea. It was an inauspicious start to what became an illustrious career as a fast and fine racing and cruising yacht, then strangely enough, later on as an excellent fishing boat based in Western Australian waters.

In addition to the hype surrounding *Era*, the year 1887 ended with Reeks attending meetings as a committee member of the Centennial Regatta, to be held the following year.[507] He was also back at the helm of *Assegai*, achieving much success that season.[508]

The New South Wales Centennial Regatta was held on 27 January 1888, one of many recreational and social activities organised to mark the centenary of European settlement in Australia. The weather shone brightly and amongst the many sailing and rowing events that day, the showcase event was the Challenge Cup Yacht Race for the One Hundred Years' Challenge Cup, valued at £500.[509] Featuring Milson's newly-launched *Era*, J. R. Fairfax's *Magic*, W. Buckingham's *Mistral* and *Waitangi*, now owned by Dr Fred Milford, the race was considered the highlight of the day, arousing a great deal of interest, including from 4,000 passengers packed on board local steamers.[510] The outcome was a deserved victory for Reeks. *Era* led from start to finish, winning by a margin of 23 minutes and 58 seconds from *Magic*. Both *Mistral* and *Waitangi* did not finish.[511] All told, it was a busy day with Reeks receiving unbounded praise as part of the committee's efforts to organise and stage the event.[512]

Social engagements continued too. In May of 1888 Reeks helped organise a picnic for employees of the Mort's Dock and Engineering Company, including overseeing various running races.[513] He also continued to publish pieces in the Sydney press. For example, in May of 1888 the *Illustrated Sydney News* carried a five-page article written by Reeks titled *Yachting and Sailing* whereby he discussed at length the history of yachting, then delved into the types of yachts present in Sydney, with 25 corresponding illustrations.[514] Tables were also presented summarising the outcome of the recently-completed sailing season.

In terms of new vessel designs from Reeks, next came *Elf*, described as a small half-decker, and built by George Ellis to the order of J. O. Fairfax.[515] The vessel was a 2-rater and was completed in February of 1888.[516] *Elf* was also very likely the next boat Ellis built after launching *Era*. J. O. Fairfax was the son of James Reading Fairfax (later Sir James), then commodore of the Royal Sydney Yacht Squadron; a surname well-known to generations of Sydneysiders since that period. Although launched early in 1888, *Elf* does not feature in local racing fixtures until December of 1889 when it participated in one of the first races for a new class with the Prince Alfred Yacht Club: for keel yachts known as the 2½-raters.[517] Though *Elf* and another entry both rated below 2½ tons, with handicap it performed well. Racing in a Double Bay regatta sailed a week earlier *Elf* had won the class after time allowances were applied. Its involvement here suggests the craft was a small keel yacht, and the half-decked description may have been incorrect as it implies a more open centreboard class vessel. Unfortunately there are few records of *Elf* for us to compare the vessel's shape and characteristics, and also contrast it with the much larger craft Reeks was designing at the time, all to the new rule. Such an assessment would be beneficial, particularly in terms of *Elf*'s proportions. Register entries

503 *The Sydney Mail and New South Wales Advertiser*, 29 October 1887.
504 *The Sydney Mail and New South Wales Advertiser*, 23 July 1887.
505 *The Sydney Mail and New South Wales Advertiser*, 29 October 1887.
506 *The Sydney Mail and New South Wales Advertiser*, 7 January 1888.
507 *Australian Town and Country Journal*, 19 November 1887; *The Daily Telegraph*, 22 November 1887; *The Sydney Morning Herald*, 20 December 1887.
508 *The Sydney Morning Herald*, 21 November, 12 December 1887, 12 March 1888; *Australian Town and Country Journal*, 3 December 1887.
509 *The Sydney Morning Herald*, 28 January 1888.
510 *The Sydney Mail and New South Wales Advertiser*, 4 February 1888.
511 *The Sydney Morning Herald*, 28 January 1888.
512 *The Sydney Mail and New South Wales Advertiser*, 4 February 1888.
513 *Balmain Observer and Western Suburbs Advertiser*, 5 May 1888.
514 *Illustrated Sydney News*, 31 May 1888.
515 *Australian Town and Country Journal*, 18 February 1888; *Lloyd's Register of Yachts* (1903).
516 *Australian Town and Country Journal*, 18 February 1888; *Lloyd's Register of Yachts* (1903).
517 *The Sydney Morning Herald*, 21 December 1889.

indicate it had a billet head and elliptical stern, meaning the stem that was possibly raked forward also ended in a scroll, and an overhanging counter, and that the vessel was 19ft on the waterline.[518] Overall, *Elf* would have been a pretty sight and a fine contrast on Sydney Harbour to the vastly bigger *Era*.

Elf was a small interlude between larger vessels for boat builder George Ellis. As it was being readied for launch in early 1888, his yard was already setting up to build their third Reeks-designed craft in succession. *The Australian Town and Country Journal* on 18 February 1888 reported, '*This same builder is about to lay down the keel of the 20-tonner, designed by Mr. W. Reeks, for a Melbourne gentleman; and if he finishes her as quickly, in comparison, as he did the Era, she should not be long before she has a dip in the briny. In this builder's hands a good, wholesome job will be turned out; no bad work being allowed on his premises*'.

With dimensions of 57ft (overall), 49ft 6in (waterline) with a beam of 12ft 1in and named *Iduna,* this particular vessel was commissioned by Melbourne-based barrister John Gurner as a large cruising yacht that would also periodically race.[519] Yawl-rigged, a progress report gave a hint of what to expect. '*The Iduna is built on a similar principle to the famous Era, though on different lines, for pleasure rather than speed has been considered. The furnishings of the cabins are most elaborate.*'[520]

Iduna was launched from the same slipway as *Era* on the evening high tide of Tuesday 10 July 1888, and was christened by Miss Cunningham, a cousin of the owner. Newspaper descriptions state it was '*a very pretty boat*', and that its lines '*give the idea she is a clipper, and being set off with a long counter presenting a most graceful appearance*'.[521] *Iduna* was fitted out with a salon, foc'sle, ladies' cabin, and steerage quarters, with ample berths. A sail locker and a galley were also included; all desirable features for its intended offshore passages and cruising out of Port Phillip Bay.

Initially rigged with a cruising yawl configuration, *Iduna* had its first trial on Sydney Harbour late in August. '*The Iduna was out on Saturday, having her owner, Mr. Gurner, on board, who having arrived from Melbourne during the morning. She looked very well and gave every promise of being fast.*'[522] More trials took place the following week, with the vessel off to a promising start. '*The Iduna has had several trial spins both in and outside Sydney Harbour, and has fully come up to expectations.*'[523]

Little time was wasted and within days *Iduna* was off to Melbourne, under Captain Breslin, stopping in at Jervis Bay on 15 September. The voyage was a good examination which the vessel passed. '*Of the voyage of the yacht Iduna, hence to Melbourne, Captain Breslin speaks highly of her sailing qualities, and her weatherly behaviour in a sea way was well tested on the voyage round, when a heavy westerly gale was experienced.*'[524]

Once in Victoria, *Iduna* settled into its home waters with owner Gurner waiting in anticipation for the upcoming Melbourne Centennial Aquatic Carnival, to be held in November of 1888. However, back in Sydney the big boat building was not finished for the year. As *Iduna*

518 British Register of Ships for ON 101021; *Lloyds Register of Yachts* (1891).
519 *The Sydney Morning Herald*, 16 November 1887.
520 *Australian Star*, 11 August 1888.
521 *Sydney Mail and New South Wales Advertiser*, 14 July 1888; *Daily Telegraph*, 12 July 1888.
522 *Referee*, 30 August 1888.
523 *Australian Star*, 7 September 1888.
524 *Evening News*, 20 September 1888.

IDUNA UNDER CONSTRUCTION IN MID-1888.
SOURCE: AUSTRALIAN NATIONAL MARITIME MUSEUM (WILLIAM HALL COLLECTION).

neared completion in August, further up the Parramatta River, in a Drummoyne yard, yet another Reeks-designed vessel was launched. The centreboard yacht *Canomie,* of 56ft (overall), 49ft 2in (waterline) by 13ft 10in beam, was built by Captain Tom R. Dudley at a yard he had recently established.[525] The vessel was commissioned by Robert Ross of Taranganba, Queensland, with the expectation that it would carry 25 tons of goods on a 5ft draft between Rockhampton and Taranganba.[526] Rigged as a yawl with a lug mizzen sail, *Canomie* was launched on 11 August 1888.

At the same Drummoyne yard, another vessel designed by Reeks was just being started, and it was a race against time for its completion. The press relayed the scene. '*Yet another new racing yacht is to be added to our rapidly-growing fleet, this time of no less than 30 tons. Mr. T. R. Dudley is building the vessel at his yard on the Parramatta River from designs furnished by Mr. Walter Reeks - 55ft. l.w.l. by 13ft. beam are about the proportions. She is to be pushed on with all possible speed with a view to competing in the International race at Melbourne, and will, of course, add considerably to the interest centered in that event. The name of the owner or owners has not yet transpired.*'[527]

The vessel was being built for a group whose names seem to never surface.[528] The syndicate in question did include businessman, alderman and yachtsman William Smairl as secretary and he is often also listed as the sole owner.[529] Involved in yachting on Sydney Harbour for a decade, and in 1888 appointed vice-commodore of the Prince Alfred Yacht Club, Smairl had only the year prior purchased the yacht *Assegai* which he raced regularly.[530]

During this period of immense activity, Reeks also drew up plans for a 10-ton yacht of 43ft 10in (overall) to the order of Dr Fred Milford who previously owned *Sao* and was current owner of *Waitangi.*[531] Milford was then also commodore of the Prince Alfred Yacht Club. Named *Isea*, the vessel was built by Ford at Berrys Bay and launched on 9 October 1888.[532]

Isea went on to become a very successful yacht for Milford and a later owner, W. M. Marks. The vessel also helped cement Reeks' business relationship with Ford. The craft shared a hull shape and design similar to the much larger *Era* and, as is later noted, was the first 10-rater built anywhere in the world to the new 'Length & Sail Area Rule'; another large feather in Reeks' cap.[533]

Just five days after the launch of *Isea*, another 'crack' yacht was launched from the same yard, also to a design by Reeks. This particular vessel, named *Jess,* a 15-ton beamy centreboarder of 46ft 6in (overall), was built to the order of Ned King under the superintendence of Ernest Thompson, a pupil of Reeks who would later travel to Ireland to continue his training.[534] Intended for cruising, its first voyage was to Lord Howe Island.[535]

ISEA ON THE STOCKS.
SOURCE: MATT COLLISS COLLECTION.

Reeks' ability to work with his client's preferences and be adaptable is definitely on show here. *Jess* was lug rigged rather than gaff rigged. Whilst this could be viewed as a throw-back to an older, simpler but less efficient sail plan, it was a configuration well suited to a cruising or commercial vessel.

JESS.
SOURCE: AUSTRALIAN NATIONAL MARITIME MUSEUM (WILLIAM HALL COLLECTION).

Though shrouded in secrecy during its rapid building process, Smairl's mystery yacht was launched two days after *Jess* and three days before *Isea.*[536] Measuring 63ft (overall) by 56ft 6in, with a beam of 13ft 3in and a depth of hold of 8ft, the 30-rater was named *Volunteer* and was soon to become an obsession for the Sydney press.

'*THE NEW 30 TONNER. The new yacht which has been building at Dudley's yard ... for a syndicate of which Mr. W. P. Smairl is the secretary, was successfully launched on Saturday morning, and was christened by Miss Smairl, in the usual*

[525] *Referee*, 31 May 1888; *The Daily Telegraph*, 13 August 1888.
[526] *The Daily Telegraph*, 13 August 1888.
[527] *The Sydney Morning Herald*, 4 August 1888.
[528] *Evening News*, 8 October 1888.
[529] *Australian Town and Country Journal*, 14 December 1889.
[530] *The Sydney Morning Herald*, 3 May 1887.
[531] *The Sydney Morning Herald*, 10 October 1888.
[532] *Referee*, 21 June 1888; *The Daily Telegraph*, 10 July 1888; *Evening News*, 1 October 1888.
[533] *The Anchor*, 22 February 1912.
[534] *The Daily Telegraph*, 5 October 1888; *The Sydney Morning Herald*, 14 May 1892.
[535] *Evening News*, 1 October 1888.
[536] *The Daily Telegraph*, 8 October 1888.

manner, the "Volunteer," after the famous American yacht which lately appropriated the America Cup against British model Thistle.'[537]

The mid-October launch date did not give *Volunteer* much time to be fitted out, trialled and raced before having to leave for Melbourne to compete at the Victorian Centennial Aquatic Carnival. In fact the vessel was given a handicap before it even left Sydney for the event. Furthermore, in this period almost all of Sydney's first-class yacht owners preferred to have their sails made by the premier sailmaker Ratsey and Lapthorne of the Isle of Wight. This was commonly noted in the launch reports of new yachts and, in particular, some of Reeks' craft. But since sails had to be ordered and shipped out in advance, in *Volunteer*'s case it was already decided to use a set of *Magic*'s sails to fit the new vessel's spars, with modifications made to the sails locally.[538] Not ideal but practical considering the time frame.

Volunteer's first race, on 28 October 1888 at the Double Bay Regatta, included being matched against *Era*. 'Backstay', writing at great length in the *Referee,* had taken the ferry to watch the event and began with a grumble about the usual lateness of the ferry's departure. He then noted that *Volunteer* was still hoisting sails as they passed it on their way to the start line, everyone already running late for the advertised start time of 1.45 pm. '*I presume it was the same old cry, "O, they will wait for us," but in this case it was a mistaken one, for the other three yachts (Era, Sirocco and Iolanthe) were sent away at 1.50, the Volunteer not crossing the line until 6min. 5sec. later. I was indeed very sorry for Mr Smairl, who deserves everybody's thanks for his share in adding to our fleet such a yacht as Volunteer, but on the other hand I like punctuality. I know, it being the Volunteer's first race, everything was upside down, and I understand the carpenters were still on board until the last moment before leaving Farm Cove. Yet I cannot blame the starter, who I think did the correct thing.*'[539]

Both *Era* and *Volunteer* were yawl rigged, their cruising arrangement, and whilst the former won the race, *Volunteer* showed it had speed, making up considerable time on many legs, but was hampered when its spinnaker pole dropped and broke early in the second leg. Though the Sydney crowds could not wait for the pair to resume racing, including with their best sails set, the pair's next meeting was to be in Melbourne. Both craft needed to leave promptly, but this passage was to be a difficult test and, once again, a reflection of the challenging circumstances of the times.

For Reeks, the upcoming Victorian Centennial Aquatic Carnival, as well as the launch of *Jess, Isea* and *Volunteer,* would have been events not to be missed. Also completed during this period was the 17-ton centreboard yacht *Bettina,* launched from Ford's yard on 6 December 1888.[540] The 48ft (overall) craft was intended for cruising and built to the order of P. W. Creagh to a design by Reeks.[541] As yet another centreboard yacht, the vessel once again gave Reeks experience with designing something more unique and unconventional. *Bettina* went on to become a successful cruiser in this configuration, with its owner later quoted as saying he would build a bigger version if he could only sell the yacht.[542]

BETTINA.
SOURCE: AUSTRALIAN NATIONAL MARITIME MUSEUM (WILLIAM HALL COLLECTION).

As a postscript, it can be reported that *Bettina* became a bigger success under a subsequent Melbourne-based owner, Edgar Newland. In 1907 and renamed *Thistle,* the vessel was entered in Australia's first ever annual ocean race, sailing a course from Port Phillip Heads, Victoria, across Bass Strait to Low Head at the entrance to the River Tamar, Tasmania. The fleet competed for the inaugural Rudder Cup, donated by Thomas Fleming Day, editor of the American *Rudder* magazine, valued at sixty guineas.[543] Starting on Boxing Day 1907 with four competitors, *Thistle,* by then converted to a keel yacht, won easily to take the trophy.[544] The race, however, was so rough, with the fleet plagued by heavy gales, persistent rain and mist, such that Newland's wife, who had been a crew member on board during the race, refused to return the trophy, '*so as to dissuade other yachtsmen from attempting such a dangerous race*'.[545] No one heeded this warning and, over 110 years later, the Melbourne to Devonport Yacht Race remains on the yachting calendar, claiming the title of Australia's oldest ocean race.

Also noteworthy, *Bettina* was built under the superintendence of Ernest Thompson, Reeks' capable assistant, rather than Reeks himself.[546] The latter's absence was significant. Reeks had departed Sydney late on the afternoon of 5 September 1888 as a passenger on board the 3,000-ton RMS *Zealandia* for the mission of a life time.[547] He was heading to New York.

[537] *Australian Town and Country Journal*, 13 October 1888.
[538] *The Sydney Morning Herald*, 4 August 1888.
[539] *Referee*, 31 October 1888.
[540] *Evening News*, 7 December 1888.
[541] *Evening News*, 1 October 1888.
[542] *Sydney Mail*, 20 July 1889.
[543] www.orcv.org.au/deveonport-about-the-cup.
[544] *Examiner*, 30 December 1907.
[545] www.orcv.org.au/deveonport-about-the-cup.
[546] *The Daily Telegraph*, 7 December 1888; *The Daily Telegraph*, 27 August 1889.
[547] *The Sydney Morning Herald*, 6 September 1888.

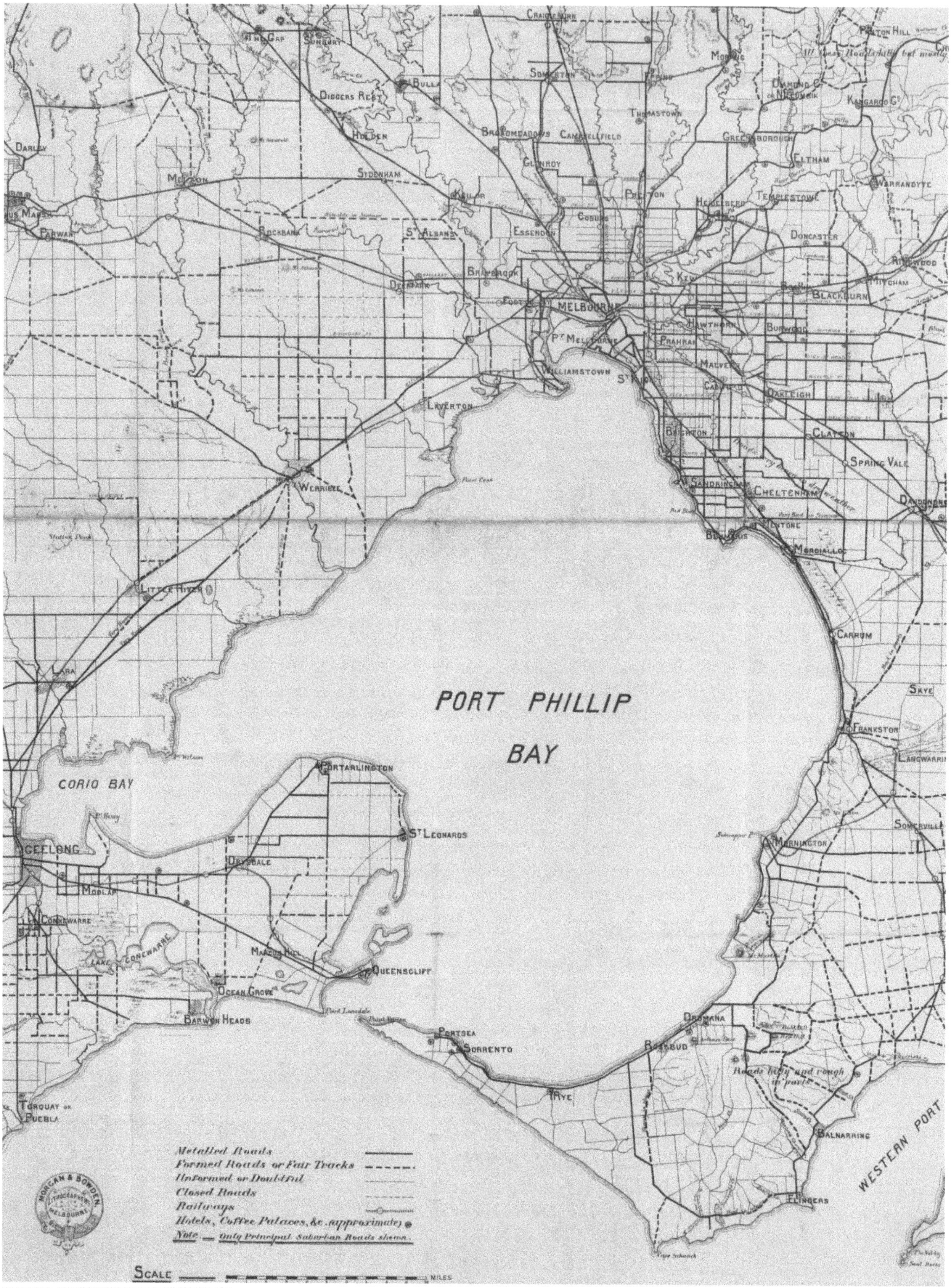

1896 MAP OF PORT PHILLIP BAY, VICTORIA.
SOURCE: NATIONAL LIBRARY OF AUSTRALIA (NLA.GOV.AU/NLA.OBJ-232389321).

CHAPTER FIVE - MELBOURNE

Before we journey with Walter Reeks to the United States, we must first pivot to the Melbourne Centennial Aquatic Carnival, held on 23-24 November 1888. Significantly, it marked the first time that representatives from multiple colonies, encompassing Victoria, New South Wales, South Australia, Tasmania and New Zealand, competed against one another across a broad spectrum of sailing events.[548]

The event promised to live up to its 'carnival' title. It was a progression from smaller intercolonial races which had demonstrated that yachtsmen from neighbouring colonies were keen to compete against one another despite the obstacle presented by distance. Such an event had been held in Hobart barely 12 months prior and, as previously noted, the Victorian yacht *Janet* had been sailed from Melbourne to Sydney to compete against *Magic* and *Waitangi* in January of 1887, with victory going to the latter.[549] That series of races had been instigated by Sir William Clarke, owner of *Janet* and commodore of the Royal Yacht Club of Victoria.

A Tasmanian-born landowner and cattle and sheep breeder, Clarke (1831 - 1897) had inherited large pastoral estates in Tasmania and Victoria from his late father in 1874, worth tens of millions of dollars in today's values, becoming the largest landowner in the colonies.[550] He was also a major philanthropist, supporting hospitals, churches and the University of Melbourne, amongst various organisations and causes.[551] A keen sportsman with a penchant for a lavish lifestyle, which he could well afford, Clarke was eager to redeem himself and his vessel at the Melbourne Centennial Aquatic Carnival, to be held in hopefully more favourable home waters.

The carnival was a multi-class event and fleet pageant, with a full spectrum of yacht racing on display. The showcase event was the race for first-class yachts of 20 tons or more, vying for prestigious bragging rights as well as £500 in prize money, plus a gold anchor valued at £100 (donated by Clarke). The entrants were *Era, Miranda* and *Volunteer* (representing New South Wales), *Taniwha, Janet* and *Iduna* (representing Victoria), and *Wanderer* (representing South Australia).[552] Notably, out of these seven vessels, Reeks had designed the three Sydney-based boats, as well as *Iduna*. In contrast, the 33-ton rating *Janet* had been designed and built locally by Henry R. Murray in 1880 for Clarke, while *Taniwha* was from New Zealand (built by Thomas Nicol of Auckland in 1880) and had come across for a previous regatta held on Port Phillip Bay. Meanwhile, the South Australian entrant *Wanderer*, built at Adelaide and launched in 1882, was actually owned by H. P. Fergie of the Royal Yacht Club of Victoria. The 46-ton yacht had been re-ballasted and fitted with new Ratsey and Lapthorne sails to bring it up to racing trim.[553] *Wanderer* sailed with the flag of the South Australian Yacht Club of which H. P. Fergie was also a member.[554]

Competing in the smaller keel yacht classes were the Victorian boats *Cooneeana* and *Secret,* as well as *Vesta* and *Tasma* from Tasmania, *Akarana* from New Zealand and *Assegai,* entered under William Smairl's ownership and thus representing New South Wales. The latter added to the mix another craft related to Reeks, through his significant design modifications.

[548] *Referee,* 21 November 1888.
[549] *The Sydney Mail and New South Wales Advertiser*, 22 January 1887.
[550] adb.anu.edu.au/biography/clarke-sir-william-john-3229.
[551] adb.anu.edu.au/biography/clarke-sir-william-john-3229.
[552] *Sportsman,* 21 November 1888; *The Age*, 23 November 1888.
[553] *Leader,* 31 August 1889.
[554] *South Australian Weekly*, 3 July 1886.

In addition, there were classes for centreboarders and open half-decked boats featuring vessels from Victoria and New South Wales, pitting their various configurations against one other to see who had the better concept. These boats included *Genesta, Ellie, Varuna* and *Ionia*, all from New South Wales.

The premier events, the race for the first-class yachts, as well as the one for the much smaller 5 and 10 tonners, were scheduled to take place on Friday 23 November 1888 with the race for yachts 10 to 20 tons taking place the following day. Finally, a special one-off challenge between the 24ft open boat *Aileen* from Sydney and the slightly larger lugger *Mayflower* of Queenscliff, Victoria, was also organised, providing yet another test of different vessel types and concepts.[555]

Never before in Australia had there been so much hype and build-up surrounding a yachting regatta. The event had been written about regularly in the months preceding, with details and commentary published in major and minor newspapers throughout the colonies. As the start of the Melbourne Centennial Aquatic Carnival neared closer, with the fleet racing scheduled to take place over only two days, plans began to unravel, with the schedule adjusted and extended beyond the initial program to account for both misunderstanding and misfortune.

There were also significant issues regarding rules. For example, a ruling by the Victorians about the maximum number of crew for the open half-decked yachts upset members of the Sydney Amateur Sailing Club primarily, as they were intending to enter some of their fleet for the regatta and would be taking their full complement of crew. '*On the programme adopted at the meeting on August 21, it is distinctly stated that there shall be no restrictions as to crews. Now, according to the latest programme issued, in the race for boats 20ft and under the limit is five hands, and in the race for boats 20ft and not over 23ft, and 23ft and not over 26ft, six hands. This seems hardly fair to our men, for not only have berths been taken in some instances, but what is perhaps more important, the boats will not be seen at their best with such small crews.*'[556]

In terms of misfortunes impacting the event, this was largely due to the difficulties experienced by the Sydney yachts not only in trying to get to Melbourne but also to arrive on time. It was typical for the period. Whilst the half-decked open boats went safely by steamer, the larger yachts made their own way to Port Phillip Bay. Their progress, or lack thereof, and the inevitable bad weather they encountered was followed closely by the press.

Iduna left Sydney first, given it had been built to the order of John Gurner of Melbourne. After the vessel's launch on 10 July 1888 and subsequent trials, no time was wasted in getting the craft sea-ready. Departing in late September, the rough weather *Iduna* encountered was simply a good opportunity to test its qualities. The vessel arrived with weeks in hand before the carnival.

For the rest of the Sydney-based fleet, it was a different story. *Era* and *Miranda* were next to depart, but not until early November. In what seemed like ample time, they left on Tuesday 6 November. However, things did not go well and the two vessels met with problems from the outset.

Era turned back soon after leaving Sydney Heads. Although clearing out to sea with a favourable north-east wind from behind, within 45 minutes and only off Coogee, '*the iron work of the gaff carried away, so she came round and returned to port*'.[557] This was repaired by taking on a spar from *Waitangi*, just to enable the craft to get to Melbourne, with the iron work fixed separately. A second departure the following day got no further; the southerly conditions were so uncomfortable that *Era* again returned to the harbour. With the weather abating the vessel finally left for Melbourne on 9 November.[558]

Miranda left Sydney late on the afternoon of 7 November and encountered the first of what became a series of southerlies. The vessel ended up sheltering south of Wollongong before heading out again, to meet worse weather further south. '*On the narration of one of her crew who knows quite a thing or two about the sea and its wonders, it was indeed a perilous trip. Making her departure from Sydney in all "the pride, pomp, and circumstance of a battleship," she met with gale after gale and ultimately crawled into Twofold Bay, a sorry weather racked barge, ultimately returning to Sydney.*'[559]

In fact, *Miranda* was still in Twofold Bay as the carnival took place. It had been quite an ordeal. '*The Miranda left for Melbourne about the same time as the Era and the two yachts met in Twofold Bay, the latter having made the faster run down the coast. She was able shortly afterwards to take refuge in Waterloo Bay, whilst the Miranda met the full force of a violent south-east gale, and after three days' combat with it, during which she lost her supply of fresh water, and sprang a serious leak, it was decided to bear away for Eden. For a couple of days, the ship's company had only soda water and whisky to drink, a state of affairs that was really worse than might appear to some, for there were several teetotallers in the crew.*'[560]

MIRANDA.
SOURCE: AUSTRALIAN NATIONAL MARITIME MUSEUM (WILLIAM HALL COLLECTION).

[555] *The Herald*, 26 November 1888.
[556] *The Sydney Mail and New South Wales Advertiser*, 10 November 1888.
[557] *The Sydney Mail and New South Wales Advertiser*, 10 November 1888.
[558] *The Sydney Morning Herald*, 9 November 1888.
[559] *The Sydney Mail and New South Wales Advertiser*, 6 November 1907.
[560] *Evening News*, 9 November 1907.

CREW OF ERA, SHOWING WALTER REEKS IN CENTRE WITH WHITE CAP AND PIPE. LIKELY TAKEN ON SYDNEY HARBOUR IN THE MONTHS BEFORE THE VESSEL WAS SAILED TO MELBOURNE IN NOVEMBER OF 1888.
SOURCE: ROYAL SYDNEY YACHT SQUADRON.

ERA.
SOURCE: AUSTRALIAN NATIONAL MARITIME MUSEUM (WILLIAM HALL COLLECTION).

VOLUNTEER.
SOURCE: NEW ZEALAND WAR MEMORIAL MUSEUM (HENRY WINKELMANN COLLECTION).

On 11 November *Volunteer* scampered away from Sydney Heads, chasing the others, with its builder Captain Tom R. Dudley as skipper and in rather messy circumstances. '*It was expected she would put underway at 10.00 o'clock on Saturday morning, but the carpenters being busy fixing up the interior arrangement, and as several other matters in connection with the yacht having to be attended to, her departure was delayed till the afternoon; and to make things worse, the cook made himself scarce early in the morning, and it was not until the last moment that one was found to take his place.*'[561] The story continues with owner, William Smairl, deciding to have *Volunteer* towed to Sydney Heads in order to waste no additional time. Setting out under racing main, with the trysail ready if needed, the passage did not last long. A strong southerly came up, and the craft returned to the harbour, finally leaving the following day. Eventually *Volunteer* arrived off Melbourne's waterfront on 20 November, with only a few days to spare.

Assegai was also one of the passage-makers, but being the smallest and almost last to leave Sydney, this proved another recipe for disaster. The vessel met southerlies and failed to reach Melbourne in time for its event on 24 November, despite concerned efforts at the Victorian end to help it along. As the deadline approached, *Assegai* was reported to have lost both anchors and was off the Victorian coast where it had taken on fresh water from the SS *Koonawarra*, but was otherwise considered okay.[562] A piece published in *The Age* provided more detail. A pilot vessel had been sent to wait at Port Phillip Heads on 22 November to tow the little yacht to Melbourne.[563] However, *Assegai* did not arrive until two days later, i.e., on the Saturday that its race was to take place at the opposite end of the bay.

All told *Era* was the better of the Sydney yachts to make the passage to Melbourne. It was also in the best shape racing-wise. By the end of its first season, *Era* was in a class of its own. Competing on Sydney Harbour, the vessel had won six important events on the calendar, including the Centennial Cup, the Brassey Cup, and the recently-created Gascoigne Cup (a prestigious trophy still today). *Era* had also cruised as a yawl with davits on the side for the tender, showing off its adaptability. The craft was very much ready to race at the Melbourne Centennial Aquatic Carnival, as were most of the other contenders, excepting perhaps *Volunteer*.

[561] *The Sydney Morning Herald*, 12 November 1888.

[562] *The Age*, 23 November 1888.

[563] *The Age*, 23 November 1888.

The day of the big event for the first-class yachts finally arrived, in hindsight proving a little anti-climactic compared to the lead up. The weather did not cooperate. Instead of the hoped-for consistent winds, the conditions offered only a fluky wind with no sea. The course had been fixed at three laps of 16 miles, but owing to a late start was reduced to two laps. This decrease also reduced handicaps. *Miranda* was of course a late withdrawal, still sheltering at Eden.

There were other hiccups. The Sydney press reported, '*There were not many spectators, and the arrangements seemed thoroughly confused, the management being very bad. Only one boat, a private steamer, hired by a number of Sydney gentlemen followed the race*'.[564]

The race started late in the morning of 23 November and was completed at 6.30 pm, surprisingly with only 12 minutes separating the finish times for the first and last boats. *Era* had lead for most of the race and went on to claim victory by a margin of just over 5 minutes. *Janet* was second and *Volunteer* third. Handicaps were then applied resulting in a win for *Iduna* with *Era* second and *Volunteer* third.[565] The yachts were not officially placed, however, owing to confusion at the finish line as to which side of the flagship the yachts were to finish on. It was claimed *Iduna* had finished on the wrong side. The vessel's owner, John Gurner, stated that he had crossed '*the originally planned winning line*', then sailed around the finishing boat to cross what he thought to be the '*last ordained winning line*'.[566] A separate report of the finish stated that '*Era and Volunteer, two New South Wales boats, both sailed round the flag ship and crossed the starting line, whilst the other four competing boats crossed without having rounded the vessel. A dispute accordingly arose as to which was the course that ought to have been pursued*'.[567]

The event's committee met later that night to discuss what seems, at least on face-value, to be a technical matter, as *Iduna* had complied with the sailing instructions that discouraged efforts '*to win the race by other methods than fair sailing and superior speed and skill*'.[568] However, in circumstances such as a very close race, this could also be a critical decision. It had to be taken seriously, even if it came down to a simple misunderstanding by some of the competing yachts. After much deliberation the committee was unable to resolve which vessel had crossed the line correctly. Instead, it was decided that the race should be re-sailed. And so on 26 November the fleet met again. Under more cooperative weather, though still very little public interest, the race was re-contested. *Era,* sailed by Alfred Milson and under protest, went on to claim victory once again, this time by a margin of 20 minutes from *Volunteer, Wanderer* was third.[569] The allocation of handicaps was not enough to deny *Era* the title of the fastest yacht in Australia, thus earning the prize money and gold anchor.

[564] *The Sydney Morning Herald*, 24 November 1888.
[565] *The Sydney Morning Herald*, 24 November 1888.
[566] R. P. Neale (1984). *Jolly Dogs are We: The History of Yachting in Victoria, 1838-1894.*
[567] *The Age*, 24 November 1888.
[568] R. P. Neale (1984). *Jolly Dogs are We: The History of Yachting in Victoria, 1838-1894.*
[569] *The Daily Telegraph*, 27 November 1888.

With regards to other events of the Melbourne Centennial Aquatic Carnival, on 24 November the fleet of smaller yachts raced without *Assegai*. On handicap *Tasma* won from *Cooeeana,* with *Vesta* third.[570] With the late arrival of *Assegai,* Sir William Clarke showed his benefactor side and proposed a race for this vessel against one other yacht. This promptly brought a response from the public, including two letters supporting the generous offer, but also asking that it be open to any of the yachts that the craft would have raced against.[571]

The new race, with four starters, was subsequently sailed. In a strong south-east wind and a nasty sea, it was won by *Galatea*, followed by *Madge, Assegai* and *Akarana*.[572]

The Sydney press followed the events of the Melbourne Centennial Aquatic Carnival closely, and provided a wrap up of the regatta from a Sydney viewpoint. '*To sum up results, many of our yachts proved their superiority as they took first and second place in the big race, but our smaller craft, which are built for the smooth waters of Port Jackson were, with limited crew, overwhelmed by the more powerful boats on the rough waters of Hobsons Bay*.'[573]

One of these smaller craft was the half-decked open boat *Varuna*, owned and sailed by W. Paul of the Sydney Amateur Sailing Club. The overall experience of his time in Melbourne may have had an influence on his decision to build a new vessel two years later, also named *Varuna,* this time to a design by Reeks.

As well as successfully providing the opportunity for vessels from many colonies to come together and compete, the Melbourne Centennial Aquatic Carnival was remarkable in that it was a meeting of the old and new from a design perspective. The racing was held under the still new Length & Sail Area Rule, and all yachts were measured and rated to this rule. Progress was still in a transition phase for this rule, however, with Reeks leading the way in Australia. His designs for *Era, Iduna* and *Volunteer* all took advantage of the new hull proportions, and were the only craft to compete in the regatta that were built to the new rule. On the other hand, *Akarana* from New Zealand was probably one of the last yachts built anywhere to the older tonnage rule, and provides an excellent comparison for the changes offered by the new rule.

Akarana was built in 1888 in Auckland by Scottish boat builder Robert Logan. He was keen to see the craft cross the Tasman Sea, building on the start made by other New Zealand boats, including *Taniwha* and *Waitangi* that had made a name for themselves in Australian colonial waters. Logan decided to build an entry for the Melbourne Centennial Aquatic Carnival and sell it after the event, not only relying on his first-class craftsmanship as a builder, a guaranteed outcome, but also on a good race result, not something that could be guaranteed. He started to build the boat while the old rule was still in use. However, by

[570] *The Age*, 26 November 1888.
[571] *The Argus*, 28, 29 November 1888.
[572] *Geelong Advertiser*, 3 December 1888.
[573] *The Sydney Morning Herald*, 31 December 1888.

the time *Akarana* was ready for launch and to be sent across by steamer, Logan knew it would have to sail under the new rule, to a higher rating that would be a disadvantage. The yacht was sent anyway, and was well received in Port Phillip Bay. It ended up winning a race held prior to the regatta, but in the main event, running rigging was carried away. It lost significant time effecting repairs and was unable to recover to gain a good place.

AKARANA ON SYDNEY HARBOUR (CIRCA 2010).
SOURCE: AUSTRALIAN NATIONAL MARITIME MUSEUM COLLECTION.

Regardless, *Akarana*'s presence was noted favourably, and although later sold to buyers from Sydney where it enjoyed success, Logan did not gain any further orders in Australia for yachts. Just over a decade later, however, his boat building sons would reap the benefits of trans-Tasman trade, selling a number of their yachts to New South Wales owners.

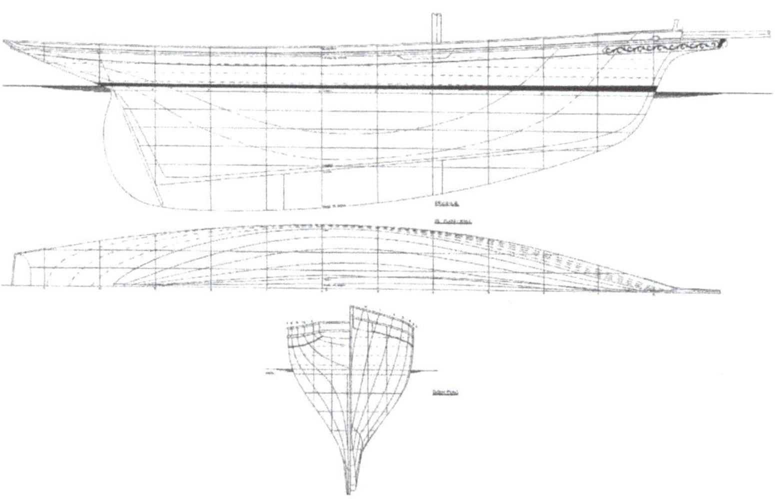

LINES OF THE AKARANA. BY DAVID PAYNE.
SOURCE: AUSTRALIAN NATIONAL MARITIME MUSEUM.

Of note, over a 130 years later, only *Akarana* survives from all of the keel boats that competed at the 1888 Melbourne Centennial Aquatic Carnival. The vessel, and its cohort, reflect the earlier beam-dominated measurement rules, with their narrow and deep hulls. However, *Akarana* carried much of its nearly 6 tons of ballast in a 5-ton lead keel bolted to the bottom of its deep hull. It also had a clipper bow and elegant profile. In comparison the Victorian yacht *Janet*, designed in 1880, had a plumb or vertical stem that gave a ram bow appearance of raking forward at the waterline when the yacht sat in the water, and was trimmed down by the stern. *Assegai* was another deep keel, narrow vessel and although modified by Reeks, its shape was still not ideal for the new rule's direction.

On the other hand, both *Era* and *Volunteer,* which were primarily designed to race, showed what was possible with the new freedoms the Length & Sail Area Rule permitted. Especially important was how, in just a few months of working to the rule, the two yachts demonstrated how Reeks was not only evolving but also improving his designs.

In profile *Era* and *Volunteer* shared an identical curved sheerline, a plumb stem rounding into the keel with the greatest depth just forward of the rudder post, and a long elegant overhanging counter. With regard to the mid-section, which is generally a good indication of the whole shape of the hull, things were developing nicely for Reeks over these two designs, less than a year apart. *Era,* the first design, when compared to tonnage rule-designed mid-sections, had more hollow or concavity in the curve to the garboard strakes coming off the keel, and a bit more overall beam relative to depth. This provided a more pronounced round to the bilge section as it curved up to the topsides. *Volunteer* had even more beam and more concavity to the hollow garboards, showing a lovely rounded bilge shape and bringing more width to the hull at the waterline. It was in fact more powerful for its size than *Era*.

Both *Era* and *Volunteer* had a significant amount of lead ballast slung low on the base of their keels, but in the overall calculation for stability, this was complimented by the width of their hulls. In the case of *Era*, its beam was about 17 per cent of its overall length. With the *Volunteer*, Reeks had increased this to 20 per cent of overall length. Because stability increases as an exponential function of beam, the actual increase in stability this offered was significantly more. For comparison, the cruising design of *Iduna* had even more, with its beam 21 per cent of its length; good for both stability and allowing for more volume inside for a cruising accommodation. Finally, the racing yacht *Thelma*, launched much later in 1889, was shorter than the others, with a ratio of 20.5 per cent.

Although hampered by poor sails in Melbourne, the fruits of this design development were soon revealed by *Volunteer* in subsequent races back on Sydney Harbour. In Melbourne, *Era* had completed the mission Alfred Milson had set when commissioning the yacht's construction immediately following *Waitangi*'s win against *Janet* in

PLANS OF VOLUNTEER (TAKEN FROM HALF-MODEL).
BY DAVID PAYNE

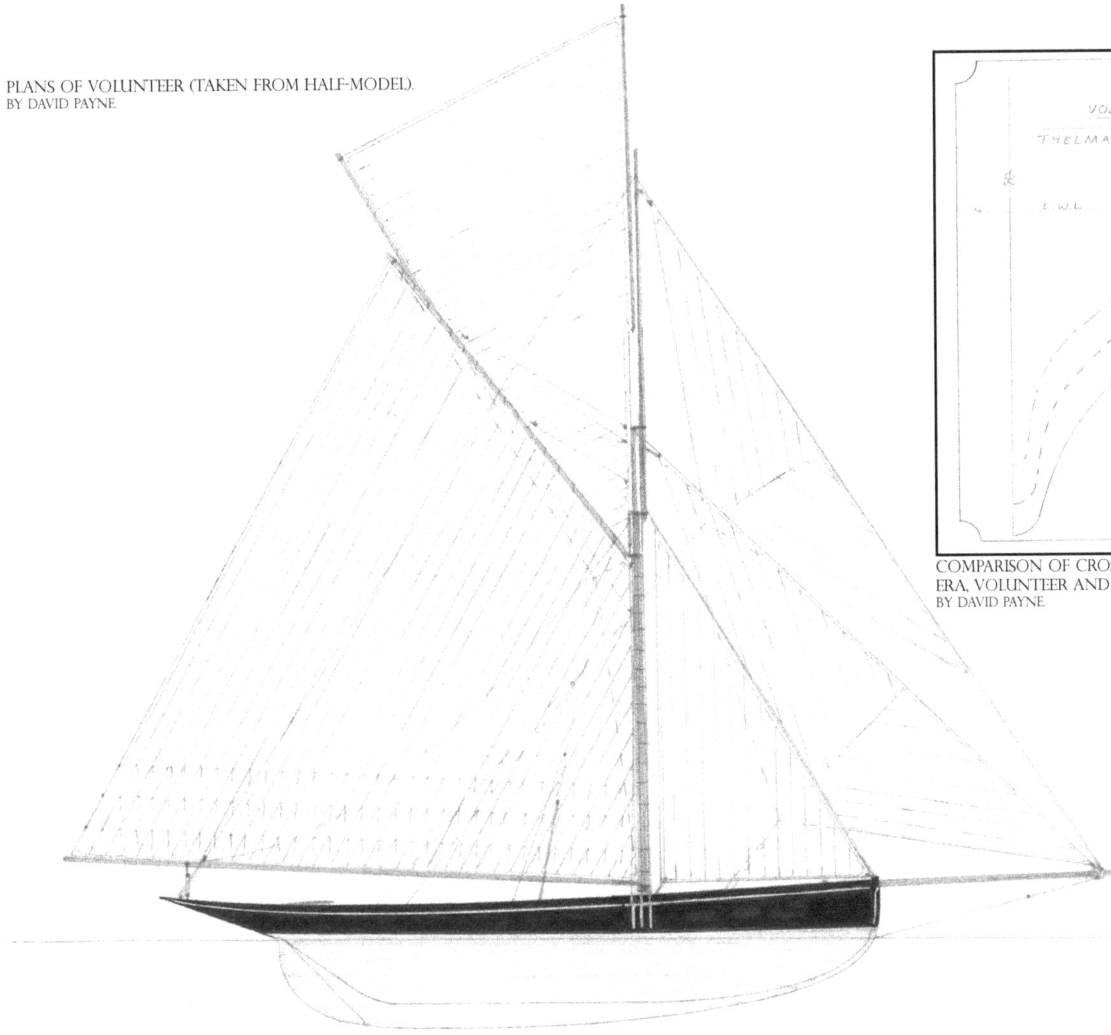

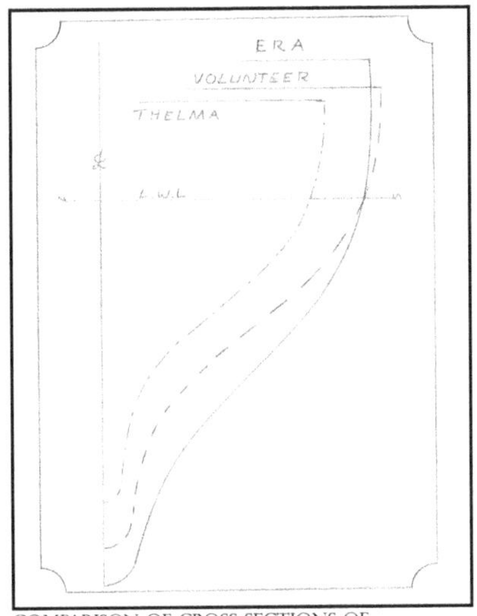

COMPARISON OF CROSS-SECTIONS OF ERA, VOLUNTEER AND THELMA (TO SCALE).
BY DAVID PAYNE

early 1887: to once again win the showcase intercolonial event. But *Volunteer* had not shown its true colours. With its Ratsey and Lapthorne set of sails hoisted, the 30-tonner *Volunteer,* although slightly smaller in all respects than the 40-tonner *Era*, was able to beat *Era* across the line on many occasions upon their return to Sydney.

As we chart the career of Reeks and take in its full perspective, it begins to look like he was already making plans for bigger things by mid-1888, including developing an understanding of what was possible in yacht design. Writing a story about the past year's sailing season he is very positive about many things, striking a very national tone at the end. '*Thus we find during our centennial year no less than forty to fifty craft ready to "shoot the line" in their respective classes. No city in the world of equal population can boast such a fleet. We are already abreast* [of] *the times in yachting and sailing, and with the desire for improvements so deep rooted in the hearts of our aquatic men and with Australia's honour as our reward, when we succeed, surely we shall soon lead the world, cease to speak of the American type and the English type, and have a type of our own, which other countries will look at with envious eyes, and call Australian.*'[574]

It appears Reeks was sowing the seeds for an endemic type of vessel before the Melbourne Centennial Aquatic Carnival but, significantly at this event, they began to sprout. This 'Australian type' was probably a reference to his own type, as this is revealed later in 1888 in an interview where he discusses different yacht types. '*Mr Reeks, with pardonable pride, called special attention to the difference between the English and Australian boats. "My boats" he said "have more beam and less draught than theirs. I have learned yacht designing in England, but I haven't built a single English boat since I went to Australia.*'[575]

Reeks was obviously ready and eager to move ahead with further development, particularly of a racing yacht to eclipse all. However, before his next big commission could be realised, the one that might make other countries envious, first he had to investigate the ground rules for the design parameters, and the conditions of the playing field. This wasn't practical from afar in Sydney, so he went to the theatre where it would all take place, New York, to assess what had to be done to win that coveted trophy, the America's Cup. Given Reeks had already demonstrated an active mind and endless bursting energy, it is highly probable that the steamer he took from Sydney to San Francisco couldn't get there quick enough.

[574] *Illustrated Sydney News*, 31 May 1888.

[575] *Evening Star* (Washington, D.C, USA), 23 October 1888.

Australian Mail Service.

FOR SAN FRANCISCO,

The new and fine A1 steel steamship

"Zealandia,"

Of the Oceanic Steamship Company, wil be due at Honolulu from Sydney and Auckland on or about

September 23, 1888.

And will leave for the above port with mails and passengers on or about that date.

For freight or passage, having SUPERIOR ACCOMMODATIONS, apply to

WM. G. IRWIN & CO., Agents

SOURCE: EVENING BULLETIN (HONOLULU, HAWAII), 1 SEPTEMBER 1888.

CHAPTER SIX - NEW YORK

In order to provide background to Walter Reeks' trip to New York in late 1888, we must rewind to early June of 1886 when a luncheon held at the Prince Alfred Yacht Club in honour of its commodore, Dr Fred Milford, resulted in a discussion regarding the merits of a local yacht challenging for the America's Cup. Enlarging upon this idea, Reeks' name was soon cast forward as a person quite capable of designing such a vessel, even though he had only been in the colony a little over a year.[576] Of note, there was not only widespread agreement amongst those present at the event, but absolute enthusiasm for such a daring proposition.

It was not the first time Reeks' name had been used in this context. One of the pet projects of 'Bobstay' during the time he wrote opinion pieces for *The Sydney Mail and New South Wales Advertiser* in the mid-1880s, was the encouragement of a Sydney-built challenger to contest for the America's Cup. '*Why should not Sydney have the cup*!' he quipped in the same piece in which he touted the virtues of the newly-arrived Reeks back in May of 1885.[577] The seed of possibility was being discussed, and Sydney's yachtsmen were listening intently.[578]

Surprisingly, there was little information given in the press in the lead up to Reeks' departure for America in September of 1888. A month after he left Sydney, however, an article published in *The Sydney Mail and New South Wales Advertiser* alluded to the secret, '*he had gone to obtain designs to build a yacht for a certain gentleman, with the object of bringing, if possible, the Americas Cup over here*'.[579]

'Mainstay' in the *Leader*, once word spread to Victoria about the mission, wrote: '*Already a Sydney yacht designer is on his way to America and England to take count of all that is worth noticing in connection with the designing and building of a 90 ton racing yacht in Port Jackson, to challenge for the possessions of the Queen's Cup, won by the America, and now in the possession of the New York Yacht Club. The project is a most laudable one, and whether we are indebted to a single individual or syndicate for the carrying out of the object for which Mr. Walter Reekes* [SIC] *is now sent on his travels, Victoria will not grudge our Sydney friends any success that may attain in the bold enterprise. It certainly would be an interesting episode in the history of yachting were Australia to secure and return to old England the much coveted trophy which their own representatives and repeatedly defeated racers have failed to take away from its American defenders*'.[580]

Travelling on board RMS *Zealandia*, Reeks departed Sydney on 5 September 1888, arriving at the vessel's first port of call, Auckland, four days later.[581] Departing after only 24 hours spent in dock, RMS *Zealandia* next stopped at American Samoa to off-load and collect mail, crossed the equator, the second time Reeks had done so in less than four years, and arrived in Hawaii on 21 September.[582] After an overnight stay in Honolulu, the vessel reached its final destination of San Francisco on 29 September; all up a total voyage duration of 24 days.[583]

Disembarking in San Francisco, Reeks immediately turned to his attention to the job at hand. Seeking out members of local yacht clubs, he spent only a few days

[576] *The Sydney Mail and New South Wales Advertiser*, 12 June 1886.
[577] *The Sydney Mail and New South Wales Advertiser*, 30 May 1885.
[578] *The Sydney Mail and New South Wales Advertiser*, 30 May 1885.
[579] *The Sydney Mail and New South Wales Advertiser*, 6 October 1888.
[580] *Leader*, 24 November 1888.
[581] *The Sydney Morning Herald*, 6 September 1888; *Thames Star*, 10 September 1888.
[582] *The San Francisco Examiner*, 11 September 1888; *The Honolulu Advertiser*, 22 September 1888; *The Hawaiian Gazette*, 25 September 1888.
[583] *Evening Post*, 1 October 1888.

in the city before he was en route to his next destination. During his time in San Francisco, however, Reeks viewed a schooner named *White*, of which he was suitably impressed.[584] Reeks later stated that the vessel was more representative of the American type of craft than those of New York, which he considered to bear more of a resemblance to English yachts.[585]

Reeks then made his way to Chicago, likely by train, where on 10 October the *Chicago Tribune* published an article associated with his arrival in the city. Rather strangely the article made no mention of his occupation, nor the reason for his visit. Perhaps Reeks did not openly divulge this information? Instead it solely discussed rabbits. Though the article incorrectly referred to him as '*William Reeks*', it offers a unique window into a topic deemed international news at the time, as well as Reeks' propensity for humour, particularly since his sole occupation was maritime-based. Or conversely, and more seriously, the article hints to his background in agriculture, growing up on a farm in Christchurch that was also subject to pests and diseases, his empathy for others and his awareness of colonial events. '*William Reeks of Sydney, New South Wales, is in the city. He said he did not leave the country to get away from the rabbit plaque, but felt that he would have to if the number of those annoying creatures did not diminish. Myriads of rabbits were driving business-men to desperation, ruining the cattle industry, impoverishing the land, and yet there was no relief. Laws had been passed to compel land-owners to rid their possessions of these little creatures, but the law was powerless to afford relief to a suffering people. They might pass all the laws on this subject they pleased, Mr. Reeks said, but the rabbit had come to stay, and thus far human ingenuity had failed to discover a device that would rid the country of the plaque.*'[586]

With all of the talk of rabbits abandoned, Reeks next found himself making front page news in Boston. Only a week later *The Boston Globe* published the below headline.

The article went on to state, '*It appears probable that a challenge for the America's Cup may come from a new quarter of the globe. The Australians, not content with the rowing championship* [earlier that year Peter Kemp of New South Wales had won the 'Championship of the World' in professional sculling, defeating Edward Hanlan of Canada], *intend to build a yacht to compete in the summer of 1890 with the then holder of this cup. Walter Reeks, naval architect of Sydney, N.S.W., has come to Boston to begin his acquaintance with our yachts and waters, and proposes on his return to design*

[584] *The Daily Telegraph*, 13 February 1889.
[585] *The Daily Telegraph*, 13 February 1889.
[586] *Chicago Tribune*, 10 October 1888.

Boston Daily Globe.

BOSTON, WEDNESDAY EVENING, OCTOBER 17, 1888—EIGHT PAGES. PRICE TWO

LATEST!

For Other Evening News See Second, Fifth and Eighth Pages.

WISDOM-PACKED.

The Cranium of Ex-Chairman Barnum.

He Thinks New York Surely Democratic.

The Old Political Chieftain Interviewed in Boston.

Politics Kicks Up a Row in the Carnegie Firms.

FROM AUSTRALIA

Will Come a Challenge for the America Cup.

Walter Reeks of Sydney in Boston

To Study the Shape of Our Sturdy Yachts.

With Mr. Burgess He Visits the Volunteer.

In 1890 the Australian Ideal Will Be Ready.

and build for a gentleman of wealth there his ideal yacht for the challenge. Mr. Reeks comes under the auspices of Messrs. Henry W. Peabody & Co. He first called upon Mr. Burgess, who received him with the utmost courtesy and frankness, and enabled him to see the Volunteer and other yachts of his design now in winter quarters. Mr. Reeks will make but a short stay in America. He will shortly visit New York and other yachting centres.'[587]

The following day, Reeks and his ulterior motives were again featured on the front page of *The Boston Globe*. The headline ran as follows.

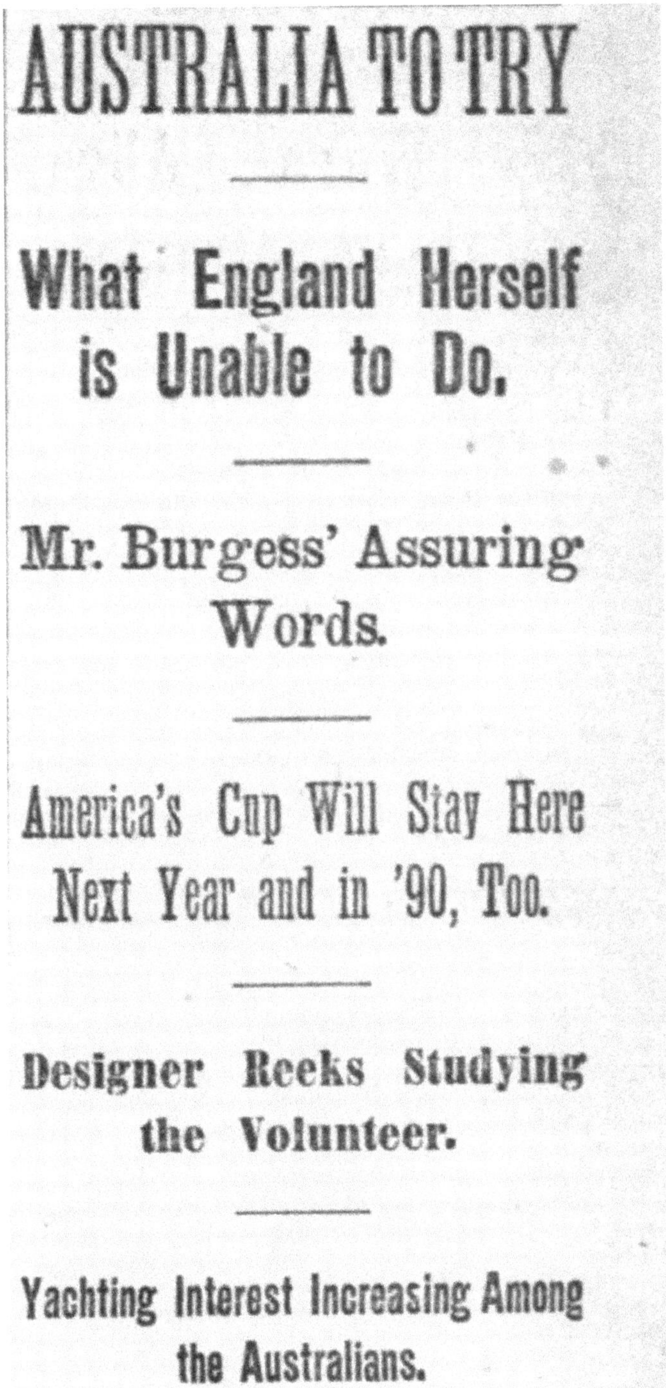
AUSTRALIA TO TRY

What England Herself is Unable to Do.

Mr. Burgess' Assuring Words.

America's Cup Will Stay Here Next Year and in '90, Too.

Designer Reeks Studying the Volunteer.

Yachting Interest Increasing Among the Australians.

SOURCE: THE BOSTON GLOBE, 18 OCTOBER 1888.

[587] *The Boston Globe*, 17 October 1888; en.wikipedia.org/wiki/World_Sculling_Championship.

The substance of the article itself provides additional information, particularly concerning the American's surprise at the possibility of a challenge. '*The presence in town of a well-known Australian naval architect, having sinister designs upon our cherished America's Cup, has caused quite a breeze of interest in yachting circles. It never occurred to the local enthusiasts that England's furthermost colony would try to do what England herself cannot accomplish, but it appears that the Australians do entertain such a notion.'*[588]

The article continued. *'A Globe reporter started today to find Walter Reeks, the Australian designer in question, but Mr. Reeks was not easily found. At the office of Henry W. Peabody & Co., in the Mason building, under whose auspices the would-be rival of our own Burgess comes to this country, no information as to Mr. Reeks's habitation here could be learned. A member of the firm who was seen was not induced to talk upon the subject, on account of the rather indefinite condition in which it still is. He said that Mr. Reeks would remain in Boston but a day or two longer, and would then probably go to New York.'*

What's interesting here is that the front page of *The Boston Globe* is printing little to no additional news since its article published the previous day. The reporter was unable to find Reeks to interview him and as a substitute reported the process that he used in failing to track him down. Instead Edward Burgess, designer of the celebrated yacht *Volunteer* that had famously won the America's Cup only the year before against *Thistle*, designed by Scotsman George Lennox Watson, was sought for comment. When asked *'What do you think of this proposed Australian yachting challenge?'* Burgess replied, *'I understand that it has been definitely decided to send one in 1890, but beyond that I'm afraid I know very little. Mr. Reeks called on me the other day and I found him a very pleasant fellow. He said it was the bona-fide intention of the Australian yachtsmen to send a challenge, but that nothing definite has yet been arranged. He is simply here to study our waters and the designs of our boats. After he has done this, he will return home and make his designs and build his yacht'*.[589]

The article concluded with a statement on Reeks himself. '*Walter Reeks is a resident of Sydney, N.S.W., where he has been for the past three years. He is about 35 years old, and before going to Australia resided with Richardson, the designer of the Irex, Stranger and other fast English boats.'*[590]

It took a few days but the reporter from *The Boston Globe* eventually succeeded in not only finding Reeks but also being granted an interview. Again, in an article published on the front page, Reeks' intentions were made clear: he was in Boston to familiarise himself with America's yachts and particularly with Burgess' ideas. The article went on to describe Reeks as a '*young, blue-eyed English-man on the sunny side of 30, who went to Australia a few years ago and revolutionized yachting there, awakening widespread interest in it, and making a science of the sport'*.[591] A direct quote from him was then given, providing an overview of yachting

[588] *The Boston Globe*, 18 October 1888.
[589] *The Boston Globe*, 18 October 1888.
[590] *The Boston Globe*, 18 October 1888.
[591] *The Boston Globe*, 23 October 1888.

in New South Wales. '*There are now more yachtsmen in Sydney, in proportion to the population, than in any other city in the world. Some years ago they used to build about one boat in five years, now they build about five boats in one year. The two leading clubs are the Royal Sydney Yacht Squadron, with 42 craft, and the Prince Alfred Yacht Club, with about 30. Our regular yachting season begins on January 26 each year, when we hold our anniversary races, and lasts till about the end of March. Regular matches are sailed every Saturday, and there were generally four or five regattas during the week*'.[592] The quote continued, '*All our boats are manned by amateur crews, and on our largest yachts, which are 75 feet overall, there are never more than two or three paid men. Our harbor at Sydney is peculiarly favorable to the development of amateur seamanship, as it is almost entirely land-locked. The harbor itself, which is only about two miles wide and five miles long, has a shore line of over 900 miles, so numerous are the bays and large and small indentations of the coast.*

All our small boats are centreboards. One style of yacht peculiar to Sydney is our small centreboards from 18 to 24 feet water line, carrying no ballast, but from 15 to 25 persons.

My views of an ideal yacht to compete for the America's cup have not been radically changed by my visit to Boston. I have learned something, of course, about the details of American yachts, but my general ideas are unchanged.'

When asked if he hoped for success in his first attempt to win the America's Cup, Reeks was reported to smile and modestly reply, '*The yacht I intend to build is for a wealthy Australian, who thinks his country can beat the world in everything when she once sets out to do it*'.[593]

Another report stated that Burgess and Reeks spent about an hour together in consultation. Reeks then travelled to South Boston to view the *Volunteer* where he apparently '*manifested the deepest interest in the study of her lines*'.[594] Other vessels he inspected were the *Mayflower* and *Puritan*.[595]

THE 1885 AMERICA'S CUP WINNER PURITAN.
SOURCE: LIBRARY OF CONGRESS
(ITEM 2016805210).

Designed by Burgess, the *Volunteer* was built by Pusey & Jones Shipbuilding Co. at Wilmington, Delaware, for Charles Paine of the New York Yacht Club. It was launched on 30 June 1887 after a construction period of only 66 days.[596] The centreboard sloop, with an all-steel frame and hull, measured 108ft (overall) by 86ft (waterline) with a beam of 23ft 2in and a draft of 10ft.[597] *Volunteer* was known as a 'compromise' sloop, i.e., characteristic of a lightweight, wide and shallow hull with centreboard, of which Burgess was deemed not only the pioneer but also the leader in designing this type of vessel.[598] His two previous efforts, the *Puritan* and *Mayflower*, were both symbolic of this type of craft. The former, Burgess' first America's Cup defender, won the event in 1885, while the later achieved the same success in 1886. Like its two predecessors, *Volunteer* went on to win the America's Cup against *Thistle* in 1887.

Born in the distinctly maritime community of Cape Cod, Massachusetts, in 1848, and growing up in Boston, Burgess' career in yacht design came after a rather successful period spent in academia, teaching entomology and natural sciences at Harvard University, his alma mater.[599] Though he grew up involved in the pastime of yachting, it was not until the age of 35, after a trip to England to study naval architecture, that Burgess decided to make the move to this vocation professionally, establishing a yacht design and brokerage firm in Boston with his brother Sidney.[600] The venture proved a success with Burgess himself designing over 130 vessels in nine years.[601] During this period, he would also be internationally recognised and celebrated for his three successive America's Cup winners, *Puritan, Mayflower* and *Volunteer.* Sadly his career was cut dramatically short. Burgess died in July of 1891 from typhoid fever. He was 43 years of age. Ironically Reeks would nearly succumb to the same disease eight years later.

—

Reeks left Boston for New York on the afternoon of 23 October 1888.[602] The press followed him. His responses to an interview published in the *Evening Star,* a Washington D.C.-based newspaper, provide an interesting insight into the situation, as well as Reeks' thoughts at that time. When asked if Australia would be challenging for the America's Cup, he responded, '*You may safely say, that a challenge will be forwarded. If I did not feel certain of it you may be rest assured I should not have abandoned my business for six months at the request of the gentlemen whom I represent. The challenge will come from the Royal Sydney Yacht Squadron, the premier club in Australia, and the only organisation bearing the title squadron. This organisation has a membership of 240 and a fleet of 42 vessels. There are a few schooners in the fleet, but most of the*

[592] *The Boston Globe*, 23 October 1888.
[593] *The Boston Globe*, 23 October 1888.
[594] *The Sportsman*, 31 October 1888.
[595] *Glasgow Evening Citizen*, 15 November 1888.
[596] en.wikipedia.org/wiki/Volunteer_(yacht).
[597] en.wikipedia.org/wiki/Volunteer_(yacht).
[598] en.wikipedia.org/wiki/America%27s_Cup.
[599] L. del Sol Knight (2005). *Encyclopedia of Yacht Designers.*
[600] L. del Sol Knight (2005). *Encyclopedia of Yacht Designers.*
[601] en.wikipedia.org/wiki/Edward_Burgess_(yacht_designer).
[602] *The Boston Globe*, 23 October 1888.

THE 1887 AMERICA'S CUP WINNER VOLUNTEER.
SOURCE: LIBRARY OF CONGRESS (ITEM 2016805225).

vessels are single-stickers, cutters or sloops. You can nowadays hardly tell the difference'.[603] When asked if there were any centreboards in the club, '*Uh, yes; indeed, all the small boats are centreboards and some of the largest ones. Our largest racing boats are 60 feet on the waterline and 75 feet overall. All but two of them have been designed by me within the last five years in Sydney, and I have sent several to Victoria and Queensland'.*[604] Next when asked about New Zealand, he replied, '*None to New Zealand. No, they build their own boats, and* [with a smile] *they think they can beat mine'.*[605] As the interview continued, Reeks alluded to the difference between the English and Australian boats. '*My boats have more beam and less draught than theirs. I learned yacht designing in England, but I haven't built a single English boat since I went to Australia'.*[606] And, perhaps the most important question of all, '*The challenger will be a keel boat of course?*' '*Why of course?*' was the quick and cryptic answer Reeks gave, once again leaving the issue unresolved.[607]

Reeks spent two weeks in New York, of which information there is surprisingly very little mentioned in the press. Perhaps he fooled them again, remaining incognito about his mission and intentions. The only available report, from the *Toronto Daily Mail* via the *New York Herald* states, *'Mr Walter Reeks, the Australian yacht designer, who came to this country a few weeks ago to inspect our*

[603] *Evening Star*, 23 October 1888.
[604] *Evening Star*, 23 October 1888.
[605] *Evening Star*, 23 October 1888.
[606] *Evening Star*, 23 October 1888.
[607] *Evening Star*, 23 October 1888.

MODEL ROOM, NEW YORK YACHT CLUB (1884).
SOURCE: NATHANIEL L. STEBBINS PHOTOGRAPHIC COLLECTION, HISTORIC NEW ENGLAND (PC047).

racing craft before building a boat to compete for the America's cup, was found yesterday in the model room of the New York Yacht Club. The aspirant for international honours is a fine looking young man, with an open, frank expression, and to all appearances may be termed a "jolly good fellow". When I was introduced to him he was engaged in looking over the Herald's accounts of the Volunteer-Thistle races, which are preserved in the club scrap book.

Upon reaching New York Mr. Reeks was made an honorary member of the New York Yacht Club, and every one has endeavoured to make his stay as pleasant as possible, and to furnish him with all the information he could desire about American yachts and waters.

When asked what he thought of America, and how he had been treated here, he said: "I am in love with the county, and the American people with whom I have come in contact have treated me as if I were a prince. I am well satisfied that, when I come over with the yacht I intend building to compete for the cup, the gentlemen of the New York Yacht Club will give me as hearty a welcome as I have received on this, my first trip."

Mr. Reeks also expressed himself as being well pleased with the Volunteer, and thought she was sufficiently large for all purposes. He believes that an 86 foot boat is large enough. "I thought," he said, "that I should build an 86 foot boat, and expressed myself in that way to one of the club men, but I find by referring to the new deed of gift that it will be safer to build at the maximum, which is 90 feet on the water line."

"Regarding the new deed of gift I must say that, although I have not studied it as carefully as I shall, the only objection I find at present is the ninety foot clause. I don't relish building a boat larger than the Volunteer; she is quite big enough for any purpose. My real object in coming to New York was to find out exactly what I had to do, and the conditions under which the races would be sailed. Every possible facility has been given me to obtain that information, and I desire to express my sincere thanks for this courtesy."

When asked if the new yacht would be a cutter or a centreboard, Mr. Reeks replied: "I shall not confine myself to either cutter or centreboard. My new boat may be a centreboard, but I am hardly prepared to answer that question yet. Any suggestion that I may receive here I will be grateful for and acknowledge with pleasure. As to my lines and dimensions, they are matters which at present are not in a matured state, and I would prefer not to give them even approximately."

"I wish to say, however," continued Mr Reeks, "that I shall build a sloop yacht and will challenge under the new deed of gift. When I send my boat, which will be in 1890, I shall comply with the rules as near as possible, and see that the challenge is in complete form. If any little details are omitted, I feel confident they can be arranged to the entire satisfaction of the club and myself."

Mr. Reeks says that when he came to America he stated frankly the object of his visit, that he had nothing to conceal, and in his preparations for designing and building the competing yacht will be open and frank.

Mr. Reeks will sail for England on Wednesday, and will begin work on the new yacht in March.'[608]

[608] *Toronto Daily Mail*, 7 November 1888.

Reeks then travelled on board the White Star Liner *Germanic* as a saloon passenger to the United Kingdom, leaving America on Wednesday 7 November and arriving in Cork, Ireland, eight days later, on 15 November 1888.[609] His arrival coincided with another exciting development that was then being played out in the English press: the possible challenge for the America's Cup by John Jameson's new 84ft composite racing cutter *Iverna*, then being built at the yard of J. G. Fay & Co. in Southampton, notably to a design by Reeks' mentor Alex Richardson.[610] This rumour was later found to be lacking any foundation. Still, speculation ran rife. Just who would challenge for the America's Cup, England, Australia or Ireland? There was also talk of *Thistle*, the Scottish yacht designed by Watson, that failed to win the cup in 1887 challenging again, after undergoing extensive alterations.[611]

Time was running out, however. The deadline with which challenges for races to be sailed in 1889 were to be received, agreed on and accepted, was only a matter of months away.[612] For Reeks, however, there was another year at least, for he was adamant that Australia would issue its challenge in 1890.

For all challengers, regardless of country or year of challenge, a revised Deed of Gift had to be adhered to. Formally issued by the New York Yacht Club on 24 October 1887, the most significant changes to the document were that the challenger provide all measurement information of the contending yacht, plus a registry certificate. Challengers were also to provide the defender with at least ten-months' notice of these vital statistics. This thus allowed the defender to know well in advance the intricacies of the vessel it would meet, such that they could potentially design and build a yacht capable of beating it. Additional discourse regarding the new Deed of Gift is discussed in detail in the following chapter. It ultimately evoked a tangled web for pending challengers.

Where Reeks spent the last few weeks of November and most of December of 1888 is not reported. He is known to have visited yacht designers and builders in Ireland, Scotland and England, likely included amongst them his mentor Alex Richardson.[613] He also would have ventured home to Christchurch, Hampshire, to visit his parents, siblings and other family members. Despite being in England for professional reasons, for Reeks personally it was an opportunity to renew ties and solidify familial relationships. It would also be the last time he would see his father. Charles Reeks died in Christchurch on 18 February 1891 at the age of 72.[614]

On 28 December 1888 Reeks left London on board the P. & O. Company's 6,500-ton steamer *Arcadia* for Australia, travelling via Gibraltar, Malta, Port Said, the Suez Canal, Aden and Colombo.[615] Considering the distance he had already ventured, it was rather a fast-paced trip around the globe.

Arriving back in Sydney via the Australian ports of Albany, Adelaide and Melbourne on 10 February 1889, 44 days later, Reeks was likely eager not only to detail the outcome of his trip to the gentleman yachtsman who had provided financial backing for his travels, but also to begin designing the yacht itself.

Two days after his return, a representative from *The Daily Telegraph* called upon Reeks at his Pitt Street office. Here he was found *'hearty and well, busily occupied in unpacking the many pictures, nautical works, samples of the latest sail cloths and metal, which he has brought for Australian yachtsmen'*.[616] Reeks then proceeded to give an interview where he discussed the interesting events that had occurred during his trip, of note his meeting with Edward Burgess and his viewing of the *Volunteer* and other famous yachts. He stated, *'One thing that struck me about Burgess's boats was that the hulls are splendidly balanced and it was impossible to find any important faults. The Volunteer is not so large as I expected, but in every respect is a clipper, and it will be a heeler that can beat her'*.

Speaking of the boat he was intending to build, Reeks said, *'she would be up to the limit allowed in the deed of gift as regards length for cutters, 90ft, but owing to the gentleman who gave him the commission being away from Sydney it has not been decided whether she will be a centreboard or a deep-keel vessel'*. He continued. *'Americans take kindly to yachting and I feel sure that every possible chance will be given to us to win the cup, as a prominent yachtsman in New York said to me, "Bring along your boat, my boy, and we will race you every day in the year, but do not act as with the Thistle; we want no more Thistles"*.'[617]

A letter to the *New York Herald* sent from Reeks and dated 16 March 1889 was published in this newspaper a month later. It contained the following direct quote. *'In a couple of weeks I shall start on designs, and in five months after the vessel will be built. I have two more matches to sail this season and then, though I have ideas pretty fully in my head at present, I'll be able to combine and formulate what I have thought of into some tangible form'*.[618]

The same article then announced that it was probable that the Sydney-built *Volunteer*, which Reeks had designed and was currently successfully helming on Port Jackson, would represent Australia in a race against *Lurline*, the 80ft San Francisco-based schooner yacht owned by John Spreckels and his brother. A wealthy businessman and member of San Francisco's Pacific Yacht Club, Spreckels had issued a challenge *'against all of Australia'* for a prize valued at £2,000.[619]

For Reeks, at only 28 years of age, there was much to do and no time to waste. First, however, there was a wedding to plan.

609 *Sheffield Independent*, 8 November 1888; *Glasgow Herald*, 16 November 1888.
610 *Evening News*, 12 November 1888.
611 *Evening News*, 12 November 1888.
612 *Evening News*, 12 November 1888.
613 *The Daily Telegraph*, 13 February 1889.
614 England & Wales, National Probate Calendar, 1892 for Charles Reeks.
615 *British Australasian*, 26 December 1888; *South Australian Register*, 5 February 1889.
616 *The Daily Telegraph*, 13 February 1889.
617 *The Daily Telegraph*, 13 February 1889.
618 *Toronto Daily Mail*, 24 April 1889.
619 *Toronto Daily Mail*, 24 April 1889.

MAP OF AUSTRALIA AND SURROUNDING COUNTRIES (1895).
SOURCE: NATIONAL LIBRARY OF AUSTRALIA (NLA.GOV.AU/NLA.OBJ-232118958).

CHAPTER SEVEN - THE OPEN OCEAN

On Monday 18 March 1889 at St. Andrew's Cathedral, Sydney, according to the rites of the Church of England, Walter Reeks married 27-year-old Mary Emma Kent. The press in announcing the union stated that, owing to the Lenten season, the ceremony was very simple and quiet with only immediate friends and a few relations attending.[620] The witnesses to the event were J. M. Manson and W. E. Broadhurst. The latter, William Edwin Broadhurst, was Mary Kent's uncle, married to her mother's younger sister Betsy (nee Ridyard).[621] The Broadhurst's and Mary Kent had only arrived in New South Wales from England five months prior, on 19 October 1888 per the *Austral*.[622] The vessel had left London on 31 August, travelling via Gibraltar, Naples, Port Said, the Suez Canal, Aden, Sri Lanka and several Australian ports before arriving in Sydney.[623]

Broadhurst was a retired miller of some wealth who was a long-term sufferer of tuberculosis. Seeking warmer climes, he and his wife had relocated to Australia with the hope of improving his condition. Having no children of their own, they travelled with their niece Mary Kent.

It must have been a whirlwind romance for Reeks and his new bride. He had only been back in Sydney for a little over a month before the marriage took place, and it is highly unlikely that, even though Mary's maiden name was Kent (also the maiden name of Reeks' mother), there was any familial or personal relationship between the pair in England. Mary hailed from Southport, Lancashire, whereas Reeks' maternal Kent line was from Hampshire.

[620] *Australian Town and Country Journal*, 23 March 1889.
[621] 1881 England Census for Betsy Broadhurst.
[622] *South Australian Register*, 12 October 1888; *The Australian Star*, 19 October 1888.
[623] *South Australian Register*, 12 October 1888.

In the early years of their marriage, Walter and Mary Reeks lived at *Blatchley*, 8 Fitzroy Street, Milsons Point with the Broadhursts. Though no longer in existence, the residence was an eight-roomed villa conveniently located, only a five minute walk to the ferry service.[624]

MAP OF MILSONS POINT, SHOWING FITZROY STREET (1880S).
SOURCE: STATE LIBRARY OF NEW SOUTH WALES (COLLECTIONS.SLNSW.GOV.AU/RECORD/74VVRW3VEBPZ).

[624] *The Daily Telegraph*, 13 January 1891.

	No.	When and where Married.	Names and Surnames of the Parties.	Conditions of the Parties, (Bachelor or Spinster.) (Widower or Widow)	Birth Place.
1335/89	698	St. Andrew's Cathedral March 2nd 1889	Stuart de Jong Miller Jane Raby	Bachelor Spinster	Faversham Kent Chili
		Married in ~~the~~ St. Andrew's Cathedral According to the rites of the Church of England			
1336/89	699	St. Andrew's Cathedral March 5th 1889	Arthur Cressy Barry & Edwyna Phillips	Bachelor Spinster	Clapham S.W. Camden
		Married in the St. Andrew's Cathedral According to the rites of the Church of England			
1425/89	700	St. Andrew's Cathedral March 18th 1889	Walter Reeks Mary Emma Kent	Bachelor Spinster	Wick, Hants, England Worsley near Manchester
		Married in the Cathedral Church of S. Andrew According to the rites of the Church of England			
1624/89	701	St. Andrews Cathedral April 3rd	William Edward Dye and Evalina Rose Church	Bachelor Spinster	Balmain Sydney Balmain Sydney

SOURCE: SYDNEY, AUSTRALIA, ANGLICAN PARISH REGISTERS, 1814-2011.

Married in the Cathedral Church of St. Andrew

According to the rites of the Church of England

in the Colony of New South Wales. Registered by

Rank or Profession.	Ages.	Usual place of Residence.	Parents. Father's Christian Name and Surname. Mother's Christian Name and Maiden Surname.	Father's Rank or Profession.
Gentleman	28	Sydney	Daniel Stuart-Millne Josephine Millne	Gentleman
			&	
—	25	Sydney	William Raby Eliza Roberts	M. Inst. C. E.

This Marriage was solemnized between us { Stuart-de Jersey Millne / Jane Raby }

In the presence of us { Lacita Hodge / Kate Chamberlain }

By (~~or before~~) me Arthur R. Rivers Officiating Minister ~~or Registrar~~.

Rank or Profession.	Ages.	Usual place of Residence.	Parents. Father's Christian Name and Surname. Mother's Christian Name and Maiden Surname.	Father's Rank or Profession.
Cane Inspector	30	Harwood Island Clarence River	James Barry Mary Barry	Brewer
—	20	Sydney	John Bede Phillips Annie Phillips	Bank Manager

This Marriage was solemnized between us { Arthur Cressy Barry / Edwyna Phillips }

In the presence of us { Wm Edw Love / Agnes MacLean }

By (~~or before~~) me Arthur R. Rivers Officiating Minister ~~or Registrar~~.

Rank or Profession.	Ages.	Usual place of Residence.	Parents. Father's Christian Name and Surname. Mother's Christian Name and Maiden Surname.	Father's Rank or Profession.
Naval Architect	28	Sydney	Charles Elizabeth } Reeks maiden Kent	Farmer
—	27	Sydney	James & Sarah } Kent maiden Kidyard	Coal Agent

This Marriage was solemnized between us { Walter Reeks / Mary Emma Kent }

In the presence of us { J. Manson / W. E. Broadhurst }

By (~~or before~~) me Ernest C. Beck Officiating Minister ~~or Registrar~~.

Rank or Profession.	Ages.	Usual place of Residence.	Parents. Father's Christian Name and Surname. Mother's Christian Name and Maiden Surname.	Father's Rank or Profession.
Banker	30	Sydney	Timothy Dye Maria Church	Comm Merchant
—	25	Sydney	Walter M. Church Annie E. Stubbs	Manager of Ins Co

This Marriage was solemnized between us { William C. Dye / E. R. Church }

In the presence of us { Walter Bradford / A. A. Dawson }

J. F. Church

Miriam Church

On a professional level, Reeks returned to Sydney from America and England with all the vigour and industriousness that he was known for. With regard to his recent overseas trip, the wealthy gentleman that funded his travels in order to glean information about a possible Australian challenge to the America's Cup was none other than Jack Want, who several years prior had commissioned Reeks to design the 40-tonner *Miranda*. Want certainly had the resources to finance such a mission, '*no other barrister of his period in Australia earned more in fees or had a greater reputation as an advocate*'.[625] He also had gumption and was known for his flamboyance and ostentatiousness.[626] Of significance, at the time of Reeks' return to Sydney, Want was in Asia where he spent the months of January through April of 1889 touring Japan and China.[627]

In May of 1889 Reeks confirmed that the yacht to challenge for the America's Cup was still going ahead and that, though he was awaiting advice from England regarding the subject, he had started designing the highly anticipated vessel with the expectation that he would then receive approval to commence its construction.[628] Perhaps in consequence of the new build, Reeks advertised *Miranda* for sale on behalf of Want, though it failed to sell and was back in commission by the start of the yachting season.[629] In August of 1889 it was noted in the press that Reeks was nearing completion of the design for the vessel to the challenge for the America's Cup.[630]

However, it remains unclear whether the challenger was designed along the lines of the English cutters, of which *Era, Iduna* and *Volunteer* are indicative of this type, or whether Reeks adopted an American centreboard type, which is represented by his designs for *Miranda* and the smaller yacht *Bettina*.

Disappointingly, though Reeks completed the design for a vessel to vie for the America's Cup, the challenge itself was never issued. The conditions of the new Deed of Gift proved too onerous to accept for both English and Australian challengers.[631] A few years later a London-based newspaper, *The Field,* summed up the situation. '*Mr. Walter Reeks, the Sydney designer, has already made a name for racing craft, and it will be remembered that two years ago his name was prominently connected with a projected challenge for the America's Cup; but the "new deed of gift" was a great deal more than the intended challenger could hoist in, and up to the present no challenge for the cup has been sent to New York.*'[632]

Want himself may have shied away from issuing the challenge, not solely because of the new Deed of Gift, but also because of his busy legal practice and looming political ambitions. A representative for Paddington and a staunch opponent of federation, he was very active in political circles during this period. These activities likely took precedence over any potential challenge for the America's Cup.

The Deed of Gift for the America's Cup has its own narrative, and has been a source of contention throughout the event's long history. As a measure of this, in recent times it has been the subject of a PhD thesis by New Zealander Hamish Ross, a sailor, legal advisor and syndicate member of several America's Cup winning teams.[633] His study relates to a fascinating legal and sporting story. Firstly, as a point of interest, nowhere in the text of the various iterations of the Deed of Gift is the event called anything but '*The Cup*', and the now registered name 'America's Cup' is simply what it has become colloquially known as.

The original Deed of Gift was a letter and covering note written in 1852 by the five surviving syndicate members of the winning schooner *America*, begging leave to present the '*The Cup*' to the New York Yacht Club. It had requirements to abide by for both the challenging and defending clubs, and these requirements had their origins in a series of horse races held in New York in 1823 involving a horse by the name of American Eclipse managed by John Cox Stevens, also a member of the *America* syndicate and a founder of the New York Yacht Club. A series of three races between American Eclipse, representing the North, and Sir Henry, the best horse representing the South, were held under a challenge arrangement determined by agreement between both parties. The three-race programme attracted a huge spectator turnout with big stakes, including the prestige of whether the North or South would prevail. Stevens' horse won 2-to-1 and in this successful event were the seeds of how Stevens and his *America* syndicate envisaged '*The Cup*' events would play out, i.e., as a match between the best yachts of the challenging and defending clubs.

The original letter was first drafted and then signed by the syndicate members, including Stevens and his brother in 1852. However, for reasons not known, Stevens did not follow through and present the letter to the New York Yacht Club before he passed away in June of 1857. At a meeting held the following month, almost six years after '*The Cup*' had been won by the schooner *America* sailing in English waters on 22 August 1851, syndicate member and donor George Schuyler presented an unsigned copy of the letter he had kept and this was accepted at the meeting as the founding Deed of Gift.

In plain English the letter had three paragraphs of conditions for a race between the challenging club and the '*Yacht Club in possession*' of the trophy. The first paragraph detailed the size of the vessel defined in tonnage, the second discussed the arrangements for the '*match*' to determine the winner, with the idea that these '*may be determined by mutual consent*', and also included what notice was required from the challenging club. The third and final paragraph stated that '*The Cup is to be the property of the Club*' who had won it at each match, and it was to be '*a Challenge Cup for friendly competition between foreign countries*'.

[625] P. Serle (1949). *Dictionary of Australian Biography.*
[626] en.wikipedia.org/wiki/Jack_Want.
[627] *Evening News,* 14 January 1889; *The Daily Telegraph,* 30 April 1889.
[628] *The Brisbane Courier,* 14 May 1889; *The Sydney Morning Herald,* 22 May 1889.
[629] *The Sydney Morning Herald,* 11 May 1889;
The Sydney Mail and New South Wales Advertiser, 2 November 1889.
[630] *The Sydney Morning Herald,* 17 August 1889.
[631] *The Sydney Morning Herald,* 22 August 1889.
[632] *The Field,* 12 December 1891.

[633] www.americascup.com/en/news/450_THE-DEED-OF-GIFT.

The deed was simple enough, but obviously open to interpretation and immediately the problems began. The New York Yacht Club held the first two matches on its own terms in 1870 and 1871, sending multiple yachts to meet a single English challenger on a New York harbour course that favoured the defender with their local knowledge. Two more challengers then came from Canada in 1876 and 1881 and both times they were completely outclassed by the United States, though the latter had relented and nominated just one boat to meet the challenger in all of the races, still taking place on a New York harbour course.

After the 1881 event, '*The Cup*' was returned to George Schuyler who wrote a new Deed of Gift and returned the trophy back to the New York Yacht Club the following year. This clarified that the holding club must nominate only one boat for the defence, that the challenging yacht had to be constructed in the country it represented and that, to prove it was seaworthy and well-constructed, the yacht had to sail to the match '*on its own bottom*' from its country of origin. Further, a defeated yacht could not challenge again until another contest was completed. Finally, to deter any Canadian challenges from clubs located on the vast freshwater Great Lakes, the challenging club had to have an annual regatta on the '*sea or an arm of the sea*', i.e., a legal term defining a site with an ebb and flow tidal pattern.

It is evident that it was not a straight-forward process for a contest to be held, and any such challenge did rely on correctly interpreting and meeting these requirements, with variations to be agreed upon by mutual consent between the challenger and defender. Three more challenges then came from Great Britain in 1885, 1886 and 1887. Played out again with races on both the New York harbour inside course and a modest offshore windward/leeward course a short distance out of the harbour, all three challengers were confidently beaten by the American defenders, though the last one became controversial. The English challenger *Thistle,* designed by George Lennox Watson and built in secrecy, alarmed members of the New York Yacht Club upon its arrival. The vessel was found to be longer than what was stipulated in the challenging document. The official umpire for the event, George Schuyler, thus adjusted *Thistle*'s handicap to account for the longer measurement, thereby significantly impacting the vessel's chance of succeeding. The length difference was a matter of about 18 inches, less than two per cent over 85ft as first advised, but the New York Yacht Club was taking no chances.

THE 1887 ENGLISH CHALLENGER THISTLE IN DRYDOCK
SOURCE: WIKIPEDIA COMMONS (EN.WIKIPEDIA.ORG/WIKI/THISTLE_(YACHT)).

Such fastidiousness demonstrated just what an Australian entry would be up against. Not only was this an event with great public interest worldwide, there was also an enormous amount of pride at stake. After only five challenges, the Americans were fixed on retaining '*The Cup*'. They were not giving away any advantages. Reeks was certainly right in going to New York to clarify and gain further information regarding the conditions of any potential challenge, especially since there was a much more detailed Deed of Gift to consider.

After the scare over the discrepancy in *Thistle*'s length, the New York Yacht Club asked George Schuyler and five others (nominated as a committee) to draft a third Deed of Gift. Of note, this document has been in effect since it was adopted in 1887 and remains in force today. As a registered trust document in the New York Supreme Court, only this governing body can arbitrate any disputes, regardless of the country of origin of the club holding '*The Cup*'.

Within this new and lengthy document there were a series of statements that gave considerably more definition to the original requirements and intent of the donors than had been delivered by previous versions. Upon its release, the revised Deed of Gift was roundly condemned by commentators on both sides of the Atlantic. Included amongst them was W. P. Stephens, an American yachting historian who today still remains widely respected for his objective and thoughtful writings.

Under the headline, '*AN ACT TO PREVENT YACHT RACING*', Stephens let forth. '*If the object of the framers was to keep The Cup they had double reason to be proud for it not only promises to accomplish that end most effectively, but it does so in such a manner that in itself it is dear to the legal heart. ... Seemingly fair and liberal on its face, the deed is far more strict in its demands than either of the others, and the concessions made in favour of racing demands are all nullified in a single short clause.*'[634]

Stephens and others were particularly put out by the demands that the challenger reveal key dimensions of the challenging yacht's hull, including the vessel's waterline length, beam at the waterline, beam overall, and draft, plus provide a registry certificate that would inevitably detail other useful information. The challenger was also to give at least ten-months' notice of these vital statistics to the defender, allowing the latter to know well in advance exactly what it would meet, and potentially allow them enough time to not only design but build a yacht with this knowledge in hand.

A positive inclusion in the new Deed of Gift, however, was that the courses had to be sailed in the open ocean, removing the option of racing in New York harbour with its local knowledge factor in favour of the defender. The mutual consent provision, allowing agreement on any of the principal conditions remained, but it still meant that both parties had to agree, and if one party did not, then the default provisions, as written, had to be adhered to.

634 E. B. Morris (1964). *Sailing for the America's Cup.*

It appears that Reeks went to New York not fully versed with the new Deed of Gift, and perhaps this was part of his reasoning for the mission. Early in 1889 he notes, '*regarding the new deed of gift, I must say that ... I have not studied it as carefully as I shall*'.[635] However, there were sharp legal minds in Australia to assist in due course, including the potential challenger Jack Want who, even without the well-reported condemnations, would likely have arrived at similar conclusions.

Reeks is further quoted as being most interested in the length of the vessel he had to build. It would appear that although he stated a preference for an 86ft waterline length hull noting, '*I don't relish building a boat larger than Volunteer (86 foot on the waterline). She is big enough for any purpose*', he did state that it would '*be safer to build at the maximum, which is 90 foot on the waterline*'.[636] He also remained confident to the end. '*I shall build a sloop yacht* [meaning a single mast], *and will challenge under the new deed of gift when I send my boat, which will be in 1890*'.[637]

It was a tricky conundrum. Reeks wasn't just landing in a new area with sometimes verbose legal interpretations complicating the new Deed of Gift. There were also new practical implications that needed to be applied to both the challenger and defender. Other events and aspects of the era, separate to the America's Cup, produced additional simmering problems. These would have likely come to the forefront during the next challenge.

All told, it was a period where the evolving America's Cup event deviated from its past regattas and shifted to a new realm. Reeks, and anyone else considering a challenge for '*The Cup*', was sailing in unchartered waters in more ways than the one.

Perhaps the primary change centred on the clause of the new Deed of Gift which stipulated how the races were to be sailed. '*All such races shall be on ocean courses, free from headlands, as follows: "The first race, twenty nautical miles to windward and return; the second race an equilateral triangular race of thirty-nine nautical miles, the first side of which shall be a beat to windward; the third race, (if necessary) twenty nautical miles to windward and return".*' This provision built upon the need for a challenging club to be one that held its '*annual regatta on an ocean water course on the sea, or on an arm of the sea, or one which combines both*'.

In effect the latter phrase excluded the option for the Canadian yacht clubs located on the Great Lakes to challenge again. The yachts of the Royal Canadian Yacht Club, based on Lake Ontario in Toronto, had been easily beaten by the New York Yacht Club's defenders in 1876 and 1881, but it had been at great expense to the latter, which those involved with were still bitter about. However, the requirement for an ocean course lost the New York Yacht Club's local knowledge advantage. It was apparently something they were willing to concede. What they had realised, however, was that if '*The Cup*' was lost, the new club could do the same and choose a home course that

635 *Evening News*, 1 January 1889.
636 *Evening News*, 1 January 1889.
637 *Evening News*, 1 January 1889.

was suited to their own local knowledge. Subsequent races were all held entirely in open waters, off Sandy Hook at the entrance to New York harbour until 1920, and this location would have also been the site for any challenges occurring in the 1890s.

The specifics of the races themselves also meant the yachts had to perform well on all points of sail over a range of conditions. Each race tested the yachts evenly, upwind and downwind, whereas other courses with fixed rounding points and sailed predominantly in the harbour could end up with the course providing just one point of sail, thereby reducing it to a 'follow the leader' type scenario and outcome. There would be only minimal tactics involved. This scenario would also only favour the boat that was best on that point of sail.

While Reeks was considering his options for a design, the same could be said of the members of the New York Yacht Club. As well as designing to a new set of sea conditions to sail over, the organisation also had a new rating rule that would affect any potential defender. As noted earlier, the Seawanhaka Corinthian Yacht Club had adopted the principle of Dixon Kemp's Length & Sail Area Rule, but had moved to modify the formula themselves, still retaining only length and sail area as the dimensions, and calling it the 'Seawanhaka Rule'. Other clubs on the east coast of the United States moved to accept this revised rule too, with the New York Yacht Club following suit in October of 1889.[638] This decision was not, however, made without considerable dissent and much prior discussion, including from Edward Burgess who stated it would significantly tax sail area. Still, when put to a formal vote the change passed, with the *Boston Globe* on 15 December 1889 running the headline, '*NEW YORK'S NEW MEASUREMENT. The Club Committee's Amendment Adopted Almost Unanimously'*.

No longer was a shallow, centreboard yacht the obvious first choice. There were other keel yacht options that might comfortably suit the new courses and new rule, particularly on the upwind legs. If nothing had changed in the Deed of Gift, the logical progression for the New York Yacht Club was to commission a new centreboard design from Burgess, otherwise they could ask General Paine, owner of the previous defender *Volunteer* to have it placed back to full racing condition for the new match. However, the latter idea was rebuked by Paine himself in March of 1890. When asked by a reporter if he intended to bring his vessel back to racing trim for a possible English challenge, he stated his preference for a new boat to be built, at the very least as a '*trial horse*'.[639] It was now recognised that things had changed; there was a new course with new sea conditions and a new rating rule that reflected aspects of yacht performance rather than the outdated tonnage rules.

All of these factors opened up the field to challenge for '*The Cup*' to new designers and new designs. The American fleet wasn't short of either. They had built very good keel boats with cutter rigs. Burgess had specifically designed additional racing yachts and enjoyed success with them in local racing. Other capable designers included Archibald Cary Smith, noted in a biography as '*one of the most successful and innovative designers in the early period of American yachting and his services were much in demand*'.[640] He had designed the *Mischief* which had successfully defended '*The Cup*' in 1881. He was also someone similar to Reeks, in terms of experimenting, taking risks and generally thinking outside of what others were doing.

William Gardner was another yacht designer entering the profession, though at that time he had only a small portfolio. His background included practical training in England, much like Reeks, and he was very active in the mid-1880s. He was included in a summary of significant American yacht designers in 1889 with a supportive comparison to Burgess. '*Mr. Gardner is a Bostonian of about 30 years age, whose talent and profession has had much the same genesis as Mr. Burgess, except that he has had the advantages of study in marine architecture at the Greenwich, England, Institute and other European schools.*'[641]

However, looming in the background was one Nathanael Greene Herreshoff. During his visit to America, Reeks had spent a day with Herreshoff, and in 1903 noted that the '*Wizard*' was then not yet well known for his yacht designs. '*At the time I was in America ... Herreshoff had no thought of sailing yacht designing ... although, mind you, I have heard how the yachting men say that Herreshoff knew more than anyone else about yachts.*'[642]

Herreshoff did know yachts and sailing craft. He had designed boats that raced against Burgess' designs and had also designed the innovative catamaran *Amaryllis*. However, even before that, Burgess was a family friend of the Herreshoffs, as well as a contemporary of Nathanael himself. The biography of Herreshoff's son L. Francis Herreshoff, also a prominent American yacht designer, notes this strong connection. '*Edward Burgess had been a yachtsman, had summered in Bristol, had sailed his own Herreshoff boats, and had become friendly with the Herreshoff family. Edward used to cross the Bristol Harbor from his summer cottage on Poppasquash Point to the Herreshoff works to talk sailboat design with Nathanael. When the Burgess fortune was wiped out in 1883, Edward took up yacht design professionally to support his family and was so successful that yachts he designed defended the America's Cup in 1885, 1886 and 1887. This trio, the Puritan, Mayflower, and Volunteer were clearly influenced by a model for a much smaller boat that Nat Herreshoff had carved in 1880 but had never built to, a model intended to be an improvement on his highly successful Shadow, and a model that Burgess, on his summer visits, had often admired as "the perfect model"*'.[643]

Nathanael Herreshoff later wrote about this to a friend Charles Foster, '*he* [Burgess] *took a house at Popasquash, just across Bristol Harbor, for the Summer and, nearly every day, he*

[638] J. Parkinson (1975). *The History of the New York Yacht Club from its Founding Through 1973.*
[639] Archival material provided to the authors by the New York Yacht Club.

[640] L. del Sol Knight (2005). *Encyclopedia of Yacht Designers.*
[641] *Commercial Advertiser*, July 22 1889.
[642] *Evening News*, 1 September 1903.
[643] R. C. Taylor (2015). *L. Francis Herreshoff, Yacht Designer.*

would come over to look around and chat. I had just made a model to be an improvement of Shadow, and he quite fell in love with it and, in coming in, would say, "I have come again to look at the perfect model." The model is quite the type of Puritan except that the stern is not carried out. I think, at this time, Ned Burgess had convinced himself of the folly of adhering to the English cutter type, and was quite ready to develop the compromise type of which Puritan and Papoose were good examples.'[644]

Despite this clear indication of Herreshoff's already long interest in yacht design, at the time of Reeks' visit in late 1888, steam yachts and other craft were dominating the Herreshoff Manufacturing Company yards. Still, Reeks' observation of Herreshoff soon proved true, and he became well known. In 1891, for example, Herreshoff had the opportunity to design a yacht for the popular 46ft waterline class, the outcome being the *Gloriana,* a vessel that marked a big step forward in yacht design with its heavily cut away forefoot, overhanging bow and high ballast ratio. '*In one season she propelled the field of yacht design into a new era.*'[645] After that season, winning every race it contested, *Gloriana* was '*withdrawn for the good of the class'.*[646] This begs the question: if Herreshoff were to have been asked to prepare a design for a 1890 or 1891 defence of '*The Cup',* would it have featured these departures?

GLORIANA.
SOURCE: LIBRARY OF CONGRESS (ITEM LC-D4-5103).

In 1893 Herreshoff designed two contenders for different America's Cup syndicates: *Colonia,* a deep keel cutter, and *Vigilant,* still very much a keel yacht in proportions, but with a shallower draft and centreboard. *Vigilant* was eventually chosen as the defender and defeated *Valkyrie II.* Further refined and more advanced than *Gloriana, Vigilant* carried 70 crew, 30 more than needed, and used live ballast on the weather rail, adding to the intrigue.[647]

VIGILANT.
SOURCE: WIKIPEDIA COMMONS (EN.WIKIPEDIA.ORG/WIKI/VIGILANT_(YACHT)).

When Reeks returned to his drawing board in Sydney to commence his design for '*The Cup'* early in February of 1889, Burgess, Smith, Gardner and Herreshoff would have all been preparing plans for other vessels, and probably were relatively unconcerned about the respectable Australian gentleman they had met briefly or read about in the American press late in 1888.[648] Their own involvement could wait. Before their services were potentially required, there first needed to be a formal challenge issued and accepted by the New York Yacht Club, and a date set for the initial race. The club itself also had one big advantage up its sleeve before they investigated and confirmed their options for a defending yacht. They would have ten-months' notice of the challenger's dimensions and likely form, and could immediately make preparations to meet the yacht, including building a new vessel, if required.

If Reeks felt he had one area where he could at least match that of a possible defender, it was his rapidly-

[644] Nathanael Greene Herreshoff letter to Charles H. W. Foster, dated March 6, 1932. Quoted in C. H. W. Foster (1932). *The Eastern Yacht Club Ditty Box, 1870-1900.*
[645] L. del Sol Knight (2005). *Encyclopedia of Yacht Designers.*
[646] L. del Sol Knight (2005). *Encyclopedia of Yacht Designers.*
[647] W. Stephens (1942). *Traditions and Memories of American Yachting.*
[648] *Evening News,* 21 January 1889

developing skill with Dixon Kemp's Length & Sail Area Rule that had been adopted in 1887. In terms of racing yachts, this is solely the rule which he had designed to and, even after only a handful of craft, he was gaining more and more confidence.

However, even if Reeks was ahead on this point, the Americans were catching up, in part thanks to their adversaries, the English. Burgess' *Papoose* lost its dominance in 1889 to *Minerva*, an English cutter type designed and built by William Fife (Jr) in Scotland and brought across the Atlantic by its American owner Charles H. Tweed. Racing *Minerva* wasn't his initial intention, but when it first fronted the race course, substituting for a disabled yacht *Liris*, *Minerva* won decisively. Given this outcome, there was no need to hold the vessel back on the race course, or for any further shyness regarding the craft's design pedigree.

Liris was another English type of yacht; an 1889 William Gardner design based on his familiarity with the traditional 'Itchen Ferry' designs hailing from Southampton. Both *Liris* and *Minerva* were true to the English cutter shape, being narrower and deeper than the wide, shallower American craft.[649]

In 1890 Burgess fired back with *Gossoon*, adopting his own adaptation of the narrower and deeper hull and some of the rig refinements too. It also had a shapely section, much comparable to designs at the time being generated by Reeks.[650] By August of 1890 the impact was noticeable. The beamy shallow centreboarder or 'skimming dish' was losing favour. The Americans were surely learning fast.

An article titled '*Vale, Skimming Dish*' published in the New York *Commercial Advertiser* noted, '*The American ideas of yacht construction have in the past few years undergone a radical change. The familiar "skimming dish" style of boat is gradually giving way to a boat that each successive season becomes narrower and deeper. … But although much has undoubtably been learned from Genesta, Galatea and Thistle, no yacht has been such a teacher as Minerva*'.[651]

In the end it probably didn't matter that the formula for calculation was different; the rules shared the same factors as variables, sail area and waterline length, and the Deed of Gift had made it clear that 90ft was the maximum length for a single-masted yacht. Reeks had realised he would need to go to the maximum length otherwise the defender would just build a longer yacht. Therefore, one attribute seemed fixed. He was going to design a yacht at 90ft on the waterline, or as close to that as he was comfortable, while leaving a margin for ballast changes. From there, Reeks could determine the sail area he felt was required, but at this point, while his parameters were pretty much final, the designer of the defender had many options available. A complicating factor was also the possibility of handicaps and time allowances being included, although Reeks would have hoped that they could agree to race on even terms with no time allowance.

[649] W. Stephens (1942). *Traditions and Memories of American Yachting*.
[650] Archival material provided to the authors by the New York Yacht Club.
[651] *Commercial Advertiser*, 19 August 1890.

The options available to the defender included:

- Build at 90ft waterline with the same sail area and race without time allowance such that first across the line was the winner.
- Build at 90ft waterline with more sail area and a higher rating than a Reeks-designed challenger, thus presenting a time allowance advantage to the Australians, if the Australians were behind.
- Build at 90ft waterline with less sail area and a lower rating than a Reeks-designed challenger, thus having a time allowance advantage in hand if they fell behind the Australian yacht.
- Adopt a shorter waterline length and add more sail area to produce the same rating.
- Adopt a shorter waterline length and a different sail area thereby providing a bigger or smaller rating, with the same time allowance possibilities as noted earlier.

There were no computer simulations to predict which, if any of these options, was going to be an acceptable course of action. Instead, pure judgement by the designer and the sailors would have to be relied upon. In Reeks' situation, setting the parameters from his side first clearly put him at a disadvantage. Although he may not have had to declare his final rating until he arrived in New York, once it was known the Americans had the home advantage of being able to quickly revise their rig to meet the possibilities noted, while the Australian boat, in all practicality, was left with what they had declared and no room to move. Technically, as the Deed of Gift stated that the boat had to be '*constructed in the country to which the Challenging Club belongs'*, the New York Yacht Club could have insisted, as it did years later, that this meant everything, including the sails and spars, had to be built in Australia. Unless they came prepared to modify sails and spars themselves, the only permission given would likely be the obvious sporting concession to effect repairs.

From here the advantage continued to favour the Americans. Another big hurdle for Reeks was construction. The Americans were building in steel or composite metal, advanced for the time and well suited to building a stiff and strong hull. The delightfully named Henry Piepgras, a German boat builder who had established himself on City Island in New York, had his yard and methods featured in an article published in the *Boston Herald* in August of 1890 titled '*SWIFT STEEL YACHTS AND HOW TO BUILD THEM'*, which explained and examined the techniques of steel yacht building.[652] Piepgras knew his trade and was building some of the top 40-footers amongst his commissions. His methods were straightforward; forming the red-hot, forged transverse frames off full-size loftings for each section, setting them up and riveting the steam-driven machine-formed plating to this skeleton of ribs, lapped in sequence. Today welding has replaced riveting and, along with CNC cutting of the steel from computer generated

[652] *Boston Herald*, 19 August 1890.

patterns, much of the fire and steam atmospherics of a shipyard's forge and forming work has been lost.

Reeks had not yet produced a steel vessel of any type, though by the end of the 1890s he was designing steel ferries with great success, indicating that perhaps he and his workers could very well have mastered the material for a substantial yacht challenger. A large wooden yacht could, however, be built but it would be less robust than any steel counterpart, and had to remain intact through a lengthy voyage in potentially stormy seas. The Deed of Gift required that challenging vessels *'proceed under sail on their own bottoms'* to the United States. Getting there may not be the problem, it was more a case of arriving there wholly intact. Reeks was not prepared to sail via the notorious Cape Horn, and instead proposed to sail via the Cape of Good Hope, a place where severe storms were very common.[653] It was later also claimed that he even considered a diversion to England to trial his yacht, to then cross the stormy North Atlantic to reach the start line.[654] Regardless of which route was chosen, the possibility of arriving in New York with a leaky hull that had opened up during a voyage halfway round the world was only adding to the problems of an already difficult situation.

It has never been confirmed what Reeks designed. In January of 1889 he is quoted as saying, *'I shall not confine myself to either cutter or centreboard'.*[655] This remained the case a month later in an interview where he stated, *'she will be up to the limit allowed in the deed of gift ... it had not been decided if it was to be a centreboard or deep keel vessel'.*[656] However, reporting on the same interview another newspaper has perhaps misunderstood or badly sub-edited the original report and instead printed, *'In all probability it would be a centreboard boat'.*[657]

The design was, however, completed. As late as 1903, when talking about the America's Cup, Reeks said, *'the scheme actually went so far as for me to complete the drawings'.*[658] This confirms a report in August of 1889, *'Little has been said or heard lately of the yacht which is to be built here to compete for the America's Cup. Mr. Reeks has been very busy lately with the designs, which are now almost complete. We have reason to hope that the actual construction will be commenced shortly. We are unable to give our readers any particulars as to the dimensions of the yacht, as they are secret at present. Side by side with this new cutter are the designs of a 37-inch model for Mr Turner, Commodore of the Moore Park Model Yacht Club. Verily extremes meet'.*[659]

The vague and mixed information in these reports does not confirm either a centreboard yacht or a deep keel cutter, while the extreme contrast in his projects is a reminder that Reeks was happy to serve a wide range of clients and design a large variety of craft. Would Burgess and Herreshoff, *et al,* have done the same?

If Reeks had decided on a shallow centreboard type along the lines of *Volunteer* there is really only *Miranda* as a guide to what he may have produced. Built for cruising and barely raced, the vessel is a poor offering for the direction he may have chosen. *Miranda*'s hull shape was ideal for cruising, and quite conservative compared to *Volunteer.* Reeks would have also needed to progress his design to have more beam, a stronger curve to the bilge and more hollow in the garboard strakes as they met the keel. In addition, he would have needed to add more sail and be confident he had created enough stability to manage the power it would generate. There was a lot to learn in a short period of time if he went down this path, and it is not possible to gauge what Reeks had already learnt or been told while inspecting the previous successful centreboard defenders in Boston.

If it was to be a deep keel yacht, it is possible to consider a few things that Reeks may have adopted, as there are examples of how his shapes were progressing from 1887 through to the early 1890s. The best guide is the next big keel racing yacht that he designed in 1889, after completing his America's Cup design. This was the 52ft cutter *Thelma,* rating 16 tons, for J. F. Hoare, built by George Ellis and launched late in 1889.

Thelma carried the same elegant sheer and keel line profiles of *Era* and *Volunteer,* but had more of the forefoot cut away than either of those two vessels. *Thelma* also had a similar curve to *Volunteer* at the midsection, and this continued aft, creating what is often called a wine-glass section where the hull sits poised on a narrow keel like the body of a glass on its delicate stem. This was a beautiful shape but also a powerful shape in terms of stability. In plan view *Thelma*'s wide beam to length ratio was accentuated by the fine tapered stern Reeks had designed. In relation to the mid-section and tapered stern he had advanced his designs further, while the bow profile was reducing gradually from the depth of his earlier designs. It was also reducing wetted surface, but not in the more substantial manner Herreshoff achieved with *Gloriana* two years later. Up until this point in time, all of Reeks' big racing designs had the plumb bow. The clipper overhanging bow was only used on two of his cruising designs and the smaller *Elf.* It did, however, make an appearance about the same time on the small 2½ rater *Louelle,* that Reeks also designed later in 1889.

In an earlier piece published in the *Illustrated Sydney News* on 31 May 1888, Reeks notes favourably of the radical yacht *Australian,* a true double-end yacht considered a freak of its time when it raced on Sydney Harbour, and was under recognised for what it had achieved until Reeks made his observations. Amateur designed by Richard Hayes Harnett, the vessel was based on the shape of a mackerel and built by Dan Sheehy at Woolloomooloo in 1857 from salvaged timbers off a shipwreck. The fore and aft lines of *Australian* are arcs of a circle, and the underwater sections are straight lines.

[653] *Evening Star*, 23 October 1888.
[654] *Boston Herald*, 6 July 1890.
[655] *Evening News*, 1 January 1889.
[656] *Daily Telegraph*, 13 February 1889.
[657] *South Australian Advertiser*, 13 February 1889.
[658] *Evening News*, 1 September 1903.
[659] *The Sydney Morning Herald*, 17 August 1889.

THELMA.
SOURCE: AUSTRALIAN NATIONAL MARITIME MUSEUM (WILLIAM HALL COLLECTION).

Australian, and two other yachts designed by Harnett, were built along these lines and as Reeks observed, '*They* [taking *Australian* as a type] *are an approximation to the best forms known to modern times. They are simply the outcome of those two well-known geometrical truths—a straight line is the shortest distance between two points, and a circle is the shortest line that will enclose a given area. Every cross section in the Australian is a right-angled triangle and every waterline and diagonal the arc of a circle, thereby bringing about this near approach to perfection: minimum of wet surface, maximum displacement to wet surface, and, above all harmony of parts. She is almost a perfect vessel*'.[660] Later in the text he stated, '*It may be well to add that the most important elements of success other than those mentioned already, are, distribution of weights, balance of canvas to hull, positions of centres of buoyancy to gravity, and good copper*'.[661]

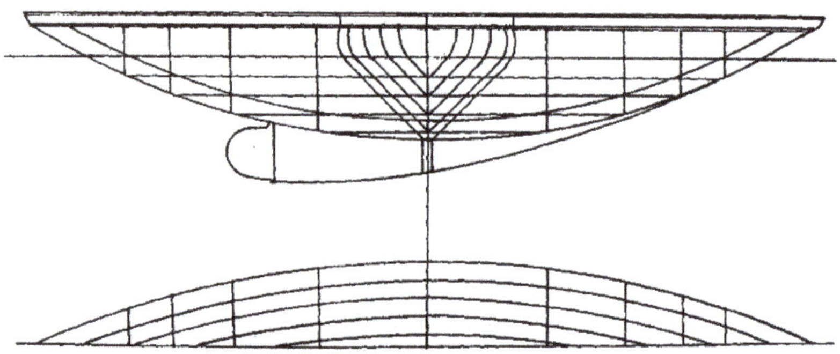

PLANS OF THE AUSTRALIAN.
SOURCE: P. R. STEPHENSEN (1962). SYDNEY SAILS: THE STORY OF THE ROYAL SYDNEY YACHT SQUADRON'S FIRST 100 YEARS (1862-1962).

Australian was not perfect enough though for Reeks to copy, but he did respect the need for reduced wetted surface and his designs all had a harmony about their lines and rig balance. Most importantly, however, it can be seen immediately that Reeks, as a naval architect with engineering training, was versed in the hydrodynamic calculations required. This set him apart from other craftsmen with their reliance on hand-carved half models and the experience of a previous design. It also gives us a window into his drawing office.

The '*distribution of weights*' and '*positions of centres of buoyancy to gravity*' are fundamental elements of basic yacht design and engineering. They are found by working through detailed tables of calculation from the lines and construction plans, followed then by further changes to the plans and recalculations, until all are satisfactory. A yacht has to have enough volume to support its overall weight, and this must be distributed such that the vessel's centre of gravity is over the hull's centre of buoyancy. To do some of this easily, Reeks probably had a planimeter amongst his drafting tools. This was a widely used instrument with a micrometer-style readout created in 1854 by Swiss engineer and mathematician Jakob Amsler-Laffon. It enabled the accurate measurement of areas bounded by irregular curves, such as those of the underwater hull cross sections; a figure needed to determine displacement.

Meanwhile, the calculation of weights was still very much a long inventory and accounting exercise. Reeks may even have had Amsler-Laffon's crowning achievement from 1855, the mechanical integrator. A quantum advance on the planimeter, this was a much bigger instrument than that used to provide data to assess the stability of the hull at different angles of heel, also working from the hull cross sections.[662] Each of these tabulations was usually time spent measured in hours, but prior to these tools being available, days could be spent trying to reach the same single step in the overall calculation. Working in pen and pencil on linen, the hand drawn and methodically calculated designs from Reeks' drawing board were a product of classical engineering practices and also works of art.

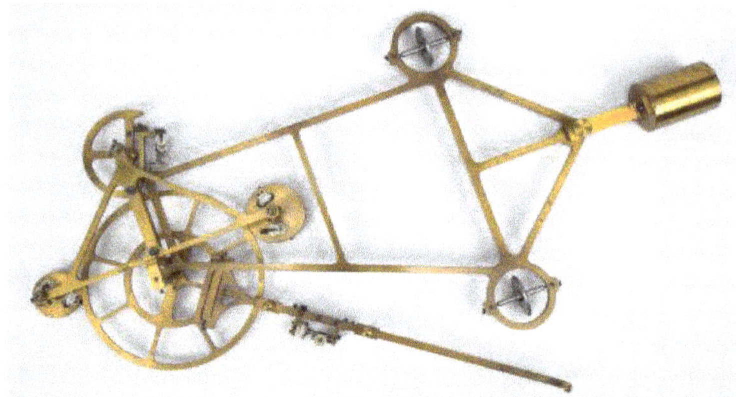

AMSLER MECHANICAL INTEGRATOR (NOTE THE PLANIMETER IS MUCH SMALLER BUT ITS 'ACTIONS' ARE LOCATED ON THE INTEGRATOR AS WELL).
SOURCE: HARVARD UNIVERSITY (WAYWISER.FAS.HARVARD.EDU/OBJECTS/167/).

Theory was important too, and this was a period of immense development in this area. Resistance was identified as something that should be reduced to improve speed, and waves were targeted as the source of the problem. In the late 1830s English naval architect John Scott Russell proposed that hull shape be determined by the shape of the waterlines in plan view, developing a mathematical way of drawing out the horizontal below-waterlines of the hull. This resulted in craft with long hollow entries parting the water gradually.[663] It was called Wave-Line Theory and had a profound influence on vessel design for a long period in the nineteenth century. It was later challenged by Scottish (then Norwegian-based) yacht designer Colin Archer who tested it with models, and was supported by American designer John Hyslop. The duo went further, seeking to improve upon the original theory.

Archer, in particular, first presented a paper about the theory in 1878 formally titled '*On the Wave Principle Applied to the Longitudinal Disposition of Immersed Volume*', a mouthful but still a neat summary of the subject. Joint discussions between Archer and Hyslop then followed and led to the pair proposing that the distribution of the volume was more critical, and the way to determine how this should be done was through the shape of the '*curve of*

[660] *Illustrated Sydney News*, 31 May 1888.
[661] *Illustrated Sydney News*, 31 May 1888.
[662] D. McGee (1998). *The Amsler Integrator and the Burden of Calculation*.
[663] L. del Sol Knight (2005). *Encyclopedia of Yacht Designers*.

areas', which is a graphical depiction of the underwater area of each section along the hull. This curve would then '*indicate how the water was actually parted and then re-joined as the boat passed* [through the water]'.[664] Referred to as Wave-Form Theory this took Russell's Wave-Line Theory and drafting methods and applied them only to the curve of areas, defining the forebody with a '*curve of versed sines*', and the after body with a '*trochoid*'.

Both are classical mathematically-generated curves favoured by engineers and draughtsmen for their precision. Ultimately these theories were very much a simplification of what is now known to be correct, but each time they were steps toward the right outcome in both thinking and methodology. Russell and Archer had both used models in a progressive series of experiments to generate their results. Burgess, Smith and others were known to favour this theory and its application to the design of hull lines.

Of course Reeks was well aware of these theories too, and those then coming from English engineer William Froude whose scientific work began in the 1860s using scale models. Froude had already established the need to reduce frictional resistance by reducing surface area and was developing laws for scaling the results from models to full-size vessels.

Reeks' preferred theories, calculations and drafting methods were all typical of the period, and they delivered results for his yacht designs over this period. *Thelma,* discussed in more detail shortly, was another success, including in its first season, and then had even greater success in 1896 when Reeks lowered the ballast further by adding depth to the wooden keel in two stages of development, and bringing it more into line with trends that had transpired in that time.[665] Pulled out of retirement in 1909 to race in the Sayonara Cup, *Thelma* nearly won the event despite being well outdated.

It is difficult to look at any other aspects of what Reeks' design for an America's Cup challenger may have looked like if it had been a keel yacht. As noted, in 1893 the Herreshoff-designed *Vigilant* featured a centreboard as well as a deep keel. This would have improved the vessel's upwind sailing, and provided an advantage over a yacht without this appendage. The extra crew brought on as moveable or live ballast weight was another trick up their sleeve. Would Reeks have considered something different that had not been thought of by the American designers? The American boats also featured a lot of detail in their sails, spars, rigging and fittings. Reeks did use the new copper alloy 'Delta metal' on some of *Thelma*'s fittings, '*in lieu of galvanised iron*' to save weight.[666] Later, his ferry designs demonstrated close attention to detail to improve performance, so he was not ignorant of the need to be as thorough as possible in his design work.

The overarching picture painted, however, is a tall order for any Australian challenger in 1890. History shows that when Australia did finally mount a challenge for the America's Cup in 1962 it was still a massive undertaking and reach, with some of the same obstacles remaining in place, along with new ones. *Gretel* surprised the American defender *Weatherly* with a closer series than the 4-to-1 victory sheet shows. Some observers even noted, '*Gretel was a very fast yacht, perhaps inherently faster than the four-year old Weatherly*'.[667] *Gretel*'s designer was the naval architect Alan Payne (the co-author's uncle) and kindly referred to as '*heir to the tradition of Walter Reeks*' who, like him, faced a similarly monumental task.[668] If history is a guide, then Reeks may have had a close series as well.

Still, despite the challenge not proceeding and the Australian vessel not being built, Reeks' reconnaissance visits to both America and England included meeting at length with several of the greatest yacht designers of the time. The opportunity to sit down one-on-one with Edward Burgess and Nathanael Greene Herreshoff, for example, and to see their vessels provided Reeks with the opportunity of a life time, and much impetus for his professional journey ahead.

Perhaps our last word on the America's Cup challenge, at least for this chapter, comes from an American report published in the *Boston Herald* in July of 1890, which quotes E. C. Holske who had spoken to Reeks while in Australia some months earlier.

Holske was a retired champion walker who arrived in Australia with other athletes from the United States for a visit in November of 1889. In his specific situation, '*His visit here is for health, recreation and business of a speculative character*'.[669] The article, using information sourced directly from Holske, then claims that an earlier and separate challenge for the America's Cup had formed in Victoria before the Melbourne Centennial Aquatic Carnival of 1888, with the prospect promoted in a Melbourne shop front, as well as in newspapers at that time. However, nothing came of it. Holske felt that a challenge would come instead in 1893 from Sydney with Jack Want involved. Unfortunately, Holske confuses the timing of the loss of Want's yacht *Mignonette* in 1884, stating that it was more recent, and reveals it as a factor for the delay of the project. This was obviously incorrect. In addition, Holske notes that Reeks' business interests '*had monopolized his attention*', also causing a delay. This may well have been true. The article, however, apart from confirming Want's close association with the possible America's Cup challenge, and Reeks' involvement unfortunately does not bring to light any of the mysteries about the design or other financial supporters involved.

—

Interestingly, around the same time that Reeks returned to Sydney from America via England in early 1889, the possibility of an international challenge race against the Californian yacht *Lurline* was gaining more

[664] L. del Sol Knight (2005). *Encyclopedia of Yacht Designers*.
[665] *Australian Yachtsman and Canoeist*, 12 December 1896.
[666] *The Sydney Morning Herald*, 7 December 1889.
[667] E. B. Morris (1964). *Sailing for the America's Cup*.
[668] P. R. Stephensen (1962). *Sydney Sails: the Story of the Royal Sydney Yacht Squadron's First 100 Years (1862-1962)*.
[669] *Sportsman*, 20 November 1889.

and more traction, helped in part by the vessel's owner (John Spreckels) offer of £2,000 in prize money.[670] The idea received even further momentum when at a dinner for those involved with the *Volunteer,* held in Sydney in mid-April of that year, the craft's owner (or part-owner) William Smairl suggested to his crew that '*they could have the yacht, and the prize money if they won, but he did not feel disposed to let the expenses of the trip fall on his shoulders. But if any of the public liked to form a syndicate to send the yacht across, he would be most happy to fall in with them*'.[671]

At the dinner, Reeks' opinion on the matter was then sought. He responded that he did not see the *Lurline* when he went to America, but he did see several boats built upon its lines, and thought that the *Volunteer* would have no difficulty in competing against the *Lurline*.[672] Captain Tom R. Dudley next spoke up, stating that he would take the *Volunteer* across to San Francisco if £400 - £500 could be raised for expenses and fitting of the boat.[673]

THE 80FT SAN FRANCISCO SCHOONER LURLINE.
SOURCE: JOURNAL OF SAN DIEGO HISTORY, FALL 1974, VOL XX, NO. 4.

This was an interesting offer from Dudley who only five years earlier, in England, had been sensationally found guilty of murder following a legal case that sparked international interest and intrigue, and still to this day is discussed and analysed in legal circles.

An experienced seaman, in July of 1884 Dudley and three others were returning to Australia from England on board the 40-ton yacht *Mignonette* which they had purchased on behalf of Jack Want.[674] After leaving Southampton in mid-May, in early July the vessel was sunk by a giant wave somewhere between Madeira and Cape Town. Though all on board managed to escape in the yacht's dinghy, they suddenly found themselves in a perilous situation; provisions comprised only two tins of turnips and no water.[675]

[670] *Toronto Daily Mail,* 24 April 1889.
[671] *The Sydney Morning Herald,* 12 April 1889.
[672] *The Sydney Morning Herald,* 12 April 1889.
[673] *The Sydney Morning Herald,* 12 April 1889.
[674] www.historyextra.com/period/cannibalism-at-sea-sailors-ate-the-cabin-boy/.
[675] www.historyextra.com/period/cannibalism-at-sea-sailors-ate-the-cabin-boy/.

For nearly two weeks the dinghy and its occupants drifted aimlessly and by 17 July all supplies had been consumed.[676] Extremely perished, 17-year-old Richard Parker could no longer resist the urge to consume salt water, drinking far in excess that his body could handle. He soon became sick with diarrhoea which only added to his dehydration. Over the next few days his condition deteriorated such that the other three on board the dinghy raised the subject of the 'Custom of the Sea', i.e., the notion of sacrificing the weakest on board to save themselves. Several days later, with Parker's medical condition much worse, potentially on the point of death, the trio made the sacrifice with the aid of a penknife. Dudley and the two others then survived up until 29 July, i.e., 24 days after the *Mignonette* had sunk, at which time a ship was sighted. Their dinghy was spotted by the German vessel *Moctzuma* and the trio were subsequently rescued.[677]

Open and honest about the events that had transpired, authorities in Great Britain stepped in and had the three men arrested. The resulting legal case sparked widespread attention and scrutiny considering those involved in past cannibalisation at sea, specifically to aid their own survival, had historically received lenient sentences, if any at all.[678] For Dudley and his fellow crew member Edwin Stephens, however, the outcome of the trial was ultimately that necessity was no justification for killing. A sentence of death was imposed in December of 1884. In-line with popular opinion at the time, it was at this juncture that the Home Office stepped in with the sentence reduced to six months of imprisonment. The term was spent at London's Holloway Prison whereupon Dudley then returned to Sydney, going on to take up yachting and, several years later, establish his own boat building yard.

As stated, Dudley's offer to captain *Volunteer* on an ocean voyage to California after enduring such hardship on board *Mignonette,* and then the subsequent trial and prison sentence, was interesting. However, like the potential America's Cup challenge, nothing came of this proposed race and any mention of it quietly dissipated from newspaper articles.

With William Smairl subsequently experiencing financial difficulties, *Volunteer* was advertised for sale in July of 1891.[679] In early 1892 Smairl was jailed for five years for embezzling funds from the company he managed, the Imperial Land, Building, and Deposit Co.

Had this particular challenge proceeded, it would have been an interesting contest. It was also a potential mismatch in terms of vessel types. *Lurline,* an 80ft two-masted schooner, would have raced against *Volunteer,* a 63ft single-masted cutter. The match would have also been held on San Francisco Bay, notorious for its strong tides. Regardless of other issues, *Volunteer* would have also needed a local expert on board to navigate the tricky harbour to its best advantage.

[676] www.historyextra.com/period/cannibalism-at-sea-sailors-ate-the-cabin-boy/.
[677] www.historyextra.com/period/cannibalism-at-sea-sailors-ate-the-cabin-boy/.
[678] www.historyextra.com/period/cannibalism-at-sea-sailors-ate-the-cabin-boy/.
[679] *The Australasian,* 11 July 1891.

As a powerful cutter designed by Reeks and built to the latest rule, *Volunteer* may have had an advantage in any upwind sailing, being more efficient on this point of sail, but would it have been enough to match the potential reaching and running advantages of the longer *Lurline,* with its great spread of sail? This is something that remains unknown.

THE ILLUSTRATED LONDON NEWS,

No. 2370.—VOL. LXXXV. SATURDAY, SEPTEMBER 20, 1884. WITH EXTRA SUPPLEMENT SIXPENCE.

THE LOSS OF THE YACHT MIGNONETTE.—FROM SKETCHES BY MR. EDWIN STEPHENS, THE MATE.

The way in which they stowed themselves in the dinghy.

Sailing before the wind: How the dinghy was managed during the last nine days.

LAVENDER BAY AND HOUSES ON MILSONS POINT WITH GARDEN ISLAND IN THE BACKGROUND (1870S).
SOURCE: MITCHELL LIBRARY, STATE LIBRARY OF NEW SOUTH WALES

CHAPTER EIGHT - MILSONS POINT

In addition to the prospect of a challenge for the America's Cup, as well as one possibly against the Californian yacht *Lurline,* and what both of these efforts entailed, Walter Reeks returned to his various professional vocations once back in Sydney in early 1889. He also resumed teaching naval architecture two nights a week at the Sydney Technical College.[680] On a personal note, Reeks was very likely enjoying the first few months of wedded bliss, he and his bride Mary building a new life together at home in Milsons Point.

In terms of vessel commission work, Reeks returned to Sydney to find small yachts in favour. He designed two 2½-raters, one for Sid Gilchrist and the other for Harry Stevens.[681] Sister-ships, both yachts were built by Tom Cubitt at Berrys Bay, late foreman to 'Watty' Ford.

This was an interesting decision by Reeks, placing his trust in someone he more than likely knew well, through previous Ford commissions, but also someone who appears to have just been starting out as a boat builder of his own standing. Cubitt's yard was located just over the fence from that of Ford, his mentor and late employer. However, the latter's yard appears not to have been in full operation at the time with very few new vessels being built. This was likely the result of Ford facing legal and financial issues due to monies owed to creditors. In May of 1889 he would enter bankruptcy with unsecured liabilities amounting to £1,500.[682] Two years later, with these issues resolved, Ford resumed working at his Berrys Bay yard, buoyed by commissions for pearling luggers and missionary schooners. Overall, however, the issue was a small blip in the Ford family's long boat building history.

[680] *The Sydney Morning Herald,* 18 May 1889.
[681] *The Sydney Morning Herald,* 22 May 1889.
[682] *The Sydney Morning Herald,* 24 May 1889, 29 November 1890, 17 February 1891.

In time, Ford's son, W. M. 'Watty' Ford (Jr), would take over the family's Berrys Bay premises, continuing its operation well into the 1930s, not only carrying on the Ford legacy but becoming a leading boat builder of Sydney in his own right. For Reeks, who already benefited from a strong working relationship with Ford, he would continue to maintain this good standing with his son; the pair going on to produce many vessels in the decades to come.

The orders for the two small 2½-raters Reeks received in early 1889 were the result of the Prince Alfred Yacht Club instituting races for the 2-ton class. The Royal Sydney Yacht Squadron soon followed suit, agreeing to implement the class at its August 1889 meeting.[683] Dimensions of the two vessels were 27ft (overall) by 24ft with a beam of 6ft and a depth of 4ft 7in. Sid Gilchrist's yacht, *Louelle,* was launched in August of 1889, while Harry Stevens' *Jenny Wren* was launched in late October of that same year.[684] Both craft carried 2 tons of lead on the keel and 740 square ft of canvas.[685]

Yachting as a pastime was changing. Smaller craft were beginning to gain traction amongst Sydney's yachtsmen, joining a pattern forming in both Great Britain and America. Fast sailing, requiring less initial investment, fewer crew members to operate and reduced management and maintenance costs and resources, it was the start of a new era in yacht racing, helped in part by the implementation of Dixon Kemp's Length & Sail Area Rule. The larger boats were being supplanted. Acting as broker, Reeks subsequently advertised *Magic, Waitangi, Electra, Sao* and *Assegai* for sale.[686]

[683] *The Sydney Morning Herald,* 8 August 1889.
[684] *The Sydney Morning Herald,* 22 August 1889; *The Daily Telegraph,* 26 October 1889.
[685] *The Sydney Morning Herald,* 22 May 1889.
[686] *The Sydney Morning Herald,* 10 August 1889.

LOUELLE.
SOURCE: AUSTRALIAN NATIONAL MARITIME MUSEUM COLLECTION (OBJECT NO. 00013777).

A third 2½-rater designed by Reeks and also built by Cubitt was launched on 13 December 1889.[687] Commissioned by G. E. Fairfax and named *Norna*, the craft measured 28ft (overall) with a beam of 6ft 6in and a draft of 4ft 6in.[688] *Norna* was straight-stemmed with a tapering counter and, like *Louelle* and *Jenny Wren*, joined the Prince Alfred Yacht Club's expanding fleet of vessels of this class.

[687] *The Sydney Morning Herald*, 14 December 1889.
[688] *The Sydney Morning Herald*, 14 December 1889.

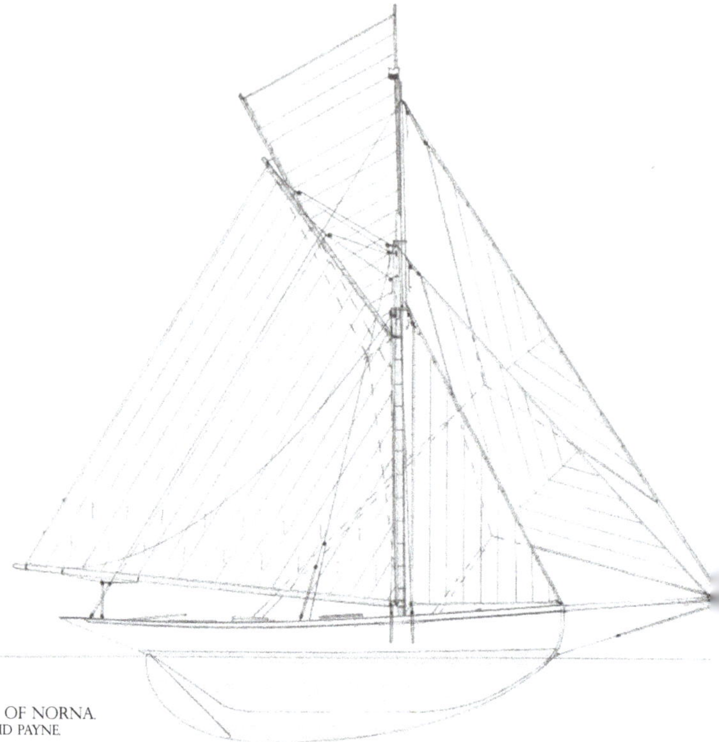

PLANS OF NORNA.
BY DAVID PAYNE.

JENNY WREN.
SOURCE: AUSTRALIAN NATIONAL MARITIME MUSEUM COLLECTION.

Meanwhile on the Parramatta River at the yard of Captain Tom. R. Dudley, another vessel designed by Reeks was noted to be underway in May of 1889, a 28ft racing yacht built to the order of a Tasmanian-based customer.[689] With a beam of 9ft and a draft of 7ft 6in, the yacht was to carry 1,400 square ft of canvas and intended to race with the 28-footers, then popular on the River Derwent.[690] Disappointingly, there is little additional information known about this vessel. Rather uncharacteristically it is not mentioned in either the New South Wales nor Tasmanian press of the time. Since such a yacht arriving in Hobart from a high-profile Sydney-based yacht designer and builder would have merited a large amount of coverage at the local level, it can be surmounted that this yacht was not completed, through one reason or another.

From Reeks' drawing board in mid-1889 also came designs for two steam yachts.[691] Again there is little further information known about these particular vessels. Around the same time, Reeks was also busy superintending alterations to the *Volunteer,* which he helmed successfully in local races later that season.[692] Rather uniquely, and showing his penchant to take on a diverse array of projects, a first class model yacht measuring 37 inches and named *Albert* was designed by Reeks for W. Turner, commodore of the Moore Park Model Yacht Club in August of 1889.[693]

Indicative of his entire career, Reeks enjoyed mixing small tasks with much bigger projects. Mid-to-late 1889, for example, saw him designing detail work as well as a steam launch for the 155ft Scottish-built steam yacht *Saide,* recently purchased by C. Millar of Melbourne, as well as making improvements to the *Iduna.*[694] The next big racing yacht designed by Reeks was also then under construction at George Ellis' Balmain yard, built to the order of J. F. Hoare, newly-elected vice-commodore of the Royal Sydney Yacht Squadron, and previous owner of *Sao.*[695] With dimensions of 52ft (overall), a beam of 10ft 8in and a draft of 7ft 1in, *Thelma* was cutter rigged, had a 15 rating and was launched on 6 December 1889.[696] The craft featured a straight stem and a long counter, similar in design to *Era* above water. However, *Thelma* was more than that and upon its launch immediately became recognised as something special.

[689] *The Sydney Morning Herald,* 22 May 1889.
[690] *The Sydney Morning Herald,* 22 May 1889.
[691] *The Sydney Morning Herald,* 22 May 1888.
[692] *The Sydney Morning Herald,* 13 January 1890, 20 January 1890.
[693] *The Sydney Morning Herald,* 17 August 1889.
[694] *The Sydney Morning Herald,* 17 August 1889; *Illustrated Australian News and Musical Times,* 1 August 1889.
[695] *The Sydney Morning Herald,* 17 August 1889.
[696] *The Sydney Morning Herald,* 17 August 1889, 7 December 1889.

THELMA ON SYDNEY HARBOUR (1890S).
SOURCE: AUSTRALIAN NATIONAL MARITIME MUSEUM COLLECTION.

'*Yesterday morning, shortly before 7 o'clock ... there was launched from the building yard of George Ellis at Waterview Bay, Balmain a splendid specimen of the modern yacht designed and built under the personal supervision of Mr. Walter Reeks, the well-known naval architect ... Thelma as she is named, is quite a beauty. Her long tapering and graceful counter (10ft) adds very much to her handsome appearance. ... Taken altogether, she reflects great credit on her designer (Mr. Reeks) also on the builder (Ellis) for the workmanship and materials used in her construction, and should, under the able guidance of her owner, secure many victories.*'[697]

These victories did eventuate over a long career, and the design accolades also continued over several decades. Nearly twenty years after launch, the press reported, '*The beautiful Thelma, of 15 tons — the best boat, perhaps, that Mr. Reeks ever designed.*'[698] In 1912 *The Anchor* claimed that the vessel was '*regarded by Reeks as his "flower of the flock"*'.[699] Similar comments were published numerous times over many decades whenever *Thelma* and Reeks were mentioned in the same sentence. The vessel also continued to feature in Reeks' personal story for the next two decades, though sadly it would end in a tragedy.

Thelma's first outing took place two weeks after launch, under a yawl cruising rig. During January of 1890, the craft went north to Broken Bay.[700] Its initial race came a little later. With Reeks at the helm, *Thelma* appeared on the start line for the 'All-Yachts Race' at the Anniversary Regatta on 26 January 1890 and began its career well by winning the event. For this achievement Reeks was rewarded handsomely at a later prize giving ceremony. '*For the all-yachts race Mr. Hoare, the owner of the winning boat, presented Mr. W. Reeks, who designed the Thelma, with a gold medal for the successful sailing of the yacht in her first contest.*'[701]

A second win followed on 8 February 1890 with *Thelma* taking first in a Prince Alfred Yacht Club race on the harbour.[702] Another cruise was undertaken at the end of the season, north to Port Stephens and then back into some heavy weather which *Thelma*'s crew and those on board *Era* thought the better of not long after heading out to sea. '*The heavy weather along the coast has prevented the yachts Era and Thelma from returning to port. On Sunday Era essayed to come down, but after standing off shore some two hours with the wind to the west of south, and a very nasty sea, she came round and returned to Broken Bay, where she will remain till the weather moderates. Thelma left Nelson's Bay at 7 a.m. last Saturday, but she found things outside unpleasant, so she returned at 9 a.m., and is still weather bound.*'[703]

In the previous chapter, *Thelma*'s shapely wine-glass sections were noted, along with its graceful tapered counter. These were key features of both the vessel's beauty and speed. However, internally *Thelma* was a showpiece too. The launch descriptions include a long and detailed paragraph of the craft's layout and high-quality fitout, best summarised by the fact that this work was given over to a specialist team. '*The internal decorations were designed by Mr. Manson, and are being carried out by Mr. Kirkaldie in a most workmanlike manner.*'[704] Its likely the former was J. M. Manson who witnessed Reeks' marriage to Mary Kent on 18 March 1889 at St. Andrew's Cathedral.

In addition to the much-anticipated launch of *Thelma*, the year 1890 began with personal joy for Reeks and his wife. The couple welcomed their first child, a son named Kent, on 8 January at No. 8 Fitzroy Street, Milsons Point, the home they shared with the Broadhursts, Mary's uncle and aunt.[705] Maybe because of the new baby or for health reasons, Reeks and William Broadhurst took a trip to the Blue Mountains a few weeks later, staying at the Carrington Hotel in Katoomba.[706]

There was more personal news, however, in the months that followed. William Broadhurst sadly died at their Milsons Point home on 26 May 1890, at the age of 53.[707] Now a widower, Betsy Broadhurst immediately returned to England, being present in Lancashire to prove her late husband's will (leaving her an estate valued at over £9,600, equivalent to more than £1.2 million in today's money) in December of that year.[708]

With the death of William Broadhurst, and the departure of Betsy Broadhurst back to England, Reeks and his wife Mary became the sole occupants of No. 8 Fitzroy Street, Milsons Point, remaining at this location with their new baby Kent until early 1891.[709]

In terms of Reeks' work commitments during this period, another design for a 2½-rater, similar to that of *Norna,* though slightly larger at 29ft 2in (overall), came from his drawing board in the early months of 1890.[710] Built by Cubitt at Berrys Bay, who so far had built the majority of the vessels for this growing class, *Astroea* was commissioned by James McCabe of Neutral Bay and carried 652 square ft of sail area. The vessel was launched in May of 1890, just after the close of the yachting season.[711]

Later that year, a fifth 2½-rater was noted to be under construction at Cubitt's yard, built to the order of J. O. Fairfax.[712] Notably designed by William Fife (Jr) of Scotland and named *Bul Bul*, construction was superintended by Reeks with the vessel launched on 21 November 1890.[713] Measuring 36ft (overall) by 25ft (waterline) with a beam of 6ft 5in and a draft of 6ft 3in, this vessel was longer, narrower and deeper than

697 *The Sydney Morning Herald*, 7 December 1889.
698 *The Sydney Mail and New South Wales Advertiser*, 6 November 1907.
699 *The Anchor*, 22 February 1912.
700 *The Sydney Mail and New South Wales Advertiser*, 21 December 1889.
701 *The Sydney Mail and New South Wales Advertiser*, 15 February 1890.
702 *The Sydney Mail and New South Wales Advertiser*, 15 February 1890.
703 *The Sydney Mail and New South Wales Advertiser*, 19 April 1890.
704 *The Sydney Mail and New South Wales Advertiser*, 14 December 1889.
705 *The Sydney Morning Herald*, 16 January 1890.
706 *Illustrated Sydney News*, 20 February 1890.
707 *The Sydney Mail and New South Wales Advertiser*, 7 June 1890; *The Daily Telegraph*, 13 January 1891.
708 England & Wales, National Probate Calendar for William Edwin Broadhurst.
709 *The Daily Telegraph*, 13 January 1891.
710 *The Daily Telegraph*, 15 February 1890; *Evening News*, 15 February 1890, 12 May 1890.
711 *The Daily Telegraph*, 15 February 1890; *Evening News*, 12 May 1890.
712 *The Sydney Mail and New South Wales Advertiser*, 27 September 1890.
713 *The Australian Star*, 13 November 1890.

BUL BUL
SOURCE: AUSTRALIAN NATIONAL MARITIME MUSEUM COLLECTION (OBJECT NO. 00013803).

those of the class previously designed by Reeks.[714] *Bul Bul* was also differently rigged, with a standing lug and single jib, and had considerably more lead on its keel.

Overall, the 2½-rater class was a short but happy diversion for Reeks, potentially offering a welcome reprieve from the frustrations of the stalled America's Cup challenge. For a season or two, the class also had its own colourful story of sailing on Port Jackson. Initially established in England around 1887, once adopted in Sydney, four boats were built and, along with a couple of existing smaller yachts, including the Reeks-designed 2-rater *Elf,* they combined to establish a fledgling fleet of 'miniature yachts', as they were referred to. The vessels bore all the trademarks of the bigger cutters in rig and appearance, but could safely race on the harbour with only three or four hands. Sometimes referred to as 'Hencoops', the 2½-raters were also popular with the few ladies who had taken to yacht racing in England.

As stated, Reeks designed three of the four yachts that initially formed the class, including the aforementioned *Louelle* and *Jenny Wren*, cutters with clipper or raked bows. His third boat, *Norna,* was a straight-stemmed cutter. It was a little bigger than its two predecessors, being 28ft (overall) with slightly less sail area (657 square ft), as opposed to 27ft (overall) with 740 square ft of sail area. His fourth boat, *Astroea*, was slightly longer still, at 29ft 2in (overall) and carried 652 square ft of sail. Of note, Reeks' former student and employee Ernest Thompson was responsible for designing another boat for the class. *Sirona* was built by Captain Dudley at Drummoyne and launched in October of 1889.[715]

The Prince Alfred Yacht Club ran a regular series of races for the 2½-raters on Sydney Harbour over the 1889/90 season. The results were largely shared by the vessels designed by Reeks, including *Elf.* On 9 November 1889 *The Sydney Mail and New South Wales Advertiser* reported, '*The 2½ raters were the novelty of the day, this being the first appearance of this class of boat. Elf won a hard race from Jennie* [SIC] *Wren by 7 seconds on her time. Mr. Stevens had a reefed mainsail, his sail being too big for her spar. Harold* [Stevens] *lost a little time through a dinghy getting in his way first time around the flagship.*' Then follows a leg-by-leg description of the race, noting that *Jenny Wren* in fact '*ran down a dinghy',* an action that seemed to cause no concern or further comment. Better luck was to follow.

'*The Jenny Wren scored her second win this season ... for a prize of 5 pounds ... the meeting of these tiny yachts under racing canvas has been looked forward to with a great deal of interest amongst aquatic men generally, they being quite a new departure from what we have been accustomed to see in Sydney Harbour.*'[716] Another leg-by-leg, tack-by-tack description followed in the same column, with sail changes and weather variations noted, and times at each buoy. It is a long read, but this reflects the level of interest in yachting at the time. The race was followed by steamer, with the Prince Alfred Yacht Club's secretary, Mr Bennett, playing host to the guests on board.

[714] *The Sydney Mail and New South Wales Advertiser,* 27 September 1890.

[715] *The Sydney Morning Herald,* 21 October 1889.

[716] *The Sydney Mail and New South Wales Advertiser*, 28 December 1889.

The 2½-rater class, however, was not without controversy. Taking advantage of the rules was apparent, even then. Sid Gilchrist, owner of *Louelle,* upset everyone by having three different ratings (and therefore time allowances) at once, depending on how much sail area he declared for each race. For example, leaving the jack yard tops'l and the jib tops'l ashore on a windy day dropped the vessel's rating to below 2, thus giving *Louelle* a significant time allowance over its competitors, who had the sails aboard but did not use them.

An off-season meeting stamped out this practice. Only one certificate was to be allowed for the 1890/91 season, but then a new problem was highlighted. The introduction of centreboarders was noted, and with it the possibility of taking the class away from the yacht basis. It was seen that if the big open boats of local design were entered, even with reduced sail, they could dominate the yachts with deep keels.

No decision by the Prince Alfred Yacht Club on an action to stall this issue is noted, but in the meantime Reeks saw potential in a longer yacht with less sail area. His design for James McCabe's *Astroea*, launched in May of 1890, was another step forward from *Norna.* Prior to launch, the press reported, '*She will have an overhanging bow more so than any other yacht in the harbour. She will have a tapering counter, longer than Norna's, and will also have a little more sheer than her*'.[717] This paragraph describes a very pretty yacht indeed. *Astroea* began the 1890/91 season with good form until the proverbial 'cat amongst the pigeons' emerged.

[717] *The Sydney Morning Herald*, 1 March 1890.

In late 1890 J. O. Fairfax sold his 2-rater *Elf* and commissioned the great Scottish designer William Fife (Jr) for a new 2½-rater design. As stated, *Bul Bul* was built by Cubitt with the construction period supervised by Reeks. Relative to *Astroea,* the craft was considered a major step up in size, being 36ft (overall) and 25ft (waterline), but rigged with only 600 square ft of sail area.[718] The key, however, was how this sail area was deployed. It abandoned the classic gaff cutter and tops'l sail plan and sported a rig new to the class in Australia, encompassing a single small headsail, and a large scotch-lug headed mainsail. The simplicity of this two-sail rig is obvious when compared with the multiple sails, spars and stays on the cutters, and it was much more efficient. *Bul Bul* romped away from the fleet with a runaway win on its first outing, just before Christmas in 1890.

'*She sailed her maiden race on the 20th December last, when she easily defeated Astroea, Norna and Jenny Wren. Her next race was against the larger class, when, though entered as a 5-rater, she saved her time on lsea, Sao, and lolanthe. On the beat to Manly she was 2 minutes ahead of these boats, which are all over 7-rating, and eventually saved her time and won the race. Her third race at the Anniversary Regatta was a run-away victory from Norna and Landair, both 2½ rating. She has thus won all the races she started in, although once meeting boats of a class above her.*'[719]

The exploits of *Bul Bul* seemed to have knocked the wind out of the 2½-rating class for the rest of the season. The 1891/92 season began without the regular mentions

[718] *The Sydney Mail and New South Wales Advertiser*, 21 February 1891.
[719] *The Sydney Mail and New South Wales Advertiser*, 21 February 1891.

PAINTING OF BRONZEWING ON SYDNEY HARBOUR BY GEORGE FREDERICK GREGORY, JR.
SOURCE: AUSTRALIAN NATIONAL MARITIME MUSEUM COLLECTION (OBJECT NO. 00004608).

of the craft in the aquatic columns, and *Bul Bul* raced with the larger yachts. Early in 1892 it was this vessel's turn to face the music, however. Samuel Hordern imported from England a lug rigged 2½-rater from the design board of George Lennox Watson. *Bronzewing* had the same sail area and waterline length as *Bul Bul* but with a straighter stem.

Bul Bul and *Bronzewing* had three match races and the latter won them all. Together the pair seem to have sent everyone else in the 2½-rating class into hiding. When the 1892/93 season opened, neither of the two yachts were certain of being commissioned to race, but they eventually both made a subdued appearance.

The freak boat is a recurring theme in yachting history. Just as a class or rule seems settled, along comes an interpretation or improvement that puts the others to pasture. While *Bul Bul* and *Bronzewing* were tormenting the Sydney fleet, back in England the fox was again in the hen house in terms of new designs. The 2½-rater class had its following in the Lakes District, and A. E. Payne was dominating the fleet with radical fin keeled designs. These new ideas were soon copied in Sydney.

In January of 1893 a 25ft one-rater was launched, *Naiad*, built by W. Goulding of Fig Tree Point to a design by Ernest Thompson.[720] Later that same year, a 0.75-rater from Neutral Bay was noted, coming from that ever-present person lurking about in Australian yachting, the owner/designer/builder, a Mr Farmer in this instance. Then in November of 1893 we see mention of a vessel (built at inner city Newtown) called *Viva Brandon* and rating about 2½.[721] All three vessels carried fin or centreboard keels, lug mains and jibs on pole masts.

JENNY WREN ON SYDNEY HARBOUR (2008).
SOURCE: JOHN JEREMY.

[720] *The Australian Star*, 14 January 1893.
[721] *The Sydney Morning Herald*, 25 November 1893.

All told, it is evident that Reeks no longer had the Sydney yacht design scene to himself. Before long other locally-based amateurs and semi-professionals would appear, then an influx from New Zealand and, finally, more professionals moving in on the turf, all adding spice to the mix. Reeks very likely welcomed the competition.

Moving forward to the present, of all the boats built to the 2½-rater class that raced regularly on Sydney Harbour, only three remain in existence. Comfortably maintained in private hands, *Jenny Wren* sails locally, along with the elegant *Kelpie*, a yacht that rarely raced after its launch in 1893.

KELPIE ON SYDNEY HARBOUR (2010S).
SOURCE: AUSTRALIAN NATIONAL MARITIME MUSEUM COLLECTION.

Kelpie also has some connections with Reeks. '*A new yacht has been launched from the yard of Messrs E. Thompson and Co. Drummoyne, built to the order of Dr. Eric Sinclair, from a design furnished by Mr Russel* [SIC] *Sinclair ... The Kelpie is built of teak ... She will be cutter rigged and in appearance will somewhat resemble the Jenny Wren.*'[722] Reeks' former student and employee Ernest Thompson was very fast becoming one of his professional rivals.

Jenny Wren and *Kelpie* join *Akarana* from the same period, now part of the Australian National Maritime Museum's fleet. All three are rare survivors of the harbour and of a glorious era in yachting.

In terms of Reeks' work independent of the 2½-rater class, by April of 1890 he had moved his office to 91 Pitt Street.[723] The following month he presented a paper on

[722] *The Sydney Mail and New South Wales Advertiser*, 23 September 1893.
[723] *The Australian Star*, 30 April 1890.

'Delta Metal' at the monthly meeting of the Engineering Association of New South Wales.[724] Here he discussed the history of the metal, explained its properties, and the uses to which it could be applied, including shipbuilding, engineering and sanitary work.[725]

There were also bigger commissions coming his way. In August of 1890 the contract for a new 17-ton yacht to be built for T. Busby of Double Bay by Tom Cubitt was signed.[726] Designed by Reeks for cruising purposes, the vessel in many aspects resembled his previous design for *Jess,* including in size being 46ft 9in (overall) with a beam of 11ft 6in and a depth of 7ft 6in.[727] With Reeks also supervising the building process, *Boronia* was launched on 15 November.[728]

In late September of 1890, a 22-footer designed by Reeks, stated to be a departure from the present type of this vessel, was launched.[729] '*Mr. Paul's new 22-footer is to be out for a trial to-day. She is designed by Mr. Walter Reeks, and is a departure from the present type of boat. It will be interesting to see how scientific principles will compare with evolution.*'[730] The latter statement was a noteworthy comment, going straight to the heart of the issue. Reeks had introduced scientific methods to yacht designing with great success. He had created his hull shapes from first principles as a plan, and was now taking on open boats at the very core of the old school way of thinking, where evolving the design from subtle changes made to the half-model or mould frames of the last boat had been the craftsman's approach for decades.

Built by George Barnett to the order of W. M. Paul, the half-decked yacht *Varuna* had a beam of 9ft 4in and a depth of 2ft 7in.[731] From the beginning the vessel proved that science worked and it sailed very well, winning many races over a number of seasons with the Sydney Amateur Sailing Club.

Another smaller vessel soon followed, built by Cubitt to a design by Reeks to the order of James Robb, Reeks also supervising the building.[732] Launched in early October

[724] *The Daily Telegraph*, 14 August 1890.
[725] *The Sydney Morning Herald,* 15 August 1890.
[726] *The Sydney Morning Herald,* 20 August 1890.
[727] *The Australian Star*, 13 November 1890.
[728] *The Sydney Morning Herald,* 20 August 1890.
[729] *The Sydney Mail and New South Wales Advertiser,* 9 August, 27 September 1890.
[730] *The Sydney Mail and New South Wales Advertiser,* 9 August, 27 September 1890.
[731] *The Sydney Mail and New South Wales Advertiser,* 27 September, 25 October 1890.
[732] *Australian Town and Country Journal*, 11 October 1890.

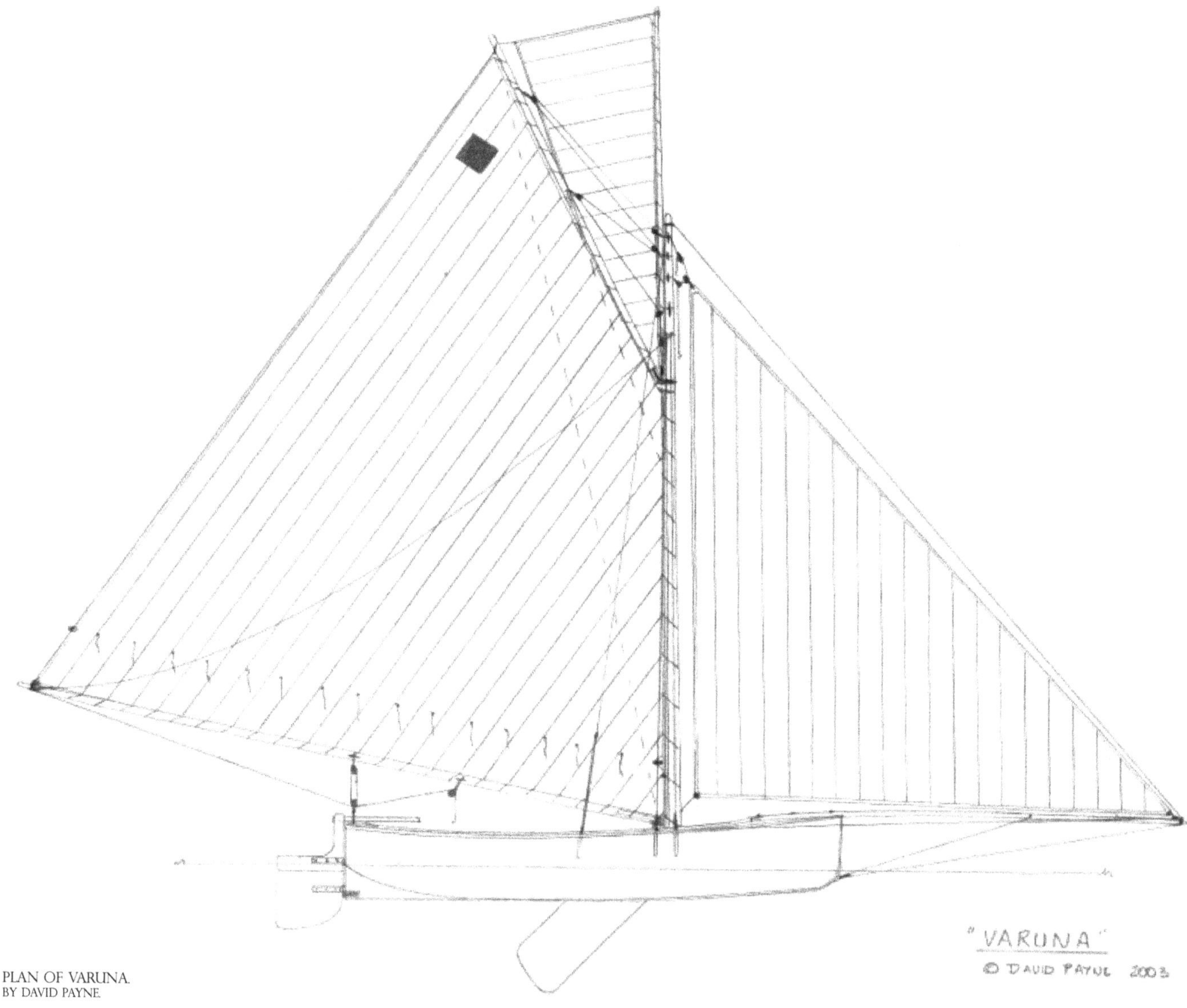

PLAN OF VARUNA.
BY DAVID PAYNE.

VARUNA.
SOURCE: AUSTRALIAN NATIONAL MARITIME MUSEUM COLLECTION (OBJECT NO. 00013837).

of 1890 and named *Viola*, the 6-ton yacht was intended purely for cruising and measured 30ft 9in (overall) by 25ft (waterline) with a beam of 7ft 6in and a depth of 4ft 6in.[733]

Continuing with the trend of designing smaller yachts, of which Port Jackson's yachtsmen were beginning to prefer more and more, Reeks next designed a '*very pretty and substantial 18-footer*'.[734] The craft was launched from the Berrys Bay yard of Cubitt on 12 November 1890, the seventh vessel he had constructed to a design by Reeks in under two years.[735] Built to the order of Dr Newmarch and named *Myee,* it measured 18ft 6in (overall) by 17ft 6in (waterline) with a beam of 7ft and a draft of 3ft 3in.[736] Half-decked, the vessel was rigged with a lug and jib and was intended to sail with the Neutral Bay Sailing Club.[737]

With little business activity occurring over the 1890 Christmas period and into the New Year, the Anniversary Regatta, held on 26 January 1891 in Sydney, saw Reeks achieve success at the helm of the 2½-rater *Astreoa,* after he had made several alterations to enhance its performance, including remoulding of the leaden keel.[738] The bulk of 1891, however, was a very quiet period for Reeks in terms of new vessel designs. Though he continued to act as a broker for the purchase and sale of used vessels, there was little ardent activity in the way of new racing yacht commissions, both small and large.

VIOLA (RENAMED LAH LOO).
SOURCE: YACHTSMAN'S GUIDE TO SYDNEY HARBOUR (1889).

The economy of New South Wales, and Australia more generally, was enduring a testing time. Heavily dependent on agricultural exports, particularly wheat and wool, prolonged and intense labour strikes by the shearers and maritime workers had both significantly stymied economic output, resulting in reduced prices for these commodities, as well as negatively impacting property

The Sydney Mail and New South Wales Advertiser, 11 October 1890.
733 *Australian Town and Country Journal*, 11 October 1890.
734 *The Australian Star*, 13 November 1890.
735 *The Australian Star*, 13 November 1890.
736 *The Sydney Morning Herald*, 15 November 1890.
737 *The Australian Star*, 13 November 1890.
738 *Referee*, 21 January 1891.

prices. These events subsequently translated into the collapse of many local banks at the peril of their depositor's funds. Unemployment and poverty rose considerably, government tax revenues decreased and public works projects were either put on hold or abandoned altogether. Adding to the collective turmoil, agriculture and associated rural industries were plagued by a long drought, further fuelling a downward trend in the Australian economy.[739] There was also great change happening at the political level fostered by the rise of socialism, the Labor Party, and a nationalistic sentiment.

Despite this uncertainty, Reeks continued to procure commissions, and consolidate his working relationship with Cubitt. The early months of 1891 were busy for the pair, two schooners intended for the Torres Strait pearl fishing industry were under construction, both designed by Reeks. Described as '*Clipper schooners*', the craft were in fact ketch rigged and built to the order of Andrew Andersen, a pearler of Prince of Wales Island, and Herbert Bowden, a pearler of Thursday Island. Both vessels measured 43ft by 11ft in beam with a draft of 6ft.[740] The first of the pair, *Freya,* was launched on 23 April 1891 while *Minerva* was launched on 15 May, several weeks later.[741]

The two vessels were aptly described in the press. '*New Pearling Fleet — Clipper Schooners. —Mr. T. Cubitt, shipbuilder, of Berry's Bay, has two fine little schooner pearlers now well in hand at his yards. Those vessels are the first of a regular fleet of specially designed pearlers for the Torres Straits fisheries. Some smart vessels are already engaged on the "grounds," and their magnificent weatherly qualities have been proved repeatedly in gales which have dispersed the fleet for hundreds of miles, not one of them being able to show a rag of canvas. Torres Straits' hurricanes are well known. The new boats have been designed by Mr. Walter Reeks to answer all the requirements of the pearling trade, and at the same time to be as speedy as yachts.*'[742]

The words '*specially designed*' and '*speedy as yachts*' are even more critical than they might appear. *Freya* and *Minerva* were probably a major first step in a project that developed into another milestone of success in Reeks' career, which will be told in greater detail in the next chapter. However, before we get there we must close in on his professional outcomes in the year 1891.

In June of that year, a '*pretty little yacht*' Reeks designed for Dr Robert Scot Skirving was launched from Cubitt's yard. Named *Hygiea* (the Greek goddess of health), the yawl-rigged craft measured 27ft (overall) by 21ft (waterline), with a beam of 7ft 3in and a draft of 4ft 6in.[743] In August of 1891 a 50ft steam launch built in Melbourne and subsequently purchased by Edward Noyes of Potts Point was re-launched. Converted into one of the prettiest steam yachts on Sydney Harbour by Reeks, the vessel was fitted with the latest torpedo engines and had been modified from '*a rather ordinary straight-*

[739] www.parliament.nsw.gov.au/about/Pages/1890-to-1900-Towards-Federation.aspx.
[740] *The Australian Star,* 25 March 1891.
[741] *The Australian Star,* 25 March 1891; *Evening News*, 24 April 1891; *Maryborough Chronicle, Wide Bay and Burnett Advertiser*, 19 May 1891.
[742] *Australian Star,* 25 March 1891.
[743] *Evening News*, 26 June 1891.

MINERVA.
SOURCE: QUEENSLAND MARITIME MUSEUM COLLECTION
NORNA PEARLING FLEET, THURSDAY ISLAND 1902 ALBUM.

stemmed launch to a pretty little clipper-bowed, schooner-rigged yacht'.[744] Fitted with '*a pretty cutwater stem, with gold scroll, her counter was run out 3ft. And with her neat schooner rig, she proved when she made her first appearance in the harbor, viz., following the R.M.S. Parramatta clear off the Heads, one of the prettiest and smartest little launches a yachtsman would care to set eyes on*'.[745] Disappointingly, the name of this particular vessel has not been determined to date.

Meanwhile in Brisbane there was launched in August of 1891 a new boat for the pearl shelling trade in Torres Strait, designed by Reeks.[746] Named *Waratah* and built by Anderson Brothers of Kangaroo Point to the order of James Clark, the vessel was also intended to compete in first class yacht races on Morton Bay.[747] *Waratah* measured 49ft (overall) with a 12ft beam and a draft of 6ft 6in, its planking and decks were of kauri pine.[748] As evident in the coming chapter, the craft was the first of many vessels, all built to the same design, and constitute Reeks' next step after *Freya* and *Minerva* toward his pivotal and profitable role in the design of pearl shellers.

For the same customer Reeks also designed a 22-footer, *Clyanthus,* solely for racing purposes that was built at James Clark's residence in New Farm on the Brisbane River by William Warbrick and launched in October of 1891.[749] In all probability *Clyanthus* was a close sister to *Varuna*. Perhaps the theoretical manner in which Reeks worked, highlighted by his use of science to direct his designs was what swayed Clark to have this particular vessel built. Alternatively, Reeks could have made a positive impression on Clark who was not only a yachtsman, but also a very canny businessman and entrepreneur. Reeks' capabilities would have immediately come to his attention. Clark also understood the pearl shell industry and the techniques involved in harvesting this natural resource. He very likely recognised in Reeks someone with the qualities he needed to further his business interests. The pair's relationship may have been cemented with just these two boats. However, over time, it would grow significantly for both parties.

Overall, during the early 1890s, Reeks continued to grow his business, particularly through the aid of non-Sydney based clients, as well as diverge away from racing yachts into commercial vessels. In late August of 1891 a steam drogher built by Rock Davis of Blackwall on the Brisbane River to a design by Reeks was launched. Commissioned by W. T. Yeager with construction superintended by Reeks, the 70ft (overall) by 16ft vessel was intended for the Richmond River drogher trade.[750] Named *Osprey,* it was a screw-steamer with a very light draft to help navigate the shallow and shoal-riddled river and estuaries. The vessel was used in conjunction with Yeager's Coraki sawmilling business for many years until sold to the North Coast Steam Navigation Company.[751]

It is at this point in time that we end this chapter on Reeks' life. Both personally and professionally, and a little bit geographically, he was about to make some big moves.

[744] *The Australian Star,* 8 August 1891.
[745] *The Australian Star*, 29 August 1891.
[746] *The Brisbane Courier,* 18 August 1891, 31 August 1891; *The Daily Telegraph*, 25 July 1892.
[747] *The Brisbane Courier,* 18 August 1891, 31 August 1891.
[748] *The Brisbane Courier,* 18 August 1891.
[749] *The Brisbane Courier,* 18 August 1891, 27 October 1891; *Lloyd's Register* (1901).
[750] *The Sydney Morning Herald*, 1 September 1891.
[751] M. Flapan (2021). *Register of Australian and New Zealand Ships and Boats.*

CLYANTHUS.
SOURCE: ROYAL QUEENSLAND YACHT SQUADRON.

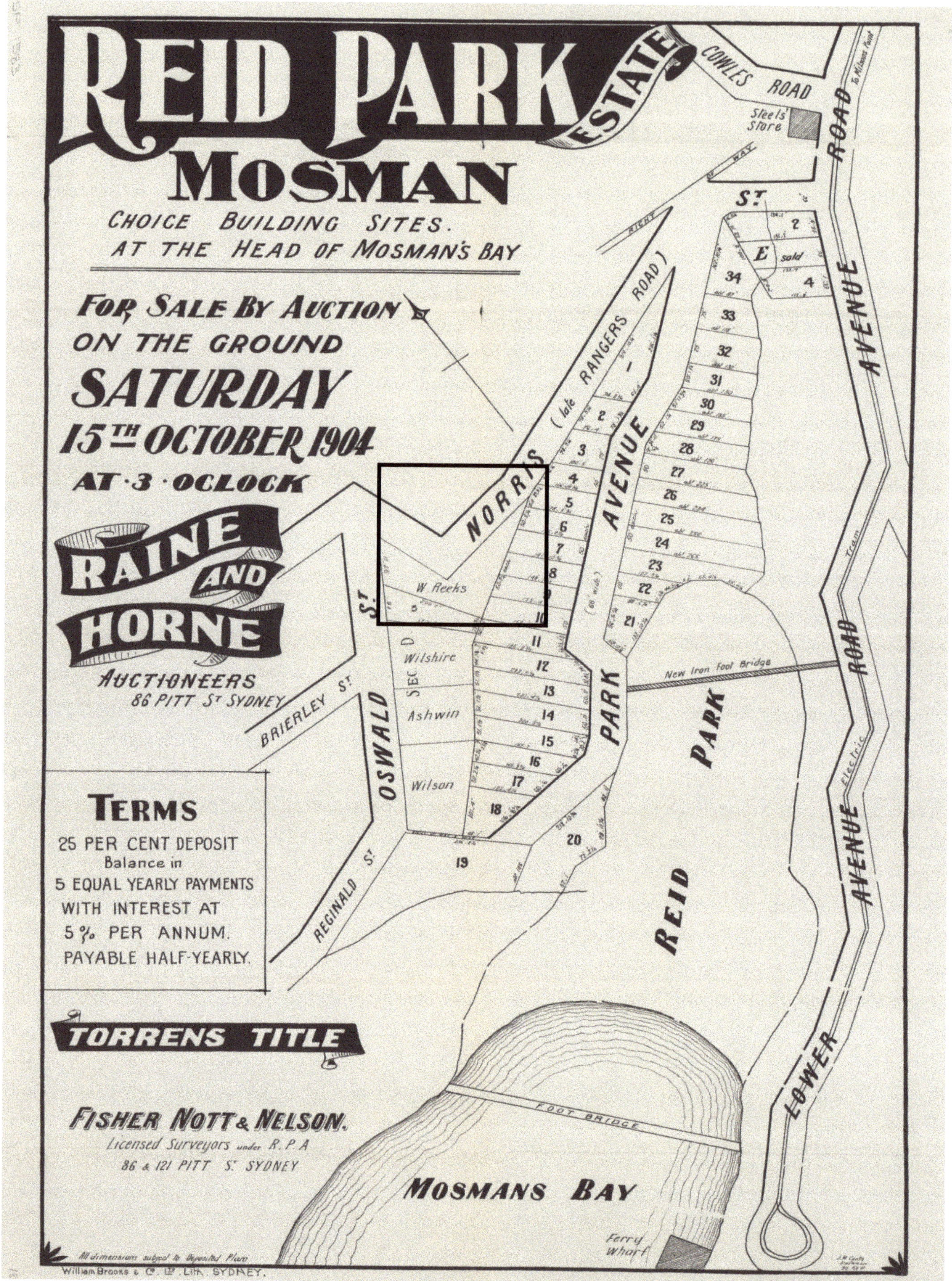

SOURCE: NATIONAL LIBRARY OF AUSTRALIA (NLA.GOV.AU/NLA.OBJ-230306155).

CHAPTER NINE - MOSMAN BAY

The first hint of what was to transpire next professionally for Walter Reeks came on 1 July 1891 when *The Sydney Morning Herald* carried an advertisement announcing the establishment of his shipbuilding business. The same advertisement was then published in this publication and several others in the months that followed. Despite the challenging economy and rather turbulent political climate Australia was then facing, establishing his own business was a risk Reeks considered well worth taking. With capacity for building vessels from 5 to 10,000 tons, it was to be a large operation allowing for the construction of a multitude of vessel types and sizes.

By September of 1891 Reeks had moved his office from 91 Pitt Street to 26 Market Street.[752] Though still in the heart of Sydney, this location offered slightly closer proximity to Darling Harbour and transport services across the harbour. Continuing his involvement with local yachting clubs, the opening day of the Prince Alfred Yacht Club saw Reeks spend the evening giving a recitation on electricity to its members.[753]

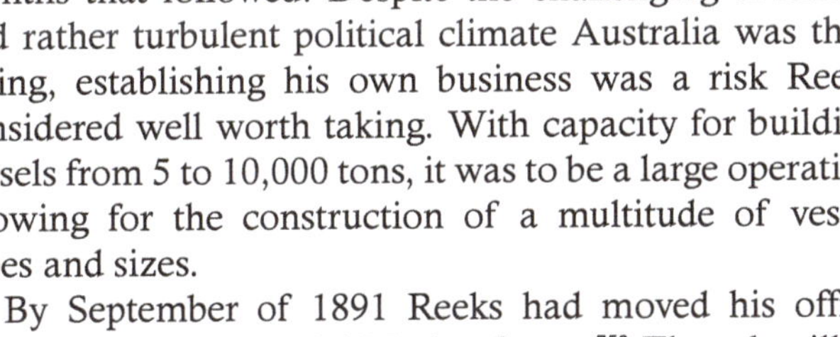

SOURCE: DAILY COMMERCIAL NEWS AND SHIPPING LIST, 31 OCTOBER 1891.

[752] *The Sydney Morning Herald*, 23 September 1891.
[753] *The Sydney Mail and New South Wales Advertiser*, 24 October 1891.

Two significant career milestones then took place near simultaneously for Reeks. These were the design and launch of specialised luggers (also referred to as pearl shellers and pearling ketches/schooners) for the growing and increasingly lucrative Torres Strait pearling industry and the development of a pioneering steamer for the newly-established Balmain New Ferry Company.

With regards to the pearling industry, on 21 January 1892 the first vessel to be sent off the ways from Reeks' new shipyard at Berrys Bay was launched, a 28-ton schooner named *Admiral* built by Reeks in conjunction with Tom Cubitt who was now acting as his foreman.[754] With construction of hardwood and New Zealand kauri, the vessel was built to the order of T. G. Kelly and measured 57ft 4in (overall), with a beam of 15ft 7in and a depth of 7ft 2in.[755] *Admiral* was intended for use in the Thursday Island region, as both a store ship and shell collector.[756] In describing the launch event, the press reported that two additional vessels were then in the course of construction, a 100ft steamer for the Balmain ferry trade (mentioned above), and a lugger for the Government of New Guinea.[757]

The lugger was launched just over a month later, on 29 February 1892, built to the order of the British Government of New Guinea. Named *Peuleule,* the 12-ton vessel measured 42ft 6in (overall), with a beam of 11ft 6in and a depth of 5ft 6in and was intended for magisterial work.[758] Following fit out, the craft left Sydney for Samarai three months after launch and would remain in service for more than 20 years.

[754] *Evening News*, 22 January 1892; *The Sydney Morning Herald*, 22 January 1892.
[755] *Evening News*, 22 January 1892; *The Sydney Morning Herald*, 22 January 1892; Register of Australian and New Zealand Shipping 1893-1894.
[756] *Evening News*, 22 January 1892; *The Sydney Morning Herald*, 22 January 1892.
[757] *The Sydney Morning Herald*, 22 January 1892.
[758] *The Sydney Morning Herald*, 1 March 1892.

By March of 1892 Reeks had once again moved offices, this time to Change Alley, Pitt Street, Circular Quay.[759] Given the developments occurring in his professional life, as well as his home situation, there was likely very little time for him to devote to yachting on Port Jackson. Reeks is not noted as being at the helm of any of the larger yachts this season.[760] Instead spare hours were directed towards building a new home for his growing family.

On 25 May 1892 Reeks and his wife Mary welcomed their daughter Thelma, born at *Twynham,* Mosman Bay, a home they had built overlooking the harbour and named for Reeks' birthplace of Christchurch, Hampshire, which was originally referred to as Twynham.[761] The house was situated on the south side of the corner where Rangers Avenue and Oswald Streets now meet, on the allotment where No. 6A Rangers Avenue currently stands. It was a busy period at the new home. Just a few months earlier, Mary had advertised for '*Lady Help*' with potential applicants asked to apply in person at *Twynham*.[762]

[759] *Daily Commercial News and Shipping List*, 7 March 1892.
[760] *Evening News*, 11 April 1892.
[761] *The Sydney Morning Herald*, 11 June 1892; *The Daily Telegraph*, 22 June 1892.
[762] *The Sydney Morning Herald*, 12 February 1892.

The couple's allotment was originally part of the Rangers Estate, a large colonial homestead built in 1844 by Oswald Bloxsome located between Rangers Avenue and Brierly Street.[763] The estate, comprising over 40 acres of land, had been subdivided and advertised for sale in smaller parcels starting from 1886 and continuing into the 1900s.[764] The Reeks' had purchased two of these allotments (lots 1 and 2 off Oswald Street, see image opposite) at auction in April of 1890, just over a year after their marriage.[765] Theirs were two of 15 allotments sold that day, with prices ranging from £2 to £5 per foot.[766] Legal documents held by the Reeks family confirm that the properties were in fact purchased exclusively by Mary Reeks, perhaps using funds given to her by the Broadhursts, her uncle and aunt.

Perched on the hillside overlooking Mosman Bay, the residence *Twynham* itself was of an unusual design, as one would expect from Reeks. For example, some of the timbers used in the building process were steam bent, a typical practice used in boat construction. In addition, its

[763] lostmosman.com/page/2/.
[764] *The Sydney Morning Herald*, 5 June 1886, 1 March 1890, 6 February 1892, 23 January 1904.
[765] *The Sydney Morning Herald*, 19 April 1890.
[766] *The Sydney Morning Herald*, 21 April 1890.

TWYNHAM (DATE UNKNOWN).
SOURCE: RAE FIDOCK COLLECTION.

SOURCE: NATIONAL LIBRARY OF AUSTRALIA (NLA.GOV.AU/NLA.OBJ-230312141).

principal feature appears to be a large two story circular or bull-nosed section on the south-east corner, topped with a conical slate or timber-slatted roof.

Located on Sydney's Lower North Shore, with a population approaching 1,600, Mosman Bay was an ideal location for Reeks and his wife Mary to raise their family. Situated on the harbour, with commanding vistas and beautiful natural charm, the area was at the time a developing minor suburb that would soon see expansion in terms of residential property development, in part helped by the new steam ferries that Reeks would be involved with. A ferry service at the time ran from Circular Quay to Mosman Bay, and there was a wharf at the bay's entrance on the end of Curraghbeena Point, as well as at the head of the bay.

Coinciding with the construction of *Twynham*, Mosman Public School was established, the local Congregational Church held its first service, sporting clubs were beginning to garner members, postal and telegraph services began, the water supply was installed and an electric tram service commenced operation.[767] Combined, these improvements were immense, benefiting Mosman's new residents and increasing their standard of living.

Reeks and his wife Mary joined Mosman's budding society; the suburb was fast becoming known as a haven for artists, writers, intellectuals and businessmen, many of whom strongly rejected proposals which could impact the natural aesthetic of their locality.[768] In September of 1892, Mary Reeks attended a dance given by the ladies of Mosman Bay, with her dress stated to be of black silk and lace.[769] The following year Mosman became a new and separate municipality, helped in part by a petition from ratepayers that Reeks himself had signed.[770]

Of note, the first elected mayor of Mosman was Richard Hayes Harnett (Jr). He was the son of a property developer who had not only pioneered the growth and expansion of Sydney's North Shore area, but also established its first transportation services, including building the Mosman and Neutral Bay wharves and running the ferry service to Mosman and several of its omnibus lines.[771] A keen sportsman, an original member of the Prince Alfred Yacht Club and a promotor of the Anniversary Regatta, Harnett's father had also designed and built the unusual yacht *Australian* (profiled in chapter six), of which Reeks thought very highly of.[772]

All told, Mosman Bay offered Reeks boundless opportunities, not only to raise his family and enjoy his recreational hours, but also to establish and build business connections. It was an idyllic location in more ways than one.

With major changes taking place on a personal level, Reeks' professional accomplishments continued apace, including the launch of several more pearling luggers from his newly-established shipyard at Berrys Bay, an easy commute by ferry from his new home. The 15-ton *Jessie*

[767] https://mosman.nsw.gov.au/community/people-culture-and-history/a-brief-history-of-mosman.
[768] https://mosman.nsw.gov.au/community/people-culture-and-history/a-brief-history-of-mosman.
[769] *Evening News*, 7 October 1892.
[770] *The Australian Star*, 21 June 1892.
[771] *The Sydney Morning Herald*, 18 November 1902, 21 September 1937.
[772] *The Sydney Morning Herald*, 18 November 1902.

MOSMAN BAY, SHOWING A FERRY LEAVING THE WHARF (CIRCA 1900S).
SOURCE: STATE LIBRARY OF NEW SOUTH WALES.

was completed on 22 July 1892, another vessel built to the order of James Clark.[773] With measurements of 45ft, with a 12ft beam and a depth of 6ft 5in and named for Clark's wife, *Jessie* was indicative of Reeks' adaptation of vessels for the purpose of pearl shelling in Torres Strait. Notably, the craft also marked the second of three identical pearlers designed by Reeks for this particular customer; the pioneer being *Waratah* which had been built in Brisbane and launched in August of the previous year.[774]

While *Jessie* left Sydney for Cooktown six days after its launch, the third boat, *Isabel,* measuring 45ft by 12ft 1in, with a depth of 6ft 2in, was completed three weeks later at Reeks' Berrys Bay yard.[775]

Overall, the pearl sheller or pearling lugger was a uniquely Australian-developed craft, helped in part by Reeks' involvement in its design. The vessel's evolution as a sail-rigged working craft spans from the 1880s through to the 1950s. For a long time the boats were more commonly known as pearling luggers, a term relating to their original lug-rigged main and mizzen sails, which gave way to the gaff rig during the 1890s.

Reeks became associated with the pearl shellers at a time when the industry was lucrative and growing. The main operators were looking to improve the profitability of pearl shelling, seeking further efficiency and economy. Reeks was once again able to apply the required design expertise that others did not have and to focus on one aspect of this, improving the craft involved in collecting the shell.

The beginnings of the industry had been humble, like so many others. Pearl shelling then evolved as needs arose. Ultimately it is a simple story of a trade often viewed romantically in novels and film, whereas the reality was a tough and dangerous workplace, where margins were sometimes slim and social inequality was rife for decades.

The associated activities of collecting mother-of-pearl shell *Pinctada maxima* and their pearls, along with beche-de-mer and trochus shell were a common indigenous practice around the northern coastlines and islands of Australia. Colonists began collecting the shell on an organised basis during the 1860s, along parts of the north-west coast, including Broome and Shark Bay, and in the Torres Strait. The shell was the principal object and had various uses, including inlay and expensive buttons, and was much sought after in Europe and North America in the late nineteenth century. Initially collectors worked the shoreline, then they moved into deeper waters, around five to six fathoms, skin diving without apparatus. The vessels they originally used were nothing special, but when the industry became more organised, a schooner or yawl rigged sailing craft was employed, primarily a mixture of craft adapted for the purpose and some built to specifically service the growing industry.

The catalyst for a significant change was the acceptance of diving suits and their associated diving gear of air pumps and hoses. After experimentation in the 1870s, diving apparatus was widely adopted from the mid-1880s.[776] This opened up grounds beyond the six-fathom limit to unassisted skin divers, and allowed for greater time to be spent on the bottom, over a larger area. It also required a boat that could support a hand pump, two operators, storage and space for the hoses, the divers and their associated dress. The dinghy and other adapted craft were no longer suitable. It had to be a moderate-sized vessel and the builders, largely based in Fremantle up until this point, responded with working craft built to order.

THREE IMAGES PORTRAYING THE PEARLING INDUSTRY, AUSTRALIA, CA. 1887.
SOURCE: NATIONAL LIBRARY OF AUSTRALIA (NLA.GOV.AU/NLA.OBJ-149702997).

[773] *The Daily Telegraph*, 25 July 1892.
[774] *The Daily Telegraph*, 25 July 1892.
[775] *The Daily Telegraph*, 29 July 1892; *The Sydney Morning Herald*, 22 August 1892.

[776] G. Kerr (1985). *Craft and Craftsmen of Australian Fishing, 1870 - 1970.*

Operators and builders had their own ideas on both design and construction details, observing and learning as the pearling trade grew further, and the craft for the north-west fields began to adopt certain shared characteristics that in the longer term became a type of its own. The cornerstone of the north-west style of vessel from the late 1880s lies in the section shapes and keel profile that were used consistently over many decades from this point in time. The craft tended to have a shallow draft, a straight keel with little rake aft and a midsection with only moderate dead rise, leading to a firm bilge below the waterline. This shape appears to be dictated by the operation of the vessels in port, and not at sea on the pearling grounds.

The north-west coast of Australia has a significant tidal range. The port areas dry out on the low tide, leaving craft to lie over on one side or the other, until refloating on the incoming tide occurs. The section shape and straight keel enabled the craft to layover and right themselves without shipping water, or sit level if propped up on their sides.

A key period in the design of the Broome-based vessels was around 1888 and 1889. Established Fremantle boat builders, including the well-known names of A. E. Brown and the Chamberlain and Mews families, were joined by Charles Walker, originally from New South Wales. He produced an improved craft that was compared to a yacht. It was also fast. One of his first customers was James Clark.

The *West Australian* on 20 April 1889 reported, '*Charles Walker ... has launched one of the prettiest models ever seen here ... The finish is equal to a yacht ... Those who take an interest in pearling should visit the boat ... she will be a fast little sailing craft'*. This was followed by another report later in the year. '*Information has been received in Fremantle that the Queen, lugger, owned by Mr. J. Clark the well-known pearler, and built by Mr. C. Walker has been proved the fastest lugger in the pearling fleet of the North West.*'[777]

In the Torres Strait things were evolving quite differently, as an 1884 description shows. '*Large vessels for depots and whaleboats for divers, have been replaced by a number of smart luggers, nearly all of which have been built on the shores of Port Jackson. These handy little vessels range from six to fourteen tons burden, are of yacht-like model, can thrash a passage against a head breeze, or run down the wind at a clipping pace under a pair of French lugs; and in a clock calm can be easily propelled by sweeps.*'[778]

Once again yachts are referred to and an interesting point to note is that since earlier times yachts had tended to be derived from working craft. In contrast, we now have this relationship going in the reverse direction.

Around 1891 it appears that a different type of craft had finally emerged in the Torres Strait fields, a craft with a long counter, low freeboard, and a deep keel being its principal features. The vessels also had a gaff ketch rig with a conventional mizzen, somewhat smaller in height and area than the mainsail on the fore mast. Most craft prior to these vessels being developed were lug-rigged, often double-ended, and smaller in length and tonnage. There were occasional large boats, and some with ketch rigs, so these elements were beginning to appear but perhaps not in the combination that was established by Reeks with his initial designs for *Freya* and *Minerva,* then *Waratah, Jessie* and *Isabel.*

The concept of this vessel type appears to be very much based around its operation at sea and on the pearling fields. An analysis of the lines plan and sail plan show a craft that is a significant contrast to the Broome style of vessel. It had a vertical stem with a forefoot that rounded down to the keel line, and the line of the keel profile had a significant rake downwards to the lowest point at the heel or rudder post. The elegant sheer tailed off with a long tapered and elliptical counter, and the section shapes showed a deep hull form, with a steep dead rise to the planking, forming a high turn of the bilge at the waterline. The plan view showed moderate beam and a well-rounded deck edge from stem to stern.

A new type of vessel was needed and one of the pioneers of the industry, James Clark, later to be referred to as the 'Pearl King', was well aware of the problems related to controlling the vessels as they acted unsupervised for a long period at sea before returning to shore with their proceeds. He devised an operation whereby the pearl shelling vessels formed a small fleet. He had already successfully trialled a scenario of this type in the north-west pearling fields. The craft still stayed at sea but every few days the live shell was taken and opened aboard a much larger mother-ship schooner, so both the take of shell, any pearls therein, and the rate of work were much more closely monitored. The results were significant increases in yield and productivity.

The yacht-like shape would have had many advantages in this situation. In the first instance it is able to perform particularly well '*thrashing to windward*', a useful attribute on passages, and useful for sailing back upwind to drift over a field again, or upwind to the mother-ship. A deeper-keeled yacht-like hull form is also a very sea-kindly shape, with no sudden motions as it rides the waves and swells (the opposite being very evident in much wider and shallower hull shapes).

The deep keel profile, with a cut away forefoot and the deepest point right aft, also gave an ideal hull profile for controlled sailing and drifting. Combined with the ketch rig and the varying sail combinations that could be set, it led to a craft that could easily be manoeuvred in tacks or gybes and set to sail in a controlled manner on all points of sail. It was also well suited to its requirement of being able to sail along slowly with the divers working the bed below. The ketch rig with a small mizzen was additionally relatively easy to set up for slow sailing or drifting, but with all sails set and sheeted in for sailing, it was generally a better rig for good all-round performance on all points of sail, as opposed to the schooner rig which is much more downwind orientated.

[777] *West Australian*, 1 October 1889.
[778] *The Illustrated Sydney News*, 25 October 1884.

The deck layout and long counter of the pearling lugger both suited the location and operation of the pump, hoses and other gear, while the low freeboard assisted the diver entering and leaving the water in a heavy suit. In terms of the yacht-like comparison, only one item was missing: the luggers with their small sail area did not have a ballasted keel, carrying internal ballast if needed, whereas a yacht of that period had a ballast piece outside of the hull on the keel to match its substantial sail area.

If we return to a report of the launch of *Jessie* from Reeks' Berrys Bay yard in July of 1892, we can garner more information on just what was happening with the development of this new industry-specific type of vessel at the time.

'*LAUNCH OF A PEARLSHELLER. On Friday evening, the 22nd instant there was launched from Mr. Walter Reeks yard, Berry's Bay, a new 15-ton ketch rigged pearlshelling craft, built to the order of James Clark of Torres Straits. As the vessel left the ways she was christened "Jessie". ... she will leave for her destination (Cooktown) this afternoon, or Tuesday morning at the latest. Captain George Cavill is in command and expresses himself highly pleased with his charge. It may be interesting to note that the Jessie is one of seven sisters. The Waratah, a boat built in Brisbane last year from a design prepared by Mr. Walter Reeks proved herself so well adapted to the purpose of shelling, and fast shown by the fact that she won the first class shellers race last 24th May (Queens Birthday Regatta), with 13 minutes to spare from the next best boat, that Mr. Clark immediately put into hand an order for six others – four in New Zealand, and two with Mr. Walter Reeks – but all from the same design. The second will be ready in about a fortnight. Mr. Cubitt, yard manager, is to be congratulated on the excellent work put into Jessie.*'[779]

As stated, obviously pleased with the prototype design Reeks had developed for *Waratah*, James Clark ordered another four pearling schooners, this time to be built in New Zealand by Messrs Lane and Brown of Whangaroa.[780] The vessels were all launched in 1892 and measured 45ft 6in (overall) with a beam of 11ft 11in and a depth of 5ft 10in.[781] Ketch rigged, in order of completion, they were *Essie, Our Boys, Mascotte* and *Waterwitch*.[782]

On 29 July 1892, a week after its launch, *Jessie* left Sydney as planned.[783] On board was someone who would go on to make his mark in another area entirely, as a professional artist and pioneering member of the Heidelberg School art movement (also known as Australian impressionism), though for now was supporting himself with ship work and writing.[784]

In an article titled '*Going North*', published in *The Argus* on 12 November 1892, Tom Roberts wrote, '*We are now a week from Sydney Heads, and have brought the little Jessie well up into Queensland waters. Only about a fortnight ago she was in the stocks of Reeks' yard at Berry's Bay, just a very elegant*

[779] *The Sydney Morning Herald*, 25 July 1892.
[780] *The Daily Telegraph*, 29 July 1892; *The Sydney Morning Herald*, 22 August 1892.
[781] New Zealand, Registered Ships & Owners, 1840-1950.
[782] New Zealand, Registered Ships & Owners, 1840-1950; Register of Australian and New Zealand Shipping 1893-1894, 1902-1903.
[783] *The Sydney Morning Herald*, 29 July 1892.
[784] en.wikipedia.org/wiki/Tom_Roberts.

PEARLING SCHOONER FOLLOWING CONSTRUCTION AT BERRYS BAY (EARLY 1890S).
SOURCE: MUSEUM OF APPLIED ARTS & SCIENCES 'POWERHOUSE MUSEUM' (HTTPS://MAAS/30717).

looking shell, with fish like shape, and after this trial we all believe she will fulfil the promise of beating any of the pearl-shelling fleet up north - as the authoritative George says "she go and steer like a weetch" and it is exhilarating to sit a whole watch through at the tiller, when in a strong west wind we slip along at eight or nine knots, the water leaving the counter so clearly the waves seem scarcely to have noticed our passing through.'

This piece described the voyage north Tom Roberts was taking on board the pearl sheller *Jessie*. He was first mate, delivering the vessel to the Torres Strait pearling grounds where the craft would join its sister-ship the *Waratah*. About four weeks later, following close in *Jessie*'s wake would come the *Isabel.* Over the next few months, more vessels of a similar style joined them in Torres Strait.

It is interesting to note that Roberts also had a connection to Mosman Bay and Reeks' new home. In 1891 he had travelled to Sydney from Victoria to stay at Curlew Camp, an artists' camp established on the eastern shore of Little Sirius Cove, the next bay to the east of Mosman Bay and Curraghbeena Point, and only a short walk from either of the two wharves.[785] Following his voyage on *Jessie*, and several other adventures in between, Roberts returned to Mosman Bay where in 1894 he painted his highly prized work '*Mosman's Bay 1894.*' The painting, significantly for the authors, shows *Twynham* on the hillside in the background. It is interesting to speculate whether Roberts intended to highlight the home, possibly because of his connection to Reeks or that its conical roof and large chimney, being quite clear amongst less distinctive details, shapes and colouring, offered a wonderful contrast with which to depict the background. Either way, the work clearly shows *Twynham.*

While it is apparent that Reeks and Roberts had become well known to one another, it is not known how they first met. It may have been at Curlew Camp, commuting on a ferry, through their connections in society, or even through Roberts' involvement with the pearling lugger *Jessie.* Unfortunately, no sketches of his time aboard this vessel have come to light.

Returning to Reeks' professional life, there was much going on. The connection to James Clark and his influence in this sudden arrival of a new design is very significant. As evident from the earlier quote about the vessels built on the north-west coast of Australia, already Clark was fostering the evolution of pioneering craft in this location, with the construction of one of the fastest luggers *Queen* from a progressive builder (Charles Walker). It seems Clark repeated this process, in both Sydney and New Zealand. He was also prepared to listen to the experts. Long-time

[785] https://en.wikipedia.org/wiki/Curlew_Camp.

'MOSMAN BAY, 1894' BY TOM ROBERTS.
SOURCE: EN.WIKIPEDIA.ORG/WIKI/TOM_ROBERTS.

pearler Ted Norman was able to recall some detail of Clark's methods, including that he was apparently rigid with his use of technical advice.

'*I understand the plans and those plans of any changes in lines, no matter how small, were referred to a naval architect for opinion before building commenced. These deep boats above referred to were counter stern.*'[786]

Clark continually showed an awareness of the need to keep abreast of the technical aspects of the pearling industry too. For example, when he established the Pilot Cultivation Co. in 1893, to develop a cultivated pearl shell operation, he employed a biologist from England on a three-year contract.

With the demand for this vessel type remaining high, Reeks continued to build pearling ketches and schooners at his Berrys Bay yard toward the latter stages of 1892. The 11-ton *Gipsey* was launched in September of that year, built to the order of Captain T. L. Grainger, and under his supervision.[787] The vessel was constructed of colonial hardwood with kauri planking and measured 44ft 9in (overall), 37ft 3in (waterline), with a beam of 11ft 4in and a depth of 5ft 6in.[788] The following month the lugger *Camoola* was launched.[789] This craft measured 42ft (overall), with a beam of 11ft and a depth of hold of 5ft 9in, and left for Thursday Island on 15 October under the command of Captain Powell.[790]

In early November of 1892 the ketch *Sana* was launched, another vessel for Clark's pearl shelling fleet based in Torres Strait.[791] Measuring 49ft 8in (overall), with a breadth of 12ft and a depth of 5ft, the craft was slightly larger than its cohort and fitted with a clipper bow and elliptical stern.[792] A few weeks later the launch of the 15-ton ketch *Ehime* took place. Also intended for Torres Strait pearling, this particular vessel measured 45ft 4in (overall), with a beam of 12ft 3in and a depth of 5ft 8in and was built to the order of Captain Grainger, representing Burns Philp & Co.[793] Its sister-ship, *Takachiho,* built for the same firm, was launched on 5 December of 1892.[794]

As stated, the second concurrent major project Reeks' Berrys Bay slipway was involved with in 1891/1892, in addition to the pearling luggers, centred on a ferry for the Balmain New Ferry Company. In late 1891 Reeks was awarded one of two contracts by this new organisation to design and build a steamer for service as part of its 'quarter-hour penny trips', calling at the wharves of Stephen and Reynolds streets, and also at Bald Rock.[795] The second contract had been awarded to George Ellis of Balmain.[796] While both craft were somewhat similar in design and features, they were also considered unique in comparison to previous harbour passenger ferries.

With an initial capital investment of over £12,000, the Balmain New Ferry Company had been established less than two years prior and subscribed to by nearly 900 shareholders, a large proportion of whom were residents of the suburb.[797] At Reeks' shipyard on 30 July 1892, the project became a reality with the launch of its pioneering vessel. In the presence of many high-profile people, including the Governor of New South Wales, Lord Jersey, and his wife Lady Jersey, Henry Copeland (Minister for Lands), Sir Julian Salomons and other members of parliament, Alderman Wheeler (Mayor of Balmain), and hundreds of interested spectators, Lady Mary Villiers, the Governor's daughter, performed the christening ceremony of *The Lady Mary*, intended to initiate the long-promised penny service to Balmain.[798]

With a guaranteed speed of 11 knots, a capacity to seat 375 passengers, and fitted with electric light, the vessel was considered far superior to its harbour predecessors in terms of passenger comfort. Constructed of hardwood with hull planking of Oregon and deck planking of kauri, the vessel's measurements were 96ft 6in (overall), with a beam of 22ft, a depth of hold of 7ft 4in and a draft of 5ft 9in.[799] In terms of its lines, the craft was considered to have evolved from some of the best American ferry boat designs.[800] *The Lady Mary* also featured nicely-furnished enclosed cabins, and a well-arranged promenade space.[801] Its engines were manufactured by Messrs Ross and Duncan of Glasgow, Scotland, specifically to suit the vessel, from specifications and drawings prepared by Reeks and under the immediate supervision of his consulting engineer, Alex Richardson of Liverpool.[802] This last fact is noteworthy. Reeks had called upon his past employer and mentor for help with such an immense and ambitious project.

The Lady Mary was also considered revolutionary in terms of its hull shape around the propellers: the vessel being driven by double screws, one at each end.[803] '*The deadwood was cut entirely away, leaving an aperture of some 7ft or 8ft, in which the shaft is exposed to the water, and this, together with the peculiar "cutting up" of the vessel at both ends is probably the greatest departure from the conventional ferry boat. The principle was introduced in the Lady Mary by Mr. Reeks with a view to give exceptional steering power, and this is without in any way weakening the structure, which is rendered rigidly strong by a judicious use of the materials.*

The engines are compound surface condensing, with cylinders of 13in and 26in, with a stroke of 16in. All the pumps are run by a separate compound engine, specially designed for the purpose, and, in addition, there is a donkey on board, so that the steamer is fitted with three distinct and separate engines. The boilers are 9ft by 8ft 6in, and have been tested to work up to 120lb. The speed of the vessel is expected to be 11 knots. An installation of the electric light has been fitted throughout the vessel by Kingsbury and Co.

[786] G. Kerr (1985). *Craft and Craftsmen of Australian Fishing, 1870 - 1970.*
[787] *The Daily Telegraph*, 16 September 1892; *The Sydney Morning Herald*, 23 September 1892.
[788] *The Daily Telegraph*, 23 September 1892.
[789] *Daily Commercial News and Shipping List*, 15 October 1892.
[790] *Evening News*, 15 October 1892.
[791] Register of Australian and New Zealand Shipping 1893-1894.
[792] *The Daily Telegraph*, 15 November 1892.
[793] *Newcastle Morning Herald and Miners' Advocate*, 28 November 1892.
[794] *The Brisbane Courier*, 6 December 1892.
[795] *The Australian Star*, 27 October 1891.
[796] *The Australian Star*, 27 October 1891.
[797] *The Sydney Morning Herald*, 1 August 1892.
[798] *The Sydney Morning Herald*, 1 August 1892; *The Daily Telegraph*, 1 August 1892.
[799] *The Australian Star*, 29 July 1892.
[800] *The Sydney Morning Herald*, 30 July 1892.
[801] *The Sydney Morning Herald*, 30 July 1892.
[802] *The Daily Telegraph*, 31 October 1892; *The Sydney Morning Herald*, 31 October 1892.
[803] *The Sydney Morning Herald*, 1 August 1892.

The steamer has attracted a considerable number of visitors during the week, and the general opinion is that the design and workmanship are of the highest character. The Lady Mary certainly is a credit to her designer and builder, and should do good service to the New Balmain Ferry Co.'[804]

The fitting out and installation of *The Lady Mary*'s machinery took some additional months and it was not until 29 October 1892 that the craft made its trial trip on the harbour in the presence of a number of shipbuilding experts, master mariners and others.[805] According to reports, '*The boat more than filled all expectations and did what work was required of her in a smart manner, covering the measured mile in 5min. 20sec. She is smart at coming round, answers her helm promptly and can be stopped and started with great celerity.*'[806]

The company's second steamer, designed by its builder with some assistance from Norman Selfe, was built by George Ellis of Balmain. The vessel was launched as the *Lady Napier* on 9 September 1892.[807] Slightly larger than the *The Lady Mary,* the craft measured 108ft 6in overall, with an extreme beam of 26ft and a depth of 7ft 6in.[808] Accommodation was provided for up to 500 passengers.

Collectively, both *The Lady Mary* and *Lady Napier,* and others that followed, were successful in helping advance the development of Balmain, historically a working-class suburb of Sydney. Much needed, they provided a short, affordable and reliable commute to and from the city centre, and other nearby locales.

In addition, there were other diversions from the slew of pearling luggers and the passenger ferry *The Lady Mary* for Reeks during 1892. Steam yachts and launches were coming into vogue. He supervised the building of the large steam yacht *Isis,* designed by William Fife (Jr) to the order of J. R. Fairfax, the craft making its first appearance at the Royal Sydney Yacht Squadron's opening day in October of that year.[809] Reeks likely accepted this commission as a way of gaining experience and keeping his yard busy, but more importantly keeping his relationship with the 'establishment' well nourished.[810] It is also noteworthy that he was prepared to act for another designer, in this case one of the most significant yacht designers of the period and indeed any time since. William Fife (Jr), and his father and grandfather before him, built their own designs and only rarely allowed another builder to work from their plans, almost exclusively reserving this honour for builders outside of Great Britain. The Australian

[804] *The Australian Star,* 29 July 1892.
[805] *The Daily Telegraph,* 31 October 1892.
[806] *The Daily Telegraph,* 31 October 1892.
[807] *The Sydney Morning Herald,* 9 September 1892.
[808] *The Daily Telegraph,* 26 July 1892.
[809] *The Sydney Morning Herald,* 24 October 1892.
[810] *The Sydney Morning Herald,* 24 October 1892.

DIVING FOR PEARLS FROM THE PEARLING LUGGER MASCOTTE AT THURSDAY ISLAND.
SOURCE: STATE LIBRARY OF QUEENSLAND (HDL.HANDLE.NET/10462/DERIV/144744).

THE LADY MARY.
SOURCE: GRAEME ANDREWS WORKING HARBOUR PHOTOGRAPH COLLECTION, CITY OF SYDNEY ARCHIVES.

LAUNCH OF THE LADY MARY FROM WALTER REEKS' BERRYS BAY YARD (30 JULY 1892). ALSO SHOWS THE STEAM YACHT ISIS BEING FITTED OUT ON THE LEFT.
SOURCE: AUSTRALIAN NATIONAL MARITIME MUSEUM COLLECTION (OBJECT NO. ANMS1092[255]).

colonies were the main beneficiary of this decision, with Fife designs built in just about all of the colonies and later states. In a number of instances, Fife's shipwrights prepared the frames in Scotland and then had them shipped out as a means of ensuring the shape would be properly maintained. Entrusting Reeks to supervise a Fife design was a great credit to him. *Isis* was also substantial. The craft was schooner rigged, of about 90 tons burden, with 20 h.p. engines. Its internal fitout radiated opulence with the salon of polished American walnut and crimson plush cushions.[811]

Isis marked the tenth vessel that had been launched from Reeks' shipyard at Berrys Bay that year. It was rather an impressive and significant feat given the yard was only in its first year of operation, as well as considering the downward state of the global economy. With his professional and personal life coalescing on positive notes, there was much for Reeks to look forward to.

[811] *The Sydney Morning Herald*, 24 October 1892.

1890 MAP OF BERRYS BAY AND SURROUNDS (NEW SOUTH WALES).
SOURCE: NATIONAL LIBRARY OF AUSTRALIA (NLA.GOV.AU/NLA.OBJ-231445915).

CHAPTER TEN - BERRYS BAY

The pattern of designing and constructing commercial vessels as opposed to substantial recreational and racing yachts looked to continue into early 1893 at Walter Reeks' Berrys Bay yard. Unfortunately, there was something on the horizon that would have a major impact on the already declining Sydney economy and thus demand for new vessels of either persuasion. The Panic of 1893 that began in the United States as a result of a stock market crash led to a 25 per cent rate of unemployment in that country and widespread social and political upheaval. Soon the aftermath shattered global economies, making its way to Australia. Compared to 1892, the year saw the colony's gross domestic product decrease by 17 per cent. It was Australia's most severe financial crisis to date, the population experiencing its first depression in 50 years. The property boom of the previous decade, financed by a rapid increase in bank and building society lending, soon turned sour, resulting in the collapse of many institutions. With it, public confidence hit an all-time low, a run on the banks being another precursor to further economic decline, many suspending transactions or closing altogether, the act of freezing accounts hitting depositors hard. The fluidity of the economy was at a standstill. There was no trust, only scepticism.

CLOSE UP OF THE PEARLING LUGGER ISABEL BEING BUILT AT WALTER REEKS' BERRYS BAY YARD.
PHOTO TAKEN AT THE LAUNCH OF THE LADY MARY ON 30 JULY 1892.
REEKS IS SHOWN NEAR THE BOW OF ISABEL, INSPECTING PREPARATIONS FOR LAUNCHING OF THE STEAMER.
SOURCE: AUSTRALIAN NATIONAL MARITIME MUSEUM COLLECTION (OBJECT NO. ANMS1092[255]).

It was not an optimistic climate to be running a business, particularly a rather new operation centred on the construction of large commercial and recreational vessels and, thereby, the transaction of great sums of money and resources needed to build them. There was no money to lend. No-one had confidence to take a risk and there was widespread hesitancy about entering new markets.

There was one industry, however, that remained somewhat sheltered from the global economic decline: the pearling industry. While many new pearl shellers had been built during the prior year, both in Sydney by Reeks and his neighbouring boat builders, as well as in New Zealand, there was still a deficiency in supply such that demand remained high. The industry had its own deterrents though, one being wild weather, with fleets and crew decimated when storms and cyclones hit the areas where they operated. It was a risky business but one that offered great reward, for the time being at least.

Before we discuss Reeks' continued involvement in this industry, we must segue to the first vessel launched from his yard in January of 1893: the mission ketch *Albert Maclaren*. Built to the order of the Australian Board of Missions under the supervision of Captain Grainger, the 15-ton vessel measured 46ft 10in in length, with a beam of 12ft and depth of 6ft 5in.[812] Its keel, stem and sternpost were constructed of the finest seasoned timber, with the planking and decks of kauri.[813] Capable of accommodating six passengers, the *Albert Maclaren* was reported to give the appearance of a fast sailer and good sea boat.[814] Intended for missionary work in New Guinea in connection with the Church of England, the craft was dedicated by the Bishop of Newcastle on 4 February 1893 and left Sydney for its new home a few weeks later under the charge of Charles Protheroe.[815] The *Albert Maclaren* remained in operation for many decades.

MISSION SCHOONER ALBERT MACLAREN IN PAPUA NEW GUINEA (1922).
SOURCE: G. WHITE (1923), FRANCIS DE SALES BUCHANAN: MISSIONARY IN NEW GUINEA.

[812] *The Sydney Morning Herald*, 20 January 1893.
[813] *The Sydney Morning Herald*, 20 January 1893.
[814] *The Sydney Morning Herald*, 20 January 1893.
[815] *The Sydney Morning Herald*, 6 February 1893.

Following the launch of the *Albert Maclaren*, Reeks' yard set to work on its next major project, laying the keel of a 100-ton schooner to be used as a store ship on the Torres Strait pearling grounds.[816] Yachting was also back on his agenda with Reeks helming *Thelma* at the Anniversary Regatta in a race that solely featured yachts he had designed. *The Australian Star* reported the event as '*without doubt the race of the day, the big yachts looking at their best in their new Lapthorne suits, with immense jackyarders* [obviously the economic downturn had not yet affected some of Sydney's yachtsmen]. *Volunteer and Thelma had the best of it during the early portion of the race, but when the yachts got outside Era's power and weight brought her to the front, and she finished with a lead of 9½min. from Volunteer, and 10½min. from Thelma. On time allowance, however, she resigned first place to Thelma. The Era was sailed by Mr. P. O. Williams, the Volunteer by Mr. A. Minnett, and the Thelma by Mr. W. Reeks.*'[817] This proved the first of several races that Reeks successfully handled *Thelma* in that season.

THELMA SAILING ON SYDNEY HARBOUR.
SOURCE: AUSTRALIAN NATIONAL MARITIME MUSEUM COLLECTION (OBJECT NO. 00012204).

On 5 April 1893 the 102-ton topsail schooner *Aladdin* was launched from Reeks' shipyard.[818] The vessel measured 90ft 7in (overall) by 23ft 3in, with a depth of 8ft 4in.[819] Built to the order of Messrs Scott, Henderson, and Co. of Sydney, and intended as a pearling mother-ship,

[816] *The Sydney Morning Herald*, 20 January 1893.
[817] *The Australian Star*, 27 January 1893.
[818] *The Sydney Morning Herald*, 6 April 1893.
[819] *The Daily Telegraph*, 5 May 1893.

the vessel was registered to George Littlejohn and left for Torres Strait a few weeks later.[820] Making the trip between Sydney and Thursday Island in 14 days, the master of the craft subsequently wired Reeks stating that the *Aladdin* behaved splendidly at sea, and that its new owner was very pleased with the craft.[821]

Unlike the new direction being taken by the vessels being used in pearl shelling, the mother-ship schooners Reeks designed and built (i.e., *Admiral* and *Aladdin*) were quite conventional, with only their layout optimised for this specific purpose. They were, however, built on the precipice of a time when they would be considered a relic of the traditional tall ships era. Steam vessels were already in wide use, but a sail-powered craft was still quite appropriate for the duties and requirements the task of a mother-ship was to undertake. Indeed, standing off and perhaps drifting with the wind for long periods of time was a typical pattern of life for these craft. Sail power made practical and economic sense. Adjusting the rig plan to suit the weather and using an inexhaustible source of power was much more economical than running a steam engine, burning coal supplies and using up precious stored water.

Reeks, as a naval architect and builder, was in an interesting position, forging a career by employing the latest methods and science to his designs, but wise enough to see that a 'back to the future' approach was what was required for the pearling industry's mother-ships. Luckily, he already had the training to design a sailing ship. He had, after all, apparently worked on the design of the tall ship *Lancaster Castle* that brought him to Australia in the first place.

Towards the end of April 1893, there was more positive business news for Reeks. The Balmain New Ferry Company had accepted his tender for the construction of another ferry steamer.[822] The cost was stated to be

[820] Register of Australian and New Zealand Shipping 1893-1894.
[821] *The Sydney Morning Herald*, 2 June 1893.
[822] *The Daily Telegraph*, 29 April 1893.

ALADDIN DURING A VISIT TO HOBART (CIRCA 1896).
SOURCE: MARITIME MUSEUM OF TASMANIA (P_OM_K_32A).

£4,250.[823] For reasons unknown the build was outsourced to Messrs Young and Scott at Balmain, instead of taking place at Reeks' own Berrys Bay shipyard.

Shortly thereafter a new 35-ton centreboard ketch designed by Reeks with a length of 58ft, width of 17ft 10in and depth of 6ft 4in was launched from the yard of Rock Davis, well-known shipbuilder of Brisbane Water on the New South Wales central coast.[824] Named *Pilot*, the craft was intended for the pearling industry in Torres Strait under management of its owner James Clark with a view of carrying live pearl shells in a first attempt to culture the shell in New Guinea waters. For this purpose *Pilot* had a well fitted, as well as incorporated two bulkheads and two sea cocks to give a constant flow of water to the well.[825] The craft was notably recorded as being the first vessel to successfully carry live shell to a lease.

On 21 October 1893 the ketch *Wallaroo* was launched from Reeks' Berrys Bay yard.[826] Built to the order of J. Talman for employment in the Torres Strait pearling industry, the vessel measured 53ft (overall), with a beam of 13ft and a draft of 7ft 6in.[827] With a long clipper bow and gracefully moulded, the vessel was described as *'probably the highest class sheller ever built'*.[828] The *Wallaroo* travelled to Thursday Island under the command of Captain S. Denton early the following month.[829] The ketch was notably the last vessel built at Reeks' Berrys Bay shipyard. He appears to have given up this location and type of work entirely, opting to return to naval architecture as his main source of income, along with other vocations and new enterprises.

One of the primary reasons for the closure of Reeks' shipyard, apart from the dismal state of the economy and its negative impact on the market for new commercial and recreational vessels, may have been related to personnel. For example, after being involved with the construction of so many vessels designed by Reeks since 1889, it is interesting to note the absence of Tom Cubitt from descriptions regarding the launch of the *Albert Maclaren* in January of 1893 and all of the vessels built by the yard after this point in time. Considering he was the yard's foreman and his workmanship was deemed of immense quality in numerous press reports, Cubitt's departure would have hindered the yard's productivity significantly.

Born in Sydney around 1850, Cubitt appears to have served the majority of his early career working for W. M. 'Watty' Ford at Berrys Bay, eventually becoming foreman. In 1889 he established his own boat yard, next door to that of his former employer. It was operating out of this same location that in mid-1891 Cubitt was to become foreman of Reeks' shipyard, though for the remainder of this year and part-way into the next, he seems to have also built his own vessels. These included the pearling lugger *Congho,* launched in September of 1891; the pearling lugger *Sedney,* launched in November of 1891 and built to the order of N. G. Anderson to a design by Captain Grainger; a steam launch, the *Langham*, built to the order of the Wesleyan Missionary Society for use in Fiji to a design also by Captain Grainger and launched in December of 1891; and the pearling lugger *Heckla*, built to the order of N. G. Anderson to a design by Captain Grainger and launched in February of 1892.[830]

After being strongly associated with the launch of a multitude of vessels from Reeks' shipyard in 1892, including the *Admiral, Peuleule, The Lady Mary, Jessie* and *Isabel*, Cubitt seems to disappear from the upper ranks of the boat building trade altogether. There are no vessels attributed to being built by him at any location in and around Sydney from this year forward, though postal directories still note that he was a resident of 31 Thomas

823 *The Daily Telegraph*, 21 July 1893.
824 *The Daily Telegraph*, 6 May 1893.
825 *The Daily Telegraph*, 6 May 1893.
826 *The Australian Star*, 24 October 1893.
827 *The Australian Star*, 24 October 1893.
828 *The Australian Star*, 24 October 1893.
829 *The Daily Telegraph*, 8 December 1893.

830 *The Daily Telegraph*, 17 September 1891; *The Sydney Morning Herald*, 17 November, 5 December 1891.

PILOT.
SOURCE: QUEENSLAND HARBOURS AND MARINE.

WALLAROO.
SOURCE: STEPHEN MULLINS COLLECTION.

Street, North Sydney, throughout this decade and up until his death in 1922.[831]

There was no choice for Cubitt not to work, unless forced not to do so by illness, injury or perhaps a falling out. He had a large family to support with ten children born to him and his wife Annie between 1881 and 1899.[832]

After such an illustrious start, including the celebrated construction of the 2½ raters and the early pearling ketches, it is a rather strange occurrence for Cubitt not to continue on this path. Family records from 1893 through to the 1920s all indicate that he remained employed in some capacity as a 'boat builder' or 'shipwright'.[833] Perhaps he made his way to one of the larger boat or shipbuilding firms, looking for greater stability and security in a time when financial risk taking was out of the question. He may have also wanted to lessen his responsibility with regards to financial and administrative tasks and the burden of a lease for the yard, opting instead to work for a larger company as opposed to himself or another individual. What is clear, however, is that Cubitt ceased working in any sort of upper management or managerial role in the boat building industry and did not became further involved in any of Reeks' later vessels.

With the unexplained departure of Cubitt, Captain Grainger appears to have become more and more involved in Reeks' shipyard activities. In January of 1893 he was noted as superintending the launch of the mission schooner *Albert Maclaren.* He was later noted as helping with the design of the *Pilot.*[834]

Thomas Lewthwaite Grainger (1850 - 1928) is another interesting player in the Sydney boat building scene beginning in the early 1890s, and for a few decades following, as are his associations with both Reeks and Cubitt. He had migrated to Australia in the late 1880s after previously being involved in intercolonial trade, including helming several ships travelling between London and Australia.[835] After a short stint in Queensland, working for Burns Philp and Co. as a marine superintendent out of Townsville, Grainger moved to Sydney, being permitted by the Marine Board of New South Wales to become a marine surveyor in 1889.[836]

[831] Sands Directories: Sydney and New South Wales, Australia, 1858 - 1933; *The Sydney Morning Herald*, 7 October 1922.
[832] Sydney, Australia, Anglican Parish Registers, 1814 - 2011 for Cubitt.
[833] Sydney, Australia, Anglican Parish Registers, 1814 - 2011 for Cubitt.

[834] *The Daily Telegraph*, 30 May 1883.
[835] *The Telegraph*, 28 October 1881, *The Brisbane Courier*, 30 August 1882; *Daily Commercial News and Shipping List*, 7 January 1929.
[836] *The Telegraph*, 29 March 1886; *The Queenslander*, 10 April 1886; *The Sydney Morning Herald*, 31 July 1889; *The Australian Star*, 31 July 1889.

Shortly thereafter Captain Grainger opened an office in Bridge Street, offering his services in marine surveying and compass adjusting.[837] He also began to design and superintend the building of vessels. For example, in 1891 Grainger superintended the construction of several vessels that he had designed, including the London Missionary Society's 50ft lugger *Niue*, built by Ford at Berrys Bay; the Australiasian Wesleyan Missionary Society ketch *Dove*, also built by Ford; the Torres Strait pearling lugger *Francisco*, built by David Drake at Balmain for Burns Philp & Co.; the pearling lugger *Congho,* launched from Cubitt's yard and also built for Burns Philp & Co.; the pearling luggers *Mabel* and *Flora*, both built by Ford to the order of T. G. Kelly; the pearling lugger *Sedney,* built to the order of N. G. Anderson by Cubitt; and finally a steam launch named *Langham*, built to the order of the Wesleyan Missionary Society by Cubitt for use in Fijian waters.[838]

1892 was another busy year for Captain Grainger. He designed and/or supervised the building of the 49ft pearling lugger *Heckla*, built by Cubitt to the order of N. G. Anderson; the pearling ketch *Nina,* built by Ford to the order of James Clark; the pearl sheller *Vida*, also built by Ford to the order of Clark; the 14-ton pearling ketch *Boomerang*, built by Ford for Burns Philp & Co.; the 11-ton pearling ketch *Gipsey,* built at Reeks' yard for Burns Philp & Co.; the 12-ton pearl luggers *Victoria* and *Alice*, built by Ford to the order of William Brown and Burns Philp & Co., respectively; and the 15-ton Torres Strait pearling ketch *Ehime,* built by Reeks to the order of Burns Philp & Co., and its sister-ship, *Takachiho,* built for the same firm.[839]

The following year, in addition to assisting with construction of the *Albert Maclaren* and *Pilot* at Reeks' yard, 1893 saw Captain Grainger supervise the building of the *Assan Ceylon*, a pearl sheller designed and built by A. Fisher of Balmain to the order of H. Bowden of Thursday Island.[840] In 1894 he supervised the building of the 20-ton ketch *Siai,* built by Rock Davis at Brisbane Water to the order of the Government of New Guinea.[841]

Indicative of the state of the economy, there were no new vessels built and/or designed by Captain Grainger after this point in time for a period of nearly five years. He remained active in marine survey and insurance work involving salvage operations, wrecks and losses, and the adjustment of compasses. In mid-1894 Grainger also chartered a vessel on a 12-month contract to bring Polynesian labour to Queensland to work in the sugar industry.[842]

New builds returned to his agenda late in the 1890s. In 1898, for example, Captain Grainger supervised the construction of a 130-ton brigatine *Wanetta* for James Clark, built by David Drake.[843] In 1899 he was noted as supervising the construction of two 12-ton pearling luggers, built by Messrs Fisher and Sons and Ford, respectively, to the order of Burns Philp & Co. for the Torres Strait pearling industry, as well as the 108ft schooner *Nugarea*, built by David Drake for the Pacific Island trade.[844]

The 1900s saw Grainger supervise the construction of a handful of schooners and motor and oil launches, including some which he designed. Several were built for use in the Pacific Islands or for customers interstate. Amongst them were the 54ft launch *Austrea,* built by William Holmes in 1901 for use by the Brisbane Port Office, and the auxiliary schooner *Malekula* which Grainger designed to the order of Burns Philp & Co. for use in the New Hebrides (now Vanuatu) island trade, built by David Drake at Balmain in

[837] *Daily Commercial News*, 9 June 1891.
[838] *The Brisbane Courier*, 27 January 1891; *The Sydney Morning Herald*, 19 June, 7 October 1891; *The Daily Telegraph*, 22 August, 17 August, 28 October 1891.
[839] *The Telegraph*, 5 November 1891; *The Sydney Morning Herald*, 4 February, 23 September, 13 October 1892; *The Daily Telegraph*, 16, 29 July 1892; *Daily Commercial News and Shipping List*, September 1892; *Newcastle Morning Herald and Miners' Advocate*, 28 November 1892; *The Brisbane Courier*, 6 December 1892; Register of Australian and New Zealand Shipping.
[840] *The Daily Telegraph*, 30 May 1893.
[841] *Evening News*, 15 November 1894.
[842] *The Sydney Morning Herald*, 17 August 1894.
[843] *The Daily Telegraph*, 27 September 1898.
[844] *The Daily Telegraph*, 21 April 1899; *The Sydney Morning Herald*, 14 August 1899.

DIVING FOR PEARLS OFF THE ASSAN CEYLON, THURSDAY ISLAND.
SOURCE: STATE LIBRARY OF VICTORIA (ID 1715968).

SCHOONER WANETTA AT THURSDAY ISLAND (1919).
SOURCE: UNIVERSITY OF QUEENSLAND (FRYER LIBRARY PHOTOGRAPH COLLECTION, UQFL477).

1903 and also under his supervision.[845] Captain Grainger died at Danneverke, New Zealand, in 1928 after retiring to this location a few years earlier.[846]

—

While vessels for the pearling industry had been keeping Reeks (and several of his colleagues, including Captain Grainger) with ongoing work during 1893, Reeks was also advancing his ideas with designs for ferries and other steam-powered vessels. In late July of 1893 he applied for a patent with the New South Wales Patent Office titled, '*Improvements in design and other ferry and steamboats*'.[847] For reasons unknown, however, the application was refused and no patent issued.[848]

The opening of the yachting season in October of 1893 again saw Reeks in attendance.[849] He did not take the helm of any yachts this season, however. There were other diversions to consume his time. On 4 December he arrived in the Manning River, in the New South Wales Tablelands region, having travelled from Sydney as a passenger on board the N.C.S.N. Company's steamer *Electra*.[850] There is no record of when he returned to Sydney, though he must have been home in time to spend the Christmas holidays with his family. Just what was Reeks doing travelling to this area?

[845] *The Telegraph*, 17 May 1901, 29 June 1903.
[846] *The Sydney Morning Herald*, 5 January 1929.
[847] *New South Wales Government Gazette*, 9 August 1893.
[848] *New South Wales Government Gazette*, 10 November 1893.
[849] *The Daily Telegraph*, 16 October 1893.
[850] *The Australian Star*, 4 December 1893.

With the Balmain New Ferry Company's new steamer well underway, and with little prospect for new vessel commissions given the disastrous state of the economy, there seems to be no commercial vessel design reasons for him to make a trip to the Manning River District. However, this short interlude could have been because of another plausible motive. Switching paths, this may have been the first stage of Reeks embarking on another engineering-related project, this time an expedition in search of gold.

Gold had been taken from Cells Creek and Mummel River in the upper Manning River since 1887. However, accessibility to these areas via the upper Manning River may have been an issue he discovered. When Reeks eventually made a move into the gold fields, it was much further north. Though before we discuss his time prospecting, we must return to Sydney for the launch of two new vessels.

—

The latest addition to the Balmain New Ferry Company's fleet, the steamer *Lady Manning* was launched on 10 February 1894, built at the shipyard of Messrs Young and Scott at Balmain and under the superintendence of Reeks.[851] A double-ended screw boat, the craft measured 117ft with a beam of 27ft and a depth of 8ft, i.e., more than 20ft longer and proportionally larger than *The Lady Mary* and *Lady Napier.* Named for the Mayor of Sydney, Sir William Manning's wife, the *Lady Manning* could accommodate 500 passengers, considerably more than its

[851] *The Sydney Morning Herald*, 10 February 1894.

LADY MANNING.
SOURCE: GRAEME ANDREWS WORKING HARBOUR PHOTOGRAPH COLLECTION, CITY OF SYDNEY ARCHIVES.

two predecessors.[852] Construction comprised a hardwood keel and timbers (bent timber), solid frame every 5ft apart, Oregon planking, hardwood top strake and garboard strake, deck frames of hardwood and decks of kauri pine.[853] At the time of launch, the vessel's engines, supplied by Messrs Ross and Duncan of Glasgow, were yet to arrive though were soon expected.[854]

The launch of the *Lady Manning* is an appropriate time to stop and delve further into Reeks' design process and thinking, in part using his own words. But first, some context.

The growth of Sydney and the expansion of ferries on Sydney Harbour and the Parramatta River were intertwined: the evolution of the vessels went hand-in-hand with the establishment and growth of suburbs around the foreshore, from Parramatta to Manly. Reeks was becoming involved in a very significant area, an enterprise that became one of the largest movements of vessels and passengers in the world. His input and designs were important to the industry's development in the 1890s, a period of quite rapid expansion because established localities were growing and the new screw propulsion promised faster and more economical services. There was a market for new ferries and with that the opportunity for new designs.

Reeks did not create the double-ended configuration; it was already well established as the appropriate shape for the side-paddle ferries then in use. These vessels could negotiate wharves in bays and other areas where it was difficult and time consuming to turn around. After pulling into the wharf, they simply reversed direction when it was time to leave. Not having to turn around also paid dividends in the busy central hub of Circular Quay and Darling Harbour wharves.

The first screw-propelled ferry *Wallaby*, using a propeller at either end, had been introduced with some scepticism in Sydney in 1878. Initially there were people who positively refused to go on it, and would on no account let their families do so either. But the mood eventually calmed. *Wallaby,* designed in collaboration by the builder William Dunn and civil engineer Norman Selfe, was a success, achieving over 9 knots, and remained in service for some decades. The next double-ender *Alathea*, built at Brisbane Water in 1881 was a *'rank failure',* only able to do 6 knots, until trialled with one propeller only when it managed over 9 knots.[855] *Kangaroo* followed a decade later, designed for 12 knots but only reaching 10.

WALLABY.
SOURCE: MORT FAMILY PHOTOGRAPHS, MITCHELL LIBRARY, STATE LIBRARY OF NEW SOUTH WALES (9YMDWZP9).

[852] *The Sydney Morning Herald*, 10 February 1894.
[853] *The Sydney Morning Herald*, 10 February 1894.
[854] *The Sydney Morning Herald*, 10 February 1894.
[855] *The Daily Telegraph*, 20 October 1894.

In 1891 the Balmain New Ferry Company came into being and, as noted earlier, Reeks successfully tendered to design and build one of their first two ferries, *The Lady Mary*. George Ellis built the other, *Lady Napier*, to his own design. Reeks' design for *The Lady Mary* proved a practical success, whereas *Lady Napier* apparently did not reach the same standards.

It is fortunate that ferry design is an area where there is good documentation to show how Reeks approached the problems of designing these types of vessels and, in hindsight, is probably the best available account that reveals his methodology. He presented two papers to the Engineering Society of New South Wales on his double-ended ferry designs: one in 1894 and the other in 1906.

The first paper was titled simply, '*Notes on the Design of Double Ended Screw Ferry Boats*' and was presented on 9 August 1894. In this paper, after a discussion on his first design, *The Lady Mary*, Reeks introduces his second design, *Lady Manning*, and then summarises the essential requirements he had to meet, i.e., '*Speed, size and price*'. He goes on to state, '*The author found, as we are constantly finding, that these three elements ... do not seem to agree at all*'. Looking back, these elements and their disagreement probably applied to all of his ferry design projects.

Speed, as highlighted, was a major obsession of the early designs, and seemed to remain the primary preoccupation of the client and thus the designer. But the potential to improve speed was constrained by the desire to increase the size of the vessel within limitations, including retaining a shallow draft, and the company's budget that only allowed them to pay for an engine and associated machinery of a certain size. '*They figured on the biggest power they could afford to pay for*'.

Speed is permanently tied to resistance, i.e., the drag that is holding the vessel back. Resistance rises as speed increases to a point of maximum speed where the full power available matches the increasing resistance at that point, and the vessel can go no faster. This drag is coming from the skin friction of the hull's surface area, the wave-making from the shape of the hull as it passes through the water, and then the drag of the propellers, one at either end.

A basic rule to start with is the longer a vessel is on the waterline, the higher the speed it can achieve. Length, however, was usually constrained by practical requirements reached from experience, and the maximum overall length was written into the contract. It was not open to the designer to create his ideal shape and length, so he had to make sure the length was used to the maximum efficiency.

Reeks had not been successful at two previous opportunities to tender for a design with other ferry companies in Sydney. *The Lady Mary* was the first opportunity he had to put his ideas into practice. The Balmain New Ferry Company wanted 10.5 knots, with 96ft in length and could afford a two-cylinder steam engine, along with sundry other details. To achieve that speed on that length and with that power limit, reductions in drag, wherever possible, were required. From observations of the existing double-enders Reeks felt that the '*column of water off the bow screw impinging on the bow caused excessive wash there, and seemed to completely disarrange the natural flow of the bow wave*'. He developed a two-fold plan. First he '*turned up the ends of the hull*'. Second he cut away as much of the deadwood as was practical. Both features were intended to get the water to '*flow under the bow instead of to either side of it*'. It seemed Reeks felt the propeller was pushing the water away instead of letting it flow under and around the hull.

Built as described, *The Lady Mary* achieved 10.25 knots. The vessel was deemed an outright success as it made up for time lost through the one-quarter of a knot shortfall because the vessel's '*starting and stopping was done exceptionally quickly, making up for the less actual speed*'.

Unfortunately no information exists on the hulls' cross sections, only a profile view of the hull, which shows the keel line was curved upwards to the stem, and the long deadwood keel on the centreline had a substantial aperture cut into it, with the propeller supported on a substantial post rather than projecting off the face of the deadwood or end of the hull. This arrangement was considered noteworthy, as reported at *The Lady Mary*'s launching.

'*The rudders are of the kind known as balanced, viz., part of the blade being on one side and part on the other, which makes the steering easier and quicker than with the old-fashioned fixed rudders. Several noticeable improvements on the usual class of ferry boats have been introduced by Mr. Reeks in the vessel, not the least of which is with regard to the fitting of the propellers, the vessel being driven by double screws, one at each end. The deadwood is cut entirely away, leaving an aperture of some seven or eight feet, in which the shaft is exposed to the water, and this, together with the peculiar "cutting up" of the vessel at both ends is probably the greatest departure from the conventional ferry boat.*'[856]

This design feature replaced what was seen on existing ferries, such as the deeper hull on *Wallaby* that rose steeply to the ends. Reeks must have considered that this deeper and somewhat bluff hull of the other designs pushed the waves away whereas his gently rising hull at the keel on the centreline and the large cut out aperture allowed the waves to flow back under and along the hull. He felt vindicated as he notes, '*this turning up of the ends with a view to avoiding one of the drawbacks of this particular method of driving power proved a success to a limited extent; the wash was reduced considerably, but still not got rid of by any means*'.

Reeks did not have to wait long for another opportunity to advance his ideas. His design for *The Lady Mary* had met the company's requirements, while *Lady Napier*, from other sources, had not. Reeks became an obvious choice for the Balmain New Ferry Company's third ferry, *Lady Manning*, launched in February of 1894. On this design he took a bold step and created an unusual if not unique shape to the hull profile of a powered vessel. In his paper, presented later that year, it is perhaps an awkward

[856] *Evening News*, 1 August 1892.

description, but his words should be stated first before clarifying the intentions.

'*The next direction the Lady Mary seemed right viz. to reduce to a minimum the obstruction to the water under the boat's bow, and though almost certainly there is no actual inducement for the water to draw under, the absence of an obstruction can be called inducement for the sake of the short term. ... he* [Reeks] *got the idea of turning the keel from end to end.*' It was noted at the launch as something to highlight. '*A feature in the construction from a naval architecture point of view is that she has been given what is known as a "sprung" keel, or camber in the keel.*'[857]

This '*turning up*' or '*camber*' resolved itself simply and elegantly. The entire keel line was one long gentle curve, almost a radius or segment of a circle. While it looked exceptionally beautiful, it had to work as well. To help with that Reeks also tackled another issue he had not fully resolved on *The Lady Mary*.

The design of this particular vessel involved the waterline being as long as the hull and, with the rudder placed under the end of the hull, the propeller was then set back some feet from it. Reeks continued, '*therefore, the wash from it comes into contact not with the actual entrance but on to a comparatively bluff part, and he* [Reeks] *concludes from that that it would always be better to keep the propeller outside the end of the load waterline, even at the expense of a little overall length to protect the rudders*'.

Thus Reeks '*was firmly convinced from the results of the Lady Mary that the outside effective length of the load waterline was from propeller to propeller, that portion of the entrance beyond the propeller being useless so far as opening up the bow wave was concerned, and as he was anxious to get rid of that heavy ugly wash off the fore end, he adopted the course shown on the drawings and kept the propeller 2ft 7 inches outside the normal load waterline, taking care however to run the boat out in a sort of snout bow, so as to get all the length possible when under way*'. As a result, the *Lady Manning*'s rudder was placed projecting beyond the end of the hull, forgoing some hull length to meet the company's overall hull length limit, and the propeller, supported by a strong bracket, turned in a plane located just forward of where the hull met the water.

The curved or cambered keel, as it was described, plus the decision to forgo some length and ensure the propeller was at the entrance to the hull were the two key features Reeks employed in the *Lady Manning;* the former to further create a flow of water around the hull rather than being pushed away, the latter to get the effective length (from his perspective) as long as possible within the parameters he was set. Two significant theories being played out in the one design was a strong act of self-confidence in the face of the risks to his career if they failed. But as Reeks wrote some four months after the vessel's launch, '*It seems to act well, for the long, clean bow wave off the Lady Manning is perhaps one of her leading characteristics, and certainly one reason why she steams as well as she does*'.

[857] *The Sydney Morning Herald*, 10 February 1894.

In the same paper presented to the Engineering Society of New South Wales in August of 1894 Reeks returns to a discussion of the hull design with almost two pages of analysis about the obvious problem that the bow becomes the stern and vice versa, and that ordinarily the aft end of a vessel should be fuller than the forward end, noting at one point, '*the enormous importance of designing the stern of such length and shape as to coincide in period with the velocity of the water due to atmospheric pressure and gravity fixed by Nature's laws, and unalterable, for neglect of that will almost assuredly result in towing big ugly seas, lifting large volumes of water above their natural height and so wickedly wasting power that might be made good use of*'. With the problem of stern drag thus described colourfully, he clarifies that the hull is only going at a displacement speed and not the higher speed of a '*torpedo boat*' with its fuller stern and bigger wash. Reeks eventually resolves that the bow of a double-ended hull throws more wash than the stern, whereas single screw vessels with a fuller stern and finer bow have an equal wash or wave at either end, meaning equal resistance. This implies then that one needs to design the ends as a finer bow rather than a fuller stern. This he did, '*fully convinced, from observing other craft, that even such apparent undue fining would not detract from the efficiency of the stern, and so it has turned out in practice*'.

What we do know, from the hull sections of the *Lady Manning*, is that Reeks designed a gently rounded vee-shaped almost bilgeless hull form with a constant deadrise angle through the middle of the hull giving way to an ever so gradual steepening as it approaches the ends. In some ways it is reminiscent of the hull of the double-ended yacht *Australian*, designed by Richard Hayes Harnett (Sr), which he described favourably in 1888 as '*almost a perfect vessel*'.[858]

Theory for the hull shape of *Lady Manning* was matched with attention to engineering and construction processes. The paper describes some of the specific detail Reeks put into supporting the propeller post on *The Lady Mary*, but he also identified a key overall issue, '*the question of keeping her shape. In more cases than one, boats of this class had been known to hump in the middle*'. To make sure this did not happen he combined a keel of 20 inches, floors of six inches and keelson of 6 inches, '*which when bolted together, made a girder of 30 inches deep, and the result was that the deflection from a straight line was certainly not more than 1/8th inch, and when the machinery was in she came back to the straight*'.

This suggests that *The Lady Mary* was built with a slight convex curve in the first place. Too much deflection and the long shaft alignment was jeopardised and bearings or bushes would wear quickly, amongst other problems. On *Lady Manning* the curved keel was solved by making it in four laminations bolted together, while the tricky detail Reeks had created, passing the shaft through the keel, was solved with '*a box rather than a chock*'. Additional supporting timbers were also described in detail, '*thus getting a complete connection everywhere ... it turned out a real good, strong piece*

[858] *Illustrated Sydney News*, 31 May 1888.

of work'. More detail was lavished on the rudders and how their stocks were supported, and the use of rudder stops with springs as a cushion in case the forward rudder suddenly swung free.

Reeks' 1894 paper and responses ran into many pages, but the majority of the words came in the discussion afterwards, where a number of those present to hear the paper seemed to push their own barrow rather than respond to the dissertation more specifically. However, we do learn one thing. Reeks had a preference for the earlier theories of J. Scott Russell and Colin Archer, rather than the more recent published tables and data including William Froude's work, and all in use in England at the time. Once more, with a colourful statement, Reeks responds to the engineer Mr Cruickshank's long treatise about using methodical trials to establish data: '*as to those English shipbuilders and engineers so well up in modern practice Mr Cruickshank's imagination has put before us, note book in hand ... we should give them just a little slack, and then in a casual way mention we were followers of the deeply-condemned Scott Russell and Colin Archer ... and they might then, be opined, go home and say "we have been to the bush, and find men producing successful boats on the very system we say is wrong. Perhaps there is something successful in poor old Scott Russell after all; let us look him up"*'.

As we close this chapter and head to the next, we must provide a short description of the other vessel designed by Reeks and launched on 17 February 1894; a small half-rater yacht built to the order of W. L. Hunt by George Barnett of Drummoyne.[859] With dimensions of 20ft (overall), 17ft (waterline), with a beam of 6ft 8in and a depth of 1ft 1in, the centreboard craft was constructed of cedar, with a keel of spotted gum and hardwood frames.[860] Noted to have a rounded bow with a tuck stern, it was launched as *Taipo*.[861]

The 'Mosquito' yacht fleet of Sydney Harbour were continuing to gain momentum, albeit not as quickly as the boat designers and builders needed. Times were generally quite sombre for this industry, recreation was taking a backseat while the worldwide economy endured its fits of bumps and losses. Still, *Taipo* won its first race, in an exciting contest, and the new class of craft was being shown great interest, combined with a little intrigue, from Port Jackson's yachtsmen.[862]

[859] *The Sydney Morning Herald*, 16 February 1894.
[860] *The Sydney Morning Herald*, 16 February 1894.
[861] *The Sydney Morning Herald*, 16 February 1894.
[862] *The Sydney Morning Herald*, 12 March 1894.

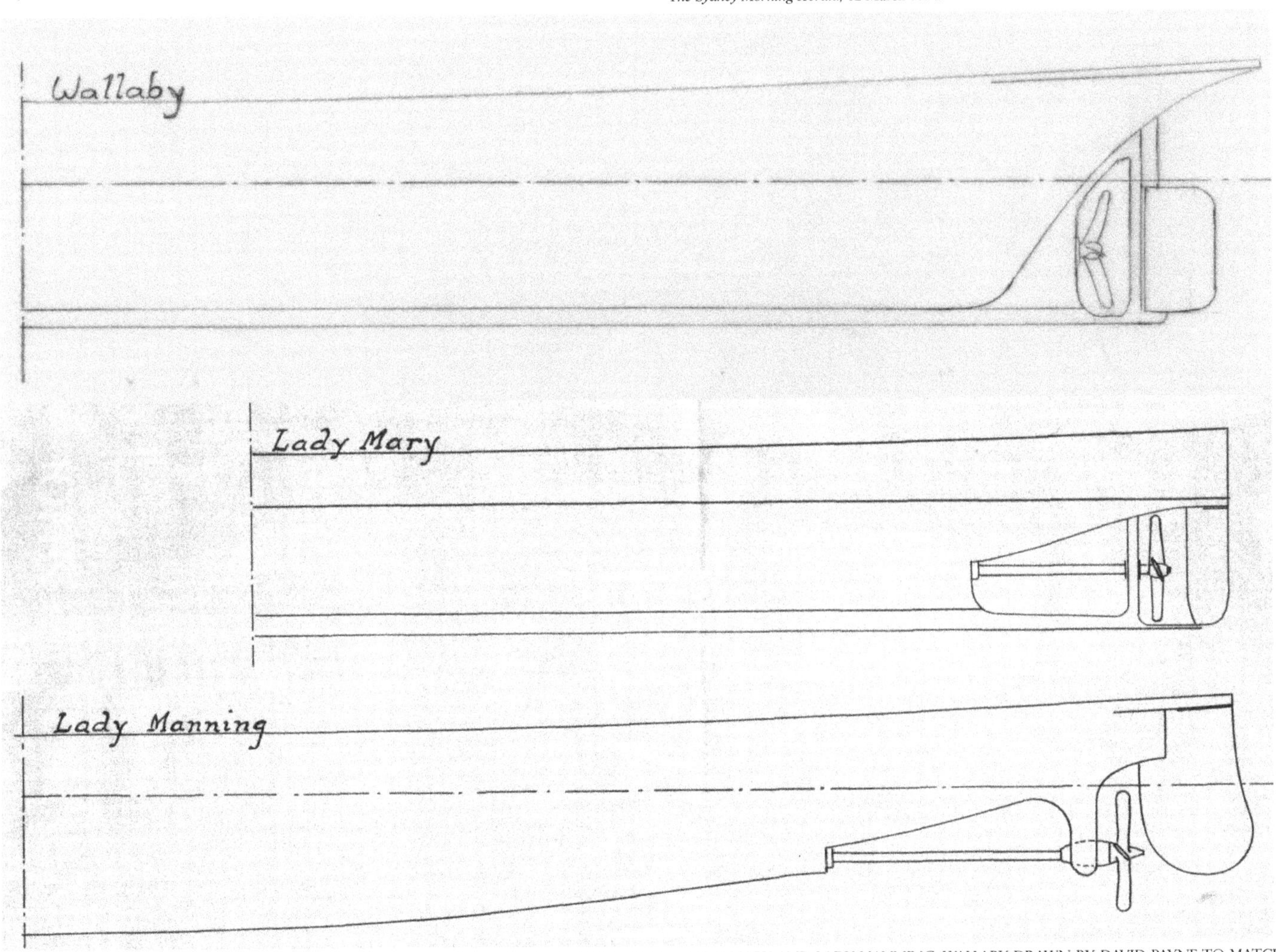

HULL SHAPES OF WALLABY, THE LADY MARY AND LADY MANNING. WALLABY DRAWN BY DAVID PAYNE TO MATCH THE DRAWING STYLE OF THE OTHER FERRIES, AS PREPARED BY WALTER REEKS FOR PUBLICATION.

PARTIAL MAP OF NEW SOUTH WALES SHOWING LOCATION OF TIMBARRA RIVER (1894).
SOURCE: STATE LIBRARY OF NEW SOUTH WALES (Z/CE 89/21).

CHAPTER ELEVEN - TIMBARRA RIVER

With two steam ferries completed, and both proving a success, Walter Reeks' opinion and expertise became more and more sought after with regard to this mode of passenger transport. For example, on 10 March 1894 he was invited to attend the inaugural meeting of the Lane Cove Co-operative Ferry Company, part of an influential collective intending to establish a new ferry company. The favourable outcome of the Balmain New Ferry Company was motivating members of other waterside communities situated around Sydney Harbour, particularly in the North Shore area and very much reliant on cheap and reliable water-based transport for their daily commutes, to band together and orchestrate positive change. Chaired by Mr T. Eardley, with those present including Captain Napier, S. German, as well as Reeks, the meeting concluded that there was no doubt as to the success of such a venture.[863] A second meeting held the following month, attended by more than 300 residents of Lane Cove, confirmed the formal establishment of Lane Cove Co-operative Ferry Company and the need to raise £10,000 through the release of 2,000 shares. It was then announced that more than 1,000 shares had already been applied for.[864]

However, instead of building new vessels to service the Lane Cove route, the company secured two existing vessels, the *Telephone* and *Thistle*, both commencing service in August of 1894.[865] While Reeks had initially been an interested party in the establishment of the company, his involvement lessened when it was decided to purchase second-hand vessels. There was very little need for his services as a naval architect given both craft proved adequate to the company's requirements. Still, the formation of co-operative ferry companies in Sydney in the mid-1890s, generally building on momentum from somewhat inconvenienced residents, provided Reeks with a new avenue of opportunity for him professionally, something that he would build on in the years to come.

The Manly Co-operative Steam Ferry Company, established in late 1893 under the same formula as the two companies just discussed, proved another worthwhile customer for Reeks. In April of 1894 he was noted as having superintended alterations to this organisation's recently-acquired steamer *Cygnet*. Built in Glasgow in 1885, the 124-ton steel screw vessel had been transferred to Sydney from Hobart where it had been employed in the Huon and Channel passenger trade for a number of years.[866] Measuring 120ft with a beam of 17ft 3in and a depth of 8ft 7in and originally rigged as a fore-and-aft schooner, the *Cygnet*'s masts had been removed and its promenade, bridge deck and saloons extensively re-worked.[867] The craft's engines had also been reconditioned. With Reeks on board, *Cygnet* took its first trial trip on Sydney Harbour on 6 April 1894.[868]

In terms of recreation, during this period the small rater yachts were continuing to dominate the waters of Sydney Harbour. Reeks was back competing too, this time taking the helm of *Bul Bul* in local races and regattas.[869]

Other opportunities came from further afield and Reeks pounced on them, possibly a combination of the need to find work and also his interest and intrigue. On 9 June 1894 the *Clarence and Richmond Examiner* announced that Rock

863 *The Australian Star*, 10 March 1894.
864 *The Sydney Morning Herald*, 4 April 1894.
865 *The Australian Star*, 1 August 1894.
866 *The Australian Star*, 7 April 1894.
867 *The Australian Star*, 7 April 1894.
868 *The Australian Star*, 7 April 1894.
869 *The Daily Telegraph*, 31 March 1894.

CYGNET AT HOBART (EARLY 1890S).
SOURCE: MARITIME MUSEUM OF TASMANIA (P_OM_C_13D).

Davis of Brisbane Water had successfully tendered for the construction of a new steamer for Messrs Henderson and Mansfield for use on the Clarence River.[870] Designed by Reeks, upon hearing of the news that the tender had been awarded, he near-immediately travelled to the Northern Rivers region of New South Wales to discuss specifics of the vessel with the clients and builder, arriving in Grafton per the *Kallatina* from Sydney on 11 June; a passage of over 350 nautical miles.[871] Reeks returned to Sydney by the same vessel just over two weeks later.[872] Given the amount of time he spent in the region, one can assume that he also ventured further afield, specifically to the Timbarra River area.

Gold! Since its initial discovery in New South Wales and Victoria in the early 1850s, the highly-prized metal had led to an influx of free settlers to Australia and the development of a robust and sometimes profitable industry, with those involved ranging from single prospectors to large mining companies. While early gold rushes had initially centred on Ballarat, Castlemaine and Bendigo in Victoria, and Orange and Bathurst in New South Wales, by the 1890s fields had opened up across much of the country, including in Western Australia, the Northern Territory, Queensland and Tasmania.

Situated at the head of McLeod's Creek, some 80 miles north-west of Grafton and 20 miles to the east of Tenterfield, gold had first been identified in the Timbarra River region of northern New South Wales in the mid-1850s, in an area that since European settlement during the decade before had primarily been used for sheep and cattle grazing. The region's Rocky River was the location of the first gold rush in 1857, bringing hundreds of parties to the area packed with provisions and claims to stake.[873] Over time other nearby rivers and creeks, many of which formed tributaries to the vast Clarence River system, were found to contain gold deposits. However, these waterways were often in flood during the winter months with heights too high for continuous mining operations, with the opposite occurring in the summer months. These fluctuations deterred many potential prospectors from investing in the area on a more permanent basis.

Even in its early establishment as a goldfield, the Timbarra River region was best accessed from Sydney via a steamer to Grafton. From here travellers had a choice of several routes, most opting to take a newly-built main road to Tenterfield, a rather circuitous passage but one that was more manageable and best suited for drays.[874] Still, it was rather a tedious experience of some 130 miles in total.[875] The terrain was forested and hilly, with ravines, creeks and steep descents common.

From its initial founding, the 20 square mile area that formed the Timbarra River goldfield produced a significant amount of the metal. Despite its somewhat inaccessible and remote location, in April of 1859 for example, it was reported that 1,032 ounces of gold had been escorted to Sydney from the region via Armidale, equivalent to nearly three standard-sized gold bars.[876] Situated on Crown land, the following month the New South Wales Government Gazette officially proclaimed the Timbarra Goldfield as

[870] *Clarence and Richmond Examiner*, 9 June 1894.
[871] *Clarence and Richmond Examiner*, 12 June 1894.
[872] *The Daily Telegraph*, 29 June 1894.
[873] *The Sydney Morning Herald*, 2 November 1857.
[874] *The Sydney Morning Herald*, 2 November 1857.
[875] *The Sydney Morning Herald*, 9 April 1859.
[876] *The Armidale Express and New England General Advertiser*, 30 April 1859.

MAP OF NORTHERN NEW SOUTH WALES, SHOWING TENTERFIELD, GRAFTON AND (IN THE RECTANGLE) THE TIMBARRA RIVER GOLDFIELDS (1897).
SOURCE: NATIONAL LIBRARY OF AUSTRALIA (NLA.GOV.AU/NLA.OBJ-231330610).

'at and in the vicinity of the Timbarra, or Northern Rocky River, and its tributaries, including McLeod's Creek, from the source of the Timbarra to its confluence with the Clarence River'.[877]

Into the 1860s and 1870s the quest to find alluvial gold in the creeks and rivers associated with the Timbarra Goldfield continued at pace. *The Clarence and Richmond Examiner and New England Advertiser* on 17 August 1880 published the following piece regarding the local industry. '*Timbarra, once the richest gold field in this district is situated on the highest granite range. The workings have all been alluvial sluicing, and have supported a considerable population during the past twenty years. Neither capital nor mechanical appliances have ever found their way there, though it is believed there are important gold deposits awaiting them. The business places comprise one European and two Chinese stores, one public-house, and a post office. Some Europeans have settled down and made comfortable homes there. The population is estimated at from 100 to 130. The quantity of gold purchased by the local storekeepers during the year from the local working and Poverty Point has amounted to 2,094 ozs.*'

By the late 1880s and into the early 1890s claims had been made on substantial tracts of land in the Timbarra Goldfield area, with leases granted. Mostly small but some formalised mining operations had also been established in this location and its surrounds, those with more resources deploying somewhat mechanised sluicing works to systematically separate the metal from the sediment, a huge advancement from the more primitive tools previously in use. During this period the population of Tenterfield had also grown to around 1,800 people; it was fast becoming one the chief townships of the region, mostly due to its fertile soil and it being a stop on the Great Northern Railway between Sydney and Brisbane.[878] In contrast, the nearby Timbarra River region remained somewhat stagnant with just a few hundred people present. The discovery of gold had resulted in a revolving door of prospectors, very few of whom were willing to move more permanently to the area.

It was in this context that Reeks first ventured up the Manning River back in December of 1893, likely undertaking some form of a reconnaissance trip. We know he then returned to the Northern Rivers region in June of 1894, in part to discuss the new steamer for use on the Clarence River out of Grafton, but also to survey the goldfields once more, this time in an area with somewhat better accessibility. It was to be the start of a new enterprise, with Reeks once again showing his entrepreneurial spirit.

Back home in Sydney later that month, Reeks and his wife Mary continued to enjoy Mosman's society. On 29 June, for example, a dance was held in '*the Barn*' at Mosman Bay, marking the first in a series of events held during the winter months of that year. With decorations of art muslin drapings, greenery, Christmas bush and variegated bamboo, amongst the guests in attendance was Mary Reeks.[879]

The second dance of the Mosman Bay Private Assembly took place at the same location on 13 July.

[877] *The Maitland Mercury and Hunter River General Advertiser*, 21 May 1859.

[878] *The Australian Handbook (incorporating New Zealand, Fiji, and New Guinea) and Shippers' and Importers' Directory* (1890).

[879] *The Daily Telegraph*, 30 June 1894.

The *Evening News* noted, '*The decorations of the hall were artistically arranged, the music well rendered, and the weather delightful for dancing. The next dance of the series is arranged to take place on Friday, August 3. Some very handsome dresses were worn by the following ladies: Mrs Martin, black velvet, white satin sleeves and front, black net and beaded trimming; Mrs Reeks, pale blue silk and black lace; Mrs Moors, white satin, white lace, and gold beaded trimming*'.[880]

During this period, Reeks continued his academic pursuits. As stated in the previous chapter, on 9 August 1894 he presented a paper titled '*Double-ended Screw Ferry Steamers*' to the Engineering Association of New South Wales at a meeting held in the Society's Rooms, 70 Pitt Street.[881] A summary of the lecture was printed in *The Sydney Morning Herald* the following day. '*Mr. W. Reeks read a paper on "Double-ended Screw Ferry Steamers." He said that double-ended ferry steamers were so few in number, both here and elsewhere, that he was safe in calling it a new field of industry. It was 16 years since the first double-ended ferry steamer was successfully launched, and since then very few had been built, those that had been being of recent date. He said that double-ended screw boats had become to be recognised as the ferry boats of the future. His first experience of a double-ended screw boat was in Liverpool in 1879, when the firm in which he was* [later] *employed constructed a steamer of this class to carry the heavy cart traffic across the Mersey. The boat proved a great success. The Brighton, which was employed by the Port Jackson Steamship Company, was built on similar lines to the boat which was built to run across the River Mersey. Screwing each end was a most economical method of propulsion; they were easy to steer and well adapted for passenger traffic. Mr. Reeks described the double-ended ferry steamers Lady Mary and Lady Manning, which he designed for the Balmain Co-operative Ferry Company, and gave a technical description of these vessels. A vote of thanks was accorded to Mr. Reeks for his paper.*'[882]

[880] *Evening News*, 20 July 1894.
[881] *The Sydney Morning Herald*, 9 August 1894.
[882] *The Sydney Morning Herald*, 10 August 1894.

On 31 July 1894 Reeks set the wheels in motion for his new mining enterprise. He applied to the New South Wales Department of Mines and Agriculture for permission to mine for gold in the Timbarra River region. '*Application No. 471; date of application, 31st July, 1894; name of applicant, Walter Reeks; description of land applied for, under Oakey Creek, from Sandy Creek to M. and S. Ogilvie's portion No. 4, parish Churchill, county Drake; extent, 8 acres, purpose for which land is to be used, to mine for gold.*'[883]

Reeks was to be involved with alluvial gold mining, i.e., extracting flakes and nuggets of the dense metal that are separated from the silt, sand, clay, and gravel material that form the river bed. At the time the fragments were best recovered by a simple sluicing and sifting process involving water. Panning was a very basic method, suitable for the lone prospector, but a slightly bigger, manned operation could make a substantial increase in the volume of material that could be sifted. The sifting device, referred to as a sluice box, featured one end raised with the water supply arranged to then flow down the box's channel or chute. The alluvial material from the river bed was washed down the sluice which had ridges across the base, called riffles, where gradually all of the rocks accumulated. The significantly-heavier gold was then picked out of the small deposits as they built up down the chute. It was a slow and laborious process, particularly by hand, and not 100 per cent effective, but at the time it was one of the best approaches available.

Receiving approval for his claim less than two weeks later Reeks, in his systematic manner and using his engineering-wired brain, took to the task of mining for gold in the Timbarra River region from a rather new perspective. On the 24 August 1894 *The Richmond River Herald and Northern Districts Advertiser* reported, '*DEPARTURE IN MINING - On 23rd instant Mr. Walter Reeks, on behalf of a small syndicate in which some local gentlemen are interested, will land at Grafton for transmission to Sandy Creek, Yulgilbar,*

[883] *Truth*, 19 August 1894.

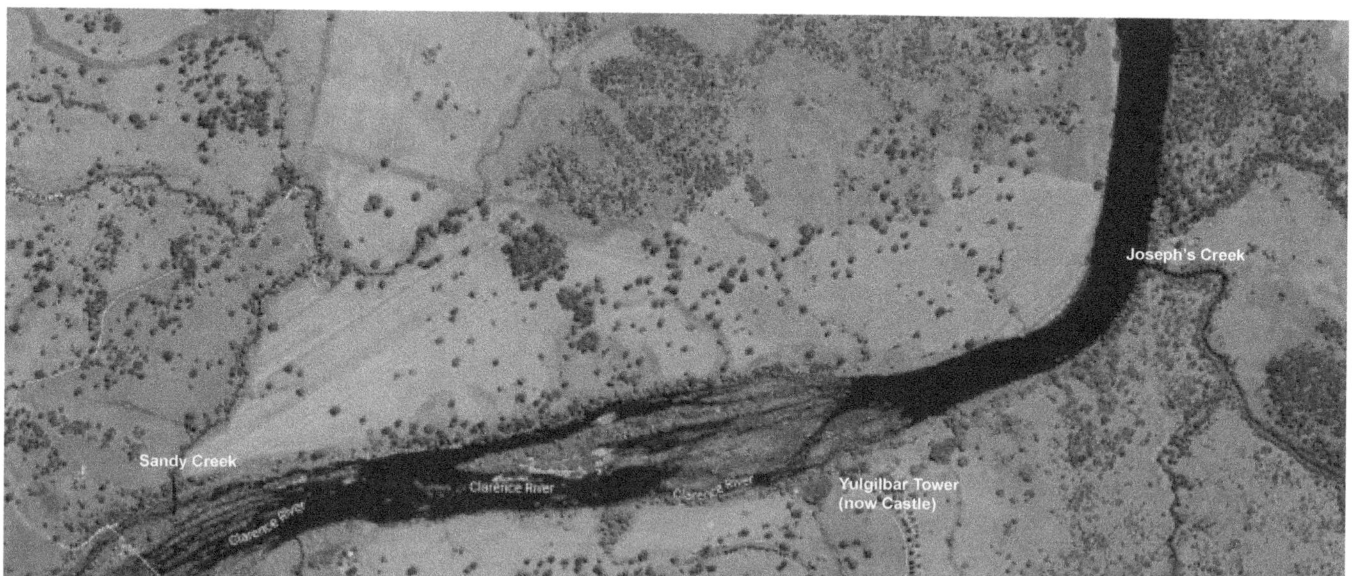

MAP OF THE CLARENCE RIVER REGION, SHOWING YULGILBAR AND SANDY AND JOSEPH'S CREEKS (NEW SOUTH WALES).
SOURCE: GOOGLE EARTH (2022).

a gold-saving apparatus which is entirely new to this colony, although it has been used with excellent result in America. The machine is constructed for working in the water, drawing up the wash on the suction principle, similar to that in use on the dredge Jupiter. When lifted the wash is passed through a set of sluice boxes, or treated in such other way as may be found necessary to save the gold. The lifting capacity of the machine is stated to be 340 tons per day, but it appears doubtful whether so large a quantity could be successfully treated in the boxes. Considerable interest will be felt in the results of the venture, as, should it prove successful, it will open up a means of working a large quantity of alluvial ground which has hithereto been inaccessible owing to the presence of too much water'.[884]

The machine for alluvial gold mining arrived in Grafton as cargo on board the *Kallatina* from Sydney on 23 August 1894. Reeks was also on board. Other passengers included his business partners, George and Thomas Allen, two Irish-born brothers both then living in Sydney and both involved in relevant vocations.[885] While George Allen was a mineralogist for the New South Wales Department of Mines, Thomas Allen was the Australian agent for commercial marine engines and propellers, including Myer's patent screw propeller which had recently been installed on the new steamer *Woolwich* at Grafton in mid-1894.[886] He later became a consulting engineer and patent attorney.[887]

With much interest centred on the mechanisation of gold mining, the local Grafton-based press continued its reporting of Reeks and his partners' efforts. '*By the Kallatina on Thursday some machinery arrived for mining on the alluvial workings on the upper Clarence. It was brought here by Messrs. Reeks, Allen and Co., and left for Yulgilbar yesterday, where it will be tried on some of the gold-bearing country in that locality. The party that introduced the machine also left for the Upper Clarence, accompanied by Mr. F. Tully* [a Grafton local], *who is also associated with the trial of the appliance, which is an American invention, and comprises engine and boiler capable of being worked up to 20 h.p., half-circular sluice box, and pumping gear. The great advantage of the machine is that it can raise auriferous sand or washdirt from a depth of 50ft. below water level, thus obviating the water difficulty which so frequently besets miners. The same is pumped into a capacious sluice box, where it is stirred by an apparatus worked by the engine, the water and earth passing off through the perforated end of the box. It is claimed that the appliance will treat 200 tons per day, and at 1dwt. per ton mining can, by this invention, be made to pay. It is proposed to start work on Sandy or Oakey Creek, but there are numerous places in the district where a machine of this description would do useful work. The engine and vertical boiler are mounted on wheels, with pole attached, so that they can be conveyed from place to place by team, without difficulty*.'[888]

On 4 September 1894 the *Clarence and Richmond Examiner* provided another update. '*We hear that the "Scweter" gold-saver, sent some weeks ago to Yulgilbar by a local syndicate, will be put into operation at Joseph's Creek, near Yulgilbar Towers, not at Sandy Creek, as at first intended*.'[889] A few weeks later came more news. '*The machine that went some weeks ago to Yulgilbar for prospecting the beds of creeks, is being removed to a different creek to that where it commenced operations. It is said to work well*.'[890]

It was an interesting time for Reeks to become involved in gold mining. Several gold-saving machines had recently been introduced to the Australian market, indicative of a period within the industry of rapid advancement and new technological developments. One of the machines, named the Jerger Gold-Saving Machine after its inventor John Jerger (an Australian watchmaker), was at the time being trialled at Goulburn, Bathurst and other parts of New South Wales with some success.[891] In addition, there were several other machines with patents pending, invented by entrepreneurial engineers and amateur gold seekers alike. These were not only intended to source gold, but also to identify gold in sluice that might otherwise be missed. The inventors and investors were all out to secure a share of the growing market. With the recent economic depression continuing, there were many prospectors trying to find a quick financial fix out of this melancholy. While gold mining was proving to be something of a low-cost though high-energy investment for single miners, a machine offered the potential for a much higher yield compared to back-breaking manual labour. It also had a higher initial investment.

In Reeks' case, his imported gold-saver had a vertical boiler and engine, which was not a new idea, but this machine appears capable of a significant increase in the amount of material that could be processed each day, by attaching the boiler and engine to a strong pump and what could be an improved, mechanised sluicing arrangement, the '*half-circular sluice box*'. Along with its portability, these were probably the features that attracted the engineering minds of both Reeks and his partners to the device. It must have had something extra for them to have it shipped from America without any examples locally to inspect before making the purchase. However, taking a considered risk and adopting new ideas was already typical of Reeks' mindset, so it should not come as a surprise that he supported this venture. The machine certainly raised some interest as its progress was reported in the press.

After spending 25 days in the Timbarra River goldfields region, Reeks returned to Sydney from Grafton on board the *Kallatina* on 16 September 1894.[892] It marked the end of a rather busy period for him. Coinciding with his efforts to mine for gold, some 350 miles further down the coast, the steamer he had designed for David Henderson of Grafton had been launched from the shipyard of Rock Davis of Brisbane Water on 18 September. Named *Orara* and intended to trade between Clarence Heads and

[884] *The Richmond River Herald and Northern Districts Advertiser*, 24 August 1894.
[885] *Clarence and Richmond Examiner*, 25 August 1894.
[886] *The Sydney Morning Herald*, 22 April 1893; *Evening News*, 28 June 1894; *The Sydney Mail and New South Wales Advertiser*, 30 June 1894.
[887] *The Daily Telegraph*, 29 July 1941.
[888] *Clarence and Richmond Examiner*, 25 August 1894.
[889] *Clarence and Richmond Examiner*, 4 September 1894.
[890] *Clarence and Richmond Examiner*, 29 September 1894.
[891] *Clarence and Richmond Examiner*, 23 January 1893.
[892] *The Australian Star*, 17 September 1894.

Copmanhurst, the 75ft vessel with an 18ft beam and 4ft depth, was designed to carry some 250 bags of maize.[893] A crane installed on deck was also capable of lifting up to two tons.[894] In addition, the vessel had seating accommodation for 30 passengers. The *Orara*'s engines, imported by Messrs Waugh and Josephson, were quadruple expansion of 50 h.p. to give the vessel a speed of 10 knots.[895] Sadly the craft was wrecked at Woolgoolga en route to Sydney in late December of the following year, all hands saved.[896]

AN OBJECT FOUND AT THE WRECK SITE OF ORARA, AND THOUGHT TO BE AN UNUSUAL PROPELLER DESIGN BY MARITIME HISTORIANS.
SOURCE: COFFS HARBOUR REGIONAL MUSEUM (MUS07-13171).

The remainder of 1894 appears to have been rather quiet for Reeks. There were no new vessel design commissions that the authors are aware of, and he likely maintained an interest in the Timbarra River gold mining operation from afar, not making any further trips to the area. He was also not involved in any yachting activities on Sydney Harbour.

While Reeks spent Christmas of 1894 with his family in Mosman Bay, his friend, artist Tom Roberts, travelled to the Timbarra River region spending some time at the gold mining site, which by now appears to have moved to the river itself. His notes and several paintings, including one titled '*On the Timbarra, Reeks' and Allen's sluicing claim*', offer a unique window into the location, the practicality of the works being carried out, the gold-saving machine and its set-up, as well as the social structure of the region. The picture of the adventure created by Roberts as he journeyed across the region, and the Timbarra River mining scene he painted at the camp are quite revealing in different ways. Reeks' social connections and adventurous spirit are both expanded a bit more from what can be found in the detail.

Roberts was certainly at home in this region and by 1894 '*had a network of patrons there, and a number of stations he could stay at. ... the outback provided him with a number of patrons and a congenial way of life. His letters of the period refer to musical evenings, picnic races, tennis and long rides on his own horse Black Bess*'.[897]

According to a biography on Roberts by Humphrey MacQueen (published in 1996), on 2 November 1894 Roberts '*set out for New England to a gold mining venture run by Reeks*'. He took the train from Sydney to Tenterfield, then a coach to Drake, above the Timbarra goldfields. Leaving Drake on horseback, Roberts rode Black Bess down a steep track well-worn from other miners.[898] It meandered through a long valley with stunning rainforest on either side, before leaving the relative shade and coming out into the hot open pastures. He was then on the flats around the Timbarra River where it made a significant right-angle bend on its journey northwards toward the upper reaches of the Clarence. This is probably where Scotsman Alexander Stewart had his property *Millera* station (it has since burnt down), a variation on the region's name *Malara*. While at *Millera* Roberts sketched three of Stewart's seven daughters. They all then ventured to '*the mining camp nearby for afternoon tea*'.[899] This was located a short distance up-stream and to the south, and most likely somewhere close to where Nicholson's Creek met the Timbarra River, trickling down the ranges from the east.

At the camp Roberts' '*fellow entrepreneurs were city gents*', colleagues of Reeks and the two Allen brothers.[900] They included 23-year-old John Le Gay Brereton, a poet and highly-regarded student of English unable to find work in that profession, an issue likely shared by others in the group.[901] Indeed they may all have met at one time or another at the artist camp at Mosman known as Curlew Camp. Le Gay Brereton, later a Sydney University professor of English, has Mosman connections too, with a park named in his honour.

TOM ROBERTS IN A BOAT AT CURLEW CAMP, C1895.
SOURCE: EN.WIKIPEDIA.ORG/WIKI/CURLEW_CAMP.

[893] *The Australian Star*, 20 September 1894.
[894] *The Australian Star*, 20 September 1894.
[895] *The Australian Star*, 20 September 1894.
[896] *The Daily Telegraph*, 31 December 1895.
[897] I. & T. Donaldson (1985). *The First Australians*.
[898] H. MacQueen (1996). *Tom Roberts*.
[899] H. MacQueen (1996). *Tom Roberts*.
[900] H. MacQueen (1996). *Tom Roberts*.
[901] H. MacQueen (1996). *Tom Roberts*.

ON THE TIMBARRA - REEK'S AND ALLEN'S SLUICING CLAIM BY TOM ROBERTS (1894).
SOURCE: ART GALLERY OF NEW SOUTH WALES (OA19.1961).

Roberts stayed at the mining camp for some weeks, probably doing some prospecting as part of the team, as well as painting. '*On the Timbarra - Reeks and Allen's sluicing claim c1894*' is a well-regarded example of his extraordinary landscape output depicting Australian scenes, and is held by the Art Gallery of New South Wales. It is also recorded as '*Dredging in the Clarence (Reek's) On the Timbarra - Reek's and Allen's Sluicing Claim Gold Digger*'. For our purposes, the painting forms an invaluable record of the camp.

The short shadows shown in the painting, cast almost directly under their hosts, suggest it is the middle of the day and the heat and stillness is apparent. With no hint of wind, smoke is rising gently from an obscured fire in the background. Roberts has set himself up on the dry river bed, with the eroded banks stepping up either side, a lone, hatted figure in the foreground. Further back, the sluice box is visible with a stream of water coming out of it and flowing down a narrow channel in the sand towards the location of the artist; one presumes he is located just downstream from the work site. Under the casuarinas in the shade is a platform for the sluice box, with three figures working away, their backs to the artist. Beside them, poking up behind a grassy ridge, like an alien feature, is a black metal vertical boiler and its upside down, spinning-top-shaped flue; a mini-portrait of the 'gold digger' and looking completely unaware of the attention it has created over the preceding months. The attention to detail is perfect.

After a period spent at the camp, Roberts decided to resume his travels and head to Yulgibar via Tabulum and the Clarence River valley, where he spent Christmas and the New Year. Settled in 1840, Yulgibar was a vast sheep and cattle station encompassing some 200,000 acres, owned by wealthy pastoralist Edward Ogilvie, then in his 80s. Roberts painted at least two portraits during his time here, one of Ogilvie (also held by the Art Gallery of New South Wales) and one of an Aboriginal woman named Maria. After some weeks, Roberts then returned to Reeks' mining camp, '*the additional baggage of canvas and clothes increased the hazards of the ride over the Great Dividing Range*'.[902]

By February of 1895 Reeks had also returned to the Timbarra River area, though it is not known if he and Roberts crossed paths. After a likely period of some weeks spent at the mining camp, Reeks returned to Sydney on the 20th of that month as a passenger on board the *Kallatina,* also on board was his prospecting partner George Allen.[903]

This was rough country. Less than 10 years earlier Reeks was still at home in England and at work in the city of Liverpool, a contrast of delightful green countryside and industrial cities with their smoke and soot. Here, he and his colleagues were engaged with a risky venture operating in 100 degrees heat, amongst poisonous animals, at risk

[902] H. MacQueen (1996). *Tom Roberts*.
[903] *The Daily Telegraph*, 22 February 1895.

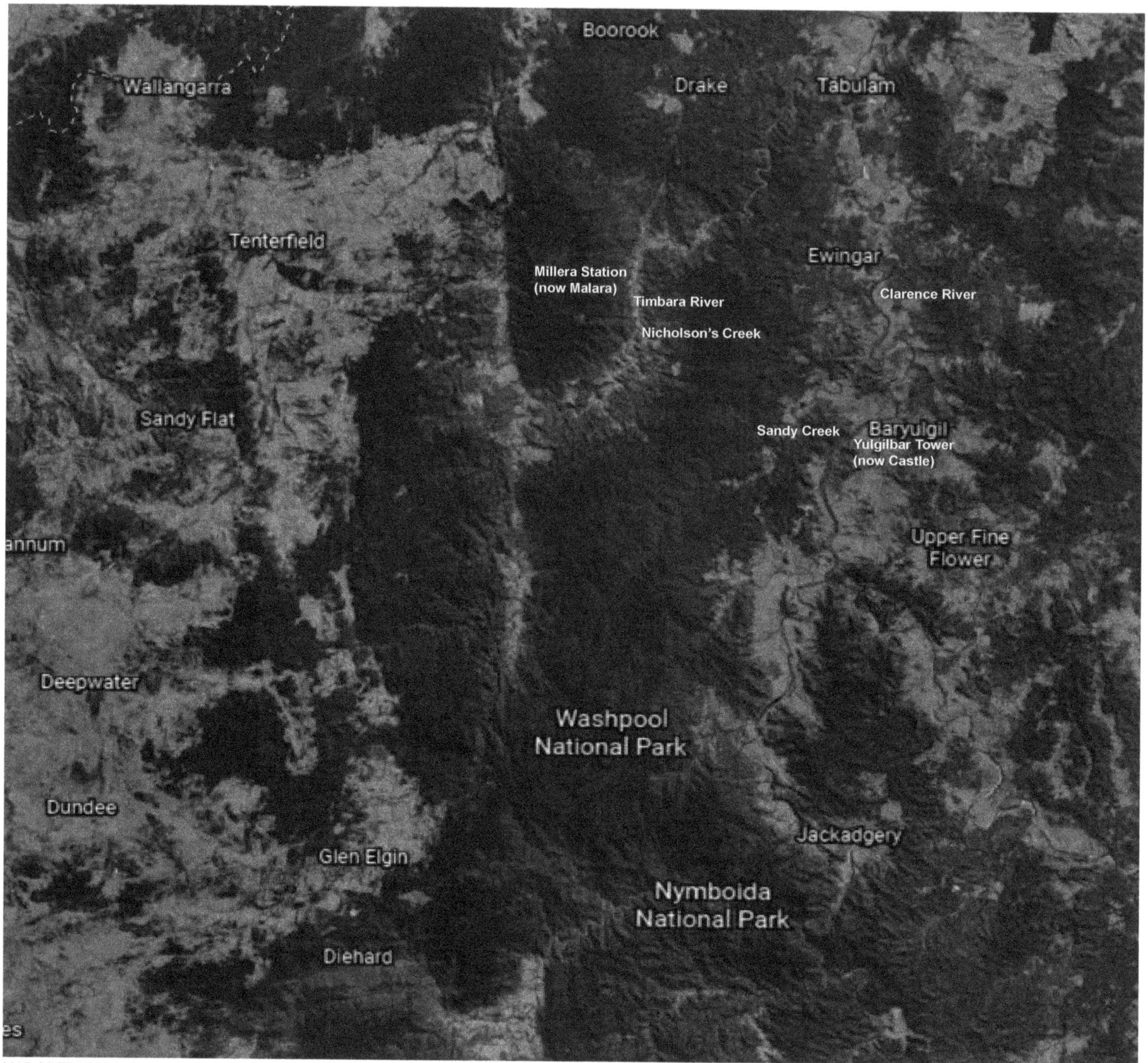

MAP OF THE TIMBARRA AND CLARENCE RIVER REGION OF NEW SOUTH WALES.
SOURCE: GOOGLE EARTH (2022).

of bushfires, and miles from anywhere. The gold digger machine he and his partners had invested in had made it to Yulgilbar easily enough, but the subsequent trip to where it was employed on the Timbarra River probably involved hauling it with a modest-sized bullock team, negotiating rocky tracks with steep rises and drops, difficult turns, and the constant danger of it all over-turning. Could Reeks have ever foreseen any of this growing up in Christchurch, and the subsequent years he spent working in Lymington and Liverpool?

Back in Sydney, in early March of 1895 Reeks was a notable guest at the annual picnic of the Shipwrights' Provident Union of New South Wales.[904] The riches and rewards offered from social events and gold mining, however, failed in comparison to what transpired next for Reeks and his family. His wife Mary died on 7 April 1895 at their home *Twynham* and was buried the next day at St Thomas Rest Park, North Sydney.[905] The cause of death was acute rheumatism pericarditis.

REEKS.—April 8, at her residence, Twynham, Rangers Estate, Mosman, Mary Emma, wife of Walter Reeks.

SOURCE: THE SYDNEY MAIL AND NEW SOUTH WALES ADVERTISER, 20 APRIL 1895.

Reeks suddenly found himself a 34-year-old widower with two young children to solely care and provide for. His son Kent was aged just five, while his daughter Thelma was just shy of three years old. A family photo, likely taken a few years after Mary's death, shows the trio sitting on a chair on the veranda at *Twynham*, a fascinating moment captured in time.

[904] *Evening News*, 5 March 1895.

[905] *The Sydney Morning Herald*, 8, 11 April 1895.

MARY REEKS (NEE KENT).
SOURCE: RAE FIDOCK COLLECTION.

WALTER, THELMA AND KENT REEKS AT TWYNHAM (LATE 1890S).
SOURCE: RAE FIDOCK COLLECTION.

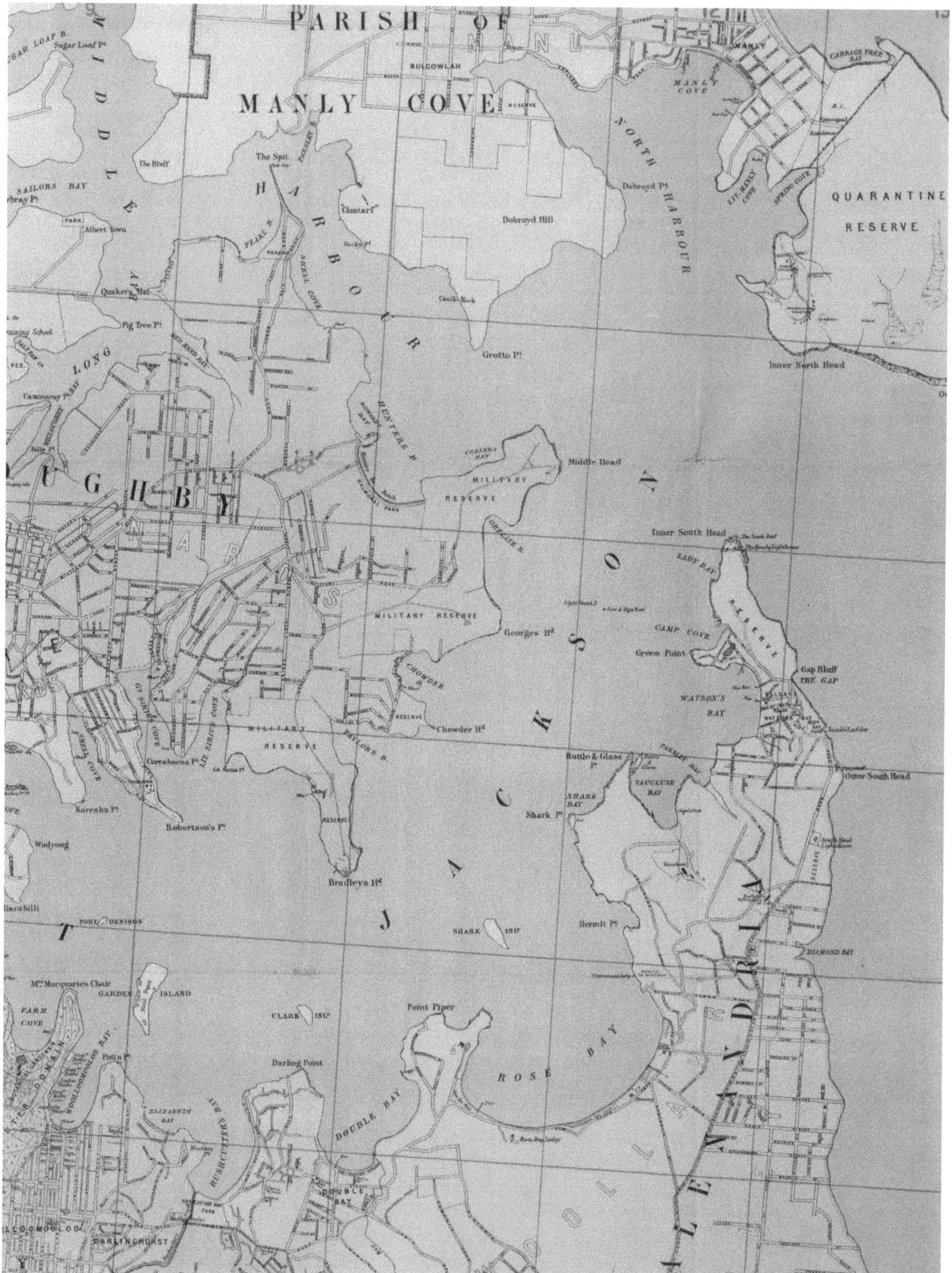

1890 MAP OF MANLY AND SURROUNDS (NEW SOUTH WALES).
SOURCE: NATIONAL LIBRARY OF AUSTRALIA (NLA.GOV.AU/NLA.OBJ-231445915).

CHAPTER TWELVE - MANLY

Mary Reeks died of inflammation of the pericardium, often a complication of rheumatic fever and other conditions such as rheumatoid arthritis. Given this diagnosis, it is likely that she had been ill for some time. Still, her condition would have worsened significantly over a short period of time and her death would have come as an absolute shock to her family and friends.

For Walter Reeks, with no family living in Australia, he was once again left to his own devices. Obviously friends would have stepped in to help with his care and that of his children, but ultimately he would have had to employ help to manage the day-to-day running of his household, particularly if he was to continue earning an income to provide for his family.

After several weeks spent grieving Reeks, either by choice or necessity, returned to work. So began another pivotal period in his professional career. On 15 June 1895 *The Sydney Morning Herald* published a prospectus of the Manly Co-operative Steam Ferry Company Limited. Having been established 18 months prior, applications were invited for the purchase of 15,000 contributing shares at £1 each. The money raised was to go towards the procurement of two steamers for the Manly ferry trade, both of which had already been designed by Reeks with the plans handed over to the company's directors.[906] The announcement was carried in the press over a period of several weeks.

The following month it was publicised that the London Missionary Society had commissioned construction of a new vessel for use in Torres Strait. Designed by Reeks, the 46-ton schooner was built in Whangaroa, New Zealand, by Messrs Lane and Brown and launched as the *Olive Branch* on 20 December 1895.[907] The craft measured just over 67ft by 16ft 10in, with a depth of 7ft 7in. Sadly, the vessel would be a total loss some six years later at Samarai, New Guinea.[908]

Back in Sydney, tenders for the construction of the hull, fitting of machinery, and installation of lighting relative to one of the two double-ended screw ferry steamers to be built for the Manly Co-operative Steam Ferry Company closed on 6 September 1895.[909] The organisation did not have enough capital to provide for the building of the two vessels concurrently, opting instead to build them one at a time. As we summarise the history of this company in the following pages, however, only one vessel would ever be built. Still, great lengths were taken to showcase the company's potential. A feature article, bordering on an advertisement, published in *The Daily Telegraph* on 27 August 1895 hyped the Manly Co-operative Steam Ferry Company's vision and objectives in terms of both its customers and its ferries.

'*Probably no better proof of the rapid strides made, the strength of the positions gained, and the bright and happy prospects of future success for those who, actuated by the powerful and wide-reaching spirit of co-operation, could be given than by the fact that in a few shorts months passage to our beautiful village Manly will be made in as a fine a steamer for her size as ever floated on "blue ocean," an illustration of which we have the privilege of her presenting to our readers. The Manly Co-operative Ferry Company Limited, fully alive to the call for safe, comfortable, seaworthy, and speedy vessels, have determined to*

[906] *The Sydney Morning Herald*, 15 June 1895.

[907] *The Daily Telegraph*, 31 July 1895; *New Zealand Herald*, 7 September 1895; *West Coast Times*, 21 December 1895.

[908] New Zealand, Registered Ships & Owners, 1840-1950.

[909] *The Daily Telegraph*, 17, 24 August 1895.

SCHOONER OLIVE BRANCH AT MAPOON.
SOURCE: STATE LIBRARY OF QUEENSLAND (HDL.HANDLE.NET/10462/DERIV/83591).

respond to that call in the shape of ferry boats in which these and many other important features shall be as conspicuous by their presence as, in many instances in the past, they have been by their absence.'

The layout of the impending craft was also a step up, particularly from the then 'Queen of the Manly' ferries, the capable and handsome paddle steamer *Brighton* coincidentally designed by Reeks' mentor Alex Richardson, that had steamed its way from Great Britain to run on the harbour. With no Virtual Reality or 3D graphics to work with, the reporter used the only ploy available at the time, a make-believe tour in words, and one still worth taking.

'*Above water, perhaps an imaginary visit to the boats will convey the best idea. Lying alongside the wharf, one has a choice of three gangways, and naturally taking the saloon or lady's end of the ship first, one finds a well-appointed cabin fitted with easy lounge seats all round, and upholstered settee in the centre, a convenient side board, and umbrella stand. Partial bulk heads and draped curtains, together with corresponding blinds to the windows, give a light, but cosy, feeling. Through and beyond these curtains every care has been taken to earn for the service the good opinion and appreciation of our fair friends. Passing out of the well-appointed apartments, the visitor is struck by the unusual (in ferry boats) feeling of being still under cover, and that in point of fact is the case, for the ends of the deck houses are completely guarded from wind, rain, and flying spray by long, gracefully curved whale or turtlebacks, which, while guarding the deck, serve also to lengthen the promenade deck, and give a striking smart, torpedo boat sort of cut to the whole craft.*

The midship portion of the deck is devoted to engine and boiler space, and to the grand staircase to upper deck, which is wide, of unusually easy grade, and with handsome polished hand-rails on each side give the air of safety and comfort.

The smokeroom, or gentlemen's cabin, together with their lavatories, occupy the rest of the main deck, and are fitted in light wood in which neatness is the prevailing idea. On to the upper or promenade deck, by either of the three stairways as conveniently placed as to make the ascent hardly appreciable, are shelter houses, in the nooks and niches of which nestle, so to speak, many easy seats, like garden chairs so admitting of the upper deck being used, even in the very worst weather. Long rows (but all being all

SOURCE: THE DAILY TELEGRAPH, 27 AUGUST 1895.

MANLY CO-OPERATIVE FERRY COMPANY'S NEW BOAT.

gracefully curved) of seats also provide for the accommodation of the many hundreds of passengers who are sure to travel by these boats. The lifeboats also find convenient resting places on this upper deck, amidships, and are so placed as to be ready at a moment's notice. It will be noticed that this company, with a view to keeping ahead of the times in the matter of life-saving appliances, are providing two lifeboats instead of one only, (which might be the wrong side of the ship in case of emergency) and also double the number of lifebuoys, which, being distributed all over the vessel, leaves nothing wanting in this all-important time in a passenger boats, life saving gear. The pilot houses, arranged to form part of the upper deck erections, certainly add much to the general composition, and are at the same time placed as most convenient for the steersman.'[910]

Such was the grace of the period, both in the outfitting of the vessel and the description. A trip was going to be an experience to remember. After this substantial discussion, the article continued on to the mechanics. '*While carefully avoiding the great outlay due to excessive horse-power, sufficient will be provided (more particularly boiler power) to ensure a speed whereby the whole of the passengers will be landed inside the half-hour, or from wharf to wharf in 26 to 27 minutes. Safety, comfort, and speed form the watchword of this enterprising company, and it is not the least exaggeration to say that the passage to New Zealand could be made in perfect safety. It need only be added that the new boats are to be lit throughout by electricity.*'

As noted, speed was once again a key issue, with the desired outcome to be able to complete the voyage to Manly (or the return one to Circular Quay) in under 30 minutes, including disembarking. In addition, Reeks had recognised another key requirement for the design: a sea-going capacity to cross Sydney Heads in rough weather to keep the route on time and operational. This design constraint remains recognised today by the fact that the ferries to Manly have to meet much higher survey requirements than their inner harbour counterparts.

Finally, the article concluded with the following paragraph. '*The designs of both boats and machinery have been prepared by Mr. Walter Reeks. The boats and all their appurtenances will be produced here, while the bare engines will be made in England. The work of construction will be put in hand forthwith, and as will be seen by reference to another part of this issue, share list and tenders close on the 6th of next month.*'[911]

The choice of Reeks as the designer for such an important craft is interesting in itself. Up until this point, only two other Sydney Harbour ferries of his design had been built: *The Lady Mary* in 1892 and *Lady Manning* in 1894, both screw propelled. They were inner harbour ferries for the Balmain New Ferry Company. Fortunately both were proving to be successful on their designated routes and the favourability of Reeks' practical designs for this vessel type was increasing.

With his home life likely settling in to a new routine, Reeks also returned to sailing, being a guest on board the *White Star* for the Royal Sydney Yacht Squadron's opening day activities on 19 October 1895.[912] Several weeks later the *Sunday Times* published a cablegram providing an update on the activities of the Manly Co-operative Steam Ferry Company. '*Everything satisfactory, tenders for engines and machinery for new steamer accepted.*'[913] Further detail was provided in *The Daily Telegraph* two days later. '*The Manly Co-operative Ferry Company has accepted tenders for the construction of the new steamer, the work to be taken in hand at once. The hull and fittings will be built by R. Young and Son, from the designs of Mr. Walter Reeks, and the engines and machinery will be constructed by Fawcett, Preston, and Co., of Liverpool, England.*'[914]

More work for Reeks soon came on hand from the Balmain New Ferry Company. On 30 October 1895 it was announced that this organisation, owing to high demand and a healthy profit of some £3,000 for the year to date, was adding another vessel to its fleet, along lines similar to the popular and practical *Lady Manning*.[915] The design and general supervision of the craft was once again placed with Reeks.[916]

910 *The Daily Telegraph*, 27 August 1895.

911 *The Daily Telegraph*, 27 August 1895.
912 *The Australian Star*, 21 October 1895.
913 *Sunday Times*, 27 October 1895.
914 *The Daily Telegraph*, 29 October 1895.
915 *The Sydney Morning Herald*, 19 November 1895.
916 *Evening News*, 30 October 1895.

The local ferry trade was growing and, happily for Reeks, he was well entwined with the companies that were at the forefront of the industry. It was welcome news after what had been a melancholy year for him personally, and a rather sparse few years professionally, owing to the depressed economy. There was gold to be found not only in the fields of Timbarra but also on Sydney Harbour.

The year 1895 ended with an interesting concert of which Reeks took part, reminiscent of his family's involvement in community activities back in Christchurch when he was growing up. While nowadays a minstrel show with blackface would be considered decidedly racist and absolutely uncouth, they were quite popular in America, England and Australia at the time, eventually becoming replaced by vaudeville. *The Sydney Morning Herald* described the event. '*The Mosman Variety Company gave their first public entertainment on Wednesday, the 18th instant, in the Congregational Schoolroom, Mosman, the net proceeds to be handed to the North Sydney Cottage Hospital. The programme was of a miscellaneous character, and was received by the audience, which was a large and enthusiastic one, with great approbation. Messrs. W. McLenehan and Joe Thiering were bones, Mr. W. Reeks interlocutor, and Messrs. Tas. Sykes and Sam. H. Martin tambos, one and all contributing to the success of the entertainment by the excellent manner in which they performed the parts alloted them.*'[917]

While, with over 125 years of hindsight, the bigotry and inferred racism in the show itself is abhorrent, the fact that Reeks was actively participating in the Mosman Amateur Variety Company, given all that had transpired in the past year, is pleasing to note. It was likely a welcome reprieve for himself as he looked to spend the first Christmas with his children without his wife Mary.

Returning to work after the New Year, Reeks advertised for tenders for the building of the hull of the Balmain New Ferry Company's new steamer.[918] The vessel was 4ft longer than the *Lady Manning* and had 2ft more beam. The tender was subsequently awarded to James Scott of Waterview Bay.[919] This vessel, along with the steamers he designed for the Manly Co-operative Steam Ferry Company, provided Reeks with another opportunity to further develop the principles of design around the keel and propellers that he had first established on the *Lady Manning*.

While the initial steamer for the Manly New Ferry Company was still under construction, the organisation was eager to begin its service and in mid-February 1896 purchased the recently completed Newcastle-built steamer *Fearless,* a vessel originally designed as a first-class ocean tug capable of 14 knots.[920] After being placed in the hands of Reeks, in only a two-week period, the *Fearless* was converted into a first-class promenade deck steamer well-suited for the Manly service.[921] The craft's deck arrangement was altered to comprise a comfortable smoking cabin at one end, and a luxuriously-furnished ladies' cabin at the other. The promenade deck was extended, with the carpentry and cabinet work carried out by Messrs Young, Son and Fletcher.[922] The *Fearless* began service on 14 March.[923]

Finally, after delays in construction and administrative hiccups, the Manly Co-operative Ferry Company's new steamer was launched at Rozelle Bay on 13 June 1896.[924] With a length of 157ft 6in (overall), a beam of 26ft, a depth of hold of 10ft 9in and a draft of 9ft 6in, the vessel was built on a laminated hardwood keel with Oregon topsides.[925] Its construction costs were high, estimated at £9,000. The craft was also the largest wooden ferry boat present on Sydney Harbour, able to accommodate up to 1,000 passengers.[926]

This ambitious project combined with a reduction in fares on its existing vessels at the expense of profit, however, appears to have over-stretched the Manly Co-operative Ferry Company's capital. A month before the launch of the new steamer, negotiations between the directors of this organisation and those of its major competitor, the Port Jackson Co-operative Steamship Company, began.[927] After much negotiation and a mostly disgruntled group of shareholders, the merging of the two organisations was finalised on 9 June.[928] There was evidently not enough money in the Manly service for two opposition companies to flourish but plenty of room for one amalgamated organisation to do well. It was in this context that the new ferry was launched only four days later, for some time afterward remaining un-named.

In the months that followed, the affairs of the Manly Co-operative Ferry Company, particularly its accounts after an audit revealed numerous financial discrepancies, continued to be the focus of an upset shareholder base. In late October of 1896 the subject reached a crescendo with the late secretary of the company, W. G. German, charged with falsifying written accounts, though he was later acquitted.[929] Several days later, the vessel which had caused so much of a financial burden for the company undertook its official trial trip. Named *Manly* and by now conclusively the property of the Port Jackson Co-operative Steamship Company, the craft left from Circular Quay and made a rapid run to Manly. Its return trip was even faster, being declared a record over the measured mile with a time of 4 minutes and 9 seconds.[930] Though designed to travel at 14 knots, *Manly* exceeded expectations by achieving 14½ knots. Those on board were suitably impressed, with Reeks receiving glowing reports in the press for his design prowess.[931] The vessel was equipped with 600 h.p. triple expansion engines, with steam supplied by a steel boiler of 14ft by 10ft, built to carry 160lb. of steam. The engines,

[917] *The Sydney Morning Herald*, 21 December 1895.
[918] *The Daily Telegraph*, 18 January 1896.
[919] *The Australian Star*, 2 June 1896.
[920] *The Australian Star*, 6 March 1896.
[921] *The Australian Star*, 6 March 1896.
[922] *The Australian Star*, 6 March 1896.
[923] *The Australian Star*, 13 March 1896.
[924] *Evening News*, 13 June 1896.
[925] *Evening News*, 13 June 1896; *The Australian Star*, 31 October 1896.
[926] *The Daily Telegraph*, 15 June 1896.
[927] *The Daily Telegraph*, 6 May 1896.
[928] *Sunday Times*, 17 May 1896; *The Daily Telegraph*, 10 July 1896.
[929] *The Daily Telegraph*, 23 October 1896.
[930] *The Australian Star*, 31 October 1896.
[931] *The Australian Star*, 31 October 1896.

boiler and associated parts were manufactured by Messrs Fawcett, Preston and Co., of Liverpool, England, under the supervision of Alex Richardson who was once again acting as Reeks' British-based agent.[932]

Meanwhile, the Balmain New Ferry Company's new steamer also continued to garner press. The craft had a novel layout worthy of generous descriptions during its construction period and then launch event.

'*The new ferry steamer, which is now being built by Mr. James Scott, at Waterview Bay, as an addition to the Balmain New Ferry Company's fleet, promises to eclipse any other of the company's vessels. Several novelties in fact, are being introduced into her construction which will make the steamer unique among the ferry boats in the harbor.*

One peculiarity is the rudder, which is above the shaft, and the propeller is outside of all, so that she can work in more solid water, as the phrase goes. The main deck houses are straight on the top.

Another specialty is a well, or open deck, cut in the main deck, 10ft. wide and 56ft. long, fitted with a stage to serve for concert purposes. The idea is that passengers shall be able to sit round the well as well as on top, and hear all the music, when concerts are given.

The whole model is different from any other ferry now running. The new boat, which has been designed by Mr. Walter Reeks, of Berry's Bay, is much larger and stronger than the Lady Manning, the finest of the company's present boats. She is expected to be ready for launching in about two months' time. Mr. Scott is also engaged on a beautiful model of the new vessel for the company, on a three-eighth scale.'[933]

The model referred to in the last sentence is priceless for the information it holds about this craft's unique underwater configuration. Significantly, it remains safely in the collection of the Powerhouse Museum, along with James Scott's equally superb sister model of *Lady Manning.*

[932] *Evening News*, 24 September 1896; *The Daily Telegraph*, 31 October 1896.
[933] *The Daily Telegraph*, 13 June 1896.

Built at a cost of £4,757, nearly half the cost of the *Manly,* the *Lady Hampden* was launched on 8 December 1896.[934] The vessel took its official trial trip on 1 March 1897 where, despite adverse weather, *Lady Hampden* travelled the measured mile in a time of 5 minutes and 11 seconds, with government officials pleased with its output.[935] Later that afternoon 400 members of the public were on board to experience their first voyage on the new vessel. The *Lady Hampden* commenced service a few weeks later.

With a length of 122ft (overall), a beam of 26ft, a moulded depth of 9ft 7in and a draft of 7ft, the *Lady Hampden* was decidedly smaller than the *Manly* with accommodation for 400 passengers.[936] Its keel was built in four thicknesses, with worked frames every 2ft 6in, its sponsons were built in the same manner, only with frames every 2ft.[937] The craft was planked with Oregon pine and hardwood, and copper fastened throughout. The *Lady Hampden*'s hull was similar in model to that of the *Lady Manning*, with the exception of the propellers working outside the rudder and its lines somewhat finer.[938] The vessel was also considered extremely unique in that its purpose was three-fold: for use as a ferry steamer, a floating concert hall, and a flagship for regattas and yacht races.[939] This multi-faceted use scenario meant that the *Lady Hampden's* passenger accommodation arrangements

[934] *The Daily Telegraph*, 5 December 1896.
[935] *The Daily Telegraph*, 2 March 1897.
[936] *The Daily Telegraph*, 13 June, 5 December 1896.
[937] *The Daily Telegraph*, 13 June 1896.
[938] *The Australian Star*, 8 December 1896.
[939] *The Australian Star*, 8 December 1896.

SS MANLY.
SOURCE: STATE LIBRARY OF WESTERN AUSTRALIA (N63A).

LADY HAMPDEN.
SOURCE: GRAEME ANDREWS WORKING HARBOUR PHOTOGRAPH COLLECTION, CITY OF SYDNEY ARCHIVES.

were somewhat peculiar. On the main deck there were two wide gangways and at the ends of the vessel, a neatly-fitted ladies' cabin and a smoking room, respectively.[940] In front of the former was a moveable platform, which contained seats when the vessel was in use as a ferry, but when converted to a concert hall were moved close to the cabin which also answered as a dressing room.[941] The *Evening News* reported, '*The general appearance of the forward part of the vessel is particularly American, and is different to anything yet seen in Sydney. The chief innovation so very noticeable in the internal arrangements, i.e., the concert room, which is really on the main deck, but around which runs a gallery or promenade deck 60 feet long on either side or about 130ft right round. From this gallery an excellent view of anything taking place on deck can be obtained. The seating on the gallery is so constructed that the backs revolve which will allow occupiers to either face in or outboard. The after end of the boat is set apart for ladies, and is suitably fitted with all conveniences. Past experience has taught the directors that to meet the requirements of harbor traffic, a steamer must be able to provide something further than that required to run a ferry service, and in this respect the Lady Hampden stands alone, for she is most suitable for moonlight concerts and other excursion purposes. She can seat 700 without crushing. It may be mentioned that in the concert room there is a small stage and piano'*.[942] Another feature was an awning deck, situated above the promenade deck, where space was provided for officials to be installed while overseeing regattas and sailing races.[943]

James Scott's builder's model of the *Lady Hampden* clearly shows the novel underwater arrangement, while other reports help add to the detail. Viewed by just focusing on the profile of the bow, the propeller was located forward under the bow, with the rudder reduced in size and sitting aft of the propeller, in the aperture formed by the shaft and the bracket supporting the propeller bearing housing. That was clearly as far forward as the propeller could go, and its positioning allowed Reeks to finally have his maximum effective load length as he had defined it in his paper of 1894, that being the distance between the propellers. On the *Manly* and the *Lady Hampden* ferries it was now equal with the allowed length of the hull.

The same location was described on the *Manly* as '*a projecting propeller head*'.[944] For this arrangement the rudders were linked and used in tandem, a fact only learnt from the *Manly*'s earlier description. '*She is steered by two rudders at each end, coupled together, which, although looking small, will doubtless answer the purpose, considering the peculiarly rockered form of the keel line*'.[945] Judged from the far forward location of the rudder quadrant on an image of *Lady Hamden*, it appears the rudders may also have been balanced, meaning the stock had a proportion of the blade area ahead of its axis, and a greater proportion aft of its axis. Engineering the rudder and stock would have been a significant challenge, but the balanced design does help share the axial twisting loads on the rudder stock onto either side.

One other point of difference clearly on display with the *Lady Hampden* was the location of the funnel, well towards one end rather than at midships. It was an unusual appearance for a double-ended ferry, where previously they had all displayed a balanced look in profile. If it was not already obvious by now, Reeks was not afraid to be different.

[940] *The Australian Star*, 8 December 1896.
[941] *The Australian Star*, 8 December 1896.
[942] *Evening News*, 2 March 1897.
[943] *The Australian Star*, 8 December 1896.
[944] *The Sydney Morning Herald*, 19 December 1896.
[945] *The Daily Telegraph*, 15 June 1896.

HALF MODELS OF THE LADY HAMPDEN (TOP) AND LADY MANNING (BOTTOM).
SOURCE: MUSEUM OF APPLIED ARTS & SCIENCES 'POWERHOUSE MUSEUM' (HTTPS://MA.AS/147684 AND HTTPS://MA.AS/147684).

While the *Manly* and *Lady Hampden* had significantly different routes on Sydney Harbour, the former expected to cross the Heads and thus be subjected to harsher and more open seas, Reeks applied the same configuration to both. This outcome is likely the culmination of his unorthodox way of thinking of how best to combine both the cambered keel along with the location of the rudders and propellers in relation to the hull's overall length, as was discussed earlier in his presentation to the Engineering Association of New South Wales in 1894.

From the dates when both craft are first mentioned in the Sydney press, and the publication of tenders calling for their construction, the design for the *Manly* was conceived first. However, both vessels were designed to have their propellers right at the ends of the hull, and their rudders back toward midships from the propeller, i.e., the reverse of what was normal. It is thus significant to note that in preparing the design associated with this unique configuration first on the *Manly,* Reeks was willing to have it initially trialled on a new type of ferry for a new company and one that he had not designed a vessel for previously. The *Manly* was also expected to service a route he was not familiar with. It was an ambitious approach for both the designer and the company who were putting their faith in him. However, there were areas that Reeks was not willing to compromise. In terms of engineering expertise, for example, he stayed with the familiar. Both the *Manly* and the *Lady Hampden*'s engines and boilers were supplied by Messrs Fawcett, Preston and Co., of Liverpool and under the supervision of Alex Richardson, his former employer and mentor.[946]

In hindsight more of the *Manly*'s significance also becomes apparent. The craft shaped the parameters required for the fleet of vessels to come. Under the flag of the Port Jackson Co-operative Steamship Company, the *Manly* sparked the evolution of a series of vessels that successfully serviced the route, culminating with the launch of the ultimate craft, the SS *South Steyne* in the late 1930s; an icon of the harbour on the world stage.

Manly was large, fast and seaworthy. The vessel had high bows, with a modern and striking look. So too was *Kuring-gai,* the second screw propelled ferry designed by Reeks for the same route and launched in 1901. This particular vessel was built in steel, thus providing a much stiffer and stronger hull. However, both *Manly* and *Kuring-gai* shared the cambered keel.

Though subsequent Manly ferries were designed by others, they too shared the same attributes that Reeks had initially designed, and developed the styling that had begun with this pair. The cambered keel was, however, abandoned in these craft, and they all resumed the straight keel line for better tracking and control.

Yachting and yachts were also back on the agenda for Reeks. In August of 1896 he was appointed honourable measurer of the Sydney Amateur Sailing Club, as well as a member of the committee.[947] That same month he was tasked with superintending extensive alterations and additions to the yacht *Isea* which had recently been purchased by A. A. Griffiths.[948] Eight years since being launched, the craft was to receive additional lead in its keel and have its depth increased to provide more stability.[949] The work was undertaken by *Isea*'s builder, Watty Ford at Berrys Bay.

Come late September, Reeks was superintending work to the yacht *Oithona,* the project being stretched out over a period of several months, included adding eight tons of lead. He was also noted as overseeing alterations to several other yachts, including *Thelma,* all at Ford's yard.[950]

[946] *The Australian Star*, 8 December 1896.

[947] *The Daily Telegraph*, 5 August 1896.

[948] *Evening News*, 28 August 1896.

[949] *The Sydney Morning Herald*, 28 August 1896.

[950] *The Sydney Morning Herald*, 30 September, 7 November 1896.

COMMITTEE OF THE SYDNEY AMATEUR SAILING CLUB. COMMODORE T. H. KELLY SHOWN IN CENTRE. WALTER REEKS IS SITTING AT THE END ON THE FAR RIGHT. SOURCE: THE SYDNEY MAIL AND NEW SOUTH WALES ADVERTISER, 30 JANUARY 1897.

After several years of negativity, Sydney's economy was on an upward trajectory and so too were the fortunes of these involved in Port Jackson's yachting scene. There were few new recreational sailing craft being built but repairs, alterations and changes were being made to existing vessels to increase their competitiveness. As the season got underway in late October of 1896, Reeks successfully took charge of *Sao* on behalf of owner F. Woolcott-Waley.[951] By the season's end he would be at the helm of *Oithona*.[952]

Momentum in the smaller yacht classes was also building, particularly in the realms of a one-design class. On 10 November 1896 a meeting of intended builders of a class 18ft on the waterline with a sail area of 260 square ft was held, with all attendees agreeing to a design already developed by Reeks. A follow-up meeting was to make a final decision on ballast requirements, as well as rules and conditions governing the proposed class which will be discussed in further detail in the following chapter.[953]

By early December it was confirmed that the class would have specifications of 22ft overall, with a 6ft 9in beam to planking and 7ft 6in beam overall. The planking would be of cedar and the boats half-decked.[954] There was to be no counter, the rudder hanging on the transom. The rig was to be gaff, mainsail and jib and, as stated, the sail area limited to 260 square ft.[955] Crews were also limited to four. Though initially suggested that the yachts were to be a one-design, the idea fell by the wayside with the only restriction agreed upon being that the boats should not be more than 18ft on the waterline.[956] As further details for the class formalised, Reeks was busy preparing workings and drawings for three boats.[957]

[951] *The Daily Telegraph*, 26 October 1896.
[952] *The Sydney Mail and New South Wales Advertiser*, 20 February 1897.
[953] *The Sydney Morning Herald*, 12 November 1896.
[954] *The Sydney Morning Herald*, 4 December 1896.
[955] *The Sydney Morning Herald*, 4 December 1896.
[956] *The Sydney Morning Herald*, 4 December 1896.
[957] *The Sydney Morning Herald*, 4 December 1896.

Overall, Reeks was heralding a period of design innovation on Sydney Harbour, in terms of both commercial passenger ferries and competitive yachts. His skills and expertise were in high demand and he was relishing the challenges offered by the two types of vessels. Socially, there were more events to attend and invitations at hand. In late November of 1896, for example, Reeks contributed to a moonlight fete and concert held at James Milson's residence, Milsons Point, to raise funds for St. John's Church. With moonlight trips on the harbour on offer, Reeks provided his services on board a small launch which was well patronised.[958]

1896 ended with Reeks assisting with efforts to plan the 'Engineering and Electrical Exhibition' for the Engineering Association of New South Wales, to be held at Prince Alfred Park in June, July and August of the following year. As a member of the executive committee, there was much to oversee, with exhibits classified into 17 groups and subdivided into 200 classes, ranging from minerals; mining; railway, tramway and vehicular appliances; and civil and military engineering to naval architecture; water supply and irrigation; and refrigerators, chilling and dairying.[959]

However, with just two weeks left before heralding in the new year, 1896 still had news to deliver and after just a short period of successful service *Manly* was making headlines for all the wrong reasons. On 18 December the vessel was involved in an accident, colliding with the coastal steamer *Merksworth* off Bradleys Head and causing the latter to sink near Cremorne Point, with fortunately no loss of life. A Marine Board of Inquiry was established and began its proceedings five days later, sitting again on 28 December at which time it found that the master of *Manly,* Thomas Capburn, was at fault for failing to keep clear, as was the obligation of an overtaking vessel.

Evidence given from those involved with both the *Manly* and *Merksworth* was published in the Sydney press over the same period that the inquiry was held, and by following *The Sydney Morning Herald*'s sequence of reports, the evidence seems to be conflicting.

Witnesses on board the *Manly* all gave evidence that claimed the craft maintained good control over its steerage, so the cambered keel was not at fault, and the vessel never altered course at the last minute, whereas they stated the *Merksworth* had steered itself into *Manly*'s path. However, those on board the *Merksworth* said their vessel held its own course and that *Manly* turned toward their ship. What was not disputed, was that the collision was a sliding, glancing blow on *Merksworth*'s port side. Nor were the facts of the diver's report contested, noting on *Merksworth* that there was '*a cut under the forepart of the bridge ... just on the round of the bilge ... it was a clean cut ... such a cut could have been caused by the blade of a propeller*'. Significantly it was also recorded that *Manly*'s damage was limited to the loss one propeller blade.

[958] *The Daily Telegraph*, 24 November 1896.
[959] *The Australian Star*, 21 December 1896.

Whilst responsibility for the collision was apportioned to the *Manly*'s master, the actual physical cause of the sinking was clearly *Manly*'s projecting propeller at the stem, unfortunately perfectly placed to deliver the damage described. An event perhaps unforeseen during the development of its design.

Reeks' activities with the Engineering Association of New South Wales continued into the new year. On 22 January 1897 he took part in a group tour of the SS *Friedrich der Grosse*, a Norddeutscher Lloyd liner only launched the previous year.[960] Measuring 523ft with a gross tonnage of over 10,500 tons, the vessel was quite substantial. The tour, conducted by the vessel's chief engineer, took in parts of the *Friedrich der Grosse's* saloons and accommodation, as well as its machinery department, including electric installation, refrigeration and hydraulics.[961] Reeks would have been positively enamoured.

As 1897 progressed, there was a renewed sense of optimism in Sydney. The economy was finally heading in the right direction, politically there was general consensus amongst representatives for federation, and free and secular education was having positive results. Roads, bridges, railways and telegraphs increased travel and communication, and technological advancements were continuing to spur all sectors of industry, from mining to agriculture, while public parks, libraries and museums were becoming more common and widespread. Drinking water was also becoming cleaner and more readily available. And of course there was sport and recreation, and lots of opportunity for both. Overall, the standard of living was rising and the population was largely looking to the new century with a heightened sense of direction and accomplishment.

For Reeks in particular, while he continued to mourn the loss of his young wife, from a work and social point of view he appears to have been more than satisfied with his lot. The vision of his purposeful vessels had become a reality, both the *Manly* and *Lady Hampden,* along with *The Lady Mary* and the *Lady Manning* were in operation and proving a success. The engineer with an artist's sense of creativity and style, could see it himself daily, heading home from his Change Alley office on the Mosman Bay ferry. One can only hope that he felt that special sense of satisfaction you get when you see your successful creation doing its task. The same can be said of his yachts. Several were back racing on the harbour, proving their worth. Reeks was also back amongst this brethren, designing, altering, improving and sailing.

[960] *The Sydney Morning Herald*, 23 January 1897.
[961] *The Sydney Morning Herald*, 23 January 1897.

GERMAN PASSENGER LINER FRIEDRICH DER GROSSE LEAVING SYDNEY COVE (1897).
SOURCE: AUCKLAND LIBRARIES HERITAGE COLLECTIONS 1729-171.

1890S MAP OF NEW ZEALAND.
SOURCE: NATIONAL LIBRARY OF AUSTRALIA (NLA.OBJ-232741332).

CHAPTER THIRTEEN - NEW ZEALAND

By early 1897 Walter Reeks had been living in Australia for 12 years. During this period he had designed somewhere in the realm of 100 vessels, small and large, commercial and recreational. Disappointingly the details and plans of many of these craft have been lost with time. He had also travelled to America and England on a mission to gather information relative to a challenge for the America's Cup, and mined for gold in New South Wales' north. His accomplishments and achievements were diverse, the outcome of his willingness to mostly always say '*Yes*', as well as his tendency for risk taking and devising unorthodox ways to overcome complicated problems.

The years leading up to the new century did not diverge from this pattern. Reeks continued to pounce on new projects and carry on working at a great pace. He also continued his involvement in Sydney's yachting community and its professional institutions, as well as its social and cultural events and activities.

Indicating his affinity with artists involved with Mosman Bay's Curlew Camp, one interesting event that Reeks was involved with in late 1896 leading into early 1897 was an art union of works created by the celebrated Australian impressionist Arthur Streeton. Whether the pair become connected through Tom Roberts is not known. What is clear, however, is that Streeton held Reeks in such high esteem that he tasked him with helping administer an art union established to pay for the former's travelling expenses for a proposed trip to Europe. The enterprise saw the issuance of 1,700 tickets at five shillings each, with the drawing of the prizes, twenty in number, taking place in February of 1897.[962]

[962] *Table Talk*, 16 October 1896; *Evening News*, 7 November 1896.

TICKET IN ARTHUR STREETON'S ART UNION OF PICTURES, 1897.
SOURCE: STATE LIBRARY OF NEW SOUTH WALES (SV/61).

Professionally, while the first half of 1897 was spent readying the *Lady Hampden* for launch and subsequent service, Reeks appears to have also been involved in a new steamer built for the North Coast S. N. Company that was launched on 18 May.[963] Though the *Euroka*'s designer was M. C. McAllister, the presence of Reeks at the vessel's launch (he was one of only a few in attendance) indicates that he was somehow involved in the craft's development.[964] Built by Messrs Foster and Minty at Balmain, the 127ft vessel had a breadth of 22ft and a depth of 8ft. The craft was specially built with a light draft to navigate the shallow bars of the Northern Rivers region; when loaded with 150 tons dead weight, the *Euroka* was expected to draw just over 5ft of water.[965]

Another project Reeks remained heavily involved with during early to mid-1897, along with other members of the Engineering Association of New South Wales, was the Engineering and Electrical Exhibition. Held at Prince Alfred Park, the event was officially opened by the Governor of New South Wales, the Hon. Henry Brand, on 26 June.[966] It was an impressive showcase with more than 1,200 exhibits, ranging from heavy machinery to models and drawings. Much thanks was given to the executive committee in organising and staging such an auspicious and all-encompassing event, with Reeks very much included amongst them.[967] Not one to miss an opportunity, he was also one of the exhibitors, having on display a half model of the new steamer *Manly*, along with reproductions of the *Lady Manning* and a longitudinal section of the yacht *Era*.[968]

With the close of the Engineering and Electrical Exhibition in late July of 1897, Reeks continued to receive invitations to social events from within this cohort. A month later, for example, he attended a smoke social and concert at *Mona*, Darling Point, the house of Colonel Rowe, President of the Institute of Architects.[969] He also maintained a vested interest in yacht administration, once again being appointed to the committee of the Sydney Amateur Sailing Club, acting as its honorary measurer.[970] In addition Reeks joined the Sydney Canoe Club, being appointed honorary measurer.[971] He also helped administer handicap races held under auspices of the Mosman Bay Sailing Club.[972]

There was also new sailing craft to design and superintend. By mid-1897 a cruising yacht, the first built at Berrys Bay by Watty Ford's son, also named Walter and also referred to as Watty, was under construction, built to

963 *Evening News*, 19 May 1897.
964 *Northern Star*, 14 July 1897.
965 *Evening News*, 19 May 1897.
966 *Evening News*, 28 June 1897.

967 *Evening News*, 28 June 1897.
968 *The Daily Telegraph*, 7 July 1897.
969 *The Sydney Morning Herald*, 9 August 1897.
970 *The Daily Telegraph*, 1 September 1897.
971 *The Australian Star*, 6 September 1897.
972 *The Daily Telegraph*, 5 March 1898.

ENGINEERING AND ELECTRICAL EXHIBITION, ALFRED PARK, SYDNEY, 1897.
SOURCE: THE SYDNEY MAIL AND NEW SOUTH WALES ADVERTISER, 17 JULY 1897.

the order of F. Woolcott-Waley. Named *Actaea,* the vessel was designed by Reeks and measured 41ft (overall) by 33ft (waterline), with a beam of 9ft and a draft of 7ft.[973] The craft was intended for sea-going cruising and thus had ample head room and as much accommodation on deck and below that could fit without impacting its strength or stability. The vessel's new owner had originally intended to alter his existing yacht *Sao* for the same purpose, by giving it more beam, but the effort was found to be too extensive such that *Sao* was broken up. Instead *Actaea* was built with as much of the material from *Sao*'s keel as could be serviceably introduced.[974] Launched on 7 October 1897, the craft resembled both *Jess* and *Thelma* in design, though with more room and less promise for speed.[975] *Actaea* also had no overhang forward and a comparatively short counter.

Actaea was yet another example of Reeks thinking a little outside of the box. Reflecting back on this project in 1907 it was said, somewhat dramatically, that '*Sao ... under Mr Reeks directions was sawn through a few feet above the keelson, and from these ashes phoenix like rose Actaea, a cruiser, later sold to Mr C Lloyd Jones*'.[976]

ACTAEA.
SOURCE: YACHTSMAN'S GUIDE TO SYDNEY HARBOUR (1898).

Concurrent with the construction of *Actaea,* two other yachts were undergoing fit out under Reeks' supervision at the same yard, *Iduna* and *Archina.*[977] Additionally, he was supervising significant alterations to the *Jess.*[978]

Oil launches were also coming into vogue, having only been introduced to New South Wales two years prior. Reeks was called upon to design one such vessel for use by the trustees of Ku-ring-Gai Chase National Park.[979] Measuring 26ft by 7ft, the craft was launched as the *Kuring Gai* in September of 1897 and could achieve a speed of eight miles per hour.

Whilst steam remained the primary source of engine power, oil was now beginning its ascendency. Reeks was nicely placed to join in with its development as the century rolled over, and future vessels to be discussed will show once again his ability to produce excellent but different craft. As it was, *Kuring Gai* was a neat, conservative beginning, as this appears to be the first of this type that he designed. Reeks' design also fitted in well with how the press was reviewing this new type of vessel. '*Oil launches are becoming pretty common on Port Jackson and some of them leave a trail of perfume equal to anything ever produced by the Cologne scent gutters. Some of the new craft are unobjectionable as to the exhaust fumes and perfectly safe and easily managed by one man or a boy, who requires no experience except common sense. Captain J. Mergill ... has a launch of this description, and one left the other day for Kuring-Gai Chase for the use of the trustees. She was designed by Mr. Walter Reeks.*'[980]

OIL LAUNCH KURING GAI (1890S).
SOURCE: MITCHELL LIBRARY, STATE LIBRARY OF NEW SOUTH WALES.

Another oil launch designed by Reeks and built by Henry Thomas Green, a boat builder of Lavender Bay, measuring 26ft and fitted with a 2 h.p. Grob's Patent Universal Oil Engine was launched in March of 1896 and advertised for sale in May of 1897 and August of 1898.[981] Though built to the order of E. Noyes and likely one of the first oil launches to be locally constructed, unfortunately little additional information is known about the vessel.[982]

Green and Reeks appear to have been very compatible and over time would collaborate as builder and designer on several more craft. The former hailed from a well-established boat building family whose patriarch George Green had first established a yard along Sydney's North Shore in the late 1820s after emigrating to the colony from Surrey, England.[983] Following four decades of steady work, he retired to New Zealand, dying in Dunedin at the home of his son, Henry Thomas Green, in 1872 aged 62.[984]

The second eldest son of 11 children born to George Green and his wife Maria (nee Bates), Henry Thomas Green (1834 - 1909) had initially worked at the family's Sydney boat yard and then set up his own business. However financial difficulties saw him move to Dunedin in the early 1860s where he became a well-known yacht designer and builder.[985] In 1882 he returned to Sydney

973 *The Daily Telegraph*, 18 September 1897.
974 *The Daily Telegraph*, 18 September 1897.
975 *The Sydney Mail and New South Wales Advertiser*, 16 October 1897.
976 *The Sydney Mail and New South Wales Advertiser*, 6 November 1907.
977 *The Australian Star*, 20 September 1897.
978 *The Sydney Morning Herald*, 6 November 1897.
979 *The Australian Star*, 20 September 1897.
980 *Australian Star*, 20 September 1897.
981 *Australian Town and Country Journal*, 29 May 1897;
The Sydney Morning Herald, 6 August 1898.
982 *Referee*, 25 March 1896.
983 *The Sydney Herald*, 18 October 1832.
984 *The Sydney Mail and New South Wales Advertiser*, 21 September 1872.
985 *The Sydney Morning Herald*, 27 June 1862; *Otago Witness*, 10 December 1870; 17 June 1871.

with his wife and family (the couple produced 16 children, though several sadly died young), establishing a boat yard at Lavender Bay.

Green became particularly noteworthy for designing and building unorthodox craft, including a remarkable tubular single scull named *Star of Australia* which his brother Richard 'Gus' Green took to England in 1863, being the first Australian champion sculler to successfully compete in that country. Though by the late 1890s Henry Thomas Green was nearing the end of his career, Reeks likely found an ally in his approach to yacht and boat designing and building.

During this period Reeks was also noted as having in progress the design for a steam yacht of nearly 50 tons for a gentleman in Melbourne.[986] Disappointingly, nothing additional is known about this craft, including whether it was actually built, which appears doubtful.

Reeks also continued to concentrate on smaller craft. Specifically, on 16 October 1897, two small yachts made their debut on Sydney Harbour. Both were designed by Reeks and both were considered somewhat unique. The first, a small rater named *Yogi,* was a development of his 18ft one-design class proposal that had been built by Green at Lavender Bay to the order of E. M. Moore.[987] The second was a shallow punt-shaped craft named *Squid* measuring 24ft (overall), 17ft on the waterline. The vessel was stated to have been built as an experiment and, like *Yogi*, was intended to compete with the one-rating class.[988]

It could be debated if 'yacht' was the right term for these fast and handy little vessels. The bulb fin ballast keel of the original design for this type, particularly those of English designer Linton Hope, had morphed into a steel centreboard, and the half-decked craft had gradually developed quite low freeboard too. They were getting closer to big dinghies and one writer noted, '*the others are centreboarders. We believe that two recently built little marvels are good at the vanishing trick, as follows: Puff comes, over goes the boat. The enthusiastic owner scrambles out onto the centreboard, and – hey presto - up she comes again, and goes on her way rejoicing. Such a boat is very convenient!*'[989]

It was a popular fleet, racing with the Sydney Amateur Sailing Club, Mosman Bay and other clubs at the time catering for the smaller classes on Port Jackson. *Yogi* was Reeks' successor to *Taipo*, which had brought success to owner W. L. Hunt. Reflecting on these times in August of 1925, the newly established *Australian Motor Boat and Yachting* magazine wrote, '*The Speedy Raters. One of these classes was the raters of 30 years ago, which varied in size from the ½ raters to the 2½ raters. One of the fundamental reasons for their popularity was the fact of their requiring extremely limited crews - from two men upwards according to size and measurements. ... The first of the "little" Raters was Daisy ... Then along came Caraid, a great boat too. But then Mr. Walter Reeks designed for W. L. Hunt the noted Taipo, which was so consistent a winner for years that she did not owe her owners anything.*'

Between *Taipo* and *Yogi*, however, sits another entrepreneurial arch of Reeks' varied career: his involvement orchestrating a one-design class on Sydney Harbour. It began late in 1896 and was a narrative that held the attention of one of the earliest of the now common yachting magazines, the short-lived but lively *Australian Yachtsman and Canoeist*. Published weekly, this title helped build momentum for the class, commencing with a report printed in its November 1896 edition. '*With regard to the open meeting called to discuss the advisability of a "One Class Fleet" much interest was shown, and plans and models exhibited ... Already six sailing men have decided to build at once, and four others expect to follow suit almost as soon.*'[990]

The Sydney Mail and New South Wales Advertiser then picked up the story and reported a few weeks later that, '*A meeting of intending builders of boats of the new one design class was held at Aarons Exchange Hotel on Tuesday evening the 10th, for the purpose of deciding on a design, plans submitted by Mr. Walter Reeks being generally approved. It was agreed that boats should be 18 feet on the waterline, with a sail area of 260 square feet and that crews should be limited to four hands. The question of the suitability of building boats to carry ballast or not was freely discussed, and the matter was left open for final decision at another meeting, at which rules and conditions governing the class will be adopted.*'[991]

The *Australian Yachtsman and Canoeist* continued to report on the class' development. On 5 December 1896 the publication confirmed that things were progressing, but why it was called '*The new 17-foot B Class*' when it was agreed that the waterline length was 18ft remains a mystery. '*Mr Walter Reeks is busy preparing working drawings and specifications. The new boats will be ready early in February.*' Basic length and beam dimensions were adopted, as was the decision to have the '*rudder hung on the transoms, no counter*'. The starting cost was estimated at £45. With sails and extras added, it was £55 to £60 pounds to sail-away. As this report went to press it also noted that *Taipo* had won yet another race, ahead of *Bronzewing*, and the event was sailed under the basic conditions of the 'B Class' in terms of sail area, crew limit and time allowance.

However, *The Daily Telegraph* on 5 December 1896 put a dampener on the one-design aspect, and although publishing much the same information on its dimensions and Reeks' own design progress, the last sentence stated the reality. '*The original suggestion that the boats should be of one design has been abandoned, and the only restriction to be imposed is that they shall not exceed 18ft. on the waterline*'.

Despite this set-back, on 12 December, in a substantial editorial, the *Australian Yachtsman and Canoeist* continued to be optimistic. '*One Class One Design - What is to be said in favour of this class? A great deal we think, because if it were possible to build a boat of exactly the same speed as another it would do away with the handicapping, and the sailing merits of*

[986] *The Australian Star*, 20 September 1897.
[987] *Referee*, 1 September 1897; *The Daily Telegraph*, 16 October 1897.
[988] *The Daily Telegraph*, 16 October 1897.
[989] *The Yachtman's Guide to Sydney Harbour and its Environs* (1898).
[990] *Australian Yachtsman and Canoeist*, 7 November 1896.
[991] *The Sydney Mail and New South Wales Advertiser*, 21 November 1896.

any boat would then rest with the skipper, accidents and flukes of course excepted.' However, the opinion piece then stated, '*We know very well that it is quite impossible to build two boats alike, even though they may be built upon the same moulds, therefore we come to the conclusion that it is an impossibility'.*[992] As a counterpoint in favour of the class it then argued that the man (i.e., the skipper) is more important than the machine (i.e., the boat). Given that, '*it is possible to produce two boats as nearly alike as human skill and science can do, and this special class is to be found in the "ONE CLASS ONE DESIGN," therefore we say it is a good class to go in for'.*[993] Another positive point then raised that it is likely to produce '*a more serviceable class'* of boat and continues at length in that vein, finishing with good intentions. '*We shall watch the ONE CLASS ONE DESIGN BOATS with much interest as soon as they appear on our waters.'*

Unfortunately, this positive affirmation was not met with the same outcome. The 2 January 1897 edition of *Australian Yachtsman and Canoeist* brings news. '*Two gentlemen from Mosmans are having boats built to sail under the Sydney Amateurs Class B. They will probably be enlarged Taipos. This is the outcome of the "one-class-one-design" movement, which seems to have merged into a one-class-different-design.'*

While Reeks' one-design class aspirations came to an end without really taking off, 'different' was an understatement in regards to comparing *Yogi* with *Squid*. By late August of 1897 the two craft were nearly ready for launch, with *Yogi* relatively conventional for the fleet. '*The latest addition to the rating class has just been completed by H. Green, of Lavender Bay, from a design furnished by Mr. Walter Reeks, the owner being E. M. Moor* [SIC] *of the Sydney Amateur Club. The little craft will rank as 17 foot Class B under the rules of the club named. Her dimensions are length overall, 22ft.; waterline, 18ft.; beam, 6ft. 9in., plank to plank; and 7ft. 6in. overall; depth, 2ft. 1¾ in.'*[994] *Yogi* was fitted with an adjustable drop, steel centreboard, and an apparent novelty being '*the arrangement of the tabernacle by which the mast step can be adjusted in different positions'.*[995]

Launched on Saturday 28 August, '*She was christened Yogi — the rising sun — and her racing colours will be a yellow sun on a blue ground. In a short spin in the light north easter on Saturday, Yogi appeared to move along in fine style, and some good racing may be expected between the new craft and the formidable little Taipo, which rates about the same. She ought to be an improvement on Mr. Hunt's boat, but she will need some handling if she is to prove her superiority over the latter craft'.*[996]

A month passes and racing begins, and we finally learn more about the second of the boats being built for the two Mosman Bay gentlemen. '*Mr. E. M. Moor's* [SIC] *small rater Yogi is to make her racing debut in the Mosman's Bay handicap this afternoon, in which event the Squid will also make her first appearance. The latter is a shallow, punt-shaped craft, 18ft. on the waterline and 24ft. overall, which was built as an experiment, and will probably be a competitor in the one-rating and under events during the year. Both craft were designed by Mr. Walter Reeks, and the "punt" is said to be a flyer in light weather.'*[997]

Squid was built for A. E. Cutler of Mosman. Although the craft features in racing reports for a while, little is known about its description other than what has been already quoted. Cutler was later to have Green build his own unorthodox-looking yacht, the self-designed *Kukuburra*, in the early 1900s, so the combination of Reeks and this owner, with their shared appreciation of something different, is probably enough to know that *Squid* was anything but conventional.

The word 'experiment' conjures some intrigue as to what the vessel may have looked like. A punt is generally a description for a flat-bottom and straight-sided single chine hull shape, a box in cross-section. Did *Squid* end at the bow with a sharp stem or a shallow blunt transom bow? These little raters were becoming very shallow for their length, the champion New Zealand one-rater *Mercia* (discussed in the next chapter) was a superb example of what could be achieved. At almost 29ft in length, this particular vessel was less than 2ft in depth from deck to keel, with a wide, shallow, skimming dish hull. *Mercia* finished with an elegant sharp stem, but this section shape easily translates to a wide flat-bottom single chine, short straight side that could either taper to a point or run out as snub, almost like an ironing board. Given Reeks' ability to adopt the unconventional, a relatively normal profile should not be assumed for *Squid.*

Judging by racing results, neither *Yogi* or *Squid* were very successful. *Yogi*'s poor form was noted by late November of 1897, '*while the scratch boat Yogi managed to get into third position. It must be confessed that the latter has not proved a success so far, but may possibly do better in a hard blow'.*[998] *Squid* started a number of events at different clubs, but either finished at the back of the fleet or in some cases, did not finish at all.

It was perhaps poor timing for this to happen. At this point in Reeks' career, his overarching history shows that these were almost certainly the last racing yachts he designed. His existing yachts, champions such as *Iduna* and *Thelma,* continued to provide excellent results, but there were new developments taking place. Dixon Kemp's Length & Sail Area Rule was short-lived, replaced after less than a decade by the more detailed formula of the Linear Rating Rule in 1896/97. But the bigger influence on the changing directions of Port Jackson's yachting scene, that happened a year or so later, was the beginning of a New Zealand influx. Federation was not far off and, as colonial yachting drew to a close and New Zealand went their separate way, yachts designed and built by Auckland's Charles Bailey and Logan Bros. were proving to be exceptional racers.

Perhaps things would have been different if *Yogi* or *Squid* had been race-winners, but Reeks had already diversified into commercial work and no longer needed

992 *Australian Yachtsman and Canoeist,* 12 December 1896.
993 *Australian Yachtsman and Canoeist,* 12 December 1896.
994 *The Daily Telegraph,* 28 August 1897.
995 *The Daily Telegraph,* 28 August 1897.
996 *The Daily Telegraph,* 30 August 1897.
997 *The Daily Telegraph,* 16 October 1897.
998 *Evening News,* 22 November 1897.

to rely on the yacht racing market which had been a foundation of his career over the previous decade. As one door closed, others opened and his yacht design work subsequently evolved into cruising vessels by and large, while his commercial design work was expanding. Around the corner there were opportunities coming from across the Tasman Sea, as well as from Queensland.

Come the end of 1897, a second moonlight fete held on the grounds of James Milson's home on 3 December, to raise funds for St. John's Church saw Reeks once again in attendance and helping with proceedings. Specifically, in partnership with A. Minnett, he provided patrons with short trips on Sydney Harbour on board the steam launch *Pacific*.[999] He would continue to be involved with this annual event for a number of years.[1000]

1898 began with Reeks calling for tenders for repairs to the Balmain New Ferry Company's steamers.[1001] A fire that began on board the *Lady Hampden* after midnight on 7 January, while the vessel was moored at the Bald Rock Ferry Wharf, had gutted the craft, also causing significant damage to *The Lady Mary* and *Lady Napier*, both moored in close proximity.[1002] While the *Lady Hampden* was burnt to the waterline and initially deemed irreparable, the other two steamers were salvageable, but required weeks of repairs.[1003] All three vessels were insured.

After further examination, however, it was decided to also repair instead of replace the *Lady Hampden*. Reeks was tasked with developing the tender and superintending the work on all three vessels.[1004] As part of this process, he initiated some tweaks to the original designs. Specifically, the *Lady Napier* was raised 15 inches, on the same lines as the *Lady Manning,* making it more serviceable.[1005] The *Lady Hampden* was refitted too, with some minor improvements.[1006] The work was undertaken by T. F. Morrison at Pyrmont.[1007] All three steamers resumed service on Port Jackson later that year.

Tenders for the construction of six pearling luggers were also called for in January of 1898, with plans and specifications developed by Reeks.[1008] Submissions for the building of an additional three pearling luggers were requested by Reeks later that year.[1009] These vessels appear to be part of a larger order from James Clark and his business partners, encompassing some 20 luggers in total, with a portion of the order constructed in New South Wales and another portion constructed in New Zealand, as had taken place in the early 1890s.[1010]

The Torres Strait pearl shelling industry was proving more and more profitable. The year 1897 alone had resulted in over 1,200 tons of shell being exported from Thursday Island, at a value £126,000; a growth on the previous year of £31,000.[1011] Buoyed by the high yielding and increasingly lucrative market, new syndicates were forming and new vessels were in high demand. Yet, Clark remained the industry's 'Pearl King', having the largest

[999] *The Sydney Morning Herald*, 4 December 1897.
[1000] *The Sydney Morning Herald*, 31 October 1898.
[1001] *The Daily Telegraph*, 8 January 1898.
[1002] *The Daily Telegraph*, 7 January 1898.
[1003] *The Australian Star*, 17 January 1898.
[1004] *The Daily Telegraph*, 31 January 1898.
[1005] *The Australian Star*, 17 January 1898.
[1006] *The Australian Star*, 17 February 1898.
[1007] *Daily Commercial News and Shipping List*, 25 March 1898.
[1008] *The Sydney Morning Herald*, 20 January 1898.
[1009] *The Sydney Morning Herald*, 30 July 1898.
[1010] *The Sydney Morning Herald*, 11 March 1898.
[1011] *Bendigo Advertiser*, 4 January 1898.

SKETCH OF THE AFTERMATH OF THE LADY HAMPDEN FIRE
SOURCE: AUSTRALIAN TOWN AND COUNTRY JOURNAL, 15 JANUARY 1898.

share of the market. Of the nine pearling luggers he commissioned to plans by Reeks, the first vessel appears to have been the ketch *Estelle*, which was launched in Sydney in early April of 1898, with Reeks also noted as supervising its construction.[1012] Next came the *Plover* and *Zoe*, both built by Rock Davis at Brisbane Water and completed the following month.[1013] Six more pearling ketches of around the same dimensions (i.e., 43ft by 12ft by 5ft) appear to have been built that same year by either Rock Davis or Watty Ford (Jr), all for Clark. These were the *Penguin, Snipe, Curlew, Teal, Heron* and *Ospray*.[1014] Two near-identical vessels, *Bittern* and *Edith*, being 43ft 4ft in length by 12ft with a depth of 5ft 1in, were additionally built for Clark by Messrs Lane and Brown of Whangaroa, New Zealand, and launched in August of 1898.[1015]

Several pearling luggers, including the *Onyx* and *Ruby*, were also built during the latter part of 1898 by Ford (Jr) at Berrys Bay.[1016] All were built for the Queensland Pearl Shell Fishery Company. It is not known if Reeks was involved in their design and/or construction, though it is a distinct possibility.

While the pearling fleet was estimated to comprise ten schooners and around ninety luggers by January of 1898, with 600 men engaged in the work offshore and another 200 onshore, the addition of these new vessels, with the goal of ushering in a new phase of the industry proved only short-term.[1017] The Far North Queensland and Thursday Island region was absolutely decimated on 5 March 1899 when the opposing cyclones *Nachon* and *Mahina* crossed tracks and combined over Cape Melville to form the category 5 cyclone *Mahina*, the largest natural disaster that had yet been recorded in Australia. All told 400 people were killed and more than 50 pearling luggers were lost in Bathurst Bay at Cape Melville, including several of those mentioned herein.[1018] At least three pearling schooners were also wrecked, but the Reeks-designed and built *Aladdin* escaped, apparently standing off shore and avoiding the worst of the storm surge.[1019]

The region still pays its respect to this event with a memorial at Bathurst Bay. As expected, however, this terrible disaster also precipitated a new cycle of pearling lugger construction.[1020]

Another project Reeks was involved with during early 1898 was something quite risky but one that also offered great commercial reward. Fortunately the press in relaying the outcome deemed it '*A successful experiment*'. The report read, '*For some time past the directors of the Newcastle and Hunter River S. N. Co. have been considering the possibility of taking the steamer Newcastle to Morpeth. For years it has been confidently asserted that such a feat was a nautical impossibility, in the first place owing to the vessel's draught, and secondly by reason of her length, which it is said would prevent her successfully navigating the river bends and swinging at Morpeth*'.[1021]

Situated some 20 miles inland from Newcastle, at the head of the navigable part of the Hunter River, Morpeth is now considered a suburb of Maitland, but in the late 1890s was a large port town, positioned as the gateway to further regional centres and towns. The possibility of navigating the steamer *Newcastle* up the Hunter River, a steel-hulled paddle steamer with dimensions of 264ft by 33ft by 16ft, was considered an absolute impossibility; it

[1012] *Daily Commercial News and Shipping List*, 7 April 1898.
[1013] *Daily Commercial News and Shipping List*, 5, 26 May 1898.
[1014] Register of Australian and New Zealand Shipping 1902-1903; *The Daily Telegraph*, 18 July, 23 August, 18 October, 30 December 1898.
[1015] *Auckland Star*, 2 September 1898; New Zealand, Registered Ships & Owners, 1840-1950.
[1016] *The Daily Telegraph*, 31 December 1898;
Register of Australian and New Zealand Shipping 1902-1903.
[1017] *Western Mail*, 14 January 1898.
[1018] www.environment.gov.au/shipwreck/public/wreck/search.do.
[1019] *The Sydney Morning Herald*, 14 March 1899.
[1020] www.environment.gov.au/shipwreck/public/wreck/search.do.
[1021] *Newcastle Morning Herald and Miners' Advocate*, 8 March 1898.

PS NEWCASTLE.
SOURCE: GRAEME ANDREWS WORKING HARBOUR PHOTOGRAPH COLLECTION, CITY OF SYDNEY ARCHIVES.

was believed that only vessels up to 240ft in length and less than 30ft in width could safely make the journey.

The *Newcastle* had been commissioned by a predecessor of the Newcastle and Hunter River S. N. Co. and built in Scotland in 1884. Though intended to ply between Newcastle and Sydney, the vessel was found quite expensive to run such that by the late 1890s the company was looking into new and more local routes to increase the craft's use and profitability. Thus they called upon Reeks to determine if it was feasible for the much larger vessel to safely negotiate the distance to Morpeth. His response being '*Yes*', it was entirely possible for the *Newcastle* to travel along the Hunter River, including navigating its sharp curves, and shallow and narrow stretches.

And so at 5:35 am on 7 March 1898, with Reeks on board, along with a small party representing the vessel's owners and several port and government officials, along with a tug in company should any assistance be required, the *Newcastle* left the wharf of its namesake city and steamed up the harbour, the tide being about three-quarters flood.[1022] After successfully negotiating the elements of the Hunter River that were considered to be difficult for the vessel's safe passage, the *Newcastle* arrived at Morpeth three hours later with a large crowd assembled to witness the event. The vessel was then swung around, with room to spare, to make the return journey, successfully traversing the run back to Newcastle to arrive at 11:45 am.[1023] The party then adjourned to the saloon where refreshments were served and toasts given, all told the experience giving a positive boost to the company and its shareholders.

The experiment proved such a success that a director of the Newcastle and Hunter River S. N. Co., J. R. Hall, immediately announced that Reeks' proposal, which included taking 15 inches off the steamer's bottom, would be seriously considered.[1024] '*It was contemplated that very extensive alterations would be made to the vessel so as to permit of her doing similar work to the Namoi, and thus increasing her value very much to the company.*'[1025] The *Newcastle* was subsequently laid up, likely in anticipation of having the stated improvements to the vessel made.

Returning to Sydney, Reeks began work preparing a more detailed plan for altering the *Newcastle*, including reducing the vessel's displacement.[1026] However, for unknown reasons these modifications would not be made and the craft remained in the Newcastle to Sydney route.

During this period Reeks also called for tenders for the building of a part-wood, part-iron steamer, about 130ft long, from plans he had developed.[1027] This specific vessel was for the Northern Union Steamship Company of New Zealand which had commissioned him to design a vessel capable of achieving a speed of 11 knots for use in its Kaipara passenger and cargo trade.[1028] It was shortly thereafter announced that the tender had been awarded to T. F. Morrison at Pyrmont with the new vessel anticipated to be built at a cost of £6,000.[1029]

The month of March 1898 ended for Reeks with another event of interest to intercolonial yachting and also connected to New Zealand: the arrival in Sydney by steamer of the Auckland-based yacht *Meteor*, owned by A. T. Pittar. A few days later, the first of a series of races was held between this craft and the crack Sydney yacht *Bronzewing.*

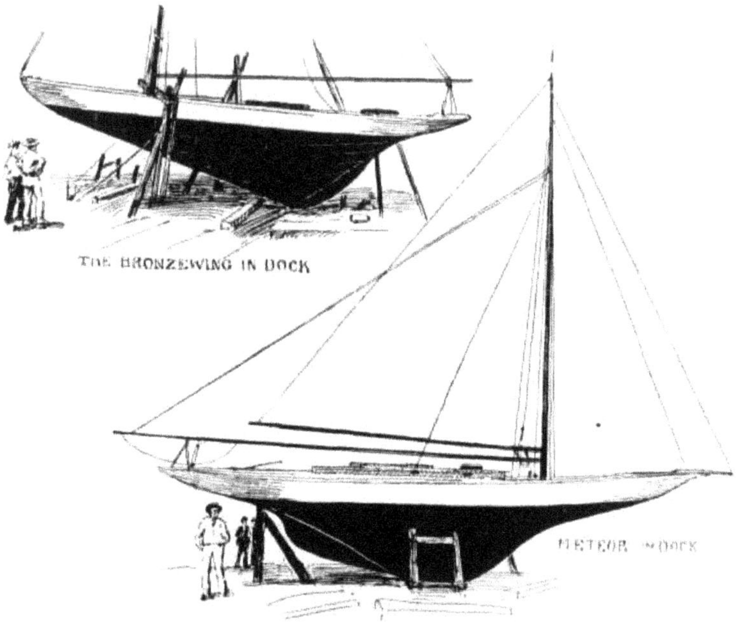

SKETCHES OF BRONZEWING AND METEOR.
SOURCE: THE SYDNEY MAIL AND NEW SOUTH WALES ADVERTISER, 9 APRIL 1898.

Whereas the challenger had only recently been launched in Auckland, built by Charles Bailey, *Bronzewing* was well known in Sydney having been built in Scotland to a design by George Lennox Watson in 1892 for Sam Hordern.[1030] This particular craft had then changed hands and at the time of the race series was owned by H. M. Cockshott.[1031] With its usual skipper at the helm of the visiting yacht, Darby Ryan, it was none other than Reeks at the helm of the local favourite.[1032]

It had been some years since the staging of an intercolonial yacht race, reminiscent of the days of *Waitangi, Janet, Era* and *Volunteer,* so it was with a near-celebrity following that the much-anticipated first race took place on 30 March on a Neutral Bay to Manly course, with £100 up for grabs.[1033] After both vessels were measured by Reeks, acting in his official capacity as measurer of the Royal Sydney Yacht Squadron, *Meteor* was given a rating of 3.8 while *Bronzewing* was given a rating of 3.[1034] The allowance conceded by the latter to the former was thus calculated at 3 minutes and 56 seconds.[1035]

1022 *Newcastle Morning Herald and Miners' Advocate*, 8 March 1898.
1023 *Newcastle Morning Herald and Miners' Advocate*, 8 March 1898.
1024 *The Daily Telegraph*, 8 March 1898.
1025 *Newcastle Morning Herald and Miners' Advocate*, 8 March 1898.
1026 *The Sydney Morning Herald*, 11 March 1898.
1027 *The Daily Telegraph*, 9 March 1898; *The Australian Star*, 12 March 1898.
1028 *The Sydney Morning Herald*, 11 March 1898.
1029 *The Daily Telegraph*, 25 March 1898.
1030 *Australian Town and Country Journal*, 2 April 1898.
1031 *The Sydney Morning Herald*, 28 March 1898.
1032 *Evening News*, 28 March 1898.
1033 *The Sydney Morning Herald*, 28, 31 March 1898.
1034 *The Sydney Morning Herald*, 31 March 1898.
1035 *The Sydney Morning Herald*, 31 March 1898.

The race, over a 17-mile course, began with light and fluky winds proving problematic for both yachts. After more than six hours, however, *Meteor* proved the faster boat on all points except in running, and pulled ahead to win by more than 10 minutes, well within the time allowance it had given to *Bronzewing*.[1036]

The second race between the pair took place the following day under the same mild conditions and on the same course. Disappointingly for Reeks, *Meteor* was once again the victor, winning the match by just under four minutes on time allowance.[1037] The prize pool was accordingly awarded to the visiting party.

While Reeks had lost the intercolonial yacht race to the New Zealanders, he did remain in favour with other members of this country's maritime community. The building of the new cargo and passenger steamer for the Northern Union Steamship Company continued.

There were other diverse projects for Reeks to be involved with too. For example, in June of 1898 he submitted a tender to the New South Wales Public Works Department for the supply of two timber steam launches, one for the Hunter River and one for the Clarence River, to be built at a cost of £4,070.[1038] The launches were to be suitable for towing and passenger ferry purposes, to be double-ended screw boats, about 65ft in length with a beam of 14ft and a shallow draft.[1039] The engines were to be non-condensing capable of achieving a speed of 10 knots.[1040] Accommodation was to be provided for 100 passengers.[1041] Reeks' submission was duly accepted one month later, with the vessels to be based out of Morpeth and Grafton, respectively.[1042] It would take another 15 months, however, before the first of these vessels was launched.

Meanwhile, a new five-rater built by Ford (Jr) at Berrys Bay to a design by William Fife (Jr) became another project that Reeks was involved with in late 1898. Tasked with supervising the vessel's construction, its principal dimensions were 48ft (overall), 31ft (waterline), with an extreme beam of 11ft 4in and a depth of 5ft.[1043] The centreboard craft was built to the order of Sam Hordern with construction of tea tree with planking of kauri pine. It also had an entirely new feature in that the runners, deck fittings and stanchions were of silveroid.[1044]

In September of 1898 Reeks requested tenders for the upper works of a houseboat.[1045] This may have been for the trustees of the Ku-ring-Gai Chase National Park for whom he had recently designed an oil launch.

OIL LAUNCH KURING GAI AND TRUSTEE'S HOUSEBOAT (1899).
SOURCE: STATE LIBRARY OF NEW SOUTH WALES (174530).

Also in September of 1898 an interesting legal case was heard in Sydney's Supreme Court over a period of several weeks, with Reeks suddenly finding himself embroiled in the controversy. The case centred on an action brought by Mahlon Clarke Lowlishaw, a merchant of Sydney, against William McLeod, printer and publisher of the Sydney newspaper *The Bulletin*. The former was claiming £2,000 as compensation for an alleged libel published in the paper. Specifically, an article printed in association with the foundering of the barque *Atacama* after springing a leak, a vessel that Mahlon Lowlishaw and his brother had only purchased a few months before, was believed to contain serious reflections on the plaintiff, whereby he had suffered in his character, reputation and circumstances.[1046]

Built in Rotterdam in 1875, on 29 January 1898 the uninsured *Atacama* was lost at sea after leaving Newcastle for San Diego with 1,700 tons of coal on board, only four of its 17 hands were saved.[1047] The vessel had first arrived in Australia from South America in October of the year prior and, upon being purchased for £700, had undergone structural work and overhaul at Sydney at a cost of £1,000.[1048] However, the *Atacama* had been condemned by Lloyds in 1890 owing to its unseaworthiness, apparently a fact that the new owners were unaware of because they had failed to check the register to source information about their new purchase.[1049]

As part of the case, over 70 witnesses and experts were called upon to give testimony. Included amongst them was Reeks who provided information on the vessel's history and condemned status based on information gleaned from Lloyds registers, his main point being that the *Atacama* had been condemned in 1890 until the day it foundered. The vessel had not been surveyed at any moment during this period to have its categorisation changed.

Other testimony was provided by the shipwrights who worked on *Atacama* prior to its departure, its captain who was extremely fortunate to survive the foundering, those involved with the subsequent Marine Board Inquiry, and experts involved in marine insurance, underwriting and surveying.

After 17 days of proceedings, the case finally came to a conclusion. The verdict was for the defendant. Of note, the lawyer representing William McLeod and *The Bulletin* was none other than Jack Want.

1036 *The Sydney Morning Herald*, 31 March 1898; *The Brisbane Courier*, 31 March 1898.
1037 *The Daily Telegraph*, 2 April 1898.
1038 *The Daily Telegraph*, 7 June 1898.
1039 *The Australian Star*, 1 April 1898.
1040 *The Australian Star*, 1 April 1898.
1041 *The Australian Star*, 1 April 1898.
1042 *The Maitland Daily Mercury*, 11 July 1898.
1043 *The Brisbane Courier*, 27 August 1898.
1044 *The Brisbane Courier*, 27 August 1898.
1045 *The Daily Telegraph*, 10 September 1898.

1046 *Truth*, 4 September 1898.
1047 *The Australian Star*, 17 February 1898.
1048 *The Daily Telegraph*, 22 September 1898.
1049 *Truth*, 4 September 1898.

WALTER REEKS GIVING TESTIMONY AT THE ACATAMA TRIAL.
SOURCE: TRUTH, 18 SEPTEMBER 1898.

OIL LAUNCH MUSA AT LAUNCHING.
SOURCE: THE SYDNEY MAIL AND NEW SOUTH WALES ADVERTISER, 12 NOVEMBER 1898.

After the sensationalism of the *Acatama* trial, Reeks returned to work. In late September of 1898 he requested tenders for the building of a 40ft launch, the construction of which was completed in mid-1899.[1050] Next came an oil launch he designed that was built under his supervision by Ford (Jr) to the order of Dr Cecil Vaughan. It was launched at Berrys Bay in mid-October of 1898.[1051] The vessel had been commissioned with the specific intention of journeying from Sydney to New Guinea.[1052] Named *Musa*, the small craft had a length of 35ft, beam of 8ft and a depth of 4ft, being rigged as a ketch.[1053] Its accommodation consisted of a forecastle for two people, a hold of cargo, space for machinery, a cabin and storeroom.

Named after the doctor's intended destination, the Musa River in south eastern Papua New Guinea, and the Musa River Plantation Company site he was setting up as their manager, the vessel had an interesting bow profile. The forefoot was angled upward and joined into a curved stem, a shape that was possibly designed to help the craft as it negotiated the shallow estuary of the Musa River at Collingwood Bay. Groundings would have been inevitable, and it would help if the craft could get close to the shoreline. This profile would have been able to nose closer toward the river bank and when accidents happened, the grounding would be more gentle, thereby allowing the vessel to reverse off the muddy bottom rather than having the stem stuck hard into the shallow obstacle.

Reeks' preparedness to take on a challenge was on show with *Musa*. Here was an unusual set of requirements by a client well out of the ordinary, destined for a region that even today is very remote. The oil engine, still new to the market, was being tested to its limits, and in many reports the novelty of the combustion engine's 'explosion' and 'explosive' power was noted.

Musa's launch and readiness for the passage was followed closely. It was an ambitious trip, considering the size and simplicity of the vessel. Still, powered by a 4 h.p. Union oil engine, *Musa* left Sydney on 15 November 1898 with Dr Cecil Vaughan on board and a crew of three.[1054]

The Daily Telegraph reported. '*A vessel that has been the object of considerable interest at the port is the oil launch Musa, recently built by Dr. Cecil Vaughan, for use in New Guinea waters. The tiny craft is designed for seagoing, and it is the intention of her owner to navigate the vessel hence to New Guinea. The venturesome voyage will be commenced in the course of a few days, and doubtless will be watched with interest, for the Musa is certainly one of the smallest craft built at the port for a long ocean trip.*

The propelling power is a four h.p. Union oil engine, made by the Union Gas Engine Company, of San Francisco, California, whose representatives in Sydney are the Union Oil and Gas Engine Agency, 29 Hunter Street. This engine is of the very latest pattern. Simplicity is a leading feature, and it is claimed that any intelligent person can see and understand the use and proper treatment of the few moving parts in a very short space of time.'[1055]

The article continued in some detail and noted that the oil used, naptha, gave it considerable range under power. A lot of thought had gone into this craft. *Musa* and its engine raised further interest as the party travelled north. Stopping at Brisbane, the vessel was described in a widely circulated report as 'a *new departure in ships, and a splendid example of the advances made with oil as a motive power ... The engine is of the latest type and its simplicity is astonishing ... The engine is small, - ridiculously small - Above all, the engine is cheap*'.[1056]

Such was the interest that *Musa*'s stopover in Cooktown was also highlighted. The craft then made a passage

[1050] *The Daily Telegraph*, 28 September 1898.
[1051] *The Sydney Morning Herald*, 12 October 1898.
[1052] *The Sydney Morning Herald*, 12 October 1898.
[1053] *The Daily Telegraph*, 15 October 1898.
[1054] *The Sydney Mail and New South Wales Advertiser*, 12 November 1898.
[1055] *The Daily Telegraph*, 15 October 1898.
[1056] *Newcastle Morning Herald and Miners's Advocate*, 26 November 1898.

MUSA LAYOUT.
SOURCE: THE DAILY TELEGRAPH, 15 OCTOBER 1898.

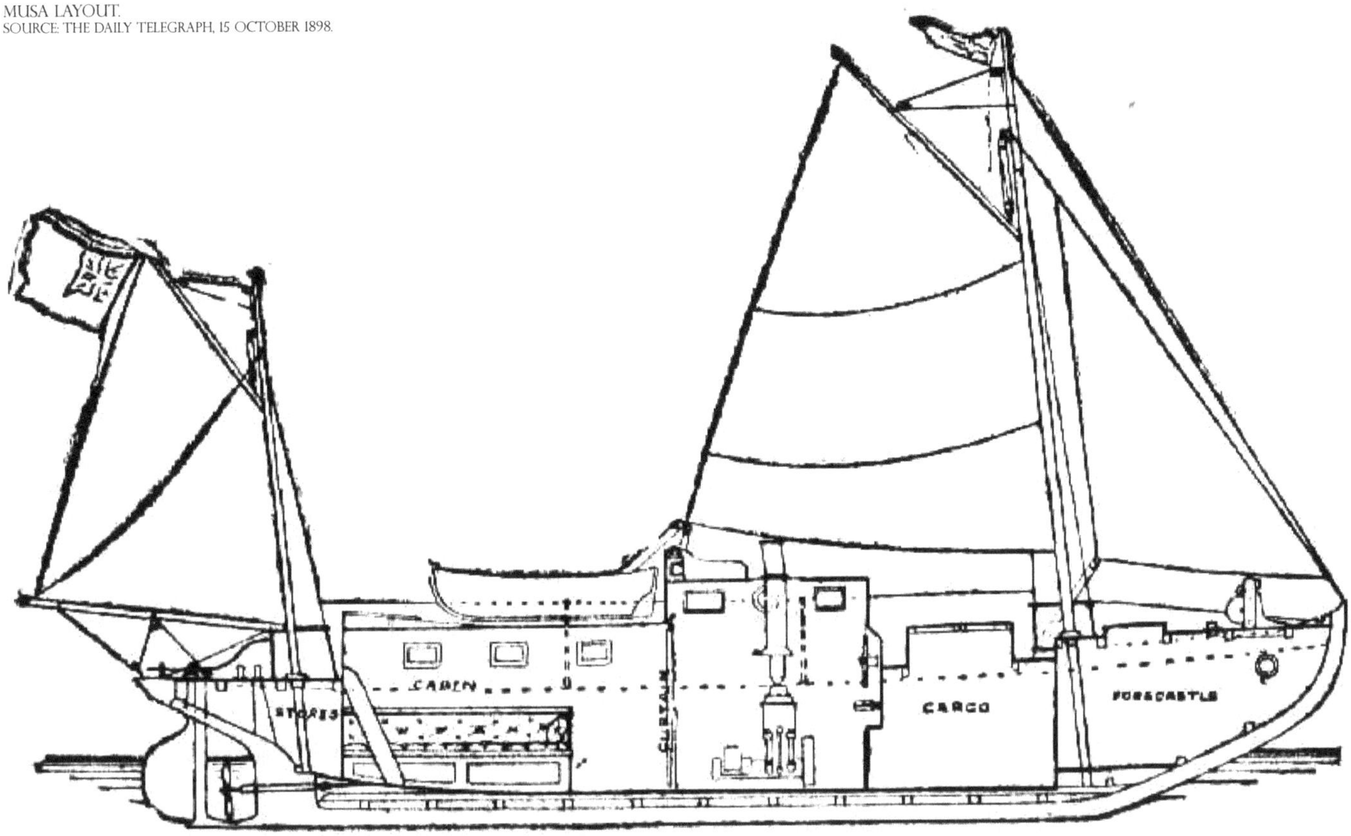

across the Coral Sea to Samarai Island, at the tip of New Guinea, a mecca and crossroads for trade at the time, now superseded by Alotau in nearby Milne Bay, where again its arrival was reported in the papers, with Dr Vaughan advising '*that the engine worked most satisfactorily – better than on the trial in Sydney*'.[1057] These reports, combined with those supplied by the Union Gas Engine Company, doubled as a promotion for the oil engine, and no doubt helped establish this as a practical alternative to steam, to the future benefit of both builders and designers.

Musa's story is worth following a little more, as it's a true adventure of the times. While the indigenous population had, from reports, proven difficult to engage with by earlier explorers, once at Musa River, Dr Vaughan's relationship with them was apparently good. There *Musa* was also put to work, proving to be an industrious little craft. '*The timber and iron for this* [a cottage at the riverhead] *were towed up from Samarai in whaleboats. When this was completed the Musa returned to Samarai for stores, and early in February was back at the Musa River again, engaged in the formidable task of carrying and towing timber and stores up to the spot selected by Dr Vaughan for the company's headquarters ... The little company of pioneers, including, besides Dr. Vaughan, two white men, and six natives from Samarai, had to be prepared for any emergency. The current of the river was so strong that it took the little vessel 14 hours travelling at seven knots, to reach the site of the company's station, whereas it only takes between three and four hours to get back to the mouth of the river again.*'[1058]

[1057] *The Sydney Morning Herald*, 26 January 1899.
[1058] *The Daily Telegraph*, 31 May 1899.

Musa certainly earned its keep for both Dr Vaughan's plantation company and the Union Gas Engine Company. The craft was an absolute credit to Reeks' design abilities too.

The cargo and passenger steamer for New Zealand that Reeks had designed for the Northern Union Steamship Company, after some delays, was finally launched on 29 October 1898, a few weeks after the *Musa*. The vessel measured 126ft 6in with a beam of 24ft 7in and a depth of hold of 13ft 3in.[1059] It was named *Aotea* and notably marked the first vessel built at the newly-established Pyrmont shipyard of T. F. Morrison.

The press coverage of the new vessel was extensive. '*A new steamer, which embodies a peculiar and interesting departure in construction, was launched on Saturday with great success. She was built by Mr. Morrison, shipbuilder of Pyrmont, to the order of the Northern Union Steamboat Company of New Zealand. There was a large gathering present to witness the launch, including representatives of shipping and commercial houses, the Marine Board, and the Legislature. The proceedings went off without a hitch, and as the Aotea glided gracefully off the ways she was christened in the orthodox manner by Miss Evelyn Lamb, daughter of Mr. Lamb, the company's representative in Sydney.*'[1060]

The article continued. '*The vessel is 126ft long overall, 24ft 7in beam, and 13ft 3in depth of hold, and she is guaranteed to*

[1059] *The Daily Telegraph*, 12 April, 28 October 1898.
[1060] *Australian Town and Country Journal*, 5 November 1898.

run 70 miles at a speed equal to 11 knots. Her draught, when fully loaded, will not exceed 6½ft, but in order to bring about this state of affairs a slight detour from the ordinary method adopted in vessels of this class has been found expedient, and Mr. Walter Reeks, her designer, has adopted a modification of the construction employed in the light draught steamers used as dispatch boats, gun boats, and the like, which do duty on the Nile and on the shallow water rivers of India.'[1061]

Another article provided further detail on this design attribute. '*The feature of this plan is the formation of the stern, which takes the shape of a long gradually rising tunnel, highest immediately over the propeller, and gently falling to the load water mark at the extreme after end. This tunnel is in the shape of half a cylinder, the edges of which are always under water, but the top of the arch inside is a foot above water when the boat is fully loaded. The propeller runs through the centre of this tunnel, and when the boat is stationary, the top of the screw is some distance out of the water. As soon as the engines are started, however, air in the tunnel is expelled and water immediately takes its place, and as the air cannot get back the propeller works in solid water, and is as efficient as it would be in a deep draught vessel. The loss of efficiency in light draught craft is mainly due to the points of the propeller taking down air, but in the tunnel-stern-steamers it cannot do so, and higher speed is the direct result.*'[1062]

This was a twin-engine vessel (developing 250 h.p. in total), but they were coupled to drive a single shaft and propeller. Thankfully a clearer view of this arrangement was given by an illustration published in *The Daily Telegraph* on 5 November 1898.

ILLUSTRATION OF THE LAUNCH OF AOTEA.
SOURCE: THE DAILY TELEGRAPH, 5 NOVEMBER 1898.

As noted, the tunnel stern was already in use in other parts of the world to allow a shallow draft, but the description and picture suggests there could be something different in the shape Reeks had designed for this particular tunnel. Of note, after this design he found it appropriate to repeat the tunnel in other craft where needed, but Reeks soon found a way that certainly made it unique to him, when he modified the configuration on a twin-screw steam ferry design for use in Brisbane.

The *Aotea* cleared Sydney Heads en route to New Zealand on 15 December 1898.[1063] Successfully navigating the Tasman Sea, the vessel arrived in Kaipara five days later.[1064]

For Reeks, the year 1898 concluded with him travelling as a passenger on board RMS *Mariposa,* leaving Sydney on 19 December bound for Auckland.[1065] He was headed to Kaipara to oversee trials of the *Aotea* in its new home waters. Proving a success, the vessel commenced service soon thereafter.[1066]

In addition, the trip was an opportunity to experience a new country, to inspect its ship and boat building yards and to see its vessels. The Sydney press reported. '*Mr. Walter Reeks is over in Auckland on a visit, and is greatly enjoying the trip, he was away for four days as a guest of Mr. Henderson in the ex-Sydney yawl Volunteer, visiting Mahurangi, Coromandel, and Waihiki, and has also been cruising with Commodore Wiseman in his big yawl* [Matangi]. *Mr. Reeks is at present at Rotorua, doing the Hot Lakes, and proposes to extend his trip down the famous Wanganui River.*'[1067]

The New Zealand-based press also provided updates of his movements. On 31 December 1898 the *New Zealand Herald* reported, '*Mr. Reekes* [SIC], *the Sydney yacht designer, has visited the various yacht-building yards during the week, and has also inspected a number of the locally-built yachts, and I understand has been much impressed with the workmanship and up-to-dateness of our local builders. It is Mr. Reekes' opinion that Auckland is a long way in advance of any of the Australian colonies regarding yachting matters generally*'.

A similar report in the same paper was published three weeks later. '*Mr. W. H. Reekes* [SIC], *the well-known Sydney naval architect and engineer, who has been on a month's visit to Auckland, returned to Sydney on Wednesday. During his stay in the city, Mr. Reekes visited the various ship and yacht building yards, and closely examined the workmanship, etc., of our builders. He is a very keen observer, and has gone away with his head fairly well stored with matters appertaining to the craft. He expresses himself as delighted with his trip, and the cordial manner in which he has been treated by everybody.*'[1068]

Interestingly, the first intercolonial yacht race to ever take place in New Zealand, was held in Auckland coinciding with Reeks' visit. The matches were sailed on 28 and 31 December 1898, and a third race was held on 4 January 1899, featuring two Australian one-raters, the *Geisha* (owned by Rymill Bros. of Adelaide), and the *Bronzewing IV* (owned by Sam Hordern, Jr, of Sydney). Both craft were competing against several local yachts of the same class. Summarising the event, which was sailed on Auckland's famed Waitamata Harbour with its strong tides, the press perhaps stated the obvious, '*all three races*

[1061] *Australian Town and Country Journal*, 5 November 1898.
[1062] *The Daily Telegraph*, 28 October 1898.
[1063] *Evening News*, 16 December 1898.
[1064] *The Daily Telegraph*, 23 December 1898.
[1065] *Evening News*, 19 December 1898.
[1066] *New Zealand Herald*, 16 January 1899.
[1067] *Referee*, 18 January 1899.
[1068] *New Zealand Herald*, 21 January 1899.

LAUNCH OF THE AOTEA.
SOURCE: GRAEME ANDREWS PHOTOGRAPH COLLECTION, CITY OF SYDNEY ARCHIVES.

for the Championship were marred by adverse weather and probably should not have been started at all. The first two were "drfiters" and the last sailed in a screaming northerly against an ebb tide'.[1069]

In the first race, the contest was won by Logan Bros.' *Mercia* (a vessel they had designed and built in 1898), with Charles Bailey's *Laurel* second, while *Geisha* finished sixth and *Bronzewing IV* seventh.[1070] The next race was won by *Laurel*, with *Mercia* fourth, *Bronzewing IV* fifth and *Geisha* sixth.[1071] The final race, sailed under trying conditions and a heavy sea, saw many of the local yachts retire or come to grief. Only three vessels remained in the contest with *Bronzewing IV* claiming first place, *Laurel* second and *Geisha* third. Accumulated points gave the prize of £100 to *Laurel*.

In relaying the outcome of the race series, Reeks was quoted as saying, '*Auckland is fully ten years ahead of Australia in the matter of yachts, both as regards to designing and workmanship*', going on to assert that the visit had been quite an '*eye-opener*' for him.[1072]

What Reeks witnessed first-hand was that the New Zealand designers and builders were well acquainted with the Linear Rule. Combined with their exceptionally strong and somewhat unique double and triple-skin kauri construction, the New Zealand yachts stood out from the pack. In just a few short years they would make their mark in Sydney yachting as well.

MERCIA COMPETING IN AUCKLAND.
SOURCE: AUCKLAND WEEKLY NEWS, 6 JANUARY 1899.

[1069] R. Elliot & H. Kidd (2001). *The Logans, New Zealand's Greatest Boatbuilding Family.*
[1070] *The Australasian*, 21 January 1899.
[1071] *The Australasian*, 21 January 1899.
[1072] *The Australasian*, 21 January 1899.

1889 MAP OF PORT JACKSON (NEW SOUTH WALES) AND SURROUNDS.
SOURCE: NATIONAL LIBRARY OF AUSTRALIA (NLA.OBJ-231707720).

CHAPTER FOURTEEN - PORT JACKSON

On 19 January 1899 Reeks left Auckland for Sydney as a passenger on board RMS *Alameda*.[1073] The month-long trip to New Zealand had proved extremely beneficial in many ways. From a commercial standpoint, he had been showered with praise, particularly for his involvement with the *Aotea*. While on board the vessel for an excursion on 14 January from Helensville to South Head, Kaipara, Reeks was the recipient of numerous toasts and accolades. There was also the prospect of more business opportunities, with the directors of the Northern Union Steamboat Company indicating their intention to commission more steamers.[1074]

AOTEA (1899).
SOURCE: AUCKLAND WEEKLY NEWS, 20 JANUARY 1899.

[1073] *Auckland Star*, 19 January 1899.
[1074] *New Zealand Herald*, 16 January 1899.

Reeks arrived back in Sydney on 22 January 1899, missing the ceremonial launch of Sam Hordern's new Fife-designed yacht by a mere three days.[1075] *White Wings* had been released from the stays at the Berrys Bay yard of Watty Ford (Jr) on 19 January, immediately hyped as a worthy addition to the under 10 rating class.[1076]

WHITE WINGS.
SOURCE: AUSTRALIAN NATIONAL MARITIME MUSEUM COLLECTION (ANMS0529[008]).

Continuing to expand his customer base, while in New Zealand Reeks had received two orders for new designs from clients based in Brisbane. The first was for a yacht commissioned by Percy Outridge, with the vessel to be built at Henry Thomas Green's yard at Lavender Bay.[1077] Outridge was another significant player in the Torres

[1075] *The Sydney Morning Herald*, 21 January 1899; *The Daily Telegraph*, 23 January 1899.
[1076] *The Sydney Morning Herald*, 21 January 1899; *The Daily Telegraph*, 23 January 1899.
[1077] *The Sydney Morning Herald*, 4 January 1899; *The Brisbane Courier*, 5 June 1899.

Strait pearling industry and a very close associate of his Brisbane neighbour, the 'Pearl King' James Clark. This specific craft was to be a centreboarder, '*she will be a 2½ rater, and built more for comfort than speed ... altogether a nice, roomy, handy little boat*'.[1078] The second order was from the Humpybong Steamship Company who tasked Reeks with designing a new steamer to trade between the Humpybong Peninsula and Brisbane, to be built under Reeks' personal supervision.[1079]

On 11 February 1899 another new order was announced. The Port Jackson Co-operative Steamship Company had commissioned Reeks with designing a steamer for its Manly service, to be built at a cost approaching £18,000.[1080] The vessel was to be based on the same principle as *Manly*, still part of the company's fleet, though was to be constructed of steel and 20ft longer.

That same day an event of interest to Reeks took place at the Rozelle shipyard of Messrs Young, Son and Fletcher. The Parramatta River Steamers and Tramway Company's new vessel *Bronzewing* was launched, gracefully entering the water after receiving the customary dousing of champagne. Immediately afterwards a small number of toasts were drunk, including one to Reeks who was involved in drawing up the lines of the new vessel, based on those taken from the company's existing steamer *Pheasant*.[1081]

Measuring 115ft (overall), with a beam of 22ft and a depth of 7ft 15in, *Bronzewing* was to be fitted with triple expansion engines supplied by Messrs Ross and Duncan of Glasgow, Scotland.[1082] However, an engineers' strike in Great Britain had caused significant delay to the arrival of the machinery such that the vessel's launch was postponed some eight months, and its trial trip not taken until 23 March 1899.[1083]

With Reeks on board that day out on Port Jackson, *Bronzewing* performed well, proving itself capable of attaining 12 knots an hour and negotiating the measured mile in a time of 5 minutes and 4 seconds.[1084] A week later the craft began service on the Parramatta River.

[1078] *Australian Canoeist and Yachting*, 24 December 1898.
[1079] *The Brisbane Courier*, 19 January 1899.
[1080] *Evening News*, 11 February 1899.
[1081] *Evening News*, 11 February 1899; *The Daily Telegraph*, 24 March 1899.
[1082] *Evening News*, 11 February 1899.
[1083] *The Daily Telegraph*, 24 March 1899.
[1084] *The Daily Telegraph*, 24 March 1899.

STEAMER BRONZEWING.
SOURCE: GRAEME ANDREWS WORKING HARBOUR PHOTOGRAPH COLLECTION, CITY OF SYDNEY ARCHIVES.

Continuing his association with Auckland's yachting community, on 27 February 1899 Reeks entertained two crew members from the one-rater *Mercia* on a moonlight bicycle tour starting in the centre of Sydney and ending at Tollemache's Hotel in Sandringham.[1085] Arch Logan and R. H. McCallum had both arrived in Sydney on board the *Westralia* along with their vessel earlier that month.[1086] Occupying the head of the table, Reeks made several toasts to the New Zealand guests. The party then adjourned to the pavilion where songs and recitations were rendered, including one performed by Reeks.[1087]

Despite being launched only the year before, *Mercia* was a Logan Bros. legend in New Zealand and would build on this reputation in Australia and later in South Africa as the craft moved through different owners. In Sydney it was later stated, '*No boat of the type* [i.e., the small raters] *ever performed so wondrously as Mercia*'.[1088] The vessel had proved just how seriously the Kiwis took their racing. Back in Auckland, *Mercia* had been designed and built as a response to Charles Bailey's *Laurel*. Immediately after the intercolonial yacht race, held over the 1899 new year period, the latter was sent to Sydney to compete at the Anniversary Regatta, held in late January.

[1085] *The Sydney Morning Herald*, 3 March 1899.
[1086] *The Daily Telegraph*, 8 February 1899; *Colonist*, 8 February 1899.
[1087] *The Sydney Morning Herald*, 3 March 1899.
[1088] *Australian Motor Boat and Yachting*, August 1925.

NEW STEAMER FOR THE MANLY FERRY.

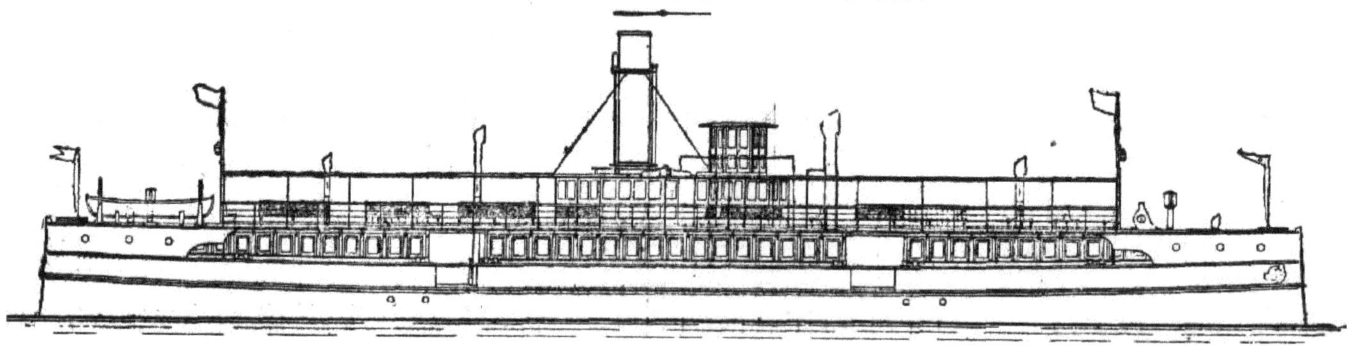

SOURCE: THE DAILY TELEGRAPH, 25 MARCH 1899.

Emboldened by *Mercia*'s own performance on Auckland Harbour against *Laurel,* and indicative of his competitive nature, Arch Logan decided to continue the duo's battle abroad and also shipped *Mercia* to Sydney. Upon arrival, however, he immediately sold the celebrated craft to local yachtsman Fred Doran.[1089]

MERCIA COMPETING IN AUCKLAND.
SOURCE: AUCKLAND WEEKLY NEWS, 6 JANUARY 1899.

While their first match up on Port Jackson went in *Laurel*'s favour, *Mercia* '*then proceeded to demolish the Sydney boats. By the end of the season Mercia was giving Laurel six minutes handicap and honour was restored*'.[1090]

This period of New Zealand yacht migration to Australia started with the small raters and grew into a number of larger yachts. The next blow was soon to be struck by another Charles Bailey yacht called *Meteor,* but thereafter it was the craft from Logan Bros. that dominated events on Sydney Harbour, and helped develop strong club racing into the new century.

Fortunately for Reeks his commercial business interests were flourishing. He had been to Auckland and witnessed first-hand the high level of craft being developed there. With the benefit of hindsight, it appears that about this time he gracefully opted out of racing yacht design, but kept involved in other ways. He did undertake some projects modifying existing yachts, but he also put a lot of time and energy into administrative efforts associated with Sydney's yacht clubs. This high level of volunteerism continued into the years and decades ahead. As an example, during 1899 alone Reeks was elected to the committees of the Royal Sydney Yacht Squadron, the Sydney Canoe Club, the Johnstone's Bay Sailing Club, and the Sydney Dinghy Sailing Club.[1091]

Reeks' efforts were noticed. A special general meeting of members of the Sydney Amateur Sailing Club held on 25 April 1899 saw him thanked for his involvement with the club's committee.[1092] He was also noted as assisting with efforts to revise a clause in the organisation's constitution dealing with 'unfair boats' with the altered wording stating '*No double-hulled or other boat whose form shall in the opinion of the committee give her an unfair advantage shall be allowed to compete in the 16ft and 30ft rating classes*'.[1093]

This change was most likely brought about by the 24ft catamaran or double-hulled 'skiff' *Flying Fish* that had been on the harbour since 1894. It had been built for the flamboyant entrepreneur and wealthy department store owner Mark Foy, founder of the Sydney Flying Squadron in 1891. It was the first catamaran seen on Sydney Harbour and showed its heels to any other boat when reaching, but was less successful on a beat to windward or running square. Initially this unusual craft confined its sailing to Foy's own club, but in 1897 with 2ft cut off both hulls *Flying Fish* appeared amongst the 22-footer open boats, and its presence was not well received. The Johnstone's Bay Sailing Club tried to ban the craft that season and it seems the Sydney Amateur Sailing Club also put this directive into effect for their classes when they revised their rules with this change. The specific language adopted also allowed discretion to review any other strange design that may appear in the future.

FLYING FISH.
SOURCE: AUSTRALIAN NATIONAL MARITIME MUSEUM COLLECTION (00013804).

For Reeks more personally, there was also recognition for his efforts associated with Sydney's yachting history. In honour of his service to the pastime on Port Jackson over several decades, a September 1899 issue of the *Town and Country Journal* published his portrait, along with those of several leading local yachtsmen and yachts.

1089 *The Sydney Morning Herald*, 23 February 1899.
1090 R. Elliot & H. Kidd (2001). *The Logans, New Zealand's Greatest Boatbuilding Family.*
1091 *The Daily Telegraph*, 29 April, 28 September 1899; *The Sydney Morning Herald*, 11 August 1899.
1092 *The Daily Telegraph*, 8 September 1899.
1093 *The Daily Telegraph*, 29 April 1899.

THE TOWN AND COUNTRY JOURNAL. 21

1.—Mr. T. W. Bremner, owner of the "Fairlie." 2.—Mr. W. M. Marks, owner of the "Isea." 3.—Mr. W. A. Cameron's "Iduna." 4.—Mr. Walter Reeks. 5.—Mr. F. J. Jackson, owner of the "Violet." 6.—Mr. Sam. Hordern, Commodore F.A.Y.C. 7.—Dr. M'Cormick, owner of the "Thelma." 8.—Mr. C. E. Newman, owner of the "Archina." 9.—Mr. T. A. Dibbs's steam yacht "Ena." 10.—"White Wings."

(See letterpress on page 24.)

PORTRAIT OF WALTER REEKS (NO. 4 ON RIGHT).
SOURCE: THE TOWN AND COUNTRY JOURNAL, 30 SEPTEMBER 1899.

With plans finalised for the Port Jackson Co-operative Steamship Company's new steel double-ended screw steamer, in April of 1899 Reeks called for tenders for the vessel's construction. Measuring 175ft in length with accommodation for over 1,000 passengers, the craft was by far the largest steamer he had designed to date.[1094] The vessel was also to be the fastest, intended to travel at a speed of 16 knots, supplied with triple expansion engines.[1095]

In addition, that month saw Reeks call for tenders for the building of two more steamers to plans he had developed, one measuring 130ft by 25ft and the other measuring 50ft by 12ft.[1096] This mode of transport was proving extremely popular across all regions of Australia and New Zealand, with Reeks certainly relishing in the high workload on offer.

Not one to decline an invitation, on 29 April 1899 Reeks took part in a trial trip of an auxiliary schooner of 100 tons register on Sydney Harbour. On board were several high-profile yachtsmen and maritime professionals. The vessel, Allan Taylor's new craft *Annandale,* had been built by Drake's shipyard at Balmain. The craft was set up with sail and auxiliary power from a 50 h.p. Union oil engine and was intended to ply the coastal trade out of Macleay River.[1097]

The *Annandale* was an absolute novelty for those on board, and for all that it represented. The combination of steam and sail had existed for some time, but auxiliary power from an oil engine in combination with sail power had the potential to further revolutionise commercial sea and river trade. While smaller launches had been fitted with oil engines in the years immediately prior, such independence gained by the fitting of engines in larger commercial craft would not only increase speed and reliability of the trade, but also profits. It was a calm morning and '*The Annandale was certainly the envy of a number of small craft at anchor waiting for a slant of wind. Mr. Allan Taylor, who was on board, states that the schooner is now independent in that she can start her voyage notwithstanding headwinds or calms. The amount of interest displayed on Saturday was remarkable*'.[1098] Exceeding expectations, the *Annandale* achieved the measured mile in a time of 10 minutes, running at a speed of six knots.[1099] The vessel was a pioneer in a new era for sea transport and Reeks, accustomed to his own interest in doing something new, was only too happy to be along for the journey.

Unfortunately, the pace at which Reeks attacked his professional and personal commitments soon caught up with him. On 10 June 1899 *The Sydney Morning Herald* published the following note. '*His numerous friends will learn with regret that Mr. Walter Reeks, the well-known naval architect, is lying dangerously ill at his residence at Mosman's Bay. Yesterday was the eighteenth day he has been confined to bed with typhoid fever, and upon inquiry at his house it was stated that he was in a critical condition.*'

Those writing for the *Australian Canoeist and Yachtsman* were just as concerned, with an initial reference on 3 June, '*typhoid fever, we understand is the cause. We join with his many friends in sincerely hoping that his case is not a bad one*'. The next week, 10 June, an update was published. Reeks was improving, '*we are glad to add that Dr Newmarch reports his patient is progressing very favourably, and that he is very hopeful that the attack will not be a serious one*'. A fellow yachtsman, Newmarch was also a previous client of Reeks', being the owner and skipper of *Myee.*

Caused by the pathogenic bacteria *Salmonella typhi*, typhoid was generally contracted by the consumption of contaminated food or water. A serious disease that could often be fatal, patients with typhoid experienced many symptoms over a period of several weeks, including prolonged fever, sweating, headaches, abdominal pain and stomach issues, as well as lack of appetite and weight loss.[1100] While the bacteria that caused typhoid, and others associated with cholera, diphtheria and tuberculosis, had first been identified in the early 1880s through advances in medical technology, a vaccination for typhoid was not developed until the late 1890s and it would be some decades before it was in widespread use. In the meantime, efforts to control the disease centred on reporting of outbreaks and increasing access to clean and safe drinking water and installing separate systems for sewerage.

A report by Dr Mailler Kendall, medical adviser to the Sydney-based Water and Sewerage Board published in June of 1899, the same month that Reeks contracted the disease, discussed the status of typhoid cases in the city and its surrounding suburbs.[1101] Overall, in the first quarter of 1899 cases averaged between 4 and 10 per 10,000 people, depending on location.[1102] These numbers were similar to the previous year but it was only since 1898 that mandatory reporting had been required.[1103] The report surmised that more than half of the reported typhoid cases in and around Sydney came from districts where sewers did not exist, and that providing districts with a proper system of sewerage and a good water supply could successfully lower the cases.[1104]

Fortunately for Reeks he did recover from typhoid. But it was a slow process as he is not mentioned until late July as attending any meetings or events. However, it is curious to note that seven months earlier, on 12 November 1898, the *Australian Canoeist and Yachtsman* had noted that Reeks was ill, and the ominous word 'typhoid' had been mentioned, so perhaps this was not his first brush with the illness.

While he lay seriously ill, work on Reeks' many projects continued. Building of the 40ft launch that he had designed and called tenders for its construction in September of the prior year was completed in June of 1899, though it is doubtful that Reeks was present to witness the

[1094] *The Australian Star*, 10 March 1899.
[1095] *The Daily Telegraph*, 25 March 1899.
[1096] *The Sydney Morning Herald*, 14 April 1899.
[1097] *The Sydney Morning Herald*, 2 May 1899.
[1098] *The Sydney Morning Herald*, 2 May 1899.
[1099] *The Sydney Morning Herald*, 2 May 1899.

[1100] www.health.nsw.gov.au/Infectious/factsheets/Pages/typhoid.aspx.
[1101] *Evening News*, 9 June 1899.
[1102] *Evening News*, 9 June 1899.
[1103] *Evening News*, 9 June 1899.
[1104] *Evening News*, 9 June 1899.

launch itself given his convalescence.[1105] Named *Amy,* the pleasure craft was built by James Hayes at Milsons Point to the order of J. S. Meiklejohn, manager of the Commercial Bank in Bundaberg, Queensland.[1106] To be based on the Burnett River, with occasional trips out to sea, the vessel was driven by a 4 h.p. '*oil-electric*' engine and measured 39ft overall, with a beam of 7ft 3in and a draft of 2ft 6in when loaded; its carrying capacity was stated to be about 20 passengers.[1107] The term 'oil-electric' used at the time, reflects the still novel method of power. It simply refers to the electric spark-plug ignition of a typical petrol engine.

Returning to work, on 22 July 1899 Reeks called for tenders for the building and fitout of a 100ft by 22ft wooden steamer, with submissions due in three days.[1108] This may have been the order placed by the Northern Union Steamship Company of New Zealand for a second, slightly smaller steamer.[1109] The craft was intended for service on the Otamatea River. However, for reasons unknown, but possibly owing to the quick three-day turnaround time, it appears that this vessel was not built. Instead, in 1903 the company would commission Reeks to design a much larger and more substantial vessel.[1110]

Of note, also in July of 1899, the board of directors of the Port Jackson Co-operative Steamship Company announced that Mort's Dock Company was awarded the contract for building their new steamer. It was a local coup and one that certainly challenged the mindset of English-built vessels being considered superior to those built in Sydney. A statement was issued, '*The whole of the work is to be done in Balmain. It is, perhaps, our first effort at federation, and at all events our new ship will be entirely from Mort's, and will be the largest and the most expensive vessel that has been built there'*.[1111] The piece countered, '*Why not follow some precedents and go to the old country?*' However, this option had been thoroughly considered but owing to the costs involved was decided against. It was a pivotal display of trust and confidence in Sydney's shipbuilders. Suddenly they found themselves not only competing against their older and more established brethren in Great Britain, but also winning important contracts.

1105 *The Daily Telegraph*, 28 September 1898.
1106 *The Bundaberg Mail and Burnett Advertiser*, 7 July 1899.
1107 *The Bundaberg Mail and Burnett Advertiser*, 7 July 1899.
1108 *The Daily Telegraph*, 22 July 1899.
1109 *Daily Commercial News and Shipping List*, 3 August 1899.
1110 *Auckland Star*, 23 November 1903.
1111 *The Sydney Morning Herald*, 28 July 1899.

MORT'S DOCK, BALMAIN (CIRCA 1890S).
SOURCE: STATE LIBRARY OF VICTORIA (H91300/369).

With so many diverging projects underway for a variety of different customers, it is hard to imagine how Reeks managed them all. For example, in September of 1899 he would have been supervising the construction of the large steamer for use on the Port Jackson Co-operative Steamship Company's Manly service (the laying down of this particular vessel commenced on 23 August), and finalising the tender for the Northern Union Steamship Company's new steamer.[1112] There were also then under construction two additional steamers that he had designed, one 130ft by 25ft and the other 50ft by 12ft. The first craft was for the Brisbane-based Humpybong Steamship Company, the second was for a customer from the Northern Rivers region.

Also nearing completion was Percy Outridge's 2½ rater yacht. And remember the two steamers that Reeks had successfully tendered the New South Wales Public Works Department for the previous year, to be used on the Clarence and Hunter rivers, respectively? Well, the first was about to be launched. While contracting typhoid had certainly resulted in a month or two hiatus, Reeks appears to have returned to work with the same energy and vigour that he was known for.

Built for use by the Roads and Bridges Branch of the Public Works Department to be based on the Clarence River, the double-ended screw steam launch *Una* took its trial trip on Port Jackson on 20 September 1899.[1113] Measuring 70ft in length, with a beam of 16ft and depth of 7ft, the multi-purpose vessel had seating accommodation for 100 passengers, but was also able to undertake towing duties. *Una* was arranged with pilot houses located at each end in which patent steering gear was fitted, such that the wheel could be locked while the helmsman managed the gangways.[1114] The craft was propelled by compound surface condensing engines, of 60 h.p., and during the trial trip, its contract speed was easily obtained.[1115] *Una* conveyed the measured mile in a time of 6 minutes and 15 seconds.[1116]

Una's sister-ship *Helen*, intended for use on the Hunter River based out of Morpeth, was then at the time getting its machinery installed. The craft was launched only a matter of weeks later.[1117] Compared to the larger double-ended ferries Reeks had designed for harbour routes, this pair of twin vessels were petite versions in the same style, with excellent proportions and proper attention to detail. As noted at the end of the report detailing *Una*'s trial, '*the Una adds one more to his already long list of successes*'.[1118]

[1112] *Evening News*, 5 July 1900.

[1113] *Evening News*, 21 September 1899.
[1114] *Evening News*, 21 September 1899.
[1115] *The Sydney Morning Herald*, 21 September 1899.
[1116] *The Australian Star*, 21 September 1899.
[1117] *The Sydney Morning Herald*, 21 September 1899.
[1118] *The Sydney Morning Herald*, 21 September 1899.

STEAM LAUNCH UNA.
SOURCE: STATE RECORDS AUTHORITY OF NEW SOUTH WALES (NRS-4481-2-[4/8610]-547).

STEAM LAUNCH HELEN.
SOURCE: GRAEME ANDREWS WORKING HARBOUR PHOTOGRAPH COLLECTION, CITY OF SYDNEY ARCHIVES.

On 18 November 1899 a substantial oil launch was completed, this one constructed by Thomas King, well known boat builder of the Northern Rivers region of New South Wales. Designed by Reeks, the vessel was built to the order of S. McNaughtan of Maclean and intended for trading purposes.[1119] The craft was aptly named *Fairtrader* and measured 50ft in length with a beam of 12ft and a draft of 3ft 6in.[1120] It was in fact one of a classic type, the 'dealers boats' or 'cream boats' that were common to the Northern Rivers over many decades. They were akin to a travelling shop and carried goods and mail to the folk living on the rivers and took their produce, including milk, on the return journey, destined for markets in the larger cities and towns. Once again the reports regarding this particular vessel note that it had '*a light draught*' and '*being intended for dealing purposes she is provided with a roomy cabin and storeroom amidships, extending the full width of the hull*'.[1121] A detailed description of the oil engine's workings and the method of reversing the shaft to go astern '*by moving a lever*' were included, plus the all-important point that it was christened with '*a bottle of whisky*' by Miss Kate Cameron.[1122]

In later years the *Fairtrader* was owned by Tom Henderson of the Clarence River. His son, Tom (Jr), also worked on the boat, and his daughter Julie passed on the following story, indicative of the lifestyle. '*He told me how delighted the farm ladies would be to be helped aboard from their little jetties or wharves to buy cottons and laces etc., while the eggs and chickens etc. were loaded on top.*'[1123]

Tom Henderson (Jr) also painted the *Fairtrader*, a copy of which is provided below, complete with flying ducks and paddling pelicans. The work captures the curved bow which Reeks gave the vessel, a different profile to the normal straight stem, according to Lois Carrington, who was another riverboat operator relative. '*The curved stem, always a talking point, out of iron bark, a hook.*'[1124] This probably refers to the stem being made from a grown crook of iron bark with the curved shape already present. *Fairtrader* would have been strong but also, like Reeks' launch *Musa*, gave a shape that suited manoeuvring the boat close into the river banks.

Socially, 1899 ended with Reeks attending a smoking concert at Aaron's Exchange Hotel for members and friends of the Electrical Association of New South Wales.[1125] On a professional level, the year ended with the arrival in Brisbane of the *Evelyn*, Percy Outridge's new 2½ rater that had been built by Green at Lavender Bay. Intended more for cruising than racing, its principal dimensions were 37ft length (overall), 24ft (waterline), with a beam of 9ft and a draft of 2ft 3in.[1126] The craft had overhangs at both ends and was notably fitted with a brass centreboard and case. The *Evelyn* was sailed from Sydney

[1119] *The Richmond River Herald and Northern Districts Advertiser*, 17 November 1899.
[1120] *The Richmond River Herald and Northern Districts Advertiser*, 17 November 1899.
[1121] *Clarence River Advocate*, 21 November 1899.
[1122] *Clarence River Advocate*, 21 November 1899.
[1123] Julie Duell letter to David Payne, 24 September 2004.
[1124] Lois Carrington letters/emails to David Payne, November 2004.
[1125] *Sunday Times*, 19 November 1899.
[1126] *The Brisbane Courier*, 11 January 1900.

COPY OF PAINTING OF OIL LAUNCH FAIRTRADER BY TOM HENDERSON.
SOURCE: JULIE DEULL COLLECTION.

to Brisbane, completing the trip in 10 days, four days of which were spent in Trial Bay awaiting better weather.[1127]

The new century dawned and Australia was well on the way to becoming a federation, a nation governing itself. The apron strings of the British Empire were less tight and there was more independence. With it came changes: politically, economically, socially, all generally positive. For Reeks, it was an exciting time to be in Sydney and to be part of the movement, and the continued development of Australia as a nation. He was nearing 39 years of age. He had spent the last two years primarily involved in the design of commercial steamers and oil launches, the pace at which he took on projects sending him to near-death in mid-1899 after being struck down by typhoid. Thankfully he had recovered but it had been a slow process.

Returning to work after the New Year, Reeks' first project for 1900 was for the Balmain New Ferry Company. Plans for a new iron ferry steamer had been developed with Reeks requesting tender submissions for building of the hull in mid-January of 1900.[1128] The vessel's principal dimensions were 130ft by 30ft by 10ft 2in.[1129] In May of 1900 he invited tenders for the building of a double-ended screw ferry steamer for the same company.[1130] However, it is doubtful that either of these vessels were built.

[1127] *The Brisbane Courier*, 11 January 1900.
[1128] *The Daily Telegraph*, 13 January 1900.
[1129] *The Daily Telegraph*, 13 January 1900.
[1130] *The Sydney Morning Herald*, 19 May 1900.

January of 1900 also saw Reeks assisting with organisation efforts for the Anniversary Regatta, acting in the capacity of measurer and helping out on the day, particularly with the intercolonial yacht race, a race once again won by a New Zealand craft named *Rainbow*, a 36ft linear rater designed and built by Logan Bros.[1131] At a follow-up meeting held a few weeks after the event, Reeks was given a vote of thanks by the committee for his efforts.[1132] He would continue to be involved with the Anniversary Regatta and its committee in the decades ahead.[1133]

The steamer Reeks had designed for the Humpybong Steamship Company was also nearing completion by February of 1900 at the Balmain shipyard of George Ellis. *The Brisbane Courier* reported on 12 February 1900. '*At the yards in Balmain the vessel was found to be well forward, planked, and in course of decking. From above the yards, she looks bluff-bowed, but this idea dissolves as one approaches the stern, and looking at her from the forefoot she is seen to have smart if not graceful lines. No one would suppose that the Queenslander is built with an almost flat floor amidships looking from forward at the entrance, but such is the case. Underneath the hull is as saucer-like as one can imagine. When she is in the water this buoyancy quality would be in evidence, but she is designed to carry some cargo in the fore hold and 450 excursionists on her main and upper decks. Mr. Reeks it seems was placed with an*

[1131] *Evening News*, 18 January 1900; *The Daily Telegraph*, 13 January 1900; *The Sydney Morning Herald*, 15 February 1900.
[1132] *The Sydney Morning Herald*, 15 February 1900.
[1133] *The Sydney Morning Herald*, 11 May 1900; *Evening News*, 6 December 1900.

YACHT RAINBOW COMPETING ON AUCKLAND HARBOUR.
SOURCE: AUCKLAND LIBRARIES HENRY WINKELMANN COLLECTION (755-ALB2-32).

order for a steamer capable of steaming eleven knots per hour, and possessing large deck space for the enjoyment of the passengers, and in the new boat he has certainly come as near as possible to realising the ideal. The steam trial has yet to come. The engines (imported) are not to hand, but the contract for the machinery work was taken by Messrs. Begg and Greig, and yesterday Mr. Begg, speaking of the time the ship would leave, hoped to get her away early in April. In her inner works the steamer is built with iron frames, and her knees and frames are of natural hardwood. Mr. Reeks's design embraces a full promenade deck, with awning, and on the main deck a large saloon with sliding ports. Aft is the ladies' cabin, and in the main saloon are the usual setees and tables. The vessel takes the place of the steamer Garnet.'

On 2 April 1900 Reeks was amongst a large assembly of invited guests and interested parties to witness the launch of this new steamer, the *Emerald.* Dr Creed, chairman of the Humpybong Steamship Company, gave a speech of which an extract was published in *The Daily Telegraph* the following day, '*the Emerald was the second vessel of the kind built in the Southern Hemisphere. She was specially designed for high speed in shallow waters. The first was built to Mr. Reeks's designs for a New Zealand firm. By the configuration of the after part of the steamer, provision was made for her propellers working in a tunnel under the vessel in such a position as to give the blades a much stronger hold upon the water than is usually obtained with shallow draft vessels. The steamer had been built conjointly, Dr. Creed said, by the company, who found the money; Mr. Reeks, who found the brains; and Mr. Ellis, who found the labor; and the happy results must be apparent to them all. He congratulated the company on their enterprise and Mr. Reeks on the success of his design*'.

The article continued. '*The Emerald is a light draft twin-screw vessel, 130ft. in length overall, with an extreme breadth of 27ft., and 5½ft. draught of water, with a speed of 11 to 11½ knots per hour. In all her equipment and appointments she is thoroughly up to date, the special object kept in view being to provide ample accommodation for about 400 passengers in such a manner as to give everyone the sense of freedom, space and air, with choice of seating, standing and promenade; and to afford all the comfort and enjoyment required for pleasure traffic in a hot climate. The vessel is to be lighted with upwards of 70 electric lights. In the fore part of the ship, excellent quarters are provided for the officers, as well as a commodious smoking saloon and bar, with table, sets and lounges. The vessel is constructed of hardwood keel, kelson, frames, stringers, beams, etc., kauri or Oregon planking between the several hardwood planks, kauri decks, housewalls, etc.; saloons and cabins are panelled in teak, blackwood, and silky oak, with mirrors and pier glasses, and lifeboat, dingy, lifebuoys, and bulkheads are provided to Board of Trade and Marine Board requirements. Thus stability, seagoing qualities, roominess, and comfort are the leading features of the vessel, and are exceptionally good for a boat of her limited tonnage, her register not exceeding 300 tons. The Emerald will be propelled by two pairs of compound surface condensing engines, supplied with steam from a single marine type boiler, 12ft. diameter by 10ft. long, having three furnaces, with a working pressure of 120lb. per square inch, giving about 300 i.h.p. The condensor, air circulating, bilge, and feed pumps are all run separately, each by its own engine.*'[1134]

Notably, there was one unique feature regarding the *Emerald*'s tunnel hull arrangement compared to the configuration of the earlier tunnel sterned *Aotea.* This detail is perhaps best explained in a description of the *Emerald* given by an employee of the firm that later operated the craft in Western Australian waters. Writing in his autobiography, published 97 years after the launch, Hugh Cameron noted, '*The most unusual feature about*

[1134] *The Daily Telegraph*, 3 April 1900.

A NEW STEAMER FOR QUEENSLAND.

LAUNCH OF THE EMERALD.

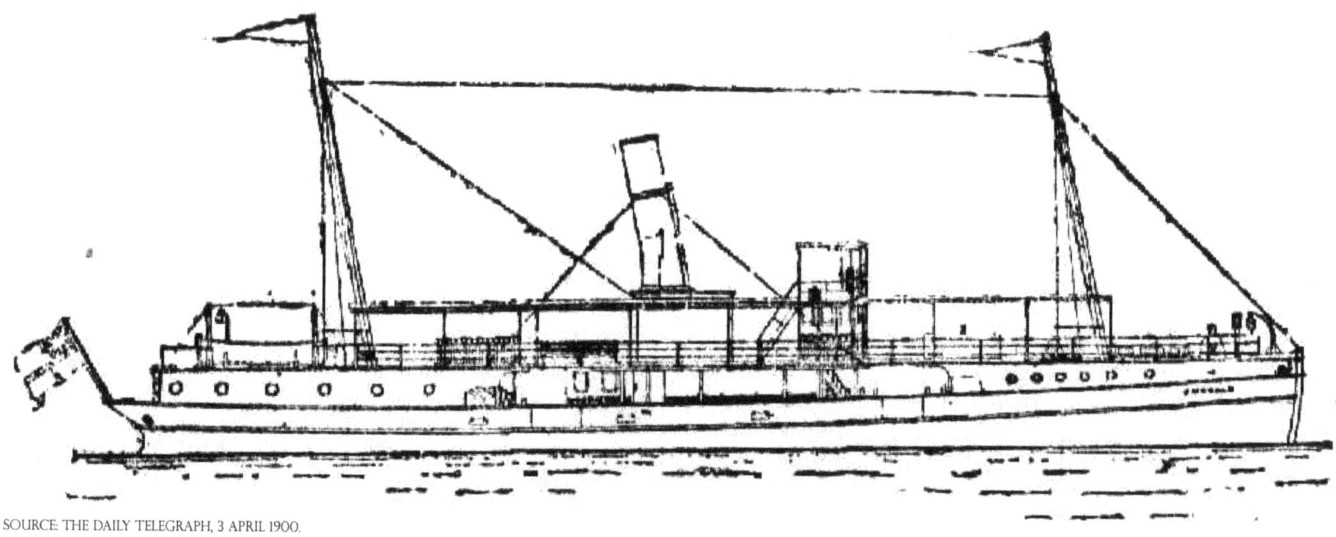

SOURCE: THE DAILY TELEGRAPH, 3 APRIL 1900.

Emerald was her tunnel stern. Both propellers were in a tunnel, but because of the restricted area, they overlapped. To overcome the problem of the blades striking each other, one shaft had been shortened by about 16 inches so that one propeller was slightly ahead of the other. To me this seemed like a bad design feature, but the efficiency seemed unimpaired'.[1135]

This was obviously quite unusual, and the author was right to question it, as the second propeller was always operating partially in the wash from the first. Yet overall, the *Emerald* seemed to perform to its requirements adequately on trials and then in service over many years. The vessel's styling was also a matter of interest. At launch another report noted, '*Looking at the vessel on the ways, and also as she floated, she appeared to be a very desirable craft, for her lines are very graceful, making an almost perfect model'.*[1136]

Upon arrival in Brisbane, however, the press were choosing their words a little differently. '*Admiration for the appearance of the new steamer may be varied, but from the accounts of the master and the owners, the Emerald has been designed in such a way that she is bound to make herself a favourite with her passengers.*'[1137]

What may be at the heart of the matter here was the combination of the features Reeks had employed, i.e., a raked funnel and spars against the vertical superstructure, plus the rounded stern and curved stem profile. Perhaps there was too much happening for people to take it all in comfortably, but the overall result was striking and drew the public's attention as much as the curious propeller arrangement interested those from the maritime sector.

The trial trip of the *Emerald* took place on Port Jackson on 13 August 1900. A large number of people, including those interested and associated with the shipbuilding industry in Sydney, assembled on board at the invitation of the Humpybong Steamship Company.[1138] Decked gaily in bunting, *Emerald* left Circular Quay and proceeded as far as Sydney Heads. The trip back saw the vessel average 11½ knots per hour. All parties on board were reportedly quite impressed with its output.

The *Emerald* arrived in Brisbane on 26 August 1900 after a three days passage from Sydney.[1139] On board was Reeks who then attended the vessel's successful trial trip to Humpybong on 28 August where it attained a speed of 10.95 knots against the tide and was generally deemed as quite satisfactory to all on board.[1140] Reeks then left Brisbane on 31 August as a passenger on board SS *Wodonga*, arriving back in Sydney on 2 September.[1141]

1135 H. Cameron (1996). *Fifty Years Afloat: The Autobiography of a Ferryman.*
1136 *Evening News*, 4 April 1900.
1137 *Telegraph*, 27 August 1900.
1138 *The Daily Telegraph*, 14 August 1900.
1139 *The Telegraph*, 27 August 1900.
1140 *The Telegraph*, 29 August 1900; *The Brisbane Courier*, 29 August 1900.
1141 *The Brisbane Courier*, 3 September 1900; *Evening News*, 3 September 1900.

FERRY STEAMER EMERALD.
SOURCE STATE LIBRARY OF QUEENSLAND (HDL.HANDLE.NET/10462/DERIV/64110).

Continuing to juggle multiple large-scale projects and important clients, on 23 February 1900 Reeks published an advertisement in *The Daily Telegraph* inviting tenders for the '*Building a Steam Yacht for T. A. Dibbs, Esq, 100ft. overall, 16ft. beam (wood), must be built under a shed. Specifications and plans may be seen on and after Monday next, at the office of Walter Reeks, Naval Architect and Engineer, Change-alley; open till Mar. 12th. The lowest of any tender not necessarily accepted*'. As we reflect back on Reeks' work, this particular vessel today stands at the forefront of his many accomplishments.

During this period there continued to be numerous social events and activities, some more high profile than others that Reeks attended. For example, he was a guest at the society wedding of Samuel Hordern (Jr) to Miss Lottie See, eldest daughter of the Hon. John See, Colonial Secretary and Minister of Defence, at St. Jude's Church, Randwick, on 14 March 1900, giving the couple an oval hand glass as a gift.[1142] The reception for 400 guests took place at *Urara*, the residence of the bride's parents.

Just over a week later Reeks, along with other members of the Engineering Association of New South Wales, made an official visit to the South Head lighthouse, including receiving a lecture on the workings of the machinery and lamps.[1143] Another excursion by the same group took place on 8 September 1900 with Reeks amongst members who paid a visit to the Australian Ropeworks at Waterloo, specifically to view the company's machinery.[1144]

Patriotism was also a priority. Australian contingents were being sent to South Africa as part of the British Empire's involvement with the Boer War (1899 - 1902). To assist with these efforts, Reeks donated to the Mosman Branch of the Patriotic Fund, a campaign tied to the larger N.S.W. Patriotic Fund.[1145]

Likely as a way to benefit existing customers and industry relationships, establish new ones and perhaps garner more business opportunities, Reeks also started to focus on publicity and marketing. On 28 May he invited some 40 gentlemen interested in or associated with Sydney's maritime enterprises to view two steamers then in the course of construction. Mort's Dock was first visited, the party arriving by way of the steam launch *Edina*. Here the hull of the steamer then in the course of construction for the Port Jackson Cooperative Steamship Company was inspected. The vessel was stated to be just under half complete. Next the party proceeded to Browntree's Dock to view the *Emerald*, then receiving its copper sheathing. It was a fruitful exercise. '*The visitors expressed themselves as more than pleased with the vessel, and called upon Mr. J. A. Paterson (director of the P.J.C.S.S. Co.) to propose the health and a successful career to Mr. Walter Reeks, whose energy and enterprise, it was contended, had done so much for the steamboat traffic of New South Wales and the sister colonies. Mr. Reeks briefly responded. A pleasant afternoon's outing terminated just at sunset, when the Edina reached Circular Quay.*'[1146]

In terms of his commercial obligations, on 2 April 1900 Reeks advertised for tender submissions for contractors willing to build two large hardwood punts about 185ft by 40ft by 15ft deep, one for delivery in July, the other in September.[1147] The following month Reeks was in the process of designing a double-ended screw steamer for Sydney Ferries Ltd. for use on its Neutral Bay service.[1148] The vessel was stated to be 100ft long, with a beam of 26ft, and to have accommodation for 500 passengers.[1149] Like the old *Wallaby,* there were to be no seats outside on the vessel's main deck. The craft was also to be capable of travelling at 12 knots.[1150] The tender for its construction, as well as another for the same company, was then announced in July of 1900 and continued to be active throughout September, closing on 1 October.[1151] Both vessels were to be double-ended steamers with plenty of beam but in some respects were considered a new departure upon the type of boats which had lately been built for the company, i.e., *Kurraba* and *Kirribilli.*[1152] There were no overhanging sponsons, the bulwarks running flush with the covering board.[1153] This alteration was considered advantageous to passengers by giving them additional deck space.[1154] As things transpired, however, both of these ferries were not built.

Also announced in July of 1900 was the tender for the building of a new double-ended screw steamer for the Balmain New Ferry Company, likely to plans modified from those announced earlier in the year.[1155] It was certainly a busy period for Reeks' steamer design work. That same month, however, another project came up that Reeks was more than happy to be involved with and it was to be one of the more unique and perhaps historically-significant of his career. It concerned HMS *Nelson,* a 126-gun frigate then approaching 90 years of age. The vessel had been in Sydney only a matter of weeks, having been purchased by Messrs W. Laidley and Company.[1156] Prior to this HMS *Nelson* had been in Melbourne, having been sold to the Government of Victoria for use as a training vessel in 1867.[1157]

The press reported. '*THE OLD WARSHIP NELSON - One of the great "wooden walls of England" of the early days of the century, the old warship Nelson, is now undergoing demolition. As she lies at Chidgey's Wharf, near the Iron Cove Bridge, where the work of dismantling is being carried on, the vessel forms an interesting object to travellers by road or water. This obsolete specimen of naval architecture was in her day a five-decker. She was launched in 1809, and the building took 10 years. It is intended to transform the old vessel into a big coal barge, with a carrying capacity - after being cut down - of 3000 tons. It is also intended to cut her top and gun decks into large*

1142 *The Australian Star*, 15 March 1900.
1143 *The Sydney Morning Herald*, 26 March 1900.
1144 *Evening News*, 11 September 1900.
1145 *The Daily Telegraph*, 6 March 1900.
1146 *The Daily Telegraph*, 29 May 1900.

1147 *The Australian Star*, 2 April 1900.
1148 *The Daily Telegraph*, 30 June 1900.
1149 *The Daily Telegraph*, 30 June 1900.
1150 *The Daily Telegraph*, 30 June 1900.
1151 *The Daily Telegraph*, 19, 28 July, 5, 20 September 1900.
1152 *The Sydney Morning Herald*, 21 August 1900.
1153 *The Sydney Morning Herald*, 21 August 1900.
1154 *The Sydney Morning Herald*, 21 August 1900.
1155 *The Daily Telegraph*, 26 July 1900.
1156 *Evening News*, 22 October 1900.
1157 *The Tasmanian Times*, 22 October 1867.

HMS NELSON AFTER BEING CONVERTED TO A COAL HULK.
SOURCE: AUSTRALIAN NATIONAL MARITIME MUSEUM (WILLIAM HALL COLLECTION).

floating stages or pontoons. The work is being carried out under the supervision of Mr. Walter Reeks, the well-known naval architect, Mr. T. S. Peaker being in charge of the work, on which 40 men are employed. The process of demolition has proved the vessel to be as solid today as when she was built nearly 100 years ago. It has been decided by the owners to dispose of most of the parts taken off the vessel as mementoes. Some of the principal parts are to be forwarded to the Australian Museum.'[1158]

Several of these specimens were presented to the Technological Museum by Reeks a few weeks later, now known as the Museum of Arts and Applied Sciences, or colloquially the Powerhouse Museum. The pieces included a section of the mainmast, 60ft tall and 4ft in diameter that was composed of eight large spars of Oregon, keyed and dowelled together, and clamped with heavy iron bands.[1159] Portions from other parts of the vessel were also given to this organisation, all showing the mode of fastenings adopted in the construction of its wooden walls.[1160] In the meantime, the work of dismantling HMS *Nelson* continued with a crane needed to remove the stern post and rudder, weighing 24 tons.[1161]

SECTION OF THE MAST OF THE HMS NELSON.
SOURCE: MUSEUM OF APPLIED ARTS & SCIENCES 'POWERHOUSE MUSEUM' (HTTPS://MA.AS/223933).

[1158] *The Sydney Morning Herald*, 2 July 1900.
[1159] *The Daily Telegraph*, 14 July 1900.
[1160] *The Daily Telegraph*, 14 July 1900.
[1161] *The Australian Star*, 3 August 1900.

While the dismantling work progressed, the owners of HMS *Nelson* determined that there was something other than removing the ship's copper bolts that could be done with the hulk. Calling upon Reeks' ingenuity, he then developed a plan whereby the old five-decked vessel would be sliced off in sections to be turned into five large barges for the storage of coal, this type of craft then in great demand on Port Jackson. With the help of two tugs, three boats and a crane, the last piece to be converted into a barge, comprising the whole of the main deck, over 170ft in length with a beam of 53ft, which when complete would have a carrying capacity of over 2,500 tons, was lifted from HMS *Nelson* on 19 October. Quite a difficult task, but one that Reeks relished being involved with.[1162]

—

In July of 1900 an update on the construction of the Port Jackson Cooperative Steamship Company's new vessel was published in the *Evening News*. The article highlighted the changes to shipbuilding then taking place; wooden ship construction was being surpassed by steel at a fast pace, as well as the uniqueness of the steamer itself. It read: '*During the past few months any person in the vicinity of Mort's Dock, in Waterview Bay, Balmain, cannot have failed to notice the incessant, in fact, the perpetual, sound of iron beating upon iron, which, to the initiated of these latter days, indicates "the building of the ship." "Iron upon iron" has been mentioned, but if accuracy of description is to be considered the phrase should be altered to "steel upon steel," for this is the material of which the vessel is being constructed, and the hammers are all steel-faced. The sounds referred to come from the stocks outside the boiler-making shops, and the vessel, slowly rising to ride the water, is a new ferry-boat for the Port Jackson Cooperative Steamship Company, Limited. The initial work of "laying down" the vessel was commenced on August 23 last, under the supervision of Mr. W. Reeks, architect. When completed, the vessel will be up to Lloyd's rules, and will have a peculiarly curved keel, technically known as "camber," but which the men employed on the work describe as a boomerang, from its resemblance to the contour of the well-known aboriginal weapon. The vessel will have a length of 175ft, a beam of 30ft, and a depth of hold of 13ft 4in. She will have a turtle-back at each end for minimising the effects of the heavy seas which are occasionally encountered in the voyage to and from Manly. At each end of the boat there will be enclosed cabins - one for ladies, and one for smokers; while there will be a very large house amidships. Wide, double stairways will lead from each end of the lower deck to the promenade deck above, the latter will have seating accommodation for 500 passengers. The entire length of the upper deck will be beneath an awning, while there will be more than the usual number of sheltered seats. A feature of the boat will be its roominess, the passage-ways being designed to be exceptionally wide; while there will be an unusual space between the seats on the two decks. Life-saving appliances, consisting of rafts, safety seats, boats, and life-buoys and belts, have been so specified for that they will form a special feature of the new boat, while particular attention has also been directed to the lavatory accommodation. The vessel, which will be of the double-ended type, with a propeller at each end, will be fitted throughout with the electric light, and will be furnished with triple expansion engines by Mort's Dock and Engineering Company, Limited, having an indicated horse-power of 850, which it is estimated will give a speed of 15½ knots on the trial trip, which is expected will take place about the end of the year. There can be little doubt but that when finished, in its new boat the Port Jackson Company will have one of the finest, if not the finest boat in the harbor*'.[1163]

The reference to the keel as curved or cambered, and then very descriptively likened to an Aboriginal boomerang by the shipwrights who built it, is a reminder that Reeks' creations still seemed to be a novelty, even though he had now used this form consistently for some years. However, as we will later see, there was more to this vessel's looks to be appreciated than just this gently curved keel.

Concurrently, a neat looking craft designed by Reeks was launched from the boat yard of William Holmes at McMahons Point in mid-July of 1900. Built to the order of Messrs H. S. Bird and Company, the steam launch measured 47ft in length with a beam of 11ft and a depth of 4ft.[1164] Named *Moa*, the craft was later used on the Manning and Richmond rivers as a cream boat.[1165]

—

In August of 1900 another Sydney-based ferry company tasked Reeks with designing a double-ended screw boat.[1166] The Watsons Bay Ferry Company commissioned him with generating a craft capable of travelling at 11 knots and accommodating 1,000 passengers.[1167] As nautical advisor to this company, Reeks also assisted with plans for alterations to the paddle steamer *Bald Rock*, a vessel the company had purchased in mid-1900 that made its official trial trip on 20 September with Reeks on board.[1168] The tender for the building of the new double-ended screw steamer for the Watsons Bay Ferry Company was released later that year.[1169]

August of 1900 again saw Reeks appointed official measurer of the Royal Sydney Yacht Squadron, a position he had held continuously since 1886.[1170] That same month it was unanimously voted that he should be made a life member of the organisation.[1171] Reeks was also appointed official measurer of the Sydney Amateur Sailing Club.[1172] The latter position was rather short-lived, however. In September of 1900 he resigned from the committee.[1173] Perhaps all of his commitments were becoming too much? Reeks had recently been elected an officer of the Mosman Swimming Club, likely because of his son Kent's

[1162] *Evening News*, 22 October 1900.

[1163] *Evening News*, 5 July 1900.
[1164] *Evening News*, 18 July 1900.
[1165] *The Daily Telegraph*, 19 October 1912; *The Sydney Morning Herald*, 16 November 1900, 23 January 1922; *Northern Star*, 7 March 1949.
[1166] *The Sydney Morning Herald*, 18 August 1900.
[1167] *The Sydney Mail and New South Wales Advertiser*, 25 August 1900.
[1168] *The Sydney Morning Herald*, 21 September 1900.
[1169] *The Daily Telegraph*, 17 December 1900.
[1170] *Evening News*, 15 August 1900.
[1171] *The Sydney Morning Herald*, 21 August 1900.
[1172] *Australian Town and Country Journal*, 25 August 1900.
[1173] *Evening News*, 11 October 1900.

involvement in the sport.[1174] He had also been reappointed honorary measurer of the Sydney Canoe Club.[1175] Reeks certainly was carrying on his family legacy of community service and volunteer work.

Not one to miss an opportunity for something out of the ordinary, Reeks also joined the Aquatic Sectional Committee tasked with organising festivities in association with the arrival of Lord Hopetoun.

The *Commonwealth of Australia Constitution Act* had become law with Queen Victoria's Assent on 9 July 1900. The following week she appointed Lord Hopetoun, who had previously been Governor of Victoria (1889 - 1895), to serve as Australia's first Governor General. With his arrival anticipated later in 1900, it had been decided that a handsome barge should be built to convey the dignitary from the *Royal Arthur* to the landing place at Sydney, and for the vessel to be manned by 30 to 40 prominent rowers.[1176] As part of this process, in November of 1900, approximately six weeks before Lord Hopetoun's expected arrival, Reeks was one of several members to submit designs for the barge.[1177]

Despite all the fanfare, Lord Hopetoun was likely just relieved to step on solid ground upon his arrival in Sydney. He had experienced a disastrous trip. Travelling via India, he had contracted typhoid fever and his wife malaria. Still, it was a remarkable spectacle that welcomed the arrival of Lord Hopetoun and his wife to Sydney on 15 December 1900. Disappointingly for Reeks, it appears that the barge idea never came into fruition. Instead a boat manned by 12 man-of-war's men was brought to the foot of the gangway at the side of the flagship where, awaiting his Excellency the Governor General, the sailors held their oars in the air at the salute.[1178] More resources appear to have been invested in the construction of a landing pavilion at Farm Cove, as shown below. Given the significance of the event, and the spectacle it generated, one can assume that Reeks would have been one of the thousands of people present at the landing, and amongst those who lined the streets to catch a glimpse of the new Governor General.

1174 *The Daily Telegraph*, 26 September 1900.
1175 *The Sydney Morning Herald*, 6 October 1900.
1176 *Sunday Times*, 4 November 1900.
1177 *Sunday Times*, 4 November 1900.
1178 *The Sydney Morning Herald*, 17 December 1900.

LANDING PAVILION AT FARM COVE FOR THE ARRIVAL OF LORD HOPETOUN AND HIS FAMILY (15 DECEMBER 1900)
SOURCE: NATIONAL LIBRARY OF AUSTRALIA (NLA.OBJ-140802814).

Intertwined with the political, technological and social changes going on in Reeks' life, there were more administrative changes afoot at the international sailing level that would have a bearing on yacht design and racing in Sydney and throughout Australia. On 24 October 1900 a meeting of members of the Yacht Racing Association, held at the Royal London Yacht Club, adopted a revision to a new rating rule that was developed and supported by several prominent yacht designers, including George Lennox Watson, William Fife (Jr), A. E. Payne, C. C. Nicholson and C. Libbich.[1179] It was to be in place for a minimum of five years, and it was a modification of the initial version of the Linear Rule from 1896. To understand the intent and outcome more specifically, we need to go back in time a few years and review changes to the yacht rating rules that Australia had been rather slow to adopt.

The Length & Sail Area Rule from 1886/87 had produced a number of good designs and had been popular in Australia. However, from the early 1890s onwards, in other countries some of the proportions had become exaggerated, with yachts showing extreme overhangs as one example of this. To counter these developments a number of high-profile yacht designers created a new rule that included length (L), breadth (B), sail area (SA) and a new factor called girth (G) in the formula. The group also moved to have the rating expressed in feet and not tonnage, with the intent that the rating represent the anticipated waterline length of the classes to be created rather than the tonnage or volume of the yacht. The formula was:

Rating in Feet = (L + B + 3/4G + ½ sq.root SA), the sum then divided by 2.

Girth was the measurement of the perimeter of the cross section or 'skin girth', taken at maximum beam, and the intention was to penalise the hollow sections of the prevailing light displacement designs that had developed. As stated, it came into force in Great Britain in 1896, and was called the Linear Rule. It was intended to be reviewed and, if necessary, revised again after five years.

In Australia, yachting was slow to adopt this new rule, unlike the velocity with which yachtsmen had taken on the Length & Sail Area Rule in 1887. The rule was discussed, including its potential merits, but perhaps with the economy still recovering from the crisis of the early 1890s, and few new boats appearing, there was no inclination to change things. Finally in December of 1898 the Royal Sydney Yacht Squadron broke the ice and, at a meeting widely reported in the press, adopted the Linear Rule with a unanimous show of hands.[1180] '*The first motion was moved by Mr T. W. Bremner ... That from August 1 1899 (next season) the rating of yachts of the squadron for time allowances shall be ascertained by the measurement rules for the time being of the Yacht Racing Association of England'.*[1181]

[1179] *The Sydney Morning Herald*, 3 December 1900.
[1180] *The Sydney Morning Herald*, 14 December 1898.
[1181] *The Sydney Morning Herald*, 14 December 1898.

The Australian Canoeist and Yachtsman of 17 December 1898, as always, offered a more colourful narrative of the event. '*It was a great meeting at the Royals on Monday last, and solemn - most solemn. All members seemed to be impressed with the gravity of the occasion, the eyes of Europe were unquestionably on the meeting. Truth to tell, it was disappointing in a way, because most of the large crowd came expecting a row, whereas everything went off as merry as a marriage bell. Not a dissent voice to either motion.*

Eloquent speech by Mr Bremner, who developed unsuspected powers of oratory, left unsaid nothing to further his cause, so that Mr Reeks, who supported him, could really find nothing more to say'.[1182] Of note, however, Reeks had pointed out that '*the new rule would have little prejudicial effect upon any existing boats, because if we looked around the room at all the models, we would see they were all very much of the same type'.*[1183]

So, the new rule came into being with little celebration or dissension on Port Jackson, and was followed by other clubs and colonies also moving to adopt it. The existing tonnage classes were simply renamed as 'X Foot Linear Raters' and events carried on as before. Across the Tasman, New Zealand had moved much faster, adopting the Linear Rule during 1897, so it was not surprising that Reeks had found them far ahead of the Sydney yachts when he visited Auckland in late 1898.

As Sydney yachting adopted the Linear Rule through 1900 it was already under review back in England. The outcome of the 24 October 1900 Yacht Racing Association meeting was a revised rule, now referred to as the 2nd Linear Rule to differentiate it from the initial or 1st Linear Rule. The formula was revised to:

Rating in Feet = (L + B + 3/4G + 4D + 1/2 sq.root SA), the sum then divided by 2.1.

'D' was a new factor that was introduced: the difference between the skin girth (G) and taught or chain girth, measured at 60 per cent of waterline aft. The D measurement was introduced because in practice it had been found that the girth measurement alone was not a big enough penalty to force designers away from beamy light displacement hull forms. This version of the rule was also intended to be unchanged until 1907.

The late adoption of the 1st Linear Rule probably helped ease Australian yachting into the 2nd version. Its immediate impact seems to have been quite limited, the pastime of yachting was building up strength again and the next 'big thing' on Sydney Harbour were the 30ft Linear raters. Most of the earlier yachts of this type were anything but radical, so they fitted in well with the newer ones that respected the revised rule and its added 'D' measurement. However, the exception was the New Zealand-built Logan boat *Sunbeam* of 1899, a fin keeler with long overhangs, ironically characteristics that the 2nd version of the Linear Rule was designed to stamp out.

[1182] *The Australian Canoeist and Yachtsman*, 17 December 1898.
[1183] *The Australian Canoeist and Yachtsman*, 17 December 1898.

In the longer term, the 2nd Linear Rule allowed for a short but relatively settled period of vessel design and development before the introduction of the International Rule that produced the many international metre boat classes after 1907. However, the 2nd Linear Rule also spanned some significant racing events, such as the early challenges for the Sayonara Cup, a story soon to unfold and occupy Reeks' attention in different ways.

Returning to Reeks' commercial activities, in September of 1900 it was published that there was very little work going on at Sydney's boat building yards, however, there was one craft under construction that was singled out. The new steam yacht being built by Watty Ford (Jr) of Berrys Bay to the order of T. A. Dibbs was progressing well. With principal dimensions of 100ft (overall), a beam of 16ft and a draft of 6ft 6in, the vessel was stated to have '*a clipper bow and a long graceful counter, which adds very much to her appearance, thanks to Mr. Walter Reeks, her designer*'.[1184]

Two months later a momentous day for Reeks took place, in more ways than one. The Manly Ferry Company's new 175ft steamer *Kuring-gai* was launched from Mort's Dock early in the morning of 8 November 1900.[1185] *The Sydney Morning Herald* reported on the event. *'The launching of the vessel was carried out under the direction of Mr. Frank [managing director of Mort's Dock]. At a given signal the dog-shores were knocked out, and the vessel, being then under the charge of Mr. C. Michelmore, glided out into the water. Mr. Walter Reeks remarked, "it was the most successful launch he had witnessed." As the vessel left the ways, a bottle of champagne was broken across her bows by Mrs. Woods, wife of the chairman of directors of the company, and christened "Kuring-gai." There were loud cheering from the employees of the works and the spectators and whistling from several launched in the close vicinity and the ferry boats.*

After the ceremony the directors and their wives and the dock officials adjourned to the board room where refreshments were partaken of ... Mr. J. Woods, chairman of directors, responded on behalf of the company, and said that in giving the building of the Kuring-gai to Mort's Dock they were paying more than they could have got the vessel built for in England, but the directors believed in giving encouragement to local industry. He was satisfied there was not a better ferry boat this side of the line than the Kuring-gai would be when completed, and he hoped it would not be long before they would be able to give an order for another of a similar character.'[1186]

It can be said that *Kuring-gai* and its predecessor *Manly* were something of a 'tour de force' for Reeks with his ferry designs. Their styling was a bold statement of elegance, speed and purpose. The contrast between the long curves of the cambered keel and concave sheer lines with the

[1184] *The Sydney Morning Herald*, 11 September 1900.
[1185] *Evening News*, 9 November 1900.
[1186] *The Sydney Morning Herald*, 9 November 1900.

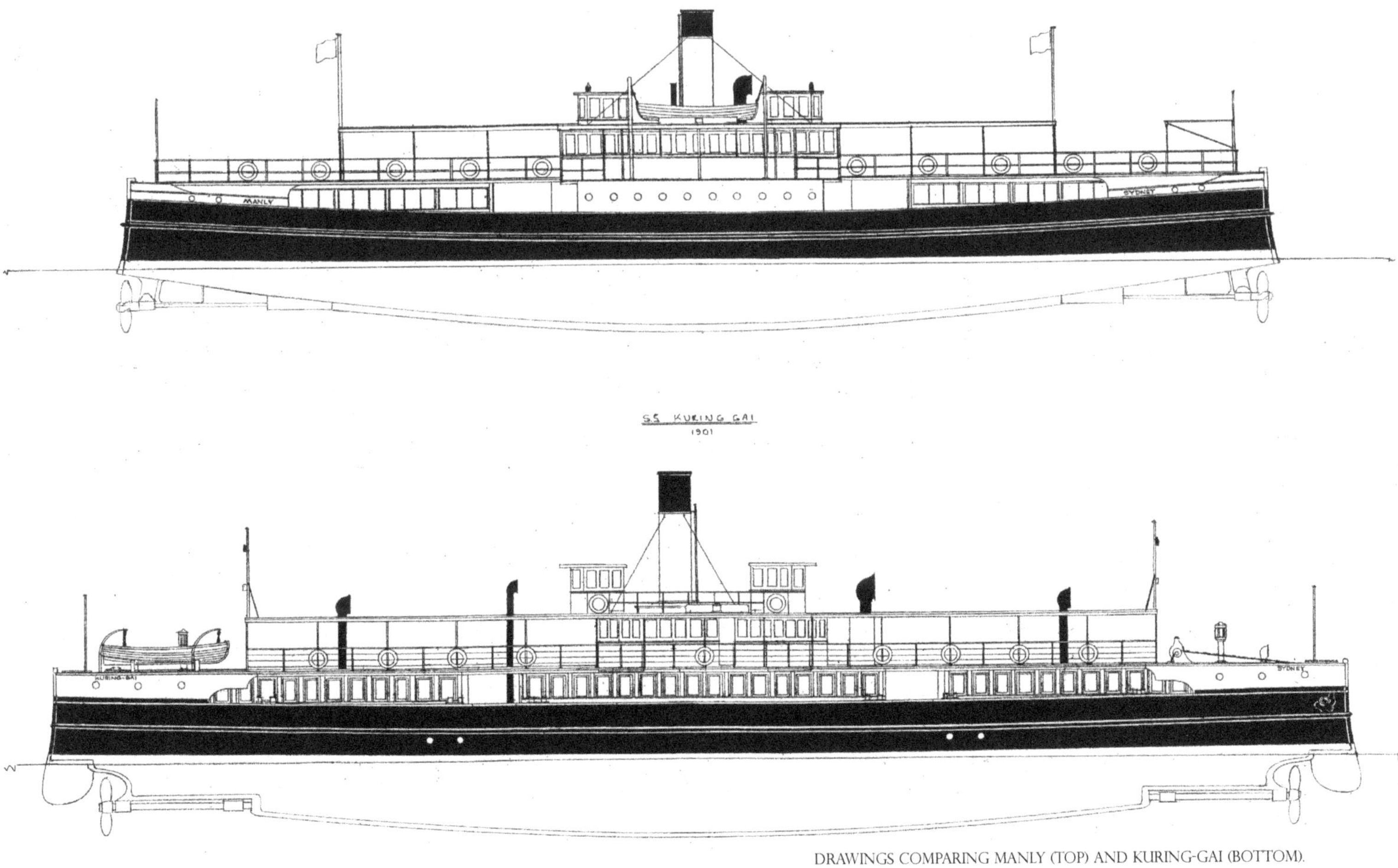

DRAWINGS COMPARING MANLY (TOP) AND KURING-GAI (BOTTOM).
BY DAVID PAYNE.

strong horizontal straight lines for the upperdeck and canopy is perhaps the key element, and this is enhanced by the ram bow profile on both craft. They were fast and they looked the part too, while the detailing of the enclosed ends reinforced the seaworthy capabilities of these two vessels. They were the largest ferries on the harbour, giving an expectation of something special, and they did not disappoint.

Historian Tony Prescott wrote a thesis on the Manly ferry service that was published in 1984. In the manuscript he summed up the collective story of these two craft, beginning with *Manly*.[1187] '*She was a revolutionary vessel, introducing a design unique to the Manly service ... she introduced the design pattern for the twentieth-century Manly boats ... Quite unlike any Manly ferry before her, she was the archetype of a design format which was to last to the present day.*'[1188]

Moving onto *Kuring-gai*, he continued. '*She was built of steel on an iron frame ... and joined Manly as a revolutionary prototype. Her metal construction made her a much sounder vessel than Manly but, unfortunately, Reeks chose to experiment with his diabolical camber keel.*'[1189] As has been explained, this feature was also apparent on *Manly*, but Prescott was able to provide some first-hand comment on its peculiarities, having interviewed a retired skipper. '*As Captain Jim Ireland, who served on her recalled, she would in certain situations, such as berthing, swing around without warning, pivoting on the lowest point of her keel. Aside for this "Achilles keel" (so to speak) she was thoroughly successful.*'[1190]

However, Reeks did not retain the projecting propeller on *Kuring-gai* as seen on *Manly*, perhaps in a response to the earlier *Merksworth* incident. Prescott discussed, '*she represented the second stage of the evolution towards the design of the twentieth century Manly steamers. She introduced broad beam, built-up steel bows with stylish curved hances, gangway exits on the promenade deck and permanent sun-deck to carry the navigation bridges*'.[1191] While draughtsman Andrew Christie produced the plans for the next six Manly ferries, only gradually improving on Reeks' basic design, '*it must be said that it took three tries before a boat was produced (Bellubra) to match Kuring Gai's 15-knot speed*'.[1192]

The wooden hulled *Manly* led to the improved steel-hulled *Kuring-gai* and its other detailed advances. They were also two of Reeks' largest designs, together giving birth to an enduring legacy on Port Jackson for him. *Manly* '*brought something new to ferry design*' while *Kuring-gai* '*was the shape of things to come*'.[1193] These were 'modern' designs, not that they would have used that word back then.

With the morning of 8 November 1900 taken up by the very successful launching of the largest and most ambitious ferry project Reeks had undertaken to date, just how did he spend his afternoon?

[1187] A. Prescott (1984). *The Manly Ferry: A History of the Service and its Operators, 1854-1974.*
[1188] A. Prescott (1984). *The Manly Ferry: A History of the Service and its Operators, 1854-1974.*
[1189] A. Prescott (1984). *The Manly Ferry: A History of the Service and its Operators, 1854-1974.*
[1190] A. Prescott (1984). *The Manly Ferry: A History of the Service and its Operators, 1854-1974.*
[1191] A. Prescott (1984). *The Manly Ferry: A History of the Service and its Operators, 1854-1974.*
[1192] A. Prescott (1984). *The Manly Ferry: A History of the Service and its Operators, 1854-1974.*
[1193] T. Mead (1988). *Manly Ferries of Sydney Harbour.*

STEAM FERRY KURING-GAI.
SOURCE: GRAEME ANDREWS WORKING HARBOUR PHOTOGRAPH COLLECTION, CITY OF SYDNEY ARCHIVES.

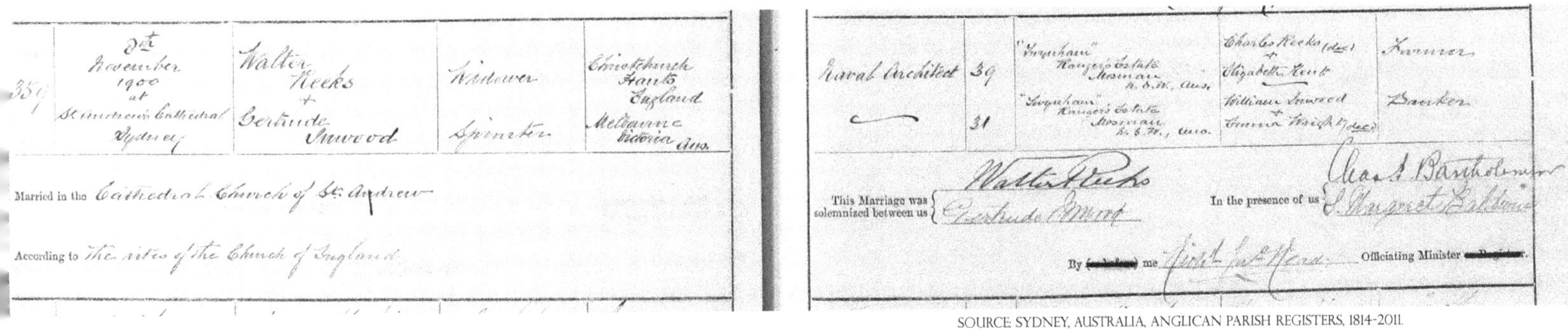

SOURCE: SYDNEY, AUSTRALIA, ANGLICAN PARISH REGISTERS, 1814-2011.

Reeks got married! Hopefully he made it to the church on time.

On 8 November 1900 at St. Andrew's Cathedral in Sydney, the principal Anglican church of the city and next door to the Town Hall, Reeks married 32-year-old Gertrude Inwood, who for some time had been living at *Twynham* and working as his housekeeper. Witnesses to the union were Charles Bartholomew and S. Margaret Baldwin. While the former will be mentioned shortly, the latter was likely a friend of the bride.

Gertrude Inwood's family history is rather interesting and deserves a few paragraphs. She was born in Melbourne in 1868 to William Inwood and Emma Annie Wright, the fifth of nine children produced by the couple, the brood comprising four daughters and five sons.[1194] Though she may not have been fully aware of the details, her father's life story would have not been out of place as a character in a Charles Dickens novel.

William Inwood had been born in London in 1836 and shortly thereafter admitted to the Foundling Hospital, an orphanage for the education and maintenance of exposed and deserted young children.[1195] After spending the early years of his childhood in the care of a foster couple, Inwood returned to the institution to receive an education, around the age of 14 becoming apprenticed to a banker.[1196] On 9 April 1859 at Saint Pancras Church, London, at the age of 23, he married Eleanor Scanlon, a 16-year-old Irish girl who had spent the majority of her childhood in England and who was the step-daughter of a well-known solicitor by the name of George Moseley Lander.[1197] She was also then pregnant with the couple's first child, a daughter, Mary Elizabeth, who was born six months later.

A second daughter, named Lillian 'Lily', was born to William Inwood and his young wife Eleanor in London late in 1861. By this time, however, Inwood was living in Australia having immigrated to Melbourne earlier in the year on the proviso that his wife and their two infants would soon follow. And so on 19 April 1862 Inwood, his wife Eleanor and their daughters were reunited, the latter trio arriving in Victoria by the *Swiftsure* from London. However, it was likely to have been a shocking situation that 19-year-old Eleanor suddenly found herself in. In the 12 months that Inwood had been living in Melbourne, he had not only formed a relationship with a local woman by the name of Emma Annie Wright, but the couple had also had a child. A second child would soon follow.

On 16 July 1863 Eleanor Inwood departed Melbourne on board the ship *Wellesley* en route to London.[1198] She left her two young daughters behind in the care of their father and his new partner. Back in London, Eleanor moved in with her step-father and his second wife, likely helping to raise their children. She returned to Australia at least one more time to see her daughters, though only for a short period. She died in London in 1895.

For William Inwood and Emma Annie Wright, there was no option for them to marry. Inwood's first wife was alive and well, and divorce was impossible. Still they would have contrived to have been 'married', allowing their children to be deemed legitimate.

Professionally, to support his ever-growing family, Inwood plied his trade as a bank teller, including working for the Oriental Bank and the Federal Bank of Australia, also becoming a respectable member of the Windsor community, an inner-city suburb of Melbourne. By 1887 he was an accountant at the Federal Bank.[1199] However, scandal seems to have followed Inwood like a bad smell and he appears to have been involved in some nefarious dealings throughout his career. One in particular can be highlighted. In the early 1890s Inwood moved to Adelaide to become manager of the Anglo-Australian Bank. He was by now widowed; Emma Annie Wright had died in a Melbourne asylum in September of 1887, and most of his children had married or gained employment.[1200] While working in this capacity, in February of 1892 Inwood committed a shameful fraud on a customer, stealing £150 from an account under false pretences. He was subsequently imprisoned for 40 days.[1201] His sentence was then extended for an additional six months, the outcome of an insolvency court verdict that was later successfully appealed.[1202] The Anglo-Australian Bank, a Melbourne-based institution, went into liquidation later that year.

For Gertrude Inwood, a few years after her father's departure to Adelaide she likely moved to Sydney to be closer to her sister Edith. Alternatively she moved to Sydney with Edith, Edith's husband (William Hassell) and their young family around 1894. Regardless of the

[1194] Australia, Birth Index 1788-1922.
[1195] https://en.wikipedia.org/wiki/Foundling_Hospital; England & Wales, Non-Conformist and Non-Parochial Registers, 1567-1936 for William Inwood.
[1196] 1841 and 1851 England Census for William Inwood.
[1197] London, England, Church of England Marriages and Banns, 1754-1932.
[1198] *The Age*, 17 July 1863; Victoria, Australia, Outward Passenger Index, 1852-1915.
[1199] *Leader*, 18 June 1887.
[1200] *The Argus*, 6 September 1887.
[1201] *Weekly Times*, 6 February 1892.
[1202] *The Herald*, 27 April 1892.

timeline, one can assume that Reeks knew nothing of his future father-in-law's history and activities, nor would they have met often, if at all. After his imprisonment Inwood remained in Adelaide continuing to work as an accountant.[1203] He died in July of 1911 at *Fernside*, the property of his eldest daughter Mary Elizabeth, located in Leongatha, Victoria.[1204]

Exactly one month after Reeks' wedding to Gertrude Inwood and the launch of the *Kuring-gai,* another highlight of his career took place. 8 December 1900 coincided with the launch of the steam yacht *Ena* from the Berrys Bay boat yard of Watty Ford (Jr). Still in existence and proudly part of the Australian National Maritime Museum's fleet based at Darling Harbour, this majestic vessel epitomises the style and grace of Reeks' design work, as well as his practicality.

Ena was built to the order of T. A. Dibbs, then commodore of the Sydney Amateur Sailing Club and manager of the Commercial Banking Company. A week after its launch he would be unanimously elected commodore of the Royal Sydney Yacht Squadron.[1205] Dibbs was a prestigious client for Reeks who no doubt understood the weight and worth of the assignment he was given; his dedication to the task, his skill and his craftsmanship very much showcased in the vessel itself.

[1203] Australia, City Directories, 1845-1948 for William Inwood.
[1204] *The Age*, 15 July 1911.
[1205] *The Sydney Morning Herald*, 13 December 1900.

Dibbs had sold his previous steam launch, also named *Ena*, to the Government of New South Wales for use by its Public Works Department in early 1900 for £2,000, in anticipation of the new build.[1206] It would be nearly a year before the second generation *Ena* was ready, but the craft was well worth the wait.

The *Evening News* wrote a glowing review of the launch event and the vessel. '*Shortly after 9 a.m. on Saturday there was launched from Mr. Ford's yard, Berry's Bay, a handsome steam yacht for Mr. T. A. Dibbs. As the vessel left the ways, she was christened Ena, by Miss Dorothy Dibbs. The launch went off without a hitch, the vessel taking the water almost without a move in her cradle. The Ena is 100ft overall, by 16ft beam, 8ft 6in deep, and she will draw 7ft 3in. On deck she will be fitted with elegant circular top skylights, with embossed glass. There is a deckhouse amidships, combining funnel casing, steering platform, stokehole-entrance, galley, and engine-room, so arranged that while giving ample room for their respective purposes, ample promenade space remains. Forward is a steam capstan, of the very latest design, and the steering gear is similar to that fitted in her Majesty's yacht. The latter has been specially imported. Below are two magnificent saloon cabins; the forward, or ladies', saloon, is 18ft long by the full width of the ship, in which are six berths, with everything arranged with a view to convenience; the after, or gentlemen's, cabin, is 16ft long, and has berth accommodation for no fewer than eight persons. Lavatory arrangements are provided for at both ends of the yacht. The Ena is schooner-rigged, with a permanent awning over the main body, and a white canvas*

[1206] *The Daily Telegraph*, 8 February 1900.

STEAM YACHT ENA (1900).
SOURCE: AUSTRALIAN NATIONAL MARITIME MUSEUM COLLECTION.

portable awning covers the quarter-deck and bow. She carries two boats in davits arranged so that they may be brought in on deck during bad weather at sea. The gangway is a special feature, giving the ease of access of a grand stairway. Steam at 120lb pressure will be supplied by a large return tube marine type boiler, to compound surface condensing engines, having cylinders 12in and 24in diameter, and 16in stroke, both with piston valves. The condenser is independent, and placed in the port wing, the air, feed, and bilge pumps being worked off the main crosshead; circulating pump being an independent Worthington, which together with the universal donkey and ejector, make up a very complete outfit. Electricity of lighting purposes will be generated by a Compton direct coupled 8-horse power plant, giving power to brilliantly illuminate the vessel fore and aft. The new yacht is from designs by Mr. Walter Reeks, and was built by Mr. W. M. Ford, Berry's Bay. The machinery is by Messrs. Chapman and Company. The inside fittings, which are all of specially selected polished oak, from the old warship Nelson (where they have been seasoning nearly 100 years), are by Messrs. W. Walker, Son, and Bartholomew. Every possible effort will be made to expedite completion, and the Ena is expected to be in commission early in the new year.'[1207]

Almost hidden amongst the design details and mechanics of the *Ena* mentioned in this article, are two points of interest not to be ignored. Firstly, Reeks had recycled part of HMS *Nelson* for use in this vessel's luxuriously-appointed fit out. Second, *Ena*'s woodwork and cabinetry were produced by the hands of Messrs. W. Walker, Son, and Bartholomew. The latter was in fact Charles Bartholomew who was a witness to Reeks' marriage to Gertrude Inwood the month prior. Bartholomew would also become a prominent member of the Royal Sydney Yacht Squadron.

Returning to the design and its fascinations, *Ena*'s handsome classic styling was the feature everyone noted first. The raked superstructure and spars gave it an air of elegance and speed, but a closer look at its lines, construction and arrangement shows considerable thought and some different features, but this time carried out by Reeks in a subtle way.

The hull shape is generally conventional but optimised to suit its use and area of operation on Sydney Harbour and nearby. The clipper bow, curved sheer and counter stern are standard classic features of the period, but the cutaway forefoot and curved keel line forward is more like a sailing yacht and a profile that Reeks used consistently on yacht designs. Many other power vessel designs from the period are more ship-like with a straight keel, parallel to the waterline, throughout the length of the hull, whereas *Ena*'s keel is straight only from about midships aft, and heavily raked down to the sternpost. The deadwood is also cutaway with a long aperture similar to his first ferry designs in 1892. A conventional approach has the deadwood and/or planking and rabbet line extending aft to the stern post.

[1207] *Evening News*, 10 December 1900.

Overall, it was an easily driven hull shape well suited to inshore operation and able to maximise its speed from the power available. The machinery installed after the launching consisted of a coal-fired return tube Scotch boiler built by Dalgleish of Glasgow, Scotland, 8ft 6in by 8ft 6in, which gave steam to a set of compound engines manufactured by Chapman & Co. of Sydney. The two cylinders were 12in and 24in by 16in stroke and ran at 160 revolutions. *Ena* could run up to 11½ knots with the ability to make 12 knots occasionally. This maximum speed was equivalent to the hull or displacement speed.

At deck level *Ena* featured a wide beam carried aft to the counter, possibly to increase the deck area for the passengers to enjoy. Consequently, the overall plan view shows a fine width forward and full shape aft, whereas other designs often had the reverse, with a full forward curve tapering to a fine tail at the counter.

The hull and superstructure followed conventional wooden boat construction methods but reference is made to the vessel being '*light in construction*' with smaller scantlings than those used in America or Europe, due to the strength of the local hardwoods.[1208] The hull was carvel planked with spotted gum to the waterline, and New Zealand kauri for the topsides to reduce weight. This all helped the design carry a modest-sized engine with less coal to be carried, thereby further reducing weight so *Ena* could achieve a high speed on the harbour.

Ena was named after Dibbs' wife Tryphena, and she may well have had a hand in some of the deliberations about the layout. The prestigious American *Rudder* magazine featured *Ena* in its March 1906 edition, a few years after the vessel's launch, and carried a lengthy description of the craft, reflecting the significance of this design.[1209]

'*The accommodation plan may strike the reader as somewhat unusual, but the arrangement meets the requirements admirably. The conditions are afternoon trips for large parties, for the genial ex-Commodore* [of the Royal Sydney Yacht Squadron] *is never so happy as when surrounded by his family and their families and friends and there seems no limit to the numbers over whom his beaming face can smile at one time and for that matter, all of the time; therefore, to accommodate these parties, large deck space was necessary and a great part of it covered with awnings. Then for sea-going purposes she was made high-wooded, and for cruising, as understood in and around Sydney, the accommodation was decided on as shown on plans, and consists of two large open cabins, that for ladies in the fore-end of the ship accommodating six, while the after or gentlemen's cabin berths eight comfortably. The climate at and around Sydney in summer is pretty hot, and experience has shown that comfort is enhanced by having big open cabins, even at the expense of some degree of privacy, for it will be seen at once that a boat of this size would divide up very badly to give everybody a cabin to him or herself.*

[1208] *Rudder* Magazine, March 1906.
[1209] *Rudder* Magazine, March 1906.

The cabin arrangements are the outcome of a great deal of consideration, and many adjustments between what "we cannot possibly do without" and what "I cannot possibly get into the space" of the ladies and architect respectively, and the compromises effected have happily turned out satisfactory, both from a utilitarian and artistic standpoint. The cabin work of both cabins is executed in English oak, with weathered and waxed finish and oxidized copper furniture. Carving of a restrained and appropriate character is used upon the companion newels and hand rail terminals. Upholstery work is covered with a tapestry cloth, having a pleasing colour effect of neutral gray, and the floor coverings are a self-coloured linoleum and Milton pile carpet. The whole effect of the cabins is pleasing in its quiet homeliness of design and colour, and the workmanship throughout is excellent being executed by Messrs. Beard Watson Ltd., from designs by Mr. Ch. Bartholomew of that firm.'[1210]

Overall *Ena* had the looks and the performance to carry it all off without disappointment, and this was undoubtably a factor in the vessel's longevity. After an extraordinary life, the craft is back and flaunting its style on Sydney Harbour over 120 years later, now in the nation's collection of the Australian National Maritime Museum.

[1210] *Rudder* Magazine, March 1906.

Following the launch of *Ena*, and with the year's end approaching, on 17 December 1900 Reeks issued two tenders. One for the building of a double-ended screw steamer for the Watsons Bay Ferry Company, and one for the building of a screw steamer, 70ft by 16ft, for Messrs Lysaghts Brothers and Co., wire netting manufacturers.[1211]

Finally, the year 1900 ended with the launch of another illustrious vessel. Designed by William Fife (Jr) and built by Ford (Jr) at Berrys Bay under the direct supervision of Reeks, the 30-rater yacht *Fleetwing* was launched on 21 December, just in time to race on Port Jackson the following day.[1212] The vessel was built to the order of Sam Hordern, commodore of the Prince Alfred Yacht Club.

[1211] *The Daily Telegraph*, 17 December 1900.
[1212] *The Sydney Morning Herald*, 22 December 1900.

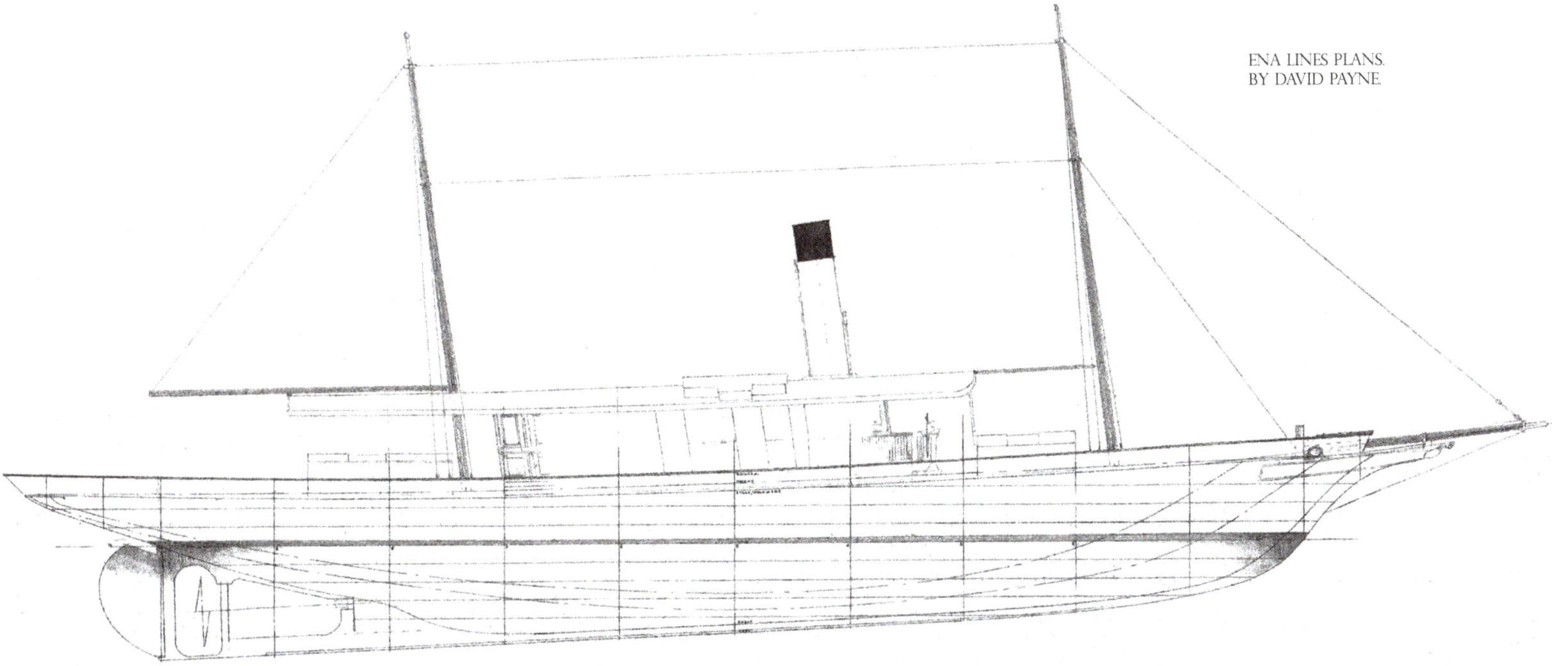

ENA LINES PLANS.
BY DAVID PAYNE.

ENA ON SYDNEY HARBOUR (2016).
SOURCE: AUSTRALIAN NATIONAL MARITIME MUSEUM.

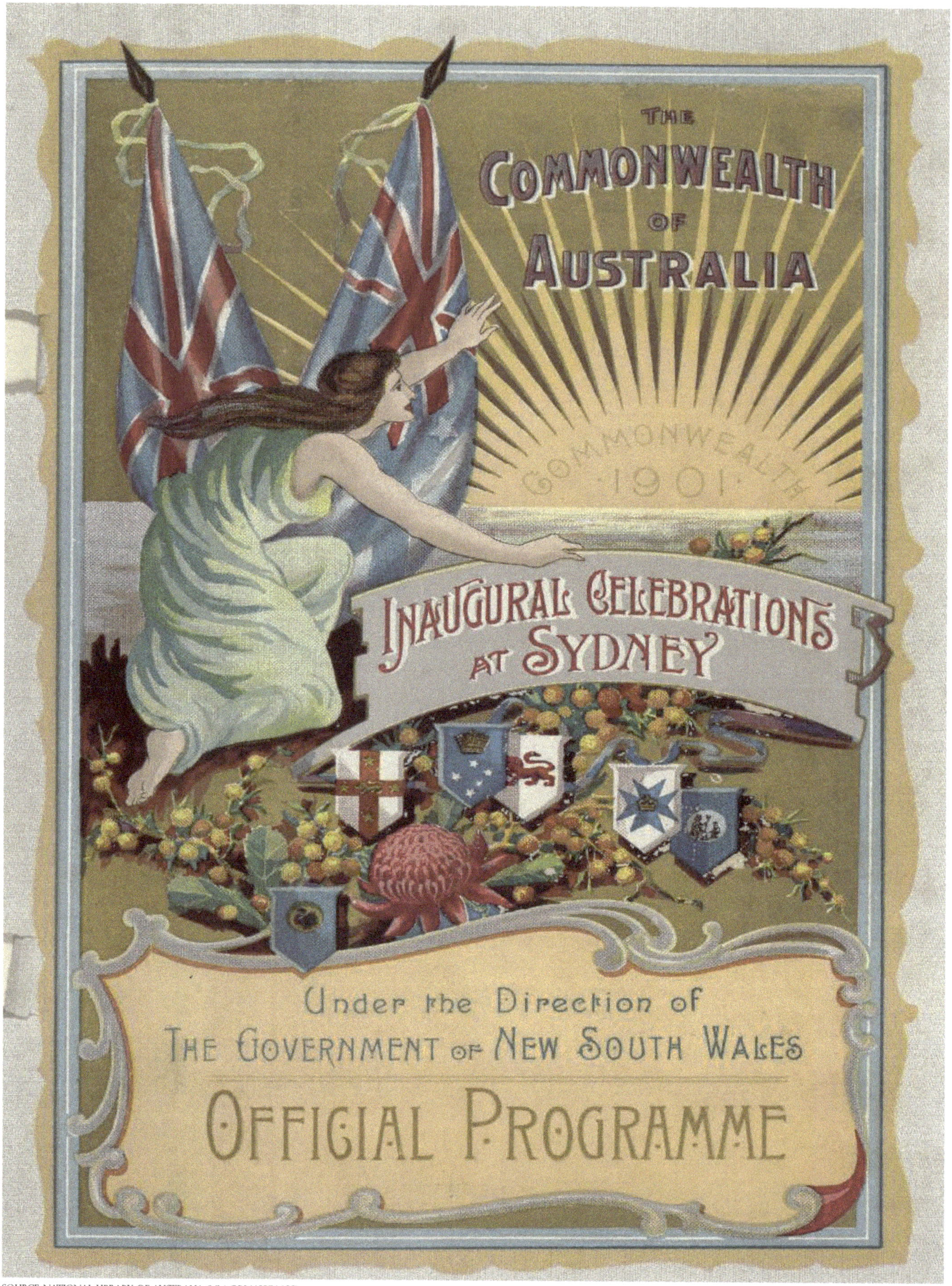

SOURCE: NATIONAL LIBRARY OF AUSTRALIA (NLA.OBJ-143374422).

CHAPTER FIFTEEN - THE COMMONWEALTH

Australia's six self-governing British colonies, Queensland, New South Wales, Victoria, Tasmania, South Australia and Western Australia, became united as the Commonwealth of Australia on 1 January 1901. It had been a rather slow and tedious process, encompassing multiple referendums, conventions and discussions held over several decades across all six colonies, and with England, before final agreement was reached. The initially-wary public had been increasingly swayed by Australian-born citizens becoming more and more active in their support for nationalism. Eventually the parties had come to agreement as to what the Commonwealth of Australia would look like in terms of its constitution (executive, legislature and judiciary structure and power), finances and trade, the role of the states, and administration. It was a historical achievement.

Though New Zealand representatives had been invited to join in discussions about becoming the seventh state of the new nation, they were hesitant to do so. There was little public interest amongst the Kiwis to be administered by a governing body situated in Australia. Several islands in the Pacific, including Papua New Guinea and Fiji, had also been referred to as part of a Commonwealth of Australia but also declined the invitation. Instead, as neighbours these island nations advocated for continued strong relationships with Australia, particularly in terms of trade.

With the Commonwealth of Australia now a reality and an interim government in place until national elections could be held, it was the start of a period of somewhat trepidation and uncertainty about how the new system of government would be implemented, including how customs, excise and other taxes would be allocated back to the six states. Up until this time the individual colonies, now states, had been responsible for their own income generation and spending. The new parliamentary system was also raising some concern, as was the High Court. Suddenly the Supreme Court in each state was no longer the highest court. Instead, matters would be referred to the High Court. In addition, sovereignty became an issue when it was realised that states could not communicate with foreign powers. This was left solely to the new federal executive. Overall, it was a period fraught with teething problems, some big and some small.

CARTOON DEPICTING NEW ZEALAND'S HESITANCY TO JOIN A FEDERATED AUSTRALIA.
SOURCE: THE FREE LANCE, 17 NOVEMBER 1900.

Though born in England, Walter Reeks likely welcomed the nationalistic approach to governance of Australia. Looking back to earlier chapters, in 1888 he was already promoting an Australian viewpoint, *'and with Australia's honour as our reward when we succeed, surely we shall soon lead the world, and cease to speak of the American type, and the English type, and have a type of our own, which other countries will look at with envious eyes, and call Australian'.*[1213]

In terms of business prospects, there were perhaps more pathways and avenues for Reeks to reach new customers. Augmented by a sense of new-found patriotism coalescing with national pride, federation had the potential to open up and streamline opportunities in states he had not previously either sought or found work, like Tasmania, South Australia and Western Australia.

With New Year's Day likely spent attending the national inauguration at Centennial Park, along with 250,000 others who had lined the route starting at Sydney's Domain, it was the beginning of an eight-day official programme of celebrations, fireworks, military parades and sporting events. For example, on 2 January Reeks and his family attended a swimming carnival at Fitzroy Dock, Cockatoo Island. On the way home his wife Gertrude unfortunately lost the cut steel belt she was wearing. Reeks offered a reward for its recovery the following day, posting an advertisement in *The Sydney Morning Herald;* likely an ambitious request.

Returning to work, on 12 January 1901 Reeks re-issued a tender for the building of Lysaghts Bros.' screw steamer. Vessel measurements were revised to 69ft by 16ft, with plans and specifications available at his office at Change Alley.[1214] Four days later he released a tender for the construction of the double-ended screw steamer he had designed for the Watsons Bay Ferry Company.[1215] A rather slow process, it would take over a year for the contract for this particular vessel to be finalised.[1216]

Despite the summer slowdown, January also coincided with another vessel being completed to a design by Reeks: a smart oil launch built by William Holmes of McMahons Point to the order of the Fijian Government. The 40ft by 8ft vessel was fitted with a Union Company oil engine capable of producing a speed of 9 knots.[1217] The craft was named *Adi Viti* and intended for medical and harbour work. The vessel arrived in Fiji in May of 1901 whereby it was subsequently christened by the young lady after whom it had been named.[1218]

Holmes is a name worth remembering. He would become a close associate of Reeks over the next decade, with the pair collaborating to produce well over 20 vessels. Born in Auckland in 1864, Holmes' father was a noted shipbuilder as were several of his uncles. Opting to build smaller craft, he initially found some success as a boat builder in Auckland before moving to Wellington. In early 1887 Holmes made an even bigger move, sailing to Sydney with Edwin Bailey, son of celebrated Auckland boat and yacht builder Charles Bailey; their translocation likely the result of the depressed state of the New Zealand economy, as well as a brimming sense of adventure. By 1896 Holmes was operating out of his own boat yard at McMahons Point in Sydney. In the decades that followed he became one of the more well-known boat builders in the Commonwealth,

[1213] *Illustrated Sydney News*, 31 May 1888.
[1214] *The Daily Telegraph*, 12 January 1901.
[1215] *The Sydney Morning Herald*, 16 January 1901.
[1216] *The Australian Star*, 8 January 1902.
[1217] *Evening News*, 22 January 1901.
[1218] *The Daily Telegraph*, 3 June 1901.

FEDERATION CELEBRATIONS AT CENTENNIAL PARK, SYDNEY, 1 JANUARY 1901.
SOURCE: NATIONAL LIBRARY OF AUSTRALIA (NLA.OBJ-141463148).

turning out a proliferation of sailing and motor craft of all classes and sizes. He died in Sydney in 1923 with his son Reginald continuing on the boat yard business until his own violent death in 1935. Reginald Holmes had become engaged in criminal fraud and smuggling. His connection to small-time criminal James 'Jim' Smith was his undoing. Smith was also a police informant but was murdered in early April of 1935. In what was a sensational discovery, Smith's body was disposed of at sea in pieces with his left arm discovered when a captured tiger shark regurgitated it in the Coogee Baths Aquarium a few weeks later.[1219]

Reginald Holmes became a person of interest in the case. He made a failed suicide attempt, was chased on the harbour by police, and then eventually gave evidence to them. Admitting that Patrick Brady, the alleged murderer of Smith, had showed the severed arm to him, Holmes said he was the one who had put it in the surf at Maroubra. Reginald Holmes was not given police protection and on 11 June 1935 was found shot dead at Dawes Point. The murder of Smith and these subsequent events became legendary, colloquially referred to as the 'Shark Arm Case'.[1220] With Reginald Holmes deceased and unable to give evidence in court, the judge ruled that an arm was insufficient evidence of death. Brady was acquitted; no one was convicted of Reginald Holmes murder. Following his sensational death, Holmes' widow Inie Parker Holmes managed the family's boat yard.

Returning to our subject at hand, in early January of 1901, after the haze of federation had subsided, Reeks had some financial reckonings to deal with, though it is not known if he was immediately made aware of them. It appears that he was still owed £325 by the Humpybong S.S. Company for work involving the *Emerald*, though the directors were hesitant to make payment and had not included the debt on the company's balance sheet, amongst several other financial issues which caused much dissension at an annual meeting held in late January of 1901.[1221]

The Brisbane Courier reported on the matter. '*Mr. Hutton said all he could say about the matter of the company's indebtedness to Mr. Reeks, was that it was arranged at a meeting of the former directors, including two of the present directors, that Mr. Reeks should receive commission on £10,000. Although that amount had not been paid to Mr. Reeks, it was none the less a liability, and should have been included in the balance-sheet as such. That the present directors recognised it as a liability, was shown by the fact that they were now paying Mr. Reeks at the rate of £50 per month. He claimed that the amount due to Mr. Reeks was not a contingent liability but an ascertained liability, and should most certainly have been included in the balance-sheet.*'

The article continued. '*Mr. Hobbs said it was impossible at present to say what the liability to Mr. Reeks really would be. When in Sydney recently he had interviewed Mr. Reeks, and had told him that until something was done to reduce the excessive vibration on the Emerald his claim for commission would not be recognised. He also mentioned some necessary repairs to the winch, and Mr. Reeks agreed to attend to both matters. Until it was known exactly what the liability would be it was not possible to include it in the balance-sheet.*'[1222]

Unless provided with information from a friend or colleague based in Queensland, or he somehow received copies of the Brisbane-based papers, Reeks would have been initially unaware that his name was being mentioned in association with the somewhat dubious financial dealings of the directors of this company. One can assume, however, that he saw to the issues with the *Emerald* and that payment was eventually made. There was no mention of the debt owed to Reeks at a special meeting of the company held in March of 1901, two months later.[1223]

For Reeks, however, more diversions soon took hold. The newly-created Commonwealth of Australia began filtering to other aspects of daily life, including sports and recreational pastimes. One such event was a conference of yachtsmen representing the various yacht clubs of the six states held on 25 January 1901 at the Prince Alfred Yacht Club's meeting rooms. Amongst the delegates present were members of the Royal Yacht Club of Victoria, the Queensland Yacht Club, the Royal Sydney Yacht Squadron and the Prince Alfred Yacht Club.[1224] It was the first time that representatives from these organisations had met to discuss yachting from a national perspective. Though not in attendance, delegates from the Auckland Yacht Club, the Royal South Australian Yacht Club and the Derwent Yacht Club had all wrote requesting to be informed of any outcomes of the meeting.

With the intention of coming together for the purpose of considering rules and regulations for interstate contests, it was a big step in the right direction, likely stimulated by a combined directive to coalesce, form and build relationships, and champion annual, organised and inclusive events on the heels of Australia's federation. It also proved a success, at least from an administrative perspective. The following motions were unanimously adopted: '*That this conference recommends that an annual series of races be instituted between the Australian states and New Zealand, with a uniform type of yacht, to be hereafter agreed upon, and that the competing yachts be built to the British Yacht Racing Association's Rule.*'[1225]

Reeks was certainly in his element, being in attendance as a representative of the Royal Sydney Yacht Squadron. He also gave an interesting dissertation on the difference between the old measurement rule of 1896 and the present rule. The new rule, he explained, should tend to do away with the bulb-fin type and bring about the building of a wholesome craft in every respect suitable to Australian waters.[1226] The following resolutions were then carried:

1219 en.wikipedia.org/wiki/Shark_Arm_case.
1220 en.wikipedia.org/wiki/Shark_Arm_case.
1221 *The Telegraph*, 31 January 1901.
1222 *The Brisbane Courier*, 31 January 1901.
1223 *The Telegraph*, 9 March 1901.
1224 *Evening News*, 28 January 1901.
1225 *Evening News*, 28 January 1901.
1226 *The Daily Telegraph*, 28 January 1901.

'That the interstate class be 30 linear-raters, under the Y.R.A. new rule, but that boats already built to the 30-rating class be allowed to compete, such boats which may be in excess of the new measurements to allow the others their rating allowance, under Y.R.A. scale.

That competing yachts shall be decked (a fixed coach top to be considered a deck), and have a companion; and if fitted with a cockpit this shall be floored and self-draining, the open area thereof not to exceed 24ft.

That races be held in each state consecutively, and that the order be decided by ballot.

That the races be held between the months of November and February in each season, the actual date to be decided by the state in which the contest is to take place, and at least three months' notice be given to the visiting states.

That competing craft be sailed by amateur members of recognised yacht clubs, and that not more than one professional hand be allowed on board each yacht.

That the various yacht clubs be invited to raise sufficient funds for the purpose of obtaining a perpetual challenge cup at a cost of not less than 50gs, and that the cash prize for each race by 100gs, with a trophy of not less than 5gs for the steersman of the winning yacht, to be in each case provided by the state in which the race is sailed.'[1227]

Despite making good progress, the last day of the conference was cancelled. Queen Victoria, who had reigned for 63 years, died on the Isle of Wight aged 81. It took some days for a more formal and united response amongst representatives of the new Australian states to be enacted, including as to how the death and subsequent mourning period should be acknowledged and observed. Her death ultimately led to businesses and commercial interests closing and the population wearing black or other tokens of mourning. The death of Queen Victoria also led to the abandonment of many meetings and events.

Disappointingly, the effort to establish a 30ft linear rater class for interstate yachts would be futile, including owing to several craft being built over the rating.[1228] Starting in 1899 the first 30ft linear rater was *Aoma*, built by Logan Bros. The same Auckland builders then delivered across the Tasman its near-sisters *Petrel* and *Heather* and the unconventional *Sunbeam* in 1900. These were followed by a return to the conventional, *Culwulla* in 1901, all for Sydney-based clients. Along with boats from other designers and builders, these craft formed a strong competitive fleet over the decade that followed. However, it was only in Sydney and the annual interstate and New Zealand races remained elusive. Still, Reeks stayed resolute in its establishment, over four years later again calling for its creation, appreciating the support from several other prominent members of the Royal Sydney Yacht Squadron.[1229]

The death of Queen Victoria additionally caused the Anniversary Regatta which was to have taken place on 26 January 1901 to be postponed until Monday 18 March.

[1227] *Evening News*, 28 January 1901.
[1228] *The Sydney Morning Herald*, 11 October 1905.
[1229] *The Sydney Morning Herald*, 11 October 1905.

Though a subdued event, Reeks was once more on hand to assist with organisation efforts.

Also in March of 1901 Reeks took part in a unique trial trip on Port Jackson of two identical government oil launches recently constructed for service on the Northern Rivers region: the *Widgi Widgi* intended for use on the Tweed River and the *Mi Mi* intended for use on the Richmond River.[1230] Propelled by a Hercules oil engine and a Union Oil Company engine, respectively, the two craft were otherwise identical and measured 40ft in length with a beam of 9ft.[1231] With considerable interest placed on a 'race' occurring during the trial trip, it was a matter of contention as to which vessel would be faster, but on the day *Widgi Widgi* travelled the measured mile 38 seconds ahead of *Mi Mi*.[1232]

Reeks was on board the *Mi Mi* along with James Orr, recently-appointed chief shipwright and naval architect for the Navigation Department.[1233] It is not known if Reeks was somehow involved in the design of these two vessels or if he had acted as a consulting engineer during their construction process; either is possible. Alternatively, not one to miss an event, he may have been just been along for the ride.

MI MI ON THE TWEED RIVER.
SOURCE: THE SYDNEY MAIL AND NEW SOUTH WALES ADVERTISER, 10 MAY 1902.

Two days later the *Kuring-gai* took its first trial trip on Port Jackson with Reeks of course on board.[1234] It would be just over a month before a second trial trip took place. This latter one, however, was more for entertainment rather than engineering purposes. The gathering comprised a large number of invited guests and interested parties, with the trip punctuated by many speeches and toasts. Amongst those receiving much congratulation was Reeks. His new bride Gertrude was also a guest on board.[1235] The *Kuring-gai* commenced service on the Manly run in May of 1901.

The trial trips continued, albeit some offering more intrigue to Reeks' inquisitive mind than others. On 15 May 1901, off Fort Macquarie, he was one of a number of men on board the launch *Nixie* to witness a trial of Lawrence's automatic syphon log, a new invention which had been fitted specifically to the vessel for the purpose of

[1230] *The Daily Telegraph*, 25 March 1901; *The Sydney Morning Herald*, 25 March 1901.
[1231] *The Daily Telegraph*, 25 March 1901.
[1232] *The Sydney Morning Herald*, 25 March 1901.
[1233] *The Sydney Morning Herald*, 25 March 1901; *The Daily Telegraph*, 12 August 1907.
[1234] *The Daily Telegraph*, 28 March 1901.
[1235] *The Sydney Morning Herald*, 30 April 1901.

demonstration. The apparatus was intended to measure a vessel's speed and was designed to replace the widely-used towed device that was not very accurate at low speeds, and had its recording instrument or gauge located at the stern of the ship.

The syphon log was very simple in principle. Placed at the bottom of the moving vessel, the variable flow of water passing through an inverted 'U' shaped tube within the ship turned an impellor, the change in revolution then reflected the speed of the vessel. It was described at the time. '*Briefly stated, two pipes, each 1in. in diameter, are carried through the bottom of the vessel. These are so arranged that one "leg" of the pipe produces a lift on the water and the other effects a suction. The circulation of the water thus produced turns a turbine in the bend of the pipe, and the spindle of the turbine being geared to a piece of short rod operates the recording instruments by means of a connecting line.*'[1236] The instruments could be placed on the bridge of the ship. On runs from Fort Macquarie to Watsons Bay and return, '*those who witnessed the trial yesterday appeared favorably impressed with the simplicity and utility of the invention*'.[1237]

The device was demonstrated in other ports including at Adelaide. However, its eventual success in the market seems to have been subdued, and despite this and other inventions, the towed log remained in favour and in use for many decades into the twentieth century

On 27 July 1901 Reeks ventured further afield, to Lockleys in the central tablelands region of New South Wales, to attend the official opening of the Fish River Gold Dredging Company's new dredge.[1238] Measuring 70ft long with a beam of 31ft and a depth of 6ft 6in, the craft was fitted with a platform providing an additional 350 square ft of deck area. Built of hardwood with framing of kauri pine, the super structure consisted of a bucket belt made up of 28 buckets and two sets of powerful grab hooks.[1239] It was described in some detail, starting with the comment that at first sight '*the dredge looks like a large house,*' and was clearly an imposing structure. Steam operated, the gravel material collected from the river bed was tipped out of each bucket and then washed through a '*revolving screen*' with the gold deposited on sloped tables below. Including the pump, engine, boiler and other items the entire structure weighed a substantial 180 tons. Though Reeks appears to have been an interested party to the event, not formally involved with the company or the dredge itself, his interest in gold mining was once again piqued and it was not long before he too joined in with additional gold mining ventures.

During this period Reeks was also tasked by the Sydney Fire Brigade with developing possible designs and specifications for floating fire-fighting apparatus.[1240] Such craft had recently been launched in New York and London, though the type used on Port Jackson would need to be designed and built to local conditions.

[1236] *The Daily Telegraph*, 16 May 1901.
[1237] *The Daily Telegraph*, 16 May 1901.
[1238] *National Advocate*, 29 July 1901.
[1239] *National Advocate*, 29 July 1901.
[1240] *The Australian Star*, 18 September 1901.

As September of 1901 came around, the days grew longer and the weather warmer. Sydney's yachtsmen began preparations for the upcoming season. In terms of administration, Reeks was again unanimously elected honorary measurer of the Royal Sydney Yacht Squadron, continuing his 15-year legacy.[1241]

The new vessels also started to make their first appearance. A few days later a motor yacht, a class of pleasure craft steadily gaining favour, was launched, designed by Reeks.[1242] Named *Lara*, the vessel was built to the order of Joseph Thompson with its kerosene-driven engines supplied by the Capitain Company of England. In appearance the craft was noted to be '*very pleasing, being buoyant and jaunty, and showing a nicely moulded stem with a graceful sheer terminating in a neat counter. The cabinhouse is not too high, and gives her the appearance of a very comfortable boat. Her dimensions are l.o.a. 36ft., l.w.l. 30ft., beam 8ft., depth 3ft. 9in., draught 2ft. 6in. She is constructed of hardwood framing with planking of kauri. A speed of seven or eight knots per hour is expected to be realised under the engine alone, but on this she is not entirely dependent, as she has been designed as a yawl with a fairly generous sail spread for a boat of this class*'.[1243]

An oil launch designed by Reeks and built by Holmes of McMahons Point to the order of Fred Rae, former owner of the *Jess*, was also then under construction.[1244] The vessel was launched on 23 December 1901 and named *Mora*.[1245] Construction was of kauri planking, the keel timbers and dead woods being of hardwood. The engine was a double cylinder, 6 h.p. Wolverine, imported from America, and expected to give a speed of about seven miles an hour. With dimensions of 31ft (overall) by 27ft (waterline), with a beam of 6ft 9in and a draft of 2ft, the comfortable and prettily modelled vessel was stated to have '*a good sheer, the stem being nicely rounded, the afterbody terminating in a neat counter*'.[1246] A brightly varnished cabin house, about 9ft long, was situated well forward, extending the full width of the boat, aft of which and opening out from the cabin was a roomy cockpit, making for a most comfortable day boat.[1247]

Late in 1901 Reeks accepted the position of vice-president of the Sydney Dinghy Sailing Club.[1248] He also continued to be part of the organising committee for the Anniversary Regatta.[1249] In addition he was noted as being a contributor to a new periodical focussing on yachting and other aquatic sports named *White Wings*.[1250] The publication was billed as a place for '*gossip ... for sailing, swimming and fishing men*'.[1251] The man's energy and willingness to be involved continued to have no bounds.

[1241] *The Sydney Morning Herald*, 10 September 1901.
[1242] *The Sydney Mail and New South Wales Advertiser*, 14 September 1901.
[1243] *The Sydney Mail and New South Wales Advertiser*, 14 September 1901.
[1244] *The Sydney Mail and New South Wales Advertiser*, 14 September 1901.
[1245] *The Sydney Morning Herald*, 21 December 1901.
[1246] *The Sydney Morning Herald*, 25 December 1901.
[1247] *The Sydney Morning Herald*, 25 December 1901.
[1248] *The Sydney Morning Herald*, 2 November 1901.
[1249] *Sydney Sportsman*, 6 November 1901.
[1250] *The Sydney Mail and New South Wales Advertiser*, 16 November 1901.
[1251] *The Sydney Morning Mail and New South Wales Advertiser*, 16 November 1901.

PLAN OF LARA.
BY DAVID PAYNE

On 25 December 1901 Reeks and his wife Gertrude sailed to Auckland as passengers on board SS *Warrimoo.*[1252] His children, Kent and Thelma, remained in Sydney. The couple's trip to New Zealand was mostly recreational, perhaps a delayed honeymoon. On arrival in Auckland, Reeks and Gertrude transferred to SS *Rotoiti* en route to Wellington, they then took SS *Rotomhana* to Lyttelton.[1253] From there the couple embarked on board SS *Waikare* for an excursion to the West Coast Sounds, including Stewart Island, Preservation Inlet, Doubtful Sounds, Smith Sound, Thomson Sound, George Sound and Milford Sound, leaving on 13 January.[1254] After two weeks spent in the area, the couple then boarded a train for Dunedin, arriving on 3 February.[1255] From here Reeks and Gertrude once more boarded SS *Warrimoo,* arriving back in Sydney on 14 February 1902, via Lyttelton and several other New Zealand ports.[1256]

Never one to miss an opportunity, the trip likely coincided with some New Zealand-based work commitments. On 14 January 1902 the *New Zealand Times* announced that Messrs W. and G. Turnbull of Wellington had '*cabled to London yesterday morning, advising their agents to accept tenders for the new steamer for their Patea trade. The vessel will be built to designs prepared by Mr W. Reeks, of Sydney, and will carry 100 tons deadweight, with seven feet draught. The steamer will average 9½ knots at sea.*'

A major milestone, this would be the first steamer that Reeks had designed that was built in Great Britain; labour conditions were apparently unsatisfactory at the time in both New Zealand and New South Wales to allow for its construction in either of these locations.[1257]

Reeks returned to Sydney to find the local boat and shipbuilding yards filled with his commissions. In particular, two steam launches he had designed for Messrs Ballande Fils Aine of Noumea were nearing completion at the yard of William Holmes at McMahons Point. The first was intended for the New Caledonia coastal trade and measured 50ft in length with a beam of 12ft and a depth of 5ft 6in.[1258] Built of kauri and hardwood, the vessel was launched on 28 February 1902 with its engines, supplied by Messrs Chapman and Co. of Sydney, capable of generating a speed of 9 knots. The second craft was a smaller launch, measuring 31ft 8in with a beam of 4ft.[1259]

At the same yard, a steamer identical to the larger vessel Reeks had designed for Ballande Fils Aine was also under construction. This particular craft was built to the order of Messrs Gilchrist, Watt and Sanderson Ltd. and intended for harbour work.[1260] The vessel, name not known, was launched in late June of 1902.[1261] It is possible that this was the second steam lighter Reeks had designed for this company; he was noted as the only naval architect in attendance at the launch of *The Admiral*, measuring 86ft 6in with a beam of 22ft 6in and a depth of hold of 7ft 4in, which had been launched from the shipyard of J. Innes of Balmain in late February of that same year.[1262]

[1252] *The Sydney Morning Herald*, 26 December 1901.
[1253] *New Zealand Herald*, 2 January 1902; *Evening Star*, 9 January 1902.
[1254] *Evening Star*, 6 January 1902.
[1255] *Evening Star*, 3 February 1902.
[1256] *The Sydney Morning Herald*, 5 February 1902.
[1257] *Auckland Star*, 29 December 1902.
[1258] *Evening News*, 3 March 1902.
[1259] *Evening News*, 3 March 1902; *The Sydney Morning Herald*, 3 March 1902.
[1260] *The Sydney Morning Herald*, 3 March 1902.
[1261] *The Australian Star*, 1 July 1902.
[1262] *The Sydney Morning Herald*, 27 February 1902.

Meanwhile another new steamer, for many decades believed to have been designed by Reeks, though this now appears doubtful, was launched from the building yard of Watty Ford (Jr) at Berrys Bay on 10 April 1902. Built to the order of the Sydney Harbour Trust Board, the *Lady Hopetoun* measured 77ft with a beam of 13ft and a depth of 6ft 4in.[1263] The craft had a straight stem, flush decks and an elliptical stern, its black hull presenting a smart appearance.[1264] The vessel's deck fittings comprised a fore cabin and deck house, covered by a large wooden awning, running from abaft the after end of the fore cabin to the taff rail. *Lady Hopetoun* was fitted with 230 h.p. engines manufactured by Strickland Co. of London, with a Thornycroft boiler, and was expected to travel at a maximum speed of 12 knots.[1265]

The launching itself was a relatively low-key affair in terms of the newspaper reports, with only a paragraph devoted to the event and vessel description, and with just one person from those attending identified, Mr Hickson, the Trust's president. However, Reeks' connection with the craft is confirmed much earlier.

On 16 July 1901 Hickson wrote to Reeks in a follow up to the Trust's previous communications regarding, '*preparation of plans and specifications for the new launch'*. In particular, the letter notes that the Trust had stated '*the original Proposal of constructing a larger boat has been abandoned*'.[1266] Instead, it was decided to reduce the size of the vessel to a '*small, handy launch'*. In turn, they requested that Reeks' fee of £50 pounds be '*somewhat reduced*' to '*whatever you think is a fair remuneration*' and it '*would be promptly paid*'.[1267] Their principal reason for this reduction seems to have been based on the fact that a smaller craft would involve less supervision. It is assumed this issue was resolved, and the new design likely was prepared or modified to suit in-house. It was then built by Ford with Reeks possibly supervising the construction period, no doubt helped by the fact that the pair were now quite experienced in working together.

Reference to the *Lady Hopetoun* is continually made to the vessel being a 'launch', suggesting a step down the ladder from a steam yacht such as *Ena*. This in no way, however, diminishes the elegance of the design. Perhaps designed with Reeks' assistance or based on his original larger design for the Sydney Harbour Trust Board (plan no. 149 that measured 87.3 by 13.6 by 7.2ft), *Lady Hopetoun* is comparable to *Ena* sharing the raked cabin house and funnel, outside or exposed steering position, a long canopy, overhanging counter, trimmed with tasteful embellishments.

Albeit, in an awkward coincidence, as *Lady Hopetoun* was undergoing trials and then put into service, the Governor General and his wife were preparing to return to England. When Lord Hopetoun's request for a greater allowance was not approved by the new parliament, he formally resigned in July of 1902 after less than two years in the role. In contrast, his namesake vessel, *Lady*

[1263] *The Australian Star*, 10 April 1902.
[1264] *The Australian Star*, 10 April 1902.
[1265] *The Australian Star*, 10 April 1902.
[1266] Letter: Harbour Trust Board to Walter Reeks, 16 July 1901.
[1267] Letter: Harbour Trust Board to Walter Reeks, 16 July 1901.

LADY HOPETOUN.
SOURCE: GRAEME ANDREWS WORKING HARBOUR PHOTOGRAPH COLLECTION, CITY OF SYDNEY ARCHIVES.

Hopetoun, steamed on as a VIP launch for decades, and at the time of writing remains operational on the harbour in immaculate working condition, a treasure of the Sydney Heritage Fleet, and is recognised widely by the familiar whistle tone that carries across the water.

Shortly after the launch of the *Lady Hopetoun*, Reeks attended an Easter yachting camp at Broken Bay, along with fellow Prince Alfred Yacht Club members.[1268] Likely itching for a vessel of his own, he partnered with Charles Bartholomew and C. W. Bennett to commission a new motor launch, *Wahine.* Built by Green at Lavender Bay, the vessel's appearance on the stocks was described in the press in great detail. '*As she rests on the stock she is seen, so far as graceful appearance is concerned, rather at a disadvantage, her flat floor and hard bilge, both of which were necessary to attain the purpose of her creation, not being nearly so pleasing to the eye as most of the sweetly turned out boats by the same designer; but afloat these features will not be noticeable, and what with her clean-cut sheer and handsome well-proportioned deckhouse she should be a very taking boat when in the water. Her extremely shallow draught calls for first notice. By a very cleverly worked out design, and by cutting away every ounce of superfluous weight, it was found possible to work on a draught of water of only nine inches, a truly remarkable achievement for this class of boat, and this should enable her to fulfil to the letter the purpose for which she was designed, that is poking about in shallow nooks and bays, both in the harbour here and further afield, such as Brisbane Water, where many delightful little creeks and cranies are open to boats that draw but a few inches of water. The problem of giving the propeller an effective grip of the water without the blades going below the line of the keel was solved by resorting to a very ingenious idea, called "a tunnel stern". This consists in shaping the run of the vessel in the form of a tunnel, by which arrangement, when the boat is stationary, nearly half the propeller is above the water, and is really in a vacuum in the tunnel chamber, but so soon as way is gathered the water is drawn into this vacuum, and the propeller has solid water to work in. A 6-h.p. De Dion marine engine is to be installed, and although a low power for a craft of this size, it is anticipated that a fair speed will be attained. Picked cedar has been used throughout for planking and so carefully has it been fitted that it looks almost like one piece. The rest of the work, too, is quite in keeping with the planking and reflects great credit upon the builder. Very noticeable features are the large amount of accommodation and the breadth of the floor. The cabin, which has over 6ft. headroom, is situated amidships, and is very nicely fitted with lockets, ice chest, lavatory, &c., polished cedar being chiefly used. Aft of the cabin and opening out from it is a roomy cockpit with an awning overhead. For'ard of the cabin is the engine-room, and for'ard of that again is the forcastle. Her general dimensions are: Length over all, 42ft. 7 in.; beam 7ft 6in.; draught of water, 9in. It is anticipated that she will be launched about the end of next week.*'[1269]

Wahine's '*extremely shallow draft*' was worthy of this attention. It was quite remarkable for a 42ft long craft. Unfortunately no images or further description clarifies the '*flat floor and hard bilge*' that were '*not so pleasing to the eye*', but this suggests Reeks created a flat bottomed, hard chine shape with vertical and probably flat topsides, forming a box in cross-section. This would help maximise displacement at that shallow draft and give the '*large amount of accommodation and breadth of floor*' that was noted as well. He could work the tunnel stern into this shape too. However, despite Reeks' previous experience with this feature, he still gave himself a challenge by ending the craft in a canoe stern, with the rudder mounted on the vertical stern frame, visible in some of the images that do exist. The shallow draft was still noteworthy in 1907,

[1268] *The Daily Telegraph*, 2 April 1902.

[1269] *The Sydney Morning Herald*, 31 May 1902.

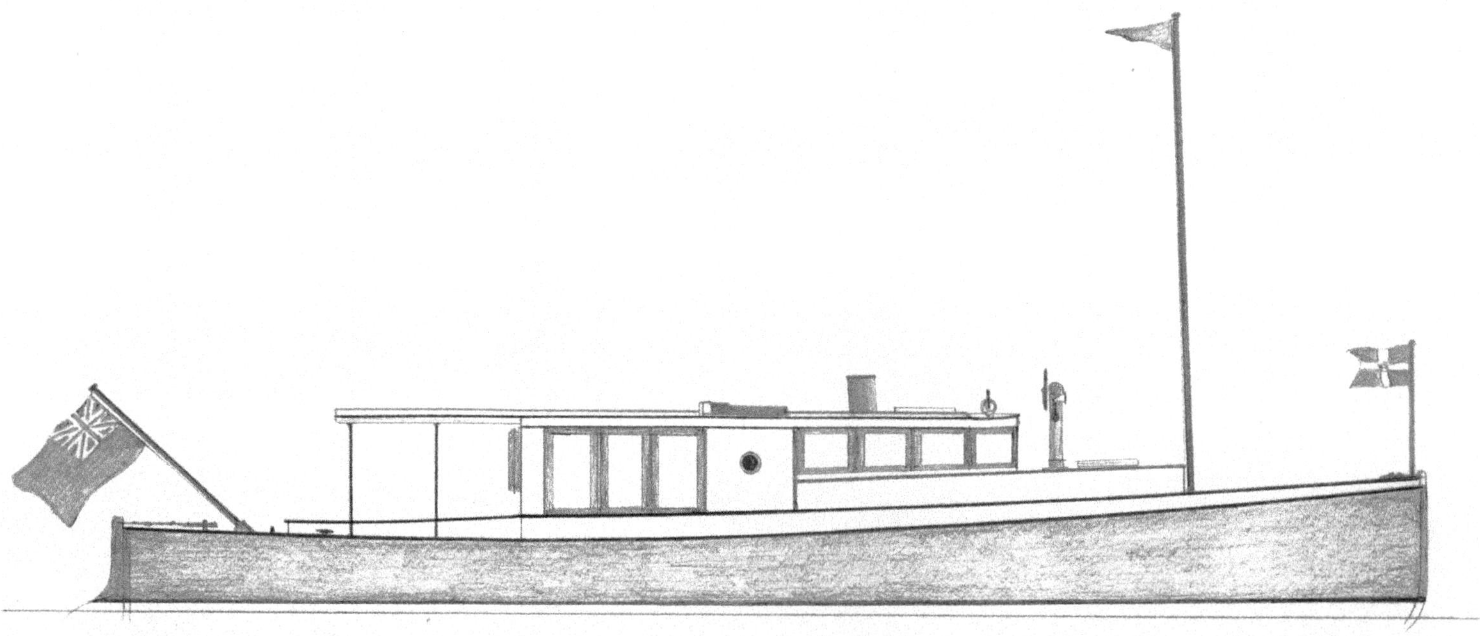

PLAN OF WAHINE
BY DAVID PAYNE

WAHINE
SOURCE: RAE FIDOCK COLLECTION.

when a review titled '*Port Jackson's Pleasure Fleets, Ancient and Modern*' by 'Lanyard' published in the *Evening News* on 14 December harked back to *Wahine*. '*Mr. Reeks' boat was designed to float on a morning dew or the perspiration of an ice box, but is a remarkably comfortable vessel, with a turn of speed.*' Long and narrow too, the 6 h.p. De Dion engine helped with this speed, but later it was to bring about a falling out with a particularly crusty client.

There were other avenues of work for Reeks' enterprising agenda too during 1902, including gold mining. The prospectus of the Belleview Gold Mining Company of North Western Australia, to be registered under the Companies Act of New South Wales, was first published on 3 April 1902. With an authorised capital of £30,000, the Board of Directors were stated to be Harry Markham Evans, managing director of the *Sunday Times* newspaper; Alexander Fraser Hill, an accountant of Sydney; Walter Reeks; and Dr J. M. McDonagh, a surgeon of Sydney. The general manager of the company was announced as Captain William Bell with W. A. McDonald appointed secretary. The intention of the organisation was to purchase an interest in an already-existing syndicate that held four gold mining leases within a 30-acre parcel of land located 215 miles from Port Hedland, Western Australia, reached via a '*moderate bush track*'.

Though a lot of effort had been invested into its establishment, disappointingly for Reeks, the Belleview Gold Mining Company of North Western Australia was not established, likely due to lack of investors. Still, it shows the lengths he was willing to go to in terms of location, as well as his persistence in the search for the prized metal.

PROSPECTUS OF

The Belleview Gold Mining Company, No Liability.

OF NORTH WESTERN AUSTRALIA.

TO BE REGISTERED UNDER THE COMPANIES' ACT. 1899, N.S. WALES.

Authorised Capital, £30,000.

In 120,000 Shares of Five Shillings Each, of which 50,000 shares are to be issued as paid up to Four Shillings per share, and allotted to the Vendors as part payment for the purchase of the mining properties, rights and appurtenances.

50,000 Shares are now Offered for Subscription (10,000 of which will be applied for by the vendors), and are payable—one shilling per share on application; one shilling per share on allotment; two shillings in calls of sixpence per share at not less intervals than one month; and as to the remaining one shilling in calls of threepence per share at similar intervals, at the discretion of the Directors. The remaining 20,000 shares will be held in reserve for future issue.

Applications for Shares must be received not later than Saturday, 5th April, 1902.

Board of Directors:
H. M. EVANS, Strathfield, and York Street, Sydney, Managing Director.
"Sunday Times" Newspaper Company, Limited.
A FRASER HILL, Waverley, Accountant.
WALTER REEKS, of Mosmans and of Pitt Street, Sydney, Naval Architect.
DR. J. M. McDONAGH, of Macquarie Street, Sydney, Surgeon.
To hold Office until the First Meeting of the Company.

Mining and General Manager in N.W.A.
CAPTAIN WILLIAM BELL, M.A.I.M.E.

Secretary:
W. A. McDONALD, (McDONALD AND WEBB).

Offices:
LONDON CHAMBERS, PITT AND MARKET STREETS, SYDNEY.

Bankers:
THE LONDON BANK OF AUSTRALIA LIMITED, PITT STREET, SYDNEY.

Solicitor:
W. G. PARISH, MUTUAL LIFE BUILDINGS, MARTIN PLACE, SYDNEY.

A PORTION OF THE PROSPECTUS OF THE BELLEVIEW GOLD MINING COMPANY.
SOURCE: DAILY COMMERCIAL NEWS AND SHIPPING LIST, 2 APRIL 1902.

Mid-1902 also coincided with some more unique opportunities and excursions for Reeks. On 23 June 1902 a ceremony of great importance, in terms of the advancement of electrical engineering in New South Wales, took place in Sydney with Reeks amongst 70 attendees to witness the historic event. The first turbine electrical engine was turned on at the Imperial Electric Light Station in Pitt Street.[1270]

Another opportunity to witness new industrial developments came a few months later, at the invitation of the Shell Transport and Trading Co. Ltd. On 31 October 1902 Reeks was one of 500 guests who were conveyed on board both the *Lady Hampden* and *The Lady Mary* to Gore Bay to inspect the firing arrangements of the SS *Clam*.[1271] The vessel, a member of the company's fleet of more than 40 craft, had recently arrived from Batoum under liquid fuel which it had been burning for the last three years.

On 9 August 1902, a smart-looking yawl-rigged cruiser, built by William Goddard at Lavender Bay to the order of C. E. Waters was launched. Named *Gulnare* and designed by Reeks, the craft measured 37ft overall, with a beam of 10ft and a depth of 4ft 9in.[1272] Construction was of kauri with spotted gum timbers and decks of narrow silky kauri planking. The press noted that every attention had been paid to space with the vessel essentially a cruiser, though from its appearance afloat, it '*should prove rather slippery for a boat of her type*'.[1273] Another report stated that it '*has all the appearance of turning out what was evidently intended, namely, a fast cruiser, combining all the modern improvements for comfort. She is a semi-fin-keel boat, with any amount of beam, and moderate overhangs. On deck she has a high coach roof surmounted by a skylight*'.[1274] A strong hint to *Gulnare*'s character comes from another statement, '*in general appearance she very much resembles the deservedly popular American knockabout type*'.[1275] The term knockabout was an American (New England region) nickname for a neat, unassuming and seaworthy cruising yacht. Simplicity was a theme, and therefore many were fitted without a bowsprit. This may have been the case here, and the moderate overhangs also suggest a shorter, rounded bow profile; a feature that would have been complimentary to the transom stern that is also noted. *Gulnare* was launched without an engine, they were still coming into vogue as an auxiliary power source, but one was added in 1904 by Pritchard Bros., with a feathering and reversing propeller designed by Reeks, again highlighting his interest in all things mechanical.[1276]

With regards to yacht racing administration, little further progress had been made on implementing an annual series of races between representatives from the Australian states and New Zealand competing in a uniform type of yacht. However, the Sydney-based clubs continued to push for greater collaboration and uniformity amongst themselves. In July of 1902 the Sydney Yacht Racing Association was formed at the recommendation of a special joint committee of the Royal Sydney Yacht Squadron and the Prince Alfred Yacht Club. The effort had great support, with Reeks at the forefront of its creation. It was a formalised attempt to align, standardise and organise yacht racing on Sydney Harbour and hopefully, by natural progression, in other states of Australia. In effect, it was another symbol of federation.

The motion adopted for the creation of the Sydney Yacht Racing Association was quite extensive. To summarise it read: '*That an association to be called the Sydney Yacht Racing Association be formed, with the following functions in connection with local yacht-racing: The framing of racing rules, the measurement and numbering of racing yachts, the decision of protests referred to it, any other matters in connection with yacht racing which may be referred to it. The membership to consist of three representatives, elected annually, and the secretary for the time being of the R. S. Y. S., P. A. Y. C., and such other clubs as may be admitted. The racing rules of the S. Y. R. A. shall be those of the British Y. R. A. for the time being, modified by such special rules as may be adopted by the S. Y. R. A. Any change in the Y. R. A. rules shall take effect six months after it takes effect in Great Britain unless otherwise decided by the association, but in no case shall the period exceed 12 months in all. Any proposed special rule may be recommended by the association, and if approved by the clubs interested shall be adopted by the association. The decision of the association on a protest shall be final, but the association may at its own instance refer a protest to the council of the Y. R. A. should it think proper to do so.*

That all races be sailed under the rules of the S. Y. R. A.

That the Y. R. A. classification of yachts be strictly adhered to, and no yacht allowed to compete in a class smaller than that to which she belongs, except as provided in S. Y. R. A. rules.

That in time by rating races under the Y. R. A. scale the time allowances should be given at the finish.

Every yacht while racing shall carry her racing number on each side of the mainsail near the peak. These numbers will be allocated and supplied by the council at a small charge. A yacht carrying its number is not required to carry a distinguishing flag.

All yachts sailing in a race at night shall comply with the Navigation Act of New South Wales for the time being as to the carrying of lights.

When time is given at the finish five flags shall be hoisted five minutes before the start ... When time is given at the start the ensign shall be hoisted five minutes before the first yacht's starting time ...

In races where time allowances are conceded at the start, a yacht running free or with the wind aft shall never alter her course by luffing or bearing away to hinder an overtaking yacht passing.

When a protest has been decided by a sailing committee the only appeal shall be to the S. Y. R. A., on the same conditions and payment of the same fees as provided in the Y. R. A. rules, and the S. Y. R. A decision shall be final.

Yachts not exceeding 36 linear rating, which have raced in Port Jackson prior to August 1, 1902, shall be allowed to compete on rating allowance in the 30ft class, providing no increase is made in the rating as shown by certificate held on November 1, 1902'.[1277]

[1270] *The Sydney Morning Herald*, 24 June 1902.
[1271] *The Australian Star*, 1 November 1902.
[1272] *The Australian Star*, 11 August 1902.
[1273] *The Australian Star*, 11 August 1902.
[1274] *Referee*, 13 August 1902.
[1275] *The Sydney Morning Herald*, 11 August 1902.
[1276] *The Sydney Morning Herald*, 17 December 1902.
[1277] *The Sydney Morning Herald*, 9 July 1902.

At a committee meeting of the Royal Sydney Yacht Squadron, held in mid-October of 1902, Reeks was notably elected as the organisation's representative to the newly-formed Sydney Yacht Racing Association. Fellow members included T. W. Bremner and Dr Gordon Craig.[1278]

Remaining steadfast to his naval architecture work involving commercial vessels, on 20 August 1902 Reeks announced a new tender for hull construction and fitting of a 122ft wooden steamer for the Balmain New Ferry Company.[1279] Names were to be submitted to him at his office, 15 Pitt Street. This vessel would be the *Lady Rawson,* completed just over a year later.

Concurrently a large steam lighter was also being built by Holmes to a design by Reeks. Commissioned by Elliott Bros., the vessel was 70ft in length with a beam of 18ft 6in and a depth of 7ft.[1280] The hull of the craft, intended for harbour use, was launched on 8 September 1902.[1281]

Nine days later, a new ferry steamer was launched from Messrs Morrison and Sinclair's yard at Pyrmont, Balmain. Built to the order of the Watsons Bay Ferry Company to a design by Reeks, the *King Edward* was a double-ended ferry steamer with dimensions of 110ft, with a breadth of 24ft and a depth of 8ft. Its construction was unique in that the vessel's deck cabins and upper works were erected before launching. The *King Edward* was described as '*nicely fitted up with accommodation for about 450 passengers, the ladies' cabin being decorated with carved panels; and the vessel will be lighted by electricity throughout. Messrs. J. Wildridge and Sinclair, of Pitt-street, Sydney, are contractors for the machinery, which will be fitted up under their supervision. The machinery will consist of one set of compound surface condensing marine engines of 250-horsepower, the diameters of the cylinders being 12in. and 24in. respectively, and 18in. stroke, and twin colonial boilers, each 4ft. 6in. diameter and 6ft. 10in. long, for a working pressure of 120lb. per square inch, complete with propellers, shafting, &c.*

The steamer has all the comfort, deck space, airiness, and stability of the best and most modern boats in the harbour, but with many points of difference with a view to making her up-to-date. The cabin has panels which are neatly and delicately carved while the midship house and forecabin are panelled in kauri, as is usual in first-class ferry boats. She is constructed of Australian hardwood and kauri, and has in addition to the ordinary construction, which is in itself unusually strong, two lattice girders running from end to end of the ship, which will make her unquestionably the strongest ferry boat of her size ever built in Sydney Harbour. She is metal-fastened throughout, and it may be worth while to note that many of her principal fastenings are of heavy, pure copper bolts obtained from dismantling the old warship Nelson, which though nearly 100 years old are better copper than can be bought to-day.'[1282] As stated, once again Reeks was recycling parts from HMS *Nelson*.

The *King Edward* made its trial trip on 3 November 1902, successfully achieving a speed of 11 knots without full pressure.[1283] The vessel's official trial trip took place a few days later with some 200 guests on board. The craft covered the measured mile in 5 minutes and 15 seconds.[1284] Reeks was heralded for his design work, being the subject of several toasts and congratulatory remarks. There was much sentiment felt amongst the directors of the Watsons Bay Ferry Company that the *King Edward* would be the pioneer of other new boats for the ferry service.[1285] In hindsight, a very true statement.

Another commercial collaboration between Reeks and Holmes saw the ketch *Kai Ora* launched from McMahons

[1278] *The Sydney Morning Herald*, 15 October 1902.
[1279] *The Australian Star*, 20 August 1902.
[1280] *The Australian Star*, 1 July 1902.
[1281] *Daily Commercial News and Shipping List*, 11 September 1902.
[1282] *The Australian Star*, 17 September 1902.
[1283] *The Australian Star*, 4 November 1902.
[1284] *The Daily Telegraph*, 7 November 1902.
[1285] *The Daily Telegraph*, 7 November 1902.

FERRY KING EDWARD.
SOURCE: GRAEME ANDREWS WORKING HARBOUR PHOTOGRAPH COLLECTION, CITY OF SYDNEY ARCHIVES.

Point on 9 December 1902.[1286] Reeks had designed this vessel to encompass some noteworthy features. Built to the order of Alex Cameron and intended for the Gippsland Lakes and Victorian coastal trade, the vessel was fitted with double headsails and a main topsail in addition to ordinary lower canvas.[1287] Two oil engines built by the Victor Motor Company were also installed, one was to act as auxiliary power and expected to drive the vessel at six knots, while the other was installed instead of a steam winch for working cargo.[1288] The *Kai Ora* featured a hold capable of carrying 30 tons, a spacious forecastle forward, and an engine room and cabin aft. While drawing only 4ft, with a full load on board, the craft featured over 6ft clear of headroom for working cargo. Overall *Kai Ora* encompassed a combination of capacity, height and light draft seldom attained. The vessel remained in service in Victorian waters for a number of years before being transferred to Papua New Guinea, like many others of its type.[1289]

With the year winding down relative to business prospects, Reeks turned his attention to his favoured pastime of yachting and administration of the sport. On 13 December 1902 an innovation in aquatic events took place with a race for motor launches held under auspices of the Neutral Bay Sailing Club, open to all such craft in Port Jackson.

With considerable interest placed on the event, more than 20 motor launches entered with a large group of spectators watching from on shore and out on the water, all eager to catch a glimpse of the contest.[1290] The handicaps, which were sealed, had been developed by Reeks in collaboration with G. Ey and G. Bredziak, in accordance with a scale recently published in *Rudder* magazine.[1291]

The race was to be the start of a new era in motor boating on Sydney Harbour. Of note, amongst the participants were several craft designed by Reeks, including *Lara* and *Mora,* both launched the previous year.[1292] Continuing to encourage this type of racing, he also devised conditions and coordinated the motor launch race at the 1903 Anniversary Regatta; the first time the event had been held at this auspicious event. This race was won by *Faa-mu*, followed by *Dart*, with *Coo-ee* third.[1293] The major yacht clubs soon followed suit, adopting motor launch races into their regular programmes. Soon motor boat clubs would be established, ushering in a new platform for the sport.

There was one last business-related event to attend to before the year concluded. The 90-ton iron steamer Reeks had designed for Messrs G. and W. Turnbull of Wellington, New Zealand, for the Patea River trade had been launched and registered with the name of *Aotea* even though the steamer of the same name Reeks had designed in 1898 was still operating in New Zealand waters. Built in Scotland by Messrs Scott of Kinghor, the vessel arrived in Sydney en route to New Zealand on 17 December 1902, after a trip encompassing the Suez Canal and Thursday Island.[1294] As stated, the craft signified the first commercial vessel designed by Reeks built in Great Britain.

Aotea measured 108ft in length, with a beam of 21ft 6in and a depth of 9ft 4in.[1295] Though the vessel was capable of carrying 180 tons of cargo, its dimensions were limited

[1286] *The Age*, 10 December 1902.
[1287] *The Daily Telegraph*, 10 December 1902;
The Snowy River Mail and Tambo and Croajingolong Gazette, 7 March 1903.
[1288] *The Daily Telegraph*, 10 December 1902.
[1289] *Daily Commercial News and Shipping List*, 24 September 1907.
[1290] *The Daily Telegraph*, 13 December 1902.
[1291] *The Daily Telegraph*, 13 December 1902.
[1292] *The Daily Telegraph*, 13 December 1902.
[1293] *Evening News*, 27, 30 January 1903.
[1294] *Auckland Star*, 29 December 1902.
[1295] *Auckland Star*, 29 December 1902.

KAI ORA AND STORMBIRD AT MARLO, VICTORIA (1905).
SOURCE: CENTRE FOR GIPPSLAND STUDIES, FEDERATION UNIVERSITY AUSTRALIA.

owing to the port of Patea having a 7ft bar, and a very narrow winding channel, that at low water was described as '*little more than a gutter*'.[1296] *Aotea* was specifically proportioned for the difficult channels it had to negotiate with a shallow draft, whilst still having a large volume for cargo. '*Good things come in small packages*' was a key quote from the newspaper article detailing its specifications, but little else is noted in this and any other reports, suggesting it was either relatively conventional, or else any unorthodox features were not recognised or considered worthy of comment.[1297] Still, Reeks was likely experiencing the same thrill many young children feel on Christmas morning when the *Aotea* docked in Sydney for the first time. Considering the location of build he had obviously been unable to personally superintend its construction. After a short stay in Sydney, owing to two plates needing to be replaced following the vessel's grounding off the Great Barrier Reef, *Aotea* proceeded to its new home port of Wellington on 6 January 1903.[1298] The vessel arrived in New Zealand on 3 February, to shortly thereafter begin shipping stock and coal in the local Wellington-based trade.[1299]

After only 18 months in the trade, however, *Aotea* sadly met its demise, a result of it touching the very narrow channel it had been designed to avoid. The craft went ashore at Patea on the southern side of the breakwater on 9 August 1904 carrying a shipment of coal, the tide believed to have been higher than actuality.[1300] Despite the best efforts of all involved to re-float the craft over a period of many weeks, *Aotea* could not be saved.[1301] After finally being freed from the bottom of the Patea on 21 September, it near-immediately was driven onto rocks, becoming a total wreck.[1302] The work of dismantling the vessel was carried out in December of that year.[1303]

With the loss of the New Zealand steamer *Aotea* well into the future, the year 1903 began with another Victorian client, C. D. Wallace of Melbourne, tasking Reeks with designing a large auxiliary yacht.[1304] It was a significant commission. Wallace was rear-commodore of the Royal Yacht Club of Victoria and current owner of the celebrated yacht *Sayonara*. Wallace also enjoyed cruising, with his favoured locations being Tasmania, Westernport and Gippsland Lakes.

The vessel Reeks designed was substantial too: a powerful auxiliary cruising schooner with a short draw to enable the craft to enter shallow and narrow bays and harbours.[1305] Built by Holmes of McMahons Point,

[1296] *Auckland Star*, 29 December 1902.
[1297] *Auckland Star*, 29 December 1902.
[1298] *The Australian Star*, 7 January 1903; *New Zealand Times*, 15 January 1903.
[1299] *Patea Mail*, 28 January, 4 February 1903.
[1300] *Wanganui Chronicle*, 11 August 1904.
[1301] *Wanganui Herald*, 13 September 1904.
[1302] *Hawera & Normanby Star*, 22 September 1904.
[1303] *Grey River Argus*, 7 December 1904.
[1304] *The Australasian*, 13 December 1902.
[1305] *The Sydney Morning Herald*, 21 March 1903.

WRECK OF THE STEAMER AOTEA AT PATEA, NEW ZEALAND, SEPTEMBER 1904.
SOURCE: AUCKLAND WEEKLY NEWS, 20 OCTOBER 1904.

another collaboration between the pair, the vessel would take some nine months to complete, and would be an extremely noteworthy addition to both the designer and its builders' resumes.

Meanwhile, the opening of the Royal Sydney Yacht Squadron's new clubhouse took place in January of 1903 with Reeks of course an attendee.[1306] The clubhouse was named *Carabella*. Following the official events, a garden party was held on the expansive grounds at Kirribilli Point. A few months later a billiard room would be built with a smoke concert taking place of which musical items or comedy sketches were delivered, including one by Reeks, ever the entertainer.[1307]

OPENING OF THE ROYAL SYDNEY YACHT SQUADRON CLUBHOUSE, CARABELLA.
SOURCE: THE SYDNEY MAIL AND NEW SOUTH WALES ADVERTISER, 28 JANUARY 1903.

Continuing to concentrate largely on commercial vessels, an auxiliary fishing ketch named *Pactolus* was the next vessel completed. The craft was launched in May of 1903, built by Holmes at McMahons Point to a design by Reeks. Fitted with twin screws and a 10 h.p. oil engine, a trial trip on Port Jackson took place on 22 May with the vessel achieving a speed of just over eight miles per hour.[1308]

Pactolus measured 52ft in length with a beam of 14ft and a draft of 4ft 6in.[1309] The craft was intended for the Northern Rivers area fishing trade with its owner, Carl Landenberger, his wife and three crew making the delivery trip to the Clarence River from Sydney in terrible weather.[1310] Despite this experience, Landenberger stated that he was more than satisfied with the new vessel.[1311] Like many others of its cohort, *Pactolus* ended up in the Pacific Island trade a few years later.[1312]

In June of 1903 Reeks called for tenders for the joinery and upper works of a passenger steamer, 77ft in length, with construction to be undertaken on board the vessel at Newcastle.[1313] This work was for the craft *Phoenix*, built to the order of W. W. Johnson by William Langford, shipbuilder of Newcastle, who had also designed the hull.[1314] With the new owner in need of a modern excursion boat, emulating those servicing Sydney Harbour, Reeks had been tasked with designing the craft's cabins and upper deck work.[1315] Capable of carrying 277 passengers, the *Phoenix* was intended for the passenger trade on Lake Macquarie and began service there late in 1903.[1316]

SS PHOENIX ON LAKE MACQUARIE (1900S).
SOURCE: IMAGE BY RALPH SNOWBALL, HELD BY THE UNIVERSITY OF NEWCASTLE LIBRARY'S SPECIAL COLLECTIONS.

Also in June of 1903 Reeks attended the launch of the new North Coast Company's 125ft steamer *Nerong,* built by Messrs Foster and Minty at Balmain.[1317] Unfortunately the designer of this large vessel is not stated, only a reference is made to a 'Mr Scott' being a part-designer.[1318] This is probably James Scott, builder of *Lady Manning* who also became a capable vessel designer and was connected to the North Coast Company Reeks could very well have been involved in some capacity.

Another tender was issued by Reeks shortly thereafter, this one for the fitting of machinery into the Balmain New Ferry Company's new steamer.[1319]

With international attention then centred on the America's Cup, about to play out once more on New York Harbour, an interesting article was published in Sydney's *Evening News* on 1 September 1903 regarding Sir Thomas Lipton's upcoming challenge in his new Fife-designed yacht *Shamrock III.* The defender, owned by a syndicate of members from the New York Yacht Club headed by Cornelius Vanderbilt III, was *Reliance,* designed by the 'Wizard' Nathanael Greene Herreshoff. The article was published far back in the paper, on page 75, but it's definitely worth revisiting.

1306 *The Sydney Mail and New South Wales Advertiser*, 28 January 1903.
1307 *The Sydney Morning Herald*, 11 July 1903.
1308 *The Sydney Morning Herald*, 23 May 1903.
1309 *The Sydney Morning Herald*, 23 May 1903.
1310 *The Grafton Argus and Clarence River General Advertiser*, 1 June 1903.
1311 *The Grafton Argus and Clarence River General Advertiser*, 1 June 1903.
1312 *The Daily Telegraph*, 24 September 1905.
1313 *Newcastle Morning Herald and Miners' Advocate*, 13 June 1903.
1314 *Newcastle Morning Herald and Miners' Advocate*, 7 November 1903.
1315 *Newcastle Morning Herald and Miners' Advocate*, 7 November 1903.
1316 *Newcastle Morning Herald and Miners' Advocate*, 7 November 1903.
1317 *The Australian Star*, 18 June 1903.
1318 *The Australian Star*, 17 June 1903;
Balmain Observer and Western Suburbs Advertiser, 20 June 1903.
1319 *The Australian Star*, 29 June 1903.

SPECTATORS WATCHING THE LAUNCH OF THE NERONG AT BALMAIN; WALTER REEKS SHOWN IN SQUARE.
SOURCE: THE SYDNEY MAIL AND NEW SOUTH WALES ADVERTISER, 24 JUNE 1903.

Nearly 15 years had passed since Reeks headed to America in consideration of an Australian challenge to the infamous trophy. He had gone on to forge a career outside of racing yacht design, with commercial steam ferries and launches becoming his niche. '*What do you know of the Wizard?*' asked the reporter. '*Just this much,*' came Reeks' immediate reply. '*He is a topnotcher all the way.*' The article continued. '*Then Mr. Reeks related some of his experiences with the great designer. He said, "On my last visit to America, about 14 years ago, I met Mr. E. Burgess, the designer of the Puritan, Mayflower, and Volunteer - three successful defenders of the America Cup, and, as was expected, I met Mr. Burgess. This gentleman had proved himself a great designer, and was at the time recognised as the leading man. His boats beat the Genista, Galatea, and Thistle. He displayed the greatest courtesy in showing me anything and everything I wanted to see, even to the lines and calculations of the Volunteer. After this, Mr. Burgess said to me, "Would you like to meet Herreshoff? He is a very busy man, and in many ways a peculiar man; you probably won't get more than 10 or 15 minutes' conversation with him altogether, but he is in my opinion the greatest man in the country, and I think in the world. It means that you will have to take a cross-country trip to Bristol* [Rhode Island], *but if you think it worth while for so short an interview, I will give you the necessary letter of introduction."*

Continuing, Mr. Reeks said, "I jumped at the chance, and proceeded to Bristol. I was at Herreshoff's yard at 7 a.m., and about half an hour later saw Herreshoff himself. He was making a survey of the great works, and seemed to take in everything at once. He then came along to me, and we conversed about yachts and steamers - for he builds steamers as well as yachts. After a few minutes, I experienced the sensation that he 'wished to goodness I would go away.' Suddenly, however, his manner changed. It may have been something I had said which he appreciated - but the result of the change was that I spent all day at the yard, and was afterwards invited to the house. During that day he showed me everything that I expressed any desire to see, and a great many things I would not have cared to ask about were also shown to me. Nathanael Herreshoff, or Nat Herreshoff, as he is more popularly known, is a quiet man, but a man of great determination and ideas. Most of his designs are started in a studio at his own house. In that studio he lays his own schemes, makes his own sketches, does all the leading computations, and then hands over to his drafting staff for completion. It is only fair to Herreshoff to say that, in my opinion, he is the only complete designer and complete engineer in one brain I have ever met. That means he does the work which is generally carried out by two men. In most cases, one man designs, the other is the engineer, and then the two confer before the scheme is laid. Herreshoff does all this work for himself".'[1320]

The interview with Reeks continued, with him offering more insight into the past and that unforgettable trip that propelled him to become the type of professional that he saw in Herreshoff. '*I have no hesitation in saying ... that by my visit to these works - extensive as they are, embracing shipbuilding, engineering, and boiler-making - and the privilege of eight or nine hours' conversation with this great man, I learned more in that one day that I had learned in any previous three years.*'[1321]

When asked about the status of the America's Cup and if it would ever be taken off America's hands, Reeks gave a rather visionary yet accurate prophecy, though it would be another eighty years before the historic and unprecedented triumph of *Australia II* resulted in the end of the New York Yacht Club's then 132-year claim on the trophy. '*My opinion is that the Cup will never be taken from*

[1320] *Evening News*, 1 September 1903.
[1321] *Evening News*, 1 September 1903.

America so long as the challengers are built on conventional lines, and while such men as Herreshoff live. The only chance of beating them is by what might be termed a "trick boat," that is to say, a type of boat unknown or untried by the Americans. This is only likely to come from such a country as, say, France, or possibly Australia or New Zealand - when the challengers must be prepared to take all chances of the untried type of boat, or, in other words, "go blind".'[1322] It's too bad that Reeks wasn't around to see Australia's victory in 1983. The wing-keel on *Australia II* providing just the element of surprise that he had predicted was necessary.

With the America's Cup occurring in October of 1903 proving to be another successful romp for the New Yorkers, Reeks' attention likely switched to a different race. Well-known Victorian Sir Rupert Clarke was at the time having a yacht built by Charles Bailey in Auckland. Named *La Carabine*, the schooner of some 60 tons was expected to be launched a few months later whereby its new owner was to sail it directly to Melbourne. From here it was arranged that *La Carabine* would then race against the Reeks-designed auxiliary schooner nearing completion at William Holmes' McMahons Point yard, built to the order of C. D. Wallace.[1323] A Melbourne to Sydney yacht race was on the cards.

It appears large auxiliary cruisers were coming into vogue. Reeks soon received an order from Dr R. Scot Skirving for a ketch-rigged yacht that was also to be built by Holmes.[1324] The craft measured 54ft (overall) by 40ft (waterline) with a beam of 13ft and a draft of 7ft.[1325] This was the second of three vessels Reeks would design for this particular customer; the first being *Hygiea* which he had launched in 1891.

There was other work on hand too, as well as two more launchings to describe, the first at Woodburn. Situated on the banks of the Richmond River, some 700km from Sydney, Reeks was likely disappointed not to have been amongst the large assemblage of 800 friends and supporters who gathered on 3 September 1903 to witness the launch of Messrs Davis Bros.' new steamer *Britannia,* a vessel he had designed.[1326] Built by Oliver Jones of Woodburn and intended to ply between this location and Lismore, the cargo and passenger steamer measured 80ft with a beam of 16ft and a draft of 5ft 2in.[1327] Its 18 h.p. engines were expected to drive the craft at a speed of 14 knots and the vessel was capable of accommodating 300 passengers.[1328]

Of perhaps considerably more interest and the reason why Reeks was unable to travel to the Richmond River for the launch of *Britannia* was the impending launch of the Balmain New Ferry Company's new ferry. The vessel, named *Lady Rawson*, was launched only a few weeks later on 24 September 1903, built by Messrs Morrison and Sinclair to a design by Reeks. This particular vessel was of a similar design to the *Lady Hampden*, though slightly larger. It also had several modern improvements.

The *Lady Rawson*'s funnel was placed well forward and there were two extra wheels, one at either end. This feature was considered a great advantage on race days when the skipper must keep well aloft to safely steer the steamer.[1329] With accommodation provided for 700 passengers, there were four stairways leading from the upper to lower decks with the entirety of the upper deck able to be closed in for concerts or in bad weather.[1330] The craft measured 122ft (overall), with a beam of 27ft 6in and a depth of just over 10ft. Heavy girders were used in construction of the hull so as to minimise vibration.[1331]

The same day that *Lady Rawson* left the stocks so too did the auxiliary schooner that Reeks had designed for

[1322] *Evening News*, 1 September 1903.
[1323] *The Australian Star*, 1 September 1903.
[1324] *The Australian Star*, 3 September 1903; *The Sydney Morning Herald*, 26 September 1903.
[1325] *The Sydney Morning Herald*, 26 September 1903.
[1326] *The Sydney Morning Herald*, 4 September 1903.
[1327] *The Richmond River Herald and Northern Districts Advertiser*, 4 September 1903.
[1328] *The Richmond River Herald and Northern Districts Advertiser*, 4 September 1903.
[1329] *The Australian Star*, 2 July 1903.
[1330] *The Australian Star*, 2 July 1903.
[1331] *The Australian Star*, 2 July 1903.

BRITANNIA.
SOURCE: LANCE LYONS.

LAUNCH OF LADY RAWSON.
SOURCE: GRAEME ANDREWS WORKING HARBOUR PHOTOGRAPH COLLECTION, CITY OF SYDNEY ARCHIVES.

C. D. Wallace of Melbourne.[1332] Named *Bona*, the craft was named after Ruby Bona Steele, Wallace's wife. Built by Holmes, the vessel was substantial. Its frames and bottom were of hardwood, while kauri was used for the topsides and decking.[1333] Fitted with a 30 h.p. oil engine, *Bona* featured a large deck with spacious accommodation below comprising saloon, ladies' cabin, gentleman's quarters, forescastle, pantry and galley, as well as a sail room.[1334] There were 20 portholes fitted to provided light and ventilation.[1335] Overall, the craft ranked amongst the finest yachts produced in Australia and with its modern appliances and machinery was to be easily worked by a moderate crew. *Bona* measured 89ft in length (overall) with a beam of 16ft 6in and a depth of 8ft. No less than 12 tons of lead had been moulded on its keel, with another four tons used for internal ballast.[1336]

[1332] *Referee*, 30 September 1903.
[1333] *Referee*, 30 September 1903.
[1334] *The Sydney Morning Herald*, 21 March 1903.
[1335] *Referee*, 30 September 1903.
[1336] *Referee*, 30 September 1903.

Bona (later renamed *Boomerang*) was in many ways a companion to *Ena* in terms of its classic design for the period and subsequent history, with both gracing Sydney Harbour over a century since they were built, and both examples of Reeks' design prowess at the top of his game.

Whilst *Bona* initially went to Port Phillip Bay, and changed hands there once, it returned to Port Jackson in the late 1920s and has been a stately figure there ever since. The new owner who brought it back to Sydney from Melbourne was Charles Lloyd Jones, who had owned a number of yachts, including the Reeks designed *Actaea* and *Thelma*.

Lloyd Jones was a businessman and patron of the arts, serving as chairman of the retail and department store chain David Jones Limited from 1920 until his death in 1958. His artistic efforts included establishing the respected journal *Art in Australia*. This flair showed through with his appreciation of *Bona*, recounted in later life. '*I watched her being built by Holmes in Lavender Bay, and always loved her. I*

BONA (LATER BOOMERANG).
SOURCE: SYDNEY HERITAGE FLEET.

did not see her many a long day after that, until, one day, I saw her moored in the Yarra. There she was, just as beautiful and delicate as ever; the passing years (25) had made no ravages on her, and she was as lovely as the day she was launched.'[1337]

In 1929 *Bona* passed into Frank Albert's hands and the Albert family became very proud but final private owners of the vessel over many decades. Under their custodianship *Bona* mostly went under power and was seen out on the harbour and at Pittwater regularly, until the family gifted the craft in 1987 to the Sydney Heritage Fleet where it is now part of their magnificent operational fleet of historic vessels.

Returning to *Bona*'s early days, its style was also noted by the prestigious American *Rudder* magazine in March of 1907, who had previously published a glowing description of *Ena*. Although not afforded the multiple pages that *Ena* was given, the shorter entry and single profile plan that

[1337] P. R. Stephensen (1962). *Sydney Sails: the Story of the Royal Sydney Yacht Squadron's First 100 Years (1862-1962).*

they published was sufficient to recognise its excellent Edwardian style and strong construction, as they focused on various details.

'*She was built for cruising pure and simple, and with a view to practically unrestricted movements was kept within a very moderate draft viz 7' 3" in full commission; to this was added a single drop board and a 30 hp engine, so Mr. Wallace could pick his ports at will.*'

Further compliments were paid to other items, including the structural timber, '*Australian hardwood, of which there is no better for shipbuilding purposes*'. In reference to the engine, the article stated that it '*drives the yacht about 8 ½ miles in smooth water; this is not remarkable when the under water body is examined; it is a very easy bottom to drive.*

Deck fittings are, of course, teak. Cabin work is of selected pine and cedar for surface finish, and all enamelled white and gold, except the facings and furniture, which are waxed oak. The upholstery, curtains, carpets, etc, are of the most beautiful materials selected with the utmost taste by Mrs Wallace.'

Summing up, the report indicated that '*Bona is a fine seaboat, and fast as cruising schooners go, for in her only race, against La Carabine, a schooner built about the same time, about the same size, and for the same class of cruising, she won easily*'.

This was the previously mentioned race against *La Carabine*. Though instead of racing between Melbourne and Sydney, the duo had competed on Hobsons Bay with *Bona* finishing the course well ahead of its rival (more detail of this race is provided in the following chapter).[1338]

The *Australian Motor Boat and Yachting* magazine, founded in 1925, also kept an eye on *Bona* and later *Boomerang* as the craft was renamed by the Alberts. The publication featured occasional images and reports of the vessel in its news sections, had it grace the cover on at least two occasions, and included substantial articles published in 1925 that were related to its then owner Harry Howard-Smith. *Bona* had cruised from Melbourne to Sydney and return and, under his guidance, had been substantially refined. The centreplate was removed, the internal arrangements optimised and new, modern fittings such as an electric capstan fitted. A deckhouse was fitted over the engine. It included '*a folding card table ... It affords a cosy and comfortable retreat at all times for two to six persons for either a smoke or a yarn or a game; or for the lesser number, a read and a possible nap, especially in cold, wet or windy weather*'. The schooner epitomised the high end of luxury yachting for the period, and still holds its own at the time of writing.

[1338] *The Argus*, 14 March 1904.

—

Always moving forward and jumping to the next project, likely leaving little time for earnest and proud reflection of his recently-completed commissions, just under two months after the launch of *Lady Rawson* and *Bona*, *The Sydney Morning Herald* carried a short article about a new steamer being built for New Zealand. '*The keel of one of the largest, if not the largest, wooden steamships ever built in this State has just been laid at the shipbuilding yards of Messrs. Morrison and Sinclair, at Balmain. She is being constructed from designs prepared by Mr. Walter Reeks of this city, to the order of the Northern Union Steamboat Company of New Zealand, and is to be employed as a passenger and cargo vessel in the Kaipara service.*'[1339]

This was the second vessel Reeks had designed for the company, the first being SS *Aotea*, built in 1898. It was an indication of the trust placed in Reeks, and the quality of his designs.

It was truly a busy time, as it always appears to be whenever Reeks was involved. Several days after this article was published, an oil launch designed by Reeks and built by H. Maumill at Lavender Bay to the order of W. J. Adams was completed. Launched on 21 November 1903, *Dawmea* was a commodious craft with a length of 35ft, a beam of 8ft and a depth of 3ft 6in.[1340] A Victor oil engine of 7½ h.p., intended to travel at a speed of eight or nine knots, was installed.[1341] The engine was placed forward and enclosed in a teak case with glass doors. Seating accommodation was provided for up to 40 people.

Dawmea was stated to differ from similar vessels of its type in that the combing was carried out to the skin of the craft, thereby allowing the full width of the boat to be available inside.[1342] Again, Reeks was showcasing his skill at improvising and adapting his designs to meet his customer's needs; in this case a demand for more internal space. He was fast solidifying his reputation as Australia's 'Herreshoff'.

[1339] *The Sydney Morning Herald*, 18 November 1903.
[1340] *The Sydney Morning Herald*, 21 November 1903; 25 February 1910.
[1341] *The Sydney Morning Herald*, 21 November 1903.
[1342] *The Sydney Morning Herald*, 21 November 1903.

1908 MAP OF TASMANIA AND SURROUNDS.
SOURCE: NATIONAL LIBRARY OF AUSTRALIA (NLA.OBJ-231122665).

CHAPTER SIXTEEN - TASMANIA

In what was nearly becoming an annual trip, on 8 December 1903 Walter Reeks arrived in Dunedin, New Zealand, per the SS *Mokoia*.[1343] This time he had travelled via Hobart.[1344] Though it is not known how many days Reeks spent in Tasmania, it was likely that he used the time wisely, touring the local boat and ship yards and meeting potential clients regarding design prospects for new commercial and recreational vessels.

In Dunedin Reeks met with representatives of the Union Steam Ship Company, surveying the steamer *Upolu* on behalf of Messrs Lever Bros. of Sydney who intended to purchase and transfer the vessel back to New South Wales.[1345] With all parties in agreement, Reeks then returned to Sydney on board SS *Waikare*, departing New Zealand's South Island on 10 December 1903 and arriving home less than a week later, via the Cook Strait.[1346]

Two days after his return from Dunedin, another Reeks-designed vessel was launched. Dr Scot Skirving's ketch-rigged yacht *Panacea* was released from the stays at William Holmes' McMahons Point yard, Mrs Scot Skirving breaking the customary bottle of wine over the bow.[1347] In appearance the craft was stated to garner both speed and seaworthiness, fully realising the design attributes Reeks had intended to embody. Constructed of hardwood and kauri, the *Panacea* was initially equipped with sails alone, with an engine to be installed at a later date. The vessel measured 54ft (overall) by 40ft 6in (waterline), with a beam of 18ft and a depth of 7ft.[1348]

Panacea continued the health-related names Scot Skirving chose for his yachts and despite any reservations he may have had from his previous experience with a yacht designed by Reeks (*Hygiea* launched in 1891), Scot Skirving had returned to him and then chosen Holmes as his preferred builder. This time, however, Scot Skirving had perhaps over-estimated his requirements and, as such, he did not keep *Panacea* for long. Writing later in his memoirs he stated, '*About 1903 I built my biggest boat to Reek's designs. She was too big and took too much room in picking up her moorings in Rushcutters Bay. I never put an engine in her*'.[1349] Being a canny Scotsman, '*I sold her while I was in England in 1904, and nearly got my money back*'.[1350]

Following Christmas and New Year celebrations, early January of 1904 coincided with Reeks being involved in another unique event about to take place, a '*great yacht race between the Victorian craft Sayonara and Sydney-owned Bona*'.[1351] As opposed to the vessel of the same name Reeks had recently designed for C. D. Wallace, this *Bona* was a 36ft (waterline length) racing yacht designed and built in Auckland in 1899 by Charles Bailey to the order of J. E. Chinnery of Sydney.[1352] Since its debut on the harbour, the craft had enjoyed several successful seasons racing against the cream of Sydney's fleet, winning many coveted titles.[1353] *Bona*'s owner was a big supporter of interstate yachting contests so when the opportunity arose for a race between his yacht and that of Alfred Gollin's Fife-designed *Sayonara*, then declared the '*speediest craft in Port*

[1343] *Evening Star*, 8 December 1903.
[1344] *Evening Star*, 8 December 1903.
[1345] *Evening Star*, 11 December 1903.
[1346] *Evening Star*, 10 December 1903; *The Daily Telegraph*, 17 December 1903.
[1347] *The Sydney Morning Herald*, 18 December 1903.
[1348] *The Sydney Morning Herald*, 18 December 1903.

[1349] A. Macintosh (1988). *Memoirs of Dr Robert Scot Skirving, 1859-1956*.
[1350] A. Macintosh (1988). *Memoirs of Dr Robert Scot Skirving, 1859-1956*.
[1351] *Evening News*, 6 January 1904.
[1352] *New Zealand Herald*, 11 November 1899; *Referee*, 10 January 1900; *Evening News*, 6 January 1904.
[1353] *New Zealand Herald*, 11 November 1899; *Referee*, 10 January 1900; *Evening News*, 6 January 1904.

Phillip', he gladly accepted the challenge, to be played out on Port Jackson.[1354] Gollin, a prominent member and ex-commodore of the Royal Victoria Yacht Club was anxious to establish an annual series of races between Sydney and Melbourne craft, and had issued the challenge to any yacht of any major Sydney-based yacht club. With Chinnery, a member of the Prince Alfred Yacht Club accepting on behalf of his craft *Bona* in mid-1903 for a race to be held that coming January, finally a state versus state contest was about to take place with Reeks ever eager to be involved.

Built at Adelaide by Alex McFarlane to the order of G. F. Garrard of Melbourne, *Sayonara* was launched late in 1897 and measured slightly longer than *Bona*, at 57ft (overall) by 38ft (waterline) with a beam of 10ft 8in.[1355] In contrast *Bona* measured 53ft (overall) by 36ft 2in (waterline) with a beam of 9ft 5in.[1356] Since its arrival in Port Phillip Bay, *Sayonara* had become Melbourne's principal racing craft, being recognised as one of the fastest and finest yachts to grace Australian waters. In mid-1899 the vessel was purchased by C. D. Wallace and two years later by Gollin, who had previously owned and successfully raced the New Zealand designed and built yacht *Uira*. With the latter intent on realistically claiming the title of owning Australia's champion yacht, he issued a challenge to the best competing yacht of Sydney to vie for what he hoped was an annual interstate trophy between yachts best representing the two locations.[1357]

With Chinnery answering this call, arrangements for the series of races, jointly agreed to by committees appointed in Victoria and New South Wales, the latter no doubt including Reeks, were complete by late October of 1903. These included '*that the races be conducted under the rules and regulations prescribed by the British Yacht Racing Association, 1903, with the following modifications: 1. Neither yacht shall have more than two paid hands on board*

[1354] *The Daily Telegraph*, 4 July 1903.
[1355] *The Australasian*, 9 October 1897; *The Daily Telegraph*, 1 January 1904.
[1356] *The Australasian*, 9 October 1897; *The Daily Telegraph*, 1 January 1904.
[1357] *Evening Journal*, 1 July 1899.

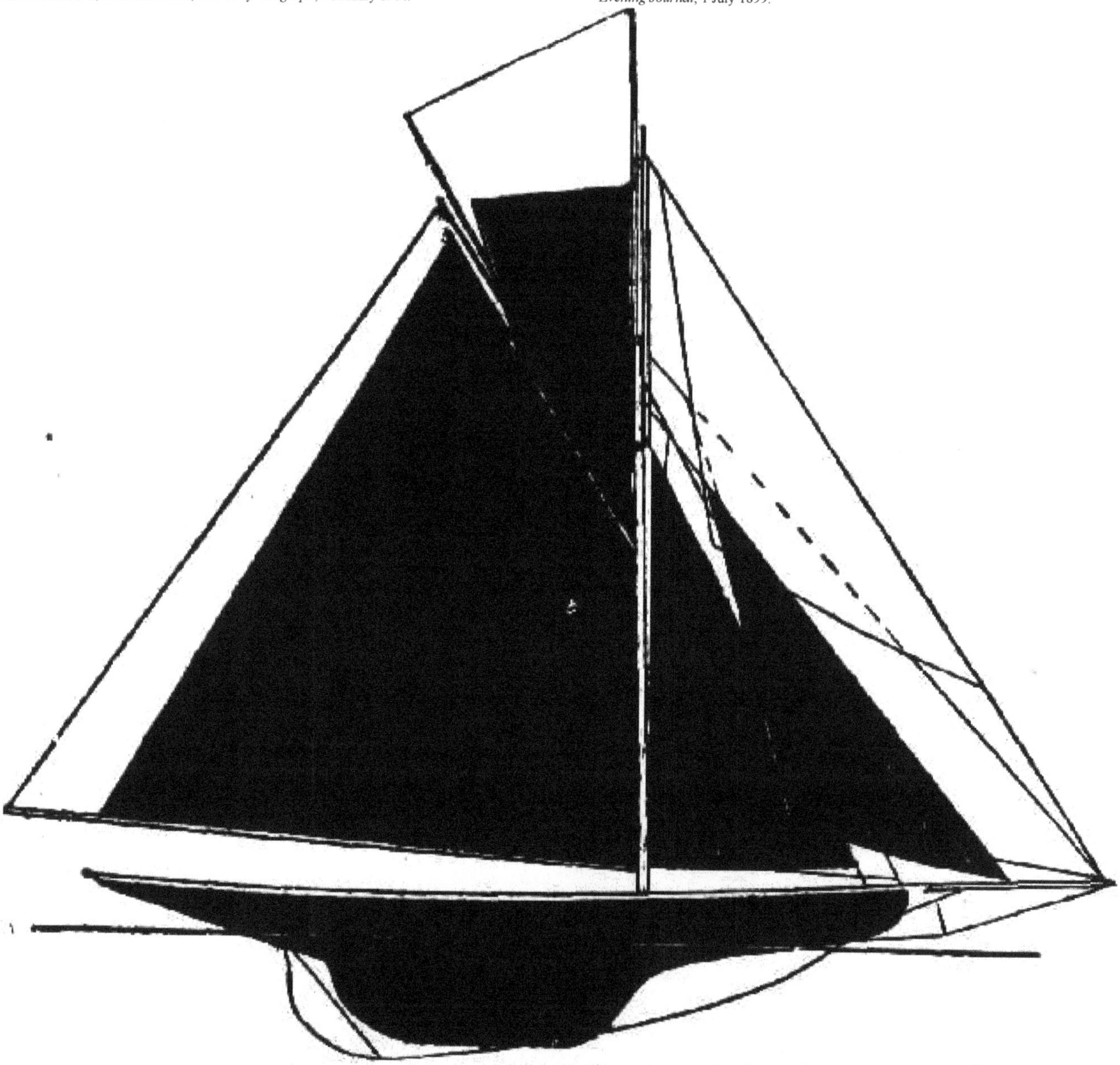

AN ILLUSTRATION OF BONA OVERLAYED ON SAYONARA DRAWN BY WALTER REEKS.
SOURCE: THE DAILY TELEGRAPH, 13 JANUARY 1904.

during a race, and they (if any) shall by included in calculating the total number of person on board. Every person on board other than the paid hands shall be a member of a yacht club of either state. No paid hand shall be allowed to touch the tiller during a race ... Courses ... shall be laid in open water, clear of headlands - the first race 10 knots to windward or leeward and return, the second a triangular race of 21 knots, and the third, if necessary, the same course as the first. The races shall be under the management and control of the sailing committee, consisting of the managing representatives of this (N.S.W) State, and in case of any dispute or protest, the matter shall be referred to the Sydney Yacht Racing Association, whose decision shall be final. Every reasonable facility shall be given to the representatives of both States for inspection of the competing yachts, and for checking measurements, marks, sail area, and such other matters as may fall within the scope of an official measurer's duty'.[1358] In addition, the Victorian representatives requested that the trophy become the property of the owner of the winning yacht, leaving it optional for the winner to offer the cup, if he chose, as a perpetual challenge cup.[1359]

With the event drawing closer, *Bona* was hurriedly readied for the race series. However, in mid-December of 1903, only weeks before the scheduled start of the event, the craft was sold to Herbert Binnie of the Royal Sydney Yacht Squadron with S. M. Dempster charged with skippering it for the races.[1360] It appears to have been a politically-motivated sale. The Sydney press reported, '*Negotiations for the purchase of the Bona have been proceeding for some time, that is, after a few members of the "Royals' failed to induce Chinnery to hand the yacht over to them, for Bona was not registered with the senior club, and their Royal Highnesses found it very galling to them to see the Bona contest the Interstate Race under the burgee of the "Alfreds". They are all pleased now that they have got her'*.[1361]

The newly-acquired owner then spent the next few weeks ensuring his craft was in perfect racing trim, making minor additions and alterations to give *Bona* the best advantage. Trial spins over the racing courses were also taken in the lead up to Christmas with Reeks on board, no doubt offering his valuable advice regarding the tune-up process.[1362]

Sayonara arrived in Sydney on New Year's Day 1904 after what was described as a '*rough trip*' taking 13 days.[1363]

[1358] *Evening News*, 23 October 1903.
[1359] *Evening News*, 23 October 1903.

[1360] *The Australian Star*, 21 December 1903.
[1361] *Sydney Sportsman*, 23 December 1903.
[1362] *The Sydney Morning Herald*, 23 December 1903.
[1363] *The Sunday Sun*, 3 January 1904.

SAYONARA.
SOURCE: AUSTRALIAN NATIONAL MARITIME MUSEUM COLLECTION (ANMS0155[004]).

It was then taken to the yard of Watty Ford (Jr) at Berrys Bay to be prepared for the race series.[1364] Meanwhile *Bona* was being overhauled at a slip at Blues Point, with Reeks superintending some last minute alterations.[1365] The differences between the two vessels were reported in the press. '*There is a vast difference between the boats. It is anticipated that the Bona will race about 40 rating, while her opponent will measure about 42. Bona will, therefore, be conceding 2min 40sec in a 20 mile contest, and 2min 48sec in 21 miles. Approximately the Bona carries 1400ft of canvas and the Sayonara 2100ft. Another difference, too, may strike yachtsmen as being significant. Both vessels are to sail cutter rigged - the Bona has a polemast while the other carries a topmast. The Sayonara's rig is described by sailing authorities as the finest fit out ever seen here. A comparison cannot be drawn in this direction as regards Bona's fit out. This is a matter which must wait until she leaves the slip at Blue's Point.*'[1366] Interestingly, and reminiscent of one of the reasons Australia's challenge to the America's Cup of 1890 was not issued, here we have Reeks making some last minute changes to *Bona* based on what he had evidently witnessed on the newly-arrived *Sayonara*.

With *Bona* now sailing under the flag of the Royal Sydney Yacht Squadron, the first race of the three-race series took place on Saturday 9 January 1904. Helmed by W. J. Robb, *Sayonara* allowed its opponent a nearly 3-minute start on size allowance.[1367] Given it had been approaching two decades since the last intercolonial sailing match had occurred, there was widespread interest in the race; steamers were packed with spectators, as were the best shore-based vantage points. The trophy, valued at 45 guineas and subscribed to by members of the Royal Sydney Yacht Squadron and the Prince Alfred Yacht Club, was well worth the effort, bragging rights being also at stake.[1368] Reeks was heavily involved, appointed official measurer of the event.

In a strong north-easterly wind, being seen as an advantage to *Sayonara,* the first race took place on an open ocean 10-mile windward course taking the duo past Manly, Long Reef and Narrabeen, and return, with the Victorian yacht showcasing its speed and capabilities to beat the local representative, *Bona,* by just over 4 minutes.[1369] The second race, held two days later in a lighter breeze on a triangular course also outside of the Heads, resulted in a win for *Bona* by nearly 2 minutes without its time allowance.[1370]

On Tuesday 12 January 1904, the third and final race of the series took place on the same course used for the first race. With much anticipated from the event, the steamers and shore-based positions were crammed with eager spectators. They were not disappointed.

In fine weather and a moderate breeze, *Sayonara* and *Bona* tussled for most of the race. The visiting yacht, however, proved the victor, winning by a nearly 5-minute

OFFICIALS INVOLVED WITH THE 1904 SAYONARA CUP. BACK (LEFT TO RIGHT): S. T. WILSON, SECRETARY OF THE ROYAL SYDNEY YACHT SQUADRON, AND S. AMES, SECRETARY OF THE ROYAL YACHT CLUB OF VICTORIA. FRONT (LEFT TO RIGHT): W. J. NEWBRIGIN, OFFICIAL MEASURER OF THE ROYAL YACHT CLUB OF VICTORIA, AND WALTER REEKS, OFFICIAL MEASURER OF THE ROYAL SYDNEY YACHT SQUADRON. SOURCE: THE SYDNEY MAIL AND NEW SOUTH WALES ADVERTISER, 13 JANUARY 19

margin, claiming the overall trophy when the former's time allowance was taken into consideration.[1371] All told it was win for the Victorians and a win for interstate yacht racing more generally. Commenting on the outcome of the race series, Reeks was quoted as saying, '*Sayonara's victory can be summed up in a very few words. She is a good cruiser as well as being a fast sailer. She, too, is certainly a better sea boat than the Bona. Bona, of course, is essentially a racing boat. Her type of yacht is neither useful for sea-going or for a cruiser.*

The races have proved that a good wholesome class of boat is quite able to hold her own in any weather against a boat built essentially for racing, and of little use for anything else.

The Sayonara is just the class of boat the new rating rule is intended to foster. She can go to sea with any boat, and is speedy as well'.[1372]

Reeks was making an interesting point. Little did he realise just how personal his involvement in what was soon to be called the Sayonara Cup would be. In five years' time he would compete for the now-acclaimed trophy on board his then 20-year creation *Thelma*, a vessel he had designed with the dual purpose of cruising and racing.

[1364] *The Australian Star*, 5 January 1904.
[1365] *Evening News*, 6 January 1904.
[1366] *Evening News*, 6 January 1904.
[1367] *The Daily Telegraph*, 18 December 1903.
[1368] *The Sydney Mail and New South Wales Advertiser,* 30 December 1903.
[1369] *Sunday Times*, 10 January 1904.
[1370] *The Australian Star*, 12 January 1904.
[1371] *The Sydney Morning Herald*, 13 January 1904.
[1372] *The Australian Star*, 13 January 1904.

Following the celebratory dinners and other events coinciding with the close of the race series, of which Reeks was a participant, he returned to the work at hand.[1373] His first task was to rearrange the internal accommodation of *Sayonara* on behalf of its owner Alfred Gollin, owing to all of the fittings being removed for the recent series of races.[1374]

As the long summer days of early 1904 dwindled, Reeks maintained his involvement in yacht racing on the harbour. In early March he steered *Magic* to victory in its maiden race of the season, competing in a general handicap race with the Royal Sydney Yacht Squadron.[1375] It was an impressive performance by Reeks at the helm of the nearly four-year-old craft, by this time owned by A. C. Saxton.[1376] *Magic* had been built to the order of J. O. Fairfax by Messrs Summers and Payne at Southampton, England, arriving in Sydney in November of 1900.[1377] Designed by the former, the '30-footer' measured 38ft (overall) by 26ft (waterline) with a beam of 8ft. It was cutter-rigged and fitted with a bulb keel.[1378] Reeks would continue to take the helm of *Magic* later in 1904 with marked success.

March of 1904 also saw Reeks (in his motor boat *Wahine*) take up official duties in a race between *Petrel, Culwulla* and *Scotia,* the three yachts having scored an equal number of points during the season for the championship of the Royal Sydney Yacht Squadron.[1379] Another sailing match took place a week later on Port Phillip Bay, arousing considerable interest amongst Victorian and New South Wales yachting circles. As previously mentioned, Sir Rupert Clarke's new schooner yacht *La Carabine* was pitted against C. D. Wallace's schooner yacht *Bona*. A recent arrival to Melbourne, the former had been built at Auckland by Bailey Brothers to a design by Captain P. W. Bull, marine superintendent of Huddart, Parker and Co.[1380] Compared to the Reeks- designed *Bona*, *La Carabine* was 2ft shorter overall (measuring 87ft) and 5ft shorter on the waterline though had 4ft more beam and over 4ft more draft.[1381] It had also been designed specifically for ocean cruising, making no pretensions that it was a racing craft. *La Carabine* was also stated to have five months' worth of weed on its bottom.[1382] Still, with no time allowance being conceded to either vessel, the match took place on a course from the breakwater pier at Williamstown to a buoy moored off Brighton, thence across the bay to Point Cook, and along the shore, keeping the Gellibrand lightship on the port hand.[1383] The total distance was about 20 miles, the winner receiving a trophy valued at 10 guineas.

With a large number of spectators and watercraft following the race, *Bona* held the upperhand for much of the duel; *La Carabine* being no match for it on a wind. By the end of the five-hour race, *Bona* was so far ahead of its rival that *La Carabin*e did not finish. Instead, the craft was sailed back to its moorings without crossing the line.[1384] Though Reeks was not present to witness the race, he no doubt read with relish of *Bona*'s triumph and all of the details of the two competitors and the race.

The first few months of 1904 were also a busy period for Sydney's shipbuilders with an increasing number of commercial commissions on hand, many of which were from customers based interstate, as well as in New Zealand and the Pacific Islands. Of relevance, Messrs Morrison, Sinclair & Co. were progressing with the steamer for the Northern Steamboat Company of New Zealand, intended for the Wairoa River trade. Designed by Reeks, the 145ft vessel was to have considerable cargo-carrying capacity, as well as first class passenger accommodation. A notable feature was a tunnel under the vessel's stern in which the propeller would work, enabling the craft to maintain buoyancy and power with shallow draft.[1385]

The commercial pearling industry was also experiencing another high with demand for new vessels filling the books of Sydney's more experienced boat builders. One such craft, a ketch of 50ft with a beam of 12ft and a depth of 6ft, *Gondolier*, was launched from William Holmes' yard in mid-March of 1904.[1386] Built to the order of James Clark, the vessel was intended for work in Torres Strait and notably marked the 169th vessel that Reeks had designed over the course of his career. Several similar pearling ketches, including *Valkyrie, Valhalla* and *Nyoola*, were launched a few weeks later in Sydney for the same owner.[1387] Considered sister-ships to *Gondolier*, we can assume these three craft were also designed by Reeks.

VALHALLA.
SOURCE: STEPHEN MULLINS COLLECTION.

Another vessel involving Reeks was also then under construction. Sydney R. Lloyd of Woy Woy had commissioned him to design a cruising launch. Built by

[1373] *The Sydney Mail and New South Wales Advertiser*, 20 January 1904.
[1374] *The Australasian*, 23 January 1904.
[1375] *The Sunday Sun*, 6 March 1904.
[1376] *The Sydney Morning Herald*, 29 November 1902.
[1377] *Australian Town and Country Journal*, 14 April 1900;
The Sydney Morning Herald, 28 July, 14 November 1900.
[1378] *The Sydney Morning Herald*, 15 November 1900.
[1379] *The Sydney Morning Herald*, 10 March 1904.
[1380] *The Argus*, 11 March 1904.
[1381] *The Argus*, 11 March 1904.
[1382] *The Australasian*, 19 March 1904.
[1383] *The Argus*, 11 March 1904.

[1384] The Australasian, 19 March 1904.
[1385] *The Daily Telegraph*, 17 February 1904.
[1386] *The Sydney Morning Herald*, 23 March 1904.
[1387] *The Sydney Morning Herald*, 23 March 1904;
Daily Commercial News and Shipping List, 29 March 1904.

CWMAVON.
SOURCE: SHIPBUILDERS OF BRISBANE WATER, NEW SOUTH WALES (GWEN DUNDON COLLECTION).

William Holmes at McMahons Point, the craft measured 40ft in length with a beam of 7ft 6in.[1388] Fitted with a 15 h.p. double cylinder Standard engine and a roomy cabin finished in polished cedar, the somewhat peculiar-looking vessel was dubbed a '*Revolution in Construction*' when it was launched on 22 April 1904 owing to the fact that an American principle of construction had been adopted whereby the lines of the hull were devoted entirely to speed.[1389] Named *Cwmavon*, the craft had a comparatively deep bow (4ft 8in), but the body then ran away flat to aft, to about half that depth.[1390] A false keel had been fitted to protect the propeller, giving the launch a depth of 6ft 6in.

Cwmavon was expected to travel at a speed of 12 knots, i.e., equal to, if not faster than the ferry boats then in operation on the harbour.[1391] Within six months of its launch, the craft was noted as the '*fastest boat at present in New South Wales*'. A press report provided some additional detail. '*This boat was designed by Mr. Walter Reeks, and a condition of the contract of sale was that she should be fast enough to beat the Standard, a launch brought here from New Zealand with a big reputation for speed. The boat Mr. Reeks built is 40ft. long, and is fitted with a duplicate engine to that of her New Zealand rival, a 15 h.p. Standard. The Cwmavon won easily in her trial race, and her speed, about 10 knots, is the fastest attainable by local boats.*' Motor boat racing was fast becoming a popular sport on Sydney Harbour with Reeks again at the forefront of an emerging type of vessel.

With so much work on hand, a long-standing industrial dispute between the Master Shipbuilders' Association, the Amalgamated Society of Carpenters and Joiners and the Shipwrights' Provident Union soon reached boiling point in Sydney, resulting in several strikes. An inquiry by a Board of Experts began on 24 March 1904 with James Orr of the Navigation Department appointed as the Arbitration Court's sole representative.[1392] It was expected to be a lengthy process, owing to the points of issue being quite technical in nature. Overall, however, the dispute centred on the shipwrights' claim against the shipbuilders, shipjoiners, and amalgamated carpenters and joiners regarding the roles and responsibilities of their specific class of tradesmen in terms of the types of work undertaken during the construction, fit-out and repair of vessels. Specifically the shipwrights, in recognising that they were paid higher wages than joiners, wanted guarantee from employers as to where and when their work started and finished, and where and when other trades could be called in to undertake work. They also proposed that there be no tonnage considered in defining a shipwrights' work. Thus, any launch, pearl boat, lugger, etc., which was carvel-built and required caulking, as well as all punts, floating stages, catamarans, etc, were to be defined as shipwrights' work, in addition to the dressing of all woodwork when not done by machinery.

With Reeks the expert appointed to represent the Master Shipbuilders and Shipjoiners, i.e., the employers, the inquiry took place over several weeks with the Board of Experts generally of the opinion that there was so much overlap between the work undertaken by the various trades that it was impossible to draw a line such that it was best left to the employers to determine which body of workers they required to complete a task.[1393] It took some

1388 *Evening News*, 21 March 1904.
1389 *The Australian Star*, 9 April 1904; *The Daily Telegraph*, 23 April 1904.
1390 *The Australian Star*, 9 April 1904.
1391 *The Australian Star*, 9 April 1904.
1392 *The Australian Star*, 23 March 1904.
1393 *The Australian Star*, 26 July 1904.

more months of deliberations, however, before the matter, including provisions for wages, penalties and working conditions, was settled with agreement by all parties.[1394]

While Reeks' involvement in the Arbitration Court's process was likely piecemeal, he would have given the matter the upmost seriousness and diplomacy that it deserved. As usual, he also had numerous other projects going on at the time, including the new steamer for New Zealand then nearing completion.

Launched on 9 June 1904 from the shipyard of Messrs Morrison, Sinclair & Co. at Balmain was the steamer for the Northern Steamboat Company of Helenvsille. Intended for the Wairoa River trade, the craft was named *Awaroa* with the christening ceremony performed by Miss Rennie Lamb, daughter of one of the senior partners representing the owner.[1395] Like its predecessor, the Reeks-designed *Aotea* launched from the same yard in 1898, the new vessel was constructed on the tunnel stern shaft principle. Its dimensions were 145ft (overall) with a breadth of 26ft and a depth of 13ft, loaded draft was 6ft. *Awaroa* was fitted with machinery amidships, a saloon aft (capable of accommodating 65 people), with the pantries, lavatories, bar and engineer's berth in the wings abreast of the machinery space.[1396] On deck was a spacious ladies' cabin and a smoking room. The machinery, manufactured by the Wallsend Slipway Company of Newcastle-on-Tyne, England, was of the triple expansion type of 500 h.p., while Howden's forced draft system was fitted to the boiler.[1397] Overall, the vessel was calculated to travel at a speed of 12 knots with room for 800 excursion passengers.[1398]

LAUNCH OF THE AWAROA.
SOURCE: AUSTRALIAN TOWN AND COUNTRY JOURNAL, 15 JUNE 1904.

Following engineer's and speed trials, the *Awaroa* was readied to make its first Tasman Sea crossing. Ironically, as the vessel was being prepared to leave Sydney for New Zealand, word came through that SS *Aotea*, designed by Reeks for Messrs G. and W. Turnbull of Wellington, New Zealand, and built in Scotland only 18 months prior, had been wrecked.[1399] Such was the precariousness of the New Zealand coastal trade. Still, demand for new passenger and cargo steamers was booming, both in Australia and New Zealand. On 28 September 1904, *The Daily Telegraph* reported, '*Mr. Walter Reeks, who designed the Awaroa, is now preparing designs for three steamers for other States, and has two building from his designs at their own ports of service.*'

One of the steamers mentioned in the last statement was designed for Messrs O'May Bros. of Hobart, Tasmania, to cope with a continually-increasing River Derwent passenger trade. The steamer was to accommodate 450 to 500 passengers and be capable of steaming from 10 to 12 knots between Hobart and Bellerive.[1400] The new vessel, being a double-ender type, would obviate the necessity of turning around after disembarking passengers, thus saving approximately three minutes of time and fuel.[1401] The boat was also to travel at a speed greater than those at present engaged in the same service, thereby resulting in a trip from the Hobart wharf to Bellerive not occupying more than eight minutes.[1402]

After visiting Sydney and meeting with Reeks, O'May Bros. had determined that the 100ft double-ended steamer *King Edward*, which had been launched in 1902 to the order of the Watsons Bay Ferry Company to a design by Reeks, embodied most of their requirements, with certain improvements, mainly to the comfort of passengers. With Reeks finalising the design, tenders calling for the vessel's construction were announced in August of 1904. Following deliberations, the submission of Frederick Moore, boat builder of Launceston, was accepted, though the vessel was to be built at Hobart's Domain shipyard and, according to the contract, was to be delivered in the very good time of eight months.[1403]

Meanwhile, another commercial vessel was completed to a design by Reeks. Built to the order of F. W. Scharer of Coraki by the local firm of Messrs Turner and Conroy, the cedar launch *Freddie*, fitted with a Victor oil engine of 4 h.p., was intended to deliver bread and parcels on the Richmond and Wilsons rivers.[1404] The 25ft craft, with a beam of 6ft 6in and a depth of 2ft 10in, was dubbed the 'Bread Boat' and launched on 16 October 1904, entering service near immediately, capable of travelling at a speed up to 9 knots.[1405] The vessel was later purchased by S. J. Fernandez of the Commercial Hotel, Coraki, for conversion into a motor boat.[1406]

1394 *The Daily Telegraph*, 30 July 1904.
1395 *Auckland Star*, 15 June 1904.
1396 *Evening News*, 10 June 1904.
1397 *Evening News*, 10 June 1904.
1398 *Evening News*, 10 June 1904.
1399 *Newcastle Morning Herald and Miners' Advocate*, 30 September 1904.
1400 *The Mercury*, 21 September 1904.
1401 *The Mercury*, 21 September 1904.
1402 *The Mercury*, 21 September 1904.
1403 *The Mercury*, 21 September 1904.
1404 *The Richmond River Herald and Northern Districts Advertiser*, 15 July, 19 August, 2 September, 14 October 1904.
1405 *The Richmond River Herald and Northern Districts Advertiser*, 21, 28 October 1904.
1406 *The Richmond River Herald and Northern Districts Advertiser*, 24 May 1907.

BLUE FLAG STORES AND BAKERY, CORAKI.

Special Notice.

STORE LAUNCH,
also carrying
BREAD.

SOUTH ARM—MONDAY or TUESDAY.
MAIN RIVER, with Creeks — WEDNESDAY, THURSDAY, FRIDAY.
NORTH ARM—SATURDAY.

BREAD LAUNCH.

MAIN RIVER & CREEKS—MONDAY.
NORTH ARM—TUESDAY.
SOUTH ARM—WEDNESDAY.
NORTH ARM—THURSDAY.
SOUTH ARM—FRIDAY.
MAIN RIVER—SATURDAY.
Serving OAKLAND every morning before Breakfast.

Intending bread customers and those living away from river bank please let me know where bread can be left.

TERMS.—Strict Monthly Cash Payments.

Parcels up to 50lbs delivered anywhere called at by the Bread Boat for 6d each.

F. W. SCHARER.

SOURCE: THE RICHMOND RIVER HERALD AND NORTHERN DISTRICTS ADVERTISER, 28 OCTOBER 1904.

With the 1904/05 yachting season about to start, Reeks continued his interest and involvement in the pastime. The Royal Sydney Yacht Squadron appointed him honorary measurer, continuing a legacy of 18 years, as well as its representative to the Sydney Yacht Racing Association.[1407] Of note, in recognition of his service, Reeks was also presented with life membership of the Sydney Amateur Sailing Club.[1408]

Motor boat racing was continuing its epic rise in popularity and new boats were being commissioned to participate in the sport. One such order was given to William Holmes of McMahons Point for a launch, designed by Reeks which was to have a 14 h.p. double cylinder Hercules engine installed. Though the craft was intended to be a very comfortable and speedy cruising launch, it was the latter distinction that earned its mark.[1409] Named *Hercules* and built to the order of E. W. Buzacott, the Sydney-based agent for the American made Hercules engine, the vessel won the motor launch race at the 1905 Anniversary Regatta with Reeks at the helm.[1410] The boat's construction had been hurried forward specifically to compete in the race.[1411]

HERCULES.
SOURCE: THE SYDNEY AND NEW SOUTH WALES ADVERTISER, 1 NOVEMBER 1905.

Hercules was probably a welcome diversion for Reeks from the many commercial commissions he had on hand at the time. During this period, in addition to launching the steamer *Awaroa* and finalising the design for the O'May Bros.' Hobart passenger ferry, Reeks was engaged in preparing plans for a new steamer for the Watsons Bay Ferry Company. The vessel was to be based on the same lines as *King Edward*, but was to be larger and faster, with engines of 500 h.p. installed.[1412] In mid-December of 1904 Reeks was also called to be an expert witness at a Marine Court of Inquiry regarding the circumstances leading to the wreck of the steamer *Maud Weston* near Terrigal on the north coast of New South Wales.[1413] The closing weeks of 1904 further coincided with a 20 h.p. Union double cylinder engine being installed in '*a fine stamp of a fishing boat designed by Mr. Walter Reeks for Mr. Jones of Williamstown, Victoria. It is the intention of Mr. Jones to use her for running either under power alone or in conjunction with a fair spread of canvas*'.[1414] Built by Rock Davis (Jr) of Brisbane Water to the order of Arthur Jones and named *Enid,* the vessel's hull was stated to be an almost exact reproduction of the Reeks-designed yacht *Bona*, though on a smaller scale.[1415] The 59ft ketch left Sydney for its home port of Melbourne in March of 1905.[1416] It then spent several years involved in the Victorian fishing industry before being sold to Gizo Plantations Co. in the Solomon Islands in 1909.[1417] *Enid* was apparently wrecked in 1925.

For Reeks, the year 1905 saw him continuing to design new steamers for commercial passenger and cargo trades. Two of these new commissions came from Tasmania, this time for customers in the state's north. First, in order to meet the growing requirements of the River Tamar

[1407] *Evening News*, 24 August 1904.
[1408] *The Australian Star*, 15 September 1904.
[1409] *The Sydney Morning Herald*, 22 October 1904.
[1410] *The Sydney Morning Herald*, 31 January 1905; *The Daily Telegraph*, 11 June 1904, 25 February 1905.
[1411] *The Daily Telegraph*, 25 February 1905.
[1412] *The Australian Star*, 29 November 1904.
[1413] *Evening News*, 14 December 1904.
[1414] *The Sydney Morning Herald*, 11 January 1905.
[1415] *The Sydney Morning Herald*, 21 March 1905; *Register of British Ships: Main Register subsequent to Merchant Shipping Act 1854*, Port Melbourne [ON 117685].
[1416] *The Sydney Morning Herald*, 21 March 1905.
[1417] *The Star*, 2 April 1909.

passenger trade, Robert Gardner has decided to replace his steamer *Agnes* with a faster and more modern vessel. Recognising that '*where the money is earned it should also be spent*', Gardner determined that everything associated with the new steamer that could be produced locally should be encouraged.[1418] However, the plans and specifications for the 130ft craft were outsourced to Reeks with Launceston-based tenders called for its construction. The contract was duly awarded to Frederick Moore who was at the time in Hobart building the Reeks-designed ferry for the O'May Bros. Moore had also been in Sydney earlier in 1905, supervising the construction of single-ended screw 130ft passenger ferry being built by Watty Ford (Jr) for Alfred Harrap and Son of Launceston. Intended for the River Tamar passenger and cargo trade, this particular craft was named *Niree* but who its designer was remains inconclusive. In one newspaper article '*W. Dunn*' is referred to as the designer while another notes that it was designed by the new owners in conjunction with William Holmes.[1419]

Concurrently Reeks finalised plans for the Balmain New Ferry Company's new double ended steamer and invited submissions for the building of the craft.[1420] The contract for the 116ft vessel was subsequently awarded to Messrs Morrison and Sinclair.[1421] Reeks also made a visit to South Woodburn on the Richmond River in March of 1905, at the request of Messrs Davis Bros., regarding the building of a new boat for the Ballina service, recently taken over from T. B. Allen.[1422] This particular craft would be built by Messrs Turner and Conroy next to the Coraki Rowing Club shed.[1423]

All up, there were five commercial vessels either under construction or about to be built to designs by Reeks in early 1905: one in Sydney (for the New Balmain Ferry Company), two in Tasmania (one for O'May Bros. of Hobart and one for Robert Gardner of Launceston), one at Coraki (for Messrs Davis Bros.) and one at Brisbane Water (for the Watsons Bay Ferry Company. Come mid-1905 Reeks had also designed a small tug for Noumea, a steam lighter for Elliott Bros. of Sydney, and another commercial fishing vessel for a Victorian customer.[1424] Later in the year he would finalise plans for a 130ft excursion steamer for McIlwraith, McEachern and Co. of Western Australia to ply between the Swan River and Rottnest Island.[1425] Reeks would also complete the design of a 65ft smart schooner, intended for Papuan Industries, Ltd. of New Guinea.[1426]

The boat building industry of Sydney was certainly gaining traction, the development of the oil engine also giving considerable impetus to the recreational component of the industry. In some instances yachts already had this auxiliary power installed aboard, while scores of other people had taken to commissioning and purchasing motor launches, many of who would have not been interested in yachting. It truly was a transitional time for the water sport industry. Not all were thrilled, however. One such comment in the Sydney press stated that while '*the true yachtsman may deprecate it, the opinion of experts is that the oil launch has come to stay*'.[1427]

Such was the extent of Sydney's ship and boat building industry that statistics at the time reveal that there were approximately 2,275 hands employed in connection with these enterprises, with the wages list totalled in the realm of £158,000 annually.[1428] The same article also stated that while the '*skillful and faithful builders have been necessary to keep Sydney in the front rank for the production of first-class boats, no less essential has been the skilled designer. Sydney has been fortunate in possessing such a skilled naval architect as Mr. Walter Reeks. "Every yacht built since his arrival in the State many years ago," says a well-known authority, "is more or less closely identified with his name." It is obvious that the designer and the builder on the spot have advantages over those outside the State. The architect knows exactly the type of boat that will best meet local requirements, and is able to produce that type, aided by the faithful work of the builder*'.[1429]

While Reeks was definitely at the forefront of both the commercial and recreational vessel design space in Sydney, by this time he was obviously becoming more and more widely recognised for the depth of his expertise outside of New South Wales. His interstate customer base would only continue to grow in the coming years.

In terms of Reeks' Sydney-based recreational vessel commissions during this period, there were at least three craft built in 1905 to his designs. These included an interesting little boat built by William Holmes' yard at McMahons Point for Dr Scot Skirving. Named *Alcyone,* the 32ft yacht was described as a '*staunch little sea-going cruiser, ketch rigged, with auxiliary power. A striking feature of the design is the carrying of a raised deck from the stem to well down aft, so as to get some 6ft headroom in the cabin without resorting to a high deck-house*'.[1430] The cockpit was also separate from the cabin.[1431] Of note, its companion and skylight were relics of HMS *Nelson;* Reeks was continuing to recycle components and pieces out of the prized ship.[1432]

Launched in June of 1905, *Alcyone* differed entirely from other boats of its class in that the cockpit was barely sufficient for the steerer to sit in comfortably, this leaving more deck room.[1433] In addition to being a smaller catchment area for water in a head sea, it also provided space for a large raised coach-house, big enough to sleep five persons.[1434] There was a 6 h.p. De Dion motor fitted, capable of developing a speed of five knots.[1435]

The vessel marked the end of what may have been a rocky relationship between Reeks and this particular client

[1418] *Examiner*, 28 February 1905.
[1419] *The Daily Telegraph,* 12 August 1905; *Daily Telegraph*, 17 November 1905.
[1420] *The Sydney Morning Herald*, 31 January 1905.
[1421] *The Daily Telegraph*, 11 May 1905.
[1422] *Casino and Kyogle Courier and North Coast Advertiser*, 25 March 1905.
[1423] *The Daily Telegraph*, 21 June 1905; *The Sydney Morning Herald*, 21 August 1905.
[1424] *The Sydney Morning Herald*, 14 June 1905.
[1425] *The Western Australian*, 25 October 1905.
[1426] *Evening News*, 12 September 1905.
[1427] *The Sydney Morning Herald*, 26 April 1905.
[1428] *The Sydney Morning Herald*, 26 April 1905.
[1429] *The Sydney Morning Herald*, 26 April 1905.
[1430] *The Sydney Morning Herald*, 8 April 1905.
[1431] *The Sydney Morning Herald*, 8 April 1905.
[1432] *The Newsletter: an Australian Paper for the Australian People*, 8 July 1905.
[1433] *The Newsletter: an Australian Paper for the Australian People*, 8 July 1905.
[1434] *The Newsletter: an Australian Paper for the Australian People*, 8 July 1905.
[1435] *The Newsletter: an Australian Paper for the Australian People*, 8 July 1905.

over three designs spanning some 15 years. Dr Robert Scot Skirving (1859-1956) was indeed a Scot, with a Calvinist upbringing, and was perhaps true to character. His biography records his early military and then merchant naval experience prior to becoming a doctor and also notes, '*he became renowned for his biting comments about his colleagues*'.[1436] Reeks seems to have been on the receiving end of some of this dialogue. Scot Skirving's great-grand daughter Ann Macintosh published his biography in 1988, with Reeks prominent in discussions regarding his yachts. For example, '*I built a four-ton yawl called Hygiea. Reeks designed her. I knew she was under ballasted and told him so.*'[1437] He also recounted being struck by a gust and knocked down soon after he had launched it, and then added an extra ton of ballast, after which '*I had a lot of pleasure out of this craft and kept her until nearly 1900*'.[1438]

Whilst Scot Skirving happily noted that *Alcyone 'was a trim, smart little ship and sailed well. I kept her for several years'*, a storm had brewed over its engine.[1439] '*Reeks sold me his own engine, a De Dion. I just got one day's work out of it and never trusted Reeks again.*'[1440] That was the original engine noted previously that came from Reeks' launch *Wahine*, which was now powered by a Thornycroft. Reeks lost a client, but given the frosty relationship, a parting of ways was probably the best course for both to take.

Returning to Reeks' commercial ventures during the year 1905, many of his designs were quickly turning into a reality, particularly in Tasmania. At Hobart's Domain shipyards on Saturday 3 June, the hull of Messrs. O'May Bros.' steamer, which had been built of Tasmanian timbers by Fred Moore from a design by Reeks specially for the passenger service between Bellerive and Hobart across the River Derwent, was successfully launched.[1441] Despite the early hour, several hundred people were present at the ceremony; forming quite a large contingent.[1442]

Named *Derwent*, the craft was a double-ender equipped with a four-bladed propeller both fore and aft. This class of vessel was quite new to Tasmania, though there had previously been in commission across the Derwent paddle steamers that could steam with equal facility either way. Messrs. O'May Bros. were prompted to build the double-ended screw steamer to increase the efficiency of the Bellerive to Hobart ferry service as opposed to a vessel that could only steam in one direction. The craft also added to the company's capacity, having been designed to carry between 500 and 600 passengers; its chief dimensions being 110ft in length with a beam of 26.5ft and depth 9ft.[1443] Though Reeks was not present for the launch of the *Derwent*'s hull, he likely took pride when reading or hearing about the event. He was also not present four months later when the vessel successfully took its trial trip, with His Excellency the Governor, Sir Gerald Strickland, on board.[1444] By this stage the *Derwent* had been fitted with its boiler and engines, built by Messrs Alex Shanks and Son of Scotland.[1445]

The *Derwent* entered service soon after, going on to have a celebrated career continually running between Hobart and Bellerive until June of 1963; a remarkable 57 years of dedicated service.[1446] During this period, there were some major structural and engineering modifications that took place, including following a severe fire that caused much damage in 1915. The vessel also experienced several changes in ownership during its tenure and was converted to diesel power in 1952.[1447]

After being pulled from the trans-Derwent ferry run, primarily due to the building of a bridge across the river, the *Derwent* was sold to Melbourne-based interests and sailed for Port Phillip Bay in September of 1965.[1448] Renamed *MV Queenscliff*, the craft was laid up a few years later and ultimately sold by the Port of Melbourne Authority to a couta fisherman. It was wrecked at Point Cook, Victoria, on 30 December 1973, nearly three-quarters of a century after its auspicious launch.[1449]

Another launch event that Reeks very likely did not attend, though the press reported he was present, was the new steamer *Vaucluse* which he had designed for the Watsons Bay ferry service. The double-ended vessel, capable of carrying 600 passengers, was built to the order of the Watsons Bay Ferry Company and successfully launched from the yard of Rock Davis (Jr) at Blackwall, Brisbane Water, on 4 July 1905.[1450] At 140ft in length with a beam of 25ft and a depth of 9ft, the craft was somewhat larger than the *Derwent*, launched the month prior.[1451] It also differed from many of the other double-ended steamers Reeks had designed in that *Vaucluse* was propelled by only one propeller, intended to drive it in one direction and pull it in another. Following its launch, the vessel was towed to Sydney where its locally-built triple expansion engines and boilers were installed, and its superstructure fitted.[1452]

Vaucluse entered service in late September of 1905 and remained on the Circular Quay to Watsons Bay run until mid-1931 whereupon the vessel was sold to the Walsh Island Dockyard for use in transporting workers.[1453] In 1938 the craft's hulk, which had been tied up at Walsh Island for a number of years, was purchased by a private salvage firm.[1454]

As Reeks had intended, *Vaucluse* was one of the fastest ferries on the harbour during its heyday. He had deliberately designed the craft with just one propeller, either pushing or pulling it. The principle had been tested enough on earlier designs, including the *Lady Manning*. This particular vessel, however, had still retained the

[1436] adb.anu.edu.au/biography/skirving-robert-scot-8448.
[1437] A. Macintosh (1988). *Memoirs of Dr Robert Scot Skirving, 1859-1956.*
[1438] A. Macintosh (1988). *Memoirs of Dr Robert Scot Skirving, 1859-1956.*
[1439] A. Macintosh (1988). *Memoirs of Dr Robert Scot Skirving, 1859-1956.*
[1440] A. Macintosh (1988). *Memoirs of Dr Robert Scot Skirving, 1859-1956.*
[1441] *The Mercury*, 5 June 1905.
[1442] *The Mercury*, 5 June 1905.
[1443] *The Mercury*, 5 June 1905.
[1444] *The Mercury*, 12 October 1905.
[1445] *The Mercury*, 12 October 1905.
[1446] A. J. (Tony) Coen (2017). *River & Coastal Vessels Trading out of Hobart 1832-2015.*
[1447] A. J. (Tony) Coen (2017). *River & Coastal Vessels Trading out of Hobart 1832-2015.*
[1448] A. J. (Tony) Coen (2017). *River & Coastal Vessels Trading out of Hobart 1832-2015.*
[1449] A. J. (Tony) Coen (2017). *River & Coastal Vessels Trading out of Hobart 1832-2015.*
[1450] *The Sydney Morning Herald*, 5 July 1905.
[1451] *The Daily Telegraph*, 5 July 1905.
[1452] *The Sydney Morning Herald*, 5 July 1905.
[1453] *The Sydney Morning Herald*, 31 August 1931.
[1454] *The Labor Daily*, 2 April 1938.

second shaft but had adopted a single propeller after being in service for a period with two propellers. *Vaucluse* was probably Reeks' first design where the second shaft was given away entirely. He also adopted a straight keel, as he had resolved that a single propeller hull had to have a straight keel for better control. Reeks went on to repeat both of these features later in the year when the *Lady Northcote* was launched for the Balmain Ferry Company, though it is not known which design he started first.

To add some contrast to Reeks' ferry designs from around this period, they include the *Derwent* with its straight keel and two propellers, and the earlier *Lady Rawson* with its cambered keel and two propellers. Comparing these two vessels with both the *Vaucluse* and *Lady Northcote*, Reeks was certainly showcasing his ability to adapt and his willingness to make changes when needed.

Vaucluse also marked another busy period for Reeks professionally. Four days following its launch, Reeks arrived in Tasmania, having travelled as a passenger on board SS *Coogee* from Melbourne.[1455] The vessel had left Victoria at 4 pm on 6 July, arriving in Launceston early afternoon of the following day.[1456] Reeks' visit to the city was primarily to view progress on Robert Gardner's new 130ft single-ended craft intended for the local river trade that

[1455] *Daily Telegraph*, 8 July 1905.
[1456] *The Mercury*, 8 July 1905.

SS DERWENT UNDER CONSTRUCTION AT HOBART'S DOMAIN.
SOURCE: THE TASMANIAN MAIL, 15 APRIL 1905.

LAUNCH OF SS DERWENT AT HOBART'S DOMAIN.
SOURCE: THE TASMANIAN MAIL, 29 JULY 1905.

was then under construction. Its builder was Fred Moore who had returned to Launceston after having launched the Reeks-designed *Derwent* at Hobart only the month prior. According to the Tasmanian press, Reeks '*expressed himself as highly satisfied with the way Mr. F. Moore is carrying out the contract*'.[1457] Despite travelling thousands of miles, Reeks stayed only a few days, apparently also travelling to Hobart to view the *Derwent*, where he was impressed with the exceptional planks of Tasmanian timbers, some over 80ft in length.[1458] He returned to Melbourne on 11 July where he wasted no time in signing a contract to design a vessel for Messrs McIlwraith, MacEachern, and Co. of Western Australia.[1459] The craft was intended to trade between Perth, Fremantle and Rottnest Island and carry 600 passengers at a speed of 12 knots.[1460]

Another outcome of Reeks' trip to Tasmania was E. P. Hart, the newly-appointed treasurer of the Tamar Yacht Club, entrusting him to design a '*handy raceabout,*' 16.5ft on the waterline and 22ft overall, with a hinged centreboard and fairly large sail area.[1461] Reeks sent the plans back to Launceston a few months later with the local press noting that the craft was to be '*right up to date - overhang fore and aft, lead on keel, not too deep draught, and could be sailed single-handed. A noticeable feature is the lack of superfluous top hamper. She is to carry no topsail. The draftsman is W. Reeks, and the order to build has been placed with Mr. Fred Moore. Yachting men will eagerly look forward to the time when this little yacht will make her debut amongst us*'.[1462] Named *Roona* and still in existence, the craft was completed early in 1906.[1463]

Robert Gardner's new river steamer was launched at Launceston on 16 November 1905 with much fanfare and ceremony. Built at a cost approaching £10,000 (equipped and fitted) and named *Togo,* the launch event took place late in the afternoon, allowing a large crowd to congregate. *The Daily Telegraph* eloquently described the scene, '*If the ship that was launched may be looked upon as a bridge, mating with the river, the wedding was an auspicious one. The order was given, the blocks released, and the vessel slowly*

[1457] *Examiner*, 8 July 1905.
[1458] *The Mercury*, 12 July 1905.
[1459] *The Mercury*, 12 July 1905.
[1460] *The Mercury*, 12 July 1905.
[1461] *Examiner*, 23 September 1905.
[1462] *Examiner*, 23 September 1905.
[1463] *Examiner*, 14 February 1906.

glided down along the chocks. At the same instant Mrs Paterson named the vessel the Togo (after the Japanese Admiral), and swung the bottle of wine, which crashed upon the bow and sent a shower of glass and liquor upon the side of the vessel. "Hoorah for the Togo," and enthusiastic cheers came from the crowd as the vessel slowly descended into the water and glided alongside the Agnes, which presently took her in tow. The ceremony was over and Mr Gardner, the owner; Mr Moore, the builder; and the representatives of the Glasgow Engineering Company were besieged with the congratulations of well wishers, who, in Mr Moore's work shed, subsequently drank to the health of the owner and success to the Togo'.[1464] The article continued, '*The building of the Togo may be said to mark a new era in the river trade. For some time past its rapid growth has demanded a larger and faster class of steamer to meet requirement; and in order to obtain the most suitable design for such a vessel, Mr Gardner instructed Mr Walter Reeks, of Sydney, who has a world-wide reputation as a successful naval architect, to prepare a design, and the result so far has justified the confidence thus placed. The vessel presents a handsome, graceful, and yachtlike appearance, which was much commented upon yesterday'.*[1465]

Togo, like many of Reeks' designs, once again set a standard and even a concept to work toward for the craft to come. One of its features was a change from the overhanging counter stern to a neat cruiser stern, then coming into vogue. This style of hull ended with a gently-rounded shape, either circular or elliptical. There was no overhang and the run of the wood planking (later steel plating) under the hull subtly curved up from the undersides toward the deck line to meet comfortably at the centreline. This remained a pleasant feature for decades, used from launches right up to huge ocean liners.

Togo's yacht-like appearance was an intended outcome of Reeks' construction plan. '*Normally heavy frames are closely spaced throughout, but Togo had no frames. Instead she had closely spaced battens all steam bent. She was built like a*

[1464] *Daily Telegraph*, 17 November 1905.

[1465] *Daily Telegraph*, 17 November 1905.

ROONA (CIRCA 1910S).
SOURCE: GRAEME BROXAM.

yacht.'[1466] The vessel was also fast and soon earned a title it became widely known for. It was for many years considered the fastest river steamer in Australia. *Togo* could very easily outrun *Niree,* letting the latter steam off ahead some distance before leaving Launceston, then sweeping past it on the way to George Town at the head of the River Tamar.

Togo's career was marked by two accidents. At 9:15 pm on 1 April 1907 on the River Tamar, it collided with a private launch *Alice* and ten lives were lost from the launch. *Togo*'s skipper was later exonerated of blame.[1467] Nearly two decades later, another collision occurred that could have been catastrophic. The vessel was then in use on the River Derwent and, vying with the *Cartela* for bragging rights, there was a regular but 'unofficial' Christmas Day race between the pair from Hobart to Green Island and return. *Togo* had participated in every race since 1920 with varying results. In 1926, however, with *Togo* in the lead and within a minute of the finish line, *Cartela*'s bow collided with *Togo*'s stern at the port quarter, swinging the latter 90 degrees and onto a course directly toward the shore at Castray Esplanade, a short distance away. All the captain could do was order full reverse and slow the inevitable grounding. Despite the cloud of steam, dust and general disturbance there was little damage and all passengers (there were hundreds on board at the time) and crew were safely disembarked, with *Togo* refloated later on.

At the time, the accident was one of the biggest sensations to have happened in the port of Hobart, with the inevitable inquiry resulting in a legal battle between the Commonwealth and the state of Tasmania as to jurisdiction rights, going all the way to the High Court.

SS TOGO ASHORE AT CASTRAY ESPLANADE.
SOURCE: WEEKLY COURIER, 29 DECEMBER 1926.

While Reeks was not present at the launch of *Togo,* he returned to Launceston two months later to oversee the vessel's trial trips where, achieving a speed of 14.5 knots though not its top speed, with hardly any vibration, Reeks

[1466] G. W. Cox (1971). *Ships in Tasmanian Waters: Riverboats, Ferries and the Floating Bridge.*
[1467] *Daily Telegraph*, 2 April 1907.

expressed '*his entire satisfaction with the manner in which the boat has been built*'.[1468]

Reeks had arrived in Tasmania on 2 February 1906 with his wife Gertrude per SS *Wakatipu* from Sydney.[1469] The couple likely combined work with pleasure, visiting several of the state's tourist attractions. They returned to Sydney via Melbourne as passengers on board *Loongana,* departing Launceston on 15 February.[1470] Less than a month later Reeks returned to Launceston, again per *Loongana*, coinciding with the *Togo*'s official trial trip.[1471] It entered service a few days later.

The 130ft *Togo* became a familiar sight on the River Tamar before being sent to Hobart for use in the trans-Derwent ferry service. Known for its graceful lines and speed, the vessel proved a popular and capable addition to the local passenger fleet. After several decades of useful service, it was ultimately supplanted in the 1940s by that ever-increasing mode of transport, motor vehicles.

Reeks arrived back in Sydney later in March of 1906. It would not be the last time he visited Tasmania, however. In 1912 the MV *Excella*, built in Hobart by John Dalgleish to the order of James Rowe and Sons, would be launched. Designed by Reeks, the vessel was another notable addition to the Hobart-based ferry trade. It was also the first of its type and size fitted with an internal combustion engine built anywhere in Australia. Despite his advancing age, Reeks was continuing to adapt and reinvent himself and his craft.

1468 *Daily Telegraph*, 15 February 1906.
1469 *The Mercury*, 3 February 1906.
1470 *Examiner*, 16 February 1906.
1471 *Tasmanian News*, 8 March 1906; *Examiner*, 10 March 1906.

LAUNCH OF SS TOGO AT LAUNCESTON.
SOURCE: THE WEEKLY COURIER, 25 NOVEMBER 1905.

SS TOGO ON THE RIVER DERWENT (CIRCA 1920S).
SOURCE: AUSTRALIAN NATIONAL MARITIME MUSEUM.

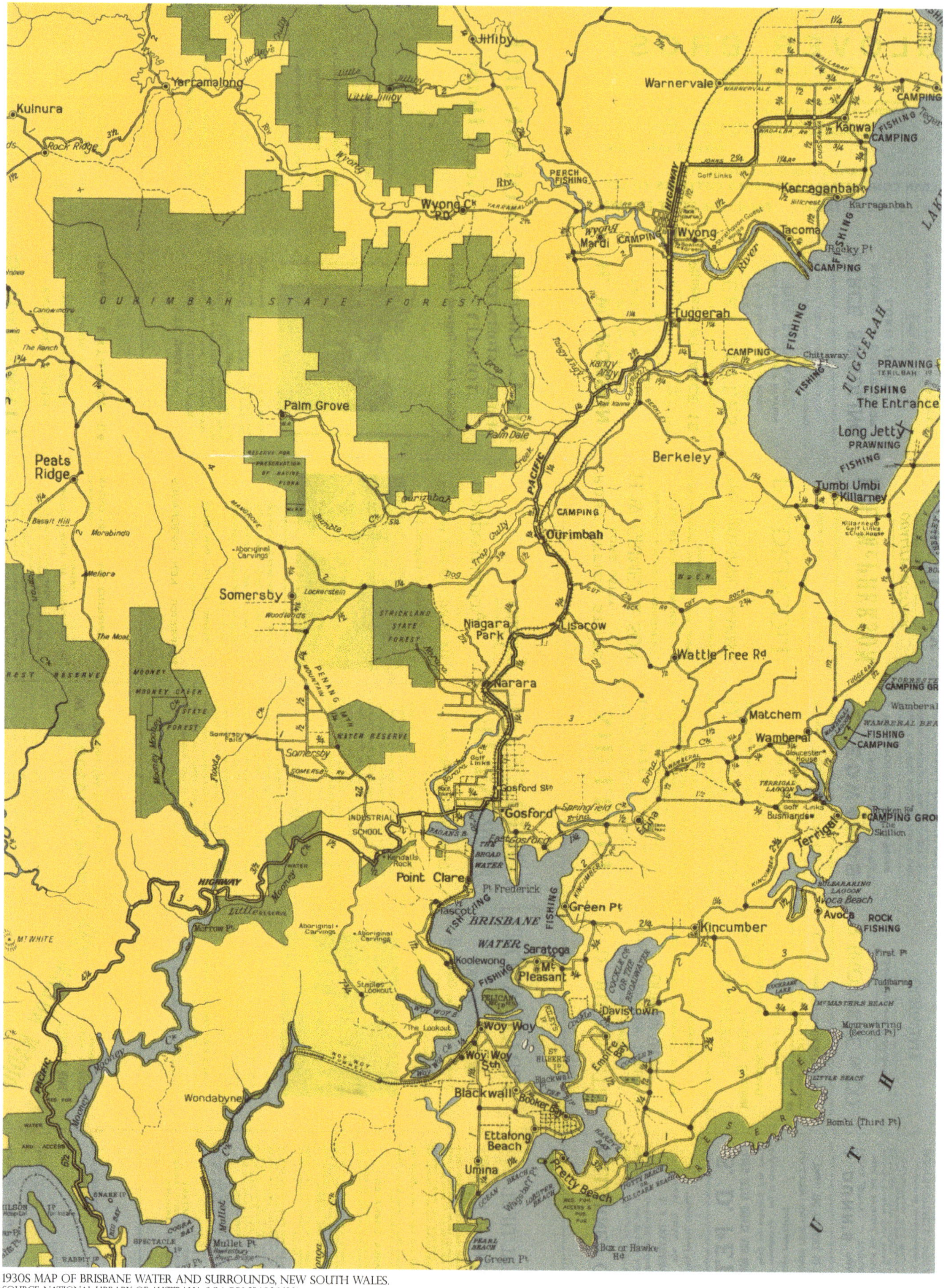

1930S MAP OF BRISBANE WATER AND SURROUNDS, NEW SOUTH WALES.
SOURCE: NATIONAL LIBRARY OF AUSTRALIA (NLA.OBJ-234026620).

CHAPTER SEVENTEEN - BRISBANE WATER

While several of Walter Reeks' projects were coming into fruition in Tasmania, also being launched were vessels he had designed for locations elsewhere. Reeks was certainly stretching his boundaries, no longer deemed solely a naval architect of Sydney, his customer base now incorporated clients in Tasmania, Victoria and Western Australia, in addition to continued patronage from clients in New Zealand and New South Wales.

The Richmond River area of New South Wales, in particular, incorporating Kyogle, Casino, Coraki and Woodburn, had long been associated with commercial boat and ship building, due to an abundance of red cedar. One such boat builder was Oliver Jones who had built the Reeks-designed steamer *Britannia* in 1903 for Davis Bros. With their expanding business enterprises, Reeks had returned to the region two years later at the request of Davis Bros. who commissioned him to design a boat for the Richmond River passenger trade.[1472] At 90ft, slightly larger than the *Britannia*, the craft was built by Messrs Turner and Conroy at Coraki over a period of 18 months and launched in February of 1907 in the presence of 2,000 spectators.[1473] Named *Frederick Davis*, the vessel had only a short career. In late 1908 it was purchased by a firm based in Melbourne and unfortunately foundered en route to its new home port in bad weather, 30 miles south of Clarence Heads.[1474] Its crew of seven were landed safely.

While the *Frederick Davis* took some time to build, other projects Reeks was involved with between 1905 and 1907 were less slow-paced. For example, in June of 1905 Reeks was noted as having the two Tasmanian-built steamers (*Derwent* and *Togo*, both mentioned in the previous chapter) under construction, as well as a fishing vessel for Victoria and a small tug for Noumea.[1475] Also under construction was the *Lady Northcote* for the Balmain New Ferry Company. In addition, Reeks was overseeing alterations to the now 17-year-old yacht *Jess* at William Holmes' yard, its owner James Cox intending to undertake a lengthy cruise to the Pacific islands.[1476] A quote for three vessels to be built for the South Perth Ferry Service was also delivered, though it is doubtful any of these craft were built.[1477]

In August of 1905 Reeks invited tenders for construction of a deckhouse on board SS *Rose,* based at Lake Macquarie.[1478] Also in hand were plans for a fast, light-draft steamer for another Australian port. As far as new work was concerned, the Sydney press reported that Reeks' '*prospects are brighter than for some time past*'.[1479]

While little is known about the fishing vessel for Victoria and the small tug for Noumea, the light-draft steamer was no doubt *Zephyr*, at the time under construction at Brisbane Water by Rock Davis (Jr) for Messrs McIlwraith, McEachern & Co. Intended for Western Australia, the vessel was to carry 600 passengers on a draft of 6ft, with a guaranteed speed of 12 knots.[1480]

Reeks and Rock Davis (Sr) had enjoyed a successful relationship spanning over two decades, beginning in 1891 with the design and construction of the wooden steam drogher *Osprey*. Other craft the pair had worked together

[1472] *The Richmond River Herald and Northern Districts Advertiser*, 2 June 1905; *The Sydney Morning Herald*, 5 July 1905.
[1473] *Northern Star*, 7 October 1905; *The Sydney Morning Herald*, 13 February 1907.
[1474] *Townsville Daily Bulletin*, 30 December 1908.

[1475] *The Sydney Morning Herald*, 14 June 1905.
[1476] *Australian Town and Country Journal*, 19 July 1905.
[1477] *Australian Town and Country Journal*, 19 July 1905.
[1478] *The Daily Telegraph*, 21 August 1905.
[1479] *The Sydney Morning Herald*, 14 June 1905.
[1480] *Evening News*, 12 September 1905.

on were the pearling luggers *Curlew, Heron, Ospray, Pilot, Penquin, Plover, Snipe, Teal* and *Zoe* (1893-1898) and the steamer *Orara* (1894). The early 1900s, however, saw Davis' health deteriorate. He died of tuberculosis in June of 1904, aged 71.[1481] By this time his yard was in the hands of his second son, also known as Rock and colloquially referred to as 'Young Rock'.

Continuing the successful legacy of his father, Young Rock soon solidified his own place within Australian ship building royalty, operating the family's yard at Brisbane Water until his own death in 1913, at the age of 49.[1482] Combined, Young Rock and his father were known to have built at least 165 vessels between the years 1863 and 1913. Of note, at least seven of the craft built by Rock Davis (Jr) between 1904 and 1913 were designed by Reeks, the first of which was the fishing boat *Enid*, built for Arthur Jones of Melbourne.[1483] Next came the double-ended steamer *Vaucluse,* built to the order of the Watsons Bay Ferry Company and launched in July of 1905.[1484]

Six months later, on 25 January 1906, the *Zephyr* was launched, designed by Reeks to the order of Messrs McIlwraith, McEacharn & Co. for use in Western Australia. Shortly thereafter, it was towed to Sydney from Brisbane Water where it was fitted with machinery, cabin and joinery, and electrics.[1485] The vessel's machinery consisted of two separate complete sets of triple expansion engines, supplied with steam from a marine type boiler, fitted with forced draft, the electric light plant being on a De Laval turbine.[1486]

Zephyr's trials were noted early in April of 1906 and once again the details were highlighted. '*Zephyr has been designed as a very light draught vessel, and it is estimated with a full complement of passengers and with coal and water on board she will draw but 5 feet.*'[1487] This would suit its intended role on the shallow Swan River. As had often been the case throughout Reeks' career, he was called upon to meet the challenge of designing a craft to otherwise awkward specifications.

Zephyr was an elegant vessel, low and long, and as well as presenting a challenge with its depth, the vessel's height was compromised by the need to fit under a fixed bridge at Fremantle. Hence the headline in an article detailing its trial, '*Collapsible Funnel, Mast and Awnings*'.[1488] Nothing actually collapsed. The funnel was designed to be telescopic, the top inner section deftly sliding into the bigger diameter lower section. The awnings were also removable and the mast stepped on a tabernacle so it could be lowered and raised easily. Simple solutions, probably not novel by themselves, but combined these features stood out for the press at the time. It is interesting to note that the same challenge would exist today. Although the bridge has been replaced, it is still fixed in height and so

[1481] *The Daily Telegraph*, 29 June 1904; en.wikipedia.org/wiki/Rock_Davis.
[1482] *The Sydney Morning Herald*, 27 August 1913.
[1483] *The Sydney Morning Herald*, 21 March 1905; *Register of British Ships: Main Register subsequent to Merchant Shipping Act 1854*, Port Melbourne [ON 117685].
[1484] *The Sydney Morning Herald*, 5 July 1905.
[1485] *The Sydney Morning Herald*, 26 January 1906.
[1486] *Evening News*, 24 January 1906.
[1487] *The Sydney Morning Herald*, 7 April 1906.
[1488] *The Sydney Morning Herald*, 7 April 1906.

ZEPHYR RETURNING TO FREMANTLE FROM ROTTNEST ISLAND (1914).
SOURCE: STATE LIBRARY OF WESTERN AUSTRALIA (SHAPCOTT COLLECTION OF PHOTOGRAPHS ; BA1104/62).

the Swan River-based yachts use a tabernacle mast to go under and head out onto Gage Roads off Fremantle.

Despite its shallow draft, *Zephyr* arrived in Western Australia late in April of 1906 after a voyage from Sydney without incident, its captain praising the vessel's sea-going qualities.[1489] While initially intended to be used on the Swan River, *Zephyr* spent only a year on this route. Instead it went on to pioneer the Rottnest Island tourist route, remaining on this service from 1907 until 1952. The vessel was ultimately demolished a decade later.

Another collaboration between the shipbuilder Rock Davis (Jr) was the Reeks-designed auxiliary schooner *Goodwill.* Launched from Brisbane Water on 17 May 1906, the 75ft craft was built to the order of the Papuan Industries Ltd. and intended to trade as a missionary boat between Badu Island and Papua New Guinea.[1490] Both Reeks and his wife Gertrude were present for the launch.

While Rock Davis (Jr) and Reeks would continue to work together over the coming years, including on the Balmain New Ferry Company's new steamer *Lady Chelmsford* (1909) and the 146ft steamer *Red Pine* (1912), at the time the largest wooden steamship built in New South Wales, Reeks was continuing to enjoy fruitful relationships with several of Sydney's more prominent boat builders. With regards to outputs from William Holmes' yard at McMahons Point, in addition to Dr Scot Skirving's *Alcyone* (mentioned in the previous chapter), Reeks designed a 58ft steam lighter for Elliott Brothers that was launched late in 1905.[1491] It marked the second such vessel Reeks and Holmes had produced for this company in the space of three years.

A smart looking cruising launch of 45ft in length was built by Holmes to a design by Reeks and completed in early March of 1906.[1492] The craft was built to the order of C. M. Johnston of Woy Woy, reportedly at a cost of £1,500, and fitted with a 25 h.p. Jersey City Standard engine.[1493] The vessel achieved an average speed of 15 miles an hour during its engineer's trials, making it one of the fastest launches in the state.[1494] A special feature of the craft, named *Meteor,* was its reversing gear, an invention of a Mr Oliver.[1495] The vessel was another light draft speedy motor launch, designed along the lines of Reeks' own *Wahine*, as well as *Cwmavon,* including the intriguing helm position out in the open on the forward cabin top.

Reeks also designed a 45ft schooner-rigged auxiliary ketch for the London Missionary Society for use in Papua New Guinea's Purari Delta.[1496] Named *Purari*, this particular vessel was launched from Holmes' yard in July of 1906 with its dedication and naming ceremony taking place the following month.[1497] The craft was specifically designed with a light draft to allow it to navigate the shallow waters of the Purari River and Delta. A few months later *Purari* was shipped to New Guinea as deck cargo on the German steamer *Willchad.*[1498]

[1489] *The Sydney Morning Herald*, 21 May 1906.
[1490] *Evening News,* 12 September 1905; *The Sydney Morning Herald*, 19 May 1906; *The Daily Telegraph*, 19 May 1906.
[1491] *Evening News*, 3 June 1905; *The Sydney Morning Herald*, 5 August 1905.
[1492] *Referee*, 20 December 1905, 7 March 1906; *Evening News*, 30 March 1906.
[1493] *The Daily Telegraph*, 25 January 1906; *Referee*, 7 March 1906.
[1494] *Evening News*, 30 March 1906.
[1495] *Evening News*, 30 March 1906.
[1496] *The Sydney Morning Herald*, 15 March 1906.
[1497] *Evening News*, 10 July 1906; *The Sydney Morning Herald*, 30 August 1906.
[1498] *The Sydney Morning Herald*, 24 August 1906; *Goulburn Herald*, 10 May 1907.

GOODWILL TIED UP AT A WHARF IN THE TORRES STRAIT ISLANDS (1928).
SOURCE: NATIONAL LIBRARY OF AUSTRALIA (NLA.OBJ-145048858).

METEOR.
SOURCE: SHIPBUILDERS OF BRISBANE WATER, NEW SOUTH WALES (GWEN DUNDON COLLECTION).

Further collaboration between Reeks and Holmes was Pearce Bros.' 42ft cruiser, intended for use as both a recreational pleasure and fishing boat. It was installed with a 16 h.p. Standard engine and launched in September of 1906.[1499] Around the same time Holmes completed a 30ft launch built to the order of T. J. Bush to a design by Reeks.[1500] The craft was intended for use at Port Hacking as an excursion vessel.[1501] Little early history is also known of a motor yacht later named *Valencia* that was built by Watty Ford (Jr) at Berrys Bay in 1906 to a design by Reeks.[1502]

Another relationship which Reeks had clearly solidified over the years was with the Balmain New Ferry Company. On 24 October 1905, with hundreds of spectators and invited guests assembled, Reeks launched the fifth vessel he had designed for this company since 1891, by now referred to as the 'Lady' class of ferry.[1503] The *Lady Northcote* was built by Messrs Morrison and Sinclair of Balmain at an estimated cost of £7,500 and at the time of launch was described as a '*handsome steamer of the combined ferry and concert boat type*' with accommodation for up to 600 passengers.[1504] Measuring 116ft the craft was somewhat smaller than both *Lady Hampden* and *Lady Rawson*, designed by Reeks for the same company and launched in 1896 and 1902, respectively. As stated previously, *Lady Northcote* was double-ended though driven by only one screw.[1505] Its surface condensing engines were supplied by Morts Dock and when put to the test a few weeks later, successfully propelled the vessel at a speed of 12 knots per hour.[1506] The *Lady Northcote* commenced service on Sydney Harbour in early January of 1906.

In addition to designs for new builds, there was also a great deal of work on hand with regards to existing vessels requiring modifications. One such craft was the 117ft fishing boat *Baroona* which was converted into a commercial steamer for the Wairoa Steamship Company of New Zealand. Built at Newcastle in 1904, the alterations were undertaken by Messrs Morrison and Sinclair under the direction of Reeks with the vessel modified to become an up-to-date passenger and cargo vessel, including the fitting of a new propeller.[1507] The *Baroona* also had its bottom strengthened and vibration reduced. Sleeping accommodation was added for 50 people, as well as a dining saloon, smokeroom, galley, and lavatories.[1508] The vessel arrived in New Zealand late in 1905 and went on to have a remarkable 105-year career, seeing service as a harbour ferry, coastal cargo ship, inter-island trader, deep-water trawler and, finally, a floating restaurant.[1509]

The following year Reeks acted as an agent for the Wairoa Steamship Company in its quest to purchase the

[1499] *Referee*, 22 August, 19 September 1906.
[1500] *Evening News*, 10 July 1906; *The Sydney Morning Herald*, 3 August 1906.
[1501] *Evening News*, 10 July 1906; *The Sydney Morning Herald*, 3 August 1906.
[1502] SYDNEY, NEW SOUTH WALES, Register of ships, single number series (II), Volume 25, 1919-1923, Folio 7.
[1503] *Evening News*, 25 October 1905.
[1504] *The Sydney Morning Herald*, 25 October 1905.
[1505] *The Sydney Morning Herald*, 5 July 1905.
[1506] *The Daily Telegraph*, 21 December 1905.
[1507] *The Sydney Morning Herald*, 5 December 1905.
[1508] *The Sydney Morning Herald*, 5 December 1905.
[1509] natlib.govt.nz/records/43654909.

LADY NORTHCOTE.
SOURCE: GRAEME ANDREWS WORKING HARBOUR PHOTOGRAPH COLLECTION, CITY OF SYDNEY ARCHIVES.

one-time, well-known steam yacht *White Star*, which had been doing duty for the Government of Queensland in Torres Strait and was previously owned by R. S. D. White, for many years commodore of the Royal Sydney Yacht Squadron. It was the company's intention to convert the vessel into a passenger boat for use on the Kaipara.[1510]

During this period Reeks maintained his association with Sydney's sailing community. He continued his decades-long legacy of being appointed honorary measurer of the Royal Sydney Yacht Squadron.[1511] In August of 1906 Reeks was also appointed vice-commodore of the squadron, a position that saw him attend garden parties, opening of the season ceremonies, dinners and other events, often having to act as host and give speeches and toasts in the absence of the club's commodore, Sir James Fairfax. In this role Reeks regularly interacted with prominent Sydney-siders, including governors and their wives, and other distinguished guests.[1512] He also frequently acted as commodore during races, often in charge of the club steamer, *Robbie Burns,* and was starter and judge for races. In addition, Reeks was the organisation's delegate to the Sydney Yacht Racing Association.[1513]

With regards to growing the sport, one of Reeks' continuing projects was advocating for the Royal Sydney Yacht Squadron to adopt class racing for 30 raters. '*Mr. Walter Reeks spoke at some length, giving the history of the failure to promote a class of 30-footers some years back, owing to many of the yachts being built over the rating. He urged the members to again take the matter up earnestly, as there was now every prospect of a successful issue.*'[1514]

Reeks had been a supporter of distinct racing classes for some time. However, as it transpired, although initially a number of individual yachtsmen expressed interest in the new 30-rater class, members of other Sydney-based clubs could not come to an agreement. A further reduction

[1510] *The Sydney Morning Herald,* 29 October 1906.
[1511] *Evening News*, 30 August 1905.
[1512] *The Sydney Morning Herald,* 29 August 1906.
[1513] *The Daily Telegraph*, 2 September 1905.
[1514] *The Sydney Morning Herald*, 11 October 1905.

Royal Sydney Yacht Squadron.

T. H. KELLY,
Vice-Commodore.

Photo. by Falk Studios.
SIR JAMES FAIRFAX,
Commodore.

Photo. by Falk Studios.
T. W. BREMNER,
Rear Commodore.

Photo. by Falk Studios.

ALFRED GEORGE MILSON,

A prominent member of "the Royals." At the first Anniversary Regatta in this State Mr. Milson's father won the Champion Yacht Race. Fifty years later the son won it. In each instance the name of the yacht was the Era. The original Era was sold to the British Government for presentation to King Thakambau when a protectorate was established over Fiji.

CHAMPION YACHT SCOTIA.

The Scotia, 30-rating, was designed by Fife (England), and built in Sydney for Mr. T. W. Bremner. For the past two seasons she has been sailing remarkably well, and is always cleverly handled. She is regarded as a very up-to-date boat, and probably the best that has been designed and built in Sydney.

Photo. by Falk Studios.
FREDERICK LASSETTER,
A Veteran of the Sport.

Photo. by Falk Studios.
E. W. KNOX,
Owner of the Sirocco, and a devoted yachtsman.

Photo. by Falk Studios.
J. A. MINNETT,
Treasurer.

Photo. by Falk Studios.
F. DORAN,
Who has many wins to his credit, the latest being the Bona over the New Zealand yacht Rawhiti on February 24.

Photo. by Falk Studios.
WALTER REEKS,
The well-known designer of yachts.

SOURCE: THE SYDNEY MAIL, 7 MARCH 1906.

of size to 24ft was recommended. A one-design proposal was also introduced, though some wanted a pure racing yacht, while others proposed one with more capability. As 1906 moved ahead a committee was appointed to further discuss options. In the end it appears that once again the idea lapsed, much to Reeks' disappointment.

Of course there were other diversions and ways to spend spare hours, as well as meet like-minded hobbyists and new clients. In 1905 Reeks joined the Motor Boat Club of New South Wales, a newly formed organisation that despite its infancy had an astonishing number of members, around 100, indicating the expanding popularity of this type of craft. At the time there was stated to be over 300 motor boats in New South Wales alone.[1515] The following year, on 6 July 1906, Reeks attended a meeting at the Aaron's Exchange Hotel to establish the Royal Motor Yacht Club.

Improvements in engine technology were considerable during this period. Holmes was noted as building a craft with a speed supposedly of 17 knots. Not to be out-done, Reeks installed a new engine in *Wahine*, along with other modifications. The press noted, '*Mr Walter Reeks' cruiser is again afloat, fitted with her new 25-h.p. Thornycroft motor. A beaver-tail has been built on to her stern to prevent "squatting" whilst running, and the hull is now painted grey, with a white stripe between that colour and the red antifouling. This gives a very smart effect to the boat. A speed of between 10 and 11 knots is expected to be obtained*'.[1516] Relishing the opportunity to further explore regional waterways, late in 1906 Reeks completed a successful five weekends' cruise of the Hawkesbury and during the time covered some 300 miles. The new engine apparently ran without a hitch.[1517]

Reeks appears to have enjoyed catering to prospective owners of these newly-found type of craft, particularly those itching for faster and more reliable vessels. Of note, there was another motor boat designed by Reeks and built by Holmes around this time. Unfortunately its original name and owner are not known. The local press, however, covered the vessel's development with interest. '*A new racing shell which we hinted at in a previous issue, is now well on the road to become an accomplished fact. Although the owner's name is not yet to be made public, we are in a position to say that from plans supplied by the well-known naval architect, Mr. Walter Reeks, Holmes of McMahon's Point, is now busily engaged laying out the lines of a 25ft speed boat. This boat will be one of the most up-to-date that has yet appeared in Australian waters, and will be a new departure from the general designs of racing boats. She is to be fitted with a set of 4 cyl. Thornycroft engines, and is expected to have a speed of at least 20 miles an hour. She will be ready during the course of six weeks, and then let the speed enthusiasts look to their laurels.*'[1518]

Details of this vessel continued to be reported in the Sydney press over several months. For example, the *Referee* of 12 September 1906 stated, '*The Balmain regatta should be productive of some splendid motor boat racing this year, as most of the new speed craft will be ready for that event, and there should be at least five boats starting in the fast event, which will be capable of doing twenty miles per hour. The new Thornycroft boat is now having her engine installed and should be out within the next week or two. She is a very pretty little model, designed by Reeks, with a beaver-tail stern and very clean entrance. A 24 h.p. Thornycroft engine will be installed in her. A somewhat similar boat in England, the Scolopendra, belonging to the Thornycroft Company, has attained a speed of nearly twenty miles per hour, with slightly less power than the Australian craft will have*'.

Interestingly, less than a month later came the statement, '*It is to be regretted that the new Thornycroft boat building at Holmes will not be used for racing*'.[1519]

It obviously took some months for the engine to be installed, perhaps it had been placed on board *Wahine* so a new one had to be imported, as the vessel only went through trials in March of 1907. '*It is reported that the Thornycroft racer, which was built some months ago by Holmes, has at last been got into running order, and a speed of 10 knots was obtained. This is considerably slower than a boat of similar design which the Thornycroft Company ran some three years ago in England, which attained a speed of 17 miles per hour.*'[1520]

It would take many more months of tweaking before the vessel finally reached a more desirable speed of 16 miles.[1521] Still, the engine must have continued to be problematic as it was replaced with a 20 h.p. Fairbanks in October of 1908. The craft was subsequently shipped to new owners, Messrs Turk and Cliff, sole agents for Fairbanks engines in Brisbane, who quickly renamed it *Fairbanks II* and advertised the vessel as '*the fastest boat on the river, attaining a speed of about 20 miles an hour*'.[1522] Competing with the Brisbane Dinghy Club, *Fairbanks II* quickly delivered on its reputation, out-racing all competitors in its class.

Despite this particular Thornycroft engine proving a nuisance, Reeks was not to be thwarted. He continued to support the installation of this brand of engine in many of his vessel designs in the years to come. One such craft whose details also remain elusive was the motor launch *Kylie,* built for F. A. Wiesener to a design by Reeks and likely launched in late 1906, builder unknown.[1523] Fitted with a 6 h.p. Thornycroft engine, the boat was advertised for sale in December of 1907.[1524]

Continuing his involvement with the Engineering Association of New South Wales, in July of 1906 Reeks delivered a second paper on ferries. Reminiscent of the presentation he had given to members of this organisation in 1894, the subsequent discussion was rather heated and further expands on Reeks' design process, as well as his delivery, thereby giving us further insight into his character. It is worthy of summation.

1515 *The Sunday Sun*, 30 July 1905; *The Australian Star*, 15 July 1905; *Southern Star*, 9 August 1905.
1516 *The Australian Star*, 16 June 1906.
1517 *The Daily Telegraph*, 17 October 1906.
1518 *Referee*, 18 July 1906.
1519 *Referee*, 10 October 1906.
1520 *Referee*, 13 March 1907.
1521 *The Daily Telegraph*, 10 July 1907.
1522 *The Australian Star*, 13 October 1908; The Brisbane Courier, 17 October 1908.
1523 *The Australian Star*, 18 February 1907.
1524 *The Sydney Morning Herald*, 14 December 1907.

Simply titled '*Double-Ended Single Screw Steamers*', the paper was presented at a meeting held on 12 July 1906.[1525] The talk was an overview of his experiences with single screw propulsion on one end of the double-ended ferries, compared to those vessels with two screws and fitted with a propeller at each end. In both configurations the propellers had to push and pull the vessel along. The discussion that followed the presentation went into quite a bit of detail, and was in fact held over to a subsequent meeting in order to allow other members to prepare their comments.

Reeks began the talk by noting the history of double-ended screw ferries, again sketching out the background of his United Kingdom experience with *Oxton* on the Mersey at Liverpool when he worked at Alex Richardson's office, and pointing out that the next successful application was on New York Harbour, closely followed by the *Wallaby* on Sydney Harbour. He was generous with his praise of *Wallaby*, stating that it had done an excellent job paving the way for further craft of its type.

Reeks then moved the discussion to what he termed '*the second generation, or double-ended, single screw boats*', first noting how they had come about, more or less by accident when *The Lady Mary* had '*doubled up one of her propellers*'. The damaged propeller was removed on the slip overnight and the vessel was run on a single prop for two weeks, including a period where a heavier propeller substituted for the still usable single one. *The Lady Mary* ran successfully in this makeshift configuration and while it was returned to the original twin screw arrangement, the notion of just one screw was not lost. '*The memory of that fortnight remained with the board of directors, particularly those of the more practical way of thinking.*'

When it came time to consider a third ferry for the Balmain New Ferry Company, '*prudence prevailed, the strong argument being advanced that it was best to provide her with two screws, and that at the first opportunity the experiment of running with one should be tried again*'. The new ferry was Reeks' *Lady Manning* with its elegant cambered keel. Later in 1894, after it was put into service, they ran the experiment more thoroughly, '*this time a propeller properly proportioned to absorb the full power of the engine was provided beforehand*', recognising a key requirement of matching engine power with the propeller size. '*The result was completely satisfactory, in fact Lady Manning has run this way ever since.*' Reeks noted its original speed with two propellers was 12 knots. With one only and pushing it, *Lady Manning* managed just over 12 knots, and when pulling it, the craft was then just under 12 knots, giving an average of 12 knots, '*which brings her speed practically the same under both sets of conditions*'.

Two more ferries were built and both were '*provided with two screws; meantime in both cases one has been removed, and the boats run with one only. The results seem to remove all doubt from the minds of the majority of the Directors. ... the Lady Northcote was built with the provision for only one propeller, likewise was the Vaucluse ... and demonstrating that to the minds of those gentlemen who have had most experience of them, the single screw double-ended ferry boat, pushing one way and pulling the other, is a practical success*'.

At this point Reeks returned to the rockered keel and stated emphatically, '*the author says right here, and without hesitation, that a double-ended, single screw ferry should not have a rocker keel*'. Citing *Lady Manning* Reeks then explained that with the combination of single screw and rocker keel, '*sometimes difficulty was experienced in making wharves ..., and this came very nearly to putting an end to the trials of obtaining a single screw double-ended boat, it being argued, ... that a ferry boat of all boats, should be under absolute control at all times*'.

The solution was to return to a straight keel and this he showed on his most recent designs with a single propeller used in both *Lady Northcote* and *Vaucluse*. This outcome, however, has an uncanny coincidence to it. The authors were surprised to find that Nathanael Greene Herreshoff also employed the rockered keel on three of his ferry designs. The twist to the tale? Herreshoff's elegant designs for North America dated from 1907 to 1911, i.e., the same time Reeks was leaving the concept behind and returning to the straight keel.

Reeks included construction considerations, and entered into a broad discussion of the value of using steel to give '*a stiffer structure than is possible in wood, but for our local requirements, and for ferry boats in particular, wood will be used as a constructive material for some time*'. The problem of stiffness connected with issues surrounding the long shafts on these ferries was noted, but Reeks was resigned to losing the battle. '*A vessel changes shape more or less in spite of you, hence vibration and humping to the detriment of the structure and inconvenience to passengers; constant attention to the propeller shaft, and frequent lining up*'.Reeks then brightens up, '*the single screw shows distinct advantages in this connection*', highlighting this by pointing out that the single propeller provides a much shorter shaft (only half the hull length) and establishes the following claims: it is easier to line up, reduces machinery costs, uses less coal, gives at least equal handling abilities, and equal average speed, resulting in '*all round a more profitable ship*'.

The presentation concluded with a return to the discussion of the shape of the bow and stern ends of the ferry, a major topic in the first paper. Again Reeks highlighted that when the propeller is pulling it will '*throw a lot of water back on to the vessel's bow at a speed greater than her passage through the water, ... and it follows that resistance will vary with the excellence or otherwise of the boat's bow*'. He suggested a form may be conceived that '*would practically not be affected by this stream of fast running water*', but fails to define what this shape might be and doubts if it would be practical. Reeks then finishes in a flourish, '*we are left to wrestle with conflicting conditions, and design, to a compromise between the practical and the ideal, and that compromise must ever remain like Sam Weller's name, depending largely on the taste and fancy of the speller*'. [Sam Weller was a character in Charles Dickens' '*Pickwick Papers*'. In his cockney accent, Weller was pronounced Veller, though if it was written as Veller, Sam would then pronounce it Weller.]

[1525] Minutes of Proceedings of the Engineering Association of New South Wales, 1905/06.

Putting that paradox aside, Reeks did note again his preference to place the propellers as far apart as possible when two are used, and if one only, to place it as far forward under the hull as possible. He brought the audience to the attention of his experience with *Lady Hamden* and *Manly* where he had '*placed them far out*', meaning right at the end of the hull, which had invited damage to them, as happened on one occasion.

Reeks finally resolves at least a rule to follow when designing the end with the propeller, accepting that will be the shape for the opposite end in a ferry boat. His rule was to design the end for the speed of the waterflow off the propeller '*for supposing the vessel's speed to be 12 knots and the column of water off the propeller is 15 knots, then it follows that the boat wants designing for 15 and not for 12*'.

Various testing methods today would help resolve this with assessments of different shapes, configurations and combinations that could make this rule look very much like another over simplification of the problem. However, in 1957, respected Sydney naval architect Jim Eken reviewed double-ended ferries on Sydney Harbour in a paper, and came to some conclusions that suggest a better hull shape could have been used, based on advances in design.[1526]

Amongst Eken's broad investigation was a review of the hull shapes of some of the earlier built ferries, including the Reeks-designed *Kuring-gai*, whose typical vee-shaped hull contrasted with the flatter floored and more box shaped sections of subsequent ferries from other designers. In Eken's focus on resistance he noted that overall, a double-ended hull shape designed to go in both directions was inherently going to have greater resistance than a hull designed to go in only one direction and being able to have the stern shape different to the bow to reduce resistance. Referring to a then well-established method of assessing a hull's resistance to seek the ideal volume distribution for the speed it was to be running at, called the prismatic co-efficient (Cp), Eken noted that the older vessels such as *Kuring-gai* with its vee-shape carried a higher Cp '*which does not appear to be conducive to lower resistance*' and that more recent ferries carried a lower Cp and were thus better optimised. Reeks' design methods focused on other co-efficients of his period which gave him the shape he desired and, perhaps without an alternative shape for comparison, he was unaware that an improvement was possible and so largely maintained the volume distribution he had always employed.

Propellers are another source of modern-day calculation and design which is highly refined. Reeks considered propellers and had made mention of this in an earlier paragraph of his 1906 paper before bringing up the hull shape question just discussed. The single propeller specifically designed for operation on *Lady Manning* was slightly bigger in all respects than the two identical propellers used initially. Reeks' discussion showed an understanding of an obvious issue: the single propeller had to work equally well in both directions of rotation, one pushing, the other pulling. It also had to have symmetry, whereas the twin propeller installations were optimised for the pushing from the stern, with a flat driving face and rounded back to maintain strength in the blades. Reeks also recognised that the propeller had to be able to impart the full power of the engine, otherwise it would '*run away at a dangerous speed*', something to be aware of when running briefly on a single propeller without a symmetrical design. The first solution for the shape of a purpose-designed single propeller was to have a slight round in both faces. Later ones were '*made quite flat with chamfer edges and ... found to answer well*'.

The design of these propellers is still a challenge today, and the authors are indebted to three retired naval architects for pooling their resources to consider the problem with some modern design tools at hand: Mori Flapan, Phil Helmore and Noel Riley. The trio recognised the same problems to be addressed, with a focus on the issue of the need to '*absorb the full power of the engine*'. With two propellers working efficiently, both share the power absorption and each one can be smaller than a single propeller, which would be larger in surface area and ideally diameter as well. The principle to be attained was nicely summed up by Reeks, '*In general, in sizing propellers you fit the largest diameter propeller that you can fit, turning at the slowest possible speed, to get maximum efficiency*'.

The added complication of 'pitch', simplified as being the distance the propeller would travel in one revolution if it were a screw thread entering a piece of wood, also needs to be considered. Pitch affects the top speed attained and acceleration, and is linked to the power output of the engine which increases until it has reached maximum rpm. Another issue of note is that if the combination of propellers was oversized, then the drag from the second propeller could exceed its additional thrust. This may arise if each propeller were sized to absorb the full power of the engine. In such circumstances the second propeller would be a hindrance and so its removal would result in improved efficiency, through the reduction in propeller losses, hull drag and shaft line.

An answer to the propeller issue was ultimately reached by employing a step-by-step process. Using *Lady Manning* as an example which was documented, our trio of modern naval architects first established what was known or fixed, such as the maximum diameter that could be fitted was 6ft, and the power of the engine. The next step assumed a small number of percentages or ratios which allowed for known losses or interferences that had to be accounted for. The conditions of power transmission to the propeller and then its ability to provide this force in the water are not perfect and suffer a series of reduced efficiencies. Modern methods have established tables and other guides to use to determine these for individual situations, but even then, in our situation, it was found necessary to move between these to find information applicable to our unique issue.

[1526] J. H. Eken (1957). *Double Ended Screw Ferries in the Harbour of Sydney.* Royal Institute of Naval Architects, Australian Branch.

The eventual answer came from a propeller series developed from systematic tests in the 1950s, and the solution was reassuring. For a single propeller absorbing the full engine power, the results came out quite similar to the pitch and diameter dimensions used by Reeks on *Lady Manning*. In addition, a comparison with two propellers, each designed to absorb half the engine power, was found to deliver slightly less pull or push power than the single propeller.

In summing up, Phil Helmore states, '*There is a lot of assumptions in the above. However, you can probably change the assumptions (including the propeller series) over quite a range without changing the basic result: Walter Reeks knew what he was talking about*'.[1527]

This contemporary review suggests there was certainly something to be said for the use of just a single propeller rather than two, but the initial discussion of the merits of Reeks' presentation and claims suggested otherwise.

Held regularly, the Engineering Association of New South Wales presentations were always well patronised by members and immediately followed by a discussion amongst those attending. In the case of Reeks' presentation in 1906, however, after two brief comments, the talk was adjourned '*so as to give members an opportunity of getting more information on the subject*', in particular, obtaining results from trials in better conditions than had so far been organised.[1528]

The Balmain New Ferry Company then kindly placed their latest ferry, *Lady Northcote*, '*at the disposal of the Association in order that trials may be held*'.[1529] The vessel had a single propeller at one end.

The trials were undertaken with great attention under the supervision of engineer Russell Sinclair, and run over the measured mile between fixed points on Pinchgut and Bradleys Head on an ebb tide immediately after it had peaked at full.[1530] An extensive series of runs were made up and down under different engine revolutions with the propeller positioned '*ahead towing*' or '*aft driving*' on 25 July 1906.[1531] The high pressure and low pressure cylinders in the twin cylinder engine were monitored. Averages were recorded for Indicated Horse Power (HIP) and speed. A co-efficient of performance was subsequently calculated. From this information a series of curves were plotted for comparisons to then be made. The group also looked at handling, i.e., '*the behaviour of the vessel when stopping, starting and steering was tested, and the results included in the table*'.[1532] In short, it was done by the book for the period.

The results were then discussed by Association members at a follow-up meeting with five representatives taking the floor, each talking extensively. They were all in agreement that the results and their experiences showed reasons to dispute the claims made by Reeks in his presentation. These included:

[1527] Phil Helmore, via email 20 August 2021.
[1528] Minutes of Proceedings of the Engineering Association of New South Wales, 1905/06.
[1529] Minutes of Proceedings of the Engineering Association of New South Wales, 1905/06.
[1530] Minutes of Proceedings of the Engineering Association of New South Wales, 1905/06.
[1531] Minutes of Proceedings of the Engineering Association of New South Wales, 1905/06.
[1532] Minutes of Proceedings of the Engineering Association of New South Wales, 1905/06.

• Where Reeks claimed that the much shorter shaft was easier to line up, it was felt there was little evidence from ferries with a long shaft that they had experienced any significant problems.

• Where Reeks claimed that less coal was used, this was disputed by the power results which showed more power was used when pulling than pushing, therefore more coal was needed to provide more steam.

• Where Reeks claimed that handling abilities were equal, this remained contentious as it was felt from the tests, in addition to experience as regular passengers, that the single propeller ferries had been difficult to manoeuvre. The speakers also provided some examples.

• Where Reeks claimed '*equal average speed*', meaning that the average of the speed for the two directions, pushing and pulling, was the same as what a twin-screw ferry would do equally in both directions, the test results fell a little short of this claim. Noting the single screw ferry was slower when pulling, it was said '*It would be poor consolation to a passenger who had just missed his train to be informed if the boat had been going the other end on, he would have made his train and some minutes to spare*'.

Reeks' claim of reduced machinery costs was accepted as valid, but perhaps overvalued, and the question was posed that if the single propeller was damaged then the vessel was disabled, whereas if it was fitted with two propellers, it could still operate with the one that was undamaged, so the margin of safety was worth the cost. Reeks' claim of '*all round a more profitable ship*' was also questioned.

Overall, Reeks' initial presentation had evolved into a robust and healthy discussion and series of practical tests amongst a cohort of engineers. The consensus from the speakers presenting the rebuttal, however, was a preference for the twin screw arrangement on a straight keel.

Reeks was then given the opportunity to respond and, far from being deflated by the somewhat negative outcome of the trials and thus poor reception of his ideas by the audience, he responded with confidence. In his words, '*the Association had an opportunity of judging, and are at full liberty to form their own conclusions; for his own part he should take, as had always been his habit, the circumstances of each individual service into account, and in future design rocker keel boats, partly or wholly straight keels, two or one screw, just as the case demanded; the day had gone by for a man to be born with one model in his head, which he ground out to order as if from a sausage factory*'.

Reeks then suggested some further considerations that the previous speakers had not accounted for in their disputes of his claims, including the suggestion that '*a perusal of the Company's coal bill would act as a strong set off against the diagrams indicating a heavy loss due to increased IHP required for the "same speed pulling"*'. He was also most emphatic that *Lady Northcote* was not lightly built, as others had said, stating '*she ranked as one of the strongest ferry steamers in the port of Sydney*'.

Reeks then had the last word in a strange way, concluding by teasing the audience with '*One question only: if the keel, to which so much importance was attached, supported the ends, what supports the keel?*' And with that missive hanging in the air, the meeting closed.

This is a good point in our story to note that Reeks had always had competitors, and some of them were present at the meeting. Reeks was not the only designer of Sydney's ferries, but he was one of the most recognised, and probably the only one providing unorthodox features in many of his vessels. There were a modest number of ship builders who had also provided designs, including Dunn; Drake; Ellis; Pashley; and Young, Son and Fletcher. James Scott, who had been foreman for building both *Lady Manning* and *Lady Hampden* also went on to design some of the 'K class' ferries, while a Mr Summerbell, the manager of the North Shore Ferry Company is credited with designs for two other 'K class' vessels. In addition, Morts Dock employee Captain Andrew Christie designed six Manly ferries after Reeks completed *Kuring-gai.*

While the presence of these competitors lessoned Reeks' opportunity for design commissions, he very likely welcomed the presence of a cohort to learn from, engage and interact with, and be inspired by. Ever the opportunist, Reeks would have been challenged to design better, faster and safer craft.

Continuing his involvement with the Engineering Association of New South Wales, in October 1906 Reeks participated in a discussion on the corrosion of metals with several other members.[1533] There was of course more income-generating work at hand too. In particular, Reeks had just signed the contract to design a substantial 110ft motor yacht to be built by Messrs Morrison and Sinclair to the order of Sam Hordern.[1534] A feature of the vessel was that its propelling power, which was to consist of '*three 100 horse-power engines - 300 h.p. in all - driving three propellers*'. [1535] What a contrast this design was from the single propeller he advocated for in the use of harbour ferries.

It was still early days for oil engines and the necessary power could only be made up by having multiple engines. To link them to a single propeller was probably an engineering feat that could have been resolved, but the likely complex result may have been impractical in terms of the resulting propeller size and ongoing running and maintenance issues. In any case, multiple engines with individual propellers offered some latitude and margin of safety if complications arose with handling or engine failure. The configuration also added to the impressive nature of this particular craft.

Other diversions for Reeks were more of a social nature. In late December of 1906 he helped organise a '*Welcome Home*' party for acclaimed Australian landscape artist Arthur Streeton, who had just returned from London after six years spent following his profession in Europe. Held at Paris House, '*In acknowledging the toast of his health, Mr. Streeton spoke in terms of appreciation of the warmth of the welcome tendered him by his Australian friends*'.[1536] Despite being in Europe for many years, Reeks and Streeton had obviously maintained their strong friendship.

1533 *The Sydney Morning Herald*, 12 October 1906.

1534 *Referee*, 21 November 1906.

1535 *Referee*, 21 November 1906.

1536 *The Sydney Morning Herald*, 28 December 1906.

ARTHUR STREETON, KENT REEKS AND TWO OTHERS ON BOARD WAHINE (1907).
SOURCE: OLIVER STREETON.

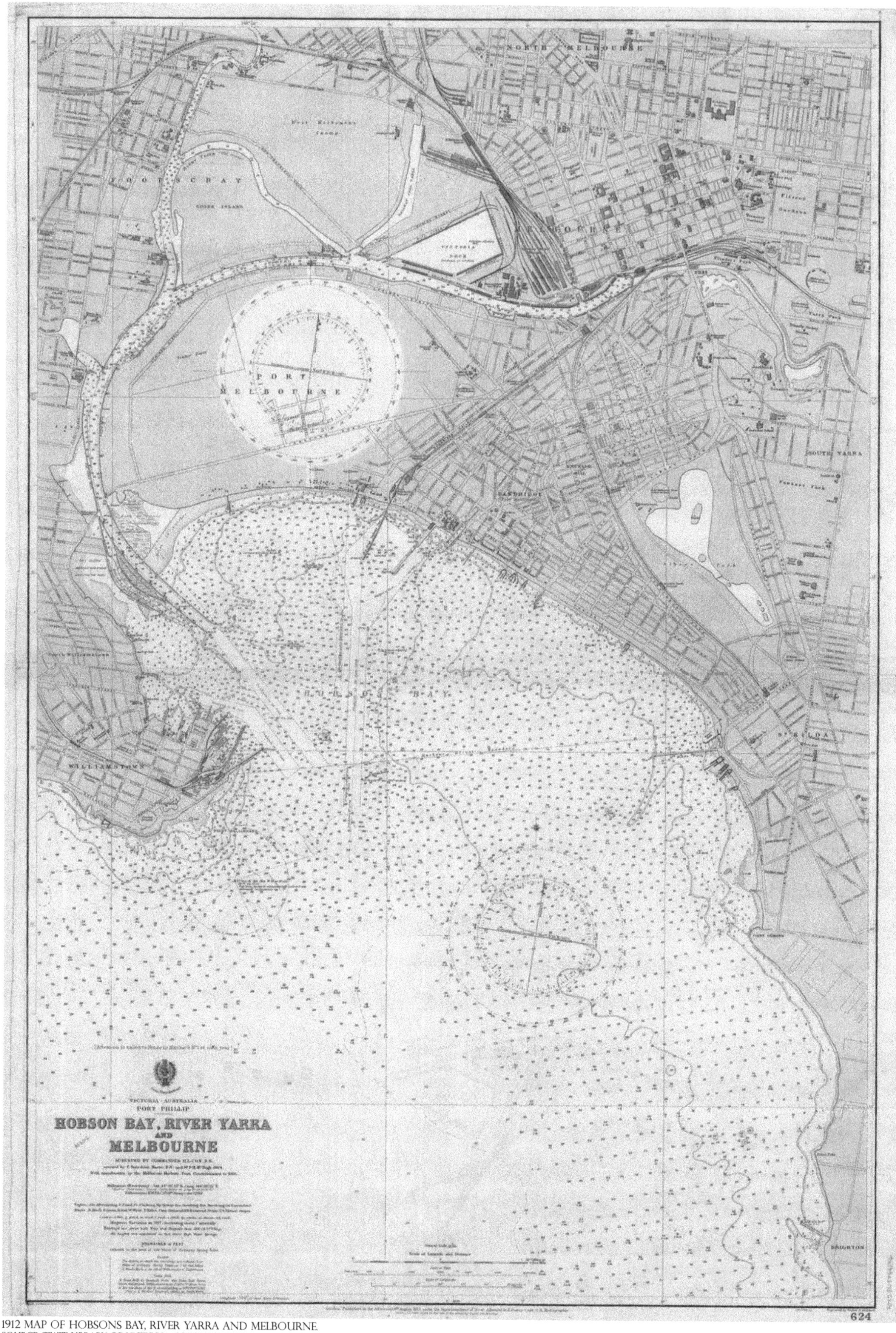

1912 MAP OF HOBSONS BAY, RIVER YARRA AND MELBOURNE.
SOURCE: STATE LIBRARY OF VICTORIA (9912989393607636).

CHAPTER EIGHTEEN - HOBSONS BAY

The origin of the Sayonara Cup began with the series of races first held on Sydney Harbour in January of 1904. It quickly became an annual event, in a few short years earning the moniker 'the America's Cup of the Australian States'.[1537] With its success, other interstate racing events, particularly those featuring yachts from New South Wales and Victoria, became established. One of these, the La Carabine Cup, was first sailed on Hobsons Bay in January of 1907, with Walter Reeks again involved.

Valued at 50 guineas, the La Carabine Cup was presented to the Royal Yacht Club of Victoria by Sir Rupert Clarke in mid-1906 for establishment of an interstate competition between yachts not exceeding 52 linear rating by the 1905 rule. The inaugural race was then confirmed as being staged on Hobsons Bay early the following year with no restrictions placed on the method of getting the yachts to Melbourne.[1538] Representatives from the Sydney-based yacht clubs then lobbied to change the conditions such that the competition be sailed over three races with the trophy going to the winner of two or more races.[1539] Other conditions included that not more than two professional hands be carried as part of the crew, that the yachts be steered by members of recognised yacht clubs, that crews be limited to 12, that mixed rigs be allowed, and that the course be 20 knots long, free from headlands.[1540]

With the conditions agreed to by all parties, entries for the 1907 event were soon received. These comprised two Sydney-based boats (*Heather* and *Rawhiti*) and nine Melbourne-based boats (*Sayonara, Uria, Eric, Fidena, S.J.S., Salacia, Kurrewa II, Fancy* and *Gitana*).

[1537] *The Sydney Mail and New South Wales Advertiser*, 9 January 1907.
[1538] *The Sydney Morning Herald*, 18 August 1906.
[1539] *Australian Town and Country Journal*, 22 August 1906.
[1540] *The Australian Star*, 3 September 1906.

With crowds numbering around 15,000, the La Carabine Cup took place on Hobsons Bay in mid-January of 1907. Reeks was in attendance, representing the Royal Sydney Yacht Squadron in an official capacity.[1541]

The first race was won by *Rawhiti* of the Royal Sydney Yacht Squadron, and the second by *S.J.S,* of the St Kilda Yacht Club.[1542] The final race was won by *Heather.*[1543] Much to the surprise of those involved, this little craft from the Royal Prince Alfred Yacht Club had also finished runner-up in both the first and second races and was thus awarded the trophy by accumulation of the most points.[1544]

With the La Carabine Cup over, New South Wales and Victorian yachtsmen immediately turned their attention to the Sayonara Cup, to be played out a few days later on Hobsons Bay between *Rawhiti* and the local yacht *Sayonara*. Reeks remained in Melbourne during this period, not only representing the Royal Sydney Yacht Squadron but also acting as an official measurer, along with Francis Davis of the Royal Yacht Club of Victoria.[1545] In the days leading up to the start of the event, the latter role proved contentious. *S.J.S.*, which had recently competed in the La Carabine Cup, was found to have sailed under an approximate rating and needed to be remeasured.[1546] The new measurement calculated its rating at 27.3 instead of 28.[1547] The reduction completely altered the position of the boats in the La Carabine Cup series. The increase in time allowance for *S.J.S.* changed the position in which it finished in one of the races, to second instead of third.[1548]

[1541] *The Sydney Mail and New South Wales Advertiser*, 16 January 1907.
[1542] *The Sydney Mail and New South Wales Advertiser*, 16 January 1907.
[1543] *Sydney Sportsman*, 23 January 1907.
[1544] *The Sydney Mail and New South Wales Advertiser*, 16 January 1907.
[1545] *The Argus*, 17 January 1907.
[1546] *Sydney Sportsman*, 23 January 1907.
[1547] *The Sydney Morning Herald*, 18 January 1907.
[1548] *The Sydney Morning Herald*, 18 January 1907.

The ultimate result was that three yachts, namely *Heather*, *S.J.S.* and *Rawhiti*, had all finished with six points.[1549] The only possible solution was to reschedule the race. It was doubtful, however, if all three vessels would compete. While *Rawhiti* was about to duel for the Sayonara Cup, *Heather* had already been dismantled and was on a cradle at Williamstown awaiting shipment back to Sydney.[1550]

The Sayonara Cup series of races took place a few days later with the Victorian yacht *Sayonara* successfully defending the championship from the New South Wales contender *Rawhiti*. On being interviewed about the result, Reeks diplomatically stated that '*Sayonara scored on every point of sailing. She is unquestionable the better boat. The conditions were perfectly fair to both*'.[1551]

Ultimately, the final race for the La Carabine Cup was re-sailed on 22 January 1907 between *Rawhiti* and *S.J.S*, with the former declared the winner.[1552] The owner of *Heather* declined to participate.

MEASURING SAYONARA.
SOURCE: THE SYDNEY MAIL AND NEW SOUTH WALES ADVERTISER, 23 JANUARY 1907.

While Reeks was in Melbourne, his wife Gertrude and two children Kent and Thelma remained home in Sydney. With Kent now seventeen years of age and looking to follow in his father's footsteps as a naval architect, Reeks would have been disappointed to miss out on the annual exhibition of students' work in connection with the Sydney School of Mechanical Drawing and Correspondence Instruction, where Kent attended. The event was opened on 11 January 1907 at 369 Pitt Street and comprised architectural, geometrical and perspective drawings, as well as furniture designs and paintings. Kent had also notably received a prize for his mechanical drawings as part of recently held annual examinations.[1553]

Reeks returned to Sydney late in January and continued his official duties with the Royal Sydney Yacht Squadron, including acting as commodore at an elaborate garden party held with 400 guests that took place on the evening of 15 February at the squadron's clubhouse, Kirribilli Point.[1554] Reek's wife Gertrude was also present at the event.[1555] The couple, along with Reeks' daughter Thelma, now approaching 15 years of age, next attended the organisation's end of season garden party a few months later.[1556]

In terms of work commitments, during the early months of 1907 Reeks had many projects on hand. A new motor boat he had designed was launched from William Holmes' yard in mid-March.[1557] After having a 12 h.p. Thornycroft engine installed and satisfactorily trialled, the craft was shipped to Adelaide. It is very likely that this vessel was named *Thornycroft* and designed and built to the order of M. S. Hawker. He used the craft to compete in motor launch races out of Milang and Goolwa for several years.[1558]

Continuing to collaborate, Reeks designed an auxiliary ketch for use by the Queensland Kanaka Mission that was built by Holmes and launched on 1 May 1907. Intended for use in connection with the mission's work in the Solomon Islands, the 60ft craft was named *Evangel* and fitted with a 20 h.p. Gardner kerosene oil engine.[1559] Designed with a draft of 6ft, there was a saloon on deck and two sleeping berths. A two-berth cabin was located aft while below featured a trade room and additional accommodation. There was also a spacious hold for the carriage of stores and a poop deck. Owing to the prevailing light winds in the Solomons, top-masts were fitted to both masts, and *Evangel* carried a lifeboat in the davits. The vessel was later sold to a commercial operation and used to transport goods between Australia and islands in the Pacific under the new name of *Ruana*. It remained active until the later years of the Second World War after which time its battered hulk was left to deteriorate on a reef to the north of Vanuatu.[1560]

In addition, Reeks designed a sea-going auxiliary cruiser for Arthur Milson that Holmes was commissioned to build.[1561] The 35ft craft was in appearance similar to *Alcyone*, Dr Scot Skirving's vessel, though more yacht-like in general outline. It was launched as *Mischief* on 24 July 1907, to be fitted with a 10 h.p. White and Poppe engine.[1562] Of note, more than 115 years after its launch, the vessel is still sailing on Pittwater north of Sydney Harbour.

Likely making good use of his time in Melbourne earlier in the year, Reeks received a contract to design a 50ft motor launch for a Victorian-based client that was also built by Holmes. The craft was lightly constructed of cedar with two skins laid longitudinally of one-quarter

1549 *The Sydney Morning Herald*, 18 January 1907.
1550 *The Sydney Morning Herald*, 18 January 1907.
1551 *The Sydney Morning Herald*, 21 January 1907.
1552 *The Argus*, 23 January 1907.
1553 *The Sydney Morning Herald*, 12 January 1907.
1554 *Punch*, 21 February 1907.
1555 *Punch*, 21 February 1907.
1556 *Sunday Times*, 14 April 1907.
1557 *Referee*, 20 March 1907.
1558 *Southern Argus*, 26 December 1907, 4 January 1912; *Observer*, 2 January 1909.
1559 *The Sydney Morning Herald*, 10 April 1907; *The Daily Telegraph*, 2 May 1907; *Evening News*, 2 May 1907.
1560 *ABC Weekly*, Vol. 7, No. 9 (3 March 1945).
1561 *The Sydney Morning Herald*, 29 March 1907.
1562 *The Daily Telegraph*, 25 July 1907.

MISCHIEF.
SOURCE MORI FLAPAN COLLECTION.

inch thickness.[1563] It was fitted with a 55 h.p. Thornycroft engine located in the forward cabin.[1564] Steering could be effected either from the engine room or the deck, a telegraph was fitted from the steering wheel on deck to the motor room.[1565] Sharing a similar design to *Meteor,* the vessel was launched as *Earlie* in September of 1907, its new owner being R. C. Brownlow of Bairnsdale, Gippsland Lakes.[1566] Following fit-out and trials, which took several months, the vessel successfully made its way to Victoria, arriving in February of 1908 after making the 384 nautical mile trip in a time of 39 hours.[1567]

Definitely the most high-profile project, however, that Reeks was involved with during this period was Sam Hordern's 110ft motor yacht, deemed to be the '*largest and finest yacht yet seen in Australia*', as well as '*the finest thing in yacht building ever attempted in Australia*'.[1568] Reeks and Hordern, a prominent Sydney businessman, had worked together on several vessels over the years and enjoyed a very cordial friendship and professional relationship.

Built by Morrison and Sinclair at Balmain, its significance was such that the vessel's plans were published in multiple newspapers, along with substantial details of its specifications and fit-out. It is worthy of reproducing, not only in terms of the design features but also its versatility and lavishness.

' *We reproduce the architect's "general arrangement plans," a reference to which will show that the yacht will have a whaleboat bow and cruiser stern; she is high-wooded, having no less than 4ft 6in freeboard at the lowest place, and stands over 10ft out of the water at the stem. Her draft of water will be 5ft, thus enabling Mr. Hordern to do all parts of the Hawkesbury River and most of its tributaries, Port Stephens, and the Myall Lakes; at the same time she will have seagoing ability capable of extended cruises, for which purpose she is fitted with a lead keel. She will be schooner rigged, the masts being hinged to make possible the passage of Ryde Bridge, Parramatta River. She will carry three boats - a dingey, a gig, and a motor boat launch. Anchors will be hove up by an electrically drive capstan, to which also the boats' davit falls can be taken if required. Steering will be by hand power, and free choice of either the pilothouse on the main deck or the bridge overhead can be had always, the gears being coupled up. Telegraph gears (electric and mechanical to the main propelling motors is also in duplicate, handy to both steering wheels. The wash deck and fire hose will be electrically pumped from below, and there will be three steel water-tight bulkheads. There will also be accommodation ladders, that on the starboard side being an easy flight of steps, the other a vertical folding ladder. Awnings fore and aft and all over the bridge deck will be on portable stanchions easily removed.*

The accommodation consists of main saloon, music saloon, owner's stateroom, two other staterooms with two bunks each and one four-berth cabin, bathrooms, and lavatory appointments; and forward for the crew of four men. There will also be pantry and galley, forecastle and pilothouse, while the bunkers for cased petrol and kerosene are in a completely insulated, oil-tight, well-ventilated compartment at the other end of the deckhouse, in

[1563] *Referee*, 17 April, 15 May, 7 August 1907.
[1564] *Referee*, 17 April 1907; *Referee*, 2 October 1907.
[1565] *Referee*, 17 April 1907.
[1566] *Referee*, 17 April, 7 August 1907; *The Daily Telegraph*, 10 August 1907; *The Sydney Morning Herald*, 14 September 1907.
[1567] *Referee*, 12 February 1908.
[1568] *Australian Town and Country Journal*, 26 June 1907; *The Sydney Mail and New South Wales Advertiser*, 26 June 1907.

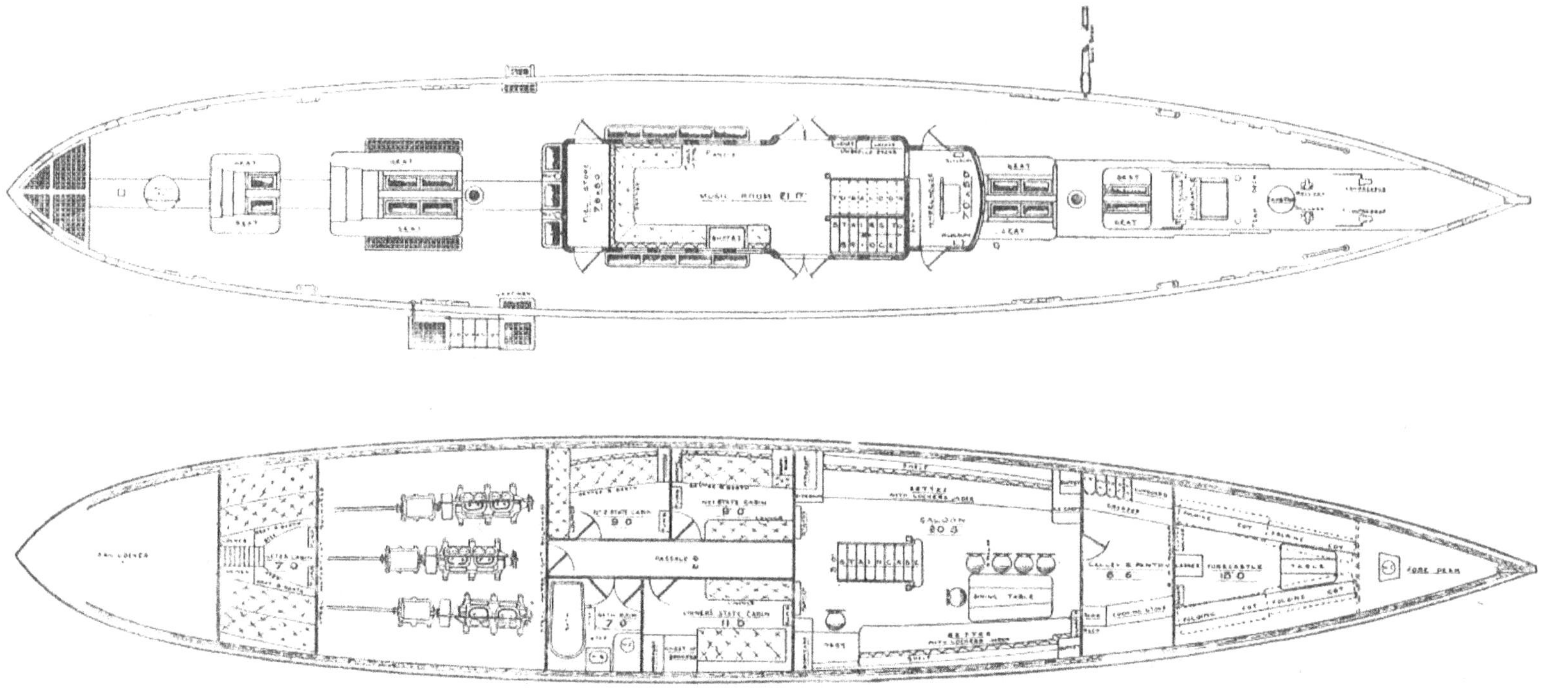

[Design by Walter Reeks

ACCOMMODATION PLAN.

110FT. TRIPLE-SCREW AUSTRALIAN BUILT MOTOR YACHT.

PLANS FOR SAM HORDERN'S 110FT MOTOR YACHT.
SOURCE: THE MOTOR BOAT, AUGUST 1907.

addition to two heavy copper tanks placed under the seats of the engine-room skylight.'[1569]

The article continued in much the same effervescence, describing in detail every room and every feature. It concluded with a note on its provenance and a statement from Reeks. '*Excepting for the engines and a few patented articles, Mr. Hordern has determined to make his yacht Australian by employing Australian skill and material whenever practicable. Mr. Reeks says she is one of the most interesting works that he has ever had to carry out. Her dimensions are: Length over all, 110ft; length on the load water line (L.W.L.), 100ft; beam, 17ft 6in; draft of water, 5ft; speed under full power, 14 knots. With tanks and stores full, she will have a range of action at full speed of 225 miles, at two-thirds power 300, at one-third power (or ordinary cruising speed) 500 miles; but for extended sea cruising double that quantity of fuel could easily be carried, and consequently double the distance run. With the centre engine running she will cruise at about 10 knots; with the two wing propellers, one right and the other left handed, she will, when required, turn in her own length.*'[1570] Another article highlighted Sam Hordern's specific brief to Reeks. '*With his later-day experience of yachting under new conditions, Mr. Hordern devoted much thought to plans for a motor craft adapted to Australian conditions and suitable for cruising on the harbour, the coast, and the rivers. He submitted his scheme to Mr. W. Reeks, with the terse instruction: "Here is the rough design of a yacht I want to build; put it into shape; build it in Australia; and specify the best that can be produced".*'[1571]

[1569] *Australian Town and Country Journal*, 26 June 1907.
[1570] *Australian Town and Country Journal*, 26 June 1907.
[1571] *The Sydney Morning Herald*, 26 June 1907.

Reeks certainly enjoyed the opportunity and freedom this project gave him. His sense of both practical and mechanical creativity, adaptability and ingenuity shined through in many aspects of the craft. Altogether, it was a revelation in completeness, size and lavish utility.

And so *Bronzewing* the '*World's largest motor yacht*', built at a cost rumoured to be between £10,000 and £12,000, was launched from Messrs Morrison and Sinclair's yard on 2 May 1908 without pomp and ceremony.[1572] Its new owner was overseas at the time and was instead represented by his son, Sam Hordern (Jr).[1573] Instead of the customary launching festivities, a monetary gift was given to all employees engaged in the vessel's construction. Less than a month later, by now fitted with its triple 108 h.p. Thornycroft engines, *Bronzewing* achieved a speed of 12 knots over the measured mile, the engines noted to be at three-quarter speed.[1574]

Concurrent with designing and supervising construction of *Bronzewing*, Reeks also designed the vessel's tender.[1575] The 20ft craft, fitted with a twin cylinder White and Poppe engine, was built by Pritchard Bros. Ltd. of Neutral Bay, its hull and seats made of cedar, its coamings of American ash, while its grating were of Canadian elm.[1576] All told, it was another display of sumptuous opulence combined with masterful design work.

[1572] *The Sydney Morning Herald*, 14 February 1908; *Australian Town and Country Journal*, 6 May 1908.
[1573] *Evening News*, 2 May 1908.
[1574] *Evening News*, 2 May 1908; *Australian Town and Country Journal*, 10 June 1908.
[1575] *The Sydney Morning Herald*, 9 March 1908.
[1576] *The Sydney Morning Herald*, 9 March 1908.

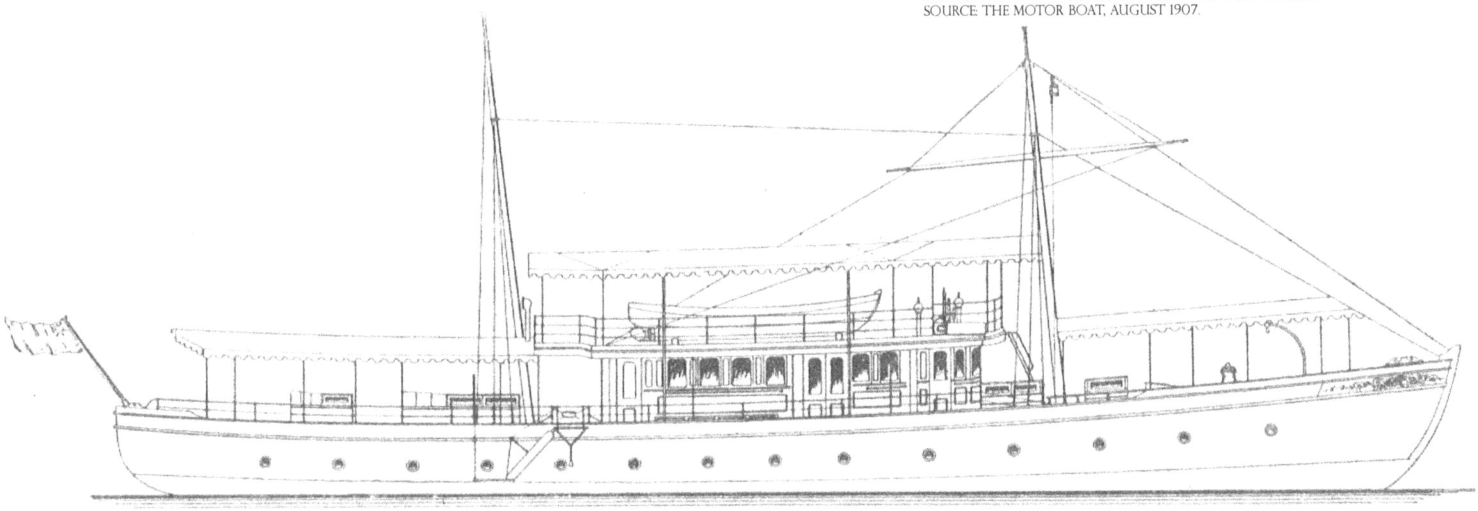

PLANS FOR SAM HORDERN'S 110FT MOTOR YACHT.
SOURCE: THE MOTOR BOAT, AUGUST 1907.

LAUNCH OF BRONZEWING ON 2 MAY 1908 AT BALMAIN.
SOURCE: AUSTRALIAN TOWN AND COUNTRY JOURNAL, 13 MAY 1908.

Sadly Sam Hordern spent little time enjoying his vessel. Just over a year following the launch of *Bronzewing,* he died at the age of 60 after a period of indifferent health.[1577] At the time of his death Hordern was '*easily the wealthiest man in Australia*'.[1578] He was also well liked. Reeks was one of 3,000 in attendance at Hordern's funeral which took place at his late residence, *Retford Hall,* Darling Point, on 15 August 1909.[1579] A week later Reeks made a sympathetic reference to the loss sustained by the Royal Sydney Yacht Squadron regarding Hordern's death.[1580] His prized vessel *Bronzewing* remained in the Hordern family's possession until 1919 when it was sold to George Sargeant and shortly thereafter onsold to J. Burke, along with its tender and dinghies, for around £3,000, by this time fitted with two 75 h.p. Jersey City Standard engines.[1581] *Bronzewing* was subsequently transferred to The Philippines where it was to be used as a pleasure craft in conjunction with Burke's business enterprises for the Philippine Manufacturing Company, based out of Manilla.[1582]

While the resplendent *Bronzewing* was under construction at Messrs Morrison and Sinclair of Balmain, another stately craft designed by Reeks was also being built at their yard. Commissioned by L. J. Davies, the 55ft motor yacht *Cambria* was a striking vessel, noted for its lines being '*most symmetrical*' and a '*very pretty model*'.[1583] Launched on 5 February 1908, after a build period of just ten weeks, the flush-decked craft, noted for its excellent accommodation, was installed with a twin-screw 25 h.p. Fairbanks.[1584]

By March of 1908 Reeks had moved his office to 9 Pitt Street. In addition to the many recreational vessels he was involved with at the time, government and commercial orders continued to take priority. Of particular note, Reeks soon became heavily involved in the design of a fishing trawler for the Australian government; a project that would likely cause him much frustration.

The first ruminations of the government's need for a fisheries trawler to assist in development of the country's

[1577] *Evening News*, 14 August 1909.
[1578] *Evening News*, 14 August 1909.
[1579] *Evening News*, 16 August 1909.
[1580] *Evening News*, 25 August 1909.
[1581] *The Sun*, 7 November 1913; *Evening News*, 4 November 1919; *The Sydney Morning Herald*, 14 January 1920; *The Daily Telegraph*, 23 June 1920.
[1582] *The Sydney Morning Herald*, 26 June 1920; *The Sun*, 4 March 1922.

[1583] *The Australian*, 4 February 1908; *The Daily Telegraph*, 27 February 1908; *The Sydney Mail and New South Wales Advertiser*, 11 March 1908.
[1584] *The Australian*, 4 February 1908; *The Daily Telegraph*, 27 February 1908; *The Sydney Mail and New South Wales Advertiser*, 11 March 1908.

CAMBRIA.
SOURCE: CHERRY JACKAMAN COLLECTION.

coastal fishing industry came at a conference that Reeks attended in early April of 1907 with representatives from the Australian and New South Wales governments present, including the Acting Minister for Customs, Mr Austin Chapman, as well as fisheries and other experts. With the intention of building a vessel for trawling purposes, Reeks was queried regarding the design of an up-to-date Norwegian trawler the group had recently received plans for, though made suitable for Australian waters.[1585] It was envisioned that if such a vessel was built and used to demonstrate that marketable fish could be caught off the Australian coastline, then state-managed and then commercially built and operated trawlers would soon follow.[1586]

A few weeks later, the design of the vessel was entrusted to Andrew Christie, a marine engineer of Sydney and member of the Institute of Naval Architects, with assistance from Reeks.[1587] Both would be paid the ordinary commission rate of 5 per cent on the cost of construction.[1588] This fee also covered supervision of its build. It was expected that the design would be finalised in a months' time. However it was soon found that a complete new design needed to be developed owing to the extensive alterations required to make the Norwegian trawler design adapted to Australian conditions.[1589]

With plans for the vessel finalised in mid-June of 1907, tenders were invited for its build.[1590] At this point in the process, criticism soon became voiced. One such article published in the Tasmanian press, copied from an article in the Melbourne-based newspaper *The Age*, opined, '*The new trawler, according to the plans, will be one of the most beautifully embellished steam yachts employed in Australian waters. It will supply accommodation for about 25 persons, including officers, crew, and passengers, and unless tanks are put on deck, will contain very little room for fish. The specifications set forth that, in certain particulars, the fittings must be equal to the best used in modern mail boats or first class railway carriages. The upholstery will be the most expensive obtainable. The specifications have been very clumsily drawn, but they tell their story none the less eloquently'*.[1591]

The article continued. '*When the hardy old sea dog who is captain of the little vessel returns to his cabin after a wild night, dripping with wet and reeking of fish, he will enter a boudoir described in the specifications thus: - "Furnishings of maple; polished enamel white walls; flat white overheads, gilt on beams, in narrow lines to approval. One iron bunk and frame of best quality, such as is used in modern mailboats. Best hair mattress, pillow, bolsters, sheets, blanks, quilts, rugs. Settee with lockers underneath divided in three. Cushions upholstered best hair moquet covers; and loose holland covers with piping red or white or blue to selections"*.'[1592] The minutiae of the detail continued for many more paragraphs, indicating the extent of time that had been invested in the internal design, layout, fit-out, furniture and decoration. Then there was a discussion of the price. '*Originally the amount placed upon the estimates to pay for the trawler was £8,000. On the estimates for 1907-1908 it is set down as £12,000, but even this sum will not nearly suffice. In Great Britain trawlers of the same kind are built for £8,000 or £9,000. The cost of the beautiful steam pleasure yacht planned here will amount to £15,000 or £20,000. Ministers assert that the vessel has been fashioned upon the simplest and most severely economical plan. But the contention that a fishing smack is incomplete without moquet upholstering, bevelled mirrors, oak panellings and decorative gildings will not be generally endorsed. Nor are these highly ornate accessories likely to contribute to the discovery of new fisheries.*'[1593]

1585 *The Sydney Morning Herald*, 2 April 1907; *The Daily Telegraph*, 2 April 1907.
1586 *The Daily Telegraph*, 2 April 1907.
1587 *The Daily Telegraph*, 27 April 1907.
1588 *The Herald*, 4 May 1907.
1589 *The Daily Telegraph*, 27 April 1907; *Examiner*, 30 April 1908.
1590 *The Sydney Morning Herald*, 14 June 1907.
1591 *Tasmanian News*, 26 August 1907.
1592 *Tasmanian News*, 26 August 1907.
1593 *Tasmanian News*, 26 August 1907.

Regardless of its detractors, the federal government's 'fishing smack' was completed and launched on 27 August 1908 from the Government Dockyard, Cockatoo Island.[1594] Named *Endeavour,* the 134ft vessel was constructed of steel to the 100 A1 class at Lloyd's and painted in Navy grey.[1595] With Reeks one of 300 people present to witness the event, as an official member of the party of guests he gave a brief speech stating simply that, '*The boat was one of his growing family - 213, he believed. It was always a pleasure to visit the dock, as the work was carried out in the most efficient manner - from the superintendent to the workmen*'.[1596]

The first trial trip of the *Endeavour* successfully took place in late December of 1908 with the vessel fitted with 450 h.p. triple-expansion engines giving it a range of 4,000 miles. The overall price tag of the craft was in the region of £15,000.[1597] Still, the vessel proved a worthwhile investment for the Commonwealth Government of Australia, spending several years investigating unexplored coastal fisheries grounds off the Australian coasts, as well as being used for valuable survey work and search and rescue operations. Sadly, the vessel and all of its passengers and crew went missing in early December of 1914 en route from Macquarie Island to Hobart after delivering stores to the meteorological station based on the island.[1598] After extensive searches, no trace was ever found of the vessel or its complement.

Another vessel built around the same time as the *Endeavour* that also caused some discourse with those involved was an oil launch the Launceston Marine Board commissioned for use at the Low Head Pilot Station. Expecting to spend around £3,000 to build the craft, the Marine Board's wardens had first tasked W. H. Jolly with developing a model. However, upon its submission, certain wardens urged for a little caution, expressing that '*Mr. Reeks of Sydney*' be communicated with and asked to supply a model and also give an estimate of the cost. Further support for the proposal came when it was realised that the two models could be compared, with the Launceston Marine Board's wardens thus able to get '*the best boat possible for the money*'.[1599] However, other wardens remarked that it was '*belittling the state to send to New South Wales for plans for such a small craft*'.[1600] Instead, it was eventually decided that prominent Launceston boat builder E. A. Jack be asked to supply a model and plans and that '*both be forwarded to Mr. Reeks to choose between them*'.[1601] A proviso was added that if neither met with his approval, '*Mr. Reeks is to be requested to forward a model most suitable*'.[1602]

A follow-up meeting took place a few weeks later where it was announced that E. A. Jack, upon being interviewed, stated that '*he was not going to send in a design to be judged by Mr. Reeks of Sydney'.*[1603] Further, '*Mr Jack has seen Mr Jolly's design, and thought it was a very good one. He did not see why they should send out of the state for one, and added that Mr. Reeks was not likely to adopt any design but his own*'.[1604] While several wardens continued to express support for sending the plans to Reeks, after more discussion, it was agreed that the plans for W. H. Jolly's model be sent to Reeks, asking his opinion about whether the boat would be suitable.[1605]

Responding to the Marine Board of Launceston's request to review the model of the proposed oil launch, Reeks stated that a consultation of the completed plans and specifications would cost a fee of £5 5s. However, if they only furnished him with a model and he then had to calculate the basis of the design, as well as displacement, etc., that an additional fee of £10 would be charged.[1606] The Marine Board soon agreed to send both the model

[1594] *The Sydney Morning Herald,* 27 August 1908; *The Australian Star*, 27 August 1908.
[1595] *The Sydney Morning Herald,* 27 August 1908.
[1596] *The Sydney Morning Herald,* 27 August 1908; *The Australian Star*, 27 August 1908.
[1597] *The Sydney Morning Herald,* 31 December 1908.
[1598] *The Mercury*, 6 January, 20 March, 21 April 1915.

[1599] *Examiner*, 9 July 1907.
[1600] *Examiner*, 9 July 1907.
[1601] *Examiner*, 9 July 1907.
[1602] *Examiner*, 9 July 1907.
[1603] *Examiner*, 23 July 1907.
[1604] *Examiner*, 23 July 1907.
[1605] *Examiner,* 23 July 1907.
[1606] *Daily Telegraph*, 6 August 1907.

FEDERAL TRAWLER ENDEAVOUR.
SOURCE: STATE LIBRARY OF VICTORIA (H91.250/1507).

and the plans to Sydney, as well as ask Reeks about recommendations for an oil engine. There followed a robust discussion amongst members regarding whether Reeks would ask for a fee for his time reviewing the engine, or if it would be included in the plans consultation.[1607]

Within a few weeks Reeks had reported back to the Marine Board with a few criticisms of the model and its specifications, the major point being an issue with displacement. Very true to character, however, he had not used the opportunity to propose a new design entirely that would be to his personal benefit. Reeks also recommended a Standard engine be fitted to the launch.[1608] At a later meeting C. F. Hutton, an engineer, reported that he had inspected various types of engines in Melbourne and Sydney and also met with Reeks to discuss options.[1609] Hutton then reported back that he recommended either a Frisco Standard or a Jersey City Standard.[1610]

With the Marine Board agreeing to Jolly's plans, with the tweaks recommended by Reeks, construction of the oil launch was put out for tender, along with the supply of an engine.[1611] The responses received for both, however, far exceeded the cost estimates that Reeks and Jolly had suggested. More meetings were held to discuss options with the wardens ultimately coming to the conclusion that the tenders be re-released.[1612] Still, there was disagreement; the end result being that the plans for the vessel were altered to an open rather than decked boat and a smaller engine installed, both changes thus reducing the costs substantially.[1613]

E. A. Jack then tendered to build the craft at a cost of £262 but this was still deemed too high such that the Board's own staff were entrusted with its construction.[1614] The half-decked vessel was thus built under the supervision of Jolly and launched at Launceston on 27 July 1908. It measured 40ft in length with a beam of 9ft and was fitted with a 16 h.p. Frisco Standard engine.[1615] Proving its utility, the vessel did its intended job so well at Low Head that it was not until mid-1925 that a larger and faster replacement was sought.[1616] The old pilot launch was subsequently advertised for sale, complete with its old engine.[1617]

While Reeks was in deliberations with the Marine Board of Launceston regarding designs for this oil launch, likely indicative of interactions he regularly had with other similar organisations throughout Australia in need of advice regarding new vessels, he was also entrenched with an abundance of other projects. In addition to those mentioned in the previous pages, the launch of another 'Lady class' of steamer that Reeks had designed for the Balmain New Ferry Company took place on 30 July 1907 at Messrs Morrison and Sinclair's Balmain shipyard.[1618] Named *Lady Carrington*, the 130ft vessel with only a 7ft 6in draft was specifically designed for the rapidly growing Lane Cove passenger trade.[1619] It was once again a double-ended steamer, similar in design to *Lady Rawson* and *Lady Northcote*, though was fitted with only one propeller; Reeks had clearly stood his ground in this instance and the directors of the Balmain New Ferry Company had full trust in his design outputs.

LADY CARRINGTON.
SOURCE: GRAEME ANDREWS WORKING HARBOUR PHOTOGRAPH COLLECTION, CITY OF SYDNEY ARCHIVES.

Another sea-going auxiliary launch was also designed by Reeks during this period. Built by William Goddard at Lavender Bay for J. T. McKittrick, well-known alderman and businessman of South Grafton, the 36ft vessel was completed in September of 1907.[1620] Named *Bosker*, it was fitted with a 16 h.p. Invincible engine and used recreationally by McKittrick, as well as occasionally employed in the Clarence River passenger trade. McKittrick also used the craft to investigate channels and inlets in the region with the prospect of having commercial steamers navigate them.

Re-elected vice-commodore of the Royal Sydney Yacht Squadron, as well as honorary measurer for the 1907/1908 season, Reeks was integral in implementation of the new international rating rule which the organisation adopted on 1 July 1907.[1621] Reeks and his wife Gertrude also continued to attend social engagements and events, with their presence often noted in the press. For example, the couple attended the Juvenile Plain and Fancy Dress Ball held at the Mosman Town Hall on 26 July 1907 in aid of the Royal North Shore Hospital.[1622] Reeks' daughter Thelma was also in attendance and won a prize for her costume.[1623] A few months later Reeks and his wife were amongst 400 guests who attended the Royal Sydney Yacht

1607 *Daily Telegraph*, 6 August 1907.
1608 *Daily Telegraph*, 3 September 1907.
1609 *Examiner*, 1 October 1907.
1610 *Examiner*, 1 October 1907.
1611 *Examiner*, 10 December 1907.
1612 *Daily Telegraph*, 4 February 1908.
1613 *Examiner*, 31 March 1908.
1614 *Examiner*, 17, 31 March 1908.
1615 *Examiner*, 29 July 1908.
1616 *Examiner*, 27 May 1925.
1617 *The Mercury*, 23 January 1926.
1618 *The Sydney Morning Herald*, 31 July 1907.
1619 *The Sydney Morning Herald*, 31 July 1907.
1620 *Evening News*, 5 August 1907; *Clarence and Richmond Examiner*, 12 October 1907.
1621 *The Australian Star*, 21 August 1907.
1622 *The Daily Telegraph*, 2 August 1907.
1623 *Mosman, Neutral and Middle Harbour Resident*, 10 August 1907.

Squadron's annual garden party held to mark the beginning of the yachting season.[1624] This was soon followed by a 'band evening' at the organisation's grounds at Kirribilli Point where Gertrude Reeks was noted as wearing a '*black lace dress with chantilly lace inlet and embroidered applique*'.[1625]

OFFICERS OF THE ROYAL SYDNEY YACHT SQUADRON, 1907/1908 SEASON, REEKS IS AT THE FRONT OF THE GROUP, LEANING ON THE STAIR POST.
SOURCE: THE SYDNEY MAIL AND NEW SOUTH WALES ADVERTISER, 30 OCTOBER 1907.

ROYAL SYDNEY YACHT SQUADRON GARDEN PARTY, OCTOBER 1907.
SOURCE: THE SYDNEY MAIL AND NEW SOUTH WALES ADVERTISER, 30 OCTOBER 1907.

[1624] *The Sunday Sun*, 27 October 1907.
[1625] *Sunday Times*, 24 November 1907.

Signifying his involvement with the organisation, on 23 November 1907 members of the Royal Sydney Yacht Squadron presented Reeks with his portrait. A statement read, '*Mr. Reeks' fame as an architect of boats of all kinds is known and appreciated through the Australian States and New Zealand. For the last 21 years Mr. Reeks has been honorary measurer to the Royal Sydney Yacht Squadron*'.[1626] Motor boating was still part of Reeks' interests, as well, with his son Kent also taking part in Motor Boat Club events and activities.[1627]

PORTRAIT OF WALTER REEKS PRESENTED TO HIM BY THE ROYAL SYDNEY YACHT SQUADRON.
SOURCE: THE SYDNEY MAIL AND NEW SOUTH WALES ADVERTISER, 30 OCTOBER 1907.

[1626] *The Arrow*, 23 November 1907.
[1627] *Referee*, 5 June 1907.

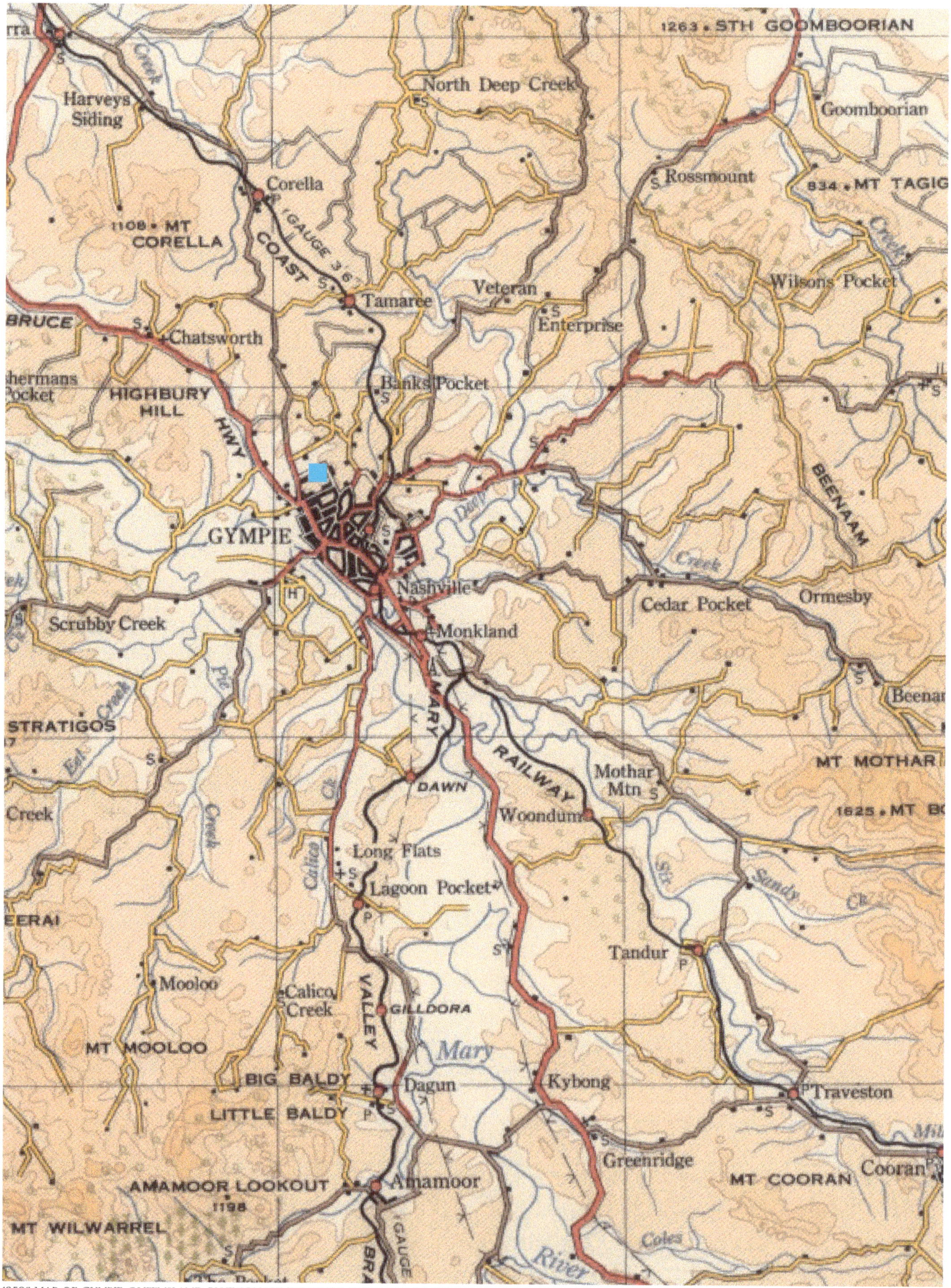

1950S MAP OF GYMPIE, QUEENSLAND. BLUE SQUARE INDICATES LIKELY SITE OF GREAT NORTHERN GOLD MINE.
SOURCE: NATIONAL LIBRARY OF AUSTRALIA (NLA.OBJ-234105259).

CHAPTER NINETEEN - GYMPIE

With such a whirlwind of professional, yachting and social commitments filling up his calendar, it's surprising that Reeks had any spare time to pursue other interests. But there appears to be one object that continued to plague him since decades before.

On 7 January 1908 a meeting of shareholders of the Great Northern Gold Mining Company, based at Gympie in Queensland, was held at Sydney's Aaron's Exchange Hotel, with Reeks notably occupying the chair.[1628] Upwards of 100 shareholders were present. The mine's manager, W. H. Reeve, was also in attendance, providing a brief outline of the work done since the company had commenced mining at a shaft sunk between Pine and Oak streets, to the north of Gympie, earlier in the previous year.[1629] Of note, this company was one of several established in 1906 to take advantage of a recently discovered gold bearing store in the area.[1630]

After studying geology at the Sydney Technical College, Reeve had relocated to Gympie in 1905 where he gained practical experience in the mining industry, as well as taught geology at the Gympie School of Arts and Mines Technical College.[1631] In May of 1906 he successfully applied for a gold mining lease on 50 acres of land in the 'Two Mile' region, just outside of Gympie, also requesting a six month exemption for the lease to allow for the raising of funds to float a company.[1632]

Situated 170km north of Brisbane, by this time Gympie had grown from a small mining and pastoral town to a city with a population approaching 12,000.[1633] Reeve was able to complete the float with a total of 23,000 shares held by shareholders living in Gympie and another 25,000 shares held by 69 shareholders based in Sydney.[1634]

With outputs from the Great Northern mine looking favourable, in February of 1908 Reeves and Reeks, along with William Morice, Charles Craig and C. A. Sussmilch, formed another company to mine at Gympie.[1635] The Peter and Paul Gold Mining Company was floated at Sydney with a capital of £15,000 in 60,000 shares worth 5s each.[1636] The mine was located adjacent to the Great Northern mine. V. L. Newberry was appointed its manager.[1637]

With his interest in gold mining still not satisfied, Reeks was soon involved in the establishment of another mining company, this time in Western Australia. A Memorandum for Registration of a No-Liability Company named The North-west Australian Mining and Exploration Company (No Liability) was published in the *Government Gazette of the State of New South Wales*, on 25 July 1908. The company's registered office was in Sydney and its intended operations in Western Australia. Its nominal capital was listed as £2,000 in 400 shares of £5 each. Reeks was one of seven founding directors.

While Reeks was investing in gold mining ventures in Queensland and Western Australia, closer to home companies continued to invest in his naval

1628 *The Sydney Morning Herald*, 7 January 1908.
1629 *Gympie Times and Mary River Mining Gazette*, 7 July 1906; *The Sydney Morning Herald*, 7 January 1908.
1630 *Gympie Times and Mary River Gazette*, 24 May 1906; *Australian Town and Country Journal*, 5 June 1907.
1631 *Gympie Times and Mary River Mining Gazette*, 6 April, 20 June, 8 July 1905; *The Daily Telegraph*, 11 September 1906; *Maryborough Chronicle, Wide Bay and Burnett Advertiser*, 30 May 1907.
1632 *Gympie Times and Mary River Mining Gazette*, 24 May 1907.
1633 *Maryborough Chronicle, Wide Bay and Burnett Advertiser*, 7 January 1905.
1634 *The Daily Telegraph*, 11 September 1906.
1635 *Australian Town and Country Journal*, 19 February 1908; *The Sunday Sun*, 18 April 1909.
1636 *The Daily Telegraph*, 9 January 1908.
1637 *The Daily Telegraph*, 7 March 1908.

architecture expertise. The directors of Sydney Ferries Ltd. commissioned him to design a new passenger steamer capable of accommodating 600 passengers and of maintaining a speed of 12 knots.[1638] Built by Messrs Morrison and Sinclair at Balmain, the *Kuranda* was launched on 9 December 1908 with principal measurements of 118ft (overall) with a beam of 27ft 3in and a moulded depth of 10ft. Its name was shortly thereafter changed to *Kaludah.*[1639] The vessel was fitted with triple expansion engines of 370 h.p.[1640] Four years later, the *Kamiri* would be launched by Messrs Morrison and Sinclair at Balmain for Sydney Ferries Ltd., built to the same plan as *Kaludah.*[1641]

Another repeat customer resulted in the construction of a new steam tug, designed by Reeks and built by Woodleys Limited, at Berrys Bay, for Messrs McIlwraith, McEacharn & Co. Pty Ltd. for use on the Swan River in Western Australia. The 70ft vessel, like the Reeks-designed *Zephyr* that was launched for the same company in 1906, was fitted out to negotiate the many bridges built over the Swan River. However, as another example of Reeks' open mind, there was a reversal of how the features reduced their height. This time the funnel was hinged and the telescopic part was the masthead.[1642] The steam tug was launched as the *Koori* on 25 March 1909.[1643]

Woodleys Limited had been established in May of 1908 as a successor to E. T. Woodley and Sons.[1644] The new enterprise had not only built the *Koori*, but in the months leading up to its launch, they had completed the cruising yacht *Arethusa*, designed by Reeks to the order of Dr J. Foreman. The 41ft yawl-rigged craft was launched on 24 February 1909.[1645] Another vessel built by Woodleys Limited to a design by Reeks was G. F. Bestie's 54ft motor cruiser *Modwena.*[1646] This particular craft was fitted with a 40 h.p. Brooke engine and launched in May of 1910.[1647] The latter was a big vessel, but also a simple, upright, handsome production of its day, neatly fitted out as a gentleman's cruiser with no pretensions to higher speed. It would satisfy many of us today, if we could afford something that size. Sadly though, it wasn't that long ago that Woodleys could have quoted on the job, but now they are no longer around to build it.

Additional vessels designed by Reeks in the years leading up to the First World War were a handsome auxiliary schooner with yacht-like features intended for cruising around the Solomon Islands that was built by Messrs Morrison and Sinclair at Balmain to the order of Captain Alexander Mackellar of Bundaberg, Queensland.[1648] Reeks also supervised the building of this particular craft with the 86ft ironbark vessel launched as the *Royal Endeavour* on 19 July of 1909.[1649] Sadly Mackellar died unexpectedly at Sydney in March of 1909 while likely viewing the vessel's construction.[1650] The order was taken over by Messrs Young Brothers, also of Bundaberg.[1651]

Designs for new passenger steamers continued to populate Reeks' desk. To ply the passenger trade between Port Melbourne and Williamstown on Port Phillip Bay, in late 1908 he was tasked with designing a steamer to the order of Messrs Baillieu and Patterson representing Melbourne Ferries Pty. Ltd.[1652] For reasons likely associated with the company's restructure, the single-ended vessel was not built for some time. In the interim, Melbourne Ferries Pty. Ltd. purchased the Reeks-designed steamer *Frederick Davis,* launched in 1907. Unfortunately the craft was completely wrecked on its delivery voyage from Sydney to Melbourne, all hands saved.[1653] The company then chartered the Gippsland Lakes steamer *Storm Bird* before their new 110ft vessel, named *Williamstown* and built by Rock Davis (Jr) at Brisbane Water, was completed and launched in October of 1909, successfully making its way to Port Phillip Bay a few months later.[1654]

WILLIAMSTOWN.
SOURCE: THE AUSTRALASIAN, 23 APRIL 1910.

Reeks also received two new orders from the Balmain New Ferry Company in 1909, marking the seventh and eighth vessels of the Lady class he had designed for this organisation.[1655] The first vessel was similar in design requirements to the *Lady Manning*, only larger at 112ft. Its principal route was to be the silted-up Lane Cove River, such that the shallow vee-hull cross sections Reeks employed were well suited. But inside that time span, Reeks had moved on from *Lady Manning*'s double-ender, cambered keel and twin propellers to his final iteration, a double-ender with a straight keel and a single propeller. It was constructed by Rock Davis (Jr) at Brisbane Water

[1638] *The Sydney Morning Herald*, 27 May 1908.
[1639] *Evening News*, 20 April 1909.
[1640] *The Daily Telegraph*, 10 December 1908.
[1641] *The Daily Telegraph*, 8 May 1912.
[1642] *The Sydney Morning Herald*, 29 October 1908.
[1643] *The Daily Telegraph*, 26 March 1909.
[1644] *Evening News*, 6 May 1908.
[1645] *Evening News*, 24 February 1909.
[1646] *Evening News*, 29 April 1910; *The Daily Telegraph*, 14 October 1912.
[1647] *Evening News*, 29 April 1910.
[1648] *The Sydney Morning Herald*, 16 January 1909.
[1649] *Evening News*, 20 April 1909.
[1650] *Maryborough Chronicle, Wide Bay and Burnett Advertiser*, 17 March 1909.
[1651] *Evening News*, 20 July 1909.
[1652] *The Argus*, 27 June 1908; *The Age*, 12 November 1908.
[1653] *Williamstown Chronicle*, 2 January 1909.
[1654] *The Age*, 15 March, 8 September 1909, 15 January 1910; *The Daily Telegraph*, 11 February 1910; *The Herald*, 19 April 1910.
[1655] *The Daily Telegraph*, 6 March 1909; *The Sydney Morning Herald*, 15 July 1909; *Evening News*, 21 January 1910.

and launched as the *Lady Chelmsford* on 14 April 1910. Its deckhouse fittings and machinery were subsequently installed in Sydney.[1656] As noted in the press, the steamer was intended for the Lane Cove service and designed to carry 500 passengers.[1657] What wasn't obvious but is something we all know now is that the *Lady Chelmsford* and its four subsequent sister-ships (soon to be recounted) would be Reeks' legacy long after he passed away. All were relatively close duplicates of the *Lady Chelmsford*, and sometimes the differences that crept in were just minor discrepancies due to the builder's preference on a detail or even his tape measure at the time. The vessels would outlive many of the other large contemporary ferries and many of their crews, operating on routes and undertaking events they were never intended for, and in locations well beyond Sydney Harbour. They were re-engined, re-painted, reinforced, re-built and even re-floated. It seems they could not be put down.

The second of these practically duplicate steamers was the *Lady Denman*. It was built at Jervis Bay by Joseph Dent and launched on 5 December 1911.[1658]

The Lady class continued on. Despite the company suffering reduced profits and higher wages, orders for two more vessel designs were soon placed with Reeks by the directors of the Balmain New Ferry Company.[1659] Both were double-enders with single screw and both were duplicates of the *Lady Chelmsford.* The former was built by George Washington at Balmain and launched as the *Lady Edeline* in October of 1913 while the latter was built by Dent's at Jervis Bay and launched less than three weeks later as the *Lady Scott*.[1660] The following year saw the completion of another similar vessel for the Balmain New Ferry Company. Built by David Drake at Balmain, it was only slightly smaller than its predecessors and launched as the *Lady Ferguson* on 25 July 1914.[1661] This craft represents the last vessel designed by Reeks for this company. Three years later the Balmain New Ferry Company went into administration, its demise precipitated by declining traffic combined with high fuel costs and labour strikes, primarily due to World War I, as well as other increased operating costs. The company and its assets were subsequently absorbed by Sydney Ferries Ltd. in late 1917.[1662]

Despite changes in administration and use, not just with these five ferries but many of the vessels Reeks designed for the Balmain New Ferry Company over his career, the bulk of these craft remained in service for several decades, including the pioneer of the fleet, *The Lady Mary,* launched in 1892 and in service until 1928.

While the *Lady Edeline* was in service until 1984, as we write this book the only surviving vessel of the Lady class of ferry is the *Lady Denman* which, after a mammoth community effort, including a stealthy tow south, is back home where it was built, preserved as part of a static display in its own gallery at the Jervis Bay Maritime Museum and Gallery.[1663]

1656 *The Daily Telegraph*, 15 July 1909; *Evening News*, 15 April 1910.
1657 *Evening News*, 21 January 1910.
1658 *Evening News*, 19 October 1911; *The Daily Telegraph*, 12 December 1911.
1659 *The Sun*, 26 February 1912; *The Daily Telegraph*, 2 November 1912.
1660 *The Daily Telegraph*, 20 October, 6 November 1913; *The Sydney Morning Herald*, 20 October 1913.
1661 *The Daily Telegraph*, 11 May 1914.
1662 *Evening News*, 2 November 1917.
1663 en.wikipedia.org/wiki/List_of_Sydney_Harbour_ferries; en.wikipedia.org/wiki/MV_Lady_Denman.

LADY EDELINE AND LADY FERGUSON AT BALMAIN FERRY BASE (1975).
SOURCE: GRAEME ANDREWS WORKING HARBOUR PHOTOGRAPH COLLECTION, CITY OF SYDNEY ARCHIVES.

GREYCLIFFE AT WATSONS BAY.
SOURCE: GRAEME ANDREWS WORKING HARBOUR PHOTOGRAPH COLLECTION, CITY OF SYDNEY ARCHIVES.

The Watsons Bay Ferry Company was another organisation to call on Reeks to design several new steamers for its harbour services in the years 1909 to 1913.[1664] Unlike the Lady class of ferries Reeks had designed for the Balmain New Ferry Company during this period, both craft were fitted with two propellers.

At 125ft in length, the first of these vessels was slightly longer than the *King Edward* but shorter than *Vaucluse,* the two steamers that Reeks had previously designed for the Watsons Bay Ferry Company and launched in 1902 and 1905, respectively.[1665] Intended to carry 600 passengers, with more outside seating accommodation than its predecessors, the vessel was built by David Drake at Balmain and launched as the *Greycliffe* on 17 October 1910.[1666] Unfortunately the craft would become famous for all of the wrong reasons.

Late in the afternoon of 3 November 1927, with 120 passengers on board, the *Greycliffe* and the Union Steamship Company mail steamer *Tahiti* collided. The former was cut in two and sank with the loss of 40 lives.[1667] The incident remains the deadliest incident to have occurred on Sydney Harbour.

Constructed along similar lines to *Greycliffe*, though slightly wider in the beam, the second Reeks-designed steamer built for the Watsons Bay Ferry Company was the *Carrara,* subsequently renamed *Woollahra*, launched from the Brisbane Water shipyard of Rock Davis (Jr) on 21 June 1913.[1668] Considered one of the most up-to-date vessels of its time, its purpose was two-fold: to meet demands for rapidly increasing traffic and to establish a service between Watsons Bay and Manly.[1669] Though the hull of the vessel was constructed by Davis (Jr), he sadly died two months after its launch, aged 49.[1670] The upperworks were then carried out by David Drake at Balmain with the new vessel entering service early in 1914.[1671]

Like the Balmain New Ferry Company, however, the longevity of the Watsons Bay Ferry Company was also fraught during the Great War. The organisation ultimately fell to the same fate as the former. It was taken over by Sydney Ferries Ltd. in 1920.[1672]

In our contemporary times it can be said that ferries have been a feature on the harbour for so long that it would be unthinkable not to see them operating. In this period of the early 20th century it was even more unthinkable, there were just so many of them, large and small, working to the ends of the harbour, from Manly to Parramatta.

1664 *The Sydney Morning Herald*, 29 October 1909.
1665 *The Sydney Morning Herald*, 29 October 1909.
1666 *The Sun*, 17 November 1910.
1667 en.wikipedia.org/wiki/Greycliffe_disaster.
1668 *The Sun*, 21 June 1913.
1669 *The Sydney Morning Herald*, 17 April 1912.
1670 *Australian Town and Country Journal*, 27 August 1913.
1671 *The Daily Telegraph*, 12 December 1913.
1672 *Smith's Weekly*, 12 June 1920.

They had their ups and downs in line with the economic flow, but were always in view; it was their patch, their heyday. They serviced both commercial and recreational passenger needs because the harbour and its waterways were still a principal highway for travel and a destination for recreation, and its foreshores and nearby areas were well established suburbs. There were trains leading inland to newer suburbs, and better roads with the beginnings of powered vehicular traffic. It was an exciting time in its own right, but to get there it was still often more efficient to go by water for all or some of the journey, and the big crossing from the city to the north side had yet to be bridged. However, already that crossing was starting to cast a growing shadow on this extraordinary ferry operation. Efforts to build that bridge had stalled for various reasons over previous decades, but in 1924 the momentum on this project built up again and, along with other progress, the harbour began to lose its grip. The opening of the famed 'coathanger' in 1932 was a milestone that brought great benefit to Sydney, but was also a collective body blow to the ferries. They were knocked down, but thankfully not knocked out.

With Reeks now considered one of, if not the leading expert on the design of ferries and other commercial craft for Sydney's waterways, he was routinely called upon to give interviews and advice regarding related topics and advances in technologies. In June of 1909, for example, he commented on the stability and adaptability of double-decked pontoons to enable the rapid embarkation and disembarkation of harbour ferry passengers, stating for the positive that '*they are the very thing required at Circular Quay*'. However, he then reiterated that '*the conversion of ferry boats to adapt them to dock berthing would mean an enormous expenditure - all the ferry boats would have to be altered at each end ... Plus the outlay by the Harbor Trust in transforming the wharves would be significant. But there is no doubt that the double-deck pontoons would facilitate matters should the ferry companies decide to adopt them*'.[1673] The ferries of Sydney were obviously doing their job well, reliably keeping to strict timetables such that the land-based infrastructure was proving a hinderance. Reeks was eager to offer a solution.

Also in June of 1909, at the general meeting of the Engineering Association of New South Wales, Reeks was elected to the committee.[1674] In addition he presented another paper, this one titled '*Notes on Some Modern Types of Steamship Construction*'.[1675] Here he discussed the use of iron in the construction of vessels, as opposed to wood, and the many changes in the methods of construction that needed to be made to accommodate this building material.[1676] Heralding the future, Reeks was obviously transitioning his own mindset to the use of metals in shipbuilding and was continuing his quest to be at the forefront of new knowledge and advances in technology. A few months later Reeks gave his opinion on marine turbines, a new American invention, stating that '*so far as he could see, there was nothing extravagant*' in the new design.[1677] This statement was indicative of Reeks' propensity to welcome new developments and not let them be over-complicated. The calibre of his notoriety amongst his peers within Sydney's engineering community resulted in Reeks being elected vice-president of the Engineering Association of New South Wales in late 1910 and re-elected to the same position the following year.[1678]

In April of 1911 Reeks gave an interview to the *Globe* regarding the status of shipbuilding in Australia lamenting the higher costs associated with building larger vessels here compared with those built in Great Britain.[1679] He added that the procurement of skilled labour was also a hurdle that needed to be overcome.[1680] Still, Reeks was optimistic. Trade was increasing and so to was the demand for new ships. He urged Australia to begin manufacturing its own steel instead of importing raw steel plates, engines and rivets from overseas.[1681] Reeks' closing statement in the article is visionary. '*Altogether I see a brilliant future for the industry. We are an island continent, and our growth depends on the volume of our overseas trade. Why should it not be carried in vessels made wholly in Australia. This trade, in its turn, will need protection. We have established the idea of an Australian owned, Australian manned Navy, and, having gone so far, why not keep the thing up and see that it is Australian built?*'.[1682] Little did he realise that a self-sufficient shipbuilding industry would become a reality in just a few short years, with World War I fuelling a boom in new vessels.

Another topic of interest for Reeks was the emerging use of oil over steam in shipbuilding, as well as the internal combustion engine. In May of 1912 he was quoted in the press as saying, '*The amount of oil required in a Diesel-engined boat is as one to five in weight compared with coal. Then there are the facts that even large combustion engines run very smoothly, and there is no smoke - two points which appeal to passengers. During the last few years tremendous strides have been made in the construction of motor vessels. Five years ago the largest internal combustion engine running was only 450-horsepower; now it is running into several thousands ... It remains for engineers to improve their construction as rapidly and successfully as they have done in the past, and fast passenger liners will follow as a natural sequence*'.[1683]

While Reeks' thoughts on the state of Australia's shipbuilding industry, the use of oil over coal, and the need to adapt to designs to accommodate advances in new technologies were likely echoed by the bulk of his professional cohort, one topic that did cause some questioning amongst his peers took place in June of 1912

1673 *The Star*, 9 June 1909.
1674 *The Daily Telegraph*, 15 October 1909.
1675 *The Daily Telegraph*, 10 June 1909.
1676 *Weekly Supplement to Building*, 15 June 1909.

1677 *The Daily Telegraph*, 6 October 1909.
1678 *The Daily Telegraph*, 15 October 1910; *Construction: Weekly Supplement to Building*, 16 October 1911.
1679 *Globe*, 12 April 1911.
1680 *Globe*, 12 April 1911.
1681 *Globe*, 12 April 1911.
1682 *Globe*, 12 April 1911.
1683 *Sunday Times*, 26 May 1912.

when he gave a presentation on '*The Use of the Spilling Staff*' to the Engineering Association of New South Wales, with many attendees questioning what exactly a spilling staff was. Soon it was revealed that 'spilling' was a method of measuring a curved shape or line with a straight edge as a datum, in order to shape a curved plank or frame to fit the hull.[1684] A specially-cut batten, i.e., a spilling staff, was often used in this process.

Reeks' propensity to fill a room and discuss such a topic shows his range of interests. He definitely grasped synthesizing and presenting technical and complicated concepts to disparate audiences, and enjoyed the resulting discussions, particularly within his professional circle. From the minute detail of a specific tool to the larger picture, Reeks was in his element no matter what the topic, relishing the debate, banter and camaraderie yet always responding in a thoughtful and courteous manner no matter which arc the discourse took.

Obviously having a lot of work on hand, other commercial craft designed by Reeks in the years leading up to World War I included the following vessels. For the Union Company he designed and supervised the building at Ford's yard of an open boat fitted with steam power capable of 10 knots.[1685] It was the largest open boat on Sydney Harbour at the time of launch in late 1911 and could carry 150 men.[1686] Also in 1911 Reeks designed a steamer for the Vacuum Oil Company to be used for towing barges and carrying cargo.[1687] Named *White Rose*, it was capable of towing a barge laden with 800 cases of oil, itself able to carry 600 cases.[1688] The *White Rose* was built at David Drake's yard in Balmain.[1689] For the same company, Reeks also designed a lighter capable of carrying 2,400 cases of oil.[1690]

Designs for a timber carrying steamer with a capacity of 450 tons were finalised by Reeks for a New Zealand-based customer in 1911.[1691] The substantial vessel was built by Rock Davis (Jr) and launched as the *Red Pine* on 26 April 1912.[1692] Measuring 153ft (overall), it was the largest vessel yet built in the Brisbane Water region of New South Wales and likely the cause of Reeks visiting the region in November of 1911.[1693] Discussions regarding this particular craft, commissioned by the Red Pine Steamship Company of Westland District, on the west coast of the South Island, may have resulted in Reeks travelling to New Zealand in late April/early May of 1909, returning to Sydney via stops in Hobart and Melbourne where he likely met with other clients.[1694]

Beginning trade as a coastal steamer operating out of Greymouth in late 1912, the *Red Pine* was completely wrecked a few months later on 7 March 1913, near Stephens Island.[1695] At the time the vessel was carrying over 850,000ft of timber and 40 tons of coal.[1696] Dense fog was blamed for the incident, all hands were saved.[1697]

Two steel lighters, each with a capacity 150 tons, were also designed by Reeks during this period for Elliott Bros. Ltd.[1698] Both were completed by early 1912 and were fitted with steam winches and cargo gear of the most up-to-date type for rapid handling.[1699]

Meanwhile, rivalry between Hobart's passenger ferry companies resulted in Reeks receiving two more orders from Tasmanian-based customers. One noteworthy craft was built at Hobart, this time at Battery Point by John Dalgleish for Messrs James Rowe and Sons and launched on 7 November 1912.[1700] As stated in a previous chapter, the 115ft *Excella* was the first of its type and size fitted with an internal combustion engine built anywhere in Australia. It was propelled by two Bolinder engines, each of 80 h.p., guaranteed to produce a speed of 12 knots. There was also the provision that if the passenger traffic increased in Hobart and the Huon and D'Entrecastreaux Channel region, where it was expected to ply, a third propeller could be added with very little alteration. Another feature was that the wheelhouse was located on the navigating bridge, above the promenade deck. As a consequence the vessel's master was able to have a commanding view ahead and on both sides; much needed on the sometimes shallow and narrow Huon River, that was also occasionally plagued by wooden logs and other debris.[1701] In addition to passengers, it was estimated that *Excella* could carry 6,000 cases of apples, a product by then proving a staple of the Tasmanian export market.[1702]

LAUNCH OF EXCELLA AT BATTERY POINT.
SOURCE: THE TASMANIAN MAIL, 14 NOVEMBER 1912.

1684 *The Sydney Morning Herald*, 14 June 1912;
Construction: Weekly Supplement to Building, 25 June 1912.
1685 *Evening News*, 19 October 1911.
1686 *Evening News*, 19 October 1911.
1687 *Evening News*, 19 October 1911.
1688 *Evening News*, 19 October 1911.
1689 *Evening News*, 19 October 1911; *Construction: Weekly Supplement to Building*, 30 October 1911.
1690 *Construction: Weekly Supplement to Building*, 30 October 1911.
1691 *Construction: Weekly Supplement to Building*, 30 October 1911.
1692 *Evening News*, 27 April 1912.
1693 *The Richmond River Herald and Northern Districts Advertiser*, 3 November 1911;
Evening News, 27 April 1912.
1694 *The Daily Telegraph*, 23 April 1909; *Southland Times*, 3 May 1909; *The Herald*, 8 May 1909.
1695 *Grey River Argus*, 10 March 1913.
1696 *Grey River Argus*, 10 March 1913.
1697 *Grey River Argus*, 10 March 1913.
1698 *Construction: Weekly Supplement to Building*, 30 October 1911;
Weekly Supplement to Building, 25 March, 29 April 1912.
1699 *Construction: Weekly Supplement to Building*, 25 March, 29 April 1912.
1700 *The Mercury*, 8 November 1912.
1701 *Huon Times*, 9 November 1912.
1702 *The Mercury*, 8 November 1912.

By mid-1912 Reeks had also nearly completed designs for a new ferry steamer for the Rosny Ferry Company, Ltd., of Hobart, largely based on the lines of the *Derwent* which he had designed for O'May Bros. in 1905.[1703] This vessel was at the time running the trans-Derwent passenger trade between Hobart and Bellerive. The client was, however, after a vessel with greater power and embracing more modern improvements. Built at Hobart's Domain shipyard, the craft would be launched the following year.

Additional vessels built to Reeks designs between 1911 and into early 1913 include the *Evangelist,* a 45ft cargo and passenger launch initially fitted with twin-screw Seal oil engines of 20 h.p. and fully equipped for long passages at sea, including passenger accommodation for six.[1704] The craft was designed with a view to providing the best possible ventilation for service in the tropics. It was built by William Holmes and launched in late August of 1912, intending to make its first trip to the Solomon Islands the following month. Due to engine failure, however, *Evangelist* limped back to Sydney where two new Fairbanks engines were installed.[1705]

Launched around the same time period was the auxiliary ketch *Miro II,* designed by Reeks and built by Holmes.[1706] The craft was fitted with 30 h.p. Wilson engines and left Sydney late in September 1912 for the Solomon Islands in charge of Captain McLauchlan.[1707] It was built for Julius Oien of Faisi, Solomon Islands, for use in conjunction with his plantation.[1708]

In mid-1912 Reeks also finalised plans to convert the three-masted auxiliary schooner *Novelty* into a full powered twin screw collier.[1709] The process was described as a '*naval shipbuilding experiment*' with '*an interesting transformation*'.[1710] The work was undertaken by David Drake at Balmain.[1711] The French steamer *St. Louis,* was also extensively repaired in 1912 under the direction of Reeks.[1712] That same year he acted as agent for the Northern Steamship Company, Ltd., Auckland, for altering and preparing the old French gunboat *Eure* to be towed across the Tasman Sea for use as a hulk.[1713]

Late 1912 coincided with Reeks finalising designs for two large towing punts for the Darling Island Stevedoring and Literage Co. Ltd., measuring 75ft each in length.[1714] The pair were built by Morrison and Sinclair and launched the following year from Balmain.[1715]

[1703] *Construction: Weekly Supplement to Building*, 26 August 1912.
[1704] SYDNEY, NEW SOUTH WALES, Register of ships, single number series (II), Volume 11, 1955-1982, ON 135521; *Construction: Weekly Supplement to Building*, 25 March, 29 April 1912; *The Daily Telegraph*, 13 September 1912.
[1705] SYDNEY, NEW SOUTH WALES, Register of ships, single number series (II), Volume 11, 1955-1982, ON 135521; *The Daily Telegraph*, 13 September 1912; *The Sydney Morning Herald*, 22 October 1912.
[1706] *The Sydney Morning Herald*, 24 September 1912.
[1707] *The Sydney Morning Herald*, 24 September 1912.
[1708] *The Sydney Morning Herald*, 24 September 1912.
[1709] *Construction: Weekly Supplement to Building*, 26 August 1912.
[1710] *The Sydney Morning Herald*, 1 November 1912.
[1711] *The Sydney Morning Herald*, 1 November 1912.
[1712] *The Daily Telegraph*, 4 November 1912.
[1713] *The Sydney Morning Herald*, 14 November 1912.
[1714] *Construction: Weekly Supplement to Building*, 26 August 1912; *The Daily Telegraph*, 31 August 1912.
[1715] *Construction: Weekly Supplement to Building*, 28 January 1913.

Commissions for new recreational vessel designs also came across Reeks' draft board. In mid-1910 the 54ft oil launch *Lotus II* was completed, built by Watty Ford (Jr) to the order of Dr J. F. Elliott of the wholesale drug and chemical manufacturing firm Elliott Bros., to a design by Reeks.[1716] In late November of 1911 a fine auxiliary yacht measuring 31ft in length with a 10ft beam and 4ft draft was launched to the order of Messrs Young and Manson.[1717] Designed by Reeks and built by Holmes, the vessel was stated to be '*something on the lines of a Scotch fishing craft*'.[1718] It was fitted with a 12 h.p. Worsley engine and named *Shona.*

A sister-ship to the Reeks-designed *Arethusa* named *Landseer III* was launched from Woodley's yard on 12 October 1912. The auxiliary yacht was built to the order of Albyn Stewart, managing director of Hydraulic Engineering and Hercules Limited.[1719] The 42ft craft was fitted with a 20 h.p. Frisco Standard engine.[1720] In addition, Reeks designed *Wyruna*, an auxiliary yacht for H. H. Dutton of South Australia.[1721] It was built by Holmes at McMahons Point and launched early in 1913.[1722] Another vessel constructed during this period was *Gypsy*, a 41ft cruising yawl scaled down by Alfred Blore of Hobart from a design by Reeks that had been published in *Rudder* magazine in 1911, though it included a centreboard that was not a feature of the original design. Built of Huon pine planking on a combination of Huon pine ribs and hardwood frames, *Gypsy* was launched in 1914 from Norfolk Bay, Tasmania.

LOTUS II.
SOURCE: AUSTRALIAN NATIONAL MARITIME MUSEUM COLLECTION.

Call it coincidence but all five of the craft just noted still exist as we compile this narrative. *Lotus II*, a stylish development moving on from *Modwena*, is still holding

[1716] SYDNEY, NEW SOUTH WALES, Register of ships, single number series (II), Volume 11, 1955-1982, ON 125243.
[1717] *Evening News*, 24 October 1911; *The Sydney Morning Herald*, 4 December 1911.
[1718] *Evening News*, 24 October 1911; *The Sydney Morning Herald*, 4 December 1911.
[1719] *The Sun*, 13 October 1912.
[1720] *The Sun*, 13 October 1912.
[1721] *Construction: Weekly Supplement to Building*, 26 August 1912.
[1722] *Construction: Weekly Supplement to Building*, 28 January 1913.

on into the twenty-first century with its vertical stem, neat cruiser stern, slight rake aft to the cabin panels, on a hull with slim, elegant proportions attune with the times. *Modwena* was upright and gentlemanly while *Lotus II* was a bit slinky, even with the exposed helm position that kept repeating on Reeks' launches and other powered vessels. The vessel is waiting on restoration in Sydney.

Landseer III and then *Wyruna* were what we now call 'wholesome cruisers': their profile was more than just pleasing, with overhangs that were not overstated, plenty of beam, a long deep keel with sufficient ballast, and not over rigged. Roomy inside, sea-kindly in their hull sections, these boats were well-suited to coastal cruising in safety, on the offshore passages undertaken between quiet anchorages. *Landseer III* was restored some years ago at Woodley's yard, whilst *Wyruna* is back in the hands of a family who once owned it, undergoing restoration in Victoria.

Shona is a little gem and always has been, likely characteristics that have kept it intact more than 110 years after its launch. All that has changed is the Bermudan rather than gaff rig. A small cruiser for Sydney Harbour and Pittwater, and with nothing pretentious, it's the choice of a canoe stern that was the feature that raised *Shona* above the rest. Perhaps influenced by the exceptional Scottish and Norwegian naval architect Colin Archer, the owners and Reeks gave the craft a tubby rounded double-ended shape with a stern hung rudder, indicative of Archer to a 'T'. The trick was getting all of the curves to be fluid as they blend together, the hardest one being the deck in plan view. No hard spots here as it moves from the bow to the middle with an even curvature, holds onto it past the midsection briefly then looks to gather more curvature evenly as it folds into an almost rounded stern from the cockpit aft. Looking back, *Shona* was a precious moment in Reeks' career, a chance to flirt a little, and back then no one was probably looking too closely as this small vessel sailed happily under the radar dominated by bigger things.

SHONA.
SOURCE: AUSTRALIAN NATIONAL MARITIME MUSEUM COLLECTION.

Of course, yachting as a pastime continued to take up the bulk of Reeks' spare time, as well as some of his professional activities. Between the years 1908 and 1910 he was again appointed vice-commodore and honorary measurer of the Royal Sydney Yacht Squadron; Sir James Fairfax was commodore.[1723] Reeks stepped down as vice-commodore in August of 1910, but remained on the committee, as well as the squadron's honorary measurer up until his death.[1724] He also continued to be part of the Anniversary Regatta's organising committee during much of this period.[1725]

Perhaps the most noteworthy event to occur in many decades regarding Reeks' sailing career was the surprise selection of the recently-resurrected *Thelma* to challenge for the Sayonara Cup to be held in January of 1909. By now nearly 20 years of age, the vessel was chosen to represent New South Wales despite more modern yachts deemed equally worthy of being up to the task, including A. C. Saxton's *Awanui*. It was a somewhat controversial decision, though multiple factors proved advantageous for *Thelma*. These included the vessel's ability to sail to Melbourne in all kinds of weather; its stability and seaworthiness were deemed superior to other craft.[1726] The races themselves were also to take place on Hobsons Bay, known for heavy winds and corresponding seas, i.e., conditions near-equal to an ocean course.

In direct comparison to *Awanui, Thelma* rated larger and carried more sail. Further, the decision was made in haste. Reeks explained, '*The Victorians would not give us time to hold the test races, and insisted upon us nominating the challenger by November 4. We sent along the necessary deposit of 50 a fortnight ago, and told them we would name the Sydney representative as soon as we held test races. The Thelma's new sails would not arrive until November, and it would be some few weeks afterwards before the tests were decided. The Victorians refused to extend the time, however, and because Thelma is considered the best sea-boat, and to have best chance of winning, we nominated her at once*'.[1727] Still, despite these justifications, it was a bold move.[1728]

In the weeks leading up to *Thelma*'s selection, Reeks had been busy helping get the craft race-ready. One important event that likely tipped the scales in its favour came during the first race of the Sydney yachting season, held on 24 October 1909. *Thelma* was matched against *Awanui* and other newer craft for the first time.[1729] With the press grandiosely stating that it was '*hardly probable*' that the former could make for a competitive race in its current rig, with Reeks at the helm *Thelma* won by a mere 57 seconds from its rival *Awanui*.[1730] The fact that it was under cruising canvas and in want of a stiffer breeze gave further merit to *Thelma*'s performance.

1723 *Sydney Sportsman*, 26 August 1908; *Evening News*, 25 August 1909.
1724 *Evening News*, 29 August 1910.
1725 *The Australian Star*, 27 October 1908; *Evening News*, 26 January 1912.
1726 *The Australian Star*, 31 October 1908.
1727 *The Australian Star*, 31 October 1908.
1728 *The Australian Star*, 31 October 1908.
1729 *The Sydney Morning Herald*, 24 October 1909.
1730 *The Sydney Morning Herald*, 24 October 1909; *Sunday Times*, 25 October 1909.

With additional wins under its belt on Sydney Harbour, and carrying a suite of new sails, *Thelma* made its way to Port Phillip Bay, leaving Sydney on 15 January 1909.[1731] Its owner C. Lloyd Jones was on board, as were Gil Henty, Walter Moore, L. Dawson, Jack Evans, J. Mulganon and Captain Feast (navigator).[1732] Reeks, along with several other crew, was not part of the complement. Instead he opted to remain in Sydney finalising last minute details and attending official trials of the federal government's fishing trawler *Endeavour.*[1733] Reeks did, however, accompany the yacht to Sydney Heads on board *Wahine,* as part of several spectator craft and well-wishers.[1734]

The voyage of *Thelma* to Melbourne was uneventful, the vessel and its crew arrived safely at Williamstown on 21 January, slipping quietly into moorings off the Royal Yacht Club of Victoria's sheds. Its competitor *Sayonara* was notably moored only a short distance away, giving interested parties a chance to eye-ball the two vessels and compare their various merits. The Melbourne press commented, '*yachtsmen who go down to Williamstown expecting to find in the Sydney boat the big overhang and long counter that marks the modern design will get something of a shock, for Thelma is straight in the stem, with a good deal of freeboard, her sheer ending with that pronounced lift of the counter that is so characteristic of the work of her designer, Mr. Walter Reeks, who will also take the helm in the Cup Race. A copper bottom, which shows little or no sign of a long ocean voyage, and port holes on her sides give Thelma more than ever all the appearance of a cruiser, and not that of a racer, to which her serviceable yawl rig at present all helps to lend colour*'.[1735]

In the days leading up to the first race of the Sayonara Cup, a comparison of the two vessels was published in newspapers throughout the country, including a very useful diagram showing the craft side by side in their cutter racing rigs. This was a good method to show the clear differences but also depict those elements where they were quite close.

These two elegant yachts were initially designed to different requirements and time periods relative to design fashions and rule developments. *Thelma,* however, had been previously modified by Reeks to improve its performance. Once if not twice, the craft was altered by lowering the ballast keel to deepen the draft, improving its stability and performance. Having been designed to the Sail Area and Waterline Rule of the late 1880s, the greater draft had given *Thelma* some of the desirable attributes of the Linear Rule that superseded the earlier rule in the late 1890s. Meanwhile, *Sayonara* had not been designed to any rule. Fife had prepared the plan as a cruising yacht for himself, which his firm then built, and *Sayonara* was the second yacht built from that plan, but constructed in South Australia.

What the diagram highlights first is the relative closeness of their sail areas: *Thelma* at 2,071 square ft was only 70 square ft less than *Sayonara* at 2,144 square ft, a minor percentage but enough to give *Thelma* a slight advantage in stronger winds, while *Sayonara* held the better cards for lighter weather. *Thelma* was also 3ft longer on the waterline, a very helpful advantage to a potential higher speed but this was then tempered by the increase in wetted surface area and subsequent added drag. However, *Thelma*'s draft was perhaps the biggest point of difference. The bulk of its lead ballast keel was placed lower than the raked keel on *Sayonara*, so *Thelma*'s stability, especially in stronger winds was going to be to its advantage.

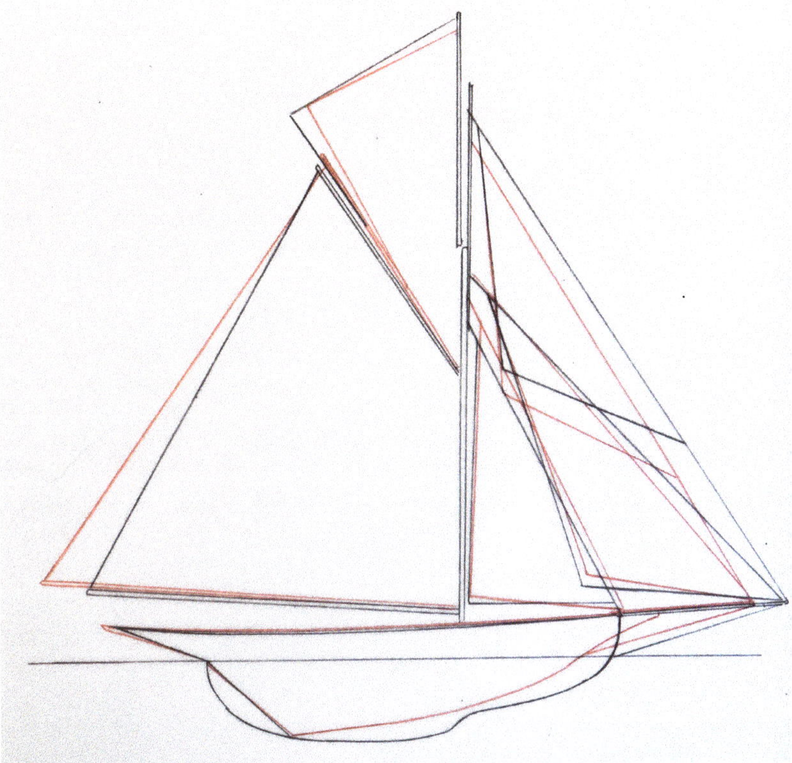

FORE AND AFT SECTION OF SAYONARA (RED LINES) VS THELMA (BLACK LINES).
SOURCE: ADAPTED FROM THE SYDNEY MAIL AND NEW SOUTH WALES ADVERTISER, 27 JANUARY 1909.

As they came to the starting line, the two craft were remarkably evenly matched despite their age and shape differences. However, yachts were now intended to be rated to the new Universal Rule, causing a potential further complication. Neither *Thelma* nor *Sayonara* had been designed to this rule, but their slightly different ratings, if measured to the rule's requirements, were both approximately equal to an International 12 metre. To avoid what was developing into a tiny handicapping difference, by mutual agreement the matches were to be sailed without any time allowance. This may well have suited Reeks, recalling that when he was interested in challenging for the America's Cup 20 years earlier, he was keen then to meet the defender on an equal basis with no time allowance required.

Perhaps the event was a chance for Reeks to make up for the disappointment of not proceeding with that America's Cup boat. The Sayonara Cup very closely matched all the design and racing arrangements for that world class event,

[1731] *The Sydney Morning Herald*, 7 December 1908.
[1732] *The Australian Star*, 15 January 1909; *Referee*, 20 January 1909; *The Argus*, 22 January 1909.
[1733] *The Australian Star*, 23 January 1909.
[1734] *Referee*, 20 January 1909.
[1735] *The Argus*, 22 January 1909.

but on a smaller, regional scale, and Reeks was going to race his design against a yacht designed by a man at the very top of the profession, William Fife (Jr). It is tempting to think that given he had probably not featured a lot in yacht racing in recent years compared to his earlier period, Reeks grabbed the chance to come out of retirement and live his dream.

With the event about to start, Reeks and the remaining members of *Thelma*'s crew travelled to Melbourne as passengers on board SS *Morea*.[1736] The group left Sydney on 30 January 1909 arriving two days later.[1737] With little time to prepare and become acclimated to both the vessel and the conditions, the first race of the Sayonara Cup took place on 5 February over a 10-mile course, starting off Williamstown. With great interest not only from those in Melbourne and Sydney but across the country, it was a duel that lived up to expectations. In the first race, *Thelma* gallantly came from behind to claim victory by a mere minute.[1738] The breeze was a strong one, which suited *Thelma,* and it was noted it had the better of things going to windward which was not what the Victorians expected. The second race, on a triangular course and in a light breeze, saw the tables turned with *Sayonara* finishing first, making a third race necessary.[1739]

With much anticipation, and on the same course as the first, the final race for the 1909 Sayonara Cup was held on 8 February. Stronger winds returned to favour *Thelma* and on the first beat the craft held its own and then went better, taking the lead, but misunderstood where the rounding mark was placed. *Sayonara* was able to round the mark first and then maintain its lead to the finish, ahead by only 56 seconds. Despite the mix up with the mark location, it was claimed to be '*one of the best races ever sailed on Hobsons Bay*'.[1740]

Disappointed by the outcome, Reeks remained stoic, thanking his crew with the remark, '*No skipper ever had a better lot of men in a boat than those working with me*'.[1741] At a celebratory dinner shared the following night between the two crews, Reeks responded to a toast thanking members of the Royal Victorian Yacht Club for their kindness and courtesies adding, '*In another way it would be idle of me to say we are not disappointed. We are, and I think you will agree we have occasion to be. We nearly lifted that Cup. But when we failed we shook hands with you, and we are grateful to you for the fight you put up to retain it. And now we congratulate you. At the same time, we'll come down again, and try to beat you, at the first opportunity*'.[1742]

Arriving back in Sydney a few days later, Reeks was met by the press. Upon being questioned about losing by the barest of margins he stated, '*Thelma did everything she was asked. She never sailed better, and I have no fault to find with her at all. She did not make a single mistake, and as for the crew they were absolutely without fault, and worked like one man*'.[1743] He added, '*Of course Thelma is 20 years old, but she surprised everybody*'.[1744] When asked if he would design another yacht to contest for the Sayonara Cup, Reeks was adamant with a '*No*', clarifying, '*I don't think the next challenger will be Sydney designed, at least, not as far as I am concerned. I've spent about five months getting the Thelma into trim, and the call on my time is too great*'.[1745]

CREW OF THELMA, WALTER REEKS SHOWN FRONT AND CENTRE.
SOURCE: WEEKLY TIMES, 13 FEBRUARY 1909.

Officially welcomed home by a series of dinners and events, Reeks and his family continued their involvement with Sydney's sailing community. While he was in Melbourne, his wife Gertrude was a guest on board the flagship SS *Morea* during the 1909 Anniversary Regatta, held late in January of that year.[1746] Reeks also presented prizes as part of the Royal Sydney Yacht Squadron's Empire Day dinner a few months later.[1747]

Another interesting sailing-related project involving Reeks occurred when well-known Sydney yachtsman Walter M. Marks commissioned Alfred Mylne of Scotland to design a 6-metre yacht for racing on Sydney waters, as well as challenging Victoria for the cup presented by the late Governor General, Lord Northcote, for interstate racing.[1748] The boat was built at the Berrys Bay yard of Watty Ford (Jr) under the supervision of Reeks and launched as *Culwulla II* on 27 October 1909.[1749] While the vessel failed to win the Northcote Cup under Marks' ownership, it did win the title in 1913 as *Rawhiti II* and

[1736] *The Arrow*, 30 January 1909.
[1737] *The Daily Telegraph*, 3 February 1909.
[1738] *The Sydney Morning Herald*, 6 February 1909.
[1739] *The Daily Telegraph*, 8 February 1909.
[1740] *The Argus*, 9 February 1909.
[1741] *The Sydney Morning Herald*, 8 February 1909.
[1742] *The Argus*, 10 February 1909.
[1743] *The Australian Star*, 12 February 1909.
[1744] *The Australian Star*, 12 February 1909.
[1745] *The Australian Star*, 12 February 1909.
[1746] *Sunday Times*, 31 January 1909.
[1747] *Evening News*, 26 May 1909.
[1748] *The Australian Star*, 9 March 1909.
[1749] *The Daily Telegraph*, 28 October 1909.

proved a successful yacht of its type under several different owners and additional name changes.[1750]

Keenly involved in interstate yachting and with the funds to support this pastime, Marks purchased the 10-metre yacht *Awanui* a month after the launch of *Culwulla II*, renaming the former *Culwulla III*.[1751] It was in this iteration that the vessel was sent to Hobsons Bay in February of 1910 to compete for the Sayonara Cup. Reeks also travelled to Melbourne to watch the series of races take place, likely enjoying the opportunity to be a spectator rather than out on the water.[1752] It proved an eventful contest between *Culwulla III* and *Sayonara* with the Sydney yacht ultimately declared the winner following a ruling by the Yacht Racing Association of Great Britain regarding time allowances that took six months to be finalised.[1753]

SAYONARA (LEFT) AND CULWULLA II (RIGHT) RACING FOR THE SAYONARA CUP.
SOURCE: WESTERN MAIL, 2 APRIL 1910.

Back in Sydney Reeks, in his role as vice-commodore of the Royal Sydney Yacht Squadron, presided over a dinner attended by 60 yachtsmen to welcome home Walter Marks and the crew of *Culwulla III*.[1754] That same month Reeks was given carte blanche to alter the New Zealand-built yacht *Rawhiti* on behalf of its new owners, E. E. Sayer and Frank Albert.[1755] Reeks set about adding lead to its keel and altering the sail plan. The improvements were a success; the craft went on to win the Sir Rupert Cup, the Jenny Trophy, the first leg of the Rawhiti Cup and first prize at the 1911 Anniversary Regatta.[1756] Later that year Reeks was noted to be back at the helm of *Thelma*, sailing in races held under the auspices of the Royal Sydney Yacht Squadron.[1757]

In March of 1912 *Thelma* won the Rawson Cup handicap. No-one was more delighted than Reeks, the press noting, '*He looked just as he felt, pleased beyond measure*'.[1758] During this same period Reeks was involved in organising races for the Sydney Amateur Club, as well as in discussions regarding establishment of a 12-metre class.[1759]

LEFT TO RIGHT: WALTER REEKS, S. M. DEMPSTER, WALTER M. MARKS, A. G. MILSON, A. J. MILSON AND NORMAN H. MURRAY DISCUSSING THE PROPOSED 12-METRE CLASS.
SOURCE: THE SYDNEY MAIL AND NEW SOUTH WALES ADVERTISER, 11 OCTOBER 1911.

Reeks continued to attend social events associated with the Royal Sydney Yacht Squadron. The opening of the 1909/1910 season, for example, coincided with a garden party held on the club's grounds at Kirribilli Point. As vice-commodore, Reeks presided over the festivities owing to Sir James Fairfax's absence. The press noted that his wife Gertrude wore a '*grey crepe-de-chine with embossed filet lace and net; grey chip hat with grey feathers and swathings of grey tulle*'.[1760] Reek's daughter Thelma was also an attendee.[1761] The trio would continue to patronise this event, held annually, in the years to come.

Throughout this period Reeks remained involved with the Engineering Association of New South Wales. In October of 1910, as part of his role as the organisation's vice-president, he travelled to Lithgow, about 150km west of Sydney.[1762] Here Reeks, along with sixty other Association members, paid a visit to the Lithgow ironworks and blast furnace.[1763] Only established a few years prior, it was Australia's first commercially-viable steel mill.[1764] Reeks was re-elected vice-president of the Engineering Association at the organisation's annual meeting in both October of 1911 and 1912.[1765]

With regards to other professional and social engagements, there were many for both Reeks and his family. A large tea party held in the Winter Garden at the Hotel Australia in early June of 1908 saw Gertrude Reeks in attendance.[1766] The event was staged as a welcome home for Mrs William Baldwin, who had just returned from overseas. A few months later Reeks attended the

1750 *Evening News*, 19 July 1910; *The Sydney Morning Herald*, 11 January 1932.
1751 *The Star*, 15 December 1909; *Referee*, 29 December 1909.
1752 *Evening News*, 12 March 1910.
1753 *Evening News*, 19 July 1910.
1754 *The Sydney Morning Herald*, 13 April 1910.
1755 *The Daily Telegraph*, 19 April 1911.
1756 *The Daily Telegraph*, 19 April 1911.
1757 *The Sydney Morning Herald*, 30 October 1911.
1758 *Evening News*, 5 March 1912.
1759 *The Sydney Mail and New South Wales Advertiser*, 11 October 1911; *The Sun*, 12 March 1912.
1760 *Sunday Times*, 10 October 1909.
1761 *Australian Town and Country Journal*, 13 October 1909.
1762 *The Daily Telegraph*, 13 October 1911.
1763 *Lithgow Mercury*, 28 October 1910.
1764 en.wikipedia.org/wiki/Lithgow,_New_South_Wales.
1765 *The Sydney Morning Herald*, 11 October 1912.
1766 *Sunday Times*, 7 June 1908.

first levee of the Governor General, the Earl of Dudley, at Government House, held on 2 October 1908.[1767]

In August of 1909 Reeks was elected vice-president of the Mosman District Cricket Club at the organisation's first annual meeting.[1768] With Allan Border now serving as its patron, the club proudly remains in existence over 100 years later.

In November of 1909 Reeks and his wife attended a dance and 'At Home' held by the commander of RMS *Otranto* where Gertrude was noted as wearing a dress of '*black crepe-de-chine and Chantilly lace*'.[1769] Late in December of that same year Reeks, his wife and daughter were amongst 250 guests to spend an evening on board the White Star training ship *Mersey*, moored at Millers Point.[1770] Based out of Liverpool, the three-masted vessel was employed as a training craft for approximately 60 cadets of the mercantile marine and was on its second voyage from England.[1771]

WHITE STAR TRAINING SHIP MERSEY.
SOURCE: STATE LIBRARY OF VICTORIA (H91.250/1489).

During April of 1910 Reeks and a party of guests on board *Wahine* were noted at Windsor on the Hawkesbury River enjoying a pleasant outing.[1772] A few months later Thelma Reeks attended the fourth annual Daffodil Dance at Baumann's Cafe in Pitt Street where she wore a white silk dress.[1773]

The early months of 1912 saw Reeks invited to lunch on board the Orient Company's RMS *Orama* at the invitation of the company's General Manager, David Anderson.[1774] Reeks and his wife also attended another party on the training ship *Mersey*, with over 250 other guests.[1775] '*Mr and Mrs Walter Reeks*' were next noted at a reception given by the Lord and Lady Mayoress of Sydney, Sir Allan and Lady Taylor, held in early April 1912 in honour of the Governor General and Lady Denman at the Town Hall.[1776] Reeks, his wife Gertrude and daughter Thelma then attended the Mosman Matrons' Ball at Warringah Hall, Neutral Bay, on 21 June 1912 where Gertrude's dress was stated to be of *'black crepe de chine, veiled in black ninon'*.[1777] A few months later Reeks was noted as attending the first private dance held in the new Rose Bay Hall as part of a farewell in honour of George Bestie's son who was travelling to England to attend Harrow College.[1778] Later that year '*Mr and Mrs Reeks and Miss Reeks*' attended an 'At Home' given by Captain Lewis and the officers of HMAS *Tingira* where '*the lads of the Australian training ship and members of the naval cadets gave a programme of athletic events at Lyne Park*'.[1779]

This juncture in time is a good opportunity to make mention of Reeks' wife Gertrude and her involvement in miniature rifle shooting. As a member of the Sydney Ladies' Miniature Rifle Club, she is first noted competing in the sport in November of 1908.[1780] Gertrude continued to attend weekly practice matches, held at the Randwick Range, during the following few years.[1781] Obviously a keen competitor and talented shooter, in 1911 she also began competing for New South Wales in interstate competitions.[1782]

MEMBERS OF THE SYDNEY LADIES' MINIATURE RIFLE CLUB,
GERTRUDE REEKS VERY LIKELY INCLUDED,
SOURCE: THE SYDNEY MAIL AND NEW SOUTH WALES ADVERTISER, 22 NOVEMBER 1911.

In March of 1912 Gertrude was elected to the committee of the Sydney Ladies' Miniature Rifle Club.[1783]

1767 *The Daily Telegraph*, 3 October 1908.
1768 *The Daily Telegraph*, 7 August 1909.
1769 *Sunday Times*, 28 November 1909.
1770 *Evening News*, 1 January 1910.
1771 *Evening News*, 1 January 1910.
1772 *Windsor and Richmond Gazette*, 16 April 1910.
1773 *Sunday Times*, 31 July 1910.
1774 *The Sun*, 4 January 1912.
1775 *Sunday Times*, 18 February 1912.
1776 *Sunday Times*, 7 April 1912.
1777 *The Sun*, 22 June 1912; *The Daily Telegraph*, 27 June 1912.
1778 *The Sun*, 1 September 1912.
1779 *Sunday Times*, 24 November 1912.
1780 *The Daily Telegraph*, 4 November 1908.
1781 *The Sydney Morning Herald*, 9 July 1909, 2 March 1911; *Sunday Times*, 7 November 1909.
1782 *Evening News*, 15 September 1911.
1783 *The Sun*, 6 March 1912.

Later that year she gave an interview to the press for an article titled, '*The Woman Behind The Gun. Lady Rifle Shots. Better Than The Men.*' The piece began, '*Through a haze of smoke the members of the Sydney Ladies' Miniature Rifle Club were found taking part in the first prize meeting of the Miniature Rifle Association last week. The basement of the Town Hall was the scene of operation, and the echo of the firing and the din made by a staff of men piling hundred of chairs close by suggesting the coming of the end of the earth. It didn't appear to trouble the fair Amazons, however, for they lay stretched upon mattresses side by side with their male opponents, and pumped lead at the tiny figure of a man appearing at the further end of the great room, as if death-dealing was their one aim*'. With the scene set, the article continued, '*Mrs. Reeks, another competitor in the various events, thinks shooting a most interesting pastime, and she remarks laughingly, "When you can handle a gun it gets you quite a lot of respect." This lady is an enthusiast, and from her it was learned that the Sydney Club is small as regards membership, but is big in courage. They have twice met and suffered defeat at the hands of the Melbourne lady shooters, but they are still ready to do further battle to recover their lost laurels. The club arranged annual matches on its own account, and a weekly practice, which is well attended, takes place each Tuesday at Randwick. Mrs Reeks will not admit any desire to shoulder a rifle if war came to Australia, but it is probably that she would prove a valuable sniper to her side if such an undesirable event came about*'.[1784]

Before we end this chapter, we must report on what became of Reeks' gold mining ventures. As previously stated, he had invested in two operations based at Gympie in Queensland: The Great Northern Gold Mining Company Limited and the Peter and Paul (Gympie) Gold Mining Company. Given he was a director of both companies, Reeks likely travelled to the region numerous times, perhaps by train from Sydney, to view operations and gain more perspective. While in the years immediately after their establishment, new leases were approved and shafts developed, both companies failed to prove long-term successes, likely a combination of lack of meaningful yields and loss of staff during the Great War. Following voluntary liquidation, the assets of both organisations were jointly auctioned off in August of 1918.[1785]

Also not proving a worthwhile investment was the North-west Australian Mining and Exploration Company (No Liability). After its registration in July of 1908, the company appears not to have been established as no further records of its existence can be found.

Persevering, Reeks continued to invest in other ventures and opportunities. On 31 May 1908 he was listed as a signature to the registration of the Australian Direct Supply Mineral Water Company, Limited. The business was stated to have a capital of £20,000 in £1 shares and intended to deal in mineral and aerated waters and other liquids. In addition to Reeks, signatures to the company's establishment were Arthur West, A. G. Milson, K. C. McCormick, A. J. Milson, W. L. Gardiner and S. W. Phillips. Renamed the Tads Mineral Water Co. Ltd. in 1912, the company established a factory at Redfern and continued to operate until it went into voluntary liquidation in October of 1920.[1786]

Another company Reeks was involved with was the Willunga Bangor Slate Quarries, Limited, based in South Australia. He was listed as a director in its initial prospectus, published in January of 1911.[1787] The enterprise proposed to acquire 26 acres of land for quarries at Willunga, south of Adelaide, with the slates reported to be of '*high quality, and have been tested by leading slate merchants, builders, and architects throughout the Commonwealth, many of who are said to have offered contracts for large consignments at highly-remunerative prices*'.[1788] The prospectus continued, '*The promoters estimate that their output should amount to 30,000 slates per week, the selling price of which at Port Adelaide should be £7 10s per thousand - a total return of £225 per week, while they consider that the cost of production should not exceed £115 per week, leaving an estimated profit of £110 per week. The company is being floated with a capital of £30,000, in 300,000 shares of 2s each. The directors are Messrs W. P. Noller, builder and contractors, Linden Court; Walter Reeks, naval architect, 15 Pitt Street; and W. G. Parish, investor. The secretary of the company is Mr. B. J. McDougall, and the registered offices are at Imperial Buildings, 16 O'Connell Street*'.[1789]

In the five months after the company's prospectus was published, the Willunga Bangor Slate Quarries, Limited, purchased land and began production. Its first load of roofing slate was ready by July of 1911. However, a scarcity of labour hindered expansion, as did the onset of World War I.[1790] It too went in to voluntary administration in August of 1917.[1791]

While Reeks was certainly not that successful at investing in companies with long and prosperous futures, one company he did hold shares in was the Sydney Ferries Ltd. In April of 1911 he was noted as owning 415 shares; the business remained in operation until 1951.[1792] If only he had invested more of his wealth in maritime-related ventures that were closer to home.

Showing his persistence for new and emerging technologies, in 1912 Reeks was named as a director of Meteor Igniter, Ltd., a new company with a capital of £30,000 in £100 shares that was established to acquire a sparking invention from J. G. Newman.[1793] The fate of this company was also not long-lasting. A series of law suits between shareholders over false representations saw it dissolved less than two years after it was originally floated.

1784 *The Sun*, 17 September 1912.

1785 *Gympie Times and Mary River Mining Gazette*, 23 May, 10 August 1918.

1786 *The Sydney Morning Herald*, 27 September 1912; *Government Gazette of the State of New South Wales*, 1 October 1920.

1787 *The Sun*, 30 January 1911.

1788 *The Sun*, 30 January 1911.

1789 *The Sun*, 30 January 1911.

1790 *The Advertiser*, 26 July 1911.

1791 *Government Gazette of the State of New South Wales*, 31 August 1917.

1792 *The Sun*, 7 April 1911.

1793 *The Daily Telegraph*, 31 October 1912.

MENU FROM SANDY'S DINNER.
SOURCE: FOX ASHWIN FAMILY COLLECTION.

CHAPTER TWENTY - EUROPE

A period of somewhat strange occurrences to absolutely devastating events began in early 1913 for Walter Reeks. Though there were successes to celebrate in terms of family unions, new vessel designs, including those mentioned in the previous chapter, new business enterprises and new technological developments, overall it was the start of a melancholic interlude for Reeks from which he never fully recovered. More hardship was just around the corner. World War I, which began in Europe in July of 1914, would have a profound impact on daily life, work and recreation for many years, well after the armistice of 11 November 1918.

For Reeks, the first such event involved his beloved yacht *Thelma*, the vessel which he had previously described as the '*flower of the flock*'.[1794] On 8 March 1913, just a few weeks after his 52nd birthday, a private sailing match took place between *Thelma* and *Sayonara*. Neither vessel was participating in club events that day as there were no races scheduled for the 10-metre class. With the consent of the squadron's vice-commodore Arthur J. Milson, the vessels' respective owners, Charles Lloyd Jones and Paul Ross, had arranged to sail for a private trophy over the Manly pile course using the Royal Sydney Yacht Squadron's start and finish lines.[1795] Obviously it was not the first occasion that these two yachts had battled for bragging rights out on the water. In 1909, with Reeks at the helm, *Thelma* had narrowly lost to *Sayonara*, representing Victoria, to claim the coveted Sayonara Cup by two wins to one on Port Phillip Bay. It was to be an interesting duel. Both yachts were getting on in years, however; *Thelma* having been built in 1889 and *Sayonara* in 1897.

Early in the week leading up to the race, on 4 March, a private dinner was held for *Thelma*'s crew to farewell Sandy Ross who was heading to Scotland on a visit back to his homeland. The dinner had many courses, including '*Poulet en Casserolle*' and the perhaps already infamous '*Peche Melba*'. It was clearly a happy occasion judging by the amusing notes on the menu, a copy of which has survived in the Reeks' family collection, and has the signatures of Reeks, Charles Lloyd Jones, and most of *Thelma*'s crew, including Walter Moore. The race had also been noted in the *Sydney Morning Herald* that same day, but no one was to know the tragedy that was just around the corner, as all present were probably thinking of the race in very different terms of expectations.

The race was sailed on Saturday 8 March with a south-easterly blowing to gale force, and driving rain. *Thelma* had a crew of ten, most if not all familiar with the yacht, but Reeks was not aboard. Its owner, a skilled sailor himself, was at the helm. Walter Moore was on the mainsheet, a position he had taken many times. The ensuing tragic story as it unfolds largely comes from the following Monday's *Sydney Morning Herald*, published on 10 March, where they found a balance between reporting the serious drama and showing genuine empathy for the tragic consequences.

'The story bristles with incidents of heroic self-sacrifice … The inevitable happened. The man who was washed overboard was lost, and the nine other members of the crew came so near to losing their lives as to make the circumstances of their rescue seem like a deliverance from death … Both rescued and rescuers agree that those who manned the whaleboat took their lives in their hands, for the sake of those who had been carried so close to death's door'.

[1794] *The Anchor*, 22 February 1912.
[1795] *The Sydney Morning Herald*, 4 March 1913.

The yachts had started off the Royal Sydney Yacht Squadron's clubhouse across from Neutral Bay with topmasts housed and then made a fast passage to Manly. Coming onto the wind to beat up the harbour they faced the full force of the gale, '*water continually came aboard, making the exercise of great care necessary*'.[1796] *Thelma* lost its aft hatch cover, so a sail bag was fitted as a makeshift alternative. '*Sayonara, having carried away the lacing of her boom, went into Middle Harbour and found shelter in Pearl Bay.*'[1797]

If Pearl Bay was indeed its sanctuary, then the seas and storm must have been ferocious. It is located a mile inland from the main harbour, behind the Spit, and indicates that even Balmoral and the lee side of Middle Head were untenable in the conditions.

Thelma remained sailing. The vessel was on port tack and on course to clear Middle Head, when a wave broke and came across the deck, cockpit and aft deck where Walter Moore was tending to the temporary cover. It carried him into the sea. '*Man-overboard*' was called immediately and the instant reaction was to tack the boat onto starboard in order to return to where he had been lost. After completing the head to wind tack, it then involved bearing away in huge seas to sail briefly on a northerly heading before going into a gybe before the wind to bring *Thelma* up close to the spot where Moore was likely to be. The all-standing gybe was always going to be a huge risk in these conditions and as they tried to come through the gybe, the mast failed.

'*The mast snapped below the deck and fell to starboard breaking away part of the deck. Almost immediately the breeze caught the sail again and lifted it with the mast and boom right across the yacht and dropped it over the port side. Fortunately, the boom lifted high enough to clear the crew, but only by a narrow margin. It was found impossible to clear away the wreckage, and the yacht drifted towards Dobroyd. Seas were breaking aboard and finding their way below, which added to the anxiety of those on board and pumping was necessary to keep the vessel afloat.*'[1798]

This appears to describe a situation where the mast failed as the yacht was almost dead before the wind but turning to port. Instead of completing that turn, the mast and rig, which had fallen on the starboard side, then brought the yacht back around to a starboard heading abeam of the wind, whereupon the wind picked up the rig and blew it over the yacht. An extraordinary, dramatic and dangerous sequence.

The situation was now perilous for all of *Thelma*'s crew, but this was not immediately apparent. This much was reported in the late afternoon paper, *The Sun* on 8 March. '*The harbor yacht race sailed this afternoon between those two yachts, Thelma and Sayonara, had an exciting ending. When the contest started, a whole gale was blowing from the south-east, accompanied by heavy rain squalls. The yachts got a severe dusting early in the contest, and before the course had been completed Thelma's mast carried away, and she was soon at the mercy of the wind. She was soon close to the rocks at North Harbor, and got foul of the rocks. It is not known to what extent she suffered. Sayonara sought shelter in Middle Harbor.*'

It now became a matter of rescuing *Thelma*'s crew and the pilot steamer SS *Buranda* came to their aid. According to the book '*Sydney Sails: The Story of the Royal Sydney Yacht Squadron's First 100 years (1862-1962)*', they had sighted the disabled *Thelma*, but were not to know Moore was lost overboard. The acting master realised he was too short-handed to undertake a rescue and chose to go at full steam to the station at Watsons Bay and embark the master Captain Sweet and additional crew. They returned to *Thelma*, by now drifting close to the dangerous bombora, a submerged rock at Dobroyd Head. In those seas the wave break would have been many metres high and incredibly powerful.

The report in the *Sydney Morning Herald* on 10 March continued with the dramatic rescue. '*A whaleboat was lowered with a line attached which was payed out as she approached Thelma, who with two anchors out, was now almost in the breakers at Dobroyd. When the whaleboat got alongside it was decided to take the crew off in one trip rather than risk a second one, and although extremely dangerous, was safely managed.*

The yachtsmen realised the imminent danger of their position and stripped off all superfluous clothing in anticipation of a struggle in the water for life. The spinnaker boom was cut in halves, with bowlines attached in order to give as much chance as possible of getting ashore at perhaps the most dangerous part of the harbour. Everyone kept calm under trying circumstances'.

The youngest member of the crew, Jack Mulganon, had told the *Herald* reporter, '*the facts of the accident recorded above*' and then continued with the following. '*The Thelma was a splendid yacht. If it had not been for Mr Moore being washed overboard we would have been absolutely safe, and would have got through the gale for sure without mishap. It was the fiercest gale I have ever experienced. We were about 800 yards off Middle Head when the cry rang out "Man overboard". I do not know who cried out. It was at once decided that even though we might risk our lives we should try to save him. We had a lifebuoy and line ready to throw as soon as we sighted him. Our skipper, Mr Jones, managed the yacht wonderfully. It is impossible to speak too highly of the acting pilot steamer Buranda. The way the coxswain managed the whaleboat in that sea was wonderful. When rescued by them we were within 20 or 30 yards of the breakers at Dobroyd, and they risked their lives to come to our aid.*'[1799]

Mulganon then corrected statements that had been made about *Thelma* capsizing, saying it was practically impossible for the yacht to capsize, and the crew were never in the water at all. But for two and a half hours '*they were a sorry plight, and were hanging on to a helpless boat, expecting death almost every minute, being completely at the mercy of the wind and waves*'.[1800]

[1796] *The Sydney Morning Herald*, 10 March 1913.
[1797] *The Sydney Morning Herald*, 10 March 1913.
[1798] *The Sydney Morning Herald*, 10 March 1913.
[1799] *The Sydney Morning Herald*, 10 March 1913.
[1800] *The Sydney Morning Herald*, 10 March 1913.

Walter Moore was paid tribute in the same report, noting his prominent membership of the Royal Prince Alfred Yacht Club, along with his extensive and respected crewing on many of Sydney's yachts, including *Thelma, Cullwulla III* and *Sayonara*. Moore had been a popular employee as a draughtsman with the Lands Department, and, most tragically, he left a wife and five children from 2 to 12 years of age.

The next day the full drama was apparent and as the week progressed, alongside continued searches on the harbour for Moore, where only traces of the boat as wreckage were recovered, a joint committee of the yacht and sailing clubs was immediately formed to raise funds for his family.

Jack Mulgonan came across some spare sails and bits of rope and rigging, another find a few days later was on Dee Why beach, a short distance up the coast, where parts of the mast, yard and mainsail were washed up.[1801] On March 28 it was reported that *Thelma*'s lead keel was found, '*The keel was located in 4ft of water off Grotto Point, about half a mile away from the spot where the Thelma was left when the crew was rescued from her. A search of the spot revealed no trace of the hull of the yacht, and so far only the leaden keel has been recovered*'.[1802] Moore's body was never recovered. He had been wearing a heavy oilskin over his full clothing and would have succumbed quite quickly in the fierce conditions.

[1801] *Evening News*, 12 March 1913; *Daily Telegraph*, 13 March 1913.
[1802] *The Sydney Morning Herald*, 28 March 1913.

As this was all unfolding, the Royal Sydney Yacht Squadron's committee began to accept subscriptions to support Moore's widow and five children, and as noted in *Sydney Sails*, it was '*subscribed to generously*'.[1803] The book also notes that the Royal Sydney Yacht Squadron added a '*monetary reward for the three men who manned the lifeboat*' in recognition of their bravery. The club also presented Captain Sweet, master of *Burnada*, with an illuminated certificate. Much later in 1936, the *Australian Boating Annual* published an article by Charles Lloyd Jones, in which he recounts some of the tragedy, in a story that comes back to the dinner. '*I remember as we were all sitting on the wind and rain swept deck ... and there seemed nothing in front of us but foundering and being dashed against the rocks. Sandy said to me, "Skipper, if we all get drowned do ye think the steamship company will refund the passage to me wife?" None but a Scotchman would think of that at such a time.*'

Somewhere, perhaps just stepped back in the shadows a little in all of this was Walter Reeks. His reactions and thoughts remain unreported or recorded, and we are left to imagine the grief and concern he must have experienced losing a close friend in a sailing accident, compounded by the loss of his wonderful yacht *Thelma*, which had brought him much success and satisfaction for over 20 years, and still had more to give.

[1803] *The Sydney Morning Herald*, 15 March 1913; P. R. Stephensen (1962). *Sydney Sails: The Story of the Royal Sydney Yacht Squadron's First 100 years (1862-1962)*.

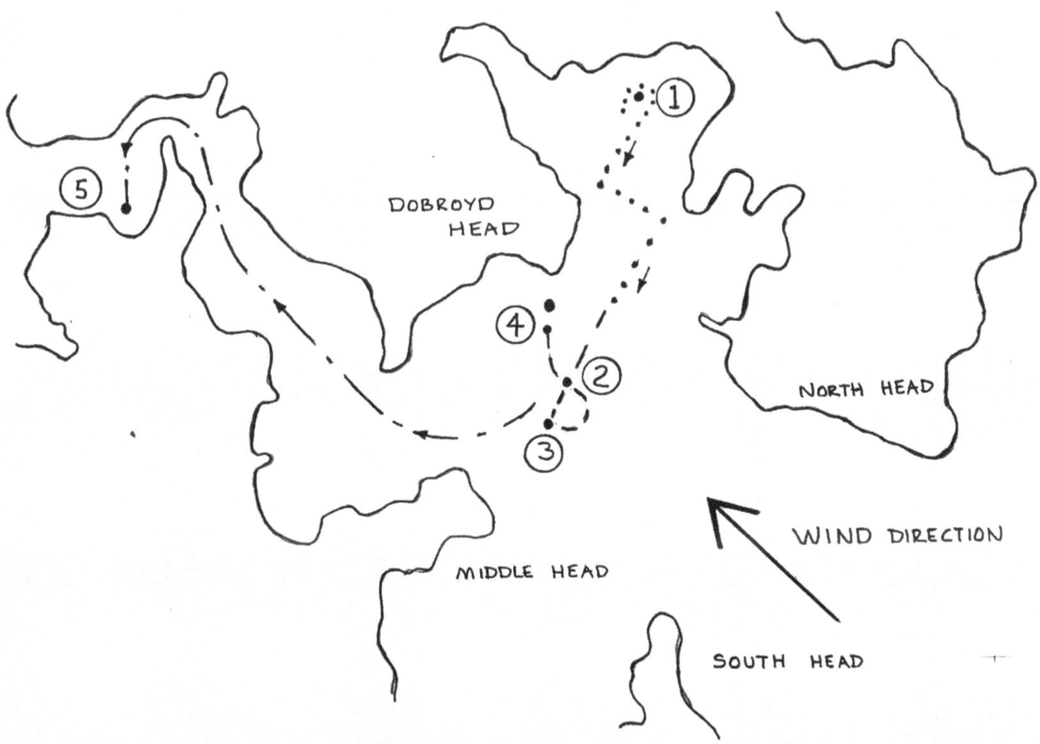

1: Manly rounding mark 2: Moore lost overboard 3: Thelma tacks to return to Moore

4: Bombora wave break and possible final location where Thelma's crew are rescued

5: Sayonara retires to Pearl Bay in Middle Harbour

DIAGRAM SHOWING THELMA'S COURSE
BY DAVID PAYNE

CONTEMPORARY PHOTO OF MIDDLE HEAD, WHERE THELMA WAS LOST.
SOURCE: DAVID PAYNE (2023).

In the months leading up to the loss of Walter Moore and *Thelma*, Reeks had remained steady in his duty to the Royal Sydney Yacht Squadron, attending and helping organise regular sailing races. He had also designed a new sail plan for *Rawhiti II*, assisting the vessel to successfully win the Northcote Cup in February of 1913.[1804] There were also commissions for new vessel designs, including the *Lady Edeline* and *Lady Scott,* both launched for the Balmain New Ferry Company later that year, and the Watsons Bay and South Shore Ferry Company's *Woollahra*, launched from the Brisbane Water shipyard of Rock Davis (Jr) in June of 1913.[1805]

Meanwhile, in Western Australia, Messrs McIlwraith, McEacharn & Co. had tasked Reeks with designing another vessel, this time a 56ft motor tug to be built by the Perth-based firm of Messrs Lawrence Bros.[1806] The scow-type craft was fitted with an 80 h.p. Gardner kerosene engine and launched as the *Kylie* in April of 1913.[1807] It was similar to the *White Rose* that Reeks had designed for the Vacuum Oil Company several years prior.[1808]

Concurrently at Battery Point, Hobart, the Rosny Estates & Ferry Co., Ltd. had contracted Fred Moore to build them a ferry. Designed by Reeks, the craft was slightly larger and more powerful than the *Derwent*, which was proving a popular addition to the Lindisfarne trans-Derwent service.[1809] Named *Rosny* the double-ended vessel was launched on 27 July 1913.[1810]

[1804] *Sunday Times*, 23 February 1913.
[1805] *The Sun*, 26 February 1912; *The Daily Telegraph*, 2 November 1912.
[1806] *Weekly Supplement to Building*, 31 March 1913.
[1807] Register of British Ships: Main Register, Port of Fremantle, Roll 3, 1856-1982, ON 131642.
[1808] *The Sydney Morning Herald*, 26 June 1913.
[1809] *The Mercury*, 10 January 1913.
[1810] *Examiner*, 28 July 1913.

Additional vessels designed by Reeks that were launched in 1913 included a 41ft motor cruiser constructed by Messrs Pritchard Bros. Ltd. of Careening Cove. The craft was commissioned by Septimus Macpherson and completed towards the end of 1913.[1811] It featured a high deck forward and large cockpit aft.[1812]

LAUNCH OF ROSNY AT BATTERY POINT, TASMANIA.
SOURCE: THE TASMANIAN MAIL, 31 JULY 1913.

Also late in 1913 Reeks completed plans for a steel steamer for the State Brickworks. The 135ft craft was

[1811] *Weekly Supplement to Building*, 28 January 1913.
[1812] *The Sun*, 3 September 1913.

intended to run at a speed of 9 knots between Sydney and the Manning River, taking bricks on the outward voyage and lime on the return trip.[1813]

An ocean-going cruiser designed by Reeks and launched in October of 1913 was built by Ford (Jr).[1814] Named *Mistral* and built to the order of E. P. Simpson, the 58ft ketch-rigged vessel was fitted with a clipper bow and at 35 tons was one of the largest auxiliary yachts amongst the Sydney fleet.[1815] A beautiful craft, it was advertised for sale in April of 1921 and over two years later subsequently sold to J. M. Hardie who renamed it *Capella*.[1816] The vessel was unfortunately lost on the rocks at North Head in December of that year after being stolen from its mooring at Rose Bay.[1817]

MISTRAL SHORTLY AFTER LAUNCH.
SOURCE: SUNDAY TIMES, 28 DECEMBER 1913.

A despatch steamer for the Papuan Government measuring 174ft in length was another craft Reeks designed during late 1913 for which he received £625 as initial payment.[1818] This particular vessel was intended to replace the large steam yacht *Merrie England* which had been wrecked some months prior off the coast of Papua.[1819] The narrative that leads eventually to a new *Merrie England,* launched in 1917 remains unclear, but it starts with this impressive vessel, which was designed to be built of steel and iron. Although tenders were called and the State Shipbuilding Yard in Williamstown, Victoria, provided specifications for a steel vessel for the Papuan Government in 1916, it is uncertain if this was the same design or a newer one. What is known, however, is that a steel vessel never went further than the drawing board. The huge vessel we know of from a profile published in 1913 was perhaps a step too far in size and even opulence for what was intended. It was a grand sight as a drawing (refer to page 356), but subsequent events more than likely put this project to replace the lost *Merrie England* aside, until it was revived more modestly in 1916.

During the interim Reeks designed a 67ft auxiliary ketch for the Papuan Government, as a temporary replacement for the yet to be completed vessel. Constructed of Kauri pine, it was built by Holmes at McMahons Point and launched as the *Elevala* on 6 November of 1913.[1820] Intended to accommodate the island's governor and his staff, the craft encompassed a large deckhouse with ventilation prioritised given the tropical waters.[1821] The *Elevala* was fitted with a 50 h.p. Standard engine.

ELEVALA AT SYDNEY SHORTLY AFTER LAUNCH.
SOURCE: SUNDAY TIMES, 28 DECEMBER 1913.

The year 1913 was certainly a busy one for Sydney's boat and shipbuilders. In fact an article published in *The Sun* on 3 September details the extent of the industry at the time. '*Never in the history of shipbuilding in Sydney has business been so brisk as it is now*'. The piece continued, '*Every yard is being deluged with orders, and builders are at their wits' end to fulfil contracts. In every instance additional employees have been engaged, without, however, being able to stem the rush. Everybody seems to want a boat at once, with the necessary consequences that many have had to wait*'.[1822]

Of course Reeks was capitalising on this peak, enjoying the opportunities provided by the increased demand in his expertise, as well as the diversity in requests for varying types of vessel designs encompassing new and emerging technologies, features, and methods of marine construction. He also sought to have a new motor launch built for himself, entrusting Holmes with its construction.[1823] The dimensions of the craft were noted as 20ft by 6ft by 2.5ft.[1824] It was to be fitted with a 3.5 h.p. Rotor, a New Zealand-made engine, that was supposedly

1813 *The Sydney Morning Herald*, 4 September 1913.
1814 *Sunday Times*, 28 December 1913.
1815 *Sunday Times*, 28 December 1913.
1816 *The Sydney Morning Herald*, 23 April 1921; SYDNEY, NEW SOUTH WALES, Register of ships, single number series (II), Volume 12, 1955-1982, ON 131573; *The Daily Telegraph*, 13 September 1912; *The Sydney Morning Herald*, 22 October 1912.
1817 *The Daily Telegraph*, 24 December 1925; *Australian Motor Boat and Yachting*, February 1926.
1818 *The Sydney Morning Herald*, 4 September 1913; *Commonwealth of Australia Gazette*, 3 January 1914.
1819 *The Argus*, 1 November 1913.
1820 *Construction: Weekly Supplement to Building*, 3 November 1913; *The Daily Telegraph*, 7 November 1913; *The Sydney Morning Herald*, 7 November 1913.
1821 *The Daily Telegraph*, 7 November 1913.
1822 *The Sun*, 3 September 1913.
1823 *The Sun*, 3 September 1913.
1824 *The Sun*, 3 September 1913.

silent when running.[1825] The estimated speed of the vessel was 6.5 knots.[1826] It is not known if the craft was actually built.

——

Continuing his involvement and investment in new enterprises, in mid-1913 Reeks took part in '*interesting experiments*' as part of his association with the Meteor Igniter, Ltd. Specifically, a special launch equipped with a 10 h.p. two-cylinder Niagara engine, was trialled in connection with a new form of sparking device.[1827] The *Sydney Morning Herald* reported, '*This invention consists essentially of a meteoric pencil, which extends for a certain distance within the combustion chamber, and a revolving disc, also in the cylinder, which is operated by a trip gear driven from the main shaft of the engine, and periodically strikes the pencil, throwing a shower of sparks among the compressed gas. It is claimed by the inventors that a quicker and more spontaneous combustion of the gas is obtained by this means than by either of the ordinary methods. The experiments, up to the present, are regarded as satisfactory*'.[1828] However, as stated in the previous chapter, the company was not viable owing to a series of lawsuits. Further the invention itself, by today's standards, seems rather cumbersome and may not have lasted very long either. At the time it seems to have been a contender along with the spark plug for internal combustion engine ignition, but the method of creating the spark appears to have had its complications.

On 22 October 1913 Reeks travelled to Brisbane as a passenger on board the *Wyreema*.[1829] This trip was at the request of the Brisbane City Council who sought Reeks' advice regarding alterations to the *Hetherington*, a steel paddle steamer which had been built to run across the Brisbane River to Kangaroo Point, though had never proved satisfactory for this purpose.[1830] Reeks was also tasked with providing advice on improvements that could be made to the Council's ferry steamer *Ena*.[1831]

Commitments to yacht administration also filled Reeks' calendar. In addition to continuing as honorary measurer of the Royal Sydney Yacht Squadron, as well as a member of the organisation's committee, there was much happening at the international level that Reeks would have been abreast of.[1832] The Yacht Racing Association's International Rule, adopted in 1906, was causing some dissatisfaction in Great Britain particularly in regards to the smaller size classes such that a new association was proposed to deal with them, the Boat Racing Association.[1833] One-design classes were also increasing in popularity with William Fife (Jr) tasked with sending plans for an 8-metre class to the Royal Sydney Yacht Squadron and Royal Prince Alfred Yacht Club for their joint consideration and possible adoption. The plans were then handed over to Reeks who was to make the necessary alterations and calculations with respect to the timbers to be used.[1834] A smaller one-design class had also been adopted in Hobart which had caught the idea of Sydney's yachtsmen.

On a personal note, Reeks' two children were entering their formative adult years. In 1913 Kent turned 23 years of age, while Thelma turned 21. Reeks' wife Gertrude remained involved in competitive shooting with the couple often noted in the press as attending various events and functions.[1835] Because life is not always kind, there were of course hurdles and grievances to endure, including a robbery in April of 1913 whereby 14 sovereigns and £50 worth of jewelry was stolen from under a flower pot located in the backyard of the family's house, *Twynham,* Mosman Bay.[1836] For reasons not known, the items had been placed there by Gertrude.

Other commitments during this period in Reeks' life included his association with the Engineering Association of New South Wales. In October of 1913 he was notably elected the organisation's president.[1837] A week prior Reeks was one of several well-known Sydney-based architects, surveyors, engineers and professionals who entertained Walter Burley Griffin, the Chicago architect whose design for the federal capital city, Canberra, had recently been accepted by the Commonwealth Government.[1838] The event took place at Paris House with Reeks noted as giving a speech.[1839]

Town planning was obviously a topic that interested Reeks, as indicated by his attendance at the inaugural meeting of the Town Planning Association of Australia which was held in Sydney's Town Hall on 17 October 1913.[1840] With the chair occupied by Colonel Vernon, the organisation was formed, '*To advocate proper planning of Australian cities and towns, modern building legislation, efficient municipal government; up-to-date water, sewerage, and lighting service; to watch Federal City developments, and other matters in keeping with the objects of the association. And that sections be established in each State*'.[1841] On the motion of Colonel Vernon, a provisional council to formulate rules and regulations was also appointed, of which Reeks was a member.[1842] The first meeting of the Town Planning Association of Australia was held on 17 November 1913 in the Royal Art Society's Rooms, 76 Pitt Street.[1843]

Following the festivities of Christmas and the New Year, Reeks was on board the flagship *Gosford II* for the Pittwater Regatta, held on 3 January 1914.[1844] Overall, he was obviously enjoying the fruits of his professional and personal efforts and achievements.

[1825] *The Sun*, 3 September 1913.
[1826] *The Sun*, 3 September 1913.
[1827] *The Sydney Morning Herald*, 26 June 1913.
[1828] *The Sydney Morning Herald*, 26 June 1913.
[1829] *The Brisbane Courier*, 17 October 1913.
[1830] *The Sydney Morning Herald*, 2 April 1914.
[1831] *Daily Standard*, 9 December 1913.
[1832] *The Sydney Morning Herald*, 27 August 1913.
[1833] *Referee*, 9 April 1913.
[1834] *The Daily Telegraph*, 3 September 1913.
[1835] *The Daily Telegraph*, 1 March 1913.
[1836] *The Herald*, 16 April 1913.
[1837] *The Sydney Morning Herald*, 11 October 1913.
[1838] *The Daily Telegraph*, 3 October 1913.
[1839] *The Daily Telegraph*, 3 October 1913.
[1840] *The Sydney Morning Herald*, 18 October 1913.
[1841] *The Sydney Morning Herald*, 18 October 1913.
[1842] *The Sydney Morning Herald*, 18 October 1913.
[1843] *Construction: Weekly Supplement to Building*, 10 November 1913.
[1844] *Evening News*, 5 January 1914.

Suddenly everything changed. On Tuesday 20 January 1914 a tragedy occurred that was to deeply and permanently impact Reeks and his family for the rest of their lives. In the small market town of Bilston, just outside of the city of Wolverhampton in the West Midlands of England, his 24-year-old son Kent was brutally murdered. It took two days for the news to reach Australia. Though the initial press reports were somewhat convoluted and erroneous, based on limited information and cable messages, Reeks knew it was his son Kent. *The Sun* newspaper of Sydney published the following article on 22 January.

'MURDER OR SUICIDE? SYDNEY MAN'S FATE. FOUND SHOT IN DISUSED SHAFT. WOLVERHAMPTON MYSTERY. London. Thursday morning.

What looks very much like a murder, of which Thomas Reeks, a young man, formerly of Sydney, was the victim, was brought to light yesterday at Bilston, near Wolverhampton.

With two bullet wounds in the forehead, one over each eye, Reeks's body was discovered lying in a disused pit shaft. The head was covered with an overcoat. Four empty cartridges were found near the corpse, but no weapon was to be seen nor were there any signs of a struggle in the vicinity of the shaft.

The deceased was 24 years of age and was recently employed as fourth engineer on the steamer Secolia, belonging to the United Fruit Company of Orleans. He arrived from Nova Scotia by the steamship Empress of Britain on Saturday, and on the same day he journeyed to Manchester on a visit to his uncle and aunt, Mr. and Mrs. Kent.

As his relatives in Manchester were unable to put him up for the night, he returned to Liverpool, his intention being to stay at the Temperance Hotel, near the Lime Street railway station.

The body was identified by his uncle, who has given the police some information which, assuming that Reeks was murdered, suggests the probable motive for the crime. Mr. Kent states that his nephew showed him a 100 dollar bill, which he had inside a wallet, and remarked that he had several others in his possession. Both the wallet and the bills were missing when the body was discovered, but a sum of £9 2s 7d and a gold watch and silver chain were found in the pockets of the dead man's clothes.

The police have ascertained that Reeks was seen in the company of another man in Liverpool on Monday night, and inquiries in Wolverhampton had elicited that two respectably dressed men, both strangers, arrived there early on Tuesday morning.

The case is generally believed to be one of murder, Reeks's friends discrediting any theory of suicide.'

The same newspaper, published later in the day carried a further piece. *'Mr Walter Reeks, the well-known Sydney naval architect, has absolutely no doubt that the Thomas Reeks mentioned in the cable message is his son, who went away from Sydney about three years ago. "The whole of the description points to my son," said Mr. Reeks, when spoken to this morning. "The only discrepancy is in the Christian name. My boy's name is Kent, not Thomas, but the fact that the names of his uncle and aunt mentioned are correct makes me certain that it is my son."*

"The name of his ship is the Saxonia, not Secotia, as mentioned in the cable."

Great sympathy is felt for Mr. Reeks by such of his many friends as have up to the present heard of the shocking tragedy.'

Two days later another statement from Reeks was published in the Melbourne press. *'Everything points to the fact that is it my son. I had a letter from him only last week. He wrote from New Orleans. It was his intention to go as far as Boston with his ship, where his contract with the company was to have ended. He then contemplated going to London to get his certificate.'*[1845]

Though the city and regional press throughout Australia continued to report extensively on Kent Reeks' movements leading up to his murder, as well as possible suspects, along with the coronal inquest and subsequent police investigation in the days, weeks and months ahead, it was the British press that published more detailed and accurate information. Still, the surrounding events, the act itself, the discovery of Kent's body, possible motives, and the people involved made for some extremely sensational reporting. For Reeks and his family in Sydney, however, it was very much a waiting game as they received spasmodic parcels of information about Kent and his movements from cablegrams, telegrams and press reports.

For Reeks, he appears to have mourned this great loss in private, at home with his wife and daughter. He made only two public statements after receiving the initial news of Kent's death. It was very uncharacteristic of him not to give interviews and be accessible to the press, but it was also an unprecedented time of grief, sorrow and many, many unanswered questions. Of course he received an untold number of messages and expressions of sympathy from his friends, colleagues and associates, it was all-told a shocking time.

Kent Reeks was described as, *'a bright boy ... well-known in Sydney engineering, sailing and social circles. He was his father's pride, and had adopted the engineering profession, in order to take part in the far-famed business conducted by Mr. Walter Reeks. He was to have sat in London for his second engineer's certificate, and then to have returned to Sydney. The letters which Kent wrote to his family were always cheery, full of interesting stories, at the same time instructive so far as the places visited by him, while at sea, were concerned. He knew Panama like a book, as well as the main ports in America'.*[1846]

After completing his studies in engineering at the Sydney Technical College in 1910, Kent Reeks had initially found work as an assistant engineer on board the White Star liner *Georgic,* travelling the east coast of Australia.[1847] The ship's home port was Liverpool and it was to this destination that the vessel sailed, departing Sydney on 6 April 1911 with Kent on board.[1848] It is not known if he then remained working on the *Georgic*, as it made several more trips between Liverpool and Australia that year, or if he found work on another vessel upon

1845 *Leader,* 24 January 1914.
1846 *Tweed Daily,* 27 January 1914.
1847 New South Wales, Australia, Unassisted Immigrant Passenger Lists, 1826-1922.
1848 *The Sydney Morning Herald*, 6 April 1911.

his arrival in Great Britain. What is known, however, is that Kent departed Liverpool as a passenger on board the Cunard liner *Lusitania* on 9 March 1912, headed for New York.[1849] He then found work as a fourth engineer on board the *Sixaola*, operating out of New Orleans, a vessel owned by the Tropical Fruit SS Co. Ltd. The *Sixaola* traded between the United States and central American ports, including Bocas del Toro and Colon in Panama, Bluefields in Nicaragua, and Port Limon in Costa Rica, primarily transporting passengers and fruit.[1850]

In early January of 1914 the *Sixaola* began a new route, sailing out of Boston for Jamaica and Port Limon.[1851] Kent's employment on board was completed during this changeover at Boston and he subsequently made his way further up the North American coast, departing Halifax, Nova Scotia, as a passenger on board the *Empress of Ireland*.

SS SIXAOLA.
SOURCE: NAVAL HISTORY AND HERITAGE COMMAND (NH 103269).

Kent arrived in Liverpool on 17 January 1914.[1852] He was in England specifically to undertake exams in association with his engineer's certificate. He also sought to seek out relations, first visiting his 75-year-old maternal grandmother Sarah Kent, and his maternal uncle Thomas Ridyard Kent, as well as his wife May and their daughters Gertrude and Muriel. The family lived at 58 Chorley Road in Swinton, some 25 miles from Liverpool. It was expected that Kent would also visit more of his mother's family who lived in Stoke on Trent, just over 40 miles from Liverpool, including his uncle Frank Kent, his aunt Gertrude and their daughter Doris.[1853]

Tragically, however, just three days after his arrival Kent was dead, the details of which are really quite gruesome. His murder case was never concluded. Though police continued to investigate his death for several months, chasing leads locally and overseas, no arrests were ever made. Still to this day the case remains an unsolved.

Back in Sydney, while Reeks mourned the devastating loss of his son in the days, weeks, months and years ahead, he remained resolute in his duties to work, pastimes and family. For example, he is noted as assisting with organisational efforts associated with races for the Northcote Cup, held in early February of 1914 which featured several yachts from Victoria.[1854]

In all of this sorrow and solemnness, however, there was some joy. On 11 February 1914 at St. Andrew's Cathedral in Sydney, Thelma Reeks, aged 21, married a close Mosman neighbour, Alfred Fox Ashwin, who at the time of the union was listed as a 28-year-old grazier then living in Baulkham Hills. Witnesses to the nuptials were listed as Winifred and Frederick Fox, the groom's sister and brother. Given the circumstances of Kent Reeks' very recent death, the celebration would have been markedly subdued.

THELMA FOX ASHWIN (NEE REEKS) (C1912).
SOURCE: FOX ASHWIN FAMILY COLLECTION.

Returning to his professional duties, Reeks gave his first speech as president of the Engineering Association of New South Wales at a meeting held on 12 March 1914, also delivering an address titled '*The Engineering of Shipbuilding*'.[1855] Though he spoke in great detail of

1849 UK and Ireland, Outward Passenger Lists, 1890-1960 for Kent Reeks.
1850 *The Times-Democrat*, 10 September 1913.
1851 *The Boston Globe*, 4 January 1914.
1852 UK and Ireland, Incoming Passenger Lists, 1878-1960 for Kent Reeks.
1853 1911 England Census.
1854 *Evening News*, 13 February 1914.
1855 *The Daily Telegraph*, 13 March 1914.

IS IT MURDER?

Ship's Officer Is Found Dead

With Bullet Wounds on His Head,

But No Revolver Is Near the Body.

The most mysterious tragedy that has occurred in the Wolverhampton district for many years has come to light by the finding of the body of a young man in a pool of blood near a disused pitshaft at Bilston, a village a few miles from the town.

An envelope found in his possession bore the name of Mr Kent Reeks, and contained a letter sent by someone from 58 Chorley Road, Swinton. This was the only means by which it was possible to ascertain who the man was, and a press representative was able to establish beyond all doubt that the victim is Kent Reeks, aged 24, a senior fourth engineer, lately engaged on the Seco-lia, one of the United Fruit Company's vessels plying between Boston, U.S.A., and the West Indies.

He was a native of Sydney, Australia, and had only once previously been in England. He had relatives in Chorley Road, Swinton, hence the letter which was found in his possession.

The discovery of the young man's body was made by two workmen who were passing along a piece of waste ground about 300 yards from the main thoroughfare. It was only a few yards away from the pitshaft, and had evidently been there for some hours.

There were two bullet wounds in the forehead, a bullet wound in each eye, and in the man's possession was £9 in gold, a gold watch, a match box, several photographs of himself and relatives, and a number of English and foreign picture cards.

Had Gone to Liverpool.

How the man came to be in that condition is a complete mystery. He was last seen by his relatives at Swinton on Saturday night when he took train to Liverpool. They understood he was to stay at an hotel there, and that he went to Liverpool and remained there until some time on Monday is proved by the fact that he called at the Liverpool Post Office and received a letter dated December 24 which had been addressed to him at Boston, U.S.A., and redirected to Liverpool.

The letter was written to him by his grandmother at Swinton, and concluded, "Your loving granny, Kent."

What his movements were after he left the Liverpool Post Office no one seems to know. Apparently he went to Wolverhampton the same day, but what caused him to go to the peaceful suburb of Bilston at present remains a mystery. It is a barren district, and leads into the heart of the Black Country.

Cartridges, But No Weapon.

Another singular part of the affair is that at the time of the discovery no weapon was found, nor have the police been able to trace one. Their theory is that it might have been thrown down the pit shaft, which contains a great deal of water, or picked up by someone who failed to notice the body. The police did, however, find near by four spent cartridges of the "pin-fire" type and three live ones, as well as a spent gun cartridge.

Of the two bullet wounds on the forehead, one had penetrated to the back of the head, and, in addition to a bullet wound in each eye, there was a bruise on the nose.

There were no signs of any struggle having taken place, and the question is—How did Reeks come to receive the bullet wounds? He had no troubles as far as his relatives knew, and they say he was exceedingly bright and cheerful when they left him.

This would scarcely lead one to presume that he committed suicide.

It would be almost impossible for him to fire four times and inflict a wound on each occasion. But, supposing that did occur, where is the revolver?

It would, in the ordinary way, have been found by his side had he taken his own life.

Dollar Bills Missing.

Do the circumstances suggest, as an alternative, murder? The theory is that had he been murdered there would be little likelihood of his being found with £9 in gold and other valuables in his possession. There is, however, a probable answer, even to that objection.

It was known when he left Manchester to go to Liverpool, on Saturday, that he had a number of dollar bills. These were not found on his body, and if he has been the victim of a dastardly crime it is quite likely that the authors of it were content to rob him of these.

They might then leave the remainder of his money and belongings alone.

At present the police are entirely in the dark as to what has occurred, and unless someone comes forward and throws a little light on the affair it is possible it may remain one of the many tragedies which have never been elucidated.

Relatives' Story.

Far from being lifted, the mystery becomes even deeper as the result of the story told to a press representative who made inquiries at the house of Mr and Mrs James Richard Kent, in Chorley Road, Swinton. The young man visited them as late as Saturday last, immediately on reaching Liverpool from America. They were unable to accommodate him overnight, and he returned to Liverpool. They saw him off from Manchester Station in the evening, and he then told them that he intended staying at a temperance hotel near Lime Street Station. They were utterly at a loss to account for his presence in the neighbourhood of Wolverhampton.

So far as they knew he was going back to Liverpool to study for an examination, with the object of getting an English engineer's certificate, as his Australian certificate was not so valuable. The examination would have taken place in a few weeks' time, and his intention then was to return home to Australia.

"He was not at all worried about this examination," said his aunt, Mrs Kent. "He was not the sort to trouble about a thing like that. He was bright and cheerful. In fact, we commented upon the improvement in him generally since we last saw him three years ago."

She emphasised his cheeriness of disposition in discussing the possibility of suicide. She said he had absolutely no troubles so far as they knew. There was no girl in the case; he had plenty of money and good prospects.

"I had chaffed him," she remarked, "as to how many hearts he had broken since we saw him last. He replied in the same strain, 'I'll get married when I'm 90.' He had plenty of money, because he showed us a 100 dollar bill which he had with others in a wallet in his pocket."

It has been ascertained that Kent Reeks was a passenger by the Empress of Britain from Halifax. He had intended to enter for an examination for an engineer's certificate.

The Mystery Deepens.

A later message says the mystery has deepened by the discovery that the three alleged live cartridges found near the four spent ones had misfired, and were ejected from a revolver.

Victim Identified.

This afternoon Mr Thomas Kent, of Manchester, identified the victim as his nephew, Thos. Kent Reeks, aged 24, a native of Sydney, Australia, and lately engaged as fourth engineer on the Scotia, a vessel belonging to the United Fruit Company, New Orleans.

The post-mortem examination of the body revealed two bullet wounds, one in the left eye and the other on the right side of the forehead, and two abrasions at first thought to be bullet wounds.

Reeks was in company with another man at Liverpool on Monday night, and inquiries at W'hampton show that two respectably-dressed men travelled on the 11.55 train from Liverpool on Monday night, arriving at W'hampton early on Tuesday morning. No particular notice, however, was taken of them.

A remarkable fact is that one of the spent cartridges found near the body was fired months ago, as the case is filled with fluffy dust, whilst the other shots have not a convincing appearance of being recently discharged.

MIDLAND MURDER MYSTERY.

REEKS' MOVEMENTS TRACED.

A Liverpool correspondent, telegraphing early this morning, states:

As the result of unceasing inquiries, the Liverpool police have traced the movements of Reeks, the young ship's engineer, who was found murdered last Tuesday beside a disused pit-shaft near Wolverhampton, and early developments are anticipated.

When Reeks left Canada on the Empress of Ireland for Liverpool he had his valuables in a special envelope provided for the purpose and gave this envelope to the purser, in whose charge it remained throughout the voyage. On the journey across the Atlantic he made friends with a girl who was returning from Canada to her home in Liverpool. He told her he had friends in Manchester, and that it was his intention to see them on the day he landed. It was afterwards ascertained that, in fact, he did so. He also told her that he had friends in Bournemouth and he would have to see them before he returned to Sydney, but he did not say when he would see them.

The girl was with him when he got his valuables from the purser. He opened the envelope in her presence on the boat, and she saw that it contained a bundle of five or six "yellow backs"—American notes of \$100 or more. That was on Friday, Jan. 16.

When she saw these notes there was no one on the deck but Reeks and herself. On Saturday morning, just before they landed, Reeks told her he had been very foolish the night before, and that he had taken off his coat and thrown it on his bunk, forgetting that he had this money in his possession. "I will just see if they are all right now," he added, and pulling out his wallet he counted the notes and observed, "They are right." There was a large number of people around at the time waiting to get off the boat.

AN AMERICAN VISITOR.

Having landed, he put his luggage in the left luggage office and, carrying a suit-case, walked with the girl to St. George's-crescent at the top of Lord-street, where she got into a taxi-cab and went home. Reeks also took a taxi-cab and drove to Lord Nelson-street. He got off at an hotel and engaged a room for three weeks, saying he might not be back that night because he was going to Manchester. He went away, and about an hour afterwards a man drove up, who said he came from Chicago. He took a room for two days. He went away, and about half-past one Reeks came back and said he would return that night. Shortly afterwards he went away and returned about eleven o'clock. The other man returned about the same time.

The two men did not see each other at all that night. Next morning (Sunday) they met at the breakfast table, and, judging from the way they spoke, they did not appear to have known each other previously. At the table, however, they appeared to become very friendly, and Reeks showed the other his discharge-book. During the whole of that day they were in and out of the hotel, always together, and in the evening they sat reading together. They asked to be called at eight o'clock next morning (Monday) and to have breakfast half an hour afterwards.

Reeks went upstairs after breakfast and the American went out. On Reeks coming down again he expressed surprise that his companion had gone and went out after him, but whether they met is not known. At ten o'clock Reeks visited an engineering academy in Castle-street, and at 10.30 went to a solicitor's office opposite, where he swore an affidavit. He was backwards and forwards between the two places until 12.45, when he left a note at the academy saying he was going to Birmingham and would be back on Thursday or Friday. About one o'clock the American came back to the hotel, paid his bill, and said he might be back in four or five days.

MYSTERIOUS PAPER.

At two o'clock Reeks returned to the hotel, obtained a piece of brown paper, went upstairs and came down with a brown paper parcel, around which there was a brown leather strap. As he went out he observed, "Neither my friend nor I will be back for tea." He was then told that the other man had gone to Leeds, and he said, "I may be back to-night, but I am not sure how long my business will take me."

That is the last known of Reeks's movements. On a search being made of the bed-room he occupied a piece of hotel paper was found on which was scribbled several times the name "J. G. Parsons," but the meaning of the paper or to whom the name referred has not been discovered.

MURDER OF KENT REEKS.

ANOTHER DISCOVERY MADE ON THE SCENE OF THE CRIME.

Yesterday thousands of people visited the scene of the Ettingshall murder. Among them was a pensioner of the Metropolitan police, who picked up in the soft ground between Millfields and the disused mine shaft where the body of Reeks was found three live revolver centre fire cartridges, marked with the letters U and L, quite a different type from the cartridges lying near the murdered man, which bore the number 32. They were handed to the local police.

The funeral of Mr. Kent Reeks, the young victim of the murder, took place quietly at Worsley, Manchester, on Saturday. Efforts had been made to keep the funeral as private as possible, and these were to a large extent successful. There were several family mourners, and the wreaths included one composed of lilies of the valley, arum lilies and evergreens. The inscription read, "From his loved ones at Tynham, Sydney, Australia."

WHO MURDERED KENT REEKS,

And Left His Body at Pitmouth?

Inquiry Into Black Country Mystery,

Which Still Baffles Solution.

Mr T. A. Stokes, coroner for South Staffordshire, to-day held the adjourned inquiry at the Town Hall, Bilston, into the circumstances of the death of Mr Kent Reeks, the young Australian ship engineer, whose body was found with bullet wounds in the head close to the top of a disused pit shaft at Ettingshall, near Wolverhampton, on 20th January last.

Three days before Reeks landed at Liverpool from Halifax (Nova Scotia). He stayed in a hotel in Liverpool, and on the 19th ult. suddenly disappeared, the next known of him being the finding of the body. Whilst at the hotel Reeks made the acquaintance and was in the company of a man, who signed the hotel book as J. H. Ramsden, Chicago, Illinois. He has since disappeared, and the police have not the slightest trace of him. When the body was found gold and silver and a gold watch were in the pockets, but a wallet supposed to contain dollar notes for a considerable amount was missing.

The pit shaft was as far as possible dragged yesterday to see if anything bearing on the crime could be discovered, but without result.

What induced Reeks to go to the Black Country is an absolute mystery, and all the efforts of the police to throw light on where the murder was actually committed, and who perpetrated it, have been unavailing.

There were very few people in Court when the proceedings opened.

Mr Thomas Kent, of Chorley Road, Swinton, Manchester, uncle of the deceased, who had given evidence of identification at the opening of the inquest a fortnight ago, was again called. He stated that upon the deceased arriving at Liverpool on the 20th he came to Manchester to visit witness. He had money in his possession. He showed witness a 100 dollar note, and showed his wife and daughter other notes the value of which was not known.

The Coroner—Did he mention the names of anybody; of any of his friends?—No, he only referred to his relatives, some of whom lived in London and Bournemouth.

Had he any relatives, friends or connections of any sort down here in this neighbourhood?—No; none whatever.

Or in the immediate neighbourhood?—No; not so far as I know.

Do you know of any reason whatever that would have brought him into this district?—I can't make the reason out at all from what he said to us.

Did he tell you whether he had made any acquaintances on the ship when he came over?—He said he spoke to two young ladies only, and that he had made their acquaintance on board. He stated they were English ladies who had been in America. But he did not say where they came from or were going to. He did not say he had made acquaintances in Liverpool.

Did he ever mention the name of a man named Ramsden?—No, he never mentioned anybody except these two girls. He had a pin in his tie which he said he had mounted on board ship.

The Tie Pin.

The Coroner asked for this and was informed by Superintendent Rowbotham of Bilston that by the instructions of the Chief Constable of Staffordshire it was handed over and was now in the possession of the Liverpool police.

The Coroner remarked that the inquiry was being held there, and not at Liverpool. Everything bearing upon it ought to be before him.

Witness further questioned, said the pin was a Brazilian green beetle mounted on silver. The deceased never expressed any intention of visiting work in the neighbourhood of the Black Country. He appeared to be most anxious to get through his marine examination and get to Australia.

A Juror—Do you know whether either of the ladies had any male friends?—Not that I know of.

Catherine Stanton, the keeper of a hotel in Lord Nelson Street, Liverpool, told the Court that Reeks came to her hotel at about ten o'clock on the morning of the 17th January, and engaged a room for three weeks. He explained to her that he had just arrived on the Empress of Ireland. He left the hotel in about ten minutes, and came back in an hour and a half. Whilst he was away a man, who gave the name of Ramsden, arrived, and engaged a room for two or three nights. He stated that he was from Chicago, but witness did not notice whether or not he had an American accent. They never noticed such things in Liverpool. Reeks and Ramsden met for the first time at breakfast on the next morning (which was Sunday). Their demeanour looked as though they had not met before. The couple were about together on the Sunday, and the next morning they had breakfast in each other's company. Afterwards Reeks announced to witness that he was going away for a few days, but he did not say where he was going to. Ramsden went out, and returning at twelve o'clock or thereabouts, asked for his bill, saying that he was going to Leeds. Reeks got back to the hotel early in the afternoon. He asked where Ramsden was, and when witness told him Ramsden had gone, he said, "Oh."

Tied-Up Parcel.

Reeks then tied up a brown paper parcel, and after stating that he would be back in a few days left the hotel about an hour after Ramsden. He did not say where he was going, and witness was not aware whether he saw Ramsden after breakfast that day.

The Coroner—Did you see any dollar bills or anything of that sort in Reeks' possession?—No; he left without paying his bill. He did not take his luggage, and consequently she did not trouble him about paying the bill.

Evidence was given of a little girl first seeing Reeks' body as she was returning from taking her father's breakfast. She thought it was a man asleep.

The police showed that when the tragedy was discovered Reeks was lying on his right side, with the collar of his overcoat partly covering his head. A police constable stated that he found close to the body four spent cartridges and three live. One of the spent cartridges was under the neck and another under the feet. The others, together with the live cartridges, were on the bank on which deceased was lying. The body was farthest away from the road, and he found traces of blood nearer the road. There was no trace of a struggle having taken place where the blood was.

No Information.

The Coroner—Have you been able to obtain any information whatever as to this man's movements?—No sir.

And you cannot find that he was in anybody's company or seen by anybody?—No.

Dr Ashley Smith, who made the post-mortem examination, told of three bullet wounds having been inflicted in the head, two penetrating the forehead, and one at the back of the skull. The shots had penetrated the brain and caused death. The top part of one of the eye cavities had been destroyed. There was no scorching on the forehead, and he argued from these signs that the shots must have been fired from a distance.

Most Mysterious Case.

A SELECTION OF ARTICLES FROM THE BRITISH PRESS COVERING THE MURDER.

PHOTO OF KENT REEKS (CIRCA 1911).
SOURCE: MIKE LONGWORTH VIA ANCESTRY.COM.

POLICE INVESTIGATING AT THE LOCATION WHERE KENT REEKS' BODY WAS FOUND.
SOURCE: UK, HISTORICAL PHOTOGRAPHS AND PRINTS, 1704-1989.

the large number of branches of scientific engineering necessary for modern shipbuilding, with reference also made to labour-saving appliances, including riveting machines, it was his opening remarks that sparked more humility and a sense of relatable humour amongst the audience, given the tendency for boisterous discussion and divergent and animated opinion of the membership. '*Already I begin to appreciate the privileges a president enjoys in reading an address, as compared with the member who reads a paper. It this were a paper, I feel sure I should see notes being made to fire back at me in the course of discussion. You would, perhaps, be inclined to just the opposite opinion, and say that the preparation of an address was hard work, and that it can scarcely be work and play at the same time. Well, rather than argue the point, I will admit that, in the words of the old Grace at meat: "For what we are about to receive, the Lord knows we have worked hard enough"*.'[1856] This comment was followed by an interesting anecdote, providing us with a snippet of Reeks' mindset and jovial relationships with his peers.

'*Some of us here are, as we say, "in the Trade." Some of us are consulting men; some of us design only; and some of us do some of all those things. Consequently scarcely two of us take precisely the same view of the things and doings round about us.*

As an example of that, on one occasion I was chatting with a friend who sold "Machines of Kinds" to his very considerable financial benefit, and discussing some matters of recent improvements, I said, "Yes, old man, but you must pay for brains." He said, "Whose brains?" I said, "why, the man who invents and perfects the improvement, of course." "Oh," he answered, "those are cheap. I thought you meant those of the salesman!"

So one sees that the point of view taken makes all the difference, and if one could only see as "God of All," to quote a line from Pope, one would probably be brought to a better understanding of the true relation of things, one to the other.'[1857] Reeks was certainly making a point, that perspective was key to understanding, and in making that statement, he was using it as a way to coalesce a membership of disparate professionals representing many different engineering and technical related professions. It was yet another indication of his personality, to build bridges amongst his peers and foster supportive discussions and knowledge sharing.

A few weeks later Reeks was in charge of the Royal Sydney Yacht Squadron's steamer for fleet races starting at Neutral Bay.[1858] Continuing his role as president of the Engineering Association of New South Wales, he next attended a dinner at Farmer's Restaurant celebrating the Electrical Association of New South Wales' 23rd anniversary.[1859]

1856 *Construction and Local Government Journal*, 6 April 1914.

1857 *Construction: Weekly Supplement to Building*, 6 April 1914.

1858 *The Sydney Morning Herald*, 23 March 1914.

1859 *The Sydney Morning Herald*, 25 April 1914.

With regards to his work outputs, April of 1914 coincided with the newly-completed ferry *Lady Scott,* designed by Reeks, having its trial runs. After being built at Jervis Bay the vessel had been towed to Sydney where its engines were fitted by Messrs Begg and Greig, and the upper works installed by Mort's Dock and Engineering Company Ltd.[1860] Named in perpetuation of the Antarctic explorer's legacy, the *Lady Scott* was to be placed in the Lane Cove River service.[1861]

Meanwhile, Reeks continued to liase with the Brisbane City Council regarding alterations to its steamers *Hetherington* and *Ena.* Following consultation, he proposed a remarkable surgery and transformation, stating that it was cheaper to build a new hull for the *Hetherington* than to make alterations to the existing one. Reeks had thus designed the new hull, with the existing engines and fittings to still be used.[1862] Regarding the *Ena*, a steel vessel, there was less work required. To increase its speed, Reeks designed an entirely new stern which was to be built into the original boat.[1863] It was proposed that this alteration would enable the vessel to make many more trips per day than it was presently capable.[1864] However, some four years later, members of the Brisbane City Council had failed to reach agreement. The two vessels remained unaltered, at great cost to the organisation.[1865]

In May of 1914 another collaboration built by William Holmes and Reeks was launched: the 47ft auxiliary schooner *Lilihina* was built to the order of Fred Erikson of Sydney who was also a large landholder in the Solomon Islands.[1866] The vessel was fitted with a 24 h.p. crude oil engine and was intended for work in association with its new owner's plantation.[1867] *Lilihina* left Sydney for Tulagi in late July with Captain Cunninghame at the helm.[1868]

More design commissions came through in May of 1914 with Messrs Davis Bros. and Burgess instructing Reeks to prepare plans and specifications for another passenger steamer for the Richmond River service.[1869] The craft was to attain a speed of 15 knots and be built in the Coraki shipyard, with work starting in only a few months.[1870]

On 2 July 1914 Reeks attended the annual dinner of the New South Wales Institute of Architects, with Prime Minister Joseph Cook and the Governor of New South Wales Sir Gerald Strickland honoured guests, the former giving a speech where he called for '*Useful and Practical Architecture*'.[1871] Following an interesting round of speeches, Reeks then gave a toast to all attendees on behalf of the Engineering Association of New South Wales.[1872] He additionally remarked, '*I feel that almost every day the professions are being brought closer together. There is a reciprocity going on that makes it almost impossible to now say where architecture ends and where civil engineering commences'*.[1873] Little did Reeks realise that his sentiments regarding cross-industry respect, collaboration and the exchange of ideas and practices would come into stark reality on a far greater scale, as Australia designed, generated and built equipment and resources in the years ahead.

World War I officially began on 4 August 1914 when Great Britain declared war on Germany. However, before we diverge into a discussion of this major international conflict, we must first mention the launch of another vessel. On 25 July 1914 the Balmain New Ferry Company Ltd.'s new steamer *Lady Ferguson* was launched from the shipyard of Messrs David Drake, Ltd. of Balmain, the customary christening ceremony performed by Miss M. Macdougall, daughter of the company's chairman.[1874] Of note, the vessel represented the fifth and final of the Lady class of ferry designed by Reeks in recent years for the Balmain New Ferry Company's Lane Cove service, the others being *Lady Denman, Lady Chelmsford, Lady Edeline* and *Lady Scott.* At 110ft long with a beam of 24ft, *Lady Ferguson* was intended to carry 500 passengers and travel at a speed of 11 knots.[1875]

LADY FERGUSON ON THE SLIP (1960S).
SOURCE: GRAEME ANDREWS WORKING HARBOUR PHOTOGRAPH COLLECTION, CITY OF SYDNEY ARCHIVES.

[1860] *The Sydney Morning Herald*, 2 April 1914.
[1861] *The Daily Telegraph*, 5 May 1914.
[1862] *The Sydney Morning Herald*, 2 April 1914.
[1863] *The Sydney Morning Herald*, 2 April 1914.
[1864] *The Sydney Morning Herald*, 2 April 1914.
[1865] *The Brisbane Courier*, 10 May 1916.
[1866] *The Sydney Morning Herald*, 21 May 1914.
[1867] *The Sydney Morning Herald*, 21 May 1914.
[1868] *The Sydney Morning Herald*, 23 July 1914.
[1869] *Northern Star*, 26 May 1914; *Truth*, 7 July 1918.
[1870] *Northern Star*, 26 May 1914.
[1871] *Construction: Weekly Supplement to Building*, 6 July 1914.
[1872] *Construction: Weekly Supplement to Building*, 6 July 1914.
[1873] *Construction: Weekly Supplement to Building*, 13 July 1914.
[1874] *The Sun*, 26 July 1914.
[1875] *The Sun*, 26 July 1914.

With Australia pledging full and complete support to Great Britain in its conflict with Germany, the 'call to arms' soon came with troops and equipment immediately deployed to Europe. In addition to the human plight, from a federal policy and resource standpoint, there was much to consider. An interstate commission was held in Sydney on 20 August 1914 in connection with proposed amendments to tariffs, including increasing tariffs on goods imported from overseas that could be manufactured in Australia.[1876] With regard to engines and boilers, Reeks testified that, '*he had seen as good marine engines and boilers made in Australia as could be made anywhere. Up to 1,500 h.p. engines were made here, and the local industry got a fair share of the trade. There were instances where local makers had been beaten by overseas competitors*'.[1877] As the fighting and hostilities increased, Australia found itself becoming more and more self-reliant.

For Reeks the war years were justifiably lean in terms of new commercial and recreational vessel design commissions. Pastimes, including yachting, were also furloughed due to the war effort. Though Reeks was again appointed honorary official measurer and to the committee of the Royal Sydney Yacht Squadron for the 1914/1915 season, the racing fixture itself was soon truncated as were associated social engagements.

With reports of soldier deaths and serious injuries filtering back to Australia, the atmosphere became duly serious with frivolities and social engagements postponed, unless they were benefitting the cause. For Reeks and his wife Gertrude, they are known to have attended the squadron's annual garden party at Kirribilli in December of 1914, but their social calendar was otherwise empty.[1878] During the early years of the war, Gertrude continued her involvement in shooting. For example, she competed in a rifle competition against a team from Hobart in March of 1915, as part of her association with the Sydney Ladies' Miniature Rifle Club.[1879] However, because of the hostilities neither team travelled. The Sydney team shot at Randwick while the Hobart team shot in Hobart; the former team claimed victory.[1880] The club would postpone its activities shortly thereafter, given the conflict.

On a personal level, there was welcome family news with the birth of Reeks' first grandchild. On 29 January 1915, a year and nine days after the death of her brother, Thelma Fox Ashwin gave birth to a boy at Mosman which she named Kent.

Despite there being less racing, yacht administration continued for Reeks, at least initially. He helped organise the Anniversary Regatta, held on 26 January 1915, and also attended the official luncheon on board the North Coast Company's steamer *Wollonbar.*[1881] In mid-February of 1915, along with several other prominent members of the Royal Sydney Yacht Squadron, Reeks welcomed the skipper of the Victorian 6-metre yacht *Killara*, J. A. Douglas, and his crew to Sydney, the contingent arriving to contest the Northcote Cup.[1882] He was also part of the sailing committee in charge of the event.[1883] The race series was won by *Killara*, the vessel far superior to the New South Wales challenger *Rawhiti II.*[1884]

By mid-1915 Reeks was part of the Royal Sydney Yacht Squadron's fundraising efforts on behalf of the Red Cross to raise money to equip *Graythwaite* as a hospital for returning sick and wounded soldiers. Many of the squadron's young and up-and-coming sailors had left for Europe to fight for their country. *Graythwaite*, a property located in Edward Street, North Sydney, had been donated to the Crown by T. A. Dibbs, former commodore of the Royal Sydney Yacht Squadron and also owner of the Reeks-designed *Ena*, to be converted into a convalescent home for sick and wounded soldiers and sailors.[1885] Such fundraising efforts involved a '*Monster Garden Fete*', held on 17 July 1915 with patronage from the Governor of New South Wales, Sir Gerard Strickland, as well as the Lord Mayor of Sydney, Sir William Cullen, and his wife, and the Mayor and Mayoress of North Sydney.[1886] This particular event included a concert by leading artists, convened under the direction of Reeks, amongst them '*The Dandies and pupils of Sara Collins*'.[1887] There was also an exhibition of deep-sea diving, and races for 12-oared cutters, gigs and dinghies.[1888] Admission to the grounds was charged at 2/-, half-price for children.[1889] During the event itself Reeks managed affairs from the squadron's pier, including organising trips to Fort Denison to view musketry.[1890]

The annual general meeting of the Royal Sydney Yacht Squadron was held on 23 August 1915 with Reeks again elected honorary official measurer, as well as to the general committee. Owing to the war, the club's usual racing programme was not held; only the Northcote Cup contest had been sailed. It was also noted that 16 of the squadron's 301 members had volunteered for service abroad. The list included Sir Alexander MacCormick (commodore), Dr Gordon Craig, Walter M. Marks and Owen B. Dibbs.[1891] Before commencing the ordinary business, however, the chairman, T. W. Bremner, moved that members of the squadron record with profound regret the death of their fellow member, Captain Stewart Milson, who was killed in action in Gallipoli, the son of vice-commodore A. J. Milson.[1892]

Australia had been at war for a year but already the casualties were high. Men were returning home with varying degrees of injuries, some permanent. Others had been killed. Their plight revealed the conflict's true cost.

[1876] *The Daily Telegraph*, 20 August 1914.
[1877] *The Advertiser*, 20 August 1914.
[1878] *The Daily Telegraph*, 26 August 1914; *Table Talk*, 10 December 1914.
[1879] *The Mercury*, 30 March 1915; *Saturday Referee and the Arrow*, 3 June 1916.
[1880] *Australian Town and Country Journal*, 21 April 1915.
[1881] *Referee*, 13 January 1915; *The Sun*, 26 January 1915.
[1882] *The Sydney Morning Herald*, 13 February 1915.
[1883] *The Daily Telegraph*, 17 February 1915.
[1884] *The Sun*, 21 February 1915.
[1885] en.wikipedia.org/wiki/Graythwaite.
[1886] *The Daily Telegraph*, 10 July 1915.
[1887] *The Daily Telegraph*, 10 July 1915.
[1888] *The Daily Telegraph*, 17 July 1915.
[1889] *The Daily Telegraph*, 10 July 1915.
[1890] *Punch*, 22 July 1915.
[1891] *The Sydney Morning Herald*, 25 August 1915.
[1892] *The Sydney Morning Herald*, 25 August 1915.

On a professional level, Reeks continued his naval architecture and marine engineering business, including consulting to the various Sydney ferry companies and other commercial enterprises. He very likely assisted the Commonwealth and various state governments who were not only frantically expanding their naval fleets with new vessels but also commandeering existing commercial vessels for use as part of a coastal auxiliary patrol fleet. Reeks' expertise and experience would have been too extensive and his business interests too great not to have been called upon to help with Australia's war efforts. And knowing his personality to date, he would have been more than happy to assist. Reeks also retained a position on the committee of the Engineering Association of New South Wales, though stepped down as the organisation's president.[1893]

There were other business commitments too. April of 1915 saw Reeks providing evidence to the Court of Marine Inquiry in Melbourne regarding the wreck of the Commonwealth fisheries steamer *Endeavour* which he had designed in 1908 in association with Andrew Christie.[1894] The vessel had been lost in December of 1914, with all on board, en route from Macquarie Island to Hobart.[1895]

In mid-May of 1915 Reeks gave an interview to the *Sunday Times* regarding the possibility of building a ship with bulkheads so strong that the vessel could remain afloat after being torpedoed, as had occurred to the *Lusitania*. Sailing from New York to Liverpool, the Cunard Liner had been sunk just a week prior by a German U-boat 11 miles off the southern coast of Ireland, resulting in the death of 1,198 passengers and crew.[1896] Reeks' response was interesting and indicates not only the wealth of his engineering and practical knowledge but just how much he knew of vessels being built around the world at the time. He stated, '*there are at the present time only three really unsinkable vessels afloat, and these are ice breakers, employed by the Russian Government to keep some of the far North ports of the Russian Empire free of ice. These three vessels are subdivided into so many compartments, that, to all intents and purposes, they are absolutely unsinkable, save by the use of high explosives. But for any other purpose they would be entirely useless. They are not called upon to carry either cargo or passengers, and the few hands they require to work them, and the coal they consume, do not call for much spaces.*

But with a modern merchant vessel the question is quite different. The cost of such a ship is enormous, and it has to be built with sufficient passenger and cargo carrying capacity to be a commercially sound proposition. It would be quite impossible to divide a ship for practical commercial purposes into a sufficient number of compartments to make her unsinkable. If that were done there would be no room for cargo or passengers, and the vessel would be as useless to her owners as she would be worthless as a utility concern.

Since the Titanic disaster the large liners have been built with two hulls, one inside the other, and divided by a space of a few feet. That space is not altogether wasted, because that portion of it below the water-line is available for water ballast. The general run of people have no idea of the enormous bulk of some of the articles ships have to find room for in their holds. It is not an uncommon occurrence for five or six locomotives to be accommodated in one hold. Obviously, then, the holds must be of large dimensions.

Let me also point out that all the great ships in the mercantile marine now afloat were designed merely to safeguard against the ordinary dangers incidental to the sea. They never contemplated having to withstand the shock of modern explosives, such as torpedoes and floating mines. Against such attacks they are powerless. Even the strongest battleship is impotent against such form of attack, and for the same reason, though in a lesser scale than is the merchantman - because the battleship as well as the trader is compelled to find accommodation for a very large crew, vast quantities of ammunition, to say nothing of coal or oil, as the case may be. Of course, it may be possible to build a battleship that would withstand the shock of modern explosives, but it is practically certain that such a ship would be useless as a fighting unit.'[1897]

In October of 1915 Reeks contributed to a special lecture series on the impact of the war, in association with experts in engineering, as well as medicine, statistics and the weather. His specific piece was published in the *Sunday Times* and gave an account on the effect the conflict was having on the building of merchant vessels. In particular Reeks noted that, '*One effect of the war has been the great benefiting of shipowners, who have been able, as a consequence of the difficulty in providing bottoms to meet the increased demands of the trade, to dispose of many of the craft laid-up in the marine "Rotten Row"*'.[1898] He then discussed the status of British and American shipbuilding, with a note that Australia could soon become more involved in the industry globally through the construction of smaller military craft, including for the defence of rivers.[1899]

Demand for new passenger and light cargo steamers in regional centres was also increasing. For Davis Bros. and Burgess Ltd. to ply on the Richmond River, Reeks designed a new 91ft craft.[1900] The vessel was built by Oliver Jones at Oakland. However, it would be some years before it was launched.

While tensions in Europe continued to escalate, the conflict played out at home too. In February of 1916, less than 700 metres from Reeks' home, a cake supply shop located at 69 Belmont Road, Mosman, was bombed, the subject of an arson attack directed at the owners that were believed to have been of German origin.[1901] None of the occupants, including small children, were injured, though the building was destroyed.

[1893] *Construction and Local Government Journal*, 1 November 1915.
[1894] *The Mercury*, 24 April 1915.
[1895] *The Mercury*, 6 January, 20 March, 21 April 1915.
[1896] https://en.wikipedia.org/wiki/RMS_Lusitania.
[1897] *Sunday Times*, 16 May 1915.
[1898] *Sunday Times*, 24 October 1915.
[1899] *Sunday Times*, 24 October 1915.
[1900] *The Richmond River Herald and Northern Districts Advertiser*, 10 December 1915.
[1901] *The Daily Telegraph*, 29 February 1916.

The grim reality of the world's events were also showcased on Sydney Harbour by the presence of troop ships and gunboats at the Anniversary Regatta held on 26 January 1917, by now the oldest continuously held regatta in the English-speaking world, having first been staged in 1837. The event was held with markedly fewer races, as well as no flagship. Reeks again served on the organising committee and attended the official luncheon on board the Governor's launch *Premier.*[1902] The chief topic of conversation was of course the war, with Sir Gerard Strickland giving a patriotic speech.[1903]

While the Great War entered its third year, come February of 1917 Sydney's shipyards were showing considerable activity. The Government of New South Wales had also established the Walsh Island Dockyard and Engineering Works at Newcastle, with multiple vessels being constructed beginning in 1914.[1904] New government and commercial vessels were increasingly in demand. Messrs Morrison and Sinclair of Balmain, for example, had in the course of construction and almost ready for launching a *'fish carrier'* for the Government of New South Wales.[1905] Designed by Reeks, the craft was to be used in connection with fishing in the state's estuaries and rivers whereby local fishermen could supply the government with fish, the latter cleaning and freezing the catch at shore-based depots that were at the time being established.[1906] The fish would then be transferred to the vessel, equipped with insulated holds to carry the fish to cold stores based in Newcastle and Sydney. It was also of light draft, for bar harbour work, and was to be driven by Skandia crude oil twin engines of the hot bulb type, while electric power was to be generated by a kerosene motor, for freezing, windlass and light.[1907] Cargo was to be handled by oil power winches. The carrier was 133ft in length overall, with a beam of 26ft 6in and a draft of 6ft loaded and was expected to reach a speed of 10½ knots.[1908]

Named *Colliboi,* the vessel was launched on 23 February 1917.[1909] After some delay, owing to its engines being destroyed while under transport to Australia and it instead being installed with conventional steam engines, the craft was not ready until October of the following year.[1910] It then worked along the New South Wales coast for several years. However, the *Colliboi*'s usefulness was plagued by mismanagement and an inadequate supply of fish. It was advertised for sale by tender in April of 1921, ultimately being purchased by the Government of Tasmania a few months later for service between Launceston and the Bass Strait islands carrying passengers, stock and cargo.[1911] Sold to William Holyman & Sons Pty. Ltd. in 1929 though remaining in the same service, the *Colliboi* struck a reef near Cape Grim on Tasmania's north-west coast in August of 1932. Badly holed, the vessel sunk and was subsequently declared a total wreck.[1912]

In addition to *Colliboi,* at least five vessels designed by Reeks were also under construction in 1917. These included a light-draft auxiliary ketch being built at Balmain by George Washington for use in the Northern Territory.[1913] The vessel was 120ft in length with a beam of 26ft 6in and a draw, when loaded, of 4ft 6in.[1914] It was to be propelled by twin screws actuated by Skandia crude oil engines, and was to have electric plant and deck equipment. In addition to designing the vessel, Reeks was also tasked with supervising its construction on behalf of the Federal

1902 *The Daily Telegraph*, 27 January 1917.
1903 *Punch*, 1 February 1917.
1904 en.wikipedia.org/wiki/Walsh_Island_Dockyard_and_Engineering_Works.
1905 *The Daily Telegraph*, 7 February 1917.
1906 *The Daily Telegraph*, 7 February 1917.
1907 *The Daily Telegraph*, 7 February 1917.
1908 *The Daily Telegraph*, 7 February 1917.
1909 *The Sun*, 23 February 1917.
1910 G. Broxam and M. Nash (2013). *Tasmanian Shipwrecks. Volume 2. 1900-2012.*
1911 *Daily Commercial News and Shipping List*, 20 April 1921; World, 5 July 1921.
1912 G. Broxam and M. Nash (2013). *Tasmanian Shipwrecks. Volume 2. 1900-2012.*
1913 *The Daily Telegraph*, 7 February 1917.
1914 *The Daily Telegraph*, 7 February 1917.

COLLIBOI AT MELBOURNE (C1920S).
SOURCE: STATE LIBRARY OF VICTORIA (H91.108/590).

SIR JOHN FORREST.
SOURCE: LIBRARY & ARCHIVES NT (PH0155/0013).

Government.[1915] He was paid £203 18s. 7d. for this work. The vessel was named *Sir John Forrest* and launched early in 1918.[1916]

Meanwhile David Drake Ltd. of Balmain was constructing a cargo lighter of 175 tons capacity for Messrs Elliott Brothers, Ltd., of a special type designed by Reeks.[1917] A similar lighter, for the same owners, was also being built by Arthur Davis of Brisbane Water, very likely to a design by Reeks.[1918] Unfortunately the names of these two vessels are not known.

The wooden hulled motor vessel that became the new *Merrie England* was design No. 288 for Reeks, completed in 1916. It replaced the impressive craft he had prepared plans for in 1913. Intended for use by the Lieutenant-Governor of Papua (New Guinea) for visiting outlying stations, it was now slowly nearing completion at Messrs Fisher and Fisher's yard at Careening Cove.[1919] With dimensions of 113ft overall, with a 25ft beam and a draw of 6ft but capable of loading 7ft with a full cargo of stores, all accommodation was on deck, except the main saloon and worker's quarters.[1920] Provision was also made for six passengers, in addition to the Lieutenant-Governor and his staff. A spoon bow and canoe stern gave the appearance of good sea-going qualities and the craft was fitted with Union oil engines.[1921] It was to be ketch-rigged, with storm or steadying sails only.[1922] The *Merrie England* was launched on 11 September 1917, built at a cost of £5,860 for hull and fittings, with the engines an additional £2,242.[1923] Reeks also acted as supervising engineer for the vessel's construction.[1924]

The *Merrie England*'s story though is an unhappy one, and records held by the National Archives of Australia help tell the tale. Once trials were completed it went to Port Moresby and entered service toward the end of 1917. Soon faults were found and the subsequent repairs and responsibilities entered into an insurance dispute. In 1918 the builders defaulted and closed their yard, partly a victim of over-runs on the quoted cost of building. Then on the 17 March 1919 a fiasco with tragic consequences saw the vessel destroyed in the harbour at Port Moresby.

Merrie England was tied up to the wharf in Port Moresby Harbour at the time, and a fuel tank aft was found to be leaking benzine (petrol). The company engineer who was asked to do the repair was not the first choice for the task and his inexperience probably led to a fatal mistake, trying to solder the leak with a blow torch in hand. The tank exploded and he was killed instantly, then fire broke out. On shore it was soon decided that the only course of action was to scuttle the vessel to put the fire out. It was a dramatic option, as the only means of sinking the *Merrie England* was to place a gelignite charge up forward to blow a hole in the hull. The charge was possibly too much, and the ensuing explosion apparently destroyed the craft. Its remains settled on the seabed near the wharf. Only the engines and other machinery were salvaged.

Rahu Doura, a native of Elevara in Papua New Guinea, and an 'engine boy' on the *Merrie England* gave the following description of the events at the inquiry held soon after. He was lucky to survive unharmed.

Initially he helped the company engineer, Mr Gregory, with soldering irons but they failed to do the repair, and a blow torch was tried. He stated, '*After dinner* (i.e., lunch) *Mr Gregory told me to put one case of benzine in the tank, Mr Gregory took down one big blow lamp and the smaller blow lamp. He told me to tip up the tank, and saw where leak come out again. He told me to put the tank down square which I did. Then he unscrewed the valve on top. Then he put the blow lamp and solder to the leak, when he did, this one he blew up, and engine room full of smoke. I was knocked down and silly through the explosion and felt my way out of the engine room*'.[1925]

The telegram from the Acting Administrator Papua dated 17/3/1919 completes the picture.

'*KEPPIE [the ship's engineer] WAS ASHORE AT THE TIME STOP TWO NATIVE CREW IN ENGINE ROOM WITH GREGORY WHO WAS REPAIRING BENZINE TANK WITH BLOW LAMP BOTH NATIVES ESCAPED GREGORY PROBABLY KILLED BY EXPLOSION STOP REMAINS OF SHIP FINALLY SUNK WITH DYNAMITE IN FIFTY TWO FEET OF WATER STOP PLEASE INFORM BLOOMFIELD AS REPRESENTATIVE INSURANCE STOP NOT CLEAR TO ME WHETHER LOSS BY FIRE COVERED BY POLICY*'.[1926]

How that detail was resolved is not known, but a note on a copy of the telegram and other aspects of the ensuing legal actions mirrors many of our own insurance claim experiences. It starts with the note that says '*Mr Bloomfield has absolutely nothing to do with this insurance*' and moves on without resolution.[1927] Subsequent documentation from the National Archives of Australia shows that there was an

[1915] *The Daily Telegraph*, 7 February 1917.
[1916] *The Advertiser*, 18 April 1918.
[1917] *The Daily Telegraph*, 7 February 1917.
[1918] *The Daily Telegraph*, 7 February 1917.
[1919] *The Daily Telegraph*, 7 February 1917.
[1920] *The Daily Telegraph*, 7 February 1917.
[1921] *The Daily Telegraph*, 7 February 1917; *Evening News*, 13 September 1917.
[1922] *The Daily Telegraph*, 7 February 1917; *Evening News*, 13 September 1917.
[1923] *The Sun*, 12 September 1917; *Papuan Courier*, 26 October 1917.
[1924] *Evening News*, 13 September 1917.
[1925] National Archives of Australia, NAA: A1 1923, 20664.
[1926] National Archives of Australia, NAA: A1 1923, 20664.
[1927] National Archives of Australia, NAA: A1 1923, 20664.

insurance dispute over earlier repairs that were underway at the time of the accident and loss of the *Merrie England*, then moved into a further dispute about the details of how the fire happened, who made the decision to sink the vessel, why did they needed to use gelignite, and how they came to their conclusions.[1928] The final outcome of all this is not recorded, but the *Elevala* that had already served as an interim vessel for the service was once again called in, until 1924 when a new ship from Morrison and Sinclair's own design and build, the *Laurabada*, finally took on the role, and served until the Second World War.

Another interesting design by Reeks completed around 1917 was that of a tug having a range of action at sea of upwards of 2,000 miles, yet capable of passing under a bridge 15ft high by the simple process of lowering the funnel, every other part being kept below that limit.[1929] Despite repeated searches of multiple sources, there are no additional details known about this craft.

In June of 1918 the Government of Queensland announced that it had purchased a trawler which was at the time under construction at Morrison and Sinclair's yard, Balmain.[1930] Designed by Reeks, the vessel had a length of 135ft, a beam of 23ft and a coal capacity of 100 tons.[1931] There were insulated fish holds, refrigerating machinery, and separate chilled departments for frozen fish. Intended for trawling and general fishing, as well as fisheries investigation work along the Queensland coast, it was named *Bar-ea-Mul* and launched at high tide on 23 August 1918, beginning service the following year.[1932]

Later in 1918 another launching ceremony involving a vessel designed by Reeks took place. Having been under construction for several years at Oliver Jones' yard at Oakland, the passenger and cargo steamer intended for the Richmond River trade, *Banksia,* finally left the stocks on 21 December.[1933] The craft was built to the order of Davis Bros. and Burgess Ltd.[1934] It would go on to have an interesting history; its hull being moored for another 20 years before it was finally completed in August of 1938.[1935]

Also in 1918 Reeks became involved in education, the first line of employment he had undertaken upon his arrival in Sydney back in 1885. On 22 March 1918 Reeks was appointed by the New South Wales Department of Education to be part of an advisory committee tasked with improving and making recommendations regarding trade education associated with shipbuilding.[1936] It was a two-year appointment that significantly saw Reeks part of a committee with representatives from the Commonwealth Naval Dockyard at Cockatoo Island, the Government Dockyard at Walsh Island in Newcastle, several of Sydney's shipbuilders and union representatives.[1937]

While this position, as well as orders for new government and commercial vessels certainly kept Reeks busy during the war years, there was very little work for him in association with recreational craft. Reeks, however, did remain involved in the sailing community, as well as continued to promote and champion benevolent causes. In April of 1917 he attended a garden fete held at *Malahadie*, the home of John Mackay in Kirribilli, to raise funds for

[1928] National Archives of Australia, NAA: A1 1923, 20664.
[1929] *The Daily Telegraph*, 7 February 1917.
[1930] *The Telegraph*, 13 June 1918.
[1931] *The Telegraph*, 13 June 1918; *The Sun*, 24 August 1918.
[1932] *The Sun*, 24 August 1918.

[1933] *The Richmond River Herald and Northern Districts Advertiser*, 10 December 1915, 24 December 1918.
[1934] *The Richmond River Herald and Northern Districts Advertiser*, 10 December 1915, 24 December 1918.
[1935] *Northern Star*, 8 August 1938.
[1936] *Government Gazette of the State of New South Wales*, 22 March 1918.
[1937] *Government Gazette of the State of New South Wales*, 22 March 1918.

BAR-EA-MUL AT SEA, NEW SOUTH WALES (C1930S).
SOURCE: NATIONAL LIBRARY OF AUSTRALIA (NLA.OBJ-162020227).

the Golden Fleece Free Kindergarten. The event saw Reeks volunteering, with R. Sinclair, to provide motor launch trips on the harbour to Fort Denison.[1938]

The annual general meeting of the Royal Sydney Yacht Squadron took place in early September of 1917. The number of members now comprised 270 while the roll of honour stood at 30. Along with Captain Stewart Milson, Lieutenant R. D. Burns had also been killed in the conflict.[1939] While no races were held that season, Reeks was once again appointed honorary official measurer, as well as a member of the committee.[1940] Re-appointed to both roles the following year, by mid-1918 the club's honour roll comprised 32 members of which two more members had sadly been killed: Captain Owen B. Dibbs, son of T. A. Dibbs, and Captain R. L. Johnson.[1941]

The Anniversary Regatta of 1918 saw Reeks on board A. G. Milson's motor yacht *Mackerel*.[1942] Though the event continued to be held, there were no speeches, no luncheon, and no prizes, conforming with the spirit of the times. Later that year Reeks helped organise a patriotic aquatic carnival.[1943] The program featured sailing and rowing events and was one of the largest aquatic functions held on Port Jackson for many years. Reeks also donated £10 to event which was held on 26 October 1918.[1944]

In December of that same year Reeks helped organise and facilitate a rowing contest for blinded soldiers, the first to take place in the Commonwealth. The event was held at Mosman Bay on 21 December 1918 in association with the Sydney Industrial Blind Institution and the Mosman Rowing Club.[1945]

1938 *The Daily Telegraph*, 2 April 1917.
1939 P. R. Stephensen (1962). *Sydney Sails: the Story of the Royal Sydney Yacht Squadron's First 100 Years (1862-1962)*.
1940 *The Daily Telegraph*, 12 September 1917.
1941 *The Daily Telegraph*, 28 August 1918.
1942 *The Daily Telegraph*, 28 January 1918.
1943 *Referee*, 11 September 1918.
1944 *The Sydney Morning Herald*, 18 October 1918; *The Daily Mail*, 2 November 1918.
1945 *Sunday Times*, 22 December 1918.

By this point in time the end of the hostilities was approaching. To this day the First World War stands as Australia's costliest conflict.[1946] The statistics are devastating. From a population of approximately five million people, nearly 420,000 had enlisted of whom more than 60,000 were killed and 156,000 wounded, gassed, or taken prisoner.[1947]

While there was much relief that the conflict had ended, a lengthy recovery process followed from multiple perspectives; globally, at the Commonwealth and state level, from a business and organisational point of view, as well as personally, both physically and mentally.

With much rebuilding to take place Reeks felt it important to honour those that had served and sacrificed. He designed and presented to the Royal Sydney Yacht Squadron a Roll of Honour. It was duly placed in the clubhouse at the end of the war and lists the names of the 33 members who enlisted and served abroad, including the four who lost their lives during active service.

It was very much a symbol of Reeks' empathy and sympathy with his fellow clubmen. Acknowledging and sharing in their loss, he was also still in mourning, for he too had lost a son well before his time.

Overall, it was the end of a tumultuous period for Reeks. While he had had many successes from a professional standpoint, personally, his son had been murdered and the weight of the First World War had been immense. Yet, there had been joy. His daughter Thelma had married and given birth to two sons, Kent in 1915 and Lindsay in 1917.[1948] Though both children were born in Sydney, Thelma, her husband Alfred Fox Ashwin and their children were soon established at Millridge, a cattle and sheep farming property near Tottenham in the central west of New South Wales, some 500 kms from Sydney.[1949]

1946 www.awm.gov.au/articles/atwar/first-world-war.
1947 www.awm.gov.au/articles/atwar/first-world-war.
1948 National Archives of Australia, A9301, 20968 and B884, N199198.
1949 *Lachlander and Condobolin and Western Districts Recorder*, 24 April 1918.

THELMA FOX ASHWIN (NEE REEKS) AND HER SON KENT (C1916).
SOURCE: FOX ASHWIN FAMILY COLLECTION.

DRAWINGS OF WALTER REEKS' HOME, TWYNHAM, MOSMAN BAY.
BY DAVID PAYNE.

CHAPTER TWENTY ONE - HOME

Following an absence of four years, yachts first appeared on Sydney Harbour on 18 January 1919, with the yachting season officially opened in ideal sailing conditions. Yachts, cruisers, motor yachts and open boats, numbering over 100, assembled at Farm Cove and, as they manoeuvred for position, presented a picture that had not been witnessed on Port Jackson since pre-war times. The arrangements were carried out by Walter Reeks, T. W. Bremner, A. G. Milson and J. C. Penrice, representing the Royal Sydney Yacht Squadron, as well as members from the Royal Prince Alfred Yacht Club.[1950]

The Anniversary Regatta of 1919 was also revived to its pre-war program. It was a large event, Sydneysiders by and large giving a collective sigh of relief to the post-conflict days and coming out in droves to enjoy the holiday and fine weather. Reeks was again involved in the event's organisation and attended the official luncheon on board the flagship *Somerset*.[1951]

The first yacht race sailed under the joint management of the Royal Sydney Yacht Squadron and the Royal Prince Alfred Yacht Club took place on 8 February 1919. The event was termed the 'Victory Yacht Race' in honour of the cessation of hostilities. The officials were Reeks and A. G. Milson.[1952] The showcase race was won by the veteran *Magic*, skippered by Stan Stevens, with second place going to *Eun-na-mara* (A. Wilson), and third to *Yeulba* (E. S. Heath).[1953]

While organised sport and pastimes were making a comeback and life was returning to somewhat normalcy, there was widespread concern regarding the Spanish Flu, at the time threatening the health of much of the world's population. The translocation of the virus to Australia was helped significantly by the return from Europe of the country's servicemen and women. While those arriving by ship were forced into quarantine, the disease was still not curtailed. *The Daily Mail* reported on the situation on 8 February 1919. The text is eerily reminiscent of the situation we found ourselves in recently regarding the COVID pandemic.

'Woe betide the man, woman, or child who appears in the streets today without a mask! It is a Government mandate and must be obeyed. The man in the streets submits cheerfully to the muzzling, putting up with optimistic outlook, to the inconvenience of the gauze that veils his nose and mouth. The Minister for Health has been greatly commended for his foresight since the outbreak of the "Spanish flu". It is a very mild outbreak, but as many more transports have to arrive through the gates of our harbour, it is best to be prepared ... If the mask keeps away the deadly germ, Sydney will wear it as cheerfully as if it were a laurel wreath.'[1954]

Reeks himself was at the time nearing 58 years of age. He had survived a serious case of typhoid decades earlier and was now faced with another potential threat to his health and well-being. It was an era without antibiotics, with only rudimentary medical technologies in use and where most illnesses were treated at home. Reeks likely took an optimistic approach to the threat of the Spanish Flu, however. With all that had transpired during the past few years, including the murder of his son and the war itself, he was likely just thankful to be returning to his projects, to administration of yacht racing and to life in general, including social and community engagements.

[1950] *The Sun*, 19 January 1919.
[1951] *The Sydney Morning Herald*, 28 January 1919.
[1952] *The Sun*, 9 February 1919.
[1953] *The Sun*, 9 February 1919.
[1954] *The Daily Mail*, 8 February 1918.

'Bob Backstay' reported on one such event that took place a few weeks after Reeks' 58th birthday. '*Among the Ancients. On Saturday last I went afloat with the Ancient Mariners and fraternised once more with many old pals. These included Ex-Commodore S. M. Dempster, Captain J. D. S. Phillips, Walter Reeks, Captain Bayldon, Captain Sam Mortimer, Captains Hammond and Anderson. The five skippers mentioned were at one time or another all on the old R.M.S. Aorangi, a craft so well known in these waters. The five had a sort of reunion, thus proving that the founders of the league had done something. If it had not been for them, men who used to be shipmates would not meet again in such happy circumstances.*'[1955]

The vessel in question, RMS *Aorangi,* was a 4,000-ton mail and passenger steamer that plied between Australian, Pacific and Canadian ports in the early twentieth century. It is interesting to note that Reeks was part of the '*Ancient Mariners*' set. More than likely he was a part-instigator of the reunion, content to be amongst a group who shared an affinity for their maritime history and enjoyed the discussions that the reunion of these stalwarts generated.

While much was going on for Reeks in the maritime realm, during this period he was still clinging to the idea that one day he would strike gold. Though his previous efforts to fund and be involved with gold mining ventures had been wound up during the First World War, there were other avenues to pursue. One such opportunity came by way of a prospectus for the Gundagai Big Reef Gold Mine N. L. which was issued in August of 1919.[1956] The company was to have a nominal capital of £100,000, divided into 200,000 shares of 10/ each, comprising 80,000 fully paid, 80,000 contributing, and 40,000 shares in reserve.[1957] It was proposed that the new company acquire certain private and other gold leases, situated some nine miles from Gundagai railway station, in New South Wales, near the main road to Tumut, the total area being 75 acres.[1958] The provisional directors, along with Reeks, were John H. King, William Brooks (MLC), Finlay McLeod and Percival Wilson.[1959] The mine was prospected and tested by January of the following year, with machinery erected.[1960] However, the venture appears not to have progressed any further and there is no mention of it past this point in time. The mine was reregistered as the Clarke-Wesley Gold Mine in October of 1920.[1961] It is not known if Reeks received a return on his investment, with this enterprise and all of the others that he poured money into.

While Reeks appears to have lessened his involvement with his various professional commitments by this stage of his life, he continued to write articles on advances in ship design. On 20 August 1919 a piece he authored was published in *The Sun* discussing the minutiae of pitch and roll, one would think a rather bland and technical topic for the general public. Still, besides giving much technical detail in the text, it is an engaging article that not only entertains the reader about this technical aspect of vessel design then undergoing development considering ships were getting larger and larger, but also keeps them amused, even distracted to the point that the reader may not even notice that they are being educated on the subject. It's worthy of some lines being copied herein, an indication of Reeks' wonderful ability to not only convey his love for the subject but also his charismatic way of breaking down a complicated subject into a engaging story. He would have been a great teacher.

'*This morning the writer* [i.e., Reeks] *received a postcard from a friend abroad with on one side a picture of a big Cunarder (name not given) and on the other "really, so far as motion is concerned, one would not know one was at sea".*'[1962] He continued, '*That is what big ships have done, and at first it would appear that there was nothing more to say about it, but let us look a little further.*

The problem of shipping is many-sided, and highly technical, but, as technical descriptions are the province of scientific journals we will leave them to their muttons and look at this matter of big ships from the point of view of these who use them. We all use them directly or indirectly, and as it is in the blood of most men to love ships, as the girls do the sailors, their raison d'etre shall forthwith be "the traffic of our stage".

All shipping, large or small, must be looked at from three main points of view, viz., that of the passengers, otherwise the public who pay, the owners who provide, and the builders who produce them.

The order of these three fundamental interests, like Sam Weller's name [again a reference to that character in Charles Dickens' *The Pickwick Papers*], *is left to the taste and fancy of the reader. Naturally each one thinks his is the most important part of it, but we will take them in sequence as above, and see first how the big ship appears and appeals to the discriminating passenger.*

Let us see what happens.

You accept an invitation to cruise with friends who have a twenty-feet launch, and a nice summer north-easter obtains. Soon you are heeling and toeing it in fine style, and the spray flies all about. One moment her stem is under, and the next the tuck seems feet below the surface. It's grand fun: but observe the common or garden ferry boat passing. She slides along with no more motion than in Circular Quay. Why?

The size (i.e., length crest to crest) of the wavelets produced by the said summer breeze is greater than that of your launch, consequently her bow is lifted by the crest of one such wave while the stern is depressed in its trough, which order of things is immediately reversed by the same wave passing aft, lifting the stern while the bow is in the hollow. And so the boat game goes on. Not so the ferry boat, the crests and troughs are the same for her as with you, but the ferry is six or seven times the length of the launch and has half a dozen crests and troughs under her at the

[1955] *Evening News*, 25 March 1919.
[1956] *The Daily Telegraph*, 8 August 1919.
[1957] *The Daily Telegraph*, 8 August 1919.
[1958] *The Daily Telegraph*, 8 August 1919.
[1959] *The Daily Telegraph*, 8 August 1919.
[1960] *The Gundagai Times and Tumut, Adelong and Murrumbidgee District Advertiser*, 20 January 1920.
[1961] *Adelong and Tumut Express and Tumbarumba Post*, 1 October 1920.
[1962] *The Sun*, 20 August 1920.

same time, she straddles the lot, and one counteracting the other, she proceeds as if in quite smooth water.

It just boils down to a simple question of length of size and size of ship. If the ship is long enough to span three complete seas there will be no pitch, but as no vessel has yet been built to do that entirely, when the longest seas obtain, it follows that pitching under extreme conditions of seas has not yet ceased.

It has been calculated with a considerable degree of accuracy that the longest seas experienced do not exceed 300 feet, and that theoretically a vessel of 900 feet length will not pitch. This appears to be very nearly true, because, as only the exceptional seas are that length, the majority are less, and therefore less than 900 feet will fill the bill.

Quite a number of well-known Atlantic liners have been built between 700 and 800 feet in length, and performed so well as to create the impression that that would be about the limit, but, as some have exceeded that already, and more are building, it is better not to prophesy limits.'[1963]

The article then continued in the same style, explaining the use of bilge keels to keep a vessel in an upright position and improve stability, as well as the introduction of '*lavish*' facilities for modern passengers, including electric lifts, writing and smoking rooms, palm courts, veranda cafes, tennis courts, gymnasiums, swimming and Turkish baths, etc. Reeks finished the article with the following statement, '*The demand for swift and luxurious travel has come to stay, and with it the need for big and yet bigger ships; so says the travelling public, so say we all*'.[1964]

An interesting response was published the following day whereby a reader cheekily stated the following.

A Long-felt Want

Mr. I. Will Soak writes: I was very much interested in Mr. Walter Reeks's remarks on the means adopted to prevent the rolling of vessels at sea. I had heard of bilge-keels before, but did not know their use. Now it seems to me, sir, that this principle might be adopted ashore. Gentlemen setting out for a convivial evening might carry with them some sort of bilge keel which could be adjusted on their boots, so that they should not roll on the way home. Such a device would save many a good fellow from bruises and contusions, and would also enable him to escape, perhaps, a humiliating appearance in the police court. The instability caused by a proper indulgence in drink has so far been the only drawback to drink, and I feel sure that if bilge keels would do away with this we should be able to escape detection by the Pussyfoot Johnsons of this fair city. I would like Mr. Reeks to communicate with me on the subject, giving me some idea of what a pair of good strong adjustable bilge keels would cost.

SOURCE: THE SUN, 21 AUGUST 1920.

It is not known if Reeks followed up with this matter.

[1963] *The Sun*, 20 August 1920.
[1964] *The Sun*, 20 August 1920.

In the weeks following, Reeks published two more articles: one on the evolution of ships crossing the Atlantic Ocean and one on the various considerations shipbuilders take into account when building big passenger steamships.[1965] While both provided ample technical information and well-researched statistics, like the first, they were engaging to the reader, taking a complex subject and making it easily digestible.

The Royal Sydney Yacht Squadron's annual general meeting was also held in late August of 1919. Membership now stood at 275 with 34 names on the Roll of Honour. Reeks was again appointed honorary measurer and to the committee. There was much discussion on the need to modernise the fleet, perhaps adopting one-design class racing, as well as increasing efforts to recruit younger members. In terms of the former, a smaller type of yacht was recommended, with details provided on the American Star class.[1966]

Reeks was also in attendance at the launch of A. C. Saxton's new six-metre yacht *Awanui III* in October of that year; one of the very first vessels built since the First World War. Intended to challenge for the Northcote Cup, the craft was constructed by James Hayes and Sons to a design by Morgan Giles of England.[1967] A few weeks later, the opening of the yachting season took place on Sydney Harbour with Reeks enjoying the festivities as an official representing the Royal Sydney Yacht Squadron.[1968] Instituting pre-war events, he also attended a garden party as part of Royal Sydney Yacht Squadron season opening events.[1969]

In January of 1920 Reeks donated a prize, in conjunction with Colonel E. W. Spain, for a race for 2ft model yachts, the first time the 'Lilliputian' class had appeared on the Anniversary Regatta programme.[1970] The prize was intended to encourage development of the class.

Reeks was also noted in the press as providing comments on the 1920 America's Cup. It was the first time since 1903 that the event had been staged. It was also the fourth time Sir Thomas Lipton had challenged for the trophy.

Held under the Universal Rule of measurement, adopted in 1914, it was a highly-anticipated event with an international following, particularly since Lipton's *Shamrock IV* had won the inaugural race of the five-race series, the first time he had won a race during his decades-long attempts.[1971] Hopes for an upset became even higher when *Shamrock IV* won the second race without incident and by a corrected time of two minutes.[1972] Interviewed about this particular race Reeks commented, '*The outstanding feature seems to me to be that the Shamrock won on her merits. By this I mean that the race was a perfectly fair one, not*

[1965] *The Sun*, 29 August, 3 September 1919.
[1966] *The Daily Telegraph*, 27 August 1919.
[1967] *Arrow*, 31 October 1919.
[1968] *The Sun*, 2 November 1919.
[1969] *The Sun*, 19 October 1919.
[1970] *Arrow*, 2 January 1920.
[1971] en.wikipedia.org/wiki/1920_America%27s_Cup.
[1972] en.wikipedia.org/wiki/1920_America%27s_Cup.

marred by any of kind of mishap'.[1973] He continued, perhaps prematurely since three races were still to be sailed, '*It is a most important thing that the cup has been brought back to the United Kingdom, and that the venue has been changed after so many years of battling. The immediate effect of this will be to create far greater interest in yachting in the future'*.[1974]

The defender, *Resolute*, designed by the 'Wizard', Nathanael Greene Herreshoff and owned by a syndicate from the New York Yacht Club headed by Henry Walters, however, went on to win the next three races, thus retaining the coveted trophy. Still, there was much optimism from the outcome, the new rule offered greater opportunity for a challenger to take the title.

Unfortunately Reeks would not be around to witness that extraordinary day when Alan Bond's *Australia II* won the America's Cup in 1983. Notably, the series was raced off Newport, Rhode Island, where it had been held after the 1920 event. However, more significantly, while Reeks did not correctly foresee the outcome in 1920, if we once more return to his interview in 1903 where he discusses Herreshoff and the Cup, Reeks did see how it would eventually be won and, as it so happened, by Australia.[1975]

Firstly, Reeks gives great credit to Herreshoff as, '*the only complete designer and complete engineer in one brain I have ever met. That means he does the work which is generally carried out by two men*'. Our review of Reeks suggests his modesty restricted him from observing what we have recounted, that Reeks too was simultaneously a gifted designer and engineer. But so was the designer of *Australia II,* Ben Lexcen (formerly Bob Miller).

The interview goes on to predict the scenario that was played out in 1983. '*Will you say whether the Cup will ever be taken from America?*' asked the reporter. Reeks responded, '*My opinion is that the Cup will never be taken from America as long as the challengers are built on conventional lines and while such men as Herreshoff live. The only chance of beating them is by what might be termed a "trick boat", that is to say, a type unknown or untried by the Americans. This is only likely to come from such a country as, say France, or possibly Australia or possibly New Zealand - when the challengers must be prepared to take all chances of the untried type of boat*'.[1976]

Reeks' prescience was uncanny here, on two points. First, *Australia II* was indeed that 'trick boat' with Lexcen's unconventional winged keel the standout feature that spooked everyone it raced against and helped give the yacht an advantage over *Liberty*, the American defender in 1983. Second, Reeks noted something else that helped. Herreshoff's mantle was eventually taken up again in the post-Second World War era by another legend, Olin Stephens, whose 12-metre class designs defended the Cup six times and only lost one race, this was *Australia (I)* designed by Ben Lexcen. By 1983 Stephens had retired and the American designs fell to protégés still sorting out

[1973] *The Sydney Morning Herald*, 22 July 1920.
[1974] *The Sydney Morning Herald*, 22 July 1920.
[1975] *Evening News*, 1 September 1903.

[1976] *Evening News*, 1 September 1903.

AUSTRALIA II ARRIVING AT FREMANTLE FROM THE BUILDER'S YARD (3 JUNE 1982).
SOURCE: STEVENSON, KINDER & SCOTT CORPORATE PHOTOGRAPHY VIA STATE LIBRARY OF WESTERN AUSTRALIA (SLWA_B3159290_6).

who would fill his seat at the top. As Reeks forecasted, the two requirements of an American challenger, i.e., meeting a radical or unconventional boat and also losing their dominate designer were present in 1983, helping to give the outcome he dreamed of, an Australian victory and, as we have seen post-1983, a big boost to yachting.

Though the *New South Wales Post Office Directory* still listed Reeks as a naval architect with an office at 232 George Street, there is little indication of what he was up to professionally during the early years of the 1920s. With regards to yachting, however, there was a great deal of activity going on. Reeks was again appointed honorary official measurer and to the committee of the Royal Sydney Yacht Squadron in August of 1920.[1977] A few months later he was presented to the Governor General, Lord Forster, at a levee at Government House, Sydney.[1978] Held on 24 October 1920 the event was an important one. It was the first levee convened by the newly-arrived Governor General, an opportunity for Sydney's esteemed politicians, businessmen and citizens to not only get a glimpse of Lord Forster, but also perhaps the chance to converse with him. Given Lord Forster was a keen yachtsman, Reeks would have been more than interested in discussing all things sailing with him.

Also in October of 1920 Reeks attended the Royal Sydney Yacht Squadron's opening day event at Kirribilli, a social occasion not to be missed.[1979] The next few months saw him busy supervising an extensive overhaul of Frank Albert's *Rawhiti II* ahead of the craft competing in the 1921 Northcote Cup.[1980] Alterations were carried out by Stewart, Sandeman, Ltd., and included a new stern and additional lead, plus a change to Marconi rig.[1981]

The Northcote Cup was held on Sydney Harbour in early February of 1921. With *Rawhiti II* as the defender, it was challenged by the newly-launched Sydney boat *Awanui III*, as well as two Victorian yachts, *The Rip* and *Killara*.[1982] Reeks was part of the event's organising committee, as well as tasked with taking care of visiting crews and arranging conditions and courses for the contest.[1983] He was also race starter and timekeeper for the three-race series which was convincingly won by *Awanui III*.[1984] At a dinner held in honour of those that participated in the Northcote Cup, Reeks wisely prophesied, '*There is never a yacht afloat that another one cannot be built to beat her*'.[1985] He continued, '*It is with this spirit inspiring them that our yachtsmen and open-boat sailors go forth to interstate matches, and I hope to see the same power sending some of them to an international match*'.[1986] Of course he was again referring to the America's Cup, that age-old race and trophy that had intrigued him for decades. Reeks described the conundrum and a not-out-of-reach solution. '*Australians have a wonderful sporting record. In competitions in almost every branch of sport they have distinguished themselves against the top-notchers of the older countries; but yachtsmen, principally through lack of opportunity, have not shared in the international honors. Now is the time, let them seize the opportunity ... Since the Panama Canal was opened Australia's path to the America Cup has been easy. Now, if our yachtsmen have any similar blood in their veins to that which flows in the body of Sir Thomas Lipton, let them, in the interest of the sport they love, move and do something.*'[1987]

Reeks continued, offering a systematic process to success. '*The thing should not be rushed; there is plenty of time ... The hardest part of the whole business will be to get anybody to take the first steps. The next difficulty would be to raise the money. Whether this can be done by a large syndicate of yachtsmen or by public subscription remains to be seen. If the latter course has to be adopted I am sure the sport-loving people of Australia would respond promptly. Now yachtsmen, it is up to you!*'[1988]

It is interesting to note that in this speech, intended to inspire and coalesce a movement, Reeks did not mention anything about the challenging yacht itself. No doubt he would have had ideas, some conventional others possibly intended to turn the yachting fraternity on its heads.

On 25 February 1921 Reeks turned 60 years old. More and more of his time was now dedicated to his favourite pastime of yachting, including increasingly greater committee and event organisation responsibilities. He also began competing in motor boat races at local regattas in *Wahine*.[1989] Another outlet for his time was the Rotary Club. He was also a prominent Freemason.[1990] In addition, there was continued involvement and investment in new and emerging technologies. Utility Machines Ltd., with a capital of £10,000, in £1 shares, was established in March of 1921 to acquire the Australian and New Zealand agencies for '*a rotary scraper machine and diagraph stencil cutting machine*'. Its directors were S. D. McLaren, D. A. Craig and Reeks.[1991] He also applied to the Commonwealth Patent Office for a patent titled, '*Adapting the force of the wind to the propulsion of sailing craft and other vehicles*'.[1992]

By far the organisation that Reeks dedicated the bulk of his time during this period was the Royal Sydney Yacht Squadron. The winter months of 1921 saw him attending a winter dance at the squadron's clubhouse.[1993] At the organisation's annual general meeting, held in August of that year, Lord Forster was elected commodore while Reeks was once more elected honorary official measurer and to the committee.[1994] The addition of Australia' Governor General to the squadron's membership was to

[1977] *The Sydney Morning Herald*, 1 September 1920.
[1978] *The Sun*, 14 October 1920.
[1979] *The Sydney Morning Herald*, 18 October 1920.
[1980] *The Sun*, 7 December 1920.
[1981] *The Sun*, 18 January 1921.
[1982] *The Daily Telegraph*, 18 January 1921; *The Sydney Morning Herald*, 31 January 1921.
[1983] *The Daily Telegraph*, 18 January 1921.
[1984] *The Sun*, 6 February 1921.
[1985] *Evening News*, 9 February 1921.
[1986] *Evening News*, 9 February 1921.
[1987] *Evening News*, 9 February 1921.
[1988] *Evening News*, 9 February 1921.
[1989] *The Sydney Morning Herald*, 29 December 1920; *The Daily Telegraph*, 24 December 1921; *The Sydney Morning Herald*, 2 January 1922.
[1990] *The Sydney Morning Herald*, 25 October 1925.
[1991] *The Sun*, 15 March 1921.
[1992] *Daily Commercial News and Shipping List*, 18 January 1922.
[1993] *The Daily Telegraph*, 16 July 1921.
[1994] *The Sun*, 30 August 1921.

have far-reaching outcomes. His sincerity and eagerness to contribute to the sport, particularly at a national level, were showcased when, presiding over his first meeting of the Royal Sydney Yacht Squadron's committee, Lord Forster announced his keenness for a healthy and inclusive interstate competition for the just-established 21ft restricted size class, going so far as to offer a deed of gift for the event.[1995]

Lord Forster had been in Australia just shy of a year. Yet in this time, despite a calendar heavy with civic and ceremonial engagements, he had purchased the yacht *Yeulba* and transferred the vessel from Sydney to Melbourne where he primarily resided, commissioned the building of a 21ft restricted class yacht to be based in Sydney, visited many of the Sydney and Melbourne-based yacht clubs, travelled to Tasmania where he participated in the 1921 Royal Hobart Regatta, accepted the position of commodore of the Royal Yacht Club of Victoria and, as stated, accepted the position of commodore of the Royal Sydney Yacht Squadron.[1996] Reeks would have been enamoured with Lord Forster's enthusiasm, relishing in a man sharing in his love not only for the sport but also its development and opportunities for success.

The official opening of the Royal Sydney Yacht Squadron's 1921/22 season took place in the format of a garden party held on Saturday 22 October within the organisation's Kirribilli premises. Despite bad weather, members took the opportunity to meet and greet Lord Forster, as well as view one of the first 21ft restricted class yachts to be built in Sydney, *E.O.J.*, belonging to J. L. Milson, which had been placed on the lawn for inspection.[1997] Reeks was of course in attendance, eager to catch up with Lord Forster and other squadron members and take part in the festivities.[1998] In fact, he was one of five members that formed a welcoming party tasked with meeting Lord Forster as he disembarked from his motor launch and chaperoning him to the grounds.[1999] Also noted as attending the event was Reeks' wife Gertrude.[2000]

With the 1921/22 yachting season underway, much of Reeks' time was soon occupied. He acted as timekeeper for the Royal Sydney Yacht Squadron's Fairfax Cup, held on 12 November 1921, the race won by *Bona*.[2001] In this capacity he was on board the steamer *Lady Carrington* which the club had hired for the day.[2002] Reeks had notably designed the vessel for the Balmain New Ferry Company in 1907. He was also timekeeper for the Rawson Cup, held on 3 December 1921, which was won by Frank Albert's *Rawhiti*.[2003] The following month Reeks was in charge of the tug *Bondi* during the Royal Sydney Yacht Squadron's Gascoigne Cup, the oldest yachting trophy in New South Wales.[2004] The short-course ocean race was also won by *Rawhiti*.[2005] In late February of 1922 Reeks was one of three officials in charge of the squadron's Peace Cup; the two others being J. D. M. Goddard and Charles Bartholomew.[2006] The latter had been a witness to Reeks' marriage in 1900 to Gertrude Inwood.

WALTER REEKS ONBOARD THE TUG BONDI AS IT PASSED THROUGH SYDNEY HEADS FOLLOWING THE OUTWARD RUN OF THE 1922 GASCOIGNE CUP RACE.
SOURCE: SYDNEY MAIL, 25 JANUARY 1922.

With the 21ft restricted class proving a boon for the sport of yachting, particularly given Lord Forster's support and patronage, there was great impetus for establishing an interstate trophy for the class.[2007] With several yachts already built in Victoria, where the class had initially been established, by December of 1921 the first of the New South Wales-built craft had been launched. These were *Inez,* designed by Fred Lomer and built for Fred Doran by Charlie Dunn of Berrys Bay; James Alderton's *Gumleaf,* which had been designed and built by William Holmes of McMahons Point; and Lord Forster's *Corella*, J. L. Milson's *E.O.J.* and Frank Albert's *Boomerang*.[2008] The latter three yachts were built at the Sandeman yard and considered exact replicates of one another.

The vessels themselves further added to the attractiveness of the 21ft restricted class. Given their moderate cost to construct, the utility of the vessels being transported between locations, and the boats being light and considered easy to handle, the medium-sized 21ft restricted class created a niche for themselves complementing the smaller and larger-sized yacht classes. Their practicality as cruising vessels also proved satisfactory.

All of these factors combined to spur discussions regarding the introduction of the 21ft restricted class in other states. Representatives from the Royal Queensland

1995 *The Sydney Morning Herald*, 7 September 1921.
1996 N. Mays, C. Grazules & D. Payne (2021). *Little Boats with Sails: The History of Australia's 21 Foot Restricted Class.*
1997 *The Sun*, 23 October 1921.
1998 *The Sun*, 23 October 1921.
1999 *Sunday Times*, 23 October 1921.
2000 *The Sydney Morning Herald*, 24 October 1921.
2001 *The Sun*, 13 November 1921.
2002 *The Sydney Morning Herald*, 14 November 1921.
2003 *The Daily Telegraph*, 5 December 1921.
2004 *The Sun*, 22 January 1922.
2005 *Sydney Mail*, 25 January 1922.
2006 *Referee*, 1 March 1922.
2007 *The Daily Telegraph*, 2 February 1922.
2008 N. Mays, C. Grazules & D. Payne (2021). *Little Boats with Sails: The History of Australia's 21 Foot Restricted Class.*

Yacht Club soon set about raising funds for a boat to be built locally.

The inaugural national championship for these '21-footers', the Forster Cup, was held on Sydney Harbour between 7 and 11 February 1922, under auspices of the Royal Sydney Yacht Squadron. It featured eight yachts: six from New South Wales and two from Queensland. Sailing into Australian yachting history, *Gumleaf* earned the distinction of being the first yacht to win the Forster Cup, gaining two victories out of three starts.[2009] Second was the Queensland representative *Lakatoi*, followed by *Inez*. Reeks was amongst those not only organising and officiating at the event, as part of its sailing committee, but there's no doubt that he would also have been carefully analysing the design and sailing qualities of each craft.[2010]

LORD FORSTER CONGRATULATING JAMES ALDERTON, GUMLEAF'S OWNER, UPON WINNING THE 1922 FORSTER CUP.
SOURCE: ORION ALDERTON.

In mid-1922 Reeks was once more elected honorary official measurer of the Royal Sydney Yacht Squadron.[2011] He also continued his involvement in education, being reappointed to the New South Wales Education Department's Advisory Committee in connection with trade and higher technical education, specifically shipbuilding. The appointment was for another two-year period.[2012]

Much interest, however, continued to be centred on developing the 21ft restricted class at the national level. In October of 1922 Lord Forster presided over a conference of interstate yachtsmen in Sydney to discuss the topic. Amongst the New South Wales delegates were S. M. Dempster, T. M. Bremner, W. D. M. Taylor, Andrew Wilson, R. Down, L. C. Waterman, W. Creagh and, of course, Reeks. Representatives from yacht clubs based in Victoria, Queensland and Tasmania were also in attendance. Supplementary to confirming that the 1923 Forster Cup would be held early in the year in Brisbane, delegates agreed that the Forster Cup was to be considered the trophy for the Australian Championship for the class. Not only was this a significant nod to the perceived benefactor of the class, it also gave the three-heat race series a level of significance that past races between yachts from the various states had been lacking, including the Northcote and Sayonara cups. The specific conditions governing interstate competition for the 21ft restricted class were unanimously agreed to and drafted. They were then formalised by Lord Forster on 25 January 1923 and distributed to the various parties.[2013]

The 21ft restricted class went on to fill an important niche in Australian sailing history, particularly between the first and second world wars. Notably the class was the first to be implemented by all six states, with more than 70 vessels were built to the class.[2014] One of these, *San Pan*, is deserving of more than a few paragraphs and will be discussed in detail shortly.

In mid-1923 Reeks travelled to Tasmania, likely to discuss specifications for a Government-owned vehicular ferry steamer which he had designed for Hobart's trans-Derwent service.[2015] The vessel was just over 137ft in length (overall), with a moulded breadth of 36ft and a depth of 10ft 10in.[2016] It was to be propelled by triple expansion service condensing engines and capable of maintaining a speed of 8 knots.[2017] The craft was expected to be built in Tasmania of local hardwood timber, with Huon pine topsides, and was to have a propeller at each end.[2018]

Though the Tasmanian Minister for Land and Works, J. Belton, stated in November of 1923 that he had made a careful and thorough investigation of the plans and specifications for the new vehicular ferry that Reeks had supplied, the vessel was not built.[2019] Tenders for its construction were called, as well as for the supply and fitting of its engines, boilers and machinery in January of 1924. Two months later, however, it was revealed that only one tender for the craft's construction had been received (from Purdon and Featherstone of Battery Point) and it was for a sum much higher than anticipated.[2020] For this reason Belton sought contact from mainland-based firms and next investigated the option of buying a second hand vessel. After months of further effort, the government contracted with Messrs J. Crighton and Son of Chester, England, to have a steel vehicular ferry constructed for the price of £34,000, i.e., £3,000 less than what Purdon and Feathersone had submitted a tender for.[2021] The resulting 550-ton vessel, *Lurgurena*, considered to be a sister-ship to one ordered by Sydney Ferries Ltd., had a length of 199ft, width of 35ft 6in and depth of 14ft.[2022] It arrived in Hobart in October of 1925, commencing service a few months

[2009] N. Mays, C. Grazules & D. Payne (2021). *Little Boats with Sails: The History of Australia's 21 Foot Restricted Class.*
[2010] *The Daily Telegraph*, 2 February 1922; *The Sydney Morning Herald*, 2 February 1922.
[2011] *The Sydney Morning Herald*, 30 August 1922.
[2012] *Government Gazette of the State of New South Wales*, 6 October 1922.
[2013] N. Mays, C. Grazules & D. Payne (2021). *Little Boats with Sails: The History of Australia's 21 Foot Restricted Class.*
[2014] N. Mays, C. Grazules & D. Payne (2021). *Little Boats with Sails: The History of Australia's 21 Foot Restricted Class.*
[2015] *The Mercury*, 18 June 1923.
[2016] *World*, 17 November 1923.
[2017] *Examiner*, 23 June 1923.
[2018] *Examiner*, 23 June 1923; *World*, 17 November 1923.
[2019] *World*, 17 November 1923.
[2020] *World*, 26 January, 4 March, 2 April 1924.
[2021] *The News*, 22 October 1924.
[2022] *Examiner*, 10 July 1924.

later.[2023] When all of the costs for its construction and delivery were factored, it was £1,000 over budget, though additional resources were spent on alterations to berthing accommodations on both sides of the River Derwent to fit the craft.[2024] It was also expected to run at a loss to the Tasmanian Government in the realm of £6,000 to $9,000 per year such that before *Lurgurena* even started crossing the Derwent, there were calls for it to be sold and a bridge built instead.[2025] Reeks would have been disappointed that his design for the trans-Derwent service was not built and likely followed the discourse surrounding the new vessel.

———

Meanwhile, Reeks continued his many ventures. In March of 1923 he was elected to the committee of the Institution of Engineers, Australia (Sydney division), also continuing in the role the following year.[2026] The national organisation had grown from an amalgamation of state engineering associations in 1919, including the Engineering Association of New South Wales.[2027]

In mid-1923 Reeks attended a winter dance at the Royal Sydney Yacht Squadron's clubhouse. The Governor General and Lady Forster were also present.[2028] A few months later, at the organisation's annual general meeting, he was once more appointed honorary official measurer.[2029]

On 12 September 1923 a new ocean-going cruising yacht named *Sea Scout* was launched, built by Messrs Morrison and Sinclair, Ltd. to the order of Harry Shelley.[2030] Designed by Reeks, the vessel measured 52ft (overall), 47ft (waterline) with a beam of 13ft 6in and a depth of 6ft.[2031] It was ketch rigged and fitted with a 100 h.p. Sterling engine.[2032]

Also in 1923 Reeks designed a large schooner for Arthur H. Davies. It was his 303rd project. Unfortunately the build did not eventuate. Plans, still in existence, reveal a classic looking 80ft schooner with a shallow draft. In an unconventional touch, it had no centreboard and no topsails. It also had an elaborate skylight over the saloon. It certainly would have added considerable grace to the fleet of yachts on the harbour had it been built.

The year 1924 saw Reeks again part of the organising committee for the Anniversary Regatta.[2033] He also acted as timekeeper for 21ft restricted class races held on Port Jackson.[2034] In addition, from 1 February 1923, Reeks was reappointed to the Education Department's Trade and Higher Technical Education Advisory Committee associated with shipbuilding. It was a two-year position.[2035]

There was also some high-profile entertainment. In early July of 1924, '*All of Sydney society was out at Kensington for the polo yesterday, and the weather was delightful*'.[2036] The Governor General, accompanied by Lady Forster, and the Governor and Lady de Chair were notable attendees. Others present included Reeks and his wife Gertrude.[2037] The following month Reeks was once more elected honorary official measurer of the Royal Sydney Yacht Squadron.[2038] In addition, he was appointed vice-president of the Sydney Yacht Racing Association.[2039] During this period Reeks continued his service to the Rotary Club, representing the organisation at various events.[2040]

In January of 1925 Reeks was noted as attending the Anniversary Regatta, including the official luncheon with the Governor of New South Wales, Sir Dudley de Chair.[2041] However, more and more, the 21ft restricted class and its continued development and introduction throughout Australia sparked Reeks' interest. Within the control of design specifications and class rules, his imagination and need to give convention a nudge saw an opportunity to push boundaries and test limits. And why not!

As his career came to a close, Reeks was not to know that his two last designs (that we can find evidence of) would more or less bookend the two contrasting characteristics of his professional work. They showcase his ability to not only produce on the one hand an unpretentious, conventional craft yet at the same time something that was totally unorthodox. *Duckwing* was the normal craft, while *San Pan*, even by Reeks' standards, was the odd one out in the fleet, indeed on Sydney Harbour. It would have also stood out on any harbour throughout the world.

DUCKWING.
SOURCE: AUSTRALIAN MOTOR BOAT AND YACHTING MAGAZINE (JULY 1925).

Duckwing has little in the way of reports to help with its description, but thankfully an image from *Australian Motor Boat and Yachting Monthly* magazine provides a complete view of this neat motor-sailer. Built by William Holmes for Edgar Perdriau, one of the sons of Henry Perdriau who founded Perdriau Rubber, this 45-footer follows on from

[2023] *Examiner*, 6 October 1925.
[2024] *Examiner*, 6 October 1925; *The News*, 13 October 1925.
[2025] *The News*, 6 November 1925.
[2026] *The Daily Telegraph*, 24 March 1923; *The Sun*, 31 March 1924.
[2027] www.eoas.info/biogs/A001029b.htm.
[2028] *The Daily Telegraph*, 24 July 1923.
[2029] *The Sydney Morning Herald*, 29 August 1923.
[2030] *The Sydney Morning Herald*, 13 September 1923.
[2031] *The Sydney Morning Herald*, 13 September 1923.
[2032] *The Sydney Morning Herald*, 13 September 1923.
[2033] *The Daily Telegraph*, 22 November 1923.
[2034] *The Daily Telegraph*, 24 January 1924.
[2035] *Government Gazette of the State of New South Wales*, 6 June 1924.
[2036] *Evening News*, 3 July 1924.
[2037] *Evening News*, 3 July 1924.
[2038] *The Sydney Morning Herald*, 26 August 1924.
[2039] *The Sydney Morning Herald*, 15 October 1924.
[2040] *The Sydney Morning Herald*, 8 September 1924.
[2041] *The Daily Telegraph*, 24, 27 January 1925.

earlier auxiliary cruising yachts that Reeks had designed, such as *Mischief* and *Wyruna* (both also built by Holmes). It featured a modest bow profile and elegant curved sheer enhancing either a raised deck or raised bulwark, balanced with a conventional rig. It went a step further toward a true motor-sailer by having a tall pilot house at the end of the trunk cabin and over the aft cockpit. With the profile showing a smaller rig set forward, it seems the engine was bigger than normal and probably the primary means of power. '*This ship is being fitted with a 30 h.p. Buffalo oil engine of special marine type*', says the report published in July 1925.[2042] *Duckwing* presents as a sturdy, capable cruising vessel ready to make passages around the coastline, and entertain guests in comfort at anchor.

At the time of *Duckwing*'s launch, the combustion engine was well and truly an accepted marine item and one which Reeks had adopted as soon as it came to Sydney, and subsequently had much success with. It was well accepted aboard yachts as an auxiliary power or, like *Duckwing*, a primary source, but there was still plenty of room to play with just sail power.

While the rig of *Duckwing* was quite modest, Reeks was ready to go out on a limb and look at a new way of configuring a rig profile for his own yacht *San Pan,* designed to the popular 21ft restricted class rule. As detailed in the previous pages, Reeks was closely involved with the group of yachtsmen around Australia who helped establish the class, particularly in Sydney, and had acted as a measurer and official in various races and regattas involving the fleet. Yet, he had not actually designed a boat to the rule which itself allowed quite a bit of individual input so that the craft were different though still fit within a set of dimensions and parameters. However, it is unlikely anyone, given these constraints, would have designed a boat as different as *San Pan* and then gone on to not only have it built, but also been brave enough to show up on the race course sailing in it. That is, anyone but Reeks.

The reports of *San Pan* after its launching in mid-February of 1925 put it clearly into perspective. The vessel was declared, '*A Freak*' and '*Not a Beauty*'. Jumping to immediate conclusions, thanks to the extraordinary mast position, rig proportions and relatively brutal hull lines, commentators reacted to the strange sight without properly considering the reasons why. Once again, it's an excellent example of Reeks applying a different approach with logic behind it, but that didn't stop the press having a good time at his expense, although the first short note was relatively benign.

'*Mr. W. Reeks had his new 21ft. yacht Sampan* [SIC] *afloat. This yacht has the mast stepped well aft, but the absence of mainsail area, is, it is claimed, more than compensated for by the sail area of the increased fore-triangle.*'[2043]

Then things stepped up a notch. '*A FREAK, Mr Reeks' yacht. Mast in Centre. Mr. Walter Reeks has designed a new type of racing 21ft class of yacht to be known as the San Pan.*

The craft was afloat on Saturday, with Mr. Reeks at the helm, and Mr. Pros. Milson on the sheet. San Pan was not built with an eye to beauty, and her design is entirely new to Australia. Her mast, instead of being carried well for'ard, is in the centre, giving her a unique appearance.

San Pan, in a light air, is not likely to trouble the present fleet, but in a stiff breeze she appears likely to take a lot of beating when properly tuned up.

Mr. Reeks does not believe in following conventions of others. He regards San Pan as an experiment and a type of boat that will lend itself to improvements. Probably the quaint one will have her first race next Saturday under the R.P.A.Y.C. flag.'[2044]

There was more criticism.

NOT A BEAUTY.

San Pan, the new 21-footer, which has been built to the restrictions laid down by the Interstate Yacht Racing Association, made her debut on Sydney Harbor on Saturday afternoon. She is a freak type and excited much curiosity. Her mast, instead of being stepped well forward, is in the centre, and she looked more like a Chinese junk than a racing yacht. She was designed by Mr. Walter Reeks, and was sailed by him on Saturday, with Mr. Pros. Milson as assistant.

SOURCE: REFEREE, 18 FEBRUARY 1925.

Finally, with tongue in cheek, the *Referee* suggested he reverse course. '*A Novel Experiment. San Pan, the new 21-footer did not put in an appearance on Saturday, and it is rumoured that she is being tried out the other way around.*'[2045]

The opportunity to give *San Pan* its due is made possible by two grainy images of the vessel sailing during its first trials and published at the time. We have included one on the following page. A couple of things are quite clear. The principal one is that the mast is actually stepped aft of midships (or the centre). The second feature to note is that Reeks adopted a simple, flat panel, hard chine hull. The unusual rig was the principal experiment. He had wisely chosen to build a simple and inexpensive hull as a test bed.

What was really going on here? The answer may very well lie in the fact that *San Pan* was designed to the rule of the 21ft restricted class which involves a set of maximum and minimum dimensions around a waterline of 21ft and length overall of 25ft. The sail area is set at a maximum of 450 square ft and how that is measured may be at the heart of Reeks' unusual design. The rule was revised in 1921 where it was established that the 450 square ft of sail area was to be measured under the Yacht Racing Association rules, where the foretriangle figure was taken at 85 per cent of its measured area, and the mainsail figure at 100 per cent of its area. Therefore, if only a headsail was carried, there was considerably more than 450 square ft of sail area available; the actual total being 529 square ft. Moreover, if only a mainsail was set then the total sail area was reduced

[2042] *Australian Motor Boat and Yachting Magazine*, July 1925.
[2043] *The Sun*, 15 February 1925.
[2044] *Evening News*, 16 February 1925.
[2045] *Referee*, 4 March 1925.

SAN PAN.
SOURCE: SYDNEY MAIL, 25 FEBRUARY 1925.

to 450 square ft. Moreover, as the area of the headsail is increased, the actual total sail area also increases.

It could well be that Reeks was looking for a way to put more sail onto this 21-footer by taking advantage of a considerably larger foretriangle with a much longer base and probably a higher hoist length. Alternately, he may have wanted to experiment with a large headsail rig, and found that the 21ft restricted class rule would suit this purpose perfectly. Whatever the reasoning, this left his sailplan with a small mainsail and required the mast to be moved aft considerably to fit this profile onto the hull without an enormous overhang forward in the form of a very long bowsprit.

What Reeks also knew was that the yacht's balance was determined by the overall centre of effort of both main and jib, relative to the underbody profile, and this would also point toward moving the mast aft to bring this centre back to where it would be on a conventional plan. As a further example of how this works, yachts that are cat-rigged with just a mainsail have their masts set well forward to bring the centre of effort into line.

The hull of *San Pan* also has merit as an experiment. The hard chine shape produced panels that would be easy to frame and plank up. In more recent times it was ideally suited to sheet materials beginning with plywood and then composite foam and balsa cored panels, but in 1925 it was not widely used, restricted mainly to commercial barge or punt hull forms. However in the USA, for example, it was the choice for very successful rowing dories and sailing sharpie hulls. Their flat bottom was planked crossways, the topsides fore and aft. *San Pan* is most likely to have had a 'vee' bottom but how it was planked up by its builders, Pritchard Bros. of Careening Cove, is not known. This gave a simple but effective hull to base the rig on and trial the experiment, without going to the expense of a fully carvel planked hull.

More evidence of Reeks' unorthodox and even lateral thinking is seen in the mainsail. The big image opposite, published in the *Sydney Mail* carries a supporting caption noting, '*A NEW SYDNEY YACHT. Mr. Walter Reeks, the Sydney yacht designer, presented this new rig to Sydney Harbour last week in the 21-footer San Pan. The mast is stepped in about the centre of the boat, and she carries two booms. The picture shows her with her smaller mainsail. The experiment has not yet been tried in a race*'.[2046]

It looks to be a windy day, possibly even one of those gusty westerly days, always uncomfortable on the harbour, and *San Pan* seems rigged down a little. The headsail is full hoist but short on the foot and leech, probably the number 2. In addition the little Bermudan mainsail, still a novel shape while the gaff rig maintained its dominance, is set off the upper of what seems to be two booms. What could be happening here is that the double boom is a combination of a normal boom set as the upper spar, sheeted to an elevated and fixed bumpkin as the lower spar, which then has a backstay fixed at its aft end to support the tall mast against the headstay loads. This spar is fixed to the mast at the gooseneck and then to the hull by a permanent triangular boom gallows at the transom. The mainsail may be reefed rather than being a smaller one, so it seems Reeks' intention all along was for a Bermudan mainsail and not a conventional gaff rig where the gaff yard would interfere with the curious backstay arrangement.

Evident in the photo though is a shortcoming that would not have helped the vessel's performance. The bumpkin and backstay idea wasn't working as well as needed. There is quite significant forestay sag, which would affect the sail shape poorly, unsettle the balance of the boat on the helm and make it difficult to point as high into the wind as it should. Forestay sag, or the inability to maintain a taut, straight as possible forestay under tension was a problem in normal rigs, and the contemporary galvanised wire and hand tensioned block and tackle methods were also still insufficient compared to the strength of modern materials and methods. Making the forestay length considerably longer was just adding more to the problem. It would have worked if the rig could have had a tight forestay, but the photo shows the mast is inverted forward and bending sideways as well, as the leeward shrouds are all very loose.

Two other things are perplexing as well, one partially evident in the photo, the other is an issue that can't be confirmed. Looking at the crew on deck, *San Pan* does not appear to have any cockpit. A valuation of the boat in late 1925 notes it had '*three small cockpits*' but they aren't easily visible, and perhaps this was a way to make it easy to build and cheap, it was an experiment after all.

In addition, there may have been a moderate width flat bottom panel off the keel giving another chine and more volume too, as the cross section drawn for a single chine and vee shape seems unlikely to give enough volume for the likely displacement.

Not much more can be said, as there are no further reports of *San Pan* sailing or racing. Perhaps this one venture out on Sydney Harbour was its only trial under this experimental rig, and we remain unclear what was learnt that day, other than the craft was seen and garnered a lot of attention. The only person the authors were able to find who had met Reeks was Brian Gale, who recalls meeting him at age about eight or nine, when he saw *San Pan* sailing. During a phone conversation Gale said, '*San Pan, about 20 to 25 foot long, plumb stem and stern. San Pan had its mast about two-thirds aft from stem with jib probably on stem head and not on a bowsprit. Big jib and Bermudian mainsail, no gaff or yard. Very slow, would not go, jib too full*'.[2047]

As revealed, *San Pan* attracted notoriety in its short life. Reeks must have talked about it too, as his descendants via his daughter Thelma and her family also had recollections to pass on in regards to the unique craft. They have stated that it was the last boat Reeks designed. It was also said to be '*modern for its time*', with a similar design used today, '*a sail and a great big jib*'. According to the family, '*People used to call out, "Hey Whiskas! You're going the wrong way"*'.[2048]

San Pan was indeed '*A craft of peculiar design and rig*'.[2049] As we have seen, that word 'peculiar' shadowed Reeks' career at times, but *San Pan* was also ahead of its time. Today we have yachts with very large foretriangles and their mast set well back toward the centre, racing at the highest level. One of the best examples is the record breaking 30-metre super-maxi *Comanche*. In addition, this vessel also shares a chine hull form. Reeks was obviously onto things with *San Pan*. He just needed more time.

Walter Reeks died at his Mosman Bay home, *Twynham*, on 2 October 1925 after a short illness.[2050] He was 64 years of age. Cause of death was exhaustion and pernicious anaemia. Reeks was survived by his second wife Gertrude and his daughter Thelma, along with her husband Alfred and their two sons, Kent and Lindsay.

Reeks was eulogised in many newspapers throughout the country. His funeral took place one day following his death and he was buried at the Macquarie Park Cemetery and Crematorium in North Ryde.[2051] Indicative of the breadth of his associations and relationships, in addition to family and friends, there was a large gathering at the funeral, including representatives from the Royal Sydney Yacht Squadron; the Royal Prince Alfred Yacht Club; the Motor Yacht Club of New South Wales; the Rotary Club of Sydney; various shipbuilders, boat builders and shipowners; the Mosman Recreation Club; the Institute

2046 *Sydney Mail*, 25 February 1925.

2047 Brian Gale, pers. comm. with David Payne, c2004.

2048 Rae Fidock, pers. comm. with David Payne, c2004.

2049 Obituary in *Australian Motor Boat and Yachting*, 1 November 1925.

2050 *The Daily Telegraph*, 3 October 1925; *The Sun*, 4 October 1925.

2051 *The Sydney Morning Herald*, 6 October 1925.

of Engineers of Australasia; and the New South Wales Department of Education.[2052]

Reeks' property, household and personal effects were left in a trust for his wife Gertrude. The remainder of his estate, including shares, was sold; the proceeds invested to provide an income to support Gertrude. Upon her death, the entire estate passed to his daughter Thelma.

Gertrude remained living at *Twynham* for a number of years but by 1930 had moved to a nearby residence at 42 Muston Street, Mosman.[2053] *Twynham* was later demolished with flats subsequently built on the property. Gertrude Reeks died on 17 November of 1933 at the age of 65. She was privately cremated.[2054]

What a life Reeks had lived! It was one of exceptional, unselfish and innovative service to naval architecture, to yacht racing and administration, and to the communities in which he lived and worked, as well as to his family. Of course he was ambitious and so were some of his projects; Australia's potential involvement in the America's Cup being one such example. There were also trials and failures, his investment in various emerging technologies and gold mining ventures were perhaps uncharacteristically not well thought out. Still, Reeks' unguarded and effervescent need to say '*Yes!*' led to great successes, particularly in his early racing yacht designs, as well as those pioneers of the Sydney ferry runs and the Torres Strait pearling industry. He appears to have approached life with gusto, juggling multiple projects with convincing commitment and conviction, and taking on opportunities with absolute determination and resolve. He fulfilled orders for clients across Australia and New Zealand, commercial and recreational, large and small; customers with limited funds to those with expansive budgets. His technical wherewithal for naval architecture also stood him at the forefront of his craft; his counsel and expertise were always sought, his advice most often taken on board. Moreover, his resounding ability to digest complex pieces of information and convey it to the general public was much appreciated, as was his involvement in education specific to his trade.

Reeks' commitment to serving others, fostered from an early age from his family life in England, and via his mentors, the Inmans of Lymington and Alex Richardson of Liverpool, is of absolute admiration. He volunteered for numerous organisations, in particular the Royal Sydney Yacht Squadron of which he was made a life member in 1900 and was notably the organisation's official measurer for 39 years, as well as its vice-commodore between 1906-10. There were other organisations too, including engineering associations, Rotary, sporting, and the masonic lodge, to which he also devoted large chunks of time and energy.

It was a life well lived and of a life's work seemingly fulfilled. Up until the last vessel, the extremely unorthodox *San Pan*, it was also a career spent chasing the perfect design for the perfect vessel, most often in the context of a client's needs and wants; the need to compromise between the practical and the ideal. In many of these projects, Reeks still couldn't get away from being bold and innovative. So often there was something unconventional coming through, maybe just a few subtle details, in other cases almost the whole project was out on a limb. As French impressionist Henri Matisse once said, '*The artist begins with a vision — a creative operation requiring effort. Creativity takes courage*'.[2055] This quote seems to sum up the dilemma that Reeks faced and is a neat description and almost an epitaph for him: a man who had the courage to find and express his own creativity.

In dramatic contrast, however, there was a burden of trials and anxieties. The death at such a young age of his first wife Mary and the murder of his son Kent casting large, grieving shadows on Reeks' family life, along with the accident to his beloved yacht *Thelma* on Sydney Harbour.

Yet, with context and clarity, as we reflect back on the life, relationships and individual and collective accomplishments of Reeks, including the more than 300 vessels he designed, through a lens almost 100 years after his death, it is a story worth relaying and one that needed to be told. There are lessons to be learned from his life for all of us. From persevering to fostering to serving to taking risks, it was definitely a life lived tenaciously well.

The legacy of Reeks remains ongoing. In addition to the numerous plans and drawings still in existence, and the biographies of the vessels he was involved with that are intertwined with local, state, national and international maritime histories, there are at least 13 vessels he designed that remain in existence, including *SY Ena*, part of the Australian National Maritime Museum's collection. Scattered throughout the country, their existence is a physical salute to Reeks and all that he achieved.

In closing this chapter on Walter Reeks, we want to convey one final story that was passed down through his daughter Thelma's family that perhaps epitomises the bold personality of Australia's first professional naval architect and his earnest need to not only pass on his knowledge but, at a deeper level, encourage others to challenge conventions.

In the summer of 1925 Thelma travelled to Sydney with her sons, Kent and Lindsay. They spent a week at *Twynham* with Reeks and his wife Gertrude, followed by a week on board *Wahine*. After the trip Reeks, at an unknown date, sent Kent a piece of wood to build a model of a sailing boat. He had, however, cheekily ordered the wood to special measurements such that it was out of balance and would not float or sail properly. Kent had trouble with the model, but soon discovered the problem and made it sail. Reeks had intentionally set a trap from which he knew his grandson would learn valuable knowledge. Kent completed the model, and looking at the image and the model's proportions it suggests he may have had to increase the draft to get it right. Whatever the case, to this day, it remains a treasured family heirloom.

2052 *The Sydney Morning Herald*, 6 October 1925; *The Daily Telegraph*, 5 October 1925.
2053 *Australia, Electoral Rolls, 1903-1980* for Gertrude Reeks.
2054 *The Sydney Morning Herald*, 18 November 1933.
2055 C. Brown (1998). *Artist to Artist: Inspiration and Advice from Visual Artists Past & Present.*

REEKS FAMILY TREE

WALTER REEKS' GRAVE.
SOURCE: NEVILLE OLLIFFE.

Charles Reeks
Born 1818
Christchurch, Hampshire, England
Died 1891
Christchurch, Hampshire, England

Elizabeth Kent
Born 1823
Ringwood, Hampshire, England
Died 1909
Christchurch, Hampshire, England

Children:

- John Reeks
  1851 - 1860
- William Reeks
  1852 - 1940
- Ellen Elizabeth Reeks
  1854 - 1930
- Marian Reeks
  1855 - 1942
- Annie Reeks
  1857 - 1883
- Emma Reeks
  1859 - 1860
- **WALTER REEKS**
  Born 1861
  Christchurch, Hampshire, England
  Died 1925
  Sydney, New South Wales
- Louis Reeks
  1863 - 1865
- Louisa Reeks
  1863 - 1933

James Kent
Born 1832
Worsley, Lancashire, England
Died 1907
Staffordshire, England

Sarah Ridyard
Born 1837
Deane, Lancashire, England
Died 1918
Swinton, Lancashire, England

married 1889

Mary Emma Kent
Born 1862
Worsley, Lancashire, England
Died 1895
Sydney, New South Wales

married 1900

Gertrude Inwood
Born 1868
Melbourne, Victoria
Died 1933
Sydney, New South Wales

Kent Reeks
Born 1890
Sydney, New South Wales
Died 1914
Bilston, Staffordshire, England

Thelma Reeks
Born 1892
Sydney, New South Wales
Died 1975
Tottenham, New South Wales

married 1914

Alfred Fox Ashwin
Born 1885
Sydney, New South Wales
Died 1950
Tottenham, New South Wales

Kent Fox Ashwin
Born 1915
Sydney, New South Wales
Married Dorothy Ford (1941)
Children: Rae (1944), Ronald (1948)
Died 1994
Tottenham, New South Wales

Lindsay Fox Ashwin
Born 1917
Sydney, New South Wales
Married Lois Lynette Gander (1945)
Children: David (1948), Judith (1949)
Died 1994
Tottenham, New South Wales

MODEL YACHT MADE BY KENT FOX-ASHWIN.
SOURCE: FOX ASHWIN FAMILY COLLECTION.

KATE AND EMMA CHARTER, WALTER REEKS' GREAT-GREAT-GRAND DAUGHTERS, DRESSED UP AS THELMA AND WALTER REEKS FOR SCHOOL HERITAGE WEEK, MUDGEE (1986).
SOURCE: FOX ASHWIN FAMILY COLLECTION.

PRITCHARD BROS. LTD.

Launch Builders & Engineers,

High Street,

NORTH SYDNEY.

9th November, 1925.

Messrs. Creagh & Creagh,
Solicitors,
S Y D N E Y.

Dear Sirs,

We herewith give a valuation and description of the experimental 21 ft Restricted class yacht "San Pan" built last season to the order of the late Walter Reeks.

"San Pan" was not successful in any way and with the present rig is practically unsaleable at any price, in our opinion £50 is a fair valuation.

"San Pan" is nearly 25ft overall with a beam nearly 8 ft built with Dorrigo pine ¾" planking over H W & Tea tree frames, fully decked with 3 small cockpits, mast is far aft with small mainsail and big jib and would have to be re-rigged and new centrecase fitted to make her at all attractive to the ordinary buyer.

We are

Yours faithfully,

Pritchard Brothers, Limited,

Henry C. Pritchard,

Director.

A VALUATION OF SAN PAN GIVEN BY PRITCHARD BROS. FOLLOWING WALTER REEKS' DEATH IN OCTOBER OF 1925.
SOURCE: WALTER REEKS PROBATE AND ESTATE PAPERS.

CHAPTER TWENTY TWO - THE VESSELS

This chapter provides a summary of the vessels known to have been designed by Walter Reeks, those he very likely designed, as well as the craft he superintended the building of. This collection of vessels is incomplete, as a number were not located in our research. The construction and/or launch of smaller boats, launches and yachts, particularly towards the latter stages of his career, was also generally not noted in the newspapers. For example, the pearling lugger *Gondolier*, launched in 1904 to a design by Reeks and built by William Holmes at McMahons Point is known to be plan no. 170. However, at this time point we have only identified 141 vessels built to designs by Reeks, indicating that there are around 30 craft missing from our list. Moreover, *Gondolier* was launched 21 years prior to Reeks' death. If the same rate he designed vessels was maintained for the next two decades, one can assume he was involved in the development of more 300 vessels during his lifespan; quite a remarkable achievement. This total number is actually confirmed by the fact that a vessel designed by Reeks for Arthur Davies in 1923, though not built, was allocated plan no. 303. So far we have only been able to account for a total of 237 vessels, meaning that about 70 craft Reeks was involved with remain elusive.

The chronological development of the vessels themselves provide a representative overview of the depth of Reeks' naval architecture expertise, as well as the progression of his career in terms of the types of vessels he was designing over time, and his use of emerging technologies. Sources used to collate this information include vessel registers, appropriation books, mercantile navy lists, newspapers, family histories, etc. We welcome additional information or corrections.

SNOWFLAKE

Year: 1882
Designer: Walter Reeks
Builder: Walter and William Reeks
Location: Christchurch, Hampshire, England
Type: Wooden yacht
Description: 16ft. Small centreboard yacht. Between 1882 and 1884 successfully competed in fortnightly sailing matches of the Christchurch Sailing Club, often helmed by William Reeks, Walter's older brother.[2056]

DOROTHY EX ABEONA

Year: 1883
Designer: Walter Reeks
Builder: George Inman & Son
Location: Lymington, Hampshire, England
Type: Wooden cutter
Description: ON 94504. 26.2 x 21 (wl) x 9.3 x 5.3ft. 6-ton cutter yacht built to the order of William Grunsell and launched in May 1883.[2057] Based at Southampton. Around 1886 sold and renamed *Dorothy*. First registered in 1888.[2058] By 1890 owned by St. John C. Crampton of London.[2059] By 1900 owned by Russell Palmer (London); by 1903 owned by R. Palmer and S. and D. Simmons (London); by 1910 owned by Norman Dallimore (London).[2060] By 1915 owned by Herbert Winder (London) who retained possession until at least 1930.[2061] Register closed in 1936.

[2056] *Christchurch Times*, 23 September 1882, 2 June 1883; *Hampshire Advertiser*, 4 July 1883; *Bournemouth Guardian*, 31 May 1884.
[2057] *Salisbury and Winchester Journal*, 19 May 1883; *Lloyd's Register of Yachts* (1903).
[2058] Appropriation Books, Official Numbers 94501 - 94550.
[2059] Mercantile Navy List, 1890, 1900; Lloyd's Yacht Register (1902-1903).
[2060] Mercantile Navy List, 1900.
[2061] Mercantile Navy Lists, 1915, 1925, 1930.

THALIA

Year: 1883
Designer: Walter Reeks (likely)
Builder: George Inman & Son
Location: Lymington, Hampshire, England
Type: Wooden yawl
Description: ON 86348. 63 x 60 (wl) x 11ft.[2062] 38-ton yawl built to the order of J. K. Soames of Blackheath and launched on 20 June 1883.[2063] Intended for cruising, carried one of Cantelo's patent capstans and was sparred for rough weather.[2064] Competed at Cowes in August 1886 in a race held on behalf of the Royal London Yacht Club, finishing third.[2065] Returned to Southampton shortly thereafter. Disappears from the records after 1886.

MILLIE

Year: 1883
Designer: Walter Reeks
Builder: Gruer McGruer
Location: Birkenhead, Cheshire, England
Type: Wooden yawl
Description: 21 (wl) x 7 x 3.8ft. Flush-decked 3½-ton yawl built by Gruer McGruer of Birkenhead to the order of Albert Coddington.[2066] Carried 1½ tons of iron on its keel, beams were of half-round iron. Raced with the Cheshire Yacht Club beginning in June 1883.[2067] Upon the sudden death of Albert Coddington at the age of 43 in May 1887, passed to his brother Francis.[2068] Competed with the Southport Corinthian Sailing Club and the Lytham Yacht Club, distinguished for being, '*a queen even among swift sailers*'.[2069] Still successfully racing in 1890 when Alex Richardson was noted to be at the helm on at least one occasion.[2070] In August 1891 won a handicap race at the Royal Anglesey Yacht Club Regatta in Wales with Francis Coddington at the helm.[2071] Disappears from the records a few years later.

ELECTRA

Year: 1885
Designer: Walter Reeks
Builder: Walter 'Watty' M. Ford
Location: Berrys Bay, New South Wales, Australia
Type: Wooden yawl
Description: ON 101143. 47.4 x 11.3 x 8.3ft. 29-ton yawl built to the order of J. F. Fitzhardinge, a solicitor of Sydney, and launched on 21 November 1885. The first vessel built in Australia to a Reeks design, also the first yawl built in the colonies. Carried 5 tons of lead on its keel and 12 tons of ballast inside. Successfully raced on Port Jackson for several years, also undertaking many coastal cruises; the purpose for which it had been originally designed and built. Sold to H. Harden, vice-commodore of the Prince Alfred Yacht Club in late 1889.[2072] Advertised for sale by auction in May 1891.[2073] Sold back to J. F. Fitzhardinge later in 1891 but re-sold in early 1892 to Herbert Lindsay Thompson who retained possession until 28 October 1895.[2074] First registered on 13 November 1895 when sold to Carleton Kelly of Double Bay. He died intestate on 20 August 1899, then transferred to his father, Thomas Hussey Kelly on 3 April 1900, commodore of the Sydney Amateur Sailing Club.[2075] The latter died in July 1901 and *Electra* was transferred to Thomas Hussey Kelly (Jr) of the Royal Sydney Yacht Squadron.[2076] Continued to race competitively for several years on Sydney Harbour, though with decreasing success. Sold to A. S. Burdekin in 1905 and sold once more in April 1907 to W. S. Tait and Company of Sydney.[2077] Transferred to New Caledonia by this firm to work as a trading vessel based out of Noumea.[2078] Sold to Dutch subjects in 1907 and British Registry closed.

ELECTRA (AND ARCHINA).
SOURCE: YACHTSMAN'S GUIDE TO SYDNEY HARBOUR (1898).

--

Year: 1885
Designer: Walter Reeks
Builder: --
Location: --
Type: Wooden yacht
Description: 10-ton racing yacht that Reeks invited quotes from Sydney's boat builders for construction of on 1 September 1885.[2079] There is no evidence of it being built.

[2062] *The Hampshire Advertiser*, 30 June 1883.
[2063] *The Hampshire Advertiser*, 23, 30 June 1883.
[2064] *The Hampshire Advertiser*, 30 June 1883.
[2065] *The Standard*, 2 August 1886.
[2066] *Field*, 31 March 1883.
[2067] *Birkenhead News*, 20 June 1883.
[2068] *Preston Herald*, 18 May 1887; *Blackburn Standard*, 29 June 1889.
[2069] *Blackburn Standard*, 29 June 1889.
[2070] *Preston Herald*, 6 August 1890.
[2071] *Liverpool Mercury*, 26 August 1891.
[2072] *The Sydney Morning Herald*, 30 November 1889.
[2073] *The Australian Star*, 16 May 1891.
[2074] *Evening News*, 21 November 1891; Registers of British Ships, Port of Sydney, single number series (II) - ROLL 9.
[2075] *The Daily Telegraph*, 26 August 1899; Registers of British Ships, Port of Sydney, single number series (II) - ROLL 9.
[2076] *Referee*, 31 July 1901, 13 August 1902.
[2077] *Evening News*, 1 May 1907.
[2078] *Evening News*, 1 May 1907.
[2079] *The Sydney Morning Herald*, 1 September 1885.

--

Year: 1886
Designer: Walter Reeks
Builder: --
Location: --
Type: Commercial fishing yacht
Description: 82 (oa) x 70 (wl) x 15 x 10ft. Yawl-rigged fishing yacht designed by Reeks as part of an article published in *The Sydney Mail and New South Wales Advertiser* on 6 March 1886 in which he stated the proposed craft could spend half of its time fishing commercially and the other half as a recreational yacht, the result being a hybridised vessel that could potentially realise a yearly profit of some £312. Featured an ice chamber, 8ft long, 14ft wide and 10ft deep, around which were shelves with fenders in front of them to store broken ice (obtained on shore). The chamber was proposed to be lined with asbestos or another non-conductor to prevent the rapid melting of the ice. The fish caught could be packed in the ice, remaining fresh for a number of days. All told, there was potential to store 400 to 500 fish. Estimated to cost £3,500 to build.

LEICHHARDT

Year: 1886
Designer: Walter Reeks (very likely)
Builder: James Pashley and Sons
Location: Balmain, New South Wales, Australia
Type: Wooden steam ferry
Description: ON 93495. 85 (oa) x 76 x 18.2 x 6.2ft. 68/46 tons (gross/net). Built to the order of the Balmain Steam Ferry Company for use between Leichhardt and Iron Cove on the Parramatta River. Launched on 8 March 1886, with its trial trip taking place on 26 May, following the boiler and machinery being fitted by Mort's Dock and Engineering Company.[2080] Transferred to the Balmain New Ferry Company Limited upon its establishment in 1892. Remained in service until destroyed by fire on 16 June 1916 while moored at the company's works, Waterview Bay, Balmain.[2081] Its remains were subsequently advertised for sale by auction.

GOSFORD

Year: 1886
Designer: Walter Reeks
Builder: James Pashley and Sons
Location: Balmain, New South Wales, Australia
Type: Wooden screw steamer
Description: ON 93536. 96 (oa) x 88.4 x 19.7 x 7.3ft. 83/57 tons (gross/net). Built to the order of Henry Jeanneret and John Pinto and launched on 1 September 1886.[2082] Intended to trade between Sydney and the New South Wales central coast. Began service on 23 December 1886 after having a second propeller installed, one capable of producing a speed of 11 knots.[2083] For the next few years, travelled daily between Sydney and Gosford, in partnership with the *Charlotte Fenwick*.[2084] Both craft were often also chartered to attend local yacht races and rowing events on Port Jackson. Another new propeller was installed in late 1887.[2085] Sold two years later to Frederick Evans of Sydney, remaining on the central coast run until April 1892 whereby it was transferred to Wollongong to carry out towage work for the Harbour Trust.[2086] Advertised for sale in September 1892 on behalf of the mortgagee and purchased by Ebenezer Waddingham of Newcastle, being used to trade between this city and Port Stephens, carrying fish and passengers.[2087] After several more changes of ownership, in October 1902 purchased by the Northern Union Steamboat Company and transferred to New Zealand where it was employed on the Kaipara.[2088] Despite being a regular on this run, it was another four years before the craft was officially registered in Auckland. Altered in 1909 after being purchased by Captain Thomas Augustus Franks of Gisborne for use as a trawler.[2089] Pioneered the local commercial trawling industry, with Franks proprietor of the Gisborne Fish Depot.[2090] By the early 1930s laid up and, following the death of Franks in 1939, remained in this state until it was dismantled and broken up in 1944.[2091]

GOSFORD AT GISBORNE, NEW ZEALAND (1909).
SOURCE: REF: 1/4-022593-G. ALEXANDER TURNBULL LIBRARY, WELLINGTON, NEW ZEALAND. (RECORDS/22631564).

WOOLWICH

Year: 1886
Designer: Walter Reeks
Builder: James Pashley and Sons
Location: Balmain, New South Wales, Australia
Type: Wooden screw steamer
Description: ON 93535. 75.5 x 17.6 x 6.8ft. 68/47 tons (gross/net). Built to the order of the Atlas Engineering

2080 *Evening News*, 8 March 1886; *The Sydney Morning Herald*, 27 May 1886.
2081 *The Daily Telegraph*, 10 June 1916.
2082 *The Sydney Morning Herald*, 2 September 1886.
2083 *The Daily Telegraph*, 25 December 1886.
2084 *The Sydney Morning Herald*, 27 July 1887.
2085 *The Sydney Morning Herald*, 28 December 1887.
2086 Sydney, New South Wales, Register of ships, single number series (II), Volume 13, 1885-1888, Folio 86; *Illawarra Mercury*, 7 April 1892.
2087 *The Sydney Morning Herald*, 7 September 1892; *Newcastle Morning Herald and Miners' Advocate*, 27 June 1893.
2088 *New Zealand Herald*, 23 October 1922.
2089 *Auckland Star*, 2 September 1909.
2090 *Gisborne Herald*, 13 September 1939.
2091 *Gisborne Herald*, 13 September 1939; New Zealand, Registered Ships & Owners, 1840-1950.

company of Woolwich, on Sydney's Lower North Shore, and launched in October 1886. Began service between the company's headquarters and Sydney in early December 1886, described as, *'one of the fastest and most comfortable ferry steamers on the river'*.[2092] Advertised for sale in September 1888, appears to have then been leased by Frank Thompson and placed in the local Chowder Bay service.[2093] In July 1889 purchased by Harry Hardwick and John Luckman of Brisbane's Queensport Aquarium Company for use conveying customers between Campbell's Wharf and the 'Queensport Aquarium'.[2094] Arrived in Brisbane on 1 August, apparently *'showing good qualities for a vessel of her class'* in *'some pretty rough weather'*.[2095] Provided service on the Brisbane River for several years, then returned to New South Wales when sold to David Henderson of Grafton for £1,900 on 1 March 1893.[2096] Used for the passenger and goods trade on the Clarence River, notably in mid-1894 having a Myers' Patent Screw Propeller installed, believed to be the first of its type in the southern hemisphere.[2097] By 1895 owned by Henry Gerard and Co. of Grafton who retained possession for several decades. In 1903 lengthened with new engines and a boiler installed.[2098] In November 1911 narrowly escaped total destruction after a fire broke out on board while moored at Chatsworth.[2099] Following extensive repairs, resumed service until November 1917 when it was withdrawn from the local passenger trade.[2100] Ultimately sank in the South Arm of the river in March 1921 after encountering a snag.[2101] The wreck was advertised for sale by auction.

YOUNG DICK (LEFT) AND WOOLWICH (RIGHT), CLARENCE RIVER.
SOURCE: THE SYDNEY MAIL AND NEW SOUTH WALES ADVERTISER, 21 JUNE 1902.

BETTINA

Year: 1887
Designer: Walter Reeks
Builder: George Ellis
Location: Balmain, New South Wales, Australia
Type: Wooden steam yacht
Description: 40 (oa) x 33.0 x 7.0 x 3.3ft. 7 tons. Built to the order of the W. C. Cooper of Rose Bay and launched early in 1887. Fitted with an overhanging bow, probably clipper style, and finished at the stern with a long counter. Timber work combined beech, rosewood and teak.[2102] Though only 40ft in length (overall), given the phrase of having a *'very light, rakish appearance'*.[2103] Took its trial trip on 19 March with as high as 200 revolutions registered by the Shipman's American engine, notably fired with kerosene instead of coal.[2104] Advertised for sale in April 1888 owing to Cooper leaving for Europe, described as *'a perfect model of symmetry and elegance'*.[2105] Very likely sold and renamed.

MAYSIE

Year: 1887
Designer: Walter Reeks
Builder: Walter 'Watty' M. Ford
Location: Berrys Bay, New South Wales, Australia
Type: Wooden yawl
Description: ON 93565. 31.3 x 9.2 x 6.4ft. Built by Ford at Berrys Bay to the order of J. M. Curtis, a Sydney solicitor, and launched in April 1887.[2106] Fitted with *'one of those "clipper" or "buttercup" bows now so much in vogue in Scotch waters ... it is the first we have seen here'*.[2107]Advertised for sale in March 1889, sold the following month to Harry Carpenter.[2108] Advertised for sale in January 1891, December 1892 and October 1893, though remained in Carpenter's hands until 1897 when purchased by Owen Tinning of Hobart, commodore of the Derwent Yacht Club, arriving on the River Derwent in November 1897 after a 14-day passage.[2109] Spent the next 15 years in Hobart, racing and cruising under Tinning's ownership. In 1912 purchased by Arthur Peck, commodore of the Port Phillip Yacht Club, making its way to Victoria in early January 1912.[2110] A decade later advertised for sale, by this time fitted with an 8 h.p. Kelvin engine.[2111] Failing to sell Peck retained ownership for two more decades, notably competing in the 1929 race from Port Phillip Bay to Low Head, Tasmania, though having to retire due to gear and rigging damage.[2112] In February 1944 sold to E. Orans, a member of the Royal Yacht Club of Tasmania.[2113] Several additional changes of ownership then followed. Register closed in 1963, noted *'Vessel reported broken up'*.[2114]

[2092] *The Daily Telegraph*, 8 December 1886.
[2093] *The Daily Telegraph*, 15 September, 19 December 1888.
[2094] *The Brisbane Courier*, 27 July 1889; *Evening News*, 30 July 1889.
[2095] *The Brisbane Courier*, 2 August 1889; *The Sydney Morning Herald*, 6 August 1889.
[2096] Register of British Ships: Main register subsequent to Merchant Shipping Act 1854, Port of Brisbane; *The Telegraph*, 25 February 1893.
[2097] *Clarence and Richmond Examiner*, 26 June 1894.
[2098] *The Richmond River Herald and Northern Districts Advertiser*, 25 September 1903.
[2099] *Clarence and Richmond Examiner*, 14 November 1911.
[2100] *The Richmond River Herald*, 6 November 1917.
[2101] *The Sydney Morning Herald*, 12 March 1921.
[2102] *Referee*, 24 March 1887.
[2103] *The Daily Telegraph*, 2 August 1886.
[2104] *The Sydney Morning Herald*, 23 March 1887.
[2105] *The Sydney Morning Herald*, 2 April 1888.
[2106] *The Sydney Morning Herald*, 21 June 1883; *The Sydney Mail and New South Wales Advertiser*, 23 April 1887.
[2107] *The Sydney Mail and New South Wales Advertiser*, 5 March 1887.
[2108] *Evening News*, 10 October 1888; *The Sydney Mail and New South Wales Advertiser*, 13 April 1889.
[2109] *The Sydney Morning Herald*, 3 January 1891, 10 December 1892, 28 October 1893; *The Mercury*, 26 November 1897.
[2110] *The Age*, 11 January 1912.
[2111] *The Australasian*, 25 February 1922.
[2112] *The Herald*, 12 December 1929; *Sunday Times*, 29 December 1929.
[2113] *The Argus*, 2 February 1944.
[2114] Register of British Ships: Main Register subsequent to Merchant Shipping Act 1854, Port Melbourne.

MAYSIE ON THE RIVER DERWENT, HOBART (CIRCA 1900S).
SOURCE: MARITIME MUSEUM OF TASMANIA (P_Y_348).

--

Year: 1887
Designer: Walter Reeks
Builder: --
Location: --
Type: Wooden screw steamer
Description: Not built. A design for a steamer with two screws, one at each end, fitted with compound surface condensing engines, estimated to cost £4,300 to build. Submitted as part of a competition initiated by the Balmain Steam Ferry Company but was not awarded the tender.[2115]

EMPRESS EX MIRANDA

Year: 1887
Designer: Walter Reeks
Builder: William Dunn
Location: Careening Cove, New South Wales, Australia
Type: Wooden centreboard yawl
Description: ON 89437. 75 (oa) x 58.0 x 13.5 x 7.5ft. 43 tons. Built by William Dunn for J. H. 'Jack' Want, a prominent Sydney solicitor and politician and launched on 24 October 1887.[2116] A '*beamy centreboarder*', the first of its type admitted to the register of the Royal Sydney Yacht Squadron. Primarily intended for ocean cruising and was fitted out with a 14ft long saloon, an owner's cabin, a ladies' cabin, an open lounge, a foc'sle for a pantry, a galley and crew accommodation, with 7ft of headroom in the main areas.[2117] All up, accommodation for 13 people. Constructed of blackbutt frames with blue gum for planking.

DRAWING OF MIRANDA ON THE STOCKS PRIOR TO LAUNCH.
SOURCE: AUSTRALIAN TOWN AND COUNTRY JOURNAL, 15 OCTOBER 1887.

Though considered a cruising yawl, showed promise as a racing craft; its first big opportunity was at the Melbourne Centennial Aquatic Carnival, held on 23-24 November 1888. Left Sydney for Port Phillip Bay late on the afternoon of 7 November and encountered a series of southerlies, forced to shelter in Twofold Bay, before returning Sydney. Made it further afield in January 1890, cruising to Hobart and returning to Sydney via many stops along Tasmania's coast, the outcome of the two-month trip published by Want in an article titled *2000 Miles in a Forty Tonner,* featured in the *The Australian Town and Country Journal* of 26 April 1890. Raced during the 1890/91 season and for several seasons thereafter. In November 1894 sold to William Burford of Adelaide and renamed *Empress.*[2118] Raced locally and was also used for cruising. In February 1897 sold to W. H. C. Lovely.[2119] Cruised to Sydney in October 1898 where it appears to have remained, often advertised for sale in the years that followed.[2120] In February 1902 sold to J. M. C. Forsyth for use trading off the coast of Papua New Guinea.[2121] Remained in Forsyth's possession, operating in these waters until 1911. Sold to German subjects likely associated with 'German New Guinea', no further history is known.[2122]

ERA

Year: 1887
Designer: Walter Reeks
Builder: George Ellis
Location: Balmain, New South Wales, Australia
Type: Wooden cutter yacht

2115 *The Daily Telegraph*, 7 June 1887.
2116 *The Sydney Morning Herald*, 25 October 1887.
2117 *Australian Town and Country Journal*, 15 October 1887.
2118 *The Daily Telegraph*, 24 November 1894.
2119 *South Australian Register*, 19 February 1897.
2120 *Evening Journal*, 3 October 1902.
2121 *The Sydney Morning Herald*, 7 February 1902.
2122 SYDNEY, NEW SOUTH WALES, Register of ships, single number series (II), Volume 18, 1899-1902, Folio 120.

EMPRESS EX MIRANDA SAILING OFF LARGS BAY, SOUTH AUSTRALIA (CIRCA 1895).
SOURCE: STATE LIBRARY OF SOUTH AUSTRALIA (B 19952).

Description: ON 93587. 72 (oa) x 61.6 x 12.7 x 8.7ft. 47 tons. Built by George Ellis for Alfred George Milson, vice-commodore of the Royal Sydney Yacht Squadron, and launched on 5 December 1887.[2123] Named in honour of several previous yachts commissioned by the Milson family. Likely '*the first boat built to the "length and sail area rule" in New South Wales, and it is believed, in any part of the world.*[2124] Elaborately fitted out, also the first yacht to carry a piano in its saloon, where there was a dining table for 14 guests.[2125] Opened its racing career with five wins and one second-place out of six races, including winning the race for the One Hundred Years' Challenge Cup, valued at £500, sailed at the 1888 Centennial Regatta in late January.[2126] Its success '*was too much for her competitors, and killed the first-class racing*'.[2127] On 9 November 1888 cleared Sydney Heads, en route to Port Phillip Bay to compete in the Melbourne Centennial Aquatic Carnival.[2128] Arrived in Melbourne five days prior to the start of the event.[2129] Won the race for 20-tonners, beating *Volunteer*, with *Wanderer* of South Australia finishing third.[2130] After a rough return voyage, arrived back in Sydney on 10 December.[2131] Continued to race successfully on Port Jackson and was used for coastal cruises and pleasure trips well into the next decade. In September 1892 had six tons of lead shorn off the 30 tons it carried on its keel. Also received new rigging and sails.[2132] All told, it won between 50 and 60 races during its career.[2133] By 1900 out of commission and remained in this condition for a number of years.[2134]

ERA AND VOLUNTEER RACING OFF MANLY (C1890S).
SOURCE: ROYAL SYDNEY YACHT SQUADRON.

In March 1905 sold to Winter, Brandt & Co. of Western Australia and arrived in Fremantle after a passage

[2123] *The Sydney Morning Herald*, 6 December 1887.
[2124] *The Sydney Mail*, 6 November 1907.
[2125] *The Australian Star*, 14 March 1905.
[2126] *The Sydney Morning Herald*, 2 February 1888; *The Sydney Mail and New South Wales Advertiser*, 2 June 1888.
[2127] *The Sydney Mail and New South Wales Advertiser*, 2 June 1888.
[2128] *Evening News*, 10 November 1888.
[2129] *The Argus*, 19 November 1888.
[2130] *South Australian Register*, 27 November 1888.
[2131] *Evening News*, 10 December 1888.
[2132] *The Sydney Morning Herald*, 15 September 1892.
[2133] *The Australian Star*, 14 March 1905.
[2134] *Australian Town and Country Journal*, 20 October 1900.

of 21 days.[2135] One of the conditions of the sale was that the vessel be taken out of eastern states' waters, as Milson could not bear to see it converted into a fishing boat.[2136] Then became involved in the shark fishing industry, operating out of Geraldton for a number of decades.[2137] In March 1918 fitted up as an auxiliary motor boat, a 30 h.p. Standard marine engine installed.[2138] At the time, capable of holding eight tons of fish in its ice chamber, making trips to and from Shark Bay every ten days.[2139] The engine only lasted three years and until 1950 worked without an engine, maintaining its fast sailing qualities.[2140] Significantly damaged in 1928 after being driven on to the breakwater at Geraldton during a storm.[2141] Successfully refloated and repaired, went on to continue in the local shark fishing industry for another three decades. Ultimately sank on 28 April 1958 in South Passage, Shark Bay, where its lead keel still remains.[2142]

ERA ON THE SLIPS AT GERALDTON.
SOURCE: CITY OF GREATER GERALDTON REGIONAL LIBRARY.

ELF

Year: 1888
Designer: Walter Reeks
Builder: George Ellis
Location: Balmain, New South Wales, Australia
Type: Wooden cutter yacht
Description: 19.5 x 7.0 x 3.5ft. 3 tons. 2-rater yacht built by George Ellis to the order of J. O. Fairfax and launched in early 1888.[2143] Advertised for sale in August 1889 and again in May 1890.[2144] By August of that year owned by Armand Kellerman of the Prince Alfred Yacht Club.[2145] He retained possession until June 1924 when it was stolen from Vaucluse Bay, taken outside the Heads and wrecked on the rocks off Dee Why.[2146]

IDUNA

Year: 1888
Designer: Walter Reeks
Builder: George Ellis
Location: Balmain, New South Wales, Australia
Type: Wooden cutter yacht
Description: ON 88965. 57 (oa) x 49.5 x 12.1 x 8.7ft. 30 tons. Built by George Ellis to the order of John Gurner, a well-known Melbourne barrister, and launched on 10 July 1888.[2147] Sailed to Port Phillip Bay two months later, going on to compete in the race for 20-tonners at the Melbourne Centennial Aquatic Carnival, held on 23-24 November.[2148] Made a circumnavigation around Tasmania early in 1889 with a crew of three; the first of several ocean-going cruises.[2149] On 30 March 1894 sold to Wallace McPherson Cameron of Sydney.[2150] Returned to Port Jackson, for many years regularly racing against the likes of *Isea* and *Thelma* (*q.v.*). Though advertised for sale in 1901, was not sold until December 1906; its new owner being Frances Snow of Adelaide.[2151] Made its way to South Australia early in 1907, going on to sail with the Royal South Australian Yacht Squadron. In December 1911 purchased by Winter, Brandt & Co. of Western Australia, the same firm that had purchased the Reeks-designed *Era* (*q.v.*) in 1905.[2152] After 21 years of successful operation in the shark fishing industry, sank during a gale in Geraldton harbour in February 1942 and was wrecked. Salvage of the craft's leaden keel, estimated to contain 17 tons of lead, was undertaken in November 1947.[2153]

CANOMIE

Year: 1888
Designer: Walter Reeks
Builder: Thomas R. Dudley
Location: Drummoyne, New South Wales, Australia
Type: Wooden sailing yacht
Description: ON 93586. 56 (oa) x 49.2 x 13.8 x 5.2ft. 24 tons. Built by Captain Thomas R. Dudley at Drummoyne on

2135 *The Australian Star*, 8 April 1905; *Geraldton Advertiser*, 21 June 1905.
2136 museum.wa.gov.au/maritime-archaeology-db/wrecks/id-1186.
2137 *Geraldton Guardian*, 6 March 1924.
2138 *Geraldton Express*, 18 March 1918.
2139 *Geraldton Express*, 18 March 1918.
2140 museum.wa.gov.au/maritime-archaeology-db/wrecks/id-1186.
2141 *The Murchison Times*, 21 July 1928.
2142 museum.wa.gov.au/maritime-archaeology-db/wrecks/id-1186.

2143 *Australian Town and Country Journal*, 18 February 1888; *Lloyd's Register of Yachts* (1903).
2144 *The Sydney Morning Herald*, 10 August 1889, 24 August 1890.
2145 *The Australian Star*, 22 August 1890.
2146 *The Sydney Morning Herald*, 7 June 1924; *The Sun*, 16 February 1935.
2147 Register of British Ships: Main Register subsequent to Merchant Shipping Act 1854, Port Melbourne [contains Volume 7 to Volume 9].
2148 *The Sydney Mail and New South Wales Advertiser*, 22 September, 1 December 1888.
2149 *The Argus*, 6 April 1889.
2150 Register of British Ships: Main Register subsequent to Merchant Shipping Act 1854, Port Melbourne [contains Volume 7 to Volume 9].
2151 *The Argus*, 12 October 1901; *The Express and Telegraph*, 10 November 1906.
2152 *Geraldton Guardian*, 12 December 1911; ADELAIDE, Register of ships subsequent to Merchant Shipping Act 1854, Volume 4, 1896-1916, Folio 120.
2153 *The West Australian*, 10 November 1947.

IDUNA ON SYDNEY HARBOUR.
SOURCE: AUSTRALIAN NATIONAL MARITIME MUSEUM COLLECTION.

the Parramatta River and launched on 11 August 1888.[2154] Commissioned by Robert Ross of Taranganba, near Rockhampton, Queensland, with the intention of carrying some 25 tons of cargo between these two locations, as well as the Keppel Islands.[2155] Centreboard craft built of hardwood and kauri and rigged as a 'Dandy'.[2156] Assigned the letters SLVM.[2157] Arrived in Rockhampton after a ten-day passage from Sydney.[2158] Then commenced in the local cargo and passenger trade, maintaining a regular service for several years.[2159] In November 1891 noted back in Sydney, competing in a race for cruisers held by the Royal Sydney Yacht Squadron.[2160] Several weeks later officially sold to Robert Macintosh of Burns Philp and Co. Ltd., Brisbane.[2161] Then readied for a 12-month cruise, encompassing Hobart, the west coast of Australia, the Java Sea, China and Japan, under the command of Captain S. G. Hooper.[2162] Left Sydney on March 1892, though was forced to return.[2163] Instead headed north for the Torres Strait via Townsville in April 1892.[2164] Several months later was advertised for sale and subsequently sold to Robert Davidson of Townsville.[2165] Two changes of ownership then took place in 1899; first in February purchased by Forbes de Lancey Hudson of Townsville, and in June by William Hamilton of Samarai, British New Guinea.[2166] In February 1901 sold to the Hamilton Pearling Co. Ltd, of Brisbane.[2167] Then commenced superintending the pearling fleet in Manning Strait.[2168] Ultimately wrecked on a reef in July 1909 while sailing between Geza and the Manning Straits, all hands saved.[2169] At the time of loss, owned by the Solomon Islands Development Company.

CANOMIE SAILING ON SYDNEY HARBOUR.
SOURCE: AUSTRALIAN NATIONAL MARITIME MUSEUM COLLECTION (00002502).

[2154] *The Daily Telegraph*, 13 August 1888; *Referee*, 16 August 1888.
[2155] *Referee*, 16 August 1888.
[2156] *The Daily Telegraph*, 13 August 1888.
[2157] *The Sydney Morning Herald*, 6 November 1888.
[2158] *The Daily Northern Argus*, 21 November 1888.
[2159] *The Daily Northern Argus*, 23 January 1889.
[2160] *The Sydney Morning Herald*, 23 November 1891.
[2161] Registers of British Ships, Port of Sydney, single number series (II) - ROLL 7.
[2162] *The Sydney Morning Herald*, 29 December 1891.
[2163] *The Sydney Morning Herald*, 18 March 1892.
[2164] *Newcastle Morning Herald and Miners' Advocate*, 8 April 1892.
[2165] *The Sydney Morning Herald*, 9 August 1893; Registers of British Ships, Port of Sydney, single number series (II) - ROLL 7.
[2166] *The Sydney Morning Herald*, 23 November 1891.
[2167] *The Sydney Morning Herald*, 23 November 1891.
[2168] *The Sydney Morning Herald*, 11 June 1901.
[2169] *The Daily Telegraph*, 24 July 1909.

JESS

Year: 1888
Designer: Walter Reeks
Builder: Walter 'Watty' M. Ford
Location: Berrys Bay, New South Wales, Australia
Type: Wooden two-masted lugger
Description: ON 93597. 46.5 (oa) x 41.0 x 12.4 x 6.4ft. 19 tons. Built by Ford, under the superintendence of Reeks' pupil Ernest Thompson, to the order of Edward 'Ned' King and launched on 4 October 1888. 15-ton beamy centreboarder intended for cruising, with its first major voyage to Lord Howe Island.[2170] Sold in June 1893 to Frederick Rae, continuing to be used primarily for cruising. In September 1897 sold to Aubrey Arthur Griffiths who had most recently owned *Isea*, undergoing a change to a cutter rig and several other alterations, supervised by Reeks.[2171] Commenced racing on Sydney Harbour with some success. Additional owners were James Cox (1899 - 1905), who took it to Queensland on a cruise in his first year of ownership, and Edward Jacobs (1905 - 1906) who intended to undertake a cruise to the South Sea Islands and possibly America, after having the centreboard removed, its bulwarks raised and some minor alterations carried out in Sydney.[2172] In April 1906 sold to the Bua Province, Fiji, for the use of Ratu Joni Madraiwiwi, Roko Tui Bua, to travel between Ra and Bua.[2173] Its subsequent history is not known.

[2170] *Evening News*, 1 October 1888.
[2171] Registers of British Ships, Port of Sydney, single number series (II) - ROLL 7; *The Sydney Morning Herald*, 6 November 1897.
[2172] Registers of British Ships, Port of Sydney, single number series (II) - ROLL 8; *The Sydney Morning Herald*, 21 October 1899, 8 July 1905; *Evening News*, 14 July 1905.
[2173] *Evening News*, 26 April 1906.

ISEA

Year: 1888
Designer: Walter Reeks
Builder: Walter 'Watty' M. Ford
Location: Berrys Bay, New South Wales, Australia
Type: Wooden sailing yacht
Description: ON 93598. 43.8 (oa) x 37.0 x 9.4 x 6.0ft. 12 tons. Built by Ford at Berrys Bay to the order of Dr Frederick Milford and launched on 9 October 1888. Believed to have been the first 10-tonner built anywhere in the world to the new 'Length & Sail Area Rule'.[2174] Was not ready to race for some time, its sails had not arrived from England.[2175] Its maiden race took place in mid-January 1889 when the press noted that it '*shaped very well*'.[2176] Went on to successfully race with the second and third class yachts for several seasons. On 1 September 1896 sold to Aubrey Arthur Griffiths, late vice-commodore of the Prince Alfred Yacht Club.[2177] Then placed in the hands of its builder with alterations and additions taking place, supervised by Reeks. One year later sold to Walter Moffat Marks under whose auspices it became an even more successful yacht on Sydney Harbour.[2178] Marks also took several coastal cruises in it. He retained possession until May of 1905 when it was sold to Dr Alfred Cooper Nathan.[2179] Just under a year later, transferred to Bernard Emerson.[2180] Subsequent owners include Sarsfield Cassidy

[2174] *The Anchor*, 22 February 1912.
[2175] *The Sydney Mail and New South Wales Advertiser*, 27 October 1888.
[2176] *Australian Town and Country Journal*, 26 January 1889.
[2177] Registers of British Ships, Port of Sydney, single number series (II) - ROLL 7.
[2178] *Evening News*, 28 August 1896.
[2179] Registers of British Ships, Port of Sydney, single number series (II) - ROLL 7.
[2180] Registers of British Ships, Port of Sydney, single number series (II) - ROLL 7.

ISEA SAILING ON SYDNEY HARBOUR.
SOURCE: AUSTRALIAN NATIONAL MARITIME MUSEUM (WILLIAM HALL COLLECTION).

(1907 - 1910), Frances Ralph (1910 - 1911) and Charles Horsfall (1911 - 1911).[2181] In May 1911 purchased by William Russell of Brisbane, by this time rigged as a schooner and fitted with an oil engine.[2182] A few months later, noted to be in Vila, Vanuatu, after a passage of 33 days from Sydney.[2183] Though returned to the possession of Horsfall, a previous owner, in 1913, its subsequent fate is unknown.[2184]

VOLUNTEER

Year: 1888
Designer: Walter Reeks
Builder: Thomas R. Dudley
Location: Drummoyne, New South Wales, Australia
Type: Wooden sailing yacht
Description: ON 93588. 63 (oa) x 56.5 x 13.3 x 8.0ft. 40 tons. 30-rater built by Dudley at Drummoyne in a hurry and in secrecy for a syndicate headed by William Smairl and launched on 6 October 1888. First raced at the Double Bay Regatta, held on 27 October, finishing second to *Era* (*q.v.*).[2185] Competed for the Intercolonial Cup at the Melbourne Centennial Aquatic Carnival, held in late November 1888, finishing second. Returned to Sydney and raced with success, particularly after its sails had been installed and rigging and masts altered.[2186] In August 1891 advertised for sale by auction on behalf of the mortgagees.[2187] Took some months to sell, ultimately purchased by Robert Ross for £500 who had previously owned *Canomie* (*q.v.*).[2188] On-sold several months later to Frederick Woolcott-Waley and resumed racing shortly.[2189] Sold in January 1893 to Henry W. Henderson of the German S. S. Company on behalf of his brother Thomas Henderson of Auckland.[2190] Left Sydney for New Zealand on 1 March 1893 arriving 16 days later.[2191] Sailed on Auckland Harbour, becoming affectionately known as 'Old Vol'. Also used for cruising well into the next century, from May 1901 owned by Alexander Alison.[2192] In August 1905 sold to Gustavus M. Hennings of Naitaba, Fiji, intended for the islands trade.[2193] Left Auckland for Suva in April 1906.[2194] Became a total loss at Levuka, Fiji, during a hurricane in early 1908 and was subsequently broken up.[2195]

THISTLE EX BETTINA

Year: 1888
Designer: Walter Reeks
Builder: Walter 'Watty' M. Ford
Location: Berrys Bay, New South Wales, Australia
Type: Wooden sailing yacht
Description: ON 93613. 48 (oa) x 42.3 x 11.2 x 5.6ft. 17 tons. Built by Ford, under the superintendence of Ernest Thompson, to the order of Patrick Creagh and launched on 6 December 1888.[2196] A beamy centreboard cruiser, embodying American ideas, also a pioneer of the large open-cockpit and coachhouse type. Went on to become a successful vessel of Sydney Harbour in this configuration. In February 1896 sold to Charles Shannon of Geelong, transferred to Victoria and renamed *Thistle*, becoming a bigger success under a subsequent Melbourne-based owner, Edgar Newland who purchased it in April 1904.[2197] Won the inaugural Rudder Cup in 1907, now known as the Melbourne to Devonport Yacht Race.[2198] Subsequently fitted with a 10-12 h.p. engine and in June 1918 sold to James Maginnis, a merchant of Melbourne.[2199] In November 1926 went ashore at Hampton during a storm though recovered.[2200] Transferred to Geoffrey Ines in December 1929 following Maginnis' death.[2201] In July 1932 sold to Dr John Bell Ferguson and primarily used

2181 Registers of British Ships, Port of Sydney, single number series (II) - ROLL 7.
2182 Registers of British Ships, Port of Sydney, single number series (II) - ROLL 7.
2183 *Evening News*, 9 September 1911.
2184 Registers of British Ships, Port of Sydney, single number series (II) - ROLL 7.
2185 *The Sydney Morning Herald*, 29 October 1888.
2186 *Evening News*, 26 August 1889; *Australian Town and Country Journal*, 14 December 1889.
2187 *The Sydney Morning Herald*, 19 August 1891.
2188 *The Australian Star*, 31 December 1891.
2189 *Evening News*, 4 June 1892.
2190 *Evening News*, 28 January 1893.
2191 *The Daily Telegraph*, 1 March 1893; *New Zealand Herald*, 18 March 1893.
2192 *New Zealand Herald*, 4 May 1901.
2193 *New Zealand Herald*, 29 August 1905.
2194 *New Zealand Herald*, 12 April 1906.
2195 *The Australian Star*, 12 March 1908.
2196 *Evening News*, 7 December 1888.
2197 Registers of British Ships, Port of Sydney, single number series (II) - ROLL 7.
2198 *The Age*, 28 December 1907.
2199 *The Age*, 13 April 1918.
2200 *The Herald*, 17 November 1926.
2201 Register of British Ships: Main Register subsequent to Merchant Shipping Act 1854, Port Melbourne [contains Volume 9 (part) to Volume 11.

VOLUNTEER OFF LEVUKA, FIJI, IN EARLY 1908.
SOURCE: NEW ZEALAND GRAPHIC, 5 MAY 1909.

for cruising, in 1932 making its way to Tasmania.[2202] Sold to Stanley Jenkins in late 1937.[2203] In 1944 sold to George Reid and converted to a shark fishing boat. Wrecked on Nepean Reef, while attempting to enter Port Phillip Bay, on 9 October 1945 with the loss of three lives, including its owner.[2204] Only one crew member managed to make it to shore.

THISTLE ASHORE AT HAMPTON, VICTORIA, 1928.
SOURCE: SANDRINGHAM & DISTRICT HISTORICAL SOCIETY.

--

Year: 1889
Designer: Walter Reeks
Builder: --
Location: --
Type: Wooden sailing yacht
Description: A design by Reeks for a vessel, likely 90ft on the waterline, intended to compete for the 1890 America's Cup, commissioned by Jack Want. No exact indication of type or rig, though suggested to be a cutter. Not built.

--

Year: 1889
Designer: Walter Reeks
Builder: --
Location: --
Type: Wooden steam yacht
Description: Not known if built. First of two vessels Reeks was noted in the press as designing in May 1889.[2205]

--

Year: 1889
Designer: Walter Reeks
Builder: --
Location: --
Type: Wooden steam yacht
Description: Not known if built. Second of two vessels Reeks was noted in the press as designing in May 1889.[2206]

--

Year: 1889
Designer: Walter Reeks
Builder: --
Location: --
Type: Wooden steam launch
Description: 27 x 5.7ft. Fitted with Kingdon's quadruple condensing engines.[2207] Designed by Reeks for C. Millar of Melbourne in conjunction with his steam yacht *Saide* ex *Star of the Sea*, a 155ft Scottish-built vessel that he had recently purchased.[2208] Though built to replace a steam launch that arrived with the vessel, it is not known if it was the one built by Simpson and Strickland of Dartmouth that was noted on board *Saide* by early 1891.[2209] Subsequent history not known.

--

Year: 1889
Designer: Walter Reeks
Builder: Captain Tom R. Dudley
Location: Drummoyne, New South Wales, Australia
Type: Wooden sailing yacht
Description: 28 x 9 x 7.6ft. Not known if completed. Was under construction by Captain Dudley at Drummoyne in May 1889, built to the order of a customer from Tasmania, assumed to be for the 28ft class then popular on the River Derwent.[2210]

ALBERT

Year: 1889
Designer: Walter Reeks
Builder: --
Location: Sydney, New South Wales, Australia
Type: Wooden model yacht
Description: 37in. First-class model yacht designed by Reeks to the order of W. Turner, commodore of the Moore Park Model Yacht Club.[2211] Competed in the first race of the season, held in mid-September 1889, finishing second.[2212] Went on to sail successfully for the remainder of the season.

LOUELLE

Year: 1889
Designer: Walter Reeks
Builder: Tom Cubitt
Location: Berrys Bay, New South Wales, Australia
Type: Wooden cutter yacht
Description: ON 106128. 27 (oa) x 24.0 x 6.0 x 4.7ft. 5 tons. 2½-rater yacht built by Tom Cubitt at Berrys Bay for

2202 *The Age*, 7 March 1932.
2203 Register of British Ships: Main Register subsequent to Merchant Shipping Act 1854, Port Melbourne [contains Volume 9 (part) to Volume 11.
2204 *The Age*, 10 October 1945.
2205 *The Sydney Morning Herald*, 22 May 1889.
2206 *The Sydney Morning Herald*, 22 May 1889.
2207 *The Brisbane Courier*, 9 October 1891.
2208 *The Sydney Morning Herald*, 17 August 1889; *Illustrated Australian News and Musical Times*, 1 August 1889.
2209 *Northern Territory Times and Gazette*, 20 April 1889; *The Sydney Mail and New South Wales Advertiser*, 24 August 1889; *The Australasian*, 7 February 1891.
2210 *The Sydney Morning Herald*, 22 May 1889.
2211 *The Sydney Morning Herald*, 17 August 1889; *Evening News*, 9 February 1891.
2212 *Evening News*, 20 September 1889.

Sid Gilchrist and launched on 21 August 1889.[2213] Carried 2 tons of lead on its keel and 740 square ft of canvas.[2214] A month after launch received 3 cwt more lead, having been found '*rather tender*'.[2215] A sister-ship to *Jenny Wren* (see next entry). Sailed with the newly-instituted 2½-rater class. Advertised for sale in February 1890 though failed to sell.[2216] Advertised again for sale in November 1891 and once more by Gilchrist in March 1895.[2217] Likely sold to S. T. Wilson who on-sold the craft in August 1901 to Arthur James Kelynack.[2218] Then immediately sold to Arthur Blix in whose name the vessel was first registered in September 1901.[2219] Subsequent history not known. Blix died in 1923 and *Louelle* was not part of his estate.[2220]

JENNY WREN

Year: 1889
Designer: Walter Reeks
Builder: Tom Cubitt
Location: Berrys Bay, New South Wales, Australia
Type: Wooden cutter yacht
Description: ON 112581. 27 (oa) x 24.3 x 6.2 x 3.1ft. 4 tons. 2½-rater yacht designed by Reeks and built by Tom Cubitt at Berrys Bay for Harry Stevens and launched on 24 October 1889.[2221] Carried 2 tons of lead on its keel and 740 square ft of canvas. A sister-ship to *Louelle* (see previous entry). Advertised for sale in mid-1891.[2222] By October 1891 owned by C. L. Garland.[2223] Advertised again for sale in late 1892.[2224] By October 1893 owned by W. S. Eaton.[2225] Advertised for sale in September 1894.[2226] By 1902 owned by Richard Old of North Sydney who first registered it in November of that year, around the same time it collided with the steamer *Woy Woy* on Sydney Harbour; thankfully only its dinghy suffered damage. A Marine Court of Inquiry subsequently found the master of *Woy Woy* at fault.[2227] In May 1906 advertised for sale.[2228] A new owner renamed it *Neaera* and it was under this name that the vessel was advertised for sale in May 1909.[2229] Its history between this date and 1922 remains unknown. Back under its original name, resumed sailing on Sydney Harbour in March 1922, by this time owned by G. S. Griffin.[2230] In September 1933 stolen from its Iron Cove mooring, though located after an extensive search in poor weather by the Water Police. At the time of the incident, owned by a Mr Hodgeson of Balmain.[2231] Subsequent owners include Paul Waites, Allan Windon, Doug Garrod and Neville Olliffe.[2232] Still in existence, located in Lane Cove, Sydney, and actively racing with other gaff-rigged cutters.

JENNY WREN IN SYDNEY (2008).
SOURCE: JOHN JEREMY.

THELMA

Year: 1889
Designer: Walter Reeks
Builder: George Ellis
Location: Balmain, New South Wales, Australia
Type: Wooden cutter yacht
Description: ON 101097. 52 (oa) x 45.3 x 10.7 x 7.1ft. 23 tons. Designed by Reeks and built by George Ellis at Balmain to the order of John F. Hoare, vice-commodore of the Royal Sydney Yacht Squadron, launched on 6 December 1889.[2233] Construction superintended by Reeks. Intended for racing and cruising. First raced at the 1890 Anniversary Regatta where it won the race for all yachts, with Reeks at the helm.[2234] Continued to win handsomely on Port Jackson for several seasons, as well as used for coastal cruises. First registered on 6 October 1893 by new owner Dr Alexander MacCormick.[2235] Under his ownership remained a standout on Sydney Harbour. In October 1896 had two tons of extra lead attached to its keel, increasing its depth and providing greater stability,

2213 *The Sydney Morning Herald*, 22 August 1889.
2214 *The Sydney Morning Herald*, 22 May 1889.
2215 *The Sydney Mail and New South Wales Advertiser*, 21 September 1889.
2216 The Sydney Morning Herald, 1 February 1890.
2217 *The Sydney Morning Herald*, 14 November 1891, 30 March 1895.
2218 *The Sydney Morning Herald*, 24 August 1901; Registers of British Ships, Port of Sydney, single number series (II) - ROLL 9.
2219 *The Sydney Morning Herald*, 24 August 1901; Registers of British Ships, Port of Sydney, single number series (II) - ROLL 9.
2220 Registers of British Ships, Port of Sydney, single number series (II) - ROLL 9.
2221 *The Sydney Morning Herald*, 25 October 1889.
2222 *The Sydney Morning Herald*, 2 May, 19 September 1891.
2223 *The Sydney Morning Herald*, 17 October 1891.
2224 *The Sydney Morning Herald*, 6 September, 19 November 1892.
2225 *Evening News*, 21 October 1893.
2226 *The Sydney Morning Herald*, 29 September 1894.
2227 *Evening News*, 20 December 1901.
2228 *The Sydney Morning Herald*, 16 May 1906.
2229 *The Sydney Morning Herald*, 15 May 1909.
2230 Evening News, 28 March 1922.
2231 *The Sun*, 30 September 1933.
2232 arhv.sea.museum/en/objects/details/126913/jenny-wren; SASC News: The Newsletter of the Sydney Amateur Sailing Club, December, 2015.
2233 *The Sydney Morning Herald*, 17 August 1889, 7 December 1889.
2234 *The Australian Star*, 14 February 1890.
2235 *The Australian Star*, 3 August 1893; Registers of British Ships, Port of Sydney, single number series (II) - ROLL 7.

the work superintended by Reeks and carried out by W. M. Ford (Jr).[2236] In early 1898 received a new mast.[2237] Sold on 18 March 1908 to Charles Lloyd Jones. Later that year selected as the New South Wales challenger for the Sayonara Cup. Despite being 20 years old, still considered the yacht best suited to do well in the heavy winds and corresponding seas Hobsons Bay was known for.[2238] Left Sydney on 15 January 1909, arriving in Melbourne seven days later.[2239] Under a new cutter rig, won the first race by a mere 15 seconds.[2240] Lost the second and third races; the latter by a narrow margin. Paid hands returned it to Sydney, though it went aground at Queenscliff (unscathed), then was buffeted by squalls.[2241] A few weeks later back to its winning ways, beating *Petrel* and *Sunbeam* in a Prince Alfred Yacht Club general handicap race.[2242] Continued to successfully race on Sydney Harbour for several more seasons, including winning the Rawson Cup in March 1912.[2243] Ultimately met its demise on 8 March 1913 when owner, Charles Lloyd Jones, agreed to race *Sayonara* (owned by Paul Ross) in a private match for a trophy over the Manly to Pile Light course.[2244] Taking place under boisterous weather conditions, later described as a cyclone, and in *'mountainous seas'*, while off Middle Harbour lost a crew member overboard. In gybing to try and pick up the disabled man, the mast snapped off and was lost. Then drifted towards Dobroyde Point where the anchor was dropped and fortunately held.[2245] In a dangerous rescue, the remaining nine crew members were transferred to a whaleboat and then to the pilot steamer *Buranda*.[2246] *Thelma* went ashore later that night and was wrecked to pieces. The drowned crew member, Walter Moore, was a prominent and popular member of the Prince Alfred Yacht Club.

2236 *The Sydney Morning Herald*, 7 November 1896.
2237 *The Sydney Morning Herald*, 22 January 1898.
2238 *The Daily Telegraph*, 30 October 1908.
2239 *Evening News*, 15 January 1909.
2240 *Daily Telegraph*, 6 February 1909.
2241 *The Herald*, 15 February 1909; *Evening News*, 17 February 1909.
2242 *The Sunday Sun*, 14 March 1909.
2243 *Evening News*, 5 March 1912.
2244 *The Sydney Morning Herald*, 4 March 1913.
2245 *Sunday Times*, 9 March 1913.
2246 *The Sydney Morning Herald*, 10 March 1913.

NORNA

Year: 1889
Designer: Walter Reeks
Builder: Tom Cubitt
Location: Berrys Bay, New South Wales, Australia
Type: Wooden cutter yacht
Description: ON 101150. 28 (oa) x 25.5 x 6.6 x 4.6ft. 4 tons. Designed by Reeks and built by Tom Cubitt to the order of G. E. Fairfax and launched on 13 December

THELMA AND IDUNA SAILING ON SYDNEY HARBOUR.
SOURCE: AUSTRALIAN NATIONAL MARITIME MUSEUM COLLECTION (ANMS1096[086]).

1889.[2247] A straight-stemmed 2½ rater with a tapering counter.[2248] Advertised for sale in April 1891.[2249] Then appears to have passed to A. Marks and, following, to Fred Doran.[2250] By 1895 owned by Joseph Ezra O'Brien who first registered it on 22 February 1896.[2251] Subsequently owned, as half-shares, by Maitland Brown and William Fonge. In August 1899 sold to Alfred Ernest Starkey who on-sold the vessel on 5 September 1901 to Frederick Probst of America, whereupon its register was closed.[2252] Subsequent history is not known.

ASTROEA

Year: 1890
Designer: Walter Reeks
Builder: Tom Cubitt
Location: Berrys Bay, New South Wales, Australia
Type: Wooden cutter yacht
Description: 29.2 (oa) x 23 (wl) x 6.8 x 5ft.[2253] 2½-rater yacht designed by Reeks and built by Tom Cubitt to the order of James McCabe and launched on 10 May 1890.[2254] Carried 652 square ft of sail area and featured a tapering counter and overhanging bow.[2255] Began racing on Port Jackson in late 1890, spending a few seasons competing then absent for several years.[2256] By 1895 racing with the Lane Cove Sailing Club and owned by W. Leigh.[2257] Stripped and recoppered in 1898, and the following year owned by E. J. Vial of the Prince Alfred Yacht Club.[2258] In 1901 owned by a Mr Marshall, and two years later by J. H. Bradley.[2259] Wrecked in July of 1910 after being swept from its moorings at Neutral Bay during a storm.[2260]

LAH LOO EX VIOLA

Year: 1890
Designer: Walter Reeks
Builder: Tom Cubitt
Location: Berrys Bay, New South Wales, Australia
Type: Wooden cutter yacht
Description: 30.8 (oa) x 25 (wl) x 7.5 x 4.7ft. 6 tons. 3-rater cruising yacht designed by Reeks and built by Tom Cubitt and launched from Berrys Bay on 2 October 1890 to the order of James Robb.[2261] Competed in the cruiser class on Sydney Harbour for one season before being advertised for sale.[2262] Purchased by W. H. Murrell of the Prince Alfred Yacht Club and renamed *Lah Loo*.[2263] Advertised for sale in early to mid-1900 and by 1901 was owned by A. P. Stewart.[2264] By 1905 owned by W. C. Heron and competing with the Royal Sydney Yacht Squadron.[2265] Last noted when advertised for sale in November 1911.[2266] Wrecked on 9 December 1920 when driven ashore at Manly, at the time owned by Clifford Henry.[2267]

VARUNA

Year: 1890
Designer: Walter Reeks
Builder: George Barnett
Location: Drummoyne, New South Wales, Australia
Type: Wooden sailing yacht
Description: 22 (wl) x 9.3 x 2.6ft. Half-decked yacht designed by Reeks and built by George Barnett for W. M. Paul at Drummoyne and launched in October 1890.[2268] Replaced an earlier 22ft half-decked boat of the same name that was successfully raced by the same owner over many seasons, including competing at the Melbourne Centennial Aquatic Carnival in November 1888.[2269] A popular and high-performing yacht on Port Jackson for many seasons, including competing for New South Wales in the Intercolonial Championship Race for 22-footers against Queensland in January 1895 where the craft finished third.[2270] Sold to W. Staunton Spain in late 1897.[2271] In March 1905 narrowly avoided a collision with the steamer *Lady Hampden* while competing in a race with the Sydney Amateur Sailing Club.[2272] Successfully competing on Sydney Harbour under this owner's management in 1909.[2273] Its subsequent history is not known.

MYEE

Year: 1890
Designer: Walter Reeks
Builder: Tom Cubitt
Location: Berrys Bay, New South Wales, Australia
Type: Wooden cutter yacht
Description: 18.5 (oa) x 17.5 (wl) x 7 x 3.3ft. Described as a '*very pretty and substantial 18-footer*' when launched from Tom Cubitt's Berrys Bay yard on 12 November 1890.[2274] A ½-decked cruiser designed by Reeks to the order of Dr Newmarch, rigged with a lug and jib and carried 14 cwt of lead on its keel, with an additional 6 cwt carried

[2247] *The Sydney Mail and New South Wales Advertiser*, 21 December 1889.
[2248] *The Sydney Mail and New South Wales Advertiser*, 21 December 1889.
[2249] *Sunday Times*, 13 October 1895.
[2250] *The Sydney Mail and New South Wales Advertiser*, 30 January 1892; *Australian Town and Country Journal*, 3 February 1894.
[2251] SYDNEY, NEW SOUTH WALES, Register of ships, single number series (II), Volume 16, 1893-1897, Folio 112.
[2252] SYDNEY, NEW SOUTH WALES, Register of ships, single number series (II), Volume 16, 1893-1897, Folio 112.
[2253] *The Sydney Morning Herald*, 1 March, 12 May 1890.
[2254] *The Sydney Morning Herald*, 12 May 1890.
[2255] *The Sydney Morning Herald*, 1 March, 12 May 1890.
[2256] *The Sydney Morning Herald*, 27 January 1892.
[2257] *Sunday Times*, 24 November 1895.
[2258] *The Daily Telegraph*, 24 August 1898; *The Sydney Morning Herald*, 14 January 1899.
[2259] *The Australian Star*, 2 September 1901; *The Daily Telegraph*, 28 February 1903; *The Sydney Morning Herald*, 25 April 1903.
[2260] *The Sydney Morning Herald*, 20 July 1910.
[2261] *The Sydney Mail and New South Wales Advertiser*, 9 August 1890; *The Sydney Morning Herald*, 3 October 1890.
[2262] *The Sydney Morning Herald*, 11 April 1891.
[2263] *The Sydney Morning Herald*, 9 October 1891
[2264] *The Sydney Morning Herald*, 17 February, 2 June 1900; *Sunday Times*, 20 October 1901.
[2265] *The Sydney Morning Herald*, 25 March 1905.
[2266] *The Sydney Morning Herald*, 18 November 1911.
[2267] www.environment.gov.au/shipwreck/public/wreck/wreck.do?key=1057.
[2268] *The Sydney Mail and New South Wales Advertiser*, 9 August 1890; *The Sydney Morning Herald*, 20 October 1890.
[2269] *The Sydney Mail and New South Wales Advertiser*, 10 November 1888; *Evening News*, 16 November 1888; *The Sydney Morning Herald*, 26 March 1890.
[2270] *The Queenslander*, 26 January 1895.
[2271] *The Daily Telegraph*, 13, 21 August 1897; *The Australian Star*, 8 November 1897.
[2272] *The Daily Telegraph*, 20 May 1905.
[2273] *The Daily Telegraph*, 27 January 1909.
[2274] *The Australian Star*, 13 November 1890; *The Sydney Mail and New South Wales Advertiser*, 22 November 1890.

inside.[2275] Appears in Sydney Flying Squadron and Sydney Amateur Sailing Club race results, sailed by its original owner, for several seasons. Advertised for sale in late 1892, then purchased by George Cannis.[2276] Advertised for sale again in December 1893, though appears not to have sold.[2277] Back competing on Sydney Harbour at the 1894 Anniversary Regatta, still owned by Cannis.[2278] Continued to race up until completion of a 23ft (overall) ½-rater yacht that Cannis had designed and built in late 1895.[2279] Advertised for sale and by late 1896 owned by A. H. Woolcott and competing with the Sydney Amateur Sailing Club.[2280] Later history is not certain, may have been the '*18ft cutter Myee*' that sailed in a private match against the 19ft yawl *Cronulla* in October 1901, at the time owned by L. C. Waterman.[2281]

BORONIA

Year: 1890
Designer: Walter Reeks
Builder: Tom Cubitt
Location: Berrys Bay, New South Wales, Australia
Type: Wooden sailing yawl
Description: ON 101005. 46.8 (oa) x 37.7 x 11.5 x 7.5ft. 17 tons. Built by Tom Cubitt at Berrys Bay to the order of T. W. F. Busby and launched on 15 November 1890.[2282] Designed by Reeks, intended for both racing and lengthy cruises. Described as '*well fitted throughout, with every comfort and convenience for cruising the coast*'.[2283] Its first trip was to the northern coast of New South Wales, undertaken a few months after launch.[2284] A more substantial trip was planned for later in 1891, encompassing New Caledonia and New Hebrides, though it returned to Sydney only a few weeks after departure having suffered damage during a storm off Lord Howe Island.[2285] Sold in April 1892 to J. Cox of Sydney and added to the fleet of the Prince Alfred Yacht Club, also undertaking coastal cruises.[2286] Sailed to Brisbane in April 1893, with Captain E. J. Tooker at the helm, having been purchased by James Clark, successful pearl sheller.[2287] Appears to have remained in Brisbane for at least a few years, its later fate is not known. Register closed in 1906 after ownership was transferred to Dutch citizens, supposedly in 1899.[2288]

BUL BUL

Year: 1890
Designer: William Fife (Jr)
Builder: Tom Cubitt (superintended by Walter Reeks)
Location: Berrys Bay, New South Wales, Australia
Type: Wooden sailing yawl
Description: ON 101127. 36 (oa) x 25 (wl) x 6.4 x 6.2ft. 7 tons. Launched from Tom Cubitt's Berrys Bay yard on 21 November 1890, built to a design by William Fife (Jr) of Scotland with construction superintended by Reeks.[2289] A 2½-rater yacht commissioned by J. O Fairfax who had previously owned the 2-rater Reeks-designed *Elf*, considered a significant departure from those of its class.[2290] Also featured a clipper bow and a long graceful counter. Dominated the 2½-rater class for several seasons before transitioning to race with the larger yacht classes. Advertised for sale in October 1892, though took some time to sell.[2291] By early 1894 owned by F. Woolcott-Waley, in February 1896 sold to J. Wrixton who advertised it for sale in October 1897.[2292] By 1900 owned by C. M. Barker.[2293] Advertised for sale in August 1906 and purchased by E. W. Barker, continuing to perform well in handicap cruiser events.[2294] Advertised for sale in late 1907, then had a series of short-term owners, including John Thornley, Charles Dibbs, and Harold and Walter Tarrant. By 1909 owned by John Hellmann of Cootamandra.[2295] Likely remained in Hellmann's possession for some time. Last noted when advertised for sale in January 1923.[2296]

FREYA

Year: 1891
Designer: Walter Reeks
Builder: Tom Cubitt
Location: Berrys Bay, New South Wales, Australia
Type: Wooden pearling lugger
Description: ON 101002. 43.0 x 11.6 x 6.0ft. 13 tons. The first pearling schooner designed by Reeks and built to the order of Andrew Andersen, a pearl sheller of Prince of Wales Island, launched on 23 April 1891.[2297] Intended for the Torres Strait pearlshelling industry, left Sydney for Queensland on 7 May, arriving at Maryborough nine days later.[2298] Continued north to the pearling grounds, with John Andersen noted at the helm in July 1892.[2299] Sold in January 1901 to Herbert Bowden, a pearler of Thursday Island.[2300] Remained in his possession until mid-1924 when sold to Hodels Limited, auctioneers of Cairns.[2301] Its history after this time point is not known.

2275 *The Australian Star*, 13 November 1890.
2276 *The Sydney Morning Herald*, 5 November, 7 December 1892; *Evening News*, 19 September 1893.
2277 *The Sydney Morning Herald*, 30 December 1893.
2278 *The Daily Telegraph*, 27 January 1894.
2279 *The Daily Telegraph*, 31 July 1895.
2280 *Sunday Times*, 1 March 1896; *The Daily Telegraph*, 19 December 1896.
2281 *Referee*, 16 October 1901.
2282 *The Sydney Morning Herald*, 15 November 1890; *The Australian Star*, 17 November 1890.
2283 *Australian Town and Country Journal*, 22 November 1890.
2284 *The Sydney Morning Herald*, 27 January 1891.
2285 *The Sydney Morning Herald*, 18 June, 4 July 1891.
2286 *The Sydney Morning Herald*, 14 October, 24 December 1892.
2287 *The Sydney Morning Herald*, 9 March 1893; *Referee*, 29 March 1893.
2288 SYDNEY, NEW SOUTH WALES, Register of ships, single number series (II), Volume 14, 1888-1891, Folio 325.
2289 *The Sydney Morning Herald*, 22 November 1890.
2290 *The Sydney Morning Herald*, 22 November 1890.
2291 *The Sydney Morning Herald*, 1 October 1892.
2292 *The Daily Telegraph*, 12 March 1894, 14 February 1896; *The Sydney Morning Herald*, 11 October 1897.
2293 *Australian Town and Country Journal*, 27 October 1900.
2294 *The Sydney Morning Herald*, 15 August 1906, 25 February 1907; *The Australian*, 30 November 1906.
2295 SYDNEY, NEW SOUTH WALES, Register of ships, single number series (II), Volume 16, 1893-1897, Folio 71; SYDNEY, NEW SOUTH WALES, Transactions register, Volume 8, 1900 onwards, Folio Tr 89.
2296 *The Sydney Morning Herald*, 3 January 1923.
2297 *The Australian Star*, 25 March 1891; *Evening News*, 24 April 1891; *Maryborough Chronicle, Wide Bay and Burnett Advertiser*, 19 May 1891.
2298 *Maryborough Chronicle, Wide Bay and Burnett Advertiser*, 19 May 1891.
2299 *The Telegraph*, 7 July 1892.
2300 SYDNEY, NEW SOUTH WALES, Register of ships, single number series (II), Volume 14, 1888-1891, Folio 311.
2301 TOWNSVILLE, Register of ships, Volume 3, 1914-1981, Folio 7.

FREYA.
SOURCE: QUEENSLAND MARITIME MUSEUM COLLECTION
NORNA PEARLING FLEET, THURSDAY ISLAND 1902 ALBUM.

MINERVA

Year: 1891
Designer: Walter Reeks
Builder: Tom Cubitt
Location: Berrys Bay, New South Wales, Australia
Type: Wooden pearling lugger
Description: ON 101009. 43.0 x 11.4 x 6.0ft. 13 tons. A similar design to *Freya*, built by Tom Cubitt to a design by Reeks and launched from Berrys Bay on 15 May 1891.[2302] Built to the order of Herbert Bowden, pearler of Thursday Island, left Sydney for Queensland four days after launch.[2303] Bowden retained ownership until October 1920 whereupon it was sold to Arthur Booker, pearler of Thursday Island.[2304] Sold back to Bowden in April 1921. In September 1925 sold to a company Bowden had established in Sydney, Bowden Pearling Co. Ltd. Wrecked a few years later, on 28 December 1928 off Tern Island, all hands saved.[2305]

HYGIEA

Year: 1891
Designer: Walter Reeks
Builder: Tom Cubitt
Location: Berrys Bay, New South Wales, Australia
Type: Wooden sailing yacht
Description: 27 (oa) x 21 (wl) x 7.2 x 4.5ft. Built by Tom Cubitt to the order of Dr Robert Scot Skirving to a design by Reeks. Described as a *'pretty little yacht'* when launched from Berrys Bay in June 1891.[2306] Named after the Greek goddess of health. Yawl-rigged, remained in its original owner's possession until November 1903 when sold to Captain Reid of Thursday Island.[2307] Subsequent history is not known.

WARATAH

Year: 1891
Designer: Walter Reeks
Builder: Anderson Brothers
Location: Kangaroo Point, Queensland, Australia
Type: Wooden pearling lugger
Description: 49 x 12 x 6.5ft. Designed by Reeks and built by Messrs Anderson Brothers of Kangaroo Point to the order of James Clark, intended to compete in first class yacht races on Morton Bay.[2308] Despite its large size, never registered and its history is somewhat hard to trace. Launched in early August 1891 and first noted later that month after having taken a cruise on Moreton Bay.[2309] Then competed in several events of the Brisbane Sailing Club of which Clark was vice-commodore, including a pearling lugger race in May of 1892 after which Clark ordered six more vessels to same design.[2310] Next headed north to the pearling grounds.[2311] Noted in March of 1893 after having suffered damage during a gale while working in the vicinity of Princess Charlotte Bay.[2312] Suffered more serious damage during a cyclone that hit the Pilbara region of Western Australia in early March 1899, with indications it sank off Cape Preston after *'drifting on*

2302 *The Australian Star*, 25 March 1891; *Evening News*, 24 April 1891; *Maryborough Chronicle, Wide Bay and Burnett Advertiser*, 19 May 1891.
2303 SYDNEY, NEW SOUTH WALES, Register of ships, single number series (II), Volume 14, 1888-1891, Folio 343; *The Daily Telegraph*, 20 May 1891.
2304 TOWNSVILLE, Register of ships, Volume 3, 1914-1981, Folio 8.
2305 *The Telegraph*, 19 September 1929.
2306 *Evening News*, 26 June 1891.
2307 *The Australian Star*, 20 November 1903.
2308 *The Brisbane Courier*, 18 August 1891, 31 August 1891.
2309 *The Brisbane Courier*, 31 August 1891.
2310 *The Brisbane Courier*, 14 September 1891; *The Sydney Morning Herald*, 25 July 1892.
2311 *The Telegraph*, 14 September 1891.
2312 *The Brisbane Courier*, 25 March 1893.

top of a boat' with the loss of all hands.[2313] Later raised, with enough temporary repairs effected to be towed to Fortescue.[2314] Likely returned to the pearling industry and may have ultimately been lost after sinking while at anchor near Cable Beach, Broome, during a cyclone in April 1908 with loss of all but two crew members.[2315]

CLYANTHUS

Year: 1891
Designer: Walter Reeks
Builder: William Warbrick
Location: New Farm, Queensland, Australia
Type: Wooden centreboard sloop
Description: ON 94174. 24.5 x 22 (wl) x 9.4 x 2.4ft. Designed by Reeks and built by William Warbrick at James Clark's residence, New Farm, Brisbane, Queensland, and launched in October 1891.[2316] Possibly a sister-ship to *Varuna*. Raced with the Brisbane Sailing Club for many years, by the mid-1890s part-owned with George Campbell.[2317] Sold to Ross Wallace in mid-1895.[2318] Competed in the Intercolonial Championship Race for 22-footers against boats from Queensland and New South Wales in late 1895, held in Brisbane.[2319] Extended into a 'rater' ahead of the 1896/1897 season, noted to look '*well in her altered state, and totally unrecognisable as a late 22-footer*'.[2320] In mid-1898 underwent additional alteration, including having its mast moved 18in forward and the centreboard case moved 18in aft.[2321] By October 1901 owned by J. B. Arthur, by 1904 owned by Bert Perry.[2322] Advertised for sale in March 1905 and '*purchased lock, stock and barrel by a Mr Barker*'.[2323] Continued to race on Moreton Bay, was noted in August 1911 when advertised for sale as a 30-foot launch with a new 8 h.p. engine.[2324] Two years later, owned by a Mr Beattie and marooned in Moreton Bay for three days, its crew having to be rescued by the Water Police.[2325] This is its last known record.

OSPREY

Year: 1891
Designer: Walter Reeks
Builder: Rock Davis
Location: Brisbane Water, New South Wales, Australia
Type: Wooden steam drogher
Description: ON 101018. 70 (oa) x 63.5 x 16.4 x 5.2ft. 51/35 tons (gross/net). Launched from Rock Davis' Brisbane Water yard on 31 August 1891, designed by Reeks and commissioned by W. T. Yeager with Reeks also superintending construction.[2326] Propelled by a high-pressure engine manufactured in Sydney by Vale and Sons, a screw-steamer with a very light draft to help navigate the shallow and shoal riddled river and estuaries.[2327] First arrived on the Richmond River in late October 1891 and used in conjunction with Yeager's Coraki sawmilling business for many years, primarily in the South Arm to Casino trade.[2328] In November 1898 replaced by a larger vessel and soon sold to the North Coast Steam Navigation Company in whose employ it remained for several decades. In early 1903 underwent an extensive overhaul and refit, including new planking, engines, boiler and beds.[2329] Watertight compartment bulkheads were also fitted, as was a coal bunker.[2330] Several months later, sunk near Woodburn after being struck by the stern of the wheeler *Trial*, though recovered.[2331] In October 1907 sank at its moorings at Coraki, though undamaged.[2332] In November 1921 listed while at the Grafton wharf with its entire top decking collapsing; the mishap caused by misloading a cargo of live pigs, the superstructure collapsed as a result of the weight of 190 bags of maize on deck.[2333] At the time of the last incident, owned by Captain Charles Pullen of the Clarence River who had only recently purchased it.[2334] Returned to service with the vessel's register closed in 1935 with comment, '*vessel dismantled*'.

ADMIRAL

Year: 1892
Designer: Walter Reeks
Builder: Walter Reeks (foreman: Tom Cubitt)
Location: Berrys Bay, New South Wales, Australia
Type: Wooden pearling schooner
Description: ON 101034. 57.3 x 15.6 x 7.2ft. 49/28 tons (gross/net). Launched from Berrys Bay on 21 January 1892, built by Reeks to his design for T. G. Kelly. Intended for use in the Thursday Island region, as both a store ship and shell collector.[2335] On 30 March 1897 sold to James Clark and its register transferred to Brisbane.[2336] Wrecked off Thursday Island on 5 March 1899 when the opposing cyclones *Nachon* and *Mahina* crossed tracks and combined over Cape Melville to form the category 5 cyclone *Mahina*.[2337]

PEULEULE

Year: 1892
Designer: Walter Reeks
Builder: Walter Reeks (foreman: Tom Cubitt)

2313 *The West Australian*, 13 March 1889; *The Brisbane Courier*, 18 April 1899.
2314 *The Daily News*, 25 March 1889.
2315 *The Daily News*, 20 April 1908.
2316 *The Brisbane Courier*, 18 August 1891, 27 October 1891; *Lloyd's Register* (1901).
2317 *The Telegraph*, 22 February 1895, 10 March 1923.
2318 *The Telegraph*, 24 December 1895.
2319 *The Daily Telegraph*, 29 December 1895.
2320 *The Telegraph*, 3 September 1896.
2321 *The Brisbane Courier*, 8 September 1898.
2322 *The Brisbane Courier*, 21 October 1901; *Queensland Figaro*, 13 October 1904.
2323 *The Telegraph*, 9 March 1905; *Queensland Figaro*, 13 April 1905.
2324 *The Telegraph*, 5 August 1911.
2325 *The Brisbane Courier*, 5 July 1913.
2326 *The Sydney Morning Herald*, 1 September 1891.
2327 *The Sydney Morning Herald*, 1 September 1891.
2328 *The Richmond River Herald and Northern Districts Advertiser*, 30 October 1891.
2329 *The Richmond River Herald and Northern Districts Advertiser*, 14 October, 18 November 1898, 8 May 1903.
2330 *The Richmond River Herald and Northern Districts Advertiser*, 8 May 1903.
2331 *The Richmond River Herald and Northern Districts Advertiser*, 2 October 1903.
2332 *The Richmond River Herald and Northern Districts Advertiser*, 11 October 1907.
2333 *The Richmond River Express and Casino Kyogle Advertiser*, 4 November 1921.
2334 *Daily Examiner*, 2 November 1921.
2335 *Evening News*, 22 January 1892; *The Sydney Morning Herald*, 22 January 1892.
2336 SYDNEY, NEW SOUTH WALES, Register of ships, single number series (II), Volume 15, 1891-1893, Folio Tr 7.
2337 BRISBANE, Main register, Volume 3, 1888-1898, Folio 77.

Location: Berrys Bay, New South Wales, Australia
Type: Wooden sailing lugger
Description: 42.5 x 11.5 x 5.5ft. 12 tons. Built by Reeks to his design, with Tom Cubitt as foreman, and launched on 29 February 1892.[2338] Built to the order of the British government of New Guinea for magisterial work.[2339] Left Sydney for Samarai in late May 1892 with Captain J. M. Hennessy at the helm.[2340] Built at a cost of £400, remained in service until September 1915 when advertised for sale by tender in Port Moresby.[2341] Its subsequent history is not known.

JESSIE

Year: 1892
Designer: Walter Reeks
Builder: Walter Reeks (foreman: Tom Cubitt)
Location: Berrys Bay, New South Wales, Australia
Type: Wooden two-masted cutter
Description: ON 101050. 45 x 12 x 6.4ft. 42/15 tons (gross/net). Designed by Reeks' and launched from his Berrys Bay yard on 22 July 1892, built to the order of James Clark.[2342] Used in the Torres Strait region for many years, later employed in Western Australian waters with its register officially transferred to Fremantle in 1917.[2343] Remained in the Clark family's ownership during this period; Victor John Clark assuming possession from 1924. In January 1926 advertised for sale in Darwin, though it appears not to have sold as 'V. J. Clark' was still noted as owner in 1939.[2344] Sequestered by the Royal Australian Navy for use as a merchant patrol vessel during World War II, fitted with an auxiliary engine, then returned to Queensland waters being used in Far North Queensland, Thursday Island and into Papua New Guinea.[2345] Last noted when advertised for sale in Townsville by the Commonwealth of Australia in May of 1945.[2346]

ISABEL

Year: 1892
Designer: Walter Reeks
Builder: Walter Reeks (foreman: Tom Cubitt)
Location: Berrys Bay, New South Wales, Australia
Type: Wooden pearling ketch
Description: ON 101054. 45 x 12.1 x 6.2ft. 40/14 tons (gross/net). Designed by Reeks and launched from his Berrys Bay yard on 18 August 1892, built to the order of James Clark, pearling magnate, for use in the Torres Strait industry.[2347] Remained in Clark's possession, along with several of his business partners, for many decades, in 1917 transferred to Western Australia and its registry from Brisbane to Fremantle. Sold in August 1924 to James Hay, pearler of Broome, substantially rebuilt in 1929/30 and an auxiliary engine installed.[2348] In December 1938 sold to Eric Akerstrom of Fremantle and in March 1947 sold to Leonard Shepheard of the Victoria Hotel, Geraldton.[2349] This is the last known record of the craft. Register cancelled in 1986 as no trace of the vessel or the owner found.

THE LADY MARY

Year: 1892
Designer: Walter Reeks
Builder: Walter Reeks (foreman: Tom Cubitt)
Location: Berrys Bay, New South Wales, Australia
Type: Wooden steam ferry
Description: ON 101083. 96.5 (oa) x 83.5 x 22.3 x 7.3ft. 79/54 tons (gross/net), 29 n.h.p. compound stem engine by Ross & Duncan, Glasgow. 400 passenger double-ended ferry designed by Reeks and launched from his Berrys Bay yard on 30 July 1892, built for the Balmain New Ferry Co. Ltd.[2350] Passed to Sydney Ferries Ltd. in 1917. Taken out of service after colliding with the Gladesville Wharf in February 1927 and sold at auction on 31 July 1928 for £150.[2351] Sold to the Tay Literage Co. and its register closed.[2352] By the mid-1930s used as a houseboat in Glades Bay.[2353]

THE LADY MARY LAID UP (C1930).
SOURCE: GRAEME ANDREWS WORKING HARBOUR PHOTOGRAPH COLLECTION, CITY OF SYDNEY ARCHIVES.

ESSIE

Year: 1892
Designer: Walter Reeks
Builder: Messrs Lane and Brown

[2338] *Evening News*, 1 March 1892.
[2339] *Evening News*, 1 March 1892.
[2340] *The Sydney Morning Herald*, 24 May 1892.
[2341] *The Australian Star*, 20 June 1892; *Papuan Times*, 8 September 1915.
[2342] *The Daily Telegraph*, 25 July 1892.
[2343] *The Daily Telegraph*, 20 February 1897; Register of British Ships: Main Register, Port of Fremantle.
[2344] *Northern Territory Times and Gazette*, 12 January 1926; Register of British Ships: Main Register, Port of Fremantle.
[2345] www.navy.gov.au/sites/default/files/documents/JAPARA_JOHN-WISE.pdf.
[2346] *Daily Mercury*, 19 May 1945.
[2347] SYDNEY, NEW SOUTH WALES, Register of ships, single number series (II), Volume 15, 1891-1893, Folio 117.
[2348] FREMANTLE, Main Register, Volume 6, 1914-1931, Folio 64.
[2349] FREMANTLE, Main Register, Volume 7, 1931-1955, Folio 56.
[2350] *The Australian Star*, 29 July 1892.
[2351] *The Sun*, 5 February 1927, 31 July 1928.
[2352] *The Sydney Morning Herald*, 1 August 1929; SYDNEY, NEW SOUTH WALES, Register of ships, single number series (II), Volume 15, 1891-1893, Folio 173.
[2353] Graeme Andrews Working Harbour Photograph Collection, City of Sydney Archives (A-00081996).

Location: Whangaroa, New Zealand
Type: Wooden pearling ketch
Description: ON 94259. 45.5 x 11.9 x 5.8ft. 13 tons (gross/net). One of several pearling luggers built in New Zealand to a Reeks design for James Clark for use in the Torres Strait pearling industry.[2354] Sailed from New Zealand to Cooktown, Queensland, three weeks after launch, arriving on 6 August 1892 and immediately put to work.[2355] Sold to Dutch owners in 1906. Repurchased by James Clark and his business partners in 1917 and register transferred to Fremantle.[2356] Subsequent history not known.

OUR BOYS

Year: 1892
Designer: Walter Reeks
Builder: Messrs Lane and Brown
Location: Whangaroa, New Zealand
Type: Wooden pearling ketch
Description: ON 94260. 45.5 x 11.9 x 5.8ft. 13 tons (gross/net). The second of several pearling luggers built in New Zealand to a Reeks design for James Clark for use in the Torres Strait pearling industry.[2357] Sailed from New Zealand to Cooktown, Queensland, a few weeks after launch and near-immediately put to work.[2358] Sold to Dutch owners in 1906. Repurchased by James Clark and his business partners in 1917 and register transferred to Fremantle.[2359] Subsequent history not known.

MASCOTTE

Year: 1892
Designer: Walter Reeks
Builder: Messrs Lane and Brown
Location: Whangaroa, New Zealand
Type: Wooden pearling ketch
Description: ON 94264. 45.5 x 11.9 x 5.85 ft. 13 tons (gross/net). Remeasured in 1917 to be 39.8 x 12.2 x 4.2ft, 16/13 tons (gross/net). The third of several pearling luggers built in New Zealand to a Reeks design for James Clark for use in the Torres Strait pearling industry.[2360] Sold to Dutch owners in 1906. Repurchased by James Clark and his business partners in 1917 and register transferred to

[2354] *The Daily Telegraph*, 13, 29 July 1892; *The Sydney Morning Herald*, 22 August 1892; SYDNEY, NEW SOUTH WALES, Register of ships, single number series (II), Volume 18, 1899-1902, Folio 66.
[2355] *Auckland Star*, 21 June, 1 July 1892.
[2356] FREMANTLE, Main Register, Volume 6, 1914-1931, Folio 57.
[2357] *The Daily Telegraph*, 13, 29 July 1892; *The Sydney Morning Herald*, 22 August 1892; SYDNEY, NEW SOUTH WALES, Register of ships, single number series (II), Volume 18, 1899-1902, Folio 66.
[2358] *Auckland Star*, 21 June, 1 July 1892.
[2359] FREMANTLE, Main Register, Volume 6, 1914-1931, Folio 57.
[2360] New Zealand, Registered Ships & Owners, 1840-1950; Register of Australian and New Zealand Shipping 1893-1894, 1902-1903.

PEARLING LUGGER ESSIE OFF BROOME (1920S).
SOURCE: NATIONAL LIBRARY OF AUSTRALIA (NLA OBJ-138406380).

Fremantle.[2361] Lost during a cyclone off Lacepede Islands on 27 March 1935, along with its entire crew.[2362] At the time, was one of 31 luggers working in the region.[2363]

WATERWITCH

Year: 1892
Designer: Walter Reeks
Builder: Messrs Lane and Brown
Location: Whangaroa, New Zealand
Type: Wooden pearling ketch
Description: ON 94265. 45.5 x 11.9 x 5.85ft. 13 tons (gross/net). Fourth of several pearling luggers built in New Zealand to a Reeks design for James Clark for use in the Torres Strait pearling industry.[2364] Sold to Dutch owners in 1906 and British registry cancelled. Returned to British registry (Fremantle #27/1917) when purchased by a syndicate of five that included James Clark.[2365] Spent the remainder of its career working as a lugger and later fishing vessel off the coast of Western Australia. Sold in September 1927 and owned by members of the Burton family of Geraldton, including Francis Charles Burton from 1941, until wrecked at Port Gregory, Western Australia, on 30 June 1957.[2366]

GIPSEY

Year: 1892
Designer: Walter Reeks
Builder: Walter Reeks (supervised by Captain T. L. Grainger)
Location: Berrys Bay, New South Wales, Australia
Type: Wooden pearling lugger
Description: ON 101059. 38.9 x 11.6 x 5.5ft (press states 44.8 (oa), 37.2 (wl), x 11.3 x 5.5ft.[2367] 11 tons (gross/net). Launched on 22 September 1892 from Reeks' Berrys Bay yard to his design, built of kauri planking with decks on hardwood frames, copper fastened.[2368] Built under the supervision of first owner Captain Thomas Lewthwaite Grainger; first registered in Sydney as #79 of 1892.[2369] Subsequent owners include Burns Philp & Co. Ltd. (1892) and James Clark, pearler of Brisbane (1895).[2370] Totally wrecked during cyclone *Mahina* at Cape Melville, Queensland, in March 1899.[2371]

CAMOOLA

Year: 1892
Designer: Walter Reeks
Builder: Walter Reeks (foreman: Tom Cubitt)
Location: Berrys Bay, New South Wales, Australia
Type: Wooden pearling lugger
Description: ON 101064. 42 (oa) x 40.2 x 11.5 x 5.9ft. 12 tons (gross/net). Built for Archibald Robert Kelly and registered in Sydney as #85/1892. Launched in October 1892 and near immediately sailed to Queensland where it operated as pearling lugger.[2372] Later owners were Patrick Joseph Doyle of Thursday Island (1894) and Simon Edwin Munro (1903).[2373] Register closed in 1953 after the registrar was informed that the vessel was beached and abandoned at Thursday Island about 1920.[2374]

SANA

Year: 1892
Designer: Walter Reeks
Builder: Walter Reeks (foreman: Tom Cubitt)
Location: Berrys Bay, New South Wales, Australia
Type: Wooden pearling lugger
Description: ON 101071. 49.7 (oa) x 42.5 x 12.2 x 5.5ft. 16 tons (gross/net). Designed by Reeks and built at his Berrys Bay yard to the order of James Clark, replacing a smaller lugger of the same name wrecked in 1890.[2375] Launched in November 1892, arriving in Brisbane a few weeks later.[2376] (Reg. Sydney #10/1877, #94/1892). Sold to Dutch subjects and register closed 1906.

EHIME

Year: 1892
Designer: Walter Reeks
Builder: Walter Reeks (foreman: Tom Cubitt)
Location: Berrys Bay, New South Wales, Australia
Type: Wooden pearling ketch
Description: ON 101074. 50 (oa) x 45.3 x 12.2 x 5.7ft. Designed and built by Reeks and launched from his Berrys Bay yard in November 1892; a sister-ship to *Takachiho* (see next entry).[2377] Built to the order of Captain T. L. Grainger, representing Burns Philp and Co.[2378] Owned by James Clark from 1895 and wrecked during cyclone *Mahina* at Cape Melville, Queensland, in March 1899.[2379]

[2361] FREMANTLE, Main Register, Volume 6, 1914-1931, Folio 57.
[2362] R. Parsons (2008). *Shipping Losses and Casualties concerning Australia and New Zealand, Second Edition; Daily Mercury, 2 April 1935.*
[2363] *The Daily News*, 30 March 1935.
[2364] *The Daily Telegraph*, 13, 29 July 1892; *The Sydney Morning Herald*, 22 August 1892; SYDNEY, NEW SOUTH WALES, Register of ships, single number series (II), Volume 18, 1899-1902, Folio 66.
[2365] FREMANTLE, Main Register, Volume 6, 1914-1931, Folio 57.
[2366] FREMANTLE, Main Register, Volume 6, 1914-1931, Folio 57.
[2367] *The Daily Telegraph*, 23 September 1892.
[2368] *The Sydney Morning Herald*, 23 September 1892.
[2369] *The Sydney Morning Herald*, 23 September 1892; SYDNEY, NEW SOUTH WALES, Register of ships, single number series (II), Volume 15, 1891-1893, Folio 126.
[2370] SYDNEY, NEW SOUTH WALES, Register of ships, single number series (II), Volume 15, 1891-1893, Folio 126.
[2371] R. Parsons (2008). Shipping Losses and Casualties concerning Australia and New Zealand, Second Edition.
[2372] *Evening News*, 15 October 1895.
[2373] SYDNEY, NEW SOUTH WALES, Register of ships, single number series (II), Volume 15, 1891-1893, Folio 132.
[2374] SYDNEY, NEW SOUTH WALES, Register of ships, single number series (II), Volume 15, 1891-1893, Folio 132.
[2375] SYDNEY, NEW SOUTH WALES, Register of ships, single number series (II), Volume 15, 1891-1893, Folio 141; R. Parsons (2008). Shipping Losses and Casualties concerning Australia and New Zealand, Second Edition.
[2376] *The Daily Telegraph*, 15 November 1892; *The Queenslander*, 3 December 1892.
[2377] *The Brisbane Courier*, 6 December 1892.
[2378] *Newcastle Morning Herald and Miners' Advocate*, 28 November 1892; SYDNEY, NEW SOUTH WALES, Register of ships, single number series (II), Volume 15, 1891-1893, Folio 144.
[2379] SYDNEY, NEW SOUTH WALES, Register of ships, single number series (II), Volume 15, 1891-1893, Folio 144.

ISIS (1920S).
SOURCE: STATE LIBRARY OF SOUTH AUSTRALIA (PRG+280/1/37/266).

TAKACHIHO

Year: 1892
Designer: Walter Reeks
Builder: Walter Reeks (foreman: Tom Cubitt)
Location: Berrys Bay, New South Wales, Australia
Type: Wooden pearling ketch
Description: ON 101075. 50 (oa) x 45.3 x 12.2 x 5.7ft. Designed and built at Berrys Bay by Reeks to the order of Captain T. L. Grainger (representing Burns Philp and Company) and launched on 5 December 1892; a sister-ship to *Ehime* (see previous entry).[2380] Sold to James Clark in April 1895 then sold to Dutch owners in 1906 and British Registry closed.[2381] Fate not located.

ISIS

Year: 1892
Designer: William Fife (Jr)
Builder: Walter Reeks (foreman: Tom Cubitt)
Location: Berrys Bay, New South Wales, Australia
Type: Steam yacht
Description: ON 101134. 97 (oa), 85.4 x 13.6 x 9.4ft. 51/34 tons (gross/net). 20 n.h.p. compound steam engine by Rankin & Blackmore (Greenock, Scotland). Built for James Reading Fairfax (Reg. Sydney #6/1895), later owned by Bernard Einerson (1918) and Charles Horace Relph (1919).[2382] Reregistered Sydney #1/1920 when converted to a motor yacht with 50/75 b.h.p. Standard Motor Construction Co. (Jersey City, USA) oil engine, and remeasured as 51/40 tons (gross/net).[2383] Reputed to have been refitted with a 100 h.p. Thornycroft engine and later owned by Samuel Perry of Adelaide (from 1923) and W. L. Buckland of Melbourne (from January 1932).[2384] Wrecked at Frankston, Victoria, on 10 March 1932 after dragging its anchors during a storm, register closed 5 April.[2385]

ALBERT MACLAREN

Year: 1893
Designer: Walter Reeks
Builder: Walter Reeks (supervised by Captain T. L. Grainger)
Location: Berrys Bay, New South Wales, Australia
Type: Ketch
Description: ON 101079. 46.8 x 12.0 x 6.4ft. 14 tons (gross/net). Designed and built by Reeks and launched from his Berrys Bay yard on 15 January 1893.[2386] Built for the Australian Board of Missions, though registered in the names of two of their executives, Alfred Garnold and Thomas Sugden Scholefield (Reg. Sydney #26/1893).[2387] Registry transferred in 1900 to Montagu John Stone-Wigg, Bishop of New Guinea. Sank at its moorings at Bwagaoia Harbour, Misima, Papua on 11 November 1930, and became a total loss.[2388]

2380 *The Sydney Morning Herald*, 6 December 1892.
2381 SYDNEY, NEW SOUTH WALES, Register of ships, single number series (II), Volume 15, 1891-1893, Folio 146.
2382 SYDNEY, NEW SOUTH WALES, Register of ships, single number series (II), Volume 16, 1893-1897, Folio 85.
2383 SYDNEY, NEW SOUTH WALES, Register of ships, single number series (II), Volume 25, 1919-1923, Folio 11.
2384 *The Sydney Morning Herald*, 22 September 1923; *The Age*, 27 January 1932; *The Herald*, 27 July 1933.
2385 *News*, 11 March 1932; *The Herald*, 27 July 1933.
2386 *Evening News*, 20 January 1893.
2387 SYDNEY, NEW SOUTH WALES, Register of ships, single number series (II), Volume 15, 1891-1893, Folio 175.
2388 SYDNEY, NEW SOUTH WALES, Register of ships, single number series (II), Volume 15, 1891-1893, Folio 175.

ALADDIN

Year: 1893
Designer: Walter Reeks
Builder: Walter Reeks (foreman: Tom Cubitt)
Location: Berrys Bay, New South Wales, Australia
Type: Pearling schooner
Description: ON 101086. 90.7 x 23.3 x 8.4ft. 102 tons (gross/net). Designed and built by Reeks for the pearling trade and launched on 5 April 1893.[2389] Built to the order of Scott, Henderson Co. but registered in the name of George Stanley Littlejohn (Reg. Sydney #31/1892).[2390] Was possibly the first mother ship or store ship for pearling luggers, an idea conceived by James Clark.[2391] From 26 March 1897 owned by Clark in conjunction with Percival Outridge and Edwin Munro, the latter becoming sole owner in 1904.[2392] Sold to Dutch owners in 1906 and British Registry closed.[2393] The sale appears to have taken place late in 1905, by which time up to six schooners, two tenders and 105 luggers employing 480 men had recently relocated from Thursday Island to the Aru Islands in the Dutch East Indies, and more were expected to follow.[2394] Its later activities have not been located.

ALADDIN.
SOURCE: DAVID PAYNE

PILOT

Year: 1893
Designer: Walter Reeks
Builder: Rock Davis (supervised by Captain T. L. Grainger)
Location: Brisbane River, New South Wales, Australia
Type: Pearling ketch
Description: ON 101089. 58.2 x 17.8 x 6.3ft. 35 tons (gross/net). Designed by Reeks and built by Rock Davis to the order of James Clark. Built especially for pearl fishing in Torres Strait as a vessel to take live shell for the first attempt at cultured pearls.[2395] Sold to the Queensland Government for £450 in August 1896 to service navigational aids, and reregistered as Brisbane #3/1896.[2396] Lost with all eight hands, including the Cooktown harbourmaster Peter Osterlund and passenger John Henry Hargreaves MLA, in a cyclone at Cape Flattery, Queensland, on 17 January 1907.[2397]

WALLAROO

Year: 1893
Designer: Walter Reeks
Builder: Walter Reeks (foreman: Tom Cubitt)
Location: Berrys Bay, New South Wales, Australia
Type: Wooden pearling lugger
Description: ON 101101. 53 (oa) x 48.5 x 12.9 x 6.5ft. 15 tons (gross/net). A pearling 'lugger' with clipper bow and elliptical counter designed and built by Reeks to the order of J. Talman of Burns Philp & Co. Ltd. (Reg. Sydney #61/1893).[2398] Launched on 21 October 1893 and sold in August 1895 to Edwin Munro of Thursday Island.[2399] British registry closed in 1907 on sale to Dutch subjects. Further details not located.

LADY MANNING

Year: 1894
Designer: Walter Reeks
Builder: Robert Young and George Scott (supervised by Walter Reeks)
Location: Roselle Bay, Balmain, New South Wales, Australia
Type: Steam ferry
Description: ON 101110, 117 (oa) x 109.0 x 24.4 x 8.2ft. 97/66 tons (gross/net). Compound steam engine, 29 n.h.p, 230 i.h.p., made by Ross & Duncan of Glasgow (Scotland). A double-ended ferry, named after the wife of the Mayor of Sydney, built with cambered keel and 475 passenger capacity for the Balmain New Ferry Co. Ltd. (Reg. Sydney #11/1894).[2400] Passed to Sydney Ferries Ltd. in November 1917.[2401] Retired and auctioned on 31 July 1928, knocked down to the Tay Lighterage Co. for £100, register closed with comment 'being broken up' a fortnight later.[2402]

2389 *The Sydney Morning Herald*, 6 April 1893.
2390 SYDNEY, NEW SOUTH WALES, Register of ships, single number series (II), Volume 16, 1893-1897, Folio 1.
2391 *The Sydney Morning Herald*, 20 January 1893, 6 April 1893.
2392 SYDNEY, NEW SOUTH WALES, Register of ships, single number series (II), Volume 16, 1893-1897, Folio 1.
2393 SYDNEY, NEW SOUTH WALES, Register of ships, single number series (II), Volume 16, 1893-1897, Folio 1.
2394 *Queensland Times*, 7 October 1905.
2395 *Queenslander*, 17 June 1893.
2396 SYDNEY, NEW SOUTH WALES, Register of ships, single number series (II), Volume 16, 1893-1897, Folio 12; BRISBANE, Main register, Volume 3, 1888-1898, Folio 70.
2397 *Brisbane Courier*, 24, 30 January, 12 February 1907.
2398 SYDNEY, NEW SOUTH WALES, Register of ships, single number series (II), Volume 16, 1893-1897, Folio 31.
2399 *The Australian Star*, 24 October 1893; SYDNEY, NEW SOUTH WALES, Register of ships, single number series (II), Volume 16, 1893-1897, Folio 31.
2400 *The Sydney Morning Herald*, 10, 12 February 1894; SYDNEY, NEW SOUTH WALES, Register of ships, single number series (II), Volume 16, 1893-1897, Folio 44.
2401 *The Daily Telegraph*, 3 November 1917.
2402 *The Sydney Morning Herald*, 1 August 1928; SYDNEY, NEW SOUTH WALES, Register of ships, single number series (II), Volume 16, 1893-1897, Folio 44.

LADY MANNING (C1908).
SOURCE: GRAEME ANDREWS WORKING HARBOUR PHOTOGRAPH COLLECTION, CITY OF SYDNEY ARCHIVES.

TAIPO

Year: 1894
Designer: Walter Reeks
Builder: George Barnett
Location: Drummoyne, New South Wales, Australia
Type: Yacht
Description: 20ft (oa) x 17 (wl) x 6.7 x 1.1ft. ½ rater designed by Reeks and launched on 17 February 1894, built to the order of W. L. Hunt by George Barnett of Drummoyne.[2403] Raced on Sydney Harbour for several seasons, by 1898 owned by Dr Craig.[2404] By 1904 owned by H. Badgery.[2405] The following year back in W. L. Hunt's hands and sailing with the Sydney Amateur Sailing Club.[2406] Advertised for sale in December 1907.[2407] Subsequent history not known.

ORARA

Year: 1894
Designer: Walter Reeks
Builder: Rock Davis (supervised by Captain T. L. Grainger)
Location: Brisbane River, New South Wales, Australia
Type: Steamer
Description: ON 101129. 75 (oa) x 70.7 x 18.2 x 4.0ft. 66/45 tons (gross/net). Quadruple-expansion steam engine 9 n.h.p., 50 i.h.p. made by Simpson & Strickland, England. Built by Rock Davis, at Brisbane Water to the order of David Henderson of Grafton (Reg. Sydney #40/1894).[2408] Launched on 18 December 1894, intended to trade between Clarence and Copmanhurst, carrying cargo and passengers.[2409] Sold to Holmes & Blain of Perth, Western Australia, in December 1895 and sailed for its new home port on 30 December, but same day lost its propeller in heavy seas and managed to limp into Woolgoolga, New South Wales, where it went onto rocks and ultimately became a total wreck, all hands saved.[2410]

--

Year: 1895
Designer: Walter Reeks
Builder: --
Location: --
Type: Steam ferry
Description: Designed by Reeks for the Manly Co-operative Steam Ferry Company.[2411] One of two vessels; the other being *Manly* (see entry below). This particular vessel was not built.

OLIVE BRANCH

Year: 1895
Designer: Walter Reeks
Builder: Messrs Lane and Brown (supervised by the Rev. F. Walker)
Location: Whangaroa, New Zealand
Type: Schooner
Description: ON 102281. 67.1 x 16.8 x 7.6ft. 46/32 (gross/net). Built to the order of the London Missionary Society, registered first in the name of their Australian agent Joseph King then from 1897 in the name of the Rev. Frederick William Walker (Reg. Auckland #14/1895, Melbourne #4/1897).[2412] Wrecked in a storm at Kwato, near Samarai, British New Guinea on 30 July 1901 and became a total loss.[2413]

MANLY

Year: 1896
Designer: Walter Reeks
Builder: R. Young and Son
Location: Rozelle Bay, New South Wales, Australia
Type: Steam ferry
Description: ON 106127. 157.6 (oa) x 147.0 x 26.0 x 10.9ft., draft 9ft 6in. 229/156 tons (gross/net), from 1897 229/139 tons. Triple-expansion steam engine 100 n.h.p., 600 i.h.p. = 14.5 knots, made by Fawcett, Preston & Co. of Liverpool (England). Built for the Port Jackson Cooperative Steam Ship Co. Ltd. (Reg. Sydney #45/1896) as a dedicated Sydney-Manly ferry, possibly the largest double-ended screw ferry then built that was capable of handling sea-going conditions. Based on a sprung three-layer laminated keel with a 9ft camber over its length.[2414] Originally intended to be named *Emancipator* and the first of a pair, but the second was not proceeded with (see above), and tenders for *Manly*'s constructed were advertised in August 1895.[2415] Owners were restructured as Port Jackson &

2403 *The Sydney Morning Herald*, 16 February 1894.
2404 *Evening News*, 28 February 1898.
2405 *The Australian Star*, 7 November 1904.
2406 *The Australian Star*, 16 January 1905.
2407 *The Sydney Morning Herald*, 7 December 1907.
2408 SYDNEY, NEW SOUTH WALES, Register of ships, single number series (II), Volume 16, 1893-1897, Folio 73.
2409 *The Sydney Morning Herald*, 18 December 1894; *The Australian Star*, 20 September 1894.
2410 *The Daily Telegraph*, 31 December 1895.
2411 *The Sydney Morning Herald*, 15 June 1895.
2412 MELBOURNE, Main Register subsequent to Merchant Shipping Act 1854, Volume 7, 1887-1898, Folio 187.
2413 *The Daily Telegraph*, 17 August 1901.
2414 *Daily Telegraph*, 31 October 1896.
2415 *Town and Country Journal*, 20 July 1895; *The Daily Telegraph*, 17, 27 August 1895.

Manly Steamship Co. Ltd. in 1907. After being laid up for some years, sold to colliery proprietor James Ruttley of Adamstown near Newcastle in February 1924 and some six weeks later again sold to shipbuilder Watty Ford (Jr) of Berrys Bay, without having left Sydney Harbour.[2416] Engines removed in mid-1926 for reuse in the former motor vessel *Durour* ex *Madal*, and register closed as being no longer required.[2417] Its hull is believed to have been demolished shortly afterwards.

MANLY.
SOURCE: GRAEME ANDREWS WORKING HARBOUR PHOTOGRAPH COLLECTION, CITY OF SYDNEY ARCHIVES.

--

Year: 1896
Designer: Walter Reeks
Builder: Henry Thomas Green
Location: Lavender Bay, New South Wales, Australia
Type: Oil launch
Description: 26 x 6.5ft. Designed by Reeks and built at Lavender Bay by Henry Thomas Green to the order of E. Noyes. Launched on 21 March 1896.[2418] Planked with ½-inch cedar with ash timbers, copper-fastened and varnished throughout; fitted with a patent steering wheel, along with 2 h.p. engines imported by Messrs Grob and Co. of Germany.[2419] Advertised for sale in May 1897 and in August 1898.[2420]

LADY HAMPDEN

Year: 1896
Designer: Walter Reeks
Builder: James Scott
Location: Balmain, New South Wales, Australia
Type: Steam ferry
Description: ON 106137. 122 (oa) x 115.6 x 26.0 x 9.7ft, 7ft draft. 135/92 tons (gross/net). Compound steam engine 42 b.h.p. made by Fawcett Preston & Co. of Liverpool (England). Double-ended screw ferry built for the Balmain New Ferry Company Ltd. (Reg. Sydney #12/1897), cambered keel with propellers working outside the rudders (see model in Powerhouse Museum collection).[2421] Launched 8 December 1896.[2422] Transferred to Sydney Ferries Ltd. in 1918, register closed 1932 as no longer required, but reregistered by same owners and details as Sydney #7/1937.[2423] Register closed 22 July 1943 as the vessel had been sold to the Government of the Commonwealth of Australia. Appears to have been sold to the Government about 1941 and converted into a motorised degaussing vessel for use on Sydney Harbour: it was badly damaged in a collision with the corvette HMAS *Deloraine* on 30 October 1941 and was beached in a sinking collision. Later returned to active service, but is reported to have been sunk as a naval practice target in 1943.

LADY HAMPDEN (C1920S).
SOURCE: GRAEME ANDREWS WORKING HARBOUR PHOTOGRAPH COLLECTION, CITY OF SYDNEY ARCHIVES.

--

Year: 1896
Designer: Walter Reeks
Builder: --
Location: --
Type: Yacht
Description: A proposal for an 18ft (wl) one-design class yacht. 24ft (oa), 260 square ft sail area, four crew, rudder on transom, and other restrictions, to sail in the 'B' class.[2424] Reeks provided plans for three identical boats to be built by one builder. No indication that they were proceeded with. Also called '17 Foot B class'. *Yogi* possibly built on dimensions or similar dimensions.

2416 *The Sun*, 1 April 1924.
2417 *Daily Telegraph*, 21 July 1926.
2418 *Referee*, 25 March 1896.
2419 *Referee*, 25 March 1896.
2420 *Australian Town and Country Journal*, 29 May 1897; *The Sydney Morning Herald*, 13 August 1898.

2421 SYDNEY, NEW SOUTH WALES, Register of ships, single number series (II), Volume 16, 1893-1897, Folio 163.
2422 *The Australian Star*, 8 December 1895.
2423 SYDNEY, NEW SOUTH WALES, Register of ships, single number series (II), Volume 27, 1928-1937, Folio 162.
2424 *The Sydney Morning Herald*, 12 November 1896.

ACTAEA

Year: 1897
Designer: Walter Reeks
Builder: Walter 'Watty' M. Ford
Location: Berrys Bay, New South Wales, Australia
Type: Cutter yacht
Description: ON 106155. 41 (oa) x 35.4 x 33 (wl) x 9.0 x 5.7ft, 7ft draft. 13 tons (gross/net). Built up from the keel of the first owner's 1883 10-ton Dixon Kemp design yacht *Sao*, which was demolished. Originally owned by Frederick George Woolcott-Waley (Reg. Sydney #46/1897); later owners were Arthur Robert Marks (April 1900), Joseph Bowen Pearce (November 1900), Allan Hammill Usher and Richard Broughton (1903), Charles Lloyd Jones (1903).[2425] Sold to a German national (destined for service in the Pacific) and register closed 24 March 1908.[2426]

KURING GAI

Year: 1897
Designer: Walter Reeks
Builder: Unknown
Location: Sydney, New South Wales, Australia
Type: Oil launch
Description: 26 x 7ft. Designed by Reeks for the trustees of Ku-ring-Gai Chase National Park and launched in September 1897.[2427] Capable of achieving a speed of eight miles per hour. Subsequent history not known.

--

Year: 1897
Designer: Walter Reeks
Builder: --
Location: --
Type: Steam yacht
Description: A nearly 50-ton steam yacht designed by Reeks for a gentleman of Melbourne.[2428] Doubtful if built.

YOGI

Year: 1897
Designer: Walter Reeks
Builder: Henry Thomas Green
Location: Lavender Bay, New South Wales, Australia
Type: Yacht
Description: 22 (oa) x 18 (wl) x 17 x 7.5 x 2.1ft. Built by Henry Thomas Green to a design by Reeks (likely modified from his '17 Foot B class' one-design) and launched on 28 August 1897.[2429] Built to the order of E. M. Moore of the Sydney Amateur Sailing Club.[2430] Competed with the ½ raters for several seasons. By 1900 R. C. Scott who retained possession until August 1902 when sold to Walter Walters of Newcastle.[2431] Competed with the Port Hunter Sailing Club for a number of seasons.[2432] Fate not known.

SQUID

Year: 1897
Designer: Walter Reeks
Builder: Henry Thomas Green
Location: Lavender Bay, New South Wales, Australia
Type: Yacht
Description: 24 (oa) x 17 (wl), 17 x 7.5 x 2.1ft. Built by Henry Thomas Green to the order of A. E. Cutler of Mosman. Described as a '*shallow, punt shaped craft ... built as an experiment*'.[2433] Like *Yogi*, competed with the ½ raters for several seasons. By 1906 up until at least 1911 owned by C. Kehlerman and sailed with the Mosman Amateur Club.[2434] Advertised for sale in October 1910 as '*sailing punt Squid*' though appears not to have sold.[2435]

ESTELLE

Year: 1898
Designer: Walter Reeks
Builder: Walter 'Watty' M. Ford, Jr (supervised by Walter Reeks)
Location: Berrys Bay, New South Wales, Australia
Type: Wooden pearling ketch
Description: ON 106168. 43.9 x 12.0 x 5.4ft. 15/11 tons (gross/net). First registered in Reeks' name in April 1898 before passing to Percival Pitman Outridge of Brisbane in July (Reg. Sydney #13/1898.[2436] Wrecked during cyclone *Mahina* at Cape Melville, Queensland, in March 1899 and register closed in August.

PLOVER

Year: 1898
Designer: Walter Reeks
Builder: Walter 'Watty' M. Ford, Jr
Location: Berrys Bay, New South Wales, Australia
Type: Wooden pearling ketch
Description: ON 106169. 43.6 x 12.2 x 5.3ft. 14/10 tons (gross/net). Altered 1934 to 19/16 tons (gross/net), 45.5 x 13.58 x 5.58ft. Built for Edwin Munro of Thursday Island (Reg. Sydney #17/1898), register closed in 1907 after sale to Dutch subjects.[2437] Reregistered at Fremantle (#6/1918) with new owner James Colin Clark of Brisbane, new tonnage 13 gross, sold to Land Stuart Ogilvie of Broome in 1920.[2438] Later owners were Bramsha Maidin (1934), Ernest Henry Chapple (1936) and Jessie Emma Edwards,

2425 SYDNEY, NEW SOUTH WALES, Register of ships, single number series (II), Volume 17, 1897-1899, Folio 20; SYDNEY, NEW SOUTH WALES, Register of ships, single number series (II), Volume 17, 1897-1899, Folio Tr 18.
2426 *The Sydney Morning Herald*, 17 March 1908.
2427 *The Australian Star*, 20 September 1897.
2428 *The Australian Star*, 20 September 1897; *The Australasian*, 23 October 1897.
2429 *The Daily Telegraph*, 28 August 1897.
2430 *The Daily Telegraph*, 28 August 1897.
2431 *Evening News*, 23 November 1900; *Evening News*, 25 August 1902.
2432 *The Australian Star*, 24 November 1902; *The Sydney Morning Herald*, 28 December 1903.
2433 *The Daily Telegraph*, 16 October 1897.
2434 *The Australian Star*, 20 March 1906; *The Daily Telegraph*, 29 October 1910; *The Arrow*, 25 March 1911.
2435 *The Sydney Morning Herald*, 29 October 1910.
2436 SYDNEY, NEW SOUTH WALES, Register of ships, single number series (II), Volume 17, 1897-1899, Folio 36.
2437 SYDNEY, NEW SOUTH WALES, Register of ships, single number series (II), Volume 17, 1897-1899, Folio 40.
2438 FREMANTLE, Main Register, Volume 6, 1914-1931, Folio 77.

widow of Darwin (1937).[2439] Register closed in 1986 when no trace of the vessel or last registered owner could be found. Fate uncertain.

ZOE

Year: 1898
Designer: Walter Reeks
Builder: Rock Davis
Location: Brisbane Water, New South Wales, Australia
Type: Wooden pearling ketch
Description: ON 106172. 43.5 x 12.2 x 5.3ft. 13/10 tons (gross/net). Built for Percival Pitman Outridge of Brisbane (Reg. Sydney #19/1898).[2440] Wrecked during cyclone *Mahina* at Cape Melville, Queensland, in March 1899 and register subsequently closed in August.

PENGUIN

Year: 1898
Designer: Walter Reeks
Builder: Walter 'Watty' M. Ford, Jr (supervised by Walter Reeks)
Location: Berrys Bay, New South Wales, Australia
Type: Wooden pearling ketch
Description: ON 106179. 44.0 x 12.2 x 5.3ft. 14/10 tons (gross/net). Built for Edwin Munro of Thursday Island (Reg. Sydney #30/1898).[2441] British registry closed in 1907 on sale to Dutch subjects. Returned to British registry (Sydney #23/1914) after purchase by Reginald Augustus Charles Hockings of Thursday Island. Later owners were Harold N. and Francis E. Hockings.[2442] Sold to Commonwealth Government and register closed in 1947, still in service mid-1950, owned by Northern Waters Pearling Co. Fate not located.

SNIPE

Year: 1898
Designer: Walter Reeks
Builder: Rock Davis
Location: Brisbane Water, New South Wales, Australia
Type: Wooden pearling ketch
Description: ON 106181. 42.9 x 12.2 x 5.2ft. 13/9 tons (gross/net). Built for James Clark (Reg. Sydney #36/1898), register closed 1906 after sale to Dutch subjects.[2443] Returned to British registry (Fremantle #8/1918) with James Colin Clark as owner.[2444] Sold to James Veich Hay of Broome in 1924 and register closed in March 1926 after the vessel was wrecked.

CURLEW

Year: 1898
Designer: Walter Reeks
Builder: Rock Davis
Location: Brisbane Water, New South Wales, Australia
Type: Wooden pearling ketch
Description: ON 106193. 43.2 x 12.2 x 4.8ft. 13/10 tons (gross/net), by 1914 47.0 x 12.44 x 4.83ft, 16/14 tons (gross/net). Built for James Clark of Brisbane, though initially registered in the builder's name (Reg. Sydney #50/1898).[2445] Register closed in 1906 after sale to foreigners. Returned to British registry (Sydney #69/1914) when owned by Reginald Augustus Charles Hockings of Thursday Island, who had rebuilt it there in 1914.[2446] From 1930 owned by Harold N. and Francis E. Hockings. Taken over by the Commonwealth Government and register closed in May 1943. Appears to have survived war service and is probably the same vessel described as the lugger *Curlew* that was briefly stolen at Fremantle in 1946.[2447]

TEAL

Year: 1898
Designer: Walter Reeks
Builder: Rock Davis
Location: Brisbane Water, New South Wales, Australia
Type: Wooden pearling ketch
Description: ON 106196. 42.6 x 12.2 x 5.1ft. 13/10 tons (gross/net). Built at Brisbane Water by Rock Davis to the order of James Clark. Rebuilt at Thursday Island in 1913 now 47.0 x 12.42 x 4.65ft, 16/14 tons (gross/net). History to 1943 identical to *Curlew*, see above. Registered in Sydney #54/1898 and #25/1914.[2448] Fate not located

HERON

Year: 1898
Designer: Walter Reeks
Builder: Rock Davis
Location: Brisbane Water, New South Wales, Australia
Type: Wooden pearling ketch
Description: ON 106202. 42.4 x 12.0 x 5.1ft. 13/9 tons (gross/net). After a survey at Thursday Island in 1920 remeasured as 47.83 x 12.58 x 5.25ft, 16/13 tons (gross/net). Launched in late December 1898, leaving New South Wales for Thursday Island on 1 January 1899.[2449] History to 1943 identical to *Curlew* and *Teal,* see above.[2450] Registered in Sydney #62/1898 and #26/1914. Fate not located.

[2439] FREMANTLE, Main Register, Volume 7, 1931-1955, Folio 13.
[2440] SYDNEY, NEW SOUTH WALES, Register of ships, single number series (II), Volume 17, 1897-1899, Folio 42.
[2441] SYDNEY, NEW SOUTH WALES, Register of ships, single number series (II), Volume 17, 1897-1899, Folio 53.
[2442] SYDNEY, NEW SOUTH WALES, Register of ships, single number series (II), Volume 23, 1912-1914, Folio 116.
[2443] SYDNEY, NEW SOUTH WALES, Register of ships, single number series (II), Volume 17, 1897-1899, Folio 59.
[2444] FREMANTLE, Main Register, Volume 6, 1914-1931, Folio 79.
[2445] SYDNEY, NEW SOUTH WALES, Register of ships, single number series (II), Volume 17, 1897-1899, Folio 73.
[2446] SYDNEY, NEW SOUTH WALES, Register of ships, single number series (II), Volume 24, 1914-1919, Folio 34.
[2447] *West Australian*, 20 April 1946.
[2448] SYDNEY, NEW SOUTH WALES, Register of ships, single number series (II), Volume 17, 1897-1899, Folio 77; SYDNEY, NEW SOUTH WALES, Register of ships, single number series (II), Volume 23, 1912-1914, Folio 118.
[2449] *The Daily Telegraph*, 30 December 1898; *The Sydney Morning Herald*, 2 January 1899.
[2450] SYDNEY, NEW SOUTH WALES, Register of ships, single number series (II), Volume 17, 1897-1899, Folio 85; SYDNEY, NEW SOUTH WALES, Register of ships, single number series (II), Volume 23, 1912-1914, Folio 119.

OSPRAY

Year: 1898
Designer: Walter Reeks
Builder: Rock Davis
Location: Brisbane Water, New South Wales, Australia
Type: Wooden pearling ketch
Description: ON 106178. 42.6 x 12.2 x 5.3ft. 13/9 tons (gross/net). Built at Brisbane Water for James Clark (Reg. Sydney #28/1898).[2451] Register closed 1906 after sale to Dutch subjects. Fate not located.

BITTERN

Year: 1898
Designer: Walter Reeks
Builder: Messrs Lane and Brown
Location: Whangaroa, New Zealand
Type: Wooden pearling ketch
Description: ON 102308. 43.3 x 12.0 x 5.1ft. 12.99 tons (gross/net). Built for William Brown but reregistered at Sydney when owned by James Clark (Reg. Auckland #13/1898, Sydney #29/1899).[2452] Register closed 1906 after sale to Dutch subjects, but returned to British Registry (Sydney #67/1912) when owned by Reginald Augustus Charles Hockings.[2453] Later owners included Cecil John Handley (1923) and Handley & Clay Ltd. (1927). Sprang a leak in bad weather and run aground at Aivei west of Orokolo Beach, Gulf of Papua, on or about 12 May 1941 and became a total loss.

EDITH

Year: 1898
Designer: Walter Reeks
Builder: Messrs Lane and Brown
Location: Whangaroa, New Zealand
Type: Wooden pearling ketch
Description: ON 102307. 43.3 x 12.0 x 5.1ft. 13 tons (gross/net). Details to 1906 as per *Bittern*, see above (Reg. Auckland #12/1898, Sydney #28/1899).[2454] Returned to British Registry (Fremantle #31/1917) now 38.9 x 12.0 x 3.4ft, 16/13 tons (gross/net), owned by James Clark, Robert Philp, Percival Pitman Outridge, Isabella Mitchell Campbell and James Theodore Clewett Mackenzie. Consolidation of owners with deaths of Isabella Campbell in 1918 and Robert Philp in 1923 with James Clark ultimately becoming sole owner in November 1922.[2455] Later owned by Montague Sydney Hadley of Derby, Western Australia (April 1923) and the Australian Aborigines Mission of Western Australia (Inc.) (November 1923). Register closed in 1928 after the vessel was broken up.

ONYX

Year: 1898
Designer: Walter Reeks (possibly)
Builder: Walter 'Watty' M. Ford, Jr
Location: Berrys Bay, New South Wales, Australia
Type: Wooden pearling ketch
Description: ON 106205. 45.9 x 12.2 x 5.4ft. 15/11 tons (gross/net). Built for The Queensland Pearl Shell Fishery Co. Ltd. (Reg. Sydney #64/1898).[2456] Later owned by

[2451] SYDNEY, NEW SOUTH WALES, Register of ships, single number series (II), Volume 17, 1897-1899, Folio 51.
[2452] SYDNEY, NEW SOUTH WALES, Register of ships, single number series (II), Volume 17, 1897-1899, Folio 116.
[2453] SYDNEY, NEW SOUTH WALES, Register of ships, single number series (II), Volume 23, 1912-1914, Folio 18.
[2454] SYDNEY, NEW SOUTH WALES, Register of ships, single number series (II), Volume 17, 1897-1899, Folio 115.
[2455] FREMANTLE, Main Register, Volume 6, 1914-1931, Folio 65.
[2456] SYDNEY, NEW SOUTH WALES, Register of ships, single number series (II), Volume 17,

PEARLING LUGGERS DAMAGED DURING A WILLY WILLY OFF BROOME, WESTERN AUSTRALIA, JANUARY 1926.
SOURCE: NATIONAL LIBRARY OF AUSTRALIA (NLA OBJ-138407088).

Thomas Joseph Farquhar of Thursday Island (1920) and Farquhar Porter Ltd. (1926). Beached and abandoned at Prince of Wales Island in 1942 and register closed in 1952.

CHARM EX RUBY

Year: 1898
Designer: Walter Reeks (possibly)
Builder: Walter 'Watty' M. Ford, Jr
Location: Berrys Bay, New South Wales, Australia
Type: Wooden pearling ketch
Description: ON 106197. 45.7 x 12.1 x 5.4ft. 14/11 tons (gross/net). First registered Sydney #36/1898, details otherwise as per *Onyx* (above) until 1926.[2457] Rebuilt by T. Furuta at Thursday Island in 1930 and renamed *Charm* (ON 153480, Reg. Townsville #4/1930, 6/1956), 47.0 x 13.75 x 5.75ft, 20/16 tons (gross/net), owner T. J. Farquhar.[2458] Passed on his death in 1931 to Sarah Farquhar, following hers in 1941 to Donald Burnett Farquhar and in 1947 to H. O. & R. N. Hockings Pty. Ltd. Remeasured and reregistered in 1956 following the installation of a 4-cylinder Gardner diesel engine (now 19/7 tons, gross/net).[2459] Sold to Geraldton fisherman William Leonard Rogers in 1961 and register closed without further comment in 1986. May have remained in service as late as 1975.

AOTEA

Year: 1898
Designer: Walter Reeks
Builder: Messrs Morrison and Sinclair
Location: Pyrmont, New South Wales, Australia
Type: Steam ferry, fore-and-aft schooner rig
Description: ON 106201. 126.5 (oa) x 120.4 x 24.1 x 13.0ft. 263/157 tons (gross/net). Compound steam engine by Ross & Duncan, Glasgow (Scotland), 35 n.h.p., 315 i.h.p. Built for the Northern Union Steamboat Co. Ltd. of Helensville, New Zealand, although registered in the name of the company's Sydney agent Robert Spence Lamb until after delivery in January 1899 (Reg. Sydney #60/1898, Auckland #4/1899).[2460] A shallow-draft steamer intended for the upper Wairoa trade, featured a screw turn in a tunnel that ensured lower draft aft while the screw was in solid water drawn from the front of the tunnel. This configuration was well-known in low-draft steamers including those used on the Nile and in India.[2461] Register closed in 1929 when reported that the vessel had been broken up, but was in fact beached at Highams Beach, Kaipara Harbour, where some timbers and the boiler remained visible into the 2000s.[2462]

SS AOTEA ON THE NORTHERN WAIROA RIVER, NEW ZEALAND (C1899) (PH-CNEG-C2312). SOURCE: AUCKLAND WAR MEMORIAL MUSEUM.

--

Year: 1898
Designer: Walter Reeks
Builder: --
Location: Sydney, New South Wales, Australia
Type: House boat
Description: Not registered. House boat designed by Reeks and built at Sydney for the trustees of the Ku-ring-Gai Chase National Park.

MUSA

Year: 1898
Designer: Walter Reeks
Builder: Walter 'Watty' M. Ford, Jr
Location: Berrys Bay, New South Wales, Australia
Type: Auxiliary oil launch, ketch rigged
Description: Not registered. 40 x 35 x 8 x 2.3ft. 4 b.h.p. engine by Union Oil Engine Co. of San Francisco. Built for Dr Cecil Vaughan who voyaged to New Guinea in it with three other passengers.[2463] Last noted in 1899 after reaching Frederick Wilhelmshaven in distress following an attack by natives and an engine breakdown.[2464]

WHITE WINGS

Year: 1899
Designer: William Fife (Jr)
Builder: Walter 'Watty' M. Ford, Jr (supervised by Walter Reeks)
Location: Berrys Bay, New South Wales, Australia

1897-1899, Folio 87; SYDNEY, NEW SOUTH WALES, Register of ships, single number series (II), Volume 17, 1897-1899, Folio Tr 30.

2457 SYDNEY, NEW SOUTH WALES, Register of ships, single number series (II), Volume 17, 1897-1899, Folio 79; SYDNEY, NEW SOUTH WALES, Register of ships, single number series (II), Volume 17, 1897-1899, Folio Tr 29.

2458 TOWNSVILLE, Register of ships, Volume 3, 1914-1981, Folio 45.

2459 TOWNSVILLE, Register of ships, Volume 3, 1914-1981, Folio 95.

2460 SYDNEY, NEW SOUTH WALES, Register of ships, single number series (II), Volume 17, 1897-1899, Folio 83.

2461 *The Daily Telegraph*, 5 November 1898.

2462 https://northlandhistory.blogspot.com/2020/06/ss-aotea-1898-1929.html.

2463 *The Queenslander*, 26 November 1898; *Sydney Mail*, 12 November 1898.

2464 *Daily News (Perth)*, 23 September 1899.

Type: Yawl
Description: ON 121154. 48 (oa) x 40.7 x 31 (wl) x 11.5 x 3.9ft. 16 tons (gross/ne). Built to the order of Sam Hordern but first British registered in 1906 when owner was John Arthur Muston (Reg. Sydney #47/1906).[2465] Sold to Arthur Muston in 1912, who requested cancellation of registry in 1918 as it was no longer required. Reregistered Sydney #2/1926, by now owned by Norman Leonard Rex Griffin.[2466] Sold in 1936 to Norman Randolph Freehill of the St Kilda Yacht Club and transferred to Melbourne.[2467] Register again closed in 1950 on sale to an American citizen.

BRONZEWING

Year: 1899
Designer: Walter Reeks
Builder: Messrs Robert Young, Son and Fletcher
Location: Rozelle, New South Wales, Australia
Type: Steam ferry
Description: ON 106216. 115 (oa), 110.0 x 22.0 x 8.1ft. 149/101 tons (gross/net). Triple expansion steam engine 35 n.h.p., 200 b.h.p. made by Ross & Duncan of Glasgow (Scotland), 12 knots. Built to the order of the Parramatta River Steamers and Tramway Company Ltd. (Reg. Sydney #13/1899) and based on the hull lines of its earlier ferry *Pheasant* and the deck fittings from the *Alathea*.[2468] Sold in 1901 to Sydney Ferries Ltd. Register closed in December 1931 when the registered owners informed that Customs that the vessel had been sold to Holyman Bros. of Hobart to be broken up by Penguin Ltd. of Sydney. Holymans presumably needed the engine and/or boiler for another vessel. *Bronzewing* had been sold at auction on 20 May 1930 for £350, but it is not known if Holymans were the purchaser at this time.

BRONZWING.
SOURCE: GRAEME ANDREWS WORKING HARBOUR PHOTOGRAPH COLLECTION, CITY OF SYDNEY ARCHIVES.

AMY

Year: 1899
Designer: Walter Reeks
Builder: James Hayes
Location: Milsons Point, New South Wales, Australia
Type: Pleasure launch
Description: Not registered. 39 x 7.2 x 2.5ft. 4 h.p. oil engine Built for J. S. Meiklejohn, manager of the Commercial Bank in Bundaberg, Queensland. History and fate not known.[2469]

--

Year: 1899
Designer: Walter Reeks
Builder: --
Location: --
Type: Steam ferry
Description: 100 x 22ft. May have been an order placed by the Northern Union Steamship Company of New Zealand for a second, slightly smaller steamer.[2470] Doubtful if built.

UNA

Year: 1899
Designer: Walter Reeks
Builder: Walter 'Watty' M. Ford, Jr
Location: Berrys Bay, New South Wales, Australia
Type: Steam ferry
Description: Not registered. 70 x 16 x 7ft depth, 5ft draft. Compound steam engine 60 i.h.p. A double-ended passenger ferry and punt tug built for the New South Wales Roads and Bridges Department for use on the Clarence River.[2471] A sister-ship to *Helen*, see next entry. Later transferred to Grafton. Sold in early 1906 to Sydney Ferries Ltd. and towed south by the steamer *Burunda,* to be employed in the Parramatta River trade.[2472] Began running on the Tarban Creek service in March 1906.[2473] After several decades of service, destroyed by fire on 4 April 1927, along with the ferry *Rose*, while moored in Five Dock Bay near the Gladesville Bridge.[2474] Hulk sold by the Fire Loss Assessors' Association for £165 to William Waugh, to be broken up.[2475]

HELEN

Year: 1899
Designer: Walter Reeks
Builder: Walter 'Watty' M. Ford, Jr
Location: Berrys Bay, New South Wales, Australia
Type: Steam ferry
Description: Not registered. 70 x 16 x 7ft depth, 5ft draft. Compound steam engine 60 i.h.p. A double-ended

2465 SYDNEY, NEW SOUTH WALES, Register of ships, single number series (II), Volume 20, 1904-1907, Folio 107.
2466 SYDNEY, NEW SOUTH WALES, Register of ships, single number series (II), Volume 26, 1923-1928, Folio 65.
2467 *The Sydney Morning Herald*, 17 October 1936.
2468 *Evening News*, 11 February 1899; SYDNEY, NEW SOUTH WALES, Register of ships, single number series (II), Volume 17, 1897-1899, Folio 100.
2469 *The Bundaberg Mail*, 7 July 1899.
2470 *Daily Commercial News and Shipping List*, 3 August 1899.
2471 *Northern Star*, 27 September 1899.
2472 *Clarence and Richmond Examiner*, 13 January, 27 January 1906.
2473 *The Cumberland Argus and Fruitgrowers Advocate*, 17 March 1906.
2474 *The Sun*, 4 April 1927.
2475 *The Sun*, 8 April 1927.

passenger ferry built for the New South Wales Roads and Bridges Department for use on the Hunter River at Morpeth, and nearing completion when sister-ship *Una* (see previous entry) was launched in September 1899.[2476] Arrived at Newcastle in early December 1899.[2477] Later based at Camden Haven and, from 1903, on the Grafton ferry service.[2478] Advertised for sale by tender in December 1904.[2479] Failed to sell, remained in service at Grafton until at least 1914.[2480] Possibly the steam launch *Helen,* belonging to Walsh Island Dockyard at Newcastle that was beached in Newcastle Harbour in August 1925 after having its bottom ripped open following collision with the Danish motor ship *Nordbo.*[2481]

UNA AT HUNTERS HILL WHARF.
SOURCE: GRAEME ANDREWS WORKING HARBOUR PHOTOGRAPH COLLECTION, CITY OF SYDNEY ARCHIVES.

FAIRTRADER

Year: 1899
Designer: Walter Reeks
Builder: Thomas King
Location: Richmond River, New South Wales, Australia
Type: Oil launch
Description: Not registered. 50 x 12 x 3.5ft. 12 tons. Built, for dealing purposes on the Clarence River, for Samuel Macnaughton, powered by a American-built naphtha engine and launched on 18 November 1899.[2482] Sold in September 1902 to T. Henderson of Brushgrove and remained in commercial service until at least 1922, when advertised for sale by now powered by a 16 h.p. heavy duty Union oil engine, installed new in 1914.[2483] Advertised for sale by Bruce Schwonberg in 1937, powered by a two-cylinder Bolinder semi-diesel.[2484] Subsequent fate unknown. Possibly later renamed *Victoria.*

EVELYN

Year: 1899
Designer: Walter Reeks
Builder: Henry Thomas Green
Location: Lavender Bay, New South Wales, Australia
Type: Cutter-rigged yacht
Description: ON 112477. 37 (oa) x 31.6 x 24 (wl) x 9.0 x 3.4ft. 7 tons (gross/net). First auxiliary engine fitted in 1906, a 12 h.p. engine fitted before 1934. Built for Percival Pitman Outridge, pearler of Brisbane (Reg. Sydney #67/1899), primarily as a cruising yacht though it had some success racing.[2485] Register closed in 1951 when vessel was noted sold about 1930 and believed lost in a cyclone off the Queensland coast in 1934. This confirms reports of it having been converted into an auxiliary fishing vessel, left Brisbane on 20 February 1934 on a trip to the northwards, but failed to return. Part-owner George Lemingo and another crewman were lost with the vessel, that was thought to have been wrecked on Fraser Island.[2486]

THE KETCH EVELYN OVERDUE FROM A FISHING CRUISE OFF THE SOUTH COAST OF QUEENSLAND.
SOURCE: THE COURIER-MAIL, 10 MARCH 1934.

--

Year: 1900
Designer: Walter Reeks
Builder: --
Location: --
Type: Iron steam ferry
Description: 130 x 30 x 10.1ft. Designed by Reeks for the Balmain New Ferry Company. Tenders for construction were called in January 1901 while new plans were available in May.[2487] The Balmain New Ferry Company absorbed the older Balmain Steam Ferry Company in mid-1900 and the iron ferry project lapsed, it being announced Reeks was now designing a new timber ferry. Ultimately the next ferry that was built for the company was the *Lady Rawson* completed in 1903.[2488]

2476 *The Australian Star*, 21 September 1899; *Clarence & Richmond Examiner*, 23 September 1899.
2477 *Newcastle Morning Herald and Miners' Advocate*, 2 December 1899.
2478 *The Clarence River Advocate*, 21 April 1903.
2479 *The Australian Star*, 9 December 1904.
2480 *Clarence and Richmond Examiner*, 26 June 1906, 18 March 1913; *The Grafton Argus and Clarence River General Advertiser*, 25 January 1911, 18 May, 10 July 1914.
2481 *The Daily Telegraph*, 6 August 1925.
2482 *Clarence and Richmond Examiner*, 21 November 1899.
2483 *The Grafton Argus and Clarence River General Advertiser*, 12 September 1902; *Clarence and Richmond Examiner*, 11 October 1902; *Daily Examiner*, 16 March 1922; S. Lee (2003). *Riverboats of the Clarence*.
2484 *Daily Examiner*, 6 February 1937.
2485 SYDNEY, NEW SOUTH WALES, Register of ships, single number series (II), Volume 18, 1899-1902, Folio 16; *Queensland Times*, 28 July 1938.
2486 *The Courier-Mail*, 10 March, 22 June 1934.
2487 *Sydney Morning Herald*, 13 January 1900; *The Daily Telegraph*, 19 May 1900.
2488 *The Daily Telegraph*, 19 May 1900; *The Sydney Morning Herald*, 19 June 1900.

--

Year: 1900
Designer: Walter Reeks
Builder: --
Location: --
Type: Steam ferry
Description: Double-ended screw ferry steamer designed by Reeks for the Balmain New Ferry Company.[2489] As with the previous entry, not built.

EMERALD

Year: 1900
Designer: Walter Reeks
Builder: George Ellis
Location: Rozelle, Balmain, New South Wales, Australia
Type: Steam ferry
Description: ON 112503. 130 (oa) x 124 x 25.2 x 7.45ft. 189/117 tons (gross/net). Two compound steam engines totalling 25 n.h., 200 i.h.p. (speed 11 knots) made by John Downey of Liverpool (England). Twin screw steam ferry, two-masted fore-and-aft schooner rig built as a ferry and excursion vessel for the Humpybong Steamship Co. Ltd. of Brisbane (Reg. Sydney #35/1900).[2490] Following financial difficulties, passed to William George Hayes and then Arthur Hobbs of Brisbane in 1907, but continued to trade under the Humpybong Steamship Co. banner.[2491] In November 1915 an unregistered sale to Melbourne interests, who purchased it for the 'Bay Trade' but soon after arrival was purchased by the Commonwealth on the behalf of the Royal Australian Navy apparently for as a tug and/or examination vessel. Soon found unsuitable for the duties expected and sold at a considerable loss.[2492] In 1924 went to Western Australia following an unregistered sale to Alfred Tilley to begin excursion work on the Swan River.[2493] In 1950 reregistered in the name of Alf. E. Tilley & Co. Pty. Ltd. and 127 tons gross. Remeasured in 1959 after superstructure alterations to 136/92 tons (gross/net) (Reg. Fremantle #1/1950).[2494] Register closed in 1963, with the comment that the vessel had been broken up.

--

Year: 1900
Designer: Walter Reeks
Builder: --
Location: --
Type: Punt
Description: First of two hardwood punts designed by Reeks, 185 by 40 x 15ft. Intended for delivery in July 1900.[2495]

[2489] *The Sydney Morning Herald*, 19 May 1900.
[2490] SYDNEY, NEW SOUTH WALES, Register of ships, single number series (II), Volume 18, 1899-1902, Folio 53; SYDNEY, NEW SOUTH WALES, Register of ships, single number series (II), Volume 18, 1899-1902, Folio Tr 24.
[2491] *Brisbane Courier*, 19 May 1913.
[2492] *Brisbane Courier*, 23 November 1915, *Herald*, 17 January 1916; *Geelong Advertiser*, 12 December 1918.
[2493] Daily News, 15 December 1924, 6 November 1925.
[2494] FREMANTLE, Main Register, Volume 7, 1931-1955, Folio 84; FREMANTLE, Main Register, Volume 7, 1931-1955, Folio 85.
[2495] *The Daily Telegraph*, 31 March 1900.

--

Year: 1900
Designer: Walter Reeks
Builder: --
Location: --
Type: Punt
Description: Second of two hardwood punts designed by Reeks, 185 by 40 x 15ft. Intended for delivery in September 1900.[2496]

--

Year: 1900
Designer: Walter Reeks
Builder: --
Location: --
Type: Steam ferry
Description: 100 x 26ft. First of two double-ended screw steamers designed by Reeks for Sydney Ferries Ltd. for use on its Neutral Bay service.[2497] Proposed to have accommodation for 500 passengers and travel at 12 knots. Tender for its construction was announced in July 1900 and continued to be active throughout September, closing on 1 October.[2498] However, was not built.

--

Year: 1900
Designer: Walter Reeks
Builder: --
Location: --
Type: Steam ferry
Description: Second of two double-ended screw steamers designed by Reeks for Sydney Ferries Ltd. for use on its Neutral Bay service.[2499] Tender for its construction was announced in July 1900 and continued to be active throughout September, closing on 1 October.[2500] However, was not built.

HMS NELSON AND PUNTS

Year: 1809-1814, refitted 1900
Designer: Redesigned by Walter Reeks
Builder: Royal Naval Dockyard
Location: Woolwich Dockyard, England. 1900 conversion by J. S. Parker
Type: Punts and hulk
Description: ON 112477. 217.0 x 56.0 x 25.0ft. 2,466/2,337 tons (gross/net), 3,036 tons displacement. 2-cylinder steam engine made by Ravenhill & Salkeld, Blackwall (England), 500 n.h.p., 2012 i.h.p. A former Royal Navy three-decked Line of Battle Ship, in 1867 was converted to a training frigate for the Colonial Navy of Victoria. Advertised for sale 28 April 1898 and sold for £2,400 to

[2496] *The Daily Telegraph*, 31 March 1900.
[2497] *The Sydney Morning Herald*, 15 August 1900.
[2498] *The Daily Telegraph*, 19, 28 July, 5, 20 September 1900.
[2499] *The Sydney Morning Herald*, 15 August 1900.
[2500] *The Daily Telegraph*, 19, 28 July, 5, 20 September 1900.

Bernard Einerson of Balmain East (Reg. Sydney #3/1898) for demolition, and arrived in Sydney on 24 May under tow by the tug *Eagle*.[2501] Sold 20 June 1900 to colliery proprietor Shepheard Edgecliff Laidley and converted into a coal hulk under plans by Reeks. Upper deck was cut off and reworked into four coal punts, while the entire main deck was rebuilt as a single punt measuring 170 x 53 x 8ft and capable of loading 2,500 tons of coal.[2502] Fate of these punts has not been located. Much timber from the dismantled vessel was souvenired for various purposes, including for fit-out of the new Reeks-designed steam yacht *Ena*. Copper bolts were also upcycled for use in the Watsons Bay Ferry Company's steamer *King Edward* that Reeks designed and launched in September 1902. Companion way and skylights were used in Dr Scot Skirving's Reeks-designed yacht *Alcyone,* built in 1905. Its lower hull was built up into a coal hulk with 2,500 tons capacity for use on Sydney Harbour.[2503] There was an unregistered sale by 1904 to the Oceanic Steamship Co., for its use as a coal hulk until the company discontinued its Trans-Pacific services in 1906. *Nelson* was offered for sale by tender in May 1907 and on 27 December formally transferred to the Union Steamship Co. of N.Z. Ltd.[2504] Employed at Beauty Point, Tasmania from 1908 until 1914, after which it was towed to Hobart for further use. Finally sold to H. Gray in 1920, towed to Limekiln (Shag) Bay on the River Derwent and demolished over a period of years.[2505]

DISMANTLING HMS NELSON.
SOURCE: AUSTRALIAN TOWN AND COUNTRY JOURNAL, 14 JULY 1900.

KURING-GAI

Year: 1900
Designer: Walter Reeks
Builder: Mort's Dock and Engineering Co.
Location: Balmain, New South Wales, Australia
Type: Steam ferry
Description: ON 112524. 175 (oa) x 171.5 x 31.3 x 14.7ft. 497/155 tons gross/net, remeasured 1906 now 497/206 tons (gross/net). Triple-expansion steam engine made by Morts Dock and Engineering Co., Balmain (Sydney). A steel-hulled, double-ended ferry for the Sydney-Manly service ordered by the Manly Co-operative Steam Ferry Co., that had been absorbed by the Port Jackson Steamship Company (Reg. Sydney #12/1901) by the time the hull was launched on 8 November 1900.[2506] At trials on 28 April 1901, reached 15.6 knots, built at a cost of £23,789 with capacity for 1,228 passengers. Owners were restructured as the Port Jackson & Manly Steamship Co. Ltd. in 1907. Vessel was re-boilered in 1922 and on 3 August 1928 sold to the Newcastle Ferries Co. Ltd. for dockyard and excursion use, making trips to Morpeth, Raymond Terrace, Broughton Island and Nelson's Bay. Reported to be out of commission and its register closed in May 1934. Sold to Morrison and Bearby of Newcastle for dismantling late in 1936.[2507] Then used as a hulk until sinking at Hexham during floods in 1952, and the Government called tenders for its removal two years later. Demolition was never completed. The remains of its hull still lie in the Hunter River near Raymond Terrace turn off, visible at low tide.

MOA

Year: 1900
Designer: Walter Reeks
Builder: William Holmes
Location: McMahons Point, New South Wales, Australia
Type: Steam launch
Description: Unregistered. 47 x 11.8 x 6ft, 4.5ft draft. Simpson and Strickland steam engine, 30 i.h.p. = 10 m.p.h. Built to the order of Messrs H. S. Bird and Company.[2508] Not named in construction reports but almost certainly the *Moa* with identical dimensions that H. S. Bird advertised for sale in 1912, capacity 40 passengers or 6 tons of cargo, and again, hull only, throughout 1913 and early 1914.[2509] The vessel (same name and dimensions) appears to have been by then operating on the Clarence River, and was offered for sale in August 1914 for £450 by F. H. Schwonberg of Maclean, kauri on hardwood construction with engines as described above.[2510] Still in service as a steam launch in 1931, owned by Thomas Turner.[2511] Probably the 'cream boat' *Moa* owned by the Norco Co-operative Ltd. ashore at Tatham in March 1949.[2512]

2501 MELBOURNE, Main Register subsequent to Merchant Shipping Act 1854, Volume 8, 1898-1911, Folio 1; *Herald*, 29 April 1898; *Daily Telegraph*, 25 May 1898.
2502 *Evening News*, 22 October 1900.
2503 *The Argus*, 26 October 1900.
2504 *Australian Star*, 28 January 1904; *The Sydney Morning Herald*, 24, 25 May 1907.
2505 *Daily Telegraph*, 27 August 1920.
2506 *The Sydney Morning Herald*, 9 November 1900; SYDNEY, NEW SOUTH WALES, Register of ships, single number series (II), Volume 18, 1899-1902, Folio 83; SYDNEY, NEW SOUTH WALES, Register of ships, single number series (II), Volume 18, 1899-1902, Folio Tr 42.
2507 *Daily Commercial News and Shipping List*, 21 October 1936; *Newcastle Sun*, 9 December 1936.
2508 *Evening News*, 18 July 1900.
2509 *The Sydney Morning Herald*, 16 November 1912, 12 August, 4 October 1913, 31 January 1914.
2510 *Clarence and Richmond Examiner*, 16 August 1913.
2511 *The Richmond River Herald*, 1 December 1931.
2512 *Northern Star*, 14 March 1949.

KURING-GAI.
SOURCE: AUSTRALIAN NATIONAL MARITIME MUSEUM COLLECTION (ANMS1092[223]).

--

Year: 1900
Designer: Walter Reeks
Builder: --
Location: --
Type: Barge
Description: Designed by Reeks to transfer Australia's first Governor General, Lord Hopetoun, from his vessel the *Royal Arthur* to the landing place at Sydney upon his arrival in Australia. Intended to be manned by 30 to 40 prominent rowers.[2513] Not built.

SY ENA EX AURORE EX ENA EX HMAS SLEUTH EX ENA

Year: 1900
Designer: Walter Reeks
Builder: Walter 'Watty' M. Ford, Jr
Location: Berrys Bay, New South Wales, Australia
Type: Steam yacht
Description: ON 112529. 100 (oa) x 88.0 x 16.5 x 8.1ft. 65/44 tons (gross/net). Compound steam engine 25 n.h.p., 180 i.h.p. by Chapman & Co. Ltd. of Sydney, 12 knots. Built for Thomas Allwright Dibbs, general manager of the Commercial Banking Company of Sydney (Reg. Sydney #21/1901).[2514] Sold to the Commonwealth Government during the First World War (renamed *HMAS Sleuth*) and although sold afterwards, no further notations till closed in 1931 and reregistered at Sydney as #4/1931, owner now Vernon Anfield.[2515] Later owned by Walter Charles James Driscoll of Hobart (1933) and (William Roche of Hobart (1940). Used briefly as a cargo steamer out of Hobart before being laid up. Refitted as a fishing vessel in the early 1940s, its steam engine replaced by a Gardner diesel and vessel renamed *Aurore* late 1940s, register closed without comment in 1986. Sank in the D'Entrecasteaux Channel in 1981, refloated, taken to Sydney and ultimately rebuilt as a steam yacht by Nick Masterman. Since 2017 proudly part of the collection of the Australian National Maritime Museum, based in Sydney, reregistered as *SY Ena*.

FLEETWING

Year: 1900
Designer: William Fife (Jr)
Builder: Walter 'Watty' M. Ford, Jr (supervised by Walter Reeks)
Location: Berrys Bay, New South Wales, Australia
Type: Yacht
Description: ON 117601. 42.5 (oa) x 28 (wl) x 33.0 x 7.8 x 3.9ft, 5.5ft draft. 5 tons (gross/net). Built under Reeks' supervision to the order of Sam Hordern, sold August 1901 to Charles Thomas Brockhoff (Reg. Sydney #13/1903).[2516] Later owned by Ernest Edward Sayer and Frank Albert (1907); E. E. Sayer (1907); Reginald Fairfax Roberts, George Gordon Woolnough, James March Hardie and Harold Richard Rabone; Percival Main Walton (1912); Clarence Linden (1914); Joseph Little (1922); Leslie James Hampton (1923). Register closed in 1926 on sale to a Spanish subject.

--

Year: 1901
Designer: Walter Reeks
Builder: --
Location: --
Type: Screw steamer
Description: 70 x 16ft. Designed by Reeks for Messrs Lysaghts Brothers and Co., wire netting manufacturers. Dimensions revised to 69 x 16ft, with plans and

2513 *Sunday Times*, 4 November 1900.
2514 SYDNEY, NEW SOUTH WALES, Register of ships, single number series (II), Volume 18, 1899-1902, Folio 92.
2515 SYDNEY, NEW SOUTH WALES, Register of ships, single number series (II), Volume 27, 1928-1937, Folio 58.
2516 SYDNEY, NEW SOUTH WALES, Register of ships, single number series (II), Volume 19, 1902-1904, Folio 52.

specifications available at his office in Change Alley. No further trace of vessel, not known if built.

ADI VITI

Year: 1901
Designer: Walter Reeks
Builder: William Holmes
Location: McMahons Point, New South Wales, Australia
Type: Oil launch
Description: Unregistered. 40 x 8ft. Fitted with a 10 h.p. Union Oil and Gas Co. engine and launched in January 1901.[2517] Built for the Fijian government and based at Suva, intended to be used for medical and harbour purposes.[2518] Additional details and fate not known.

ADI VITI.
SOURCE: SYDNEY MAIL AND NEW SOUTH WALES ADVERTISER, 12 OCTOBER 1901.

WIDGI WIDGI

Year: 1901
Designer: Walter Reeks (possibly)
Builder: Government of New South Wales
Location: Sydney, New South Wales, Australia
Type: Oil launch
Description: Unregistered. 40 x 7.2 x 3.5ft. 8 h.p. Hercules oil engine. One of two government oil launches identical other than the engine, constructed for service on the Northern Rivers region. Intended for use on the Tweed River and *Mi Mi* (see next entry) was intended for use on the Richmond River. *Widgi Widgi*'s trial trip took place on 19 March 1901, when 8½ knots was attained.[2519] The two vessels were completed together and held a trial run on 23 March 1901. Guests included Mr. Orr ('*naval architect*') and Reeks, either may have been the designer.[2520] *Widgi Widgi* proved the faster of the two. It was put up for sale by tender in December 1904. With the maximum offer of £130 being unacceptable, was transferred to the trustees of the Kuring-gai Chase.[2521] At the time the Government of New South Wales had 44 launches in its service.[2522] Last noted at Wiseman's Ferry in 1909.[2523]

MI MI

Year: 1901
Designer: Walter Reeks (possibly)
Builder: Government of New South Wales
Location: Sydney, New South Wales, Australia
Type: Oil launch
Description: Unregistered. 40 x 7.2 x 3.5ft. 8/10 h.p. 2-cylinder Union oil engine. One of two government oil launches, identical other than the engine, constructed by the Government of New South Wales for service on the Northern Rivers region; *Widgi Widgi* (see previous entry) was intended for use on the Tweed River and *Mi Mi* was intended for use on the Richmond River. Fate not been located, its Union oil engine was offered for sale by tender at Sydney in 1917.[2524]

MI MI AT STOTTS CREEK (1902).
SOURCE: TWEED REGIONAL MUSEUM.

--

Year: 1901
Designer: Walter Reeks
Builder: --
Location: --
Type: Floating fire-fighting apparatus
Description: Floating fire-fighting apparatus designed by Reeks for use in fighting water-based fires as requested by the board of the Sydney Metropolitan Fire Brigade.[2525] Not known if built.

LARA

Year: 1901
Designer: Walter Reeks
Builder: William Holmes
Location: McMahons Point, New South Wales, Australia
Type: Motor launch
Description: ON 112549. 36 (oa) x 30 (wl) x 32.0 x 8.0 x 3.35ft, 2.5ft draft. 8/5 tons (grt/net). English-made Tolch Co. 'Capitaine' Kerosene engine, replaced in 1910 by an 8 h.p. 2-cylinder oil engine made by the Union Gas Engine Co. of San Francisco (USA). Built for Joseph Thompson (Reg. Sydney #10/1902) and launched in September 1901, featuring kauri planking on hardwood frames.[2526] Later

2517 *Evening News*, 22 January 1901.
2518 *The Daily Telegraph*, 3 June 1901.
2519 *The Sydney Morning Herald*, 20 March 1901.
2520 *The Sydney Morning Herald*, 25 March 1901.
2521 *The Daily Telegraph*, 21 December 1904.
2522 *The Richmond River Herald*, 16 December 1904; *The Daily Telegraph*, 21 December 1904.
2523 *Windsor and Richmond Gazette*, 1 May 1909.
2524 *The Daily Telegraph*, 30 October 1917.
2525 *The Australian Star*, 18 September 1901.
2526 *The Sydney Morning Herald*, 7 September 1901; SYDNEY, NEW SOUTH WALES, Register of ships, single number series (II), Volume 18, 1899-1902, Folio 124.

owned by Sydney Gilchrist (1910), Errol Aubrey Gilchrist (1926), and William Thomas Albert (1933). Ran ashore at Toowong Bay near Tuggerah on 23 November 1933 and became a total wreck.

MORA

Year: 1901
Designer: Walter Reeks
Builder: William Holmes
Location: McMahons Point, New South Wales, Australia
Type: Oil launch
Description: ON 121159. 31 (oa) x 27.6 x 7.0 x 1.75ft, 2ft draft. 4/3 tons (gross/net). 2-cylinder Wolverine Motor Co. (Grand Rapid, Michigan USA) oil engine, 6½ knots. Built of kauri on hardwood for Frederick George Rae of Mosman, a member of the Sydney Amateur Sailing Club (Reg. Sydney #4/1907).[2527] Register closed in 1920 as new owner did not require it. Fate not located.

THE ADMIRAL

Year: 1902
Designer: Walter Reeks (possibly)
Builder: John Innes (press), Joseph Munro (register)
Location: Balmain, New South Wales, Australia
Type: Steam lighter
Description: ON 112573. 86.5 (oa) x 80.0 x 22.5 x 6.7ft. 117/80 tons (gross/net). Compound steam engine by Plenty & Co., Newbury (England), 20 n.h.p., 100 i.h.p., 8 knots. Built for Gilchrist, Watt and Saunderson, Ltd. (Reg. Sydney #43/1902) and launched in February 1902, sold 1934 to MacKenzie and Petersen, Ltd.[2528] Register closed 1960 on Customs being informed that the vessel had been sold for scrap. Very strongly built, primarily of ironbark, with 2½in Oregon pine decks.[2529]

--

Year: 1902
Designer: Walter Reeks
Builder: William Holmes
Location: McMahons Point, New South Wales, Australia
Type: Steam launch
Description: Not British registered. 50 x 12 x 5.5ft. Built for Messrs Ballande Fils Aine of Noumea and intended for the New Caledonia coastal trade. Launched on 28 February 1902, constructed of kauri and hardwood, with its engines supplied by Messrs Chapman and Co. of Sydney, generating a speed of 9 knots.[2530] Possibly the tug *Niaouli 'built in Sydney'* ashore at Kouaoua, New Caledonia in mid-February 1909.[2531] A coastal steamer of the same name was lost at nearby Thio in 1917.[2532]

--

Year: 1902
Designer: Walter Reeks
Builder: William Holmes
Location: McMahons Point, New South Wales, Australia
Type: Steam launch
Description: Not British registered. 31.7 x 4ft. Designed by Reeks to the order of Messrs Ballande Fils Aine of Noumea and launched in March 1902.[2533] Additional details not known.

--

Year: 1902
Designer: Walter Reeks
Builder: William Holmes
Location: McMahons Point, New South Wales, Australia
Type: Steam launch or tugboat?
Description: 50 x 12 x 5.5ft. Built for Messrs Gilchrist, Watt and Sanderson Ltd. and intended for harbour work. Identical to the larger vessel built for Messrs Ballande Fils Aine of Noumea (see previous entry).[2534] Constructed of kauri and hardwood.[2535]

LADY HOPETOUN

Year: 1902
Designer: Unknown, Walter Reeks (possibly part of)
Builder: Walter 'Watty' M. Ford, Jr
Location: Berrys Bay, New South Wales, Australia
Type: Steam launch
Description: Not registered. 77 x 13 x 6.3ft. 30 tonnes displacement. 230 i.h.p. triple-expansion steam engine made by Simpson and Strickland, London (England), speed about 12 knots. Built as VIP vessel for the Sydney Harbour Trust (later Maritime Services Board of NSW) and launched 10 April 1902, named after the wife of the Commonwealth of Australia's first Governor General, Lord Hopetoun.[2536] Designer not known though Reeks initially prepared plans and specifications for a larger 87.3 x 13.6 x 7.2ft vessel for the Sydney Harbour Trust (plan no. 149), they decided to abandon this project and build a smaller craft. However, some of Reeks ideas or components may have been incorporated into *Lady Hopetoun*. Sold by tender to the Lady Hopetoun and Port Jackson Marine Steam Museum in 1962. Restored and steaming again in 1970. Remains part of Sydney Heritage Fleet in the 2020s.

--

Year: 1902
Designer: Walter Reeks
Builder: William Holmes (supervised by Walter Reeks)
Location: McMahons Point, New South Wales, Australia
Type: Steam lighter

2527 SYDNEY, NEW SOUTH WALES, Register of ships, single number series (II), Volume 20, 1904-1907, Folio 114.
2528 *Evening News*, 26 February 1902; *The Sydney Morning Herald*, 27 February 1902; SYDNEY, NEW SOUTH WALES, Register of ships, single number series (II), Volume 19, 1902-1904, Folio 21.
2529 *The Sydney Morning Herald*, 27 February 1902.
2530 *Evening News*, 3 March 1902.
2531 *The Sydney Morning Herald*, 23 February 1909.
2532 *Queensland Times*, 7 March 1917.
2533 *Evening News*, 3 March 1902.
2534 *Daily Commercial News and Shipping List*, 3 March 1902.
2535 *Evening News*, 3 March 1902.
2536 *The Australian Star*, 10 April 1902.

Description: Not registered. 70 ft x 18 ft 6 in. x 7 ft 3 in. 40 h.p. steam engine. Built for Elliott Brothers Ltd, intended for harbour use and launched on 8 September 1902.[2537]

NATOMA EX BEACON EX GULNARE

Year: 1902
Designer: Walter Reeks
Builder: William Goddard
Location: Lavender Bay, New South Wales, Australia
Type: Yacht
Description: ON 112591. 37 (oa) x 31.1 x 25 (wl) x 10.0 x 3.8ft. 10 tons (gross/net). In 1903 fitted with a 2-cylinder engine by De Dion Bouton of Paris, (France), 8 h.p. = 5 ¾ knots; in 1907 a 4-cylinder John I. Thornycroft of London (England) oil engine, 22 b.h.p. = 12 knots. Owners were Charles Edwin Waters (Reg. Sydney #3/1903); John Spencer Brunton (1906); Charles Horace Relph (January 1912); George Henry Charles Carrett (October 1912); William Snodgrass Jacob (July 1926).[2538] Register closed in 1952 when no trace of the vessel was known and the last registered owner was reported dead in 1927. Launched 9 November 1902 as a yawl-rigged cruiser, kauri planks and decks on spotted gum frames, first engine installed the following year.[2539] Known to have been ashore at Dobroyd Point, The Spit in 1926 and 1927, but recovered on both occasions and advertised for sale in 1928.[2540] No further trace.

WAHINE

Year: 1902
Designer: Walter Reeks
Builder: Henry Thomas Green
Location: Lavender Bay, New South Wales, Australia
Type: Motor launch
Description: ON 121142. 42.4 x 7.6 x 2.7ft, 9in draft. 11/10 tons (gross/net). Reeks' own launch, initially owned in partnership with C. W. Bennett and Charles Bartholomew, hard-chine, flat-bottomed launch of cedar construction with screw tunnel.[2541] First registered with Reeks as sole owner (Sydney #29/1906).[2542] Originally fitted with a 6 h.p. De Dion engine then in 1905 with a 4-cylinder Thornycroft oil engine, 24 b.h.p. = 10 knots. Register closed in 1951 on news that the vessel had been dismantled and its remains burnt about 1926.

KING EDWARD

Year: 1902
Designer: Walter Reeks
Builder: Messrs Morrison and Sinclair (supervised by Walter Reeks)
Location: Pyrmont, New South Wales, Australia
Type: Steam ferry
Description: ON 112583. 110.0 (oa) x 102.0 x 23.0 x 8.6ft. 98/67 tons (gross/net). Compound steam engine 24 n.h.p., 200 i.h.p., 11½ knots, made by Campbell and Calderwood, Paisley (Scotland). Built for the Watsons Bay and South Shore Ferry Company Ltd. (Reg. Sydney #54/1902) and launched on 17 September 1902.[2543] Construction of Australian hardwood and kauri included a longitudinal lattice framework that made it '*the strongest ferry boat of her size ever built in Sydney Harbour*'.[2544] Sold in 1920 to Sydney Ferries Ltd., and register closed in 1932, following being auctioned off, as no longer required.[2545] One of nearly 20 ferries displaced by the Sydney Harbour Bridge, the vessel's engine was reported removed in 1935 and its hull sold for demolition in 1940.[2546]

KING EDWARD AT CIRCULAR QUAY (1920S).
SOURCE: GRAEME ANDREWS WORKING HARBOUR PHOTOGRAPH COLLECTION, CITY OF SYDNEY ARCHIVES.

KAI ORA

Year: 1902
Designer: Walter Reeks
Builder: William Holmes
Location: McMahons Point, New South Wales, Australia
Type: Auxiliary ketch
Description: ON 112595. 52.3 x 13.0 x 5.5ft. 28/15 tons (gross/net), 30 tons deadweight. 2-cylinder kerosene engine made by Victor Motor Co. of Sydney (New South Wales), 10 n.h.p. = 6 knots. Press reported as built for Alex Cameron of Gippsland, Victoria, and launched on 12 December 1902 but was first registered in the name

2537 *The Australian Star*, 1 July 1902.
2538 SYDNEY, NEW SOUTH WALES, Register of ships, single number series (II), Volume 19, 1902-1904, Folio 42.
2539 *Australian Star*, 11 August 1902; *Referee*, 8 July 1903.
2540 *Evening News*, 1 May 1926; *Dugong Chronicle*, 18 January 1927; *The Sydney Morning Herald*, 21 November 1928.
2541 *The Sydney Morning Herald*, 31 May 1902.
2542 SYDNEY, NEW SOUTH WALES, Register of ships, single number series (II), Volume 20, 1904-1907, Folio 89.
2543 SYDNEY, NEW SOUTH WALES, Register of ships, single number series (II), Volume 19, 1902-1904, Folio 32.
2544 *The Sydney Morning Herald*, 18 September 1902.
2545 *The Sydney Morning Herald*, 1 June 1932.
2546 A. M. Prescott (1984). *Sydney Ferry Fleets*.

of George Holdship (Reg. Sydney #1/1903).[2547] Sold 11 September 1903 to Albert John Ellerker of Melbourne and by him on 3 October 1906 to the Orbost Shipping Co. Pty. Ltd. (Reg. #Melbourne 13/1906).[2548] Reregistered art Samarai, British New Guinea, in 1910. A very shallow draft (4ft) auxiliary ketch built for the Victorian far east coast trade, its register was closed in 1920 with the comment that it had been '*lost in the Gulf division*' of Papua in 1920.

AOTEA

Year: 1902
Designer: Walter Reeks
Builder: Messrs Scott
Location: Kinghor, Scotland
Type: Steel steamship
Description: ON 107040, 108 (oa) x 102.4 x 21.7 x 9.7ft. 203/90 tons (gross/net). Compound steam engine, 235 n.h.p., 200 i.h.p. Built in Scotland to a Reeks design for Messrs G. and W. Turnbull of Wellington, New Zealand, for the coastal trade to and from the Patea River. Arrived at Sydney on its maiden voyage on 17 December 1902.[2549] (Reg. Kirkaldy #1/1902, Wellington #4/1903). Stranded at the mouth of the Patea River on 9 August 1904 and before salvage efforts could be completed, broke up in a gale in mid-September.[2550]

PACTOLUS

Year: 1903
Designer: Walter Reeks
Builder: William Holmes
Location: McMahons Point, New South Wales, Australia
Type: Auxiliary fishing ketch
Description: Not British registered. 52 x 14 x 4.5ft. 20 tons. 2 x 10 h.p. oil engines by the Victor Motor Company of Double Bay, Sydney (New South Wales), twin-screw = 8¼ m.p.h. Trial under power on 22 May 1903.[2551] Built for Carl Rudolph Landenberger as a fishing vessel based at Iluka, Clarence River.[2552] Sold to Jean Baptiste Octave Mouton of New Britain (a French citizen, born in Belgium, and not Naturalised until 1911, so not entitled to hold British registry for his vessel/s) and cleared with Customs at Sydney for New Britain on 21 September 1906.[2553] Subsequent fate not known.

PHOENIX

Year: 1903
Designer: Walter Reeks (joinery and upperworks only)
Builder: William Langford
Location: Newcastle, New South Wales, Australia
Type: Steam ferry
Description: ON 113147. 77 (oa) x 68.1 x 19.3 x 6.0ft. Built by William Langford at Newcastle for William Walker Johnson and intended for the Lake Macquarie passenger trade (Reg. Newcastle #7/1903).[2554] Certified to carry 277 passengers, its hull was designed by Langford while its joinery and upperworks were designed by Reeks.[2555] Sold in 1917 and again in 1918, it was ultimately destroyed by fire in May 1931 on the Hawkesbury River and its register closed a few months later.

NERONG

Year: 1903
Designer: Walter Reeks (possibly)
Builder: Foster and Minty (register states The Northern Coast Steam Navigation Co. Ltd.)
Location: Peacock Point, Balmain, New South Wales, Australia
Type: Steel twin-screw coastal steamship
Description: ON 117630. 125 (oa) x 119.8 x 22.0 x 7.9ft, 7ft draft. 219/132 tons (gross/net). Two compound steam engines totalling 27 n.h.p., 150 i.h.p., 7 knots, made by Bow and McLachlan, Paisley (Scotland) in 1901. A shallow-draft twin-screw steamer built to trade along the New South Wales north coast and enter shallow bar harbours, originally owned by the North Coast Steam Navigation Co. Ltd. (Reg. Sydney #47/1903).[2556] Launched on 17 June 1903, unfortunately the designer is not stated, only a reference is made to a 'Mr Scott' being a part-designer.[2557] This was quite possibly James Scott (builder of *Lady Manning* and *Lady Hampden*) and Reeks, who was present at the launch, could very well have been involved in some capacity. Primarily a cargo steamer with limited passenger accommodation. Half-model is in the Powerhouse Museum's collection. Vessel foundered off Norah Head, New South Wales on 19 September 1917.

BRITANNIA

Year: 1903
Designer: Walter Reeks
Builder: Oliver R. Jones
Location: Woodburn, New South Wales, Australia
Type: Steam ferry
Description: ON 121118. 80 (oa) x 74.95 x 16.1 x 5.2ft. 55/38 tons (gross/net). Compound steam engine, 12 n.h.p., 90 i.h.p. = 12 knots, made by Ross & Duncan of Glasgow, Scotland in 1899. Original owners: Frederick William Davis, Joseph George Davis and William Burgess (sawmill manager of Coraki, Reg. Sydney #3/1906), owners restructured in 1906 as Davis Bros &

[2547] *Daily Commercial News and Shipping List*, 13 December 1902; SYDNEY, NEW SOUTH WALES, Register of ships, single number series (II), Volume 19, 1902-1904, Folio 46.
[2548] MELBOURNE, Main Register subsequent to Merchant Shipping Act 1854, Volume 8, 1898-1911, Folio 113.
[2549] *The Daily Telegraph*, 20 December 1902.
[2550] *New Zealand Times*, 11 August, 23 September, 26 October 1904.
[2551] *The Sydney Morning Herald*, 23 May 1903.
[2552] *Glen Innes Examiner*, 16 March 1937.
[2553] *The Sydney Morning Herald*, 22 September 1906, *Daily Telegraph*, 24 September 1906.
[2554] NEWCASTLE, NSW, Main Register subsequent to Merchant Shipping Act 1854, Volume 5, 1888-1924, Folio 56.
[2555] *Newcastle Morning Herald*, 7 November 1903.
[2556] SYDNEY, NEW SOUTH WALES, Register of ships, single number series (II), Volume 19, 1902-1904, Folio 86.
[2557] *The Australian Star*, 17 June 1903; *Balmain Observer and Western Suburbs Advertiser*, 20 June 1903.

Burgess Ltd.[2558] Built primarily for passenger work on the Richmond River (licensed to carry 300), launched on 10 September and entered service on 21 December 1903.[2559] Unregistered sale to Richmond River Shipping Co. in 1931. Later converted to motor vessel with 120 b.h.p. National diesel, and damaged beyond repair by fire sometime after April 1945 when the vessel was sold by the mortgagee in possession, after the engine was advertised for sale. Register closed in 1950 on receipt of news that the vessel was derelict at Ballina after constructive total loss by fire.[2560]

LADY RAWSON

Year: 1903
Designer: Walter Reeks
Builder: Messrs Morrison and Sinclair
Location: Pyrmont, New South Wales, Australia
Type: Steam ferry
Description: ON 117636. 122 (oa) x 117.3 x 27.5 x 11.3ft. 172/117 tons (gross/net). Compound steam engine 40 n.h.p., 230 i.h.p. by A. G. Humford of Colchester (England), speed 12 knots. Built for the Balmain New Ferry Company Ltd. (Reg. Sydney #56/1903), a double-ended harbour ferry with 760 passenger capacity.[2561] Sold in 1915 to Sydney Ferries Ltd. Register closed 28 November 1932 as no longer required. Does not appear to have re-entered ferry service after breaking a propeller shaft on 17 January 1932, failed to bring an acceptable offer when auctioned on 31 May and loaned for use as a floating clubhouse by the Australian Motor Yacht Squadron, to be based at Beauty Point, Middle Harbour. Dedicated as such by the Lord Mayor of Sydney on 22 October 1932.[2562] Vessel finally sold for £40 for demolition in January 1934.[2563] The hulk was destroyed by fire and sank in shallow water on 28 February 1934.

BOOMERANG EX BONA

Year: 1903
Designer: Walter Reeks
Builder: William Holmes
Location: McMahons Point, New South Wales, Australia
Type: Auxiliary schooner
Description: ON 101747. 89 (oa) x 73.2 x 16.9 x 7.6ft. 42/29 tons (gross/net). 4-cylinder oil engine by Tarran Motors Co. of Melbourne (Victoria), 30 b.h.p. = 7 knots. Seagoing cruising yacht built for Charles Dunkley Wallace of St Kilda (Reg. Melbourne #4/1904).[2564] Hardwood frames and bottom, kauri topsides, copper fastened, launched 24 September 1903.[2565] Later owned by Harry Bellingham Howard-Smith of Melbourne (1918); Charles Lloyd Jones of Sydney (1927); Michel Francois Albert of Sydney (1929, reregistered Sydney #3/1929 and renamed *Boomerang*); Albert Investments Pty. Ltd. (1941).[2566] Resurveyed in 1972 following fitting of 6-cylinder diesel made by Gardner Engines (Sales) Ltd., Manchester (England) in 1969, 6.77 n.h.p., 110 b.h.p. = 8 knots and now essentially a motor yacht with rudimentary schooner rig. Donated by the Albert family to the Sydney Heritage Fleet in 1987 and remains in service as a motor yacht on Sydney Harbour.

DAWMEA

Year: 1903
Designer: Walter Reeks
Builder: H. Manmill and Wain
Location: Lavender Bay, New South Wales, Australia
Type: Oil launch
Description: ON 117694. 34.7 (oa) x 30.8 x 8.0 x 3.3ft, 2ft draft. 5/4 tons (gross/net). 2-cylinder oil engine by Victor Motor Company, Sydney (New South Wales), 8¼ n.h.p, 7 knots. Built for William John Adams (Registered Sydney #21/1905).[2567] Register closed in 1951 with comment that vessel was sold about 1909 and owner died in 1935. Built of 1in kauri pine on 1¾ x 2in bent timbers, glazed teak engine compartment forward with seating for 30-40 passengers aft. Auctioned at Sydney on 25 February 1910 and presumably renamed by its new owner.[2568]

LOLA EX PANACEA

Year: 1904
Designer: Walter Reeks
Builder: William Holmes
Location: McMahons Point, New South Wales, Australia
Type: Auxiliary ketch
Description: ON 117657. 54 (oa) x 45.6 x 40.5 (wl) x 13.0 x 6.1ft. 16/11 tons (gross/net). Built as a yacht for Dr Robert Scot Skirving, replacing his previous yacht *Hygeia,* launched 15 December 1903, first raced on 9 January 1904.[2569] Engine (2-cylinder by Sintz Gas Engine Co. of Michigan, USA, 16 n.h.p. = 8 knots) not fitted until sold to Erskine North Raymond in May 1904, destined for Islands use (Reg. Sydney #21/1904 and claimed built in that year, renamed *Lola*).[2570] Register closed in 1951, there being no trace of vessel or owner. Last noted in 1905, having survived a stranding and a cyclone in the New Hebrides, but believed to have later been used as a pilot boat in New Caledonia.[2571]

[2558] SYDNEY, NEW SOUTH WALES, Register of ships, single number series (II), Volume 20, 1904-1907, Folio 63.
[2559] *The Richmond River Herald and Northern Districts Advertiser*, 4 September 1903.
[2560] www.wrecksite.eu/wreck.aspx?50164.
[2561] SYDNEY, NEW SOUTH WALES, Register of ships, single number series (II), Volume 19, 1902-1904, Folio 95.
[2562] *Daily Telegraph*, 31 January 1934.
[2563] *The Sun*, 27 February 1934.
[2564] MELBOURNE, Main Register subsequent to Merchant Shipping Act 1854, Volume 8, 1898-1911, Folio 73.
[2565] *The Sydney Morning Herald*, 26 September 1903.
[2566] SYDNEY, NEW SOUTH WALES, Register of ships, single number series (II), Volume 27, 1928-1937, Folio 20; SYDNEY, NEW SOUTH WALES, Register of ships, single number series (II), Volume 34, 1970-1973, Folio 131.
[2567] *The Sydney Morning Herald*, 21 November 1903; SYDNEY, NEW SOUTH WALES, Register of ships, single number series (II), Volume 20, 1904-1907, Folio 35.
[2568] *The Daily Telegraph*, 12 February 1910.
[2569] *Evening News*, 16 December 1903;
The Sydney Morning Herald, 18 December 1903, 11 January 1904.
[2570] *The Sydney Morning Herald*, 12 May 1904; *The Daily Telegraph*, 29 August 1904;
SYDNEY, NEW SOUTH WALES, Register of ships, single number series (II), Volume 19, 1902-1904, Folio 122.
[2571] *The Examiner*, 29 April 1905.

BONA (LATER BOOMERANG).
SOURCE: AUSTRALIAN NATIONAL MARITIME MUSEUM COLLECTION (00011648).

CWMAVON

Year: 1904
Designer: Walter Reeks
Builder: William Holmes
Location: McMahons Point, New South Wales, Australia
Type: Speed launch
Description: Not registered. 40 x 7.5ft. 15 h.p. 2-cylinder Jersey City Standard oil engine. Built by William Holmes for Sydney R. Lloyd of Woy Woy, used on Brisbane Water and claimed to be the fasted launch in New South Wales. Torpedo-boat hull made of kauri with cedar finished cabin.[2572] No trace after 1905.[2573]

GONDOLIER

Year: 1904
Designer: Walter Reeks
Builder: William Holmes
Location: McMahons Point, New South Wales, Australia
Type: Wooden pearling ketch
Description: ON 117649. 50 (oa) x 45.2 x 12.1 x 6.5ft. 16/11 tons (gross/net). By 1917 48.6 x 13.0 x 6.1ft., same tonnage. Built by William Holmes to Plan no. 170 for James Clark, pearl sheller of Brisbane (Reg. Sydney #11/1904).[2574] Sold to Dutch owners and register closed in 1906. Reregistered Fremantle #34/1917, owners now James Clark, Robert Philp, Percival Pitman Outridge, Isabella Mitchell Campbell and James Theodore Clewett MacKenzie.[2575] Shares passed to remaining owners on deaths of I. M. Campbell in 1918 and R. Philp in 1922. Register closed without further comment in 1986. Believed to have been the *Gondolier* operating as a fishing boat out of Fremantle in 1935, with an Italian crew.[2576]

VALKYRIE

Year: 1904
Designer: Walter Reeks (possibly, same dimensions as known Reeks vessels)
Builder: Walter 'Watty' M. Ford, Jr
Location: Berrys Bay, New South Wales, Australia
Type: Wooden pearling ketch
Description: ON 117647. 45.2 x 12.1 x 6.5ft. 14/11 tons (gross/net). Built for James Clark of Brisbane (Reg. Sydney #9/1904).[2577] Register closed 1906 on sale to Dutch subjects, no further trace.

[2572] *The Sydney Morning Herald*, 21 March 1904; *The Daily Telegraph*, 19 March, 18 April 17 September 1904.
[2573] *The Sydney Mail*, 1 November 1905.
[2574] SYDNEY, NEW SOUTH WALES, Register of ships, single number series (II), Volume 19, 1902-1904, Folio 112.
[2575] FREMANTLE, Main Register, Volume 6, 1914-1931, Folio 68.
[2576] *The West Australian*, 14 January 1935.
[2577] SYDNEY, NEW SOUTH WALES, Register of ships, single number series (II), Volume 19, 1902-1904, Folio 110.

VALHALLA

Year: 1904
Designer: Walter Reeks (possibly, same dimensions as known Reeks vessels)
Builder: Walter 'Watty' M. Ford, Jr
Location: Berrys Bay, New South Wales, Australia
Type: Wooden pearling ketch
Description: ON 117648. 45.2 x 12.1 x 6.5ft. 14/11 tons (gross/net). Built for James Clark of Brisbane (Reg. Sydney #10/1904).[2578] Register closed 1906 on sale to Dutch subjects, no further trace.

NYOOLA

Year: 1904
Designer: Walter Reeks (possibly, same dimensions as known Reeks vessels)
Builder: William Holmes
Location: McMahons Point, New South Wales, Australia
Type: Wooden pearling ketch
Description: ON 117652. 45.2 x 12.1 x 6.5ft. 14/11 tons (gross/net). By 1917 48.6 x 13.0 x 6.1ft., same tonnage. Built for James Clark, pearl sheller of Brisbane (Reg. Sydney #15/1904).[2579] Sold to Dutch owners and register closed in 1906. Reregistered in Fremantle #34/1917, owners now James Clark, Robert Philp, Percival Pitman Outridge, Isabella Mitchell Campbell and James Theodore Clewett MacKenzie.[2580] Shares passed to remaining owners on deaths of I. M. Campbell in 1918 and R. Philp in 1922. Register closed 1971 on Department of Fisheries and Fauna informing Customs that the vessel had been broken up in 1923.[2581]

AWAROA

Year: 1904
Designer: Walter Reeks
Builder: Messrs Morrison and Sinclair
Location: Pyrmont, New South Wales, Australia
Type: Steam ferry
Description: ON 117670. 145 (oa) x 139.0 x 25.4 x 12.5ft, 6ft draft. 344/211 tons (gross/net). Triple expansion steam engine made by the Wallsend Slipway Company, Newcastle (England), 62 n.h.p., 450 i.h.p. = 13 knots. Built for the Northern Steamboat Company of Helenvsille, New Zealand, first registered in the name of their Sydney agent and shareholder Robert Spence Lamb (Reg. Sydney #33/1904, Reg. Auckland #13/1905).[2582] Built for the cargo, passenger and mail service on the Wairoa River, essentially a lengthened version of *Aotea* (1898, see previous entry) designed by Reeks and also with a tunnel hull for minimum draft. Launched on 9 June 1904.[2583] Went on to have many owners and trades including on the Wellington-Nelson run, Port Phillip Bay trade and ultimately the Bass Strait cargo trade for William Holyman & Sons Pty. Ltd. of Launceston.[2584] On a voyage from Launceston to Melbourne during a shipping strike with a 'volunteer' crew, sprang a leak in a severe gale and sank off Cape Liptrap in Bass Strait on 23 July 1925, most fortuitously without loss of life.[2585]

FREDDIE

Year: 1904
Designer: Walter Reeks
Builder: Messrs Turner and Conroy
Location: Coraki, New South Wales
Type: Oil launch
Description: Not registered. 25 x 6.5 x 2.8ft. An oil 'dealing launch' designed by Reeks to the order of F. W. Scharer of Coraki, Richmond River. Fitted with a 4 h.p. Victor engine, 9 knots. Survived a fire in 1905, and was noted being slipped in 1907.[2586] Subsequent fate unknown.

HERCULES

Year: 1905
Designer: Walter Reeks
Builder: William Holmes
Location: McMahons Point, New South Wales, Australia
Type: Motor launch
Description: Not registered. Dimensions not known. Designed by Reeks and built to the order of E. W. Buzacott, the Sydney-based agent for the American made Hercules engine of which a 14 h.p. model was installed.[2587] Won the motor launch race at the 1905 Anniversary Regatta with Reeks at the helm.[2588]

ENID

Year: 1905
Designer: Walter Reeks
Builder: Rock Davis (Jr)
Location: Brisbane Water, New South Wales, Australia
Type: Auxiliary fishing ketch
Description: ON 117685. 59.6 x 15.2 x 6.8ft. 26/24 tons (gross/net). 2-cylinder oil engine made by Union Oil Engine Company of San Francisco (USA), 20 n.h.p. = 8 knots. Launched from Brisbane Water yard in early 1905. Originally owned by Arthur Thomas Jones of Williamstown, Victoria (Reg. Sydney #10/1905, Melbourne #9/1905).[2589] Later owned by Samuel Philip

[2578] SYDNEY, NEW SOUTH WALES, Register of ships, single number series (II), Volume 19, 1902-1904, Folio 111.
[2579] SYDNEY, NEW SOUTH WALES, Register of ships, single number series (II), Volume 19, 1902-1904, Folio 116.
[2580] FREMANTLE, Main Register, Volume 6, 1914-1931, Folio 69.
[2581] SYDNEY, NEW SOUTH WALES, Register of ships, single number series (II), Volume 19, 1902-1904, Folio 116.
[2582] SYDNEY, NEW SOUTH WALES, Register of ships, single number series (II), Volume 20, 1904-1907, Folio 6.
[2583] *The Sydney Morning Herald*, 10 June 1904.
[2584] MELBOURNE, Main Register subsequent to Merchant Shipping Act 1854, Volume 9, 1911-1924, Folio 87; MELBOURNE, Main Register subsequent to Merchant Shipping Act 1854, Volume 9, 1911-1924, Folio 122.
[2585] *The Mercury*, 24 July 1925.
[2586] *The Richmond River Herald*, 31 May, 17 November 1907.
[2587] *The Sydney Morning Herald*, 22 October 1904.
[2588] *The Sydney Morning Herald*, 31 January 1905;
The Daily Telegraph, 11 June 1904, 25 February 1905.
[2589] SYDNEY, NEW SOUTH WALES, Register of ships, single number series (II), Volume 20, 1904-1907, Folio 24; MELBOURNE, Main Register subsequent to Merchant Shipping Act 1854, Volume

AWAROA AT LAUNCESTON (C1920S).
SOURCE: LIBRARIES TASMANIA (LMSS761-1-45).

Townsend of Mornington, Victoria (1907) and Gizo Solomons Plantations Pty. Ltd. of Melbourne (1909). Register closed in 1964 with a comment that the vessel had been wrecked in 1925. Built as a fishing vessel for use in Bass Strait but from 1906 in the Port Phillip Bay trade operating between Melbourne, Mornington and Dromana. Cleared Port Phillip Heads for the Solomon Islands on 7 April 1909.[2590]

--

Year: 1905
Designer: Walter Reeks
Builder: William Holmes
Location: McMahons Point, New South Wales, Australia
Type: Steam lighter
Description: Not registered. 58ft. Built for Elliott Bros. and launched in late 1905.[2591]

--

Year: 1905
Designer: Walter Reeks
Builder: --
Location: Sydney, New South Wales, Australia
Type: Fishing boat
Description: Commercial fishing vessel designed by Reeks and built in Sydney for a Victorian customer.[2592] Additional details not known.

--

Year: 1905
Designer: Walter Reeks
Builder: --
Location: Sydney, New South Wales, Australia
Type: Tug
Description: A small tug designed by Reeks and built in Sydney, intended for use in Noumea.[2593] Additional details not known.

ALCYONE

Year: 1905
Designer: Walter Reeks
Builder: William Holmes
Location: McMahons Point, New South Wales, Australia
Type: Cruising yacht
Description: Not registered. 33 (oa) x, 26ft (wl) x 9.3 x 5ft. Built for Robert Scot Skirving, 8 or 10 tons. Later sold to Burns Philp and Co. to carry copra in Solomons.[2594]

8, 1898-1911, Folio 98.
[2590] *Geelong Advertiser*, 8 April 1909.
[2591] Evening News, *3 June 1905; The Sydney Morning Herald*, 5 August 1905.
[2592] *The Sydney Morning Herald*, 14 June 1905.
[2593] *The Sydney Morning Herald*, 14 June 1905.
[2594] *The Sydney Morning Herald*, 31 March 1910.

QUEENSCLIFF EX DERWENT

Year: 1905
Designer: Walter Reeks
Builder: Frederick Moore
Location: Hobart, Tasmania, Australia
Type: Ferry steamer
Description: ON 119242. 110.1 x 23.8 x 9.7ft. 141/36, remeasured in 1914 to 164/74, by 1966 146/66. Compound, engine 24 n.h.p. = 10.5 knots, made by Shanks & Co., Arbroath (Scotland) made in 1904; later twin diesel engines, 225 b.h.p. = 12 knots, by Lister Blackstone, Dursley, Gloucestershire (England) made in 1951. Built for the trans-Derwent ferry service to the order of James O'May, Henry O'May, George Elvin O'May and Ernest Watson O'May, registered Hobart #30/11/1905.[2595] Reregistered Sydney in 1912 and transferred back to Hobart a few month later, now owned by the Rosny Estates & Ferry Co. Ltd. Later owners include S.S. Reemere Pty. Ltd. (1927); O'May Bros. Pty. Ltd. (1929); Hobart Bridge Co. Ltd. (1939); The Transport Commission (1947).[2596] One of the longest serving ferries on the River Derwent run. In 1965 purchased by Vasile Pantea of Queenscliff, Victoria, reregistered at Melbourne and renamed *Queenscliff*. Ultimately wrecked at Point Cook on 30 December 1973.

VAUCLUSE

Year: 1905
Designer: Walter Reeks
Builder: Rock Davis (Jr)
Location: Brisbane Water, New South Wales, Australia
Type: Ferry steamer
Description: ON 121109. 139.5 x 23.3 x 9.0ft. 121/82 tons (gross/net). Triple Expansion steam engine made by Chapman & Co., Sydney, 70 n.h.p., 500 i.h.p. = 14¼ knots. Double-ended vessel with single screw, built of kauri topsides and decks on hardwood, launched on 4 July 1905.[2597] Built for the Watson's Bay and South Shore Ferry Company (Reg. Sydney #38/1905).[2598] Said to be the fastest ferry on the Harbour. Owners reorganised on 9 July 1912 to Watsons Bay and South Shore Ferry Company Ltd. Vessel transferred 29 September 1920 to Sydney Ferries Ltd. In 1931 purchased by the Government of New South Wales to operate services to the Walsh Island dockyard and register closed.[2599] After several years laid up, sold for scrapping in 1938.[2600]

ROONA

Year: 1905
Designer: Walter Reeks
Builder: Frederick Moore
Location: Launceston, Tasmania, Australia
Type: Yacht
Description: Not registered. 22 (oa) x 16.5ft (wl). Commissioned by E. P. Hart of Launceston, to be built by Frederick Moore.[2601] Oral history via a previous owner (Howell) suggests that Reeks had based the lines on those of a boat published in *Rudder* magazine. Completed in time for the Tamar Regatta.[2602] Owned in Hobart from 1907 and later in Victoria (after decks raised). Noted at Goolwa, South Australia, in 2010s.

TOGO

Year: 1905
Designer: Walter Reeks
Builder: Frederick Moore
Location: Launceston, Tasmania, Australia
Type: Ferry steamer
Description: ON 101788. 130 (oa) x 122.0 x 22.6 x 9.5ft. Compound triple expansion engines, 46 n.h.p., 400 i.h.p. = 14 knots, made by the Glasgow Engineering Co. of Launceston (Tasmania), 1905. Built by Frederick Moore to the order of Robert Gardner and launched in November 1905 (Reg. Launceston #1/1906).[2603] Considered one of the most influential river steamers built in Tasmania, in 1907 transferred to the Launceston Shipping Co. Ltd. In 1910 arrived in Hobart to run between Hobart, Port Esperance, Southport and Port Arthur (reregistered Hobart #10/1911). Later owners include H. M. The King; James Rowe & Sons Ltd. and the Commonwealth of Australia (during World War II).[2604] Destroyed by fire in 1957 when beached alongside the hulk *Otago* at Old Beach, River Derwent.

LADY NORTHCOTE

Year: 1905
Designer: Walter Reeks
Builder: Messrs Morrison and Sinclair
Location: Pyrmont Balmain, New South Wales, Australia
Type: Ferry steamer
Description: ON 121122. 115.8 x 24.0 x 10.25ft. 128/87 tons (gross/net). Built for the Balmain New Ferry Company Ltd. Registered Sydney #5/1906 with capacity for 602 passengers.[2605] Sold 1 March 1918 to Sydney Ferries Ltd. Register closed 1932 as no longer required. Reregistered #2/1937.[2606] After being laid up, sank at Sydney Ferries' Balmain wharf in March 1941. Believed refloated and hulked, final fate unknown.

[2595] HOBART, Main Register (with continuation entries), Volume 11, 1899-1911, Folio 118.
[2596] HOBART, Main Register (with continuation entries), Volume 14, 1964-1978, Folio 13; MELBOURNE, Main Register subsequent to Merchant Shipping Act 1854, Volume 12, 1962-1975, Folio 49.
[2597] *The Daily Telegraph*, 5 July 1905.
[2598] SYDNEY, NEW SOUTH WALES, Register of ships, single number series (II), Volume 20, 1904-1907, Folio 52.
[2599] *The Sydney Morning Herald*, 31 August 1931.
[2600] *The Labor Daily*, 2 April 1938.
[2601] *Examiner*, 23 September 1905.
[2602] *Examiner*, 14 February 1906.
[2603] LAUNCESTON, TASMANIA, Register of British ships, Volume 5, 1870-1930, Folio 273.
[2604] HOBART, Main Register (with continuation entries), Volume 12, 1911-1930, Folio 3; MELBOURNE, Main Register subsequent to Merchant Shipping Act 1854, Volume 9, 1911-1924, Folio 107; HOBART, Main Register (with continuation entries), Volume 12, 1911-1930, Folio 117.
[2605] SYDNEY, NEW SOUTH WALES, Register of ships, single number series (II), Volume 20, 1904-1907, Folio 65.
[2606] SYDNEY, NEW SOUTH WALES, Register of ships, single number series (II), Volume 27, 1928-1937, Folio 157.

SOURCE: GRAEME BROXAM COLLECTION.

METEOR

Year: 1906
Designer: Walter Reeks
Builder: William Holmes
Location: McMahons Point, New South Wales, Australia
Type: Motor launch
Description: ON 121145. 45.0 x 6.6 x 3.9ft. 10/7 tons (gross/net). 4-cylinder engine by Standard Motor Construction Co. of Jersey City, New York (USA) made 1906, 25 h.p. = 13 knots. Built for Colin Macquarie Johnston of Woy Woy (Reg. Sydney #36/1906).[2607] Trials held on 31 March 1906.[2608] Register closed 1951 when noted owner died in 1935 and launch was not then part of his assets. Fate unknown. Probably renamed after being sold in 1912.[2609]

FREDERICK DAVIS

Year: 1906
Designer: Walter Reeks
Builder: Messrs Turner and Conroy
Location: Coraki, New South Wales, Australia
Type: Passenger steamer
Description: ON 121175. 90 (oa) x 86.0 x 15.6 x 6.1ft. 61/41 tons (gross/net). Compound steam engine made by Simpson, Strickland & Co. of Dartmouth (England) in 1906, 16.5 n.h.p., 150 i.h.p., 12 knots. A vessel with a long gestation but a regrettably short career. Reeks visited the Richmond River in March 1905 for discussions on a new steamer with Davis Bros., owners of SS *Britannia*, who had just taken over the Ballina service from T. B. Allen, owner of *Meteor*.[2610] It appears that Davis Bros. first considered building the vessel themselves at Woodburn, before letting out a contract for construction to Turner and Conroy of Coraki. An enlarged version of *Britannia*, the new vessel was ordered in June 1905. It was designed primarily as a passenger and light freight vessel, with cool compartments for shipping cream.[2611] The ironbark keel was finally laid on 4 October 1905, with an expected delivery within twelve months.[2612] Although the hull was reported to be almost complete in August 1906, it was not launched until 12 February 1907, and registered as *Frederick Davis* on 14 June in the names of Joseph George Davis, Frederick William Davis and William Burgess (Reg. Sydney #23/1907).[2613] In December 1908 sold to Bay Ferry Co. for the Melbourne-Williamstown service, with an expectation of being reregistered in their name on delivery.[2614] Regretfully, encountered heavy weather soon after leaving Ballina on 25 December 1908 under Captain Charles E. Young, and foundered at anchor in 5 fathoms at Wilsons Headland about 30 miles south of Clarence Heads on the following day, to become a total wreck.

2607 SYDNEY, NEW SOUTH WALES, Register of ships, single number series (II), Volume 20, 1904-1907, Folio 96.
2608 *The Sydney Morning Herald*, 2 April 1906.
2609 *The Sydney Morning Herald*, 2 March 1912.

2610 *Casino and Kyogle Courier*, 25 March 1905.
2611 *Northern Star*, 5 June 1905; *Richmond River Herald*, 7 June 1905.
2612 *The Grafton Argus*, 9 October 1905.
2613 SYDNEY, NEW SOUTH WALES, Register of ships, single number series (II), Volume 21, 1907-1909, Folio 5; *The Richmond River Herald*, 3 August 1906, 15 February 1907.
2614 *The Age*, 28 December 1908.

ZEPHYR

Year: 1906
Designer: Walter Reeks
Builder: Rock Davis (Jr)
Location: Brisbane Water, New South Wales, Australia
Type: Excursion steamer
Description: ON 120017. 126.2 x 25.5 x 7.7ft. 135/42 tons (gross/met) later revised to 178/71 (gross/net). 2 x Triple Expansion steam engines made by Campbell and Calderwood of Paisley, Glasgow (Scotland), totalling 95 n.h.p., 500 i.h.p. = 13 knots. Built for McIlwraith, McEacharn & Co. Pty. Ltd. (Reg. Fremantle #9/1906), restricted as McIlwraith, McEacharn Ltd. in 1926 and sold to Alf. E. Tilley & Co. Ltd. in 1946.[2615] Vessel broken up and register closed in 1966. A very fast excursion steamer built after displacement of the slower paddle steamer *Decoy* in response to the reconstruction of the rival *Manx Fairy* into SS *Westralian*.

GOODWILL

Year: 1906
Designer: Walter Reeks
Builder: Rock Davis (Jr)
Location: Brisbane Water, New South Wales, Australia
Type: Auxiliary steam schooner, later motor ketch
Description: ON 121148. 75 (oa) x 66.8 x 14.5 x 6.8ft. 43/29 tons (gross/net). Quadruple expansion steam engine by Simpson and Strickland, Dartmouth (England), 28 i.h.p. = 6 knots. Soon replaced by a motor engine. Auxiliary schooner ordered 19 August 1905 from Rock Davis (Jr) by Papuan Industries, Ltd., a 'Christian trading institution' formed in 1904 by interests associated with the London Missionary Society, and launched on 17 May 1906 (Reg. Sydney #38/1906, depth later remeasured as 7.3ft).[2616] Reregistered at Port Moresby on arrival there. Converted into an auxiliary motor ketch soon after arrival and sold between 1929 and 1933 to Janet Cowling of Daru, Papua.[2617] Register closed with comment 'wrecked at Cape Melville' in 1950. In fact, abandoned at Cape Melville in March 1942 following an engine break-down, and sank at anchor while manned by the Commonwealth. Mrs Cowling campaigned long and hard for compensation, and was awarded £450 late in 1944.[2618]

PURARI

Year: 1906
Designer: Walter Reeks
Builder: William Holmes
Location: McMahons Point, New South Wales, Australia
Type: Motor launch
Description: Not registered. 45 x 9.5 x 4ft, 2.2ft draft. 20 h.p. Gardner engine. Built of kauri and hardwood for the London Missionary Society for use in British New Guinea, trial trip to Paramatta on 23 August 1906, and to be shipped as deck cargo by the German steamer *Willehad*.[2619] First year of operations on the Purari River marred by engine break-downs.[2620] Still in service in 1919.[2621] Subsequent history not known.

--

Year: 1906
Designer: Walter Reeks
Builder: William Holmes
Location: McMahons Point, New South Wales, Australia
Type: Cruising yacht
Description: 42ft (oa). Built to the order of Pearce Bros. and intended for use as both a recreational pleasure and fishing boat. Installed with a 16 h.p. Standard engine and launched in September of 1906.[2622] Additional details not known.

--

Year: 1906
Designer: Walter Reeks
Builder: William Holmes
Location: McMahons Point, New South Wales, Australia
Type: Motor launch
Description: 30ft. Built to the order of T. J. Bush to a design by Reeks.[2623] Intended for use at Port Hacking as an excursion vessel.[2624] Additional details not known.

VALENCIA

Year: 1906
Designer: Walter Reeks
Builder: Walter 'Watty' M. Ford, Jr
Location: Berrys Bay, New South Wales, Australia
Type: Motor yacht
Description: ON 136455. 28.3 x 10.6 x 3.5ft. 8/7 tons (gross/net). In 1919 powered by a 2-cylinder oil engine made by Ferro Machinery and Foundry Co. of Cleveland Ohio (USA) in 1910, 11 b.h.p. = 7 m.p.h. In 1925 a new engine noted: 6-cylinder Studebaker Corporation of America, Indiana (USA) 29 b.h.p. = 9 m.p.h. Early history not known, may have been renamed *Valencia* when purchased by Arthur Henry Davies (first registered Sydney #27/1919).[2625] By 1922 owned by Harold Bradley Treacy; by 1925 owned by Norman James Frey Josephson. Advertised for sale in 1928.[2626] By 1935 owned by Arthur Neville St George Burkitt. Register closed 1951 as no longer required. Subsequent history not known.

[2615] FREMANTLE, Main Register, Volume 4, 1903-1907, Folio 94.
[2616] SYDNEY, NEW SOUTH WALES, Register of ships, single number series (II), Volume 20, 1904-1907, Folio 98; *The Sydney Morning Herald*, 17 December 1904, 20 October 1905; *The Daily Telegraph*, 19 May 1906.
[2617] *Cairns Post*, 21 November 1933.
[2618] National Archives of Australia, MP742/1, 299/4/470.
[2619] *The Sydney Morning Herald*, 24, 30 August 1906.
[2620] *Darling Downs Gazette*, 13 May 1907.
[2621] B. Mathews (1919). *Ships of Peace, London Missionary Society.*
[2622] *Referee*, 22 August, 19 September 1906.
[2623] *Evening News*, 10 July 1906; *The Sydney Morning Herald*, 3 August 1906.
[2624] *Evening News*, 10 July 1906; *The Sydney Morning Herald*, 3 August 1906.
[2625] SYDNEY, NEW SOUTH WALES, Register of ships, single number series (II), Volume 25, 1919-1923, Folio 7.
[2626] *The Sydney Morning Herald*, 19 December 1928.

FAIRBANKS II EX ?

Year: 1906
Designer: Walter Reeks
Builder: William Holmes
Location: McMahons Point, New South Wales, Australia
Type: Speed boat
Description: 25ft (oa). Originally installed with a 4-cylinder oil engine made by Thornycroft (England), 24 h.p.[2627] Replaced by a 20 h.p. Fairbanks engine in October of 1908 and subsequently shipped to new owners, Messrs Turk and Cliff, sole agents for Fairbanks engines in Brisbane, who quickly renamed the vessel *Fairbanks II*.[2628] Competed with the Brisbane Dinghy Club.

KYLIE

Year: 1906
Designer: Walter Reeks
Builder: --
Location: Sydney, New South Wales, Australia
Type: Motor launch
Description: Built for F. A. Wiesener to a design by Reeks and likely launched in late 1906, builder unknown.[2629] Fitted with a 6 h.p. Thornycroft engine, advertised for sale in December of 1907.[2630]

RUANA EX EVANGEL

Year: 1907
Designer: Walter Reeks
Builder: William Holmes
Location: McMahons Point, New South Wales, Australia
Type: Auxiliary ketch
Description: ON 121173. 60 (oa) x 55.0 x 15.5 x 5.7ft. 46/22 tons gross/net) remeasured 1914 as 48/28 tons. 20 b.h.p. 2-cylinder L. Gardner & Sons Ltd. of Patricroft, Manchester (England), kerosene engine, 6½ knots. Later replaced by a 2-cylinder oil engine 40 b.h.p. = 7 ½ knots made by Coln Deutz of Germany in 1913. Built from hardwood with kauri topsides and deck for the Queensland Kanaka Mission for use in the Solomon Islands and launched 1 May 1907.[2631] Soon sold to Florence Selina Harriett Young of Bundaberg (Reg. Sydney #20/1907, renamed in 1912 as *Ruana*).[2632] Later owned by the Malayta Co. Ltd. (1915) and the Fairymead Sugar Co. Ltd. (1936). Register closed 26 October 1943 on advice from owner that the vessel had been lost. Circumstances of the loss are not known: there is no record of impressment by Government during the Second World War.

[2627] *Referee*, 18 July 1906.
[2628] *The Australian Star*, 13 October 1908; The Brisbane Courier, 17 October 1908.
[2629] *The Australian Star*, 18 February 1907.
[2630] *The Sydney Morning Herald*, 14 December 1907.
[2631] *The Daily Telegraph*, 2 May 1907.
[2632] SYDNEY, NEW SOUTH WALES, Register of ships, single number series (II), Volume 21, 1907-1909, Folio 2.

RUANA EX EVANGEL CAREENING.
SOURCE: GRAEME BROXAM COLLECTION.

KOOKOOBURRA

Year: 1907
Designer: Walter Reeks (possibly)
Builder: Messrs Morrison and Sinclair
Location: Pyrmont Balmain, New South Wales, Australia
Type: Steam ferry
Description: ON 121169. 139.9 x 25.3 x 10.0ft. 180/123 tons (gross/net). Triple-expansion steam engine made by Campbell and Calderwood, Paisley (Scotland), 61 n.h.p., 650 i.h.p. = 12-13 knots. Owned by Sydney Ferries Ltd. (Reg. Sydney #15/1907, #23/1936).[2633] A large and fast ferry built for the Parramatta run; launch reports give the builders as the designers, but when *Kuranda* (*Kaludah*) was built for them two years later, said to be of the *Kookooburra* type (though shorter) and was designed by Reeks.[2634] The register was closed on 28 November 1932 as no longer required, but the vessel reactivated and reregistered in 1936, and remained in service as a Sydney Harbour ferry until shortly before being sold for demolition in 1948.

MISCHIEF

Year: 1907
Designer: Walter Reeks
Builder: William Holmes
Location: McMahons Point, New South Wales, Australia
Type: Cruising yacht
Description: ON 121186. 35.6 (oa) x 31.7 x 9.6 x 4.4ft, 5 ft draft. 9 tons (gross/net). White & Poppe 3-cylinder oil engine made Coventry (England), 10½ b.h.p., 6 knots. Replaced 1927 with 1918-vintage 4-cylinder oil engine made by Charles L. Seabury of Norris Heights (USA), 25 b.h.p. = 8 knots. Built for Arthur James Milson (Reg. Sydney #36/1907), who died 7 May 1926 and sold via executors to David Theodore Nicholson.[2635] Register closed 1948 as no longer required. Vessel still in existence, based in Sydney.

[2633] SYDNEY, NEW SOUTH WALES, Register of ships, single number series (II), Volume 20, 1904-1907, Folio 125; SYDNEY, NEW SOUTH WALES, Register of ships, single number series (II), Volume 27, 1928-1937, Folio 141.
[2634] *The Daily Telegraph*, 7 December 1906, 10 December 1908.
[2635] SYDNEY, NEW SOUTH WALES, Register of ships, single number series (II), Volume 21, 1907-1909, Folio 18.

LADY CARRINGTON

Year: 1907
Designer: Walter Reeks
Builder: Messrs Morrison and Sinclair
Location: Pyrmont Balmain, New South Wales, Australia
Type: Steam ferry
Description: ON 121187. 130.0 x 24.2 x 10.0ft. 146/99 tons (gross/net). Compound steam engine, 85 n.h.p., 465 i.h.p. = 12 knots, made by Morts Dock and Engineering Co. Ltd. of Balmain. Built for the Balmain New Ferry Company (Reg. Sydney #37/1907), passed to Sydney Ferries Ltd. on 1 March 1918.[2636] Register closed 29 November 1932 as no longer required. A fast 701 passenger Sydney Harbour ferry launched on 30 July 1907, reached 12¾ knots on trials on 23 September.[2637] Auctioned 30 June 1932 but no bids received.[2638] Sold for demolition in January 1934.[2639]

BOSKER

Year: 1907
Designer: Walter Reeks
Builder: William Goddard
Location: Lavender Bay, New South Wales, Australia
Type: Motor launch
Description: Not registered. 36 x 8.1 x 3.9ft. 16 h.p. Morts Dock and Engineering Co. Invincible engine. Built of kauri and Oregon pine with hardwood engine beds and stringers for J. T. McKittrick of Grafton, made delivery voyage 24-24 September 1907.[2640] Remained in commercial service until at least 1916.[2641] Fate not known.

EARLIE

Year: 1907
Designer: Walter Reeks
Builder: William Holmes
Location: McMahons Point, New South Wales, Australia
Type: Motor launch
Description: Not registered. 50 x 9ft. Cedar launch with two skins laid longitudinally fitted with a 55 h.p. Thornycroft engine built by Holmes to a design by Reeks to the order of R. C. Brownlow of Bairnsdale, Gippsland Lakes, Victoria.[2642] Following fit-out and trials, arrived in its new home port in February 1908 after making the 384 nautical mile trip in a time of 39 hours.[2643] Remained in commercial operation as a tourist ferry based as Paynesville until the 1950s. Salvaged in 1985 and trucked to Renmark, South Australia. Still in existence, based on the Murray River.

EARLIE.
SOURCE: PETER MEDLING.

PEGASUS? EX PLUME

Year: 1907
Designer: Walter Reeks
Builder: Messrs Morrison and Sinclair
Location: Pyrmont Balmain, New South Wales, Australia
Type: Steam tug/lighter
Description: Not registered. Approx. 65ft. One of two sister-ships designed by Reeks (plan no. 207) for the Colonial Oil Company (see next entry) for use on Sydney Harbour, fitted with a large capacity derrick for cargo handling. Possibly the vessel that was derelict by the 1970s and reconstructed from in the 2010s to become the motor cruiser *Pegasus*.

PEGASUS? EX SNOWFLAKE

Year: 1907
Designer: Walter Reeks
Builder: Messrs Morrison and Sinclair
Location: Pyrmont Balmain, New South Wales, Australia
Type: Steam tug/lighter
Description: Not registered. Approx. 65ft. Second of two sister-ships designed by Reeks (plan no. 207) for the Colonial Oil Company (see previous entry) for use on Sydney Harbour, fitted with a large capacity derrick for cargo handling. Possibly the vessel that was derelict by the 1970s and reconstructed in the 2010s to become the motor cruiser *Pegasus*.

THORNYCROFT?

Year: 1907
Designer: Walter Reeks

2636 SYDNEY, NEW SOUTH WALES, Register of ships, single number series (II), Volume 21, 1907-1909, Folio 19.
2637 *The Sydney Morning Herald*, 31 July, 24 September 1907.
2638 *The Sydney Morning Herald*, 1 July 1932.
2639 *Daily Commercial News*, 1 February 1934.
2640 *Clarence and Richmond River Examiner*, 12, 15 October 1907.
2641 *Daily Examiner* (Grafton), 12 January 1914.
2642 *Referee*, 17 April, 7 August 1907; *The Daily Telegraph*, 10 August 1907; *The Sydney Morning Herald*, 14 September 1907.
2643 *Referee*, 12 February 1908.

Builder: William Holmes
Location: McMahons Point, New South Wales, Australia
Type: Motor boat
Description: Not registered. Launched from William Holmes' yard in mid-March 1907.[2644] Installed with a 12 h.p. Thornycroft engine and shipped to Adelaide. Very likely named *Thornycroft* and built to the order of M. S. Hawker. Used to compete in motor launch races out of Milang and Goolwa for several years.[2645]

--

Year: 1907
Designer: Walter Reeks
Builder: --
Location: --
Type: Vehicular steam ferry
Description: Double-ended vehicular steam ferry designed by Reeks to the order of Sydney Ferries Ltd. Design no. 211. Not built. A copy of the plan is held by Bill Allen.

BRONZEWING

Year: 1908
Designer: Walter Reeks
Builder: Messrs Morrison and Sinclair
Location: Pyrmont Balmain, New South Wales, Australia
Type: Motor yacht
Description: ON 125177. 110 (oa) x 101.8 x 17.8 x 8.2ft., 6.5ft draft. 98/67 tons (gross/net). Three 4-cylinder oil engines by John I. Thornycroft of Chiswick (England), total 300 b.h.p. = 12 knots. Dubbed 'The World's Largest Motor Yacht', built by Morrison and Sinclair at Balmain to the order of Samuel Hordern (Reg. Sydney #51/1908) and launched on 2 May 1908.[2646] He died in 1909 and vessel officially passed to Lebbeus Hordern on 5 March 1914. Sold to foreign (American) subjects and register closed in 1920 when purchased by a Mr Burke of Manila and left Australia under tow from SS *Wai-Hoi*. No further record found.

[2644] *Referee*, 20 March 1907.
[2645] *Southern Argus*, 26 December 1907, 4 January 1912; *Observer*, 2 January 1909.
[2646] *The Sydney Morning Herald*, 4 May 1908; SYDNEY, NEW SOUTH WALES, Register of ships, single number series (II), Volume 21, 1907-1909, Folio 79.

--

Year: 1908
Designer: Walter Reeks
Builder: Pritchard Bros.
Location: Careening Cove, New South Wales, Australia
Type: Tender
Description: Not registered. 20ft tender built by Pritchard Bros. to the order of Samuel Hordern for use in conjunction with his motor yacht *Bronzewing*.[2647] Likely sold with the vessel in 1914, and possibly again in 1920 when towed to The Philippines. No further record found.

CAMBRIA

Year: 1908
Designer: Walter Reeks
Builder: Messrs Morrison and Sinclair
Location: Pyrmont Balmain, New South Wales, Australia
Type: Motor launch
Description: ON 125155. 55 (oa) x 51.4 x 11.0 x 4.9ft. 22/15 tons (gross/net). Two 3-cylinder oil engines by Fairbanks of New York (USA), 20 h.p. each = 14 knots. Built for Lewis James Davis and launched on 1 March 1908 (Reg. Sydney #16/1908).[2648] Sold to the Police Department of New South Wales in 1918 and register closed. Vessel remained in service for several decades. Subsequent fate not known.

ENDEAVOUR

Year: 1908
Designer: Walter Reeks (hull) and Andrew Christie of Morts Dock and Engineering Co. (machinery)[2649]
Builder: Government Dockyard of New South Wales
Location: Cockatoo Island, New South Wales, Australia
Type: Steel steam trawler/research ship
Description: ON 132436. 134 (oa) x 131.7 x 23.2 x 10.8ft. 331/80 tons (gross/net) later revised to 331/102 tons. Triple Expansion steam engine, 45 n.h.p., 362.8 i.h.p.,

[2647] *The Sydney Morning Herald*, 4 May 1908; SYDNEY, NEW SOUTH WALES, Register of ships, single number series (II), Volume 21, 1907-1909, Folio 79.
[2648] SYDNEY, NEW SOUTH WALES, Register of ships, single number series (II), Volume 21, 1907-1909, Folio 44; *The Sydney Morning Herald*, 6 March 1908.
[2649] *The Sydney Morning Herald*, 27 August 1908.

CAMBRIA AS A POLICE VESSEL
SOURCE: AUSTRALIAN NATIONAL MARITIME MUSEUM COLLECTION (00011518).

made by the hull-builders. Built for the Government of the Commonwealth of Australia (Reg. Melbourne #31/1912).[2650] Purpose-built Federal Government fisheries research ship, launched on 27 August 1908. Successfully performed work around the Australian coast before going missing on a voyage from Macquarie Island to Hobart, believed sunk in a gale on or about 5 December 1914 with the loss of all hands, register closed 4 May 1915. Reeks gave extensive evidence of the ship's seaworthiness at the inquiry held in Melbourne in April 1915.[2651]

KALUDAH EX KURANDA

Year: 1909
Designer: Walter Reeks
Builder: Messrs Morrison and Sinclair
Location: Pyrmont Balmain, New South Wales, Australia
Type: Steam ferry
Description: ON 125190. 118 (oa) x 115.0 x 25.0 x 9.7ft. 137/93 tons (gross/net). Triple expansion steam engine made by Campbell and Calderwood, Paisley (Scotland), 50 n.h.p., 370 i.h.p. = 12 knots. Built for Sydney Ferries Ltd. for the Parramatta run and launched as *Kuranda* on 9 December 1908: the Board of Trade evidently took exception to the name, which was also that of a well-known Queensland coastal steamer, and it was registered as *Kaludah* (Reg. Sydney #13/1909).[2652] Built of kauri and hardwood and said to be of the 'Kookooburra type' although that vessel was longer.[2653] '*Burned to the water's edge on the Parramatta River NSW on the 22nd March 1911… Register closed 19 May 1911.*'[2654] Wreck refloated on 5 April, engine salvaged and hull sold 16 May to Einerson Brothers of Balmain for £40, presumably for demolition.[2655] A replacement ferry *Kamiri* was built the following year to the same plans as *Kaludah* and using its engines that had been salvaged and repaired.[2656]

KOORI EX KELPIE

Year: 1909
Designer: Walter Reeks
Builder: Woodleys Ltd.
Location: North Sydney, New South Wales, Australia
Type: Steam tug
Description: ON 124996. 70 (oa) x 65.8 x 16.7 x 7.2ft. 44 tons gross, 18 net (recalculated on several occasions). Compound steam engine made by Ross and Duncan of Govan, Glasgow (Scotland) in 1908, 18 n.h.p., 150 i.h.p. = 9 knots. Built for McIlwraith McEacharn & Co. Pty. Ltd. (Reg. Fremantle #10/1909), with kauri decks, topsides and planking on hardwood frames, copper fastened and sheathed.[2657] Hinged funnel and telescoping masthead lamp to fit under bridges on Swan River. Launched as *Kelpie* on 25 March 1909, though quickly renamed *Koori*, and worked on the Swan River primarily as a tug but also did some excursion work. Owners restructured with a couple of minor name changes, including MM Line Pty. Ltd. (1913) and MM Ltd. (1926). Sold on 4 May 1927 to Bellambi Coal Co. Ltd. of Sydney and reregistered Sydney #6/1927.[2658] Record tow of 2,466 miles from Fremantle to Sydney by interstate cargo steamer *Kooringa*. Sold for breaking up and register closed 2 September 1953.

ARETHUSA

Year: 1909
Designer: Walter Reeks
Builder: Woodleys Ltd.
Location: North Sydney, New South Wales, Australia
Type: Oil screw yacht
Description: ON 125197. 41 (oa) x 35.5 x 33 (wl) x 10.4 x 4.6ft. 9/8 tons gross. Remeasured 1920 now 36.0 x 10.6 x 5.2ft. 10/9 tons (gross/net). Originally powered by a 3-cylinder Parsons of Southampton (England) oil engine 21 b.h.p. = 7 knots, in 1920 replaced with 2-cylinder Standard Motor Construction Co., Jersey City (USA), 8/10 b.h.p. = 6 knots. Built for Joseph Foreman and Alexander Jarvie Hood (Reg. Sydney #21/1909) and launched on 24 February 1909.[2659] Construction of kauri on ironbark, with copper and Muntz metal fastened. Sold in 1920 to Richard Mervyn Glasson.[2660] Register closed in 1950 when Customs could find no trace of the vessel or its owner.

RURI

Year: 1909
Designer: Walter Reeks
Builder: Charles Dunn
Location: North Sydney, New South Wales, Australia
Type: Auxiliary yawl
Description: ON 125204. 35.4 x 8.6 x 3.4ft. 11/8 tons (gross/net). 4-cylinder oil engine by J. W. Brooke, Lowestoft 18 b.h.p. = 7 knots; in 1916 noted replaced by a 2-cylinder oil engine made by Union Gas Engine Co. of San Francisco (USA) made in 1914, 12 b.h.p. = 7 knots. Built for Albert Frank Saunders (Reg. Sydney #28/1909).[2661] Later owned by George Macquarie Paine (1916); Bertie Norman Black (1919); Robert and Henry James Wild, farmers of Botany (1922); and Henry Walker Bunce of Coogee (1926). Register closed in 1950, no trace of vessel that last registered owner had sold, but believed destroyed.

[2650] MELBOURNE, Main Register subsequent to Merchant Shipping Act 1854, Volume 9, 1911-1924, Folio 31.
[2651] *Evening News*, 27 August 1908; *The Sydney Morning Herald*, 24 April 1915.
[2652] SYDNEY, NEW SOUTH WALES, Register of ships, single number series (II), Volume 21, 1907-1909, Folio 94.
[2653] *The Daily Telegraph*, 10 December 1908.
[2654] *Sydney Mail*, 29 March 1911.
[2655] *The Sydney Morning Herald*, 5 April 1911; *Daily Commercial News*, 17 May 1911.
[2656] *Construction: Weekly Supplement to Building*, 27 May 1912.
[2657] FREMANTLE, Main Register, Volume 5, 1907-1914, Folio 24.
[2658] SYDNEY, NEW SOUTH WALES, Register of ships, single number series (II), Volume 26, 1923-1928, Folio 96.
[2659] SYDNEY, NEW SOUTH WALES, Register of ships, single number series (II), Volume 21, 1907-1909, Folio 102; *Evening News*, 24 February 1909.
[2660] *Evening News*, 24 February 1909.
[2661] SYDNEY, NEW SOUTH WALES, Register of ships, single number series (II), Volume 21, 1907-1909, Folio 109.

RURI.
SOURCE: RAE FIDOCK.

ROYAL ENDEAVOUR

Year: 1909
Designer: Walter Reeks
Builder: Messrs Morrison and Sinclair
Location: Pyrmont Balmain, New South Wales, Australia
Type: Auxiliary schooner
Description: ON 125207. 86 (oa) x 76.8 x 18.4 x 8.7ft. 60/26 tons (gross/net). 3-cylinder kerosene engine by L. Gardner & Sons of Manchester (England), 55 b.h.p. = 7 knots. First owners were Young brothers of Bundaberg, Queensland, trading as the Malayta Co. Ltd. (reg. Sydney #31/1909); vessel launched 19 July 1909.[2662] Later owned by J. R. Clay & Co. Ltd. of Port Moresby (1929). Register closed 2 October 1941 as no longer required. Vessel taken over by Commonwealth of Australia and disposed of in 1946. No trace of having re-entered commercial service.

ROYAL ENDEAVOUR.
SOURCE: GRAEME BROXAM COLLECTION.

[2662] SYDNEY, NEW SOUTH WALES, Register of ships, single number series (II), Volume 21, 1907-1909, Folio 112.

LINDISFARNE EX WILLIAMSTOWN

Year: 1910
Designer: Walter Reeks
Builder: Rock Davis (Jr), W. B. Davis and J. Cameron
Location: Brisbane Water, New South Wales, Australia
Type: Steam ferry
Description: ON 125228. 110 (oa) x 98.0 x 22.0 x 7.6ft. Remeasured 1942 as to 97.4 x 22.0 x 7.5ft. 99/65 tons (gross/net), remeasured 1942 as 83/55.25 (gross/net). Compound, cylinders 12½ in. and 22½ in. diameter by 14 in. stroke, 24 n.h.p., 75 i.h.p. = 10 knots, made by Bow, McLachlan & Co., Glasgow (Scotland), 1895; 120 p.s.i. steel boiler made by Ross & Duncan (Scotland) 1904; 'Model C.O.' 4-cylinder diesel engine, 100 b.h.p. @ 340 r.p.m. = 8 knots, made by Fairbanks, Morse & Co., (USA), installed 1942. Built to the order of Melbourne Ferries Pty. Ltd. and registered Sydney on 2 March 1910, then reregistered Melbourne on 30 March that same year.[2663] Following a decade of service on Port Phillip Bay, sold to Hobart-based owners and transferred for use on the River Derwent; reregistered Hobart on 4 October 1920 by The Lindisfarne Ferry Ltd. having purchased the vessel for £3,750.[2664] Remaining in Hobart, in 1923 sold to Rosny Estates and Ferry Co. Ltd.; in 1927 to S.S. Reemere Pty. Ltd.; in 1929 to O'May Bros. Pty. Ltd.; in 1931 Henry Inkerman Featherstone; and in 1941 to Donald Gorringe. Register closed in 1942 when sold to the Government of the United States of America. Saw service around the islands of the South Pacific and was reported to be at Noumea after the Second World War. Final fate not known.

LADY CHELMSFORD

Year: 1910
Designer: Walter Reeks
Builder: Rock Davis (Jr)
Location: Brisbane Water, New South Wales, Australia
Type: Steam ferry
Description: ON 125248. 112 (oa) x 110.2 x 24.7 x 9.3ft. 98/67 tons (gross/net). Remeasured 1958 as 99/64 tons (gross/net). Compound steam engine by Bow and McLachlan, Paisley (Scotland) in 1892, 20 n.h.p., 150 i.h.p. = 10 knots. In 1937 fitted with a 5-cylinder diesel made by L. Gardner & Sons Ltd. of Manchester (England), 190 b.h.p. = 10 knots; in 1958 with a 4-cylinder diesel by Crossley Bros. Ltd. of Openshaw, Manchester (England), 300 b.h.p. = 11 knots. Single propeller operating in push-pull configuration for either direction, launched on 14 April 1910. Originally owned by the Balmain New Ferry Co. Ltd. (reg. Sydney #32/1910).[2665] Sold in 1918 to Sydney Ferries Ltd. Register closed 30 November 1932 as

[2663] SYDNEY, NEW SOUTH WALES, Register of ships, single number series (II), Volume 22, 1910-1912, Folio 9; MELBOURNE, Main Register subsequent to Merchant Shipping Act 1854, Volume 8, 1898-1911, Folio 167.
[2664] HOBART, Main Register (with continuation entries), Volume 12, 1911-1930, Folio 184.
[2665] SYDNEY, NEW SOUTH WALES, Register of ships, single number series (II), Volume 22, 1910-1912, Folio 32.

no longer required. Returned to service as motor ferry and reregistered Sydney #3/1937.[2666] Owners restructured as Harbour Lighterage and Showboat Ltd. in October 1951. In April 1952 sold to the Sydney Harbour Transport Board, on 24 May 1971 to Harold Ernest Moffatt of Woodville, South Australia, and 1 August 1971 to Showboat Pty. Ltd. of Wayville, South Australia (reregistered Port Adelaide #1/1973) and transferred to Australian registry in 1983.[2667] Relocated to Melbourne in 1984 and was owned by Keith Rankin when sank at its moorings in Victoria Harbour, Melbourne, in February 2008. Remained sunk until demolished in situ in mid-1911.[2668]

RAWHITI II EX CULWULLA II

Year: 1909
Designer: Alfred Mylne
Builder: Walter 'Watty' M. Ford, Jr (superintended by Walter Reeks)
Location: Berrys Bay, New South Wales, Australia
Type: Yacht
Description: ON 125215. 27.2 x 5.7 x 3.85ft. 3 tons (gross/net). International 6-metre racing yacht built to the order of Alfred Moffit Marks and launched by the Governor of New South Wales, Sir Richard Poore on 27 October 1909 (Reg. Sydney #40/1909).[2669] Later owned by Francois Albert and Ernest Edward Sayer, renamed *Rawhiti II* (1911); Francois Albert (1919); William Southwick Dagg of Melbourne (1922; reregistered Melbourne #11/1934); George Arthur Laird and Jack Jones of Geelong (1934), Maynard Mappin of Hampton, Victoria (1943).[2670] Register closed 1943 as no longer required. Sank in a gale at St Kilda jetty on 20 July 1946.

LOTUS II EX MARCIA EX LOTUS II

Year: 1910
Designer: Walter Reeks
Builder: Walter 'Watty' M. Ford, Jr (superintended by Walter Reeks)
Location: Berrys Bay, New South Wales, Australia
Type: Motor launch
Description: ON 125243. 54.5 x 11.2 x 5.1ft. 28/25 (gross/net), following alterations remeasured in 1934 as 35/32 tons (gross/net). 3-cylinder engine made by Standard Gas Engine Co. of San Francisco (USA) in 1909, 30 b.h.p. = 10 knots. Built for James Frederick Elliott of Balmain (reg. Sydney #34/1910). Later owned by Reginald Arthur de Tessier Prevost (1920); Robert Walder and Arthur Thomas Henry Pittar (1921); Harry Austin (1924); Edmond Albert Waldock (1928); Robert Ralston Shafto and renamed *Marcia* (1932); Herbert Shelley Phillips (1937); Alexander Neil Thomson (1938); and Edward Henry O'Brien (1941).[2671] Advertised for sale in 1949 now powered by a 6-cylinder Gray diesel 220 b.h.p. for £5,500.[2672] Register closed in 1963 as no longer required. Still in existence, reverted back to *Lotus II* and based at Pittwater, New South Wales.

MODWENA

Year: 1910
Designer: Walter Reeks
Builder: Woodleys Ltd.
Location: North Sydney, New South Wales, Australia
Type: Motor launch
Description: ON 125240. 54 (oa) x 52.7 x 11.5 x 5.25ft. 23/21 tons (gross/net). 6-cylinder oil engine by J. W. Brooke & Co. Ltd., Lowestoft (England), 40 b.h.p. = 10 Knots. Built for George Frederick Bestie of Vaucluse (reg. Sydney #21/1910).[2673] Press in 1911 gives owner as M. G. Thornton. Later owned by Walter Macfarlane Ford, Jr (April 1920), Harry Mansfield Shelley (June 1920), Arnold Crescence Cooke (1926), and Andrew Ernest Morrow (1932). Register closed in 1950 as no longer required. Fate not known.

GREYCLIFFE

Year: 1910
Designer: Walter Reeks
Builder: David Drake
Location: Balmain, New South Wales, Australia
Type: Steam ferry
Description: ON 131478. 125.0 x 24.0 x 9.9ft. 133/90 tons (gross/net). 49 n.h.p., 350 i.h.p. Campbell & Calderwood (Scotland) triple expansion screw. Built to the order of the Watsons Bay and South Shore Steam Ferry Company Ltd.; and launched on 17 October 1910 (Reg. Sydney #8/1911).[2674] In 1920 transferred to Sydney Ferries Ltd. Sunk in a collision with Royal Mail Steamer SS *Tahiti* at 4.28 p.m. on 3 November 1927 in Port Jackson, New South Wales, with the unfortunate loss of 40 lives.

--

Year: 1911
Designer: Walter Reeks
Builder: James Wilson
Location: Brisbane, Queensland
Type: Motor launch
Description: Not registered. 28 x 7 x 3.5ft. Motor tender designed by Reeks to be carried on davits on board the steam yacht *Merrie England*. Fitted with a 15 h.p. 2-cylinder kerosene engine. Built at a cost of £460 in

[2666] SYDNEY, NEW SOUTH WALES, Register of ships, single number series (II), Volume 27, 1928-1937, Folio 158.
[2667] ADELAIDE, Register of ships subsequent to Merchant Shipping Act 1854, Volume 6, 1961-1973, Folio 159.
[2668] *Docklands News*, 29 June 2011 (www.docklandsnews.com.au/vale-lady-chelmsford_6608/).
[2669] *Australian Town and Country Journal*, 3 November 1909; SYDNEY, NEW SOUTH WALES, Register of ships, single number series (II), Volume 21, 1907-1909, Folio 121.
[2670] MELBOURNE, Main Register subsequent to Merchant Shipping Act 1854, Volume 10, 1924-1943, Folio 150.
[2671] SYDNEY, NEW SOUTH WALES, Register of ships, single number series (II), Volume 22, 1910-1912, Folio 24.
[2672] *The Sydney Morning Herald*, 8 January 1949.
[2673] SYDNEY, NEW SOUTH WALES, Register of ships, single number series (II), Volume 22, 1910-1912, Folio 21.
[2674] SYDNEY, NEW SOUTH WALES, Register of ships, single number series (II), Volume 22, 1910-1912, Folio 45.

Brisbane, Queensland, by submission of a tender to the Commonwealth Home Affairs Department.[2675] Completed early in 1911 and after trials was shipped to Port Moresby by SS *Mataram.*[2676] *Merrie England* stranded on a reef at the entrance to Port Moresby on 24 October 1912 and was ultimately abandoned as a total loss: the launch was presumably saved, and may have become part of the *Merrie England* (II)'s equipment.[2677] It then may have been one of the launches lost when that vessel was destroyed by fire in 1919. Plans held by the National Archives of Australia.

--

Year: 1911
Designer: Walter Reeks
Builder: Walter 'Watty' M. Ford, Jr (superintended by Walter Reeks)
Location: Berrys Bay, New South Wales, Australia
Type: Open boat
Description: Limited details known. Fitted with steam power capable of developing a speed of 10 knots.[2678] It was the largest open boat on Sydney Harbour at the time of launch in late 1911 and could carry 150 men.[2679] Very likely the *Potiki,* ON 136936, built by Ford in 1911 for the Union SS Co. of NZ Ltd. 65.8 x 14.3 x 5.8ft. 31/14 grt/net. First Reg. Sydney #71/1914. Compound steam engine by Chapman & Co. of Sydney, 10 nhp/50 ihp = 9 knots. Sold for demolition and register closed 1953.

WHITE ROSE

Year: 1911
Designer: Walter Reeks
Builder: David Drake
Location: Balmain, New South Wales, Australia
Type: Steamer
Description: Limited details known. Designed by Reeks for the Vacuum Oil Company to be used for towing barges and carrying cargo.[2680] Operated in conjunction with the company's harbour-based business activities and was capable of towing a barge laden with 800 cases of oil, itself able to carry 600 cases.[2681]

--

Year: 1911
Designer: Walter Reeks
Builder: --
Location: Sydney, New South Wales, Australia
Type: Steam lighter
Description: Limited details known. Second vessel designed by Reeks for the Vacuum Oil Company.[2682] Capable of carrying 2,400 cases of oil.[2683]

[2675] *The Telegraph*, 30 April, 2 May, 23 June 1910.
[2676] *The Telegraph*, 4 March 1911.
[2677] *The Argus*, 4 November 1912.
[2678] *Evening News*, 19 October 1911.
[2679] *Evening News*, 19 October 1911.
[2680] *Evening News*, 19 October 1911.
[2681] *Evening News*, 19 October 1911.
[2682] *Evening News*, 19 October 1911.
[2683] *Construction: Weekly Supplement to Building*, 30 October 1911.

SHONA

Year: 1911
Designer: Walter Reeks
Builder: William Holmes
Location: McMahons Point, New South Wales, Australia
Type: Motor yacht
Description: ON 131570. 31.0 x 11.0 x 4.8ft. 9/8 tons (gross/net). 4-cylinder engine by Wolsey Motor Co. of Birmingham (England), 12 b.h.p. = 7 knots, replaced in 1967 by 4-cylinder diesel by Perkins Engines Ltd., Peterborough (England), 47 b.h.p. = 8 knots. Built for John William Manson (first Reg. Sydney #60/1913).[2684] Later owned by Robert Lee Massie (1922); Cedric Malcomson Gains Douglas Boult (1928); George Byers Boulton (1930); William Augustus Darkey (1947); Blanch Pavasour Darkey (1953); Ernest James Theodore Mashman (1954); Cecil Roy Dodd (1957); Leo Christie (1964); Thomas Alan Oxford (1973); Clyde William Clinton (1976); John Stuart Hebden (1977); Peter Lansdell Paine (1981); and Clive Arthur Atkins (1981). To Australian registry 1986. Used as prop in the film *Nim's Island* in 2007/8 Still in existence, based in Queensland.

--

Year: 1911
Designer: Walter Reeks (possibly)
Builder: Arthur White
Location: Williamstown, Victoria, Australia
Type: Tender
Description: Not registered. 22ft tender possibly designed by Reeks for use in conjunction with the steel steam trawler/research ship *Endeavour* (see page 337). Built at Williamstown in early 1911 of kauri at a cost of £350, which caused some alarm in the press.[2685] Fitted with an Invincible engine. Likely lost with the vessel in late 1914.

LADY DENMAN

Year: 1912
Designer: Walter Reeks
Builder: Joseph Dent
Location: Jervis Bay, New South Wales, Australia
Type: Steam ferry
Description: ON 131510. 112 (oa) x 110.4 x 25.0 x 9.0ft. 98/67 tons (gross/net). Remeasured 1958 as 96/65 tons (gross/net). Compound steam engine by Chapman & Co., Sydney (New South Wales) made in 1912, 38 n.h.p., 230 i.h.p. = 10 knots; in 1936 6-cylinder diesel made by L. Gardner & Sons Ltd. of Manchester (England), 228 b.h.p. = 10 ¼ knots; in 1962 4-cylinder diesel by Crossley Bros. Ltd. of Openshaw, Manchester (England), 300 b.h.p. = 10½ knots. Built to the order of the Balmain New Ferry Co. Ltd. (Reg. Sydney #31/1912).[2686] Sold in 1918

[2684] SYDNEY, NEW SOUTH WALES, Register of ships, single number series (II), Volume 23, 1912-1914, Folio 80.
[2685] *Geelong Advertiser*, 8 February 1911.
[2686] SYDNEY, NEW SOUTH WALES, Register of ships, single number series (II), Volume 22, 1910-1912, Folio 110.

to Sydney Ferries Ltd. Register closed 30 November 1932 as no longer required. Returned to service as a motor ferry (reregistered Sydney #1/1937).[2687] Sold to Harbour Lighterage and Showboat Ltd. in April 1952. Register closed September 1975 as no longer required, vessel having been sold to the Government (Public Transport Commission). Taken out of service in 1979, donated to a preservation group (Jervis Bay Maritime Museum and Gallery), based at Huskisson, Jervis Bay, New South Wales, where it was originally built. Has since been preserved ashore in a purpose-built structure.[2688]

LANDSEER III

Year: 1912
Designer: Walter Reeks
Builder: Woodleys, Ltd.
Location: North Sydney, New South Wales, Australia
Type: Auxiliary yacht
Description: ON 131536. 42 (oa) x 35.5 x 11.2 x 5.8ft, draft 6ft. 12/10 tons (gross/net). 2-cylinder oil engine by Standard Gas Engine Co. of San Francisco, California (USA), 10 b.h.p. = 7 knots. Similar in design to *Arethusa*, though not the same dimensions. Built of spotted gum with kauri topsides and decks and maple and Indian teak cabin fittings for Albyn Athol Stewart (reg. Sydney #64/1912), launched 12 October 1912.[2689] Later owned by Arthur James Hanson (1917); George Arthur Robinson (1917); Lewis James Davies (1920); David Jaffa (1937); Frederick and Gwenyth Teiffel (1965). Register closed 1968 as sold to an American citizen. Returned to Australian registry in 1981 (Sydney #122/1981).[2690] Restored by Kelly Holder and sold 2002 or thereabouts. Still in existence, based in Sydney.

LANDSEER III AT REFUGE BAY, HAWKESBURY RIVER.
SOURCE: CHERRY JACKAMAN.

--

Year: 1912
Designer: Walter Reeks
Builder: --
Location: Sydney, New South Wales, Australia
Type: Steel barge
Description: Not registered. First of two 150-ton barges built for Elliott Bros. Ltd. of Sydney. Noted as nearing completion in late March 1912 and complete a month later.[2691] History and fate not known.

--

Year: 1912
Designer: Walter Reeks
Builder: --
Location: Sydney, New South Wales, Australia
Type: Steel barge
Description: Not registered. Second of two 150-ton barges built for Elliott Bros. Ltd. of Sydney. Noted as complete in April 1912.[2692] History and fate not known.

EVANGELIST

Year: 1912
Designer: Walter Reeks
Builder: William Holmes
Location: McMahons Point, New South Wales, Australia
Type: Motor launch
Description: ON 131521. 45 x 10.3 x 4.6ft. 11/5 tons (gross/net). Cargo and passenger launch initially fitted with twin-screw Seal oil engines of 20 h.p. and fully equipped for long voyages, including passenger accommodation for six (Reg. Sydney #43/1912 to William James McGown).[2693] Designed with a view to providing the best possible ventilation for service in the tropics and launched in late August 1912, intending to make its first trip to the Solomon Island the following month. Due to engine failure, limped back to Sydney where two new Fairbanks engines were installed.[2694] Was noted en route to the Solomon Islands in January 1913. Subsequent history not known. Register closed in 1951 noting registered owned died in 1946, neither beneficial owner nor vessel could be traced.

MIRO II

Year: 1912
Designer: Walter Reeks
Builder: William Holmes
Location: McMahons Point, New South Wales, Australia
Type: Auxiliary ketch
Description: ON 131524. 51.5 x 15.5 x 6.35ft. 27/14

[2687] SYDNEY, NEW SOUTH WALES, Register of ships, single number series (II), Volume 27, 1928-1937, Folio 156.
[2688] https://en.wikipedia.org/wiki/MV_Lady_Denman.
[2689] SYDNEY, NEW SOUTH WALES, Register of ships, single number series (II), Volume 23, 1912-1914, Folio 15.
[2690] SYDNEY, NEW SOUTH WALES, Register of ships, single number series (II), Volume 39, 1980-1982, Folio 126.
[2691] *Construction: Weekly Supplement*, 25 March, 29 April 1912.
[2692] *Construction: Weekly Supplement*, 25 March, 29 April 1912.
[2693] SYDNEY, NEW SOUTH WALES, Register of ships, single number series (II), Volume 11, 1955-1982, ON 135521; *Construction: Weekly Supplement to Building*, 25 March, 29 April 1912; *The Daily Telegraph*, 13 September 1912.
[2694] SYDNEY, NEW SOUTH WALES, Register of ships, single number series (II), Volume 11, 1955-1982, ON 135521; *The Daily Telegraph*, 13 September 1912; *The Sydney Morning Herald*, 22 October 1912.

tons (gross/net). 3-cylinder oil engine by Wilsons Engineering Co. of Brisbane (Queensland), 30 b.h.p. = 7 knots. Replaced by 1937 with 3-cylinder diesel by Deutz Motorrenen Gesselschaft of Hamburg (Germany), 40/44 b.h.p. = 7 knots. Built to the order of Julius Oien of Faisi, Solomon Islands (registered Sydney #46/1912).[2695] Oien died 1921 and vessel passed via executor to Levers Pacific Plantations Ltd. In 1937 owned by Bruce McNeil Clark of Sydney; in 1938 by Edward Edison Waters (who used the vessel as a yacht); in 1939 by William Home Cameron of Mackay, Queensland; and in 1940 by Robert Bruce Jamieson of Proserpine. Register closed 12 November 1941 as no longer required. Vessel believed taken over by Commonwealth of Australia during Second World War.

RED PINE

Year: 1912
Designer: Walter Reeks
Builder: Rock Davis (Jr)
Location: Brisbane Water, New South Wales, Australia
Type: Steamship
Description: ON 131526. 153 (oa) x 146.6 x 30.1 x 10.05ft. 364/181 tons (gross/net). Ross & Duncan (Glasgow, Scotland,) compound steam engine, 45 n.h.p., 380 i.h.p. = 9 knots, made 1911. Originally owned by David Mills (Reg. Sydney #48/1912) and transferred 19 October 1912 to the Pine Steamship Co. of Westland, Ltd.. of Greymouth, New Zealand.[2696] When launched on 26 April 1912, stated to be the largest wooden steamer built in New South Wales. Built for the New Zealand coastal timber trade out of Greymouth.[2697] Wrecked at Cape Stephens, D'Urville Islands, New Zealand, on 7 March 1913 and register closed.[2698]

CAPELLA EX MISTRAL

Year: 1912
Designer: Walter Reeks
Builder: Walter 'Watty' M. Ford, Jr
Location: Berrys Bay, New South Wales, Australia
Type: Auxiliary ketch
Description: ON 131573. 58 (oa) x 52.5 x 13.5 x 6.9ft. 24/22 tons (gross/net). 2-cylinder oil engine by Jersey City Motor Construction Co. of Jersey City (USA), 18 b.h.p. = 6 knots. Built for Edward Percy Simpson (reg. Sydney #67/1913).[2699] Sold in 1923 to James March Hardie, renamed *Capella* in 1924. Stolen from moorings at Rushcutters Bay on the night of 23 December 1925 and wrecked at North Head a few hours later. Salvage, including the engine, was auctioned on 6 January 1926.[2700]

[2695] *The Sydney Morning Herald*, 11 November 1912; SYDNEY, NEW SOUTH WALES, Register of ships, single number series (II), Volume 22, 1910-1912, Folio 125.
[2696] SYDNEY, NEW SOUTH WALES, Register of ships, single number series (II), Volume 22, 1910-1912, Folio 127.
[2697] *The Sydney Morning Herald*, 27 April 1912.
[2698] *Southland Times*, 10 March 1913.
[2699] SYDNEY, NEW SOUTH WALES, Register of ships, single number series (II), Volume 23, 1912-1914, Folio 87.
[2700] *The Sydney Morning Herald*, 26 December 1925; *The Daily Telegraph*, 2 January 1926.

EXCELLA

Year: 1912
Designer: Walter Reeks
Builder: John Dalgleish
Location: Battery Point, Tasmania, Australia
Type: Motor vessel
Description: ON 124553. 115 (oa) x 107.0 x 23.9 x 8.3ft. 174/132 tons (gross/net); in 1916 174/102. 2 x semi-diesels, each two cylinders, totalling 160 b.h.p. = 12 knots, made by Messrs Bolinders of Sweden in 1912; in 1916 compound direct acting steam engine, 15 n.h.p. = 9 knots, made by R. R. Smellie & Co., Brisbane (Queensland) in 1874; in 1923 25 n.h.p., made by Chapman & Co., Sydney (New South Wales); possibly replaced 1940s/50s by a Plenty & Sons set now in *SY Ena*, and may have been ex *Ena* in the first place. Designed by Reeks as a replacement for James Rowe & Sons' *Excelsior* and launched on 7 November 1912 as the *Excel*; registered as *Excella* #1/1913.[2701] Proved a popular and successful vessel of the Huon and D'Entrecasteaux Channel, based out of Hobart, also making several voyages to the mainland. Continued in the river trade under a succession of owners, gradually used less and less, until March 1965 when its boiler was condemned. Then removed to Prince of Wales Bay. Purchased by Jim Butterworth with the intention to restore it as a tourist vessel. However sank at its moorings on three occasions, the second time being in May 1976. On 17 August 1977 towed to Hobart's Domain slip and inspected for the first time since 1952. In 1982 purchased for $2,000 by John Jacob to be based at Kettering, preserved ashore with its superstructure refitted for use as an entertainment centre. Towed to Kettering by the *Cartela* and hauled into a trench dug alongside the marina. Unfortunately its floor was not level, and *Excella* broke its back fore and aft of the boiler. Damaged beyond repair, demolished early in 1989. Some timbers were reused in the brigantine *Windeward Bound.*

WYRUNA

Year: 1912
Designer: Walter Reeks
Builder: William Holmes
Location: McMahons Point, New South Wales, Australia
Type: Auxiliary cruiser
Description: Not registered. 36 x 10 x 5ft. 18 b.h.p. Thornycroft oil engine. Based on similar lines to *Shona* and *Mischief*, built for H. H. Dutton of South Australia.[2702] Advertised for sale in 1919 described as 12½ tons, 36ft 9in x 10ft 6in. x 5ft 3in., 20 h.p. 4-cylinder Thornycroft, 8 knots, Copper-plated to waterline. £800.[2703] Various later Adelaide-based owners include Dr E. Crozier (1923), C. Vincent (1933), J. B. W. Rischbieth (1937), and C. Houlgrave (1946; this may be a sailing boat). Still in

[2701] HOBART, Main Register (with continuation entries), Volume 12, 1911-1930, Folio 28.
[2702] *Construction: Weekly Supplement to Building*, 26 August 1912.
[2703] *The Australasian*, 20 December 1920.

existence though in poor condition. Based in Melbourne where it is currently being restored by Ferdy Darley.[2704]

WYRUNA (1960S).
SOURCE: WWW.WYRUNA.COM.

KAMIRI

Year: 1912
Designer: Walter Reeks
Builder: Messrs Morrison and Sinclair
Location: Pyrmont Balmain, New South Wales, Australia
Type: Steam ferry
Description: ON 131516. 112.0 x 25.8 x 10.9ft. 144/46 tons (gross/net). Triple expansion steam engine made by Campbell and Calderwood, Paisley (Scotland), 50 n.h.p., 370 i.h.p. = 12 knots. Built for Sydney Ferries Ltd. and launched in May 1912 (Reg. Sydney #38/1912).[2705] A replacement ferry for *Kaludah* (see entry, burnt in March 1911), and built on the same plans as this vessel and using its engines that had been salvaged and repaired.[2706] Serviced the Parramatta River run until 1946 when laid up. Broken up between 1951 and 1953.

KAMIRI AT MCMAHONS POINT (1950S).
SOURCE: GRAEME ANDREWS WORKING HARBOUR COLLECTION, CITY OF SYDNEY ARCHIVES.

--

Year: 1913
Designer: Walter Reeks
Builder: Morrison and Sinclair
Location: Balmain, New South Wales, Australia
Type: Steel barge
Description: Not registered. 75ft. First of two large towing punts Reeks designed for the Darling Island Stevedoring and Literage Co. Ltd.[2707]

--

Year: 1913
Designer: Walter Reeks
Builder: Morrison and Sinclair
Location: Balmain, New South Wales, Australia
Type: Steel barge
Description: Not registered. 75ft. Second of two large towing punts Reeks designed for the Darling Island Stevedoring and Literage Co. Ltd.[2708]

GERARDO EX KYLIE

Year: 1913
Designer: Walter Reeks
Builder: W. D. Lawrence
Location: Perth, Western Australia, Australia
Type: Motor scow/tug
Description: ON 131642. 56 (oa) x 46.0 x 12.7 x 3.33ft. 19/9 tons (gross/net). 4-cylinder oil engine by L. Gardner & Sons of Manchester (England), 75 b.h.p. Built for work on the Swan River for McIlwraith, McEacharn & Co. Pty. Ltd. (reg. Fremantle #2/1913), owners reorganised shortly afterwards as McIlwraith, McEacharn Line Pty. Ltd.[2709] Sold in 1919 to John Creagh of Broome, the Vicar Apostolic of Kimberly, for the Beagle Bay Mission, and renamed *Gerardo*.[2710] Refitted as an auxiliary ketch, now 51.9 x 12.3 x 5.8ft, 25/15 tons (gross/net). Later owned by Patrick Percy, pearler of Broome (1925); Alfred Chambers Morgan (1929); and Streeter & Male Ltd. (1934).[2711] Destroyed by the Royal Australian Navy at Broome in March 1942, in expectation of an imminent Japanese invasion, and register finally closed in 1967.

--

Year: 1913
Designer: Walter Reeks
Builder: Pritchard Bros. Ltd
Location: Careening Cove, New South Wales, Australia
Type: Auxiliary cruiser
Description: 41ft. Sea-going auxiliary cruiser built by Pritchard Bros. for Septimus Macpherson to a design by

[2704] https://www.wyruna.com/.
[2705] SYDNEY, NEW SOUTH WALES, Register of ships, single number series (II), Volume 22, 1910-1912, Folio 117; *The Daily Telegraph*, 8 May 1912.
[2706] *Construction: Weekly Supplement to Building*, 27 May 1912.
[2707] *Construction: Weekly Supplement to Building*, 26 August 1912; *The Daily Telegraph*, 31 August 1912.
[2708] *Construction: Weekly Supplement to Building*, 26 August 1912; *The Daily Telegraph*, 31 August 1912.
[2709] FREMANTLE, Main Register, Volume 5, 1907-1914, Folio 119.
[2710] *The Daily News*, 14 May 1919.
[2711] FREMANTLE, Main Register, Volume 6, 1914-1931, Folio 88.

Reeks. Completed towards the end of 1913, featured a high deck forward and large cockpit aft.[2712] Advertised for sale in early 1923.[2713] Additional details, including name, not known.

MERRIE ENGLAND

Year: 1913
Designer: Walter Reeks
Builder: --
Location: --
Type: Steel steamer
Description: 175ft. Steel steamer designed by Reeks as a replacement for the Papuan Government's earlier steamer of that name wrecked at Port Moresby.[2714] Project abandoned. Reeks later designed another vessel.

--

Year: 1913
Designer: Walter Reeks
Builder: --
Location: --
Type: Steel steamer
Description: 135ft. Plans completed by Reeks for a steel steamer to achieve 9 knots for the State Brickworks, to run between Sydney and Manning River.[2715] Not built.

--

Year: 1913
Designer: Walter Reeks
Builder: William Holmes
Location: McMahons Point, New South Wales, Australia
Type: Motor launch
Description: Not registered. 20 x 6 x 2.5ft. 3½ h.p. New Zealand-made Rotor engine. 6½ knots expected. Was to be Reeks' personal boat.[2716] Not known if built.

WOOLLAHRA

Year: 1913
Designer: Walter Reeks
Builder: Rock Davis (Jr)
Location: Brisbane Water, New South Wales, Australia
Type: Steam ferry
Description: ON 131577. 125.0 x 26.5 x 10.5ft. 152/104 tons (gross/net). Triple expansion steam engine by Campbell and Calderwood, Paisley (Scotland), 49 n.h.p., 350 i.h.p. = 12 knots. Built for the Watsons Bay and South Shore Ferry Co. Ltd. (reg. Sydney #72/1913 and launched 21 June 1913 as *Carrara*, a name that for some reason seems to have been found wanting by the Board of Trade as vessel was registered as *Woollahra*.[2717] Still being referred to as *Carrara* during trials in December immediately before registration.[2718] Sold 29 September 1920 to Sydney Ferries Ltd. Register closed 25 November 1932 as no longer required. Vessel remained laid up until sold to the Commonwealth Government c1941 and employed as a hulk. Advertised for sale at Sydney in August 1945, and presumably demolished soon afterwards.[2719]

ROSNY

Year: 1913
Designer: Walter Reeks
Builder: Frederick Moore
Location: Battery Point, Tasmania, Australia
Type: Screw ferry
Description: ON 133481. 110.0 x 24.0 x 11.0ft. Compound, 24 n.h.p. = 11 knots, made by Shanks & Sons, Arbroath (Scotland); in 1964 diesel, 8-cylinders, 25 n.h.p., 320 b.h.p. = 10 knots, made by the Bergius Kelvin Co. of Glasgow (Scotland) in 1963. Built to the order of the Rosny Estates & Ferry Co. Ltd. (Registered Hobart #10/1913), in effect a slightly enlarged and improved version of the *Derwent* which Reeks designed for the O'May Bros. in 1906.[2720] Launched by the wife of Commercial Bank of Tasmania manager David Barclay on 24 July 1913. Commenced running on the River Derwent between Hobart and Lindisfarne on 29 November 1913. Sold in 1921 to the Manager, Councillors and Citizens of the City of Williamstown and reregistered Melbourne (#4/1921).[2721] In 1931 sold to Albert Edgar John Morrison of Glen Iris, Victoria. Returned to Hobart and reregistered there on 11/4/1931 to O'May Bros Pty. Ltd.[2722] In 1939 sold to the Hobart Bridge Company Ltd. for £3,800; in 1947 to the The Transport Commission; in 1963 to Tamar Cruises Pty. Ltd. and transferred to Launceston. Extensively refitted at Devonport as a diesel-powered excursion vessel for use on the Tamar out of Launceston with accommodation for 350 passengers. Owing to financial difficulties, made its final trip on 28 December 1964. Advertised for sale on 31 August 1966 by Alfred Harrap & Son Pty. Ltd. then put up for auction. Sold to local resident Ernie Park and his wife for just $750, intending to refit it as a floating restaurant, though plans were not approved. In May 1967 towed by the Marine Board's tug *Inveresk* to Kayena, Devil's Elbow, near the Batman Bridge, where it was moored. Sank less than three months later with just the top of its superstructure showing above the water. No doubt its remains are still there.

[2712] *Weekly Supplement to Building*, 28 January 1913; *The Sun*, 3 September 1913.
[2713] *The Sydney Morning Herald*, 20 January 1923.
[2714] *Papuan Courier*, 26 October 1917.
[2715] *The Sydney Morning Herald*, 4 September 1913.
[2716] *The Sun*, 3 September 1913.
[2717] *The Sydney Morning Herald*, 23 June 1913; SYDNEY, NEW SOUTH WALES, Register of ships, single number series (II), Volume 23, 1912-1914, Folio 92.
[2718] *The Sydney Morning Herald*, 12 December 1913.
[2719] *The Examiner*, 18 August 1945.
[2720] HOBART, Main Register (with continuation entries), Volume 12, 1911-1930, Folio 39.
[2721] MELBOURNE, Main Register subsequent to Merchant Shipping Act 1854, Volume 9, 1911-1924, Folio 160.
[2722] HOBART, Main Register (with continuation entries), Volume 13, 1931-1964, Folio 4, 192.

LADY EDELINE

Year: 1913
Designer: Walter Reeks
Builder: George A. Washington
Location: Balmain, New South Wales, Australia
Type: Steam ferry
Description: ON 131586. 112 (oa) x 110.7 x 24.3 x 9.0ft. 96/65 tons (gross/net). Not remeasured on conversion to a motor vessel. Compound steam engine by James Pollock of Glasgow (Scotland) in 1886, 28 n.h.p., 210 i.h.p. = 11 knots; in 1937 5-cylinder diesel made by L. Gardner & Sons Ltd. of Manchester (England), 228 b.h.p. = 10¼ knots; in 1963 4-cylinder diesel by Crossley Bros. Ltd. of Openshaw Manchester (England), 40 n.h.p., 300 b.h.p. = 10½ knots. Built to the order of the Balmain New Ferry Co. Ltd. (Reg. Sydney #10/1914), launched 18 October 1913.[2723] Sold March 1918 to Sydney Ferries Ltd. Register closed 25 November 1932 as no longer required. Returned to service as a motor ferry and reregistered Sydney #4/1937.[2724] Owners restructured as Harbour Lighterage and Showboat Ltd. in October 1951; in April 1952 sold to The Sydney Harbour Transport Board. Register closed 1975 as no longer required (owners reorganised as Public Transport Commission, from 1980 as the Public Transit Authority). Taken out of service in 1984, the longest serving of the 'Lady' class with more than 70 years of operation. Sank at Mortlake in 1988 and remains reportedly still visible there.

ELEVALA

Year: 1913
Designer: Walter Reeks
Builder: William Holmes
Location: McMahons Point, New South Wales, Australia
Type: Auxiliary ketch
Description: Not British Registered. 69 (oa) x 15 x 7ft loaded draft. 35 tons. 50 h.p. Standard oil engine = 8 knots.[2725] Built as a stop-gap measure till the replacement for the wrecked steamer *Merrie England* could be built. Launched at Sydney 6 November 1913.[2726] Left Sydney on its delivery voyage to Port Moresby on 24 December 1913.[2727] Despite an aborted attempt to sell the vessel in 1924, spent the rest of its career working gainfully for the Papuan administration until the Japanese invasion of 1942. Beached at Yalua Creek during the Battle of Milne Bay on 25 August 1942 and became a total loss.[2728]

[2723] SYDNEY, NEW SOUTH WALES, Register of ships, single number series (II), Volume 23, 1912-1914, Folio 103.
[2724] SYDNEY, NEW SOUTH WALES, Register of ships, single number series (II), Volume 27, 1928-1937, Folio 159.
[2725] *Territory of Papua Government Gazette*, 7 May 1924.
[2726] *The Sydney Morning Herald*, 7 November 1914.
[2727] *The Week*, 2 January 1914.
[2728] *Townsville Daily Bulletin*, 5 October 1946.

HARBOUR QUEEN EX JOHN CADMAN EX AQUA QUEEN EX LADY SCOTT

Year: 1913
Designer: Walter Reeks
Builder: Joseph Dent
Location: Jervis Bay, New South Wales, Australia
Type: Steam ferry
Description: ON 136379. 112 (oa) x 110.4 x 24.4 x 9.0ft. 95/65 tons (gross/net). Remeasured as 96/64 tons (gross/net) in 1953. Compound steam engine by McKie and Baxter, Govan (Scotland) in 1913, 24 n.h.p., 160 i.h.p. = 11½ knots; in 1937 diesel made 1936 by Henty & Gardner Ltd. of Manchester (England); in 1963 6-cylinder diesel by Crossley Bros. Ltd. of Openshaw, Manchester (England), 65 n.h.p. = 9 knots. c1974 refitted with 12-cylinder diesel engine made by Detroit Diesel Engine Division, Detroit, Michigan (USA) in 1968, 19.5 n.h.p., 550 b.h.p. = 10 knots. Originally owned by the Balmain New Ferry Co. Ltd. (Reg. Sydney #46/1914), launched 18 October 1913 '*practically a sister to the s.s. Lady Denman and Lady Chelmsford*'.[2729] Sold in March 1918 to Sydney Ferries Ltd. Register closed 30 November 1932 as no longer required. Returned to service as a motor ferry and reregistered Sydney #5/1937.[2730] Owners restructured as Harbour Lighterage and Showboat Ltd. in October 1951. In April 1952 sold to the Sydney Harbour Transport Board; in 1969 to John McCulloch; in 1970 to Lady Scott Cruises Pty. Ltd.; in 1972 to Aquatours Pty. Ltd. (renamed in 1974 to *Aqua Queen* and in 1975 to *John Cadman*). To Australian registry in 1983. Later renamed *Harbour Queen*. Sank at Blackwattle Bay in 2012 and demolished *in situ*.

GYPSY

Year: 1914
Designer: Walter Reeks
Builder: William Ball and Wyn Tatnall
Location: Norfolk Bay, Tasmania, Australia
Type: Auxiliary yacht
Description: Not registered. 36 (oa) x 30 (wl) x 10.5 x 4.5ft. Believed to have been designed by well-known Hobart yacht designer Alfred Blore, who apparently scaled down a plan for the 41ft cruising yawl *Arethusa* designed by Reeks that was published in *Rudder* magazine in 1911, and also included a centreboard that was not a feature of the original design. Construction of Huon pine planking on a combination of Huon pine ribs and grown hardwood frames. Launched in 1914, was not completed until after the war, when sold to brothers John and Sydney Knight. Still in the Knight family's possession, considered by many to be the most beautiful vintage yacht in Hobart.

[2729] Sunday Times, 19 October 1913; SYDNEY, NEW SOUTH WALES, Register of ships, single number series (II), Volume 24, 1914-1919, Folio 11.
[2730] SYDNEY, NEW SOUTH WALES, Register of ships, single number series (II), Volume 27, 1928-1937, Folio 160; SYDNEY, NEW SOUTH WALES, Register of ships, single number series (II), Volume 35, 1973-1975, Folio 112.

GYPSY OFF KETTERING, TASMANIA (2016).
SOURCE: COLIN GRAZULES.

LILIHINA

Year: 1914
Designer: Walter Reeks
Builder: William Holmes
Location: McMahons Point, New South Wales, Australia
Type: Auxiliary schooner
Description: ON136385. 47.6 x 14.5 x 6.45ft. 24 h.p. 2-cylinder oil engine by Forbinia (?) Engine Co. of London (England). Built to the order of Fred Erikson for use in the Solomons Islands (Reg. Sydney #56/1914).[2731] Left Sydney on its maiden voyage to Tulagi with Captain Cunninghame at the helm in late July 1914.[2732] Later owned by John Stephen and Eric Ord Mackenzie of British Solomon Islands (1917); John Stephen of Solomon Islands (1920); and Theodore John Marks though mortgaged with previous owner (1920). Stranded on the east side of Kounusugu Cove in the Solomon Islands on 10 February 1923 and became a total wreck, register closed June 1924.

LADY FERGUSON

Year: 1914
Designer: Walter Reeks
Builder: David Drake, Ltd.
Location: Balmain, New South Wales, Australia
Type: Steam ferry
Description: ON 136409. 110.0 x 24.3 x 9.0ft. Compound, 24 n.h.p., 160 b.h.p. = 11¼ knots, made by M'Kie & Baxter, Govan, (Scotland); in 1937 diesel, 6-cylinder, 228 b.h.p. = 9 knots, made by Henty & Gardner in 1936, not noted on register until 1951; in 1959 diesel, 4-cylinder, 36.69 i.h.p., 300 b.h.p. = 10 knots, made by Crossley Bros. Ltd., Manchester (England) in 1959. The last of five identical sister ferries designed by Reeks and built to the order of the Balmain New Ferry Company. Licenced to carry 470 passengers. On 31 October 1917 the assets of the Balmain New Ferry Company were taken over by Sydney Ferries Ltd., who paid £7,461 for *Lady Ferguson*. First registered at Sydney on 18/3/1918.[2733] In 1951 transferred to Harbour Lighterage and Showboat Ltd.; in 1952 to the Sydney Harbour Transport Board, register closed 5/9/1975, as no longer required. Enjoyed more than sixty years of useful service on Sydney Harbour. Then purchased by the Government of Tasmania for the trans-Derwent ferry service, owing to the Tasman Bridge disaster, arriving on 3 May 1975 after a lengthy tow by the tug *Tarpan*. Never made a commercial run on arriving in Hobart, being considered in too poor a condition to be economically repaired and soon broken up.

Year: 1917
Designer: Walter Reeks
Builder: David Drake, Ltd.
Location: Balmain, New South Wales, Australia
Type: Steel barge

[2731] SYDNEY, NEW SOUTH WALES, Register of ships, single number series (II), Volume 24, 1914-1919, Folio 21.
[2732] *Daily Commercial News and Shipping List*, 1 August 1914.
[2733] SYDNEY, NEW SOUTH WALES, Register of ships, single number series (II), Volume 24, 1914-1919, Folio 39; SYDNEY, NEW SOUTH WALES, Register of ships, single number series (II), Volume 27, 1928-1937, Folio 161.

Description: Not registered. First of two 175-ton barges built for Elliott Bros. Ltd. of Sydney. Completed in 1917.[2734]

--

Year: 1917
Designer: Walter Reeks
Builder: Arthur Davis
Location: Brisbane, Queensland, Australia
Type: Steel barge
Description: Not registered. Second of two 175-ton barges built for Elliott Bros. Ltd. of Sydney and completed in 1917.[2735]

--

Year: 1917
Designer: Walter Reeks
Builder: --
Location: Australia
Type: Tug
Description: Tug designed by Reeks with a range of 2,000 miles, yet capable of passing under a bridge 15ft high by the simple process of lowering the funnel, every other part being kept below that limit.[2736] No additional details known about this craft.

MERRIE ENGLAND

Year: 1917
Designer: Walter Reeks
Builder: Messrs Fisher and Fisher
Location: Careening Cove, New South Wales, Australia
Type: Motor vessel
Description: Not registered. 113 x 25 x 9ft. Approx. 200 tons. 2 x 80 b.h.p. Union oil engines. A replacement for the Papuan Government's earlier steamer of that name wrecked at Port Moresby: a much smaller vessel than the 175ft steel steamer that Reeks had originally drawn up in 1913 but that project had been abandoned. Construction began in October 1916.[2737] Launched on 11 September 1917.[2738] The vessel was totally destroyed by fire at Port Moresby on 17 March 1919 after a benzene tank exploded while being repaired with a blow lamp, killing the engineer doing the work. The vessel was sunk by explosives to prevent its total destruction.[2739] There was an extensive court case, the Territory of Papua vs the British New Guinea Development Co. Ltd. over the latter's negligent employee causing the explosion, during which considerable aspersions were cast against Reeks' competence in designing and supervising construction. Ultimately the owners were awarded £11,016 10/- compensation.[2740]

SIR JOHN FORREST

Year: 1918
Designer: Walter Reeks
Builder: George A. Washington (superintended by Walter Reeks)
Location: Balmain, New South Wales, Australia

[2734] *The Daily Telegraph*, 7 February 1917.
[2735] *The Daily Telegraph*, 7 February 1917.
[2736] *The Daily Telegraph*, 7 February 1917.
[2737] *Papuan Courier*, 26 October 1917.
[2738] *Evening News*, 13 September 1917.
[2739] *The Sydney Morning Herald*, 19 March 1919.
[2740] *Papuan Courier*, various dates March-April 1920; *The Sydney Morning Herald*, 26 March 1920.

SIR JOHN FORREST
(WITH ONE OF REEKS 'LADY CLASS' TWIN FUNNEL FERRIES).
SOURCE: MORI FLAPAN

Type: Motor ketch
Description: ON 136437. 117.5 x 26.1 x 5.05ft, 4.5ft draft with 100 tons of cargo. 160/122 tons (gross/net). 2 x oil engines by Skandia Engine Co. of Sweden, 200 b.h.p. = 8 knots. Commissioned by the Government of the Commonwealth of Australia (Reg. Sydney #3/1918), intended for service on the rivers of the Northern Territory, and sold on 20 October 1920 to W. R. Carpenter & Co. Ltd.[2741] Reregistered at Suva, Fiji, in 1921. Noted still in service, owned by the Fiji Shipping Co. Ltd. of Suva in 1930 and reported struck by lightning in a cyclone the following year.[2742]

COLLIBOI

Year: 1918
Designer: Walter Reeks
Builder: Morrison and Sinclair
Location: Balmain, New South Wales, Australia
Type: Refrigerated fish carrier, later coastal cargo steamer
Description: ON 101798. 133 (oa) x 130.6 x 25.8 x 9.4ft, 7.5ft loaded draft. 309/144 tons (gross/net). 2 x Compound steam engines made by the Government Dock Yard, Walsh Island (New South Wales) in 1918, 39½ n.h.p., 237 i.h.p. = 9 knots. Designed as a twin-screw motor fish transport associated with the Government of New South Wales' trawler fleet, and launched on 23 February 1917.[2743] The planned twin-Skandia diesel power plant never became available and the vessel was instead completed as a steamship. Mounting financial losses saw it sold in July 1921 to the Tasmanian Government for £10,500 (half of its construction cost) to run an improved service to the Bass Straits Islands against the established operators William Holyman & Sons. First registered in Launceston #2/1922.[2744] Ultimately the Government found mounting losses politically unpalatable and sold the ship to Holymans on the condition that it remained in the trade. Wrecked while doing so on the north-west coast of Tasmania (Woolnorth Point) on 29 August 1932 and register closed 19 March 1933.

SEA AIR EX BANKSIA

Year: 1918
Designer: Walter Reeks
Builder: Oliver Richard Jones
Location: Oakland, New South Wales, Australia
Type: River steamer, completed (?) as motor ferry
Description: Unregistered. 91 (oa) x 18.5 x 8ft, 5ft draft. Designed for triple expansion engines 425 i.h.p. by Simpson, Strickland & Co. = 17 knots. Designed in mid-1914 as a very high-speed passenger and light cargo steamer for a daily Ballina-Lismore run in the Richmond River trade, for Davis Bros. and Burgess. A very modern-looking vessel with a cruiser stern. Unfortunately the outbreak of World War I prevented the engines from being delivered and progress on the hull proceeded slowly, culminating in the vessel's launch on 21 December 1918.[2745] The original owners never completed the vessel, which remained part of the landscape at Coraki and later Ballina until 1937 when sold to Richard Watt of Brisbane, and after refitting with a small engine for its delivery voyage, relaunched in July 1938 with the name *Sea Air*.[2746] No further trace.

BAR-EA-MUL

Year: 1918
Designer: Walter Reeks
Builder: Morrison and Sinclair, Ltd.
Location: Balmain, New South Wales, Australia
Type: Auxiliary steam trawler
Description: ON 152011. 127.8 x 23.4 x 13.2ft. 237/85 tons (gross/net). Triple expansion steam engine made by the New South Wales Government Dockyard, Newcastle (New South Wales) in 1918, 51 n.h.p., 400 i.h.p. = 10 knots. Built for the Government of New South Wales as a trawler but sold before launching (which was on 23 August 1918) to the Government of Queensland, to be used for similar work.[2747] Sold in 1925 to Robert William Miller (first Reg. Sydney #15/1925).[2748] Sold on 28 July 1925 to Douglas Paul Hann; on 18 March 1926 to the Red Funnel Fisheries Ltd.; and on 19 October 1933 to the Red Funnel Trawlers Ltd. In operation until March 1948.[2749] Register closed 1 September 1948, vessel dismantled. The hull was scuttled in December 1950.[2750]

SEA SCOUT

Year: 1923
Designer: Walter Reeks
Builder: Morrison and Sinclair, Ltd.
Location: Balmain, New South Wales, Australia
Type: Auxiliary ketch
Description: ON 151986. 52 (oa) x 50.0 x 47 (wl) x 13.8 x 6.5ft. 28.10/19.10 tons (gross/net). 6-cylinder diesel, 5.4 n.h.p./72 b.h.p. = 8 ¼ knots, made by Gardner & Sons, Manchester (England) in 1955. First registered Sydney #34/1923 as 26.09/22.03 grt/net, 49.5 x 13.8 x 6.5ft.[2751] Originally owned by Henry Maxfield Shelley who died on 26/6/1923 and after probate and trusteeship sold 11/5/1933 to Robert Wild. Additional owners were 12/4/1940 Robert Thirlaway; 10/12/1941 Robert Wild; 23/8/1944 Robert Clive Bell, Thomas Joseph Bell, William Myers and John Symonds, fishermen of Eden, New South Wales. Register closed in 1950 as no longer required, then reregistered at Melbourne as #9/1957,

2741 *The Daily Telegraph*, 19 October 1917; SYDNEY, NEW SOUTH WALES, Register of ships, single number series (II), Volume 24, 1914-1919, Folio 99.
2742 *Cairns Post*, 5 March 1931.
2743 *The Daily Telegraph*, 7 February 1917; *The Sun*, 23 February 1917.
2744 LAUNCESTON, TASMANIA, Register of British ships, Volume 5, 1870-1930, Folio 328.
2745 *Richmond River Herald*, 26 May 1914, 10 February 1915, 24 December 1918.
2746 Sunday Mail, 31 July 1938.
2747 *The Sun*, 25 August 1918; *The Brisbane Courier*, 1 April 1919.
2748 SYDNEY, NEW SOUTH WALES, Register of ships, single number series (II), Volume 26, 1923-1928, Folio 51.
2749 *The Sydney Morning Herald*, 20 March 1948.
2750 *The Sun*, 19 December 1950.
2751 SYDNEY, NEW SOUTH WALES, Register of ships, single number series (II), Volume 26, 1923-1928, Folio 8.

SEA SCOUT AT TOWNSVILLE, QUEENSLAND (2018).
SOURCE: COLIN GRAZULES.

owned by Clarence Albert Noden, fisherman of Lakes Entrance.[2752] Later owned by William Harold Emile Bolleman of Paynesville (1963); Norman Irving East of Newhaven (1968); and Alan Richard Mears and Yvonne Ann Mears of Newhaven (1979). Register closed 1986. During the 1990s was in Tasmania under restoration at Kettering. In 2018 reported near Townsville, Queensland.

--

Year: 1923
Designer: Walter Reeks
Builder: --
Location: --
Type: Steam vehicular ferry
Description: Approx. 133 x 36 x 8.5ft. To be fitted with a triple-expansion steam engine, speed about 8 knots. Reeks visited Hobart in June 1923 for discussions with the Tasmanian Government over the building of a timber double-ended vehicular ferry for the Hobart-Bellerive service, to replace the *Kangaroo* built in 1855. The proposal was later abandoned for a steel ferry to be built in Britain and ultimately delivered as *Lurgurena* in 1925.[2753]

--

Year: 1923
Designer: Walter Reeks
Builder: --
Location: --
Type: Schooner
Description: 80ft. Designed by Reeks for Arthur H. Davies. Plan no. 303. Similar to *Boomerang* ex *Bona*. For reasons not known, not built.

[2752] MELBOURNE, Main Register subsequent to Merchant Shipping Act 1854, Volume 11, 1944-1961, Folio 148.
[2753] *World*, 25 June 1925.

THE RIP

Year: --
Designer: Walter Reeks
Builder: Pritchard Bros.
Location: North Sydney, New South Wales, Australia
Type: Motor boat
Description: Not registered. 22 x 6ft. Fitted with a 6 h.p. De Dion engine. Clinker-built of cedar. Advertised for sale at Alligator Creek, Cairns, Queensland, in November 1925.[2754]

DUCKWING

Year: 1925
Designer: Walter Reeks
Builder: William Holmes
Location: McMahons Point, New South Wales, Australia
Type: Motor yacht
Description: ON 152020. 45.4 x 12.575 x 5.6ft. 16/14 tons (gross/net). 2 x 4-cylinder oil engine by Buffalo Gasoline Motor Co. of Buffalo (USA), 60 b.h.p. = 10 knots. Built for Edgar Martin Perdriau (Reg. Sydney #1/1926).[2755] Register closed 1937, as no longer required. Most likely the 48ft vessel of that name used by Zane Grey in the filming of *White Death*. Sprang a leak and beached near Tyron Island off Gladstone, Queensland, in July 1949.[2756] No report of success of salvage attempts.

SAN PAN

Year: 1925
Designer: Walter Reeks
Builder: Pritchard Bros.
Location: North Sydney, New South Wales, Australia
Type: Yacht
Description: 21ft restricted class yacht, 25ft (overall). An experimental craft often referred to as 'The Freak', designed by Reeks for his own use, built by Pritchard Bros. of Careening Cove and launched in January 1925.[2757] Stated to be of a design entirely new to the class and the country, with its mast stepped in the centre of the boat and two booms carried.[2758] Did not race with the 21ft restricted class. Sold by Reeks' estate following his death in October 1925. By 1930 converted to a keeled racing boat and competed with the Sydney Amateur Sailing Club.[2759] Famously wrecked on Big Ben Reef just outside of Newcastle Harbour in April 1936 with its crew picked up by the pilot steamer *Birubi*.[2760] Was returning to Sydney from Lake Macquarie after winning the Sydney to Lake Ocean Race.

[2754] *Cairns Post*, 17, 19 November 1925.
[2755] SYDNEY, NEW SOUTH WALES, Register of ships, single number series (II), Volume 26, 1923-1928, Folio 64.
[2756] *The Manning River Times*, 6 August 1941, *Brisbane Telegraph*, 25 July 1949.
[2757] *Evening News*, 16 February 1925.
[2758] *Evening News*, 16 February 1925; *Sydney Mail*, 25 February 1925.
[2759] *Truth*, 16 February 1930; *The Sydney Morning Herald*, 14 November 1930.
[2760] *The Sydney Morning Herald*, 20 April 1936.

CHAPTER TWENTY THREE - PLANS AND DRAWINGS, PROJECT NUMBERS AND VESSELS STILL IN EXISTENCE

PLANS AND DRAWINGS

Given the size of Walter Reeks' output, it is disappointing that only a modest number of original plans exist for the craft he designed, and even then, often only as blueprint plan copies. These plans can be found in collections at public institutions, as part of private collections and within publications in the period the vessels were built. For example, the Sydney Heritage Fleet has Sydney Harbour ferry blueprints and copies donated by Bill Allen, along with some of the plans for *Boomerang* ex *Bona*, while the State Library of New South Wales has four plans for the big schooner, Project No. 303, that Reeks designed for A. H. Davies, plus two drawings of Project No. 207, the lighter Reeks designed for the Colonial Oil Company now known as *Pegasus*.

The National Archives of Australia has a plan for the MV *Merrie England,* launched in 1917, along with its power launch, plus an almost complete set for the tender for the federal trawler FRV *Endeavour*. The Australian National Maritime Museum holds plans for the lugger *Gondolier* and its sisters and another unnamed pearling lugger by Reeks, along with a couple of the Sydney Harbour ferry plans, and two cross-sections for SY *Ena*.

There are a number of Reeks plans of harbour ferries, along with some cross-sections of *Lotus II* held by private collectors. One interesting plan is the lines for *Lady Manning*, drawn up by its builder James Scott.

Magazines also published views taken from the full plans, particularly in articles describing specific vessels. For example, *Rudder* published drawings of SY *Ena*, *Boomerang* (ex *Bona)* and *Arethusa* in the early 1900s. Another American magazine published drawings of MV *Bronzewing* in 1907, ahead of its launch the following year.

Newspapers of the time provide a couple of profile drawings as well. Just how they were prepared is not known as they appear to be in scale and proportion but are drawn in an illustrative style.

In addition, Reeks' papers on the Sydney Harbour double-ended ferries included diagrams of the ferries drawn, it is assumed, by him. Copies of some of these, drawn on linen in ink but unsigned, are held by co-author David Payne, gifted to him by Graeme Andrews. These may be the only extant original drawings by Reeks, otherwise it is a tracing by someone to enable the diagrams to be copied for the publication.

Many of the complete plans still in existence have Reeks' signature and a date along with a title block and project number. We know Reeks had assistants as draughtsmen; early in his career (in 1888) one of them was Ernest Thompson, an ex naval architecture pupil of Reeks'. However, all the original plans we have located have the same style, and they cover almost all of his career. His signature and date, often bottom right on the plan, confirm he signed off on the drawings, and without further information to work with, their consistent style over this long period suggests they are all his drawings.

They are particularly neat and precise, drawn to correct drafting standards of the period, and fit that description of being able to stand as a work of art by themselves. Traditional Imperial scales are used, dimensions, scantlings, notes and text are clearly written. A couple of plans are midsections which carry all the essential information about the structure of the vessel.

Models from the period are equally scarce and more difficult to provenance. The Powerhouse Museum has two half models of the ferries *Lady Manning* and *Lady*

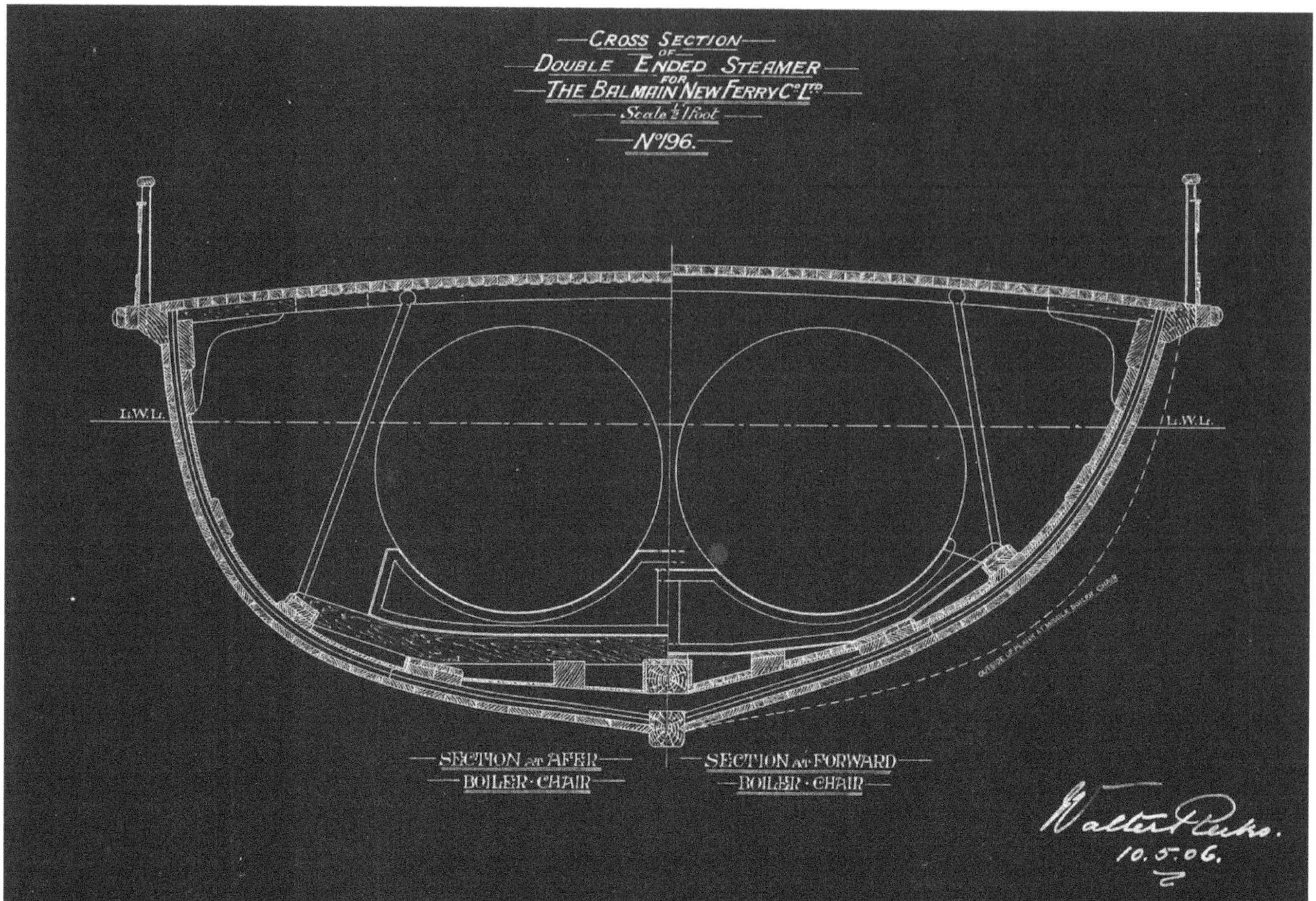

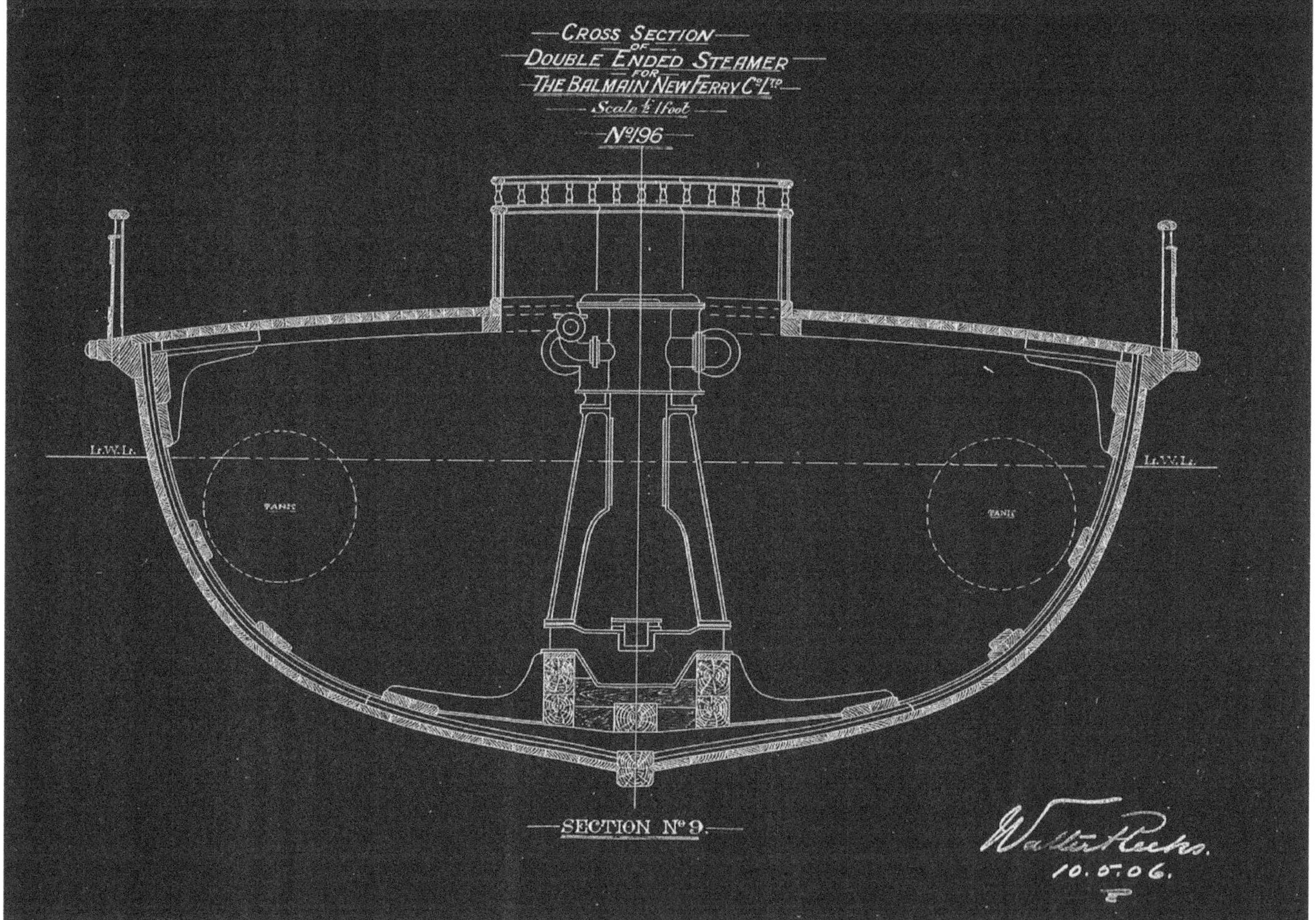

PLANS FOR PROJECT NO. 196, DOUBLE-ENDED STEAMER. NOT BUILT.
SOURCE: BILL ALLEN COLLECTION, SYDNEY HERITAGE FLEET.

Hampden, and it appears conclusive they are builders half models and both made by their builder James Scott or a tradesman from his workforce. The Royal Sydney Yacht Squadron has half models of *Era, Volunteer, Thelma, Miranda, Varuna* and one that is possibly *Iduna* but their origin is not recorded. They all give the appearance of being a builder's model for the owner or even for Reeks himself.

In recent times plans have been drawn and models made of his craft. The earliest of these is a set of plans for SS *Kuring-gai,* held by the National Archives of Australia, and drawn in the 1930s for docking at Cockatoo Island Dockyard by a draughtsman from that yard.

Reeks also provided detailed specifications as a written document. We are lucky to be able to have an excellent contrast working from an example for the yacht *Isea* in 1888 and then, from the end of his career, a set for the Arthur Davies schooner Project 303 in 1923.

Isea's specifications are handwritten, while Project 303 is typed. *Isea*'s run to about 8 pages, while Project 303 is a hefty 14 pages of detail, five covering the cabin fittings alone, and all this was commensurate with the size and fitout of this large cruising vessel compared to the smaller and simpler racing yacht that was *Isea.* The attention to detail is often very similar throughout, the scope of what they do and don't cover is clear, scantlings are stated in size and choice of material, and many small things to consider during construction are often pointed out.

Pride is shown in the finished vessel. A note about the frames on Project 303 is telling, '*Care must be taken to locate frames exactly as shown because the placing of the ports to look well from the outside and to fit accommodation is dependant upon this*'. They might be a small thing in the big picture to assemble, but Reeks was well aware that the outside appearance would be marred by misplaced port lights along the side. Even if slightly out of alignment, this would reflect badly on both himself and the builder.

Both specifications share some additional handwritten notes, corrections or revisions. They reflect the times each was written in, with the later set providing much more detail on sources and other specifics. Ballast is an interesting comparison. On *Isea*'s it noted, '*Weight about 6 tons. Owner will find lead*'. In Project No 303, over two very detailed paragraphs (Limber Ballast and Deadweight Ballast) it is very clearly stated how the internal and external ballast should be sourced and laid out inside and outside the hull, but with one minor note at the end, '*Should additional ballast be required it will be chargeable as an extra or supplied by owner as may be agreed*'. It seems, despite the passage of years, owners still had to be able to source lead.

When we pull together the depth of this practice there is no doubt Reeks was professional throughout his career. He maintained consistently high standards regardless of the nature of the project. A small tender for a vessel, drawn as an individual project, had the same attention paid to it

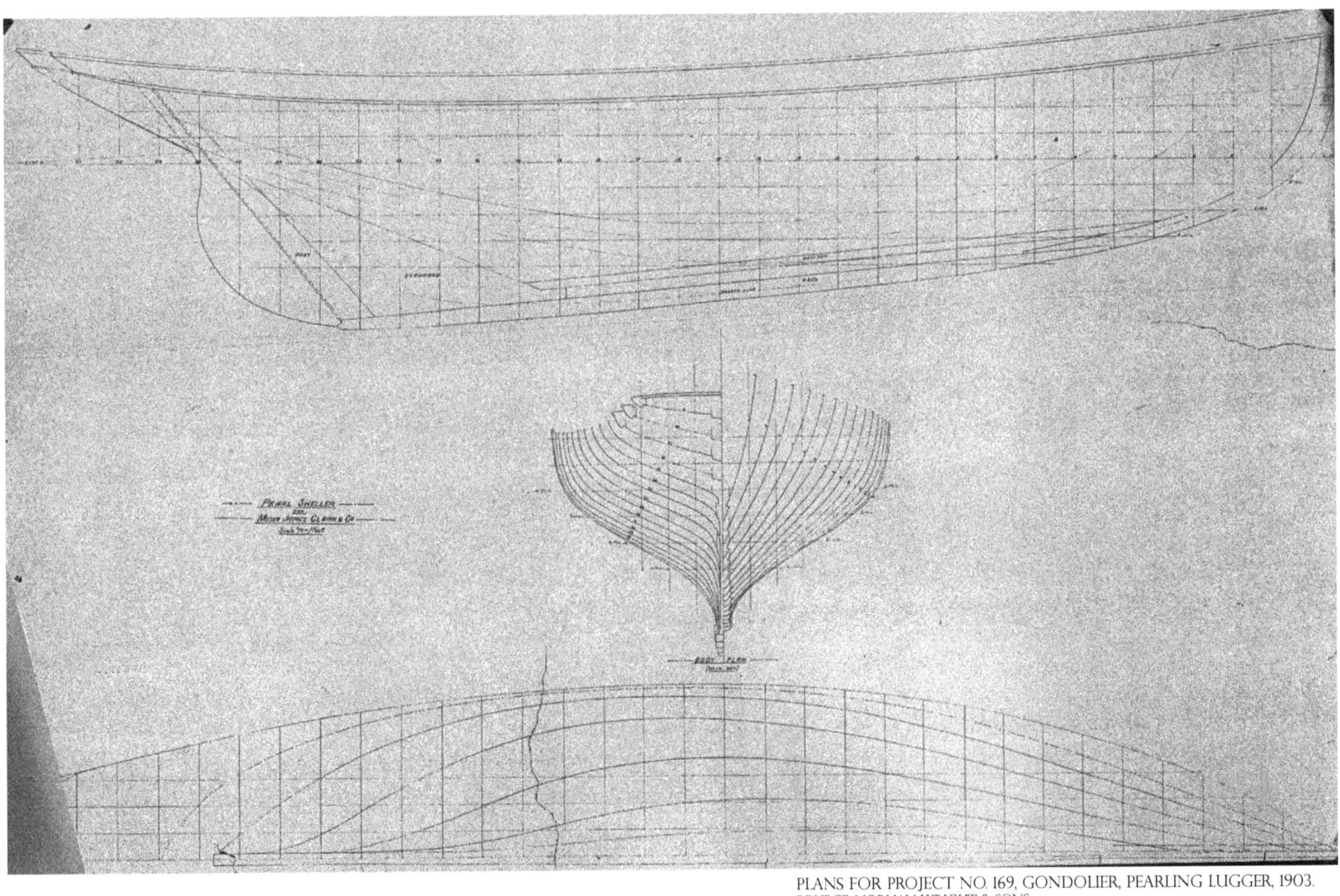

PLANS FOR PROJECT NO. 169, GONDOLIER, PEARLING LUGGER, 1903.
SOURCE: NORMAN WRIGHT & SONS.

PART OF REEKS' HAND WRITTEN SPECIFICATIONS FOR ISEA.
SOURCE: ROYAL SYDNEY YACHT SQUADRON.

Specification of a

Sailing Yacht

The vessel to be built under cover and must be carefully constructed from specifications, plans, and measurements furnished and approved by Mr Walter Reeks Naval Architect

Dimensions

Length on LWL	34' 6"
Length over all	43' 10"
Beam extreme	8' 6"
Draught extreme	5' 6"

Scantling &c.

Stem Of Blue gum, Blackbutt, or Tie kee, sided 4½" at deck and moulded to suit the shape of the vessel to be properly bolted to keel and deadwood and to be faced with a strong metal band well bolted on from the bobstay bolt well down under the forefoot.

Stern Post Of Blue or Spotted gum, in one piece sided 4½" at the deck and 3¼" keel, well morticed into the keel. The back of post to be groved out to take the fore edge of rudder

Main Keel Of Blue gum or Iron bark moulded to suit the form of the vessel and taper fore and aft into siding of stem and stern post it will be about 27" amidships and 5" deep right out

Lead Keel The builder to find labour in casting and fitting and bolts to be carefully fitted and bolted with nut and screw clenched on

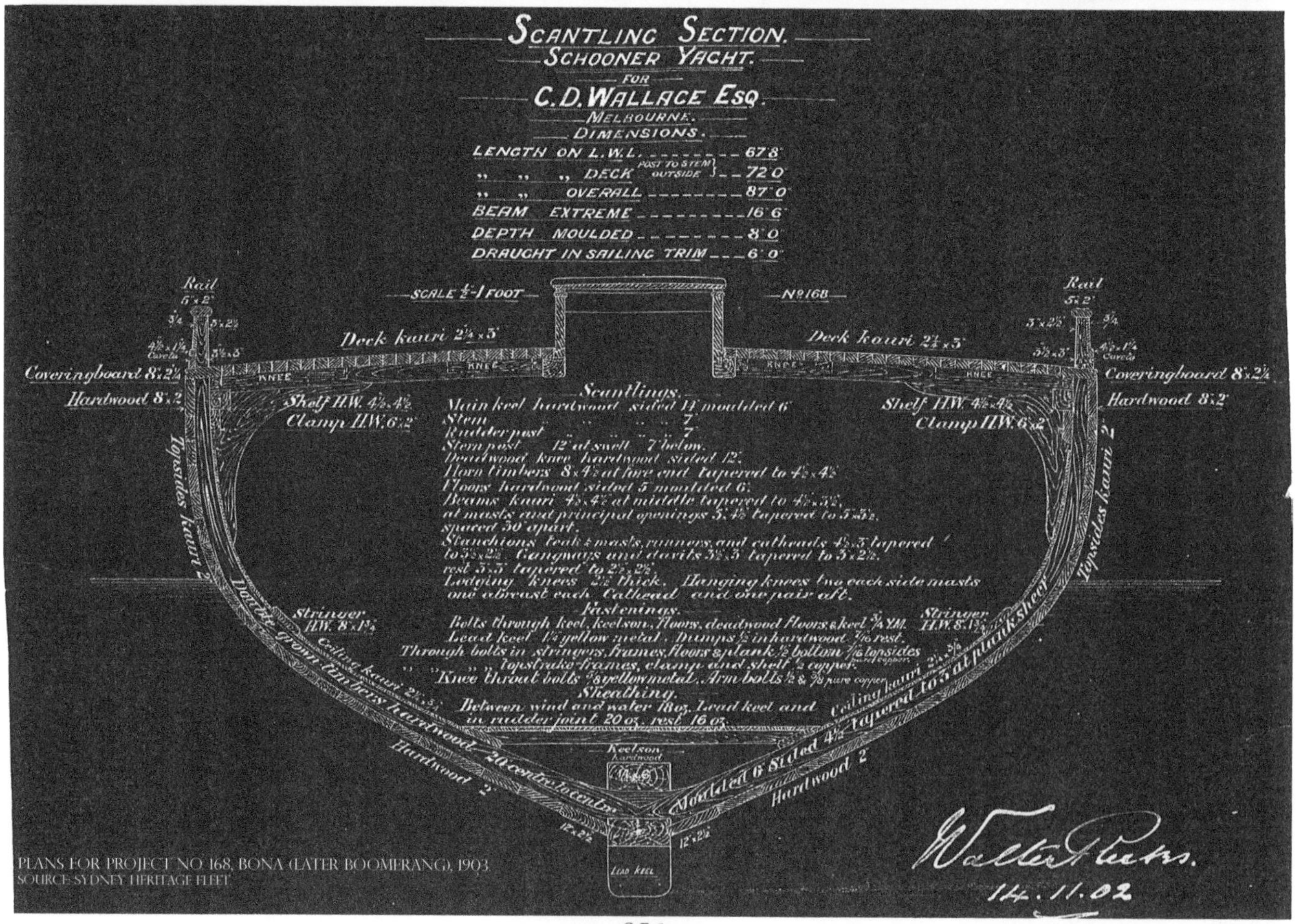

PLANS FOR PROJECT NO 168, BONA (LATER BOOMERANG), 1903
SOURCE SYDNEY HERITAGE FLEET

as a much larger vessel. It remains a pity Reeks' archive has not survived, but luckily what has been sourced has enough information to show that he, as the first full time professional naval architect in Australia, started the profession in the best way possible, setting a standard for others to follow.

For this book co-author David Payne has drawn numerous plans and illustrations, based on the half models noted, information on the blueprint plans or drawings noted, or developed as best possible from information that is available showing dimensions along with images of the vessel. These plans are included on the following pages.

PROJECT NUMBERS

Developed from the title blocks of existing plans, the list shows the rate of progress for Reeks' designs, developing gradually in his early period, picking up from early 1900 through to the First World War, and then slowing down in the years between the war ending and his death in 1925.

No. 30: *Jessie,* pearling lugger, 1892
No. 31 *Isabel*, pearling lugger, 1892
No. 149 steam launch for Sydney Harbour Trust Board
No. 168 *Bona* (later *Boomerang*), auxiliary schooner, 1902
No. 169 *Gondolier,* pearling lugger, 1903
No. 187 *Zephyr,* twin screw passenger steamer, McIlwraith and McEacharn & Co., 1905
No. 196 ferry, Balmain New Ferry Company, 1906
No. 207 steam lighter/tug, Colonial Oil Company, 1907
No. 211 vehicular ferry, Sydney Ferries, Ltd., 1907
No. 218 *Arethusa*, yacht, 1908
No. 223 tender for Federal trawler, 1909
No. 226 *Lotus II,* motor launch, 1910
No. 231 ferry, Balmain New Ferry Company, 1910
No. 232 power launch for SY *Merrie England*, 1910
No. 241 '*Lady Scott*' type ferry, Balmain New Ferry Company, 1910
No. 254 '*Lady Scott*' type ferry, Balmain New Ferry Company, 1911
No. 261 '*Lady Scott*' type ferry, Balmain New Ferry Company, 1913
No. 275 ferry, Sydney Ferries, Ltd., 1915
No. 288 MV *Merrie England*, motor vessel for Papua Government, 1916
No. 303 schooner for Arthur H. Davies, 1923

EXISTING VESSELS

The following vessels are known to be in public collections.

Boomerang ex Bona, schooner, 1902, built by William Holmes, Sydney. Owner: Sydney Heritage Fleet, Sydney. Status: operational.

SY Ena, steam yacht, 1903, built by W. M. Ford (Jr), Sydney. Owner: Australian National Maritime Museum, Sydney. Status: operational.

Lady Denman, ferry, 1911, built by Joseph Dent, Jervis Bay. Owner: Jervis Bay Maritime Museum, Huskisson, New South Wales. Status: static exhibit.

The following vessels are known to be privately owned.

Jenny Wren, 2½-rater yacht, 1889, built by Tom Cubbit, Sydney. Current owner: private, Lane Cove, New South Wales. Status: good condition.

Roona, yacht, 1906, built by Frederick Moore, Launceston, Tasmania. Current owner: private, Goolwa, South Australia. Status: good condition.

Mischief, yacht, 1907, built by William Holmes, Sydney, Current owner: private, Pittwater, New South Wales. Status: good condition.

Earlie, motor launch, 1907, built by William Holmes, Sydney, Current owner: private, Renmark, South Australia. Status: good condition.

Pegasus ex *Plume* or *Snowflake*, tug/lighter, 1907, built by Morrison and Sinclair, Ltd., Sydney. Current owner: private, Sydney, New South Wales. Status: Excellent condition.

Lotus II, motor launch, 1910, built by W. M. Ford (Jr), Sydney, Current owner: private, Pittwater, New South Wales Status: needs reconstruction.

Shona, yacht, 1911, built by William Holmes, Sydney, Current owner: private, Queensland. Status: excellent condition.

Landseer III, yacht, 1912, built by Woodleys, Sydney, Currrnt owner: private, Sydney, New South Wales Status: excellent condition.

Wyruna, yacht, 1913, built by William Holmes, Sydney, Current owner: private, Melbourne, Victoria. Status: undergoing restoration.

Gypsy, yacht, 1914, built by William Ball and Wyn Tatnall, Norfolk Bay, Tasmania. Current owner: Knight family, Hobart, Tasmania. Status: excellent condition.

Sea Scout, yacht, 1923, built by Morrison and Sinclair, Ltd., Sydney. Current owner: private, near Townsville, Queensland. Status: possibly poor condition.

DRAWINGS BY DAVID PAYNE

The following series of drawings of various craft designed by Reeks, or that he was involved with, were created by David Payne. They were produced based on an existing vessel, i.e., *Lotus* and *Lady Denman*, the existence of a half model, or based on an existing drawing developed by Reeks.

Miranda (1887) by David Payne.

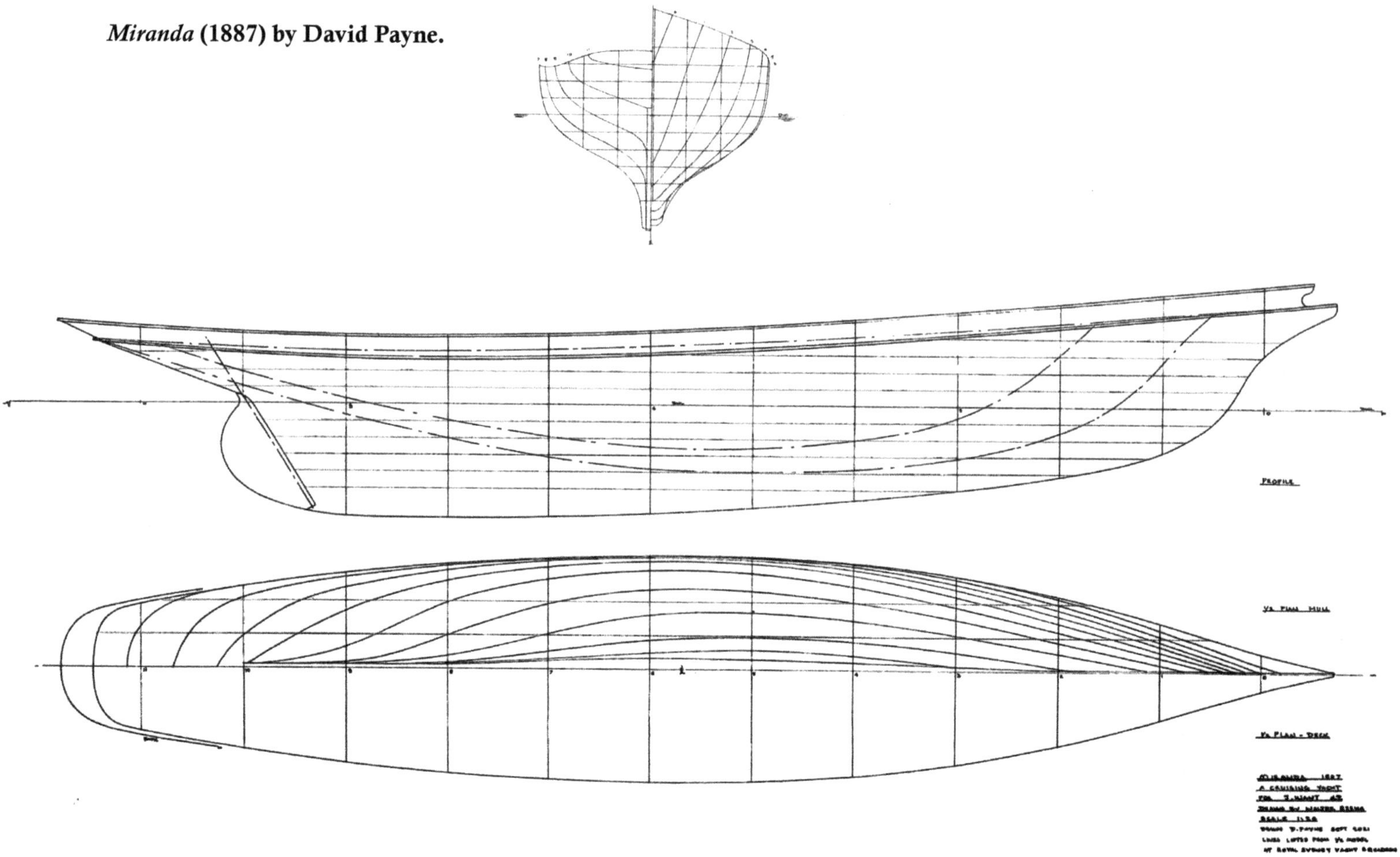

Era (1887) by David Payne.

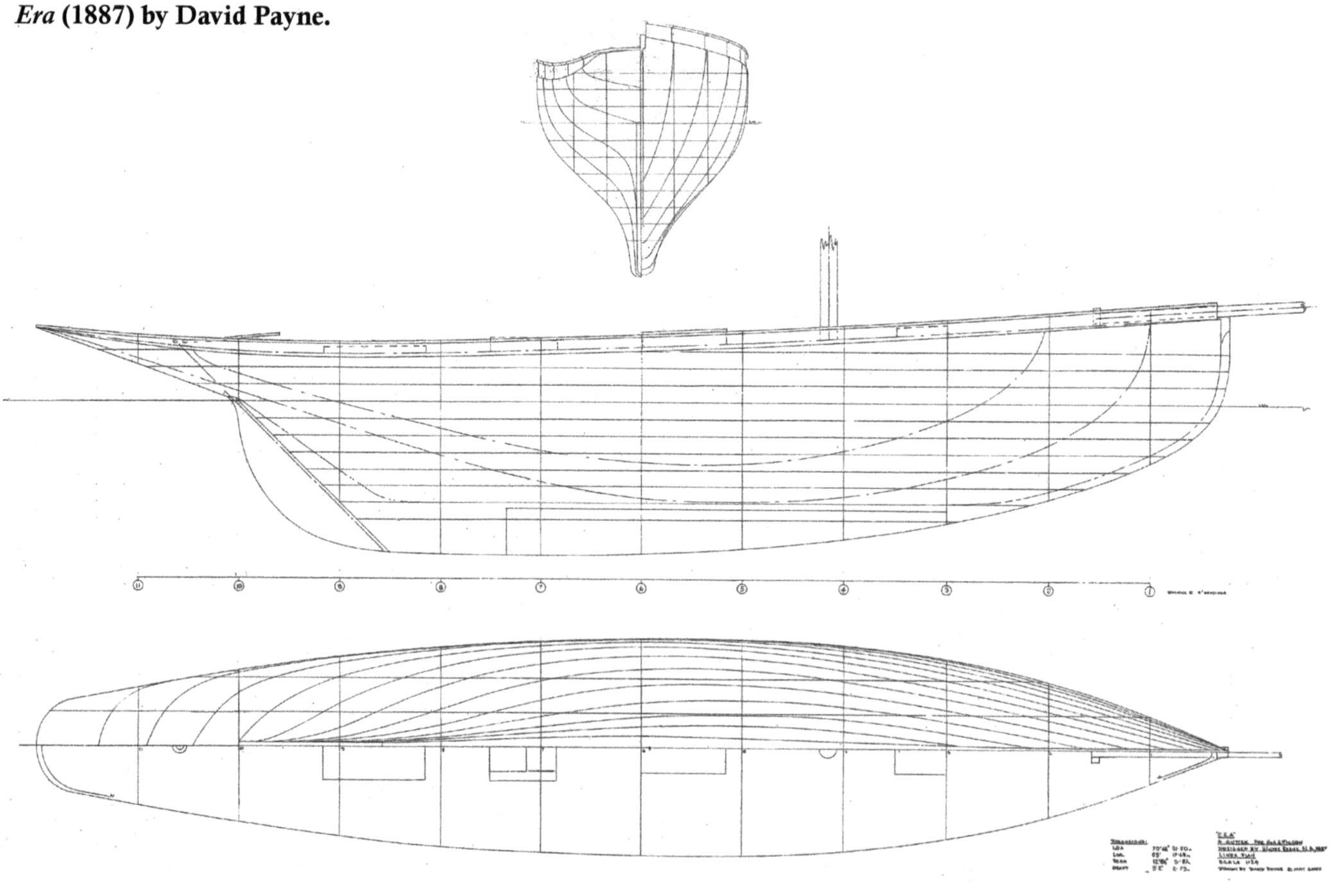

***Volunteer* (1888) by David Payne.**

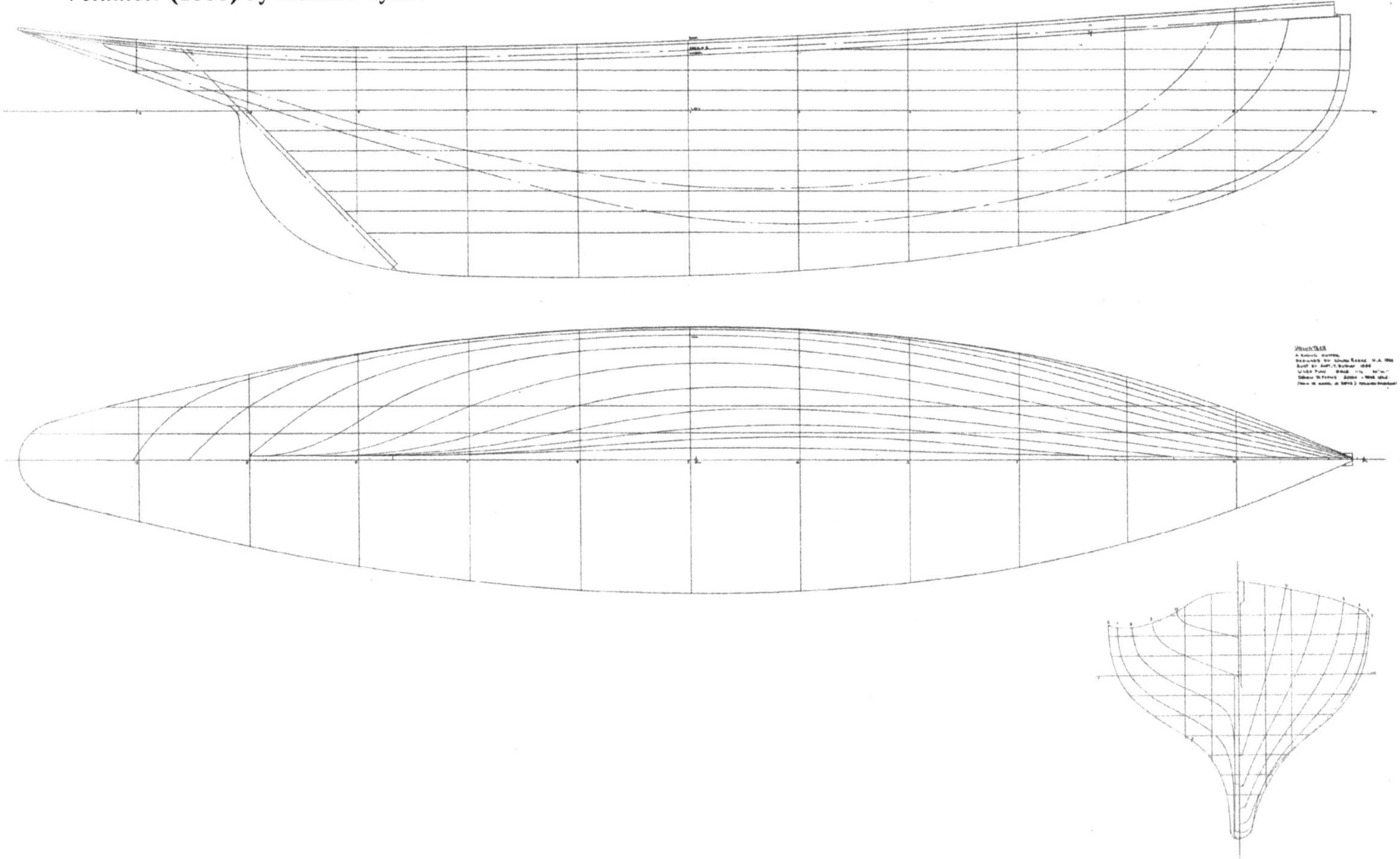

***Thelma* (1889) by David Payne.**

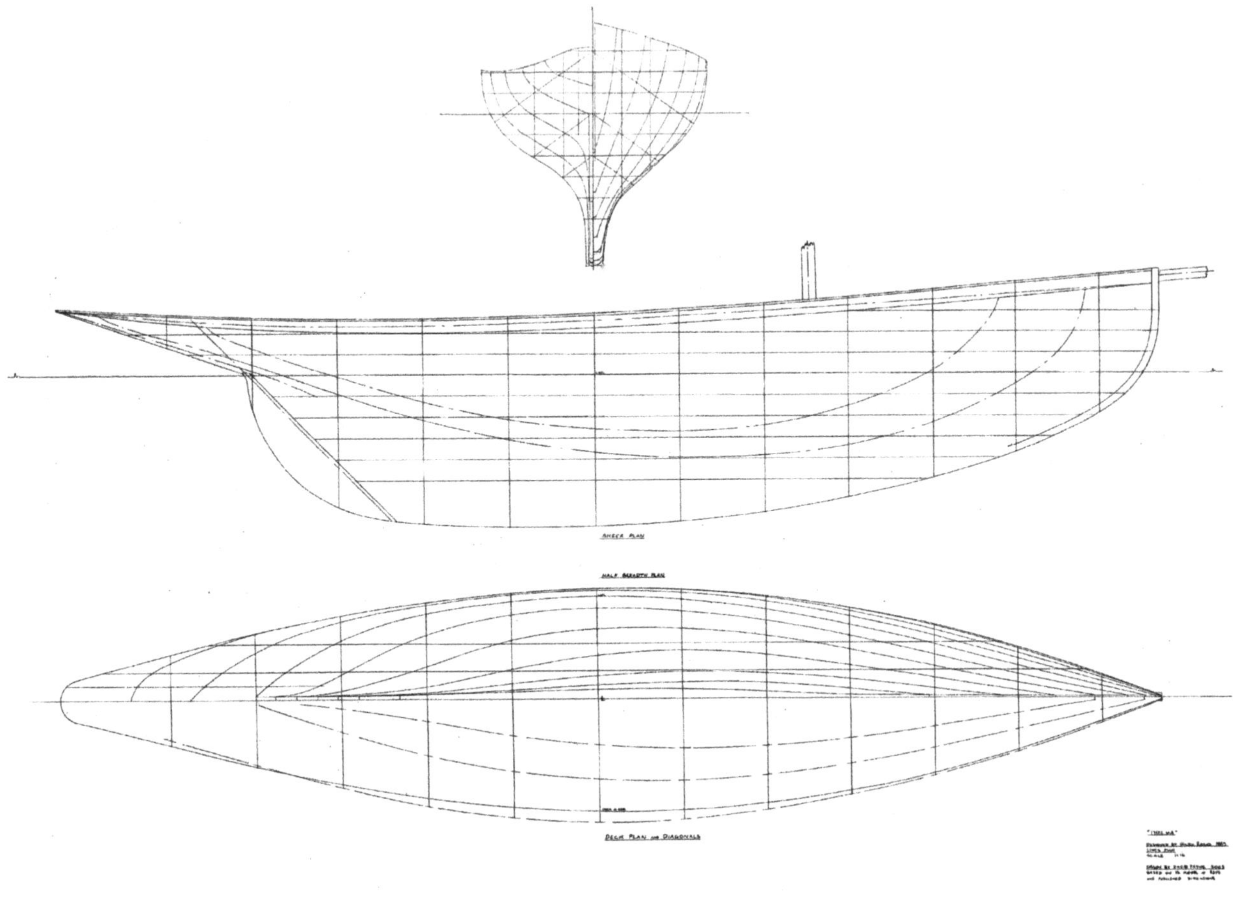

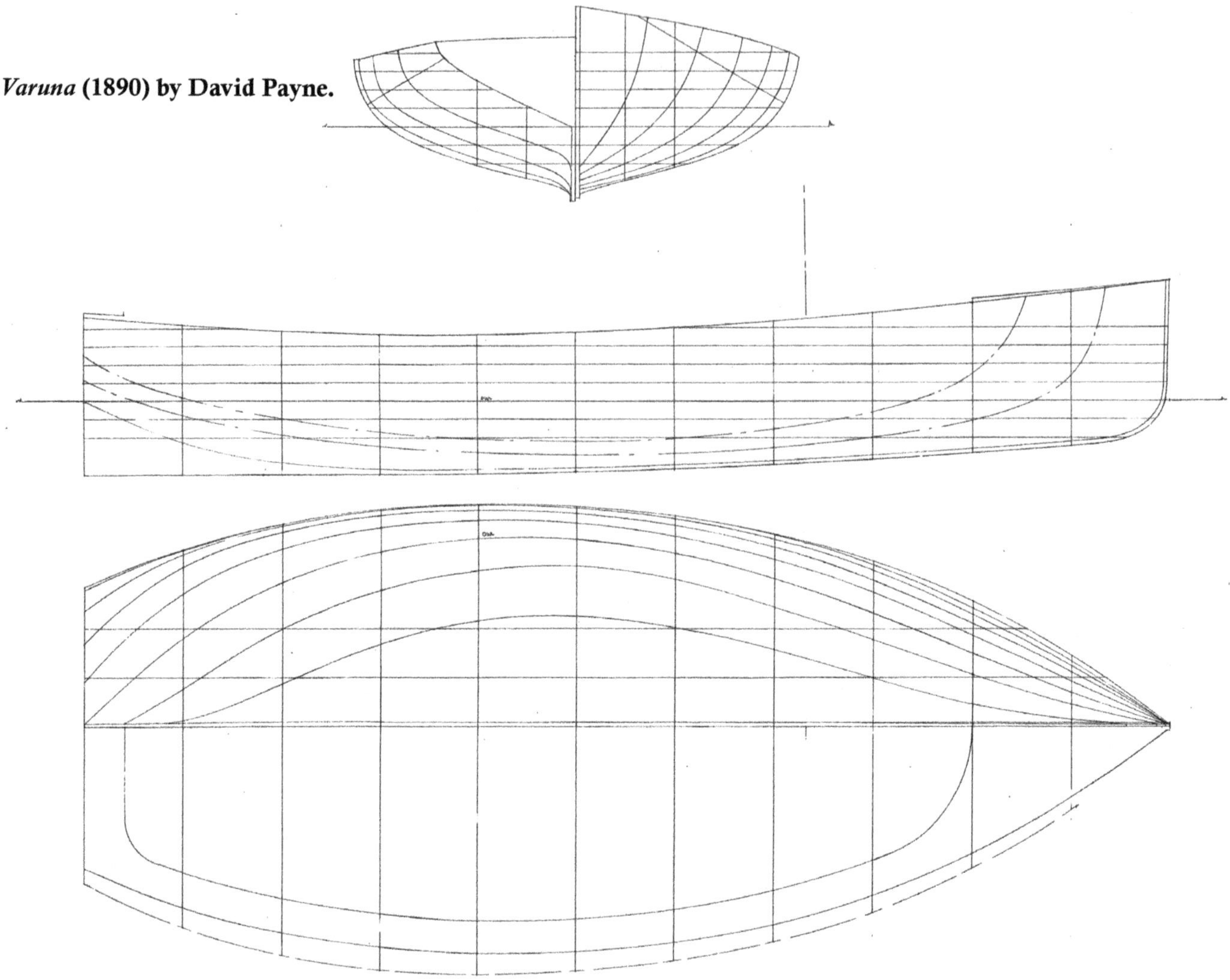

Varuna (1890) by David Payne.

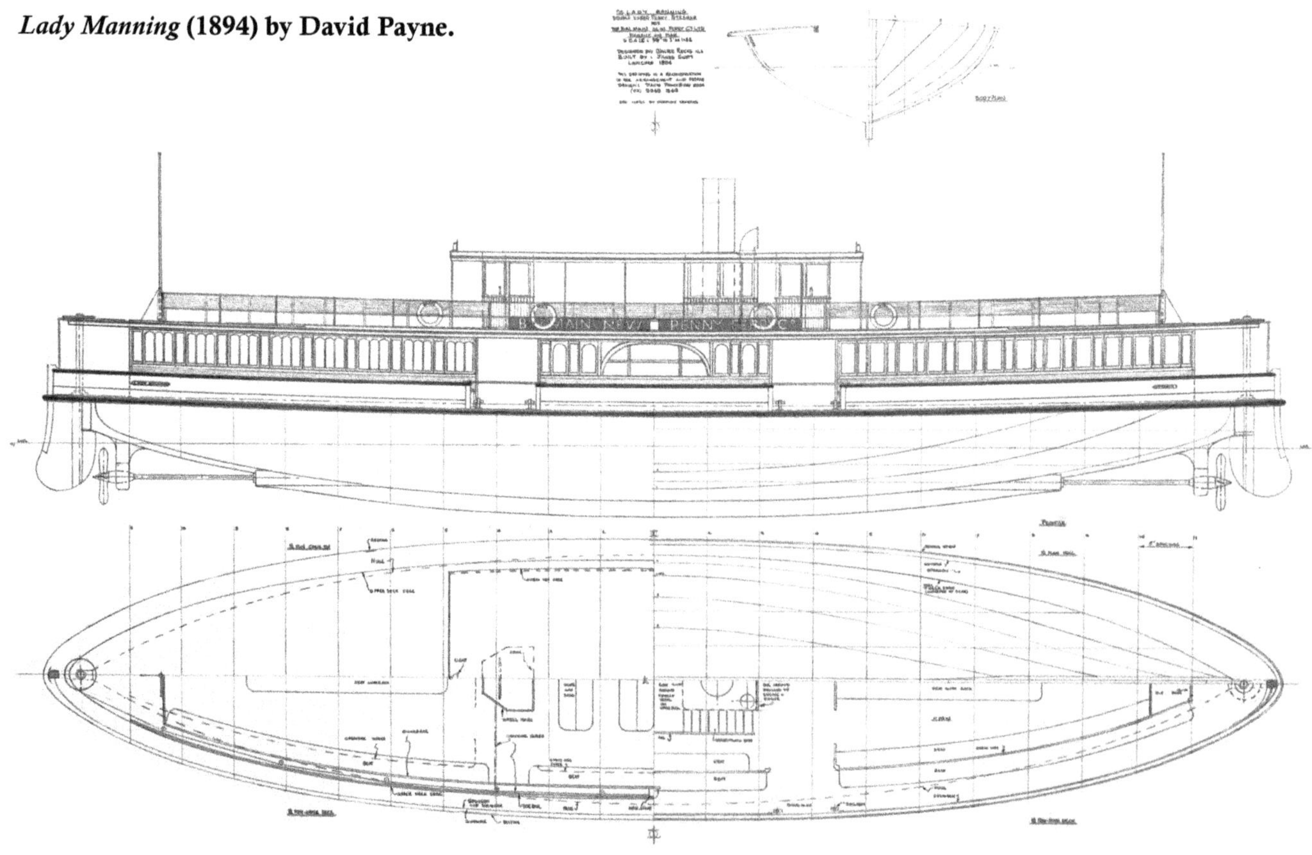

Lady Manning (1894) by David Payne.

***Twynham* (1890) by David Payne.**

SOUTH EAST ELEVATION

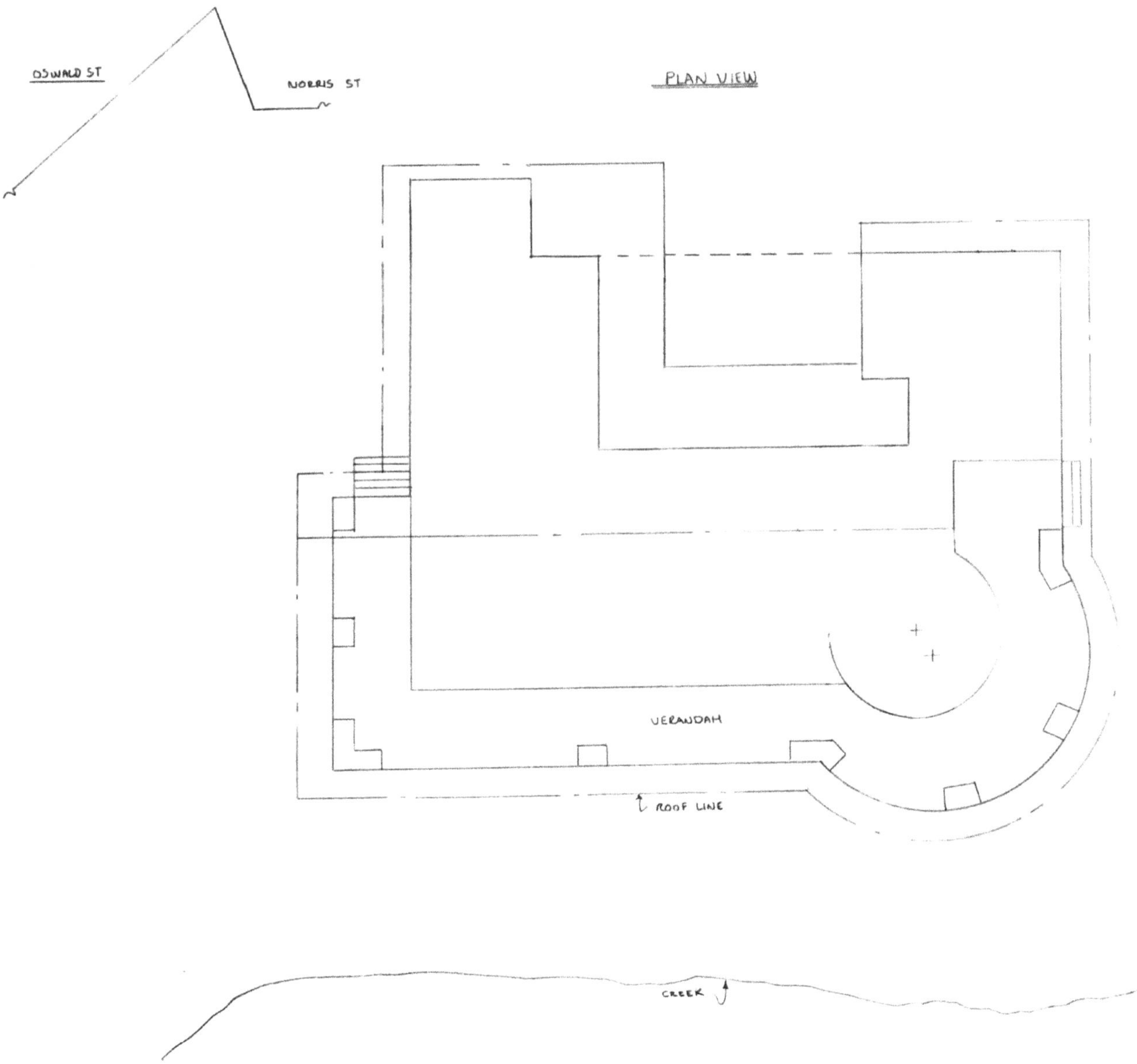

SY Ena (1900) by David Payne.

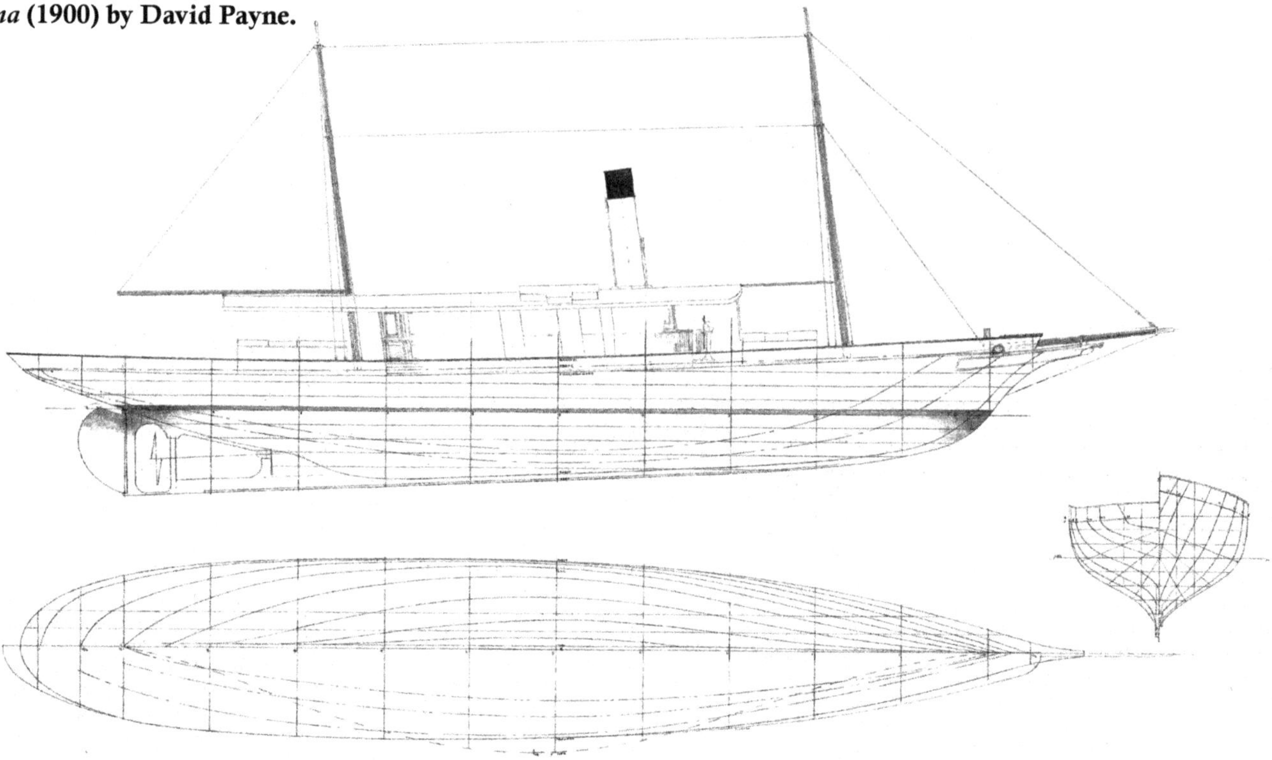

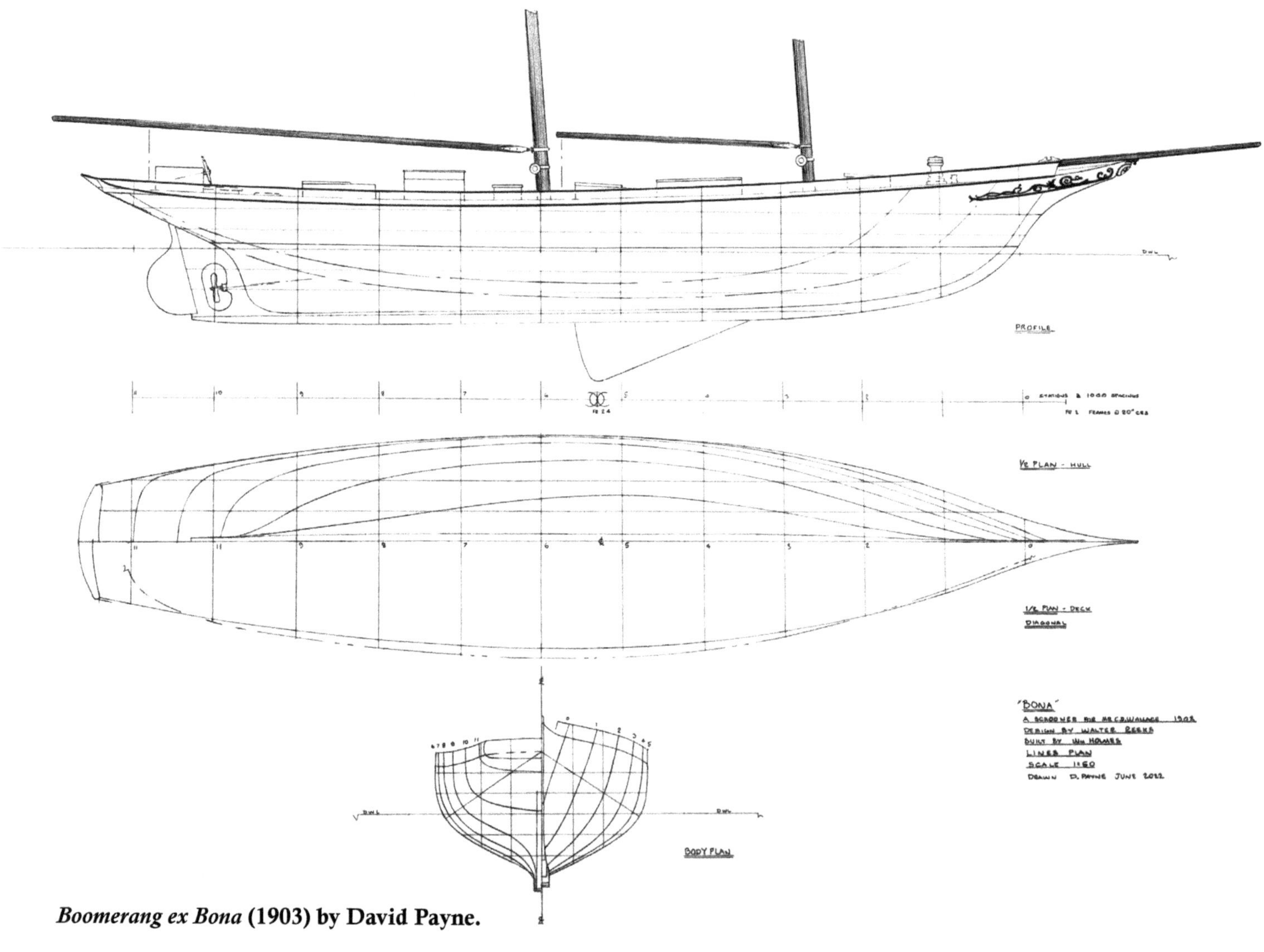

Boomerang ex Bona (1903) by David Payne.

***Pegasus* (1907) by David Payne.**

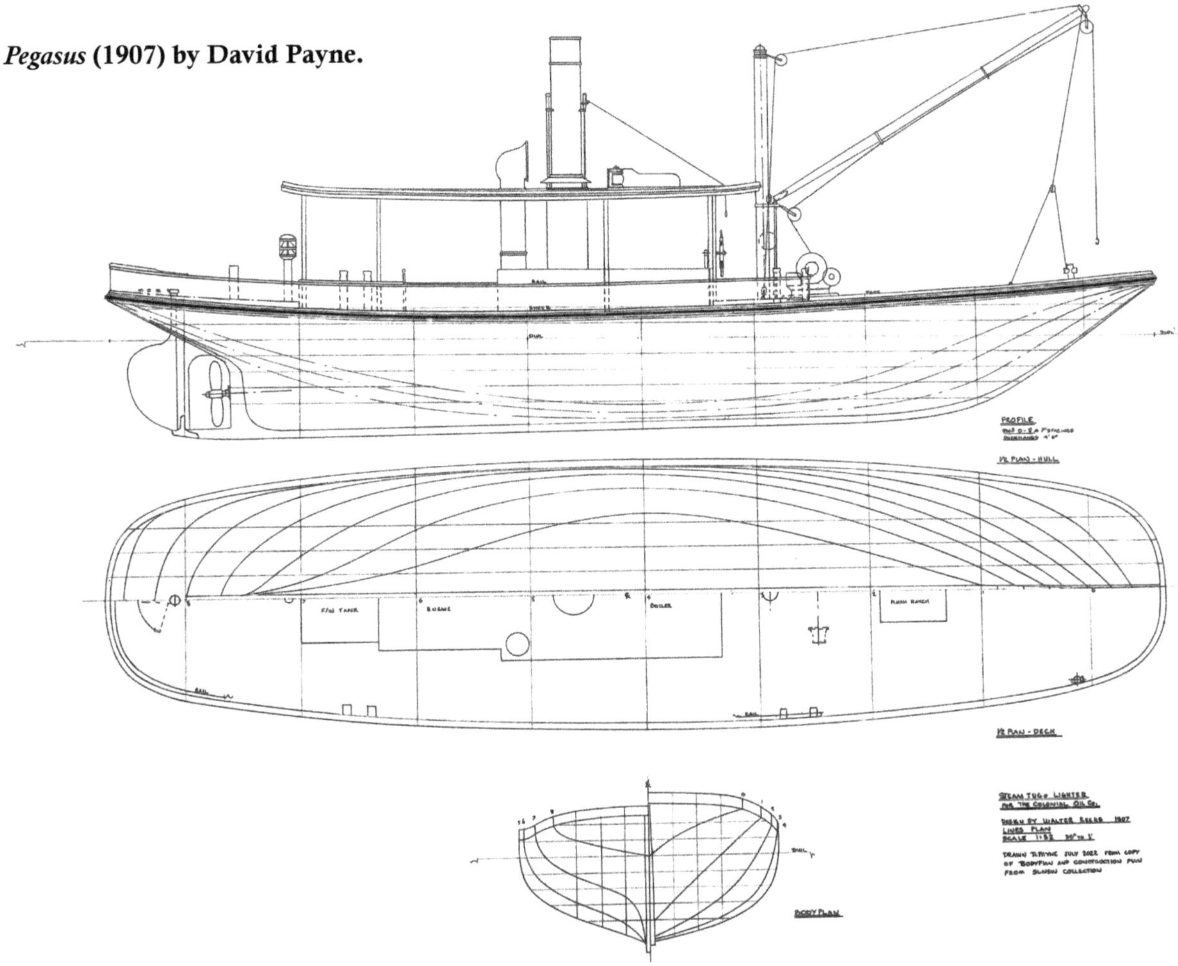

***Lotus II* (1910) by David Payne.**

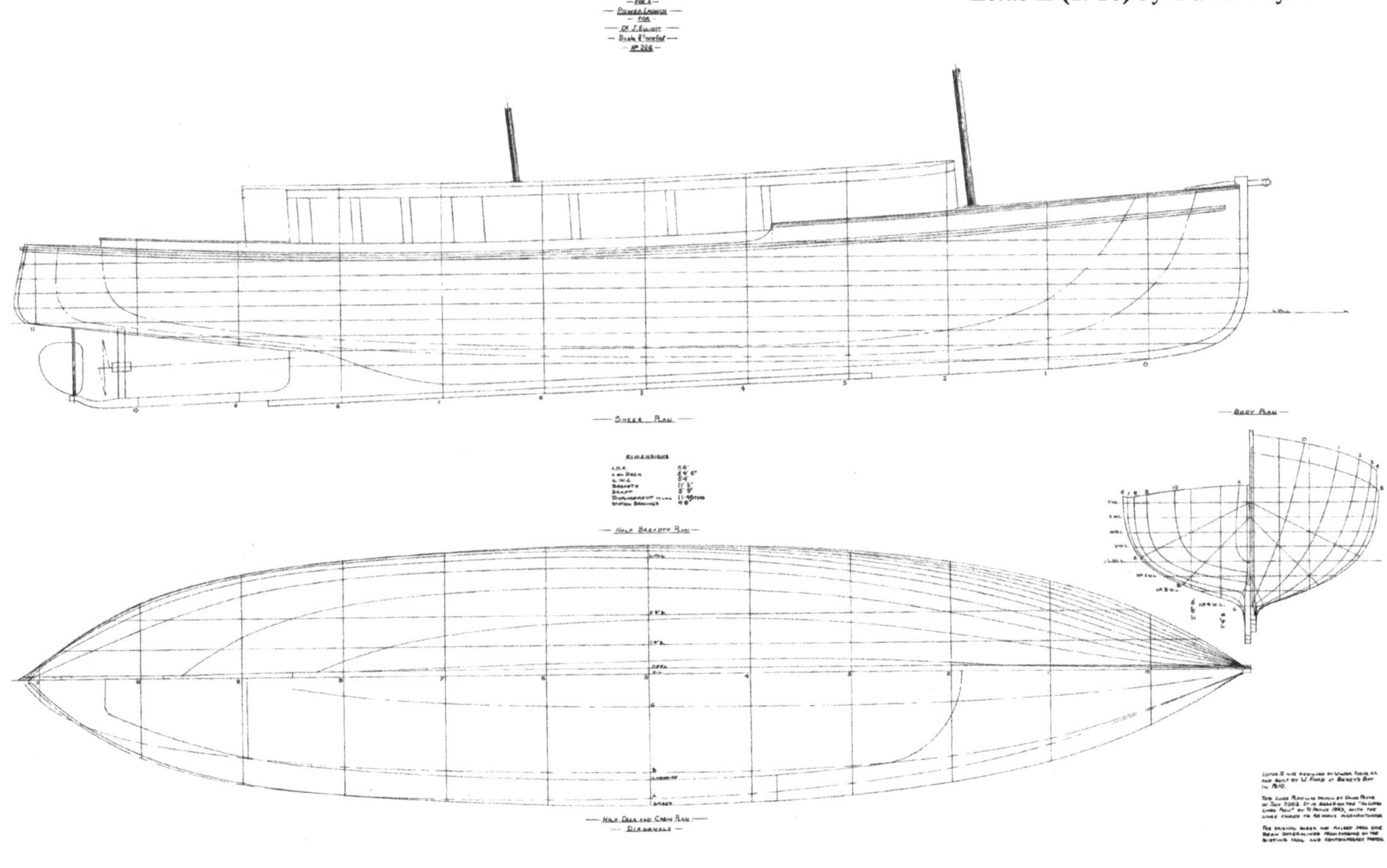

***Lady Denman* (1912) by David Payne.**

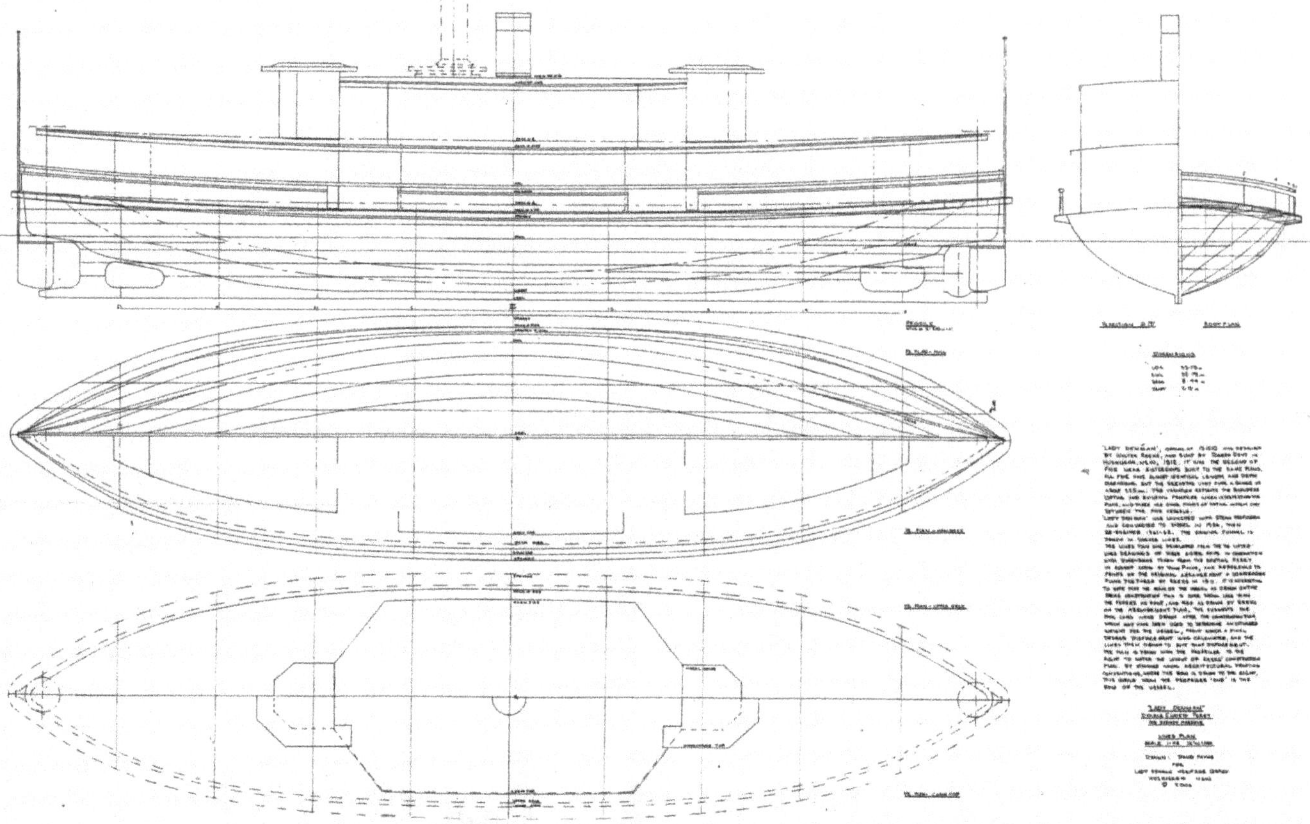

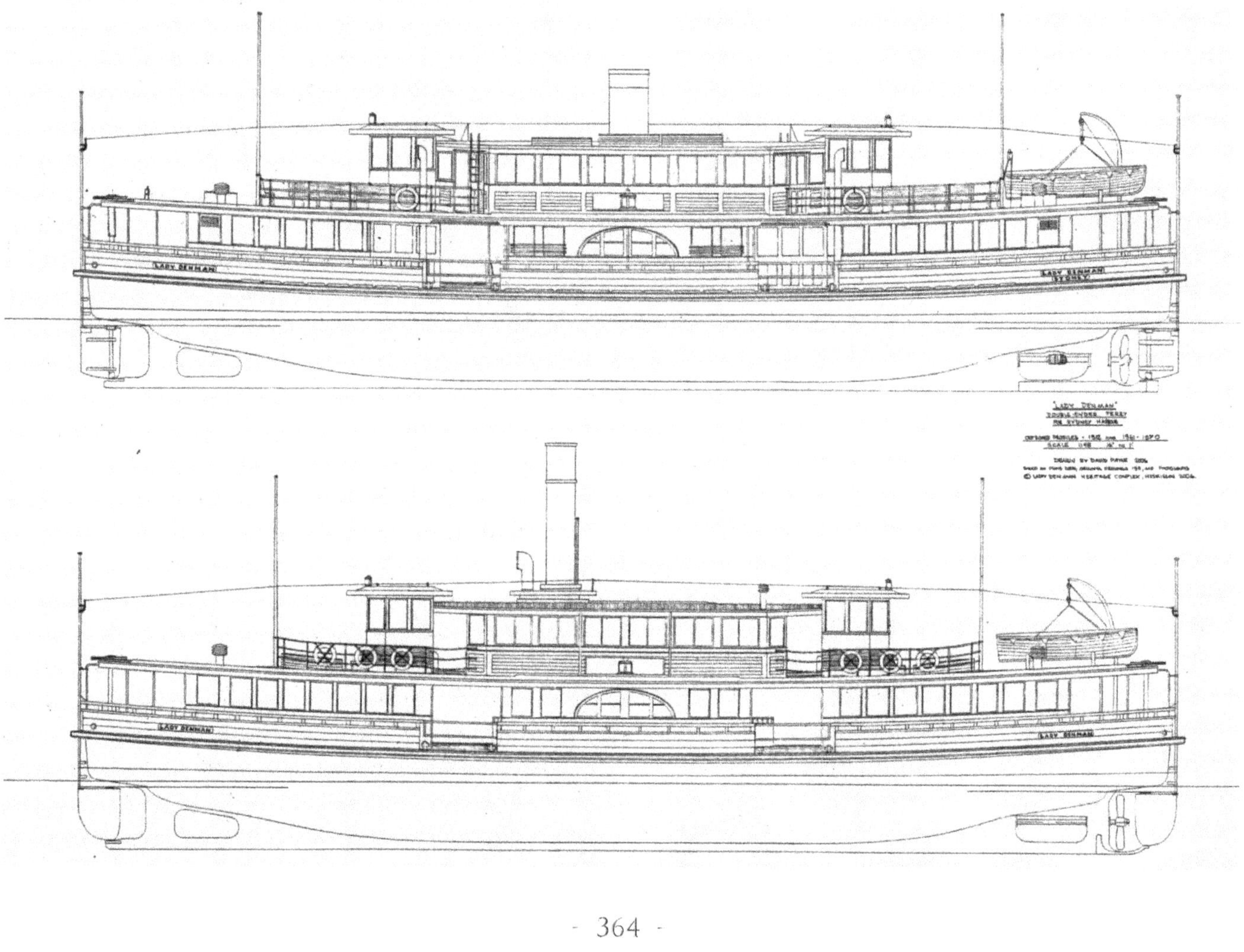

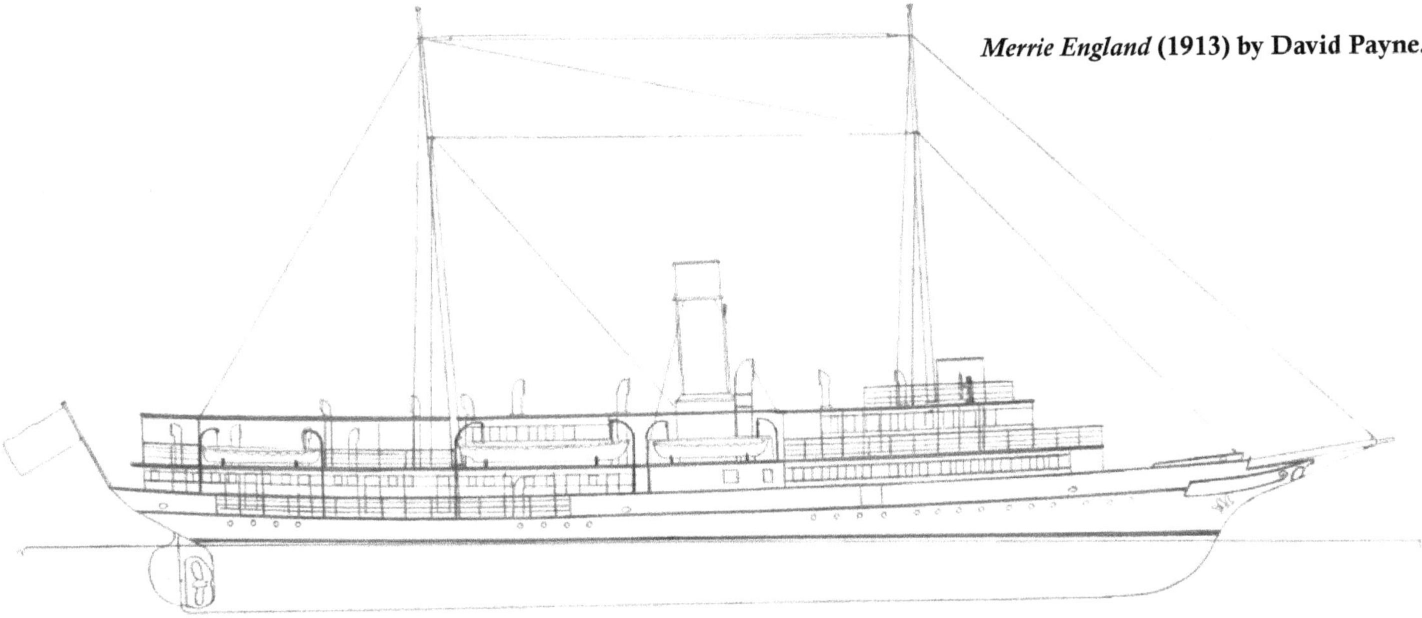

Merrie England (1913) by David Payne.

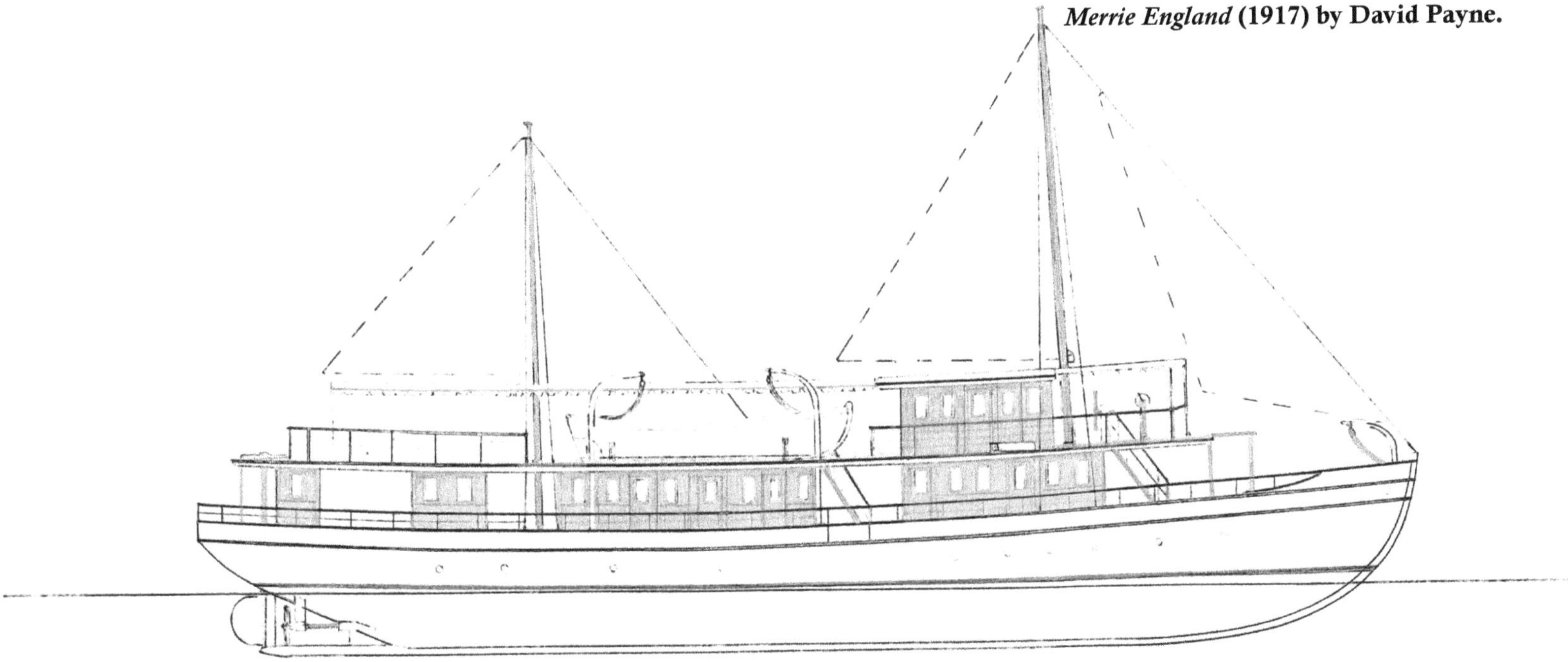

Merrie England (1917) by David Payne.

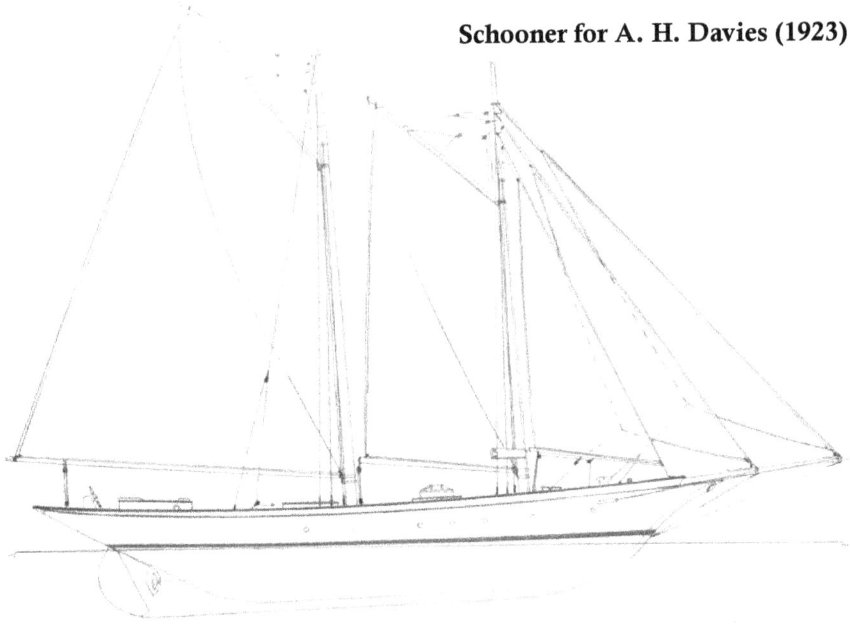

Schooner for A. H. Davies (1923) by David Payne.

SAN PAN RE-RIGGED AND SAILING ON SYDNEY HARBOUR (EARLY 1930S).
SOURCE: AUSTRALIAN NATIONAL MARITIME MUSEUM.

***San Pan* (1925) by David Payne.**

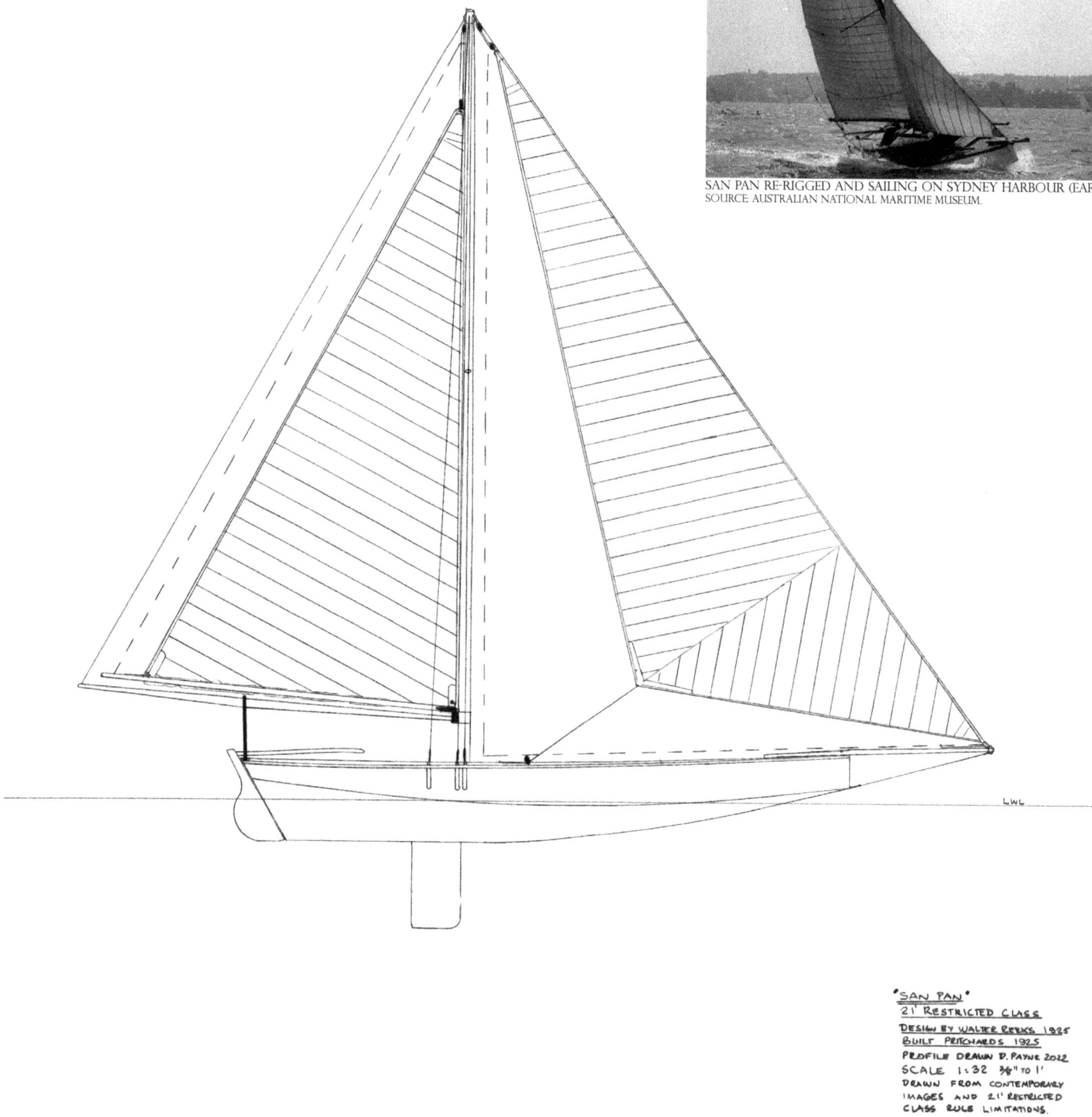

INDEX

WORLD WAR I ROLL OF HONOUR PRESENTED BY WALTER REEKS TO THE ROYAL SYDNEY YACHT SQUADRON.
SOURCE: ROYAL SYDNEY YACHT SQUADRON.

CARICATURE PORTRAIT OF WALTER REEKS
BY HERBERT BEECROFT.
SOURCE: ROYAL SYDNEY YACHT SQUADRON.

www.ingramcontent.com/pod-product-compliance
Lightning Source LLC
Chambersburg PA
CBHW042013200926
51615CB00029B/8
* 9 7 8 0 6 4 5 7 0 8 6 2 2 *